Fuzzy Modeling and Control

Studies in Fuzziness and Soft Computing

Editor-in-chief
Prof. Janusz Kacprzyk
Systems Research Institute
Polish Academy of Sciences
ul. Newelska 6
01-447 Warsaw, Poland
E-mail: kacprzyk@ibspan.waw.pl
http://www.springer.de/cgi-bin/search_book.pl?series=2941

Further volumes of this series can be found at our homepage.

Vol. 47. E. Czogała† and J. Łęski
Fuzzy and Neuro-Fuzzy Intelligent Systems, 2000
ISBN 3-7908-1289-7

Vol. 48. M. Sakawa
Large Scale Interactive Fuzzy Multiobjective Programming, 2000
ISBN 3-7908-1293-5

Vol. 49. L.I. Kuncheva
Fuzzy Classifier Design, 2000
ISBN 3-7908-1298-6

Vol. 50. F. Crestani and G. Pasi (Eds.)
Soft Computing in Information Retrieval, 2000
ISBN 3-7908-1299-4

Vol. 51. J. Fodor, B. De Baets and P. Perny (Eds.)
Preferences and Decisions under Incomplete Knowledge, 2000
ISBN 3-7908-1303-6

Vol. 52. E.E. Kerre and M. Nachtegael (Eds.)
Fuzzy Techniques in Image Processing, 2000
ISBN 3-7908-1304-4

Vol. 53. G. Bordogna and G. Pasi (Eds.)
Recent Issues on Fuzzy Databases, 2000
ISBN 3-7908-1319-2

Vol. 54. P. Sinčák and J. Vaščák (Eds.)
Quo Vadis Computational Intelligence?, 2000
ISBN 3-7908-1324-9

Vol. 55. J.N. Mordeson, D.S. Malik and S.-C. Cheng
Fuzzy Mathematics in Medicine, 2000
ISBN 3-7908-1325-7

Vol. 56. L. Polkowski, S. Tsumoto and T.Y. Lin (Eds.)
Rough Set Methods and Applications, 2000
ISBN 3-7908-1328-1

Vol. 57. V. Novák and I. Perfilieva (Eds.)
Discovering the World with Fuzzy Logic, 2001
ISBN 3-7908-1330-3

Vol. 58. D.S. Malik and J.N. Mordeson
Fuzzy Discrete Structures, 2000
ISBN 3-7908-1335-4

Vol. 59. T. Furuhashi, S. Tano and H.-A. Jacobsen (Eds.)
Deep Fusion of Computational and Symbolic Processing, 2001
ISBN 3-7908-1339-7

Vol. 60. K.J. Cios (Ed.)
Medical Data Mining and Knowledge Discovery, 2001
ISBN 3-7908-1340-0

Vol. 61. D. Driankov and A. Saffiotti (Eds.)
Fuzzy Logic Techniques for Autonomous Vehicle Navigation, 2001
ISBN 3-7908-1341-9

Vol. 62. N. Baba, L.C. Jain (Eds.)
Computational Intelligence in Games, 2001
ISBN 3-7908-1348-6

Vol. 63. O. Castillo, P. Melin
Soft Computing for Control of Non-Linear Dynamical Systems, 2001
ISBN 3-7908-1349-4

Vol. 64. I. Nishizaki, M. Sakawa
Fuzzy and Multiobjective Games for Conflict Resolution, 2001
ISBN 3-7908-1341-9

Vol. 65. E. Orłowska, A. Szalas (Eds.)
Relational Methods for Computer Science Applications, 2001
ISBN 3-7908-1365-6

Vol. 66. R.J. Howlett, L.C. Jain (Eds.)
Radial Basis Function Networks 1, 2001
ISBN 3-7908-1367-2

Vol. 67. R.J. Howlett, L.C. Jain (Eds.)
Radial Basis Function Networks 2, 2001
ISBN 3-7908-1368-0

Vol. 68. A. Kandel, M. Last and H. Bunke (Eds.)
Data Mining and Computational Intelligence, 2001
ISBN 3-7908-1371-0

Andrzej Piegat

Fuzzy Modeling and Control

With 680 Figures
and 96 Tables

Physica-Verlag
A Springer-Verlag Company

Prof. Andrzej Piegat
Faculty of Computer Science and Information Systems
Technical University of Szczecin
Zolnierska Street 49
71-210 Szczecin
Poland
Andrzej.Piegat@wi.ps.pl

ISSN 1434-9922
ISBN 978-3-7908-2486-5 e-ISBN 978-3-7908-1824-6

Cataloging-in-Publication Data applied for
Die Deutsche Bibliothek – CIP-Einheitsaufnahme
Piegat, Andrzej: Fuzzy modeling and control: with 96 tables / Andrzej Piegat. – Heidelberg; New York: Physica-Verl., 2001
(Studies in fuzziness and soft computing; Vol. 69)

Physica-Verlag Heidelberg New York
a member of BertelsmannSpringer Science+Business Media GmbH

Printed in Germany

Hardcover Design: Erich Kirchner, Heidelberg

To my Family

Foreword

Since their inception in the mid-1960s by Professor Lotfi A. Zadeh from the University of California, Berkeley, fuzzy sets have triggered mixed feeling in the scientific community. On the one hand, there has been a growing number of devotees who have early recognized potentials of fuzzy sets to model and solve many real world problems. On the other hand, however, there have been quite a considerable number of opponents – quite often prominent scholars and researchers – who have fiercely fought against this emerging new tools. One of their arguments has been a lack of applications.

The situation had changed since the mid 1980s when the so-called "fuzzy boom" occurred, primarily in Japan, but then also in Korea and Europe, much less so in the USA. Basically, the turning point was the launching on the market of fuzzy logic control based appliances and other equipment exemplified by subway trains, cranes, elevators, etc. They had primarily been successful applications of fuzzy logic control originated by Mamdani, Sugeno, Takagi and other people.

Fuzzy logic control has since then been a benchmark application of fuzzy sets, and even to many people synonymous with applications. Many excellent books have been published on this topic. However, some of them have been authored by people from outside of the control community. One of consequences of this situation has been, in my view, too high emphasis on logical, relational, etc. aspects of fuzzy control, with too little emphasis on more control specific issues and aspects.

One of those control specific aspects is *modeling.* In fact, modeling is probably much more relevant than control itself, and is of much greater applicability as a general tool and technique for problems in virtually all domains. Unfortunately, modeling has not received adequate attention in the fuzzy literature though developments in *fuzzy modeling* have been quite considerable.

The present book is probably the most comprehensive coverage of fuzzy modeling and control in the literature. First of all, it presents in depth the crucial field of fuzzy modeling, in all its aspects and flavors. The book discusses virtually all better known and more popular techniques exemplified by rule based modeling, logical models, and also hybrid models like neuro-fuzzy. The author discusses in a clear and simple yet formal and strict way issues related to fuzzy modeling that can be of much use for readers interested in both more theoretical and practical aspects.

Then, after that comprehensive discussion of fuzzy modeling, the author proceeds to the presentation of fuzzy control. He starts with a more traditional approach of what might be called a purely fuzzy logic control that is not model based, and then proceeds to modern, more promising approaches that rely on fuzzy models of plants and systems under control, and also of more sophisticated control structures like adaptive or multivariate control.

Finally, the author discusses issues related to the stability of fuzzy control systems, and again, the scope of analysis is rarely encountered in the existing literature.

To summarize, in my opinion this is an extraordinary books, like almost no ones in the literature. It gives a great, comprehensive coverage of fuzzy modeling and control, written in a manner that is both understandable and acceptable by the control community. The writing of such a book has required not only great knowledge but also a considerable scholarly and research maturity to be able to find proper, most promising tools and techniques in a vast literature.

Professor Piegat deserves thanks and appreciation of the whole fuzzy community for writing such an exceptional books that should be read by all interested in modern approaches to fuzzy modeling and control.

Warsaw, Poland, December 2000 *Janusz Kacprzyk*

Preface

Conventional mathematics enables processing of precise information, e.g.:

- temperature 39.7 (°C),
- velocity 90 (km/h),
- business pay $12317,
- sea wave amplitude 1.75 (m).

However, in the surrounding world we very often meet imprecise information such as:

- high temperature,
- high velocity,
- low business pay,
- calm sea,
- pleasant sales person,
- big customer interest,
- small turbidity of the liquid,
- high steel quality, etc.

Imprecise information has been used by people for thousands of years. However, until quite recently it has not been used at all in methods based on conventional mathematics. Therefore it has been lost. Because of this, the efficiency of many design-, control-, modeling-, forecasting- and decision-making methods was considerably limited – all the more, as in some systems imprecise information is the only accessible information. Besides, each piece of "precise" information was measured with some (often considerable) error – so it is also imprecise.

The domain of mathematics dealing with imprecise information was named fuzzy set theory. This theory in connection with conventional mathematics enables the processing and use of any information. It discloses new and very interesting possibilities and prospects for science and technology.

This book provides the reader with essential information of fuzzy set theory, fuzzy modeling and control. It is based on the latest literature and on the results of the author's investigations.

Good understanding of a theory is the main condition for its application and improvement or development of one's own ideas and concepts. To make this easier, the author illustrated the presented methods with a large number of figures and examples. The author wishes the reader many benefits from the information contained in this book.

The author would like to thank the following persons:

- Deputy Rector for Scientific Affairs of the Technical University of Szczecin, Professor **Walerian Arabczyk** for financial support of the book's preparation,
- Dean of Faculty for Computer Science and Information Systems of the Technical University of Szczecin, Professor **Jerzy Sołdek** for financial support of the book's preparation,
- Foundation for Development Support of the Technical University of Szczecin and especially its Director **Krzysztof Leszczyński** for financial support of the book's preparation,
- **Bogdan Grzywacz** Ph.D., **Stanisława Lewandowska** M.Sc., **Ewa Lisek** M.Sc., for the book's translation,
- **Richard Stark** BA, MA, CELTA (Great Britain) for language correction of the book,
- **Marcin Pluciński** Ph.D. for computer setting of the book.

Szczecin, December 2000 *Andrzej Piegat*

Contents

1. Introduction 1
- 1.1 Essence of fuzzy set theory 1
- 1.2 Development of fuzzy set theory 6

2. Basic Notions of Fuzzy Set Theory 11
- 2.1 Fuzzy sets 11
- 2.2 Characteristic parameters (indices) of a fuzzy set 22
- 2.3 Linguistic modifiers of fuzzy sets 27
- 2.4 Types of membership functions of fuzzy sets 34
- 2.5 Type 2 fuzzy sets 52
- 2.6 Fuzziness and probability: two kinds of uncertainty 56

3. Arithmetic of Fuzzy Sets 59
- 3.1 The extension principle 60
- 3.2 Addition of fuzzy numbers 67
- 3.3 Subtraction of fuzzy numbers 75
- 3.4 Multiplication of fuzzy numbers 79
- 3.5 Division of fuzzy numbers 95
- 3.6 Peculiarities of fuzzy numbers 100
- 3.7 Differences between fuzzy numbers and linguistic values 107

4. Mathematics of Fuzzy Sets 111
- 4.1 Basic operations on fuzzy sets 111
  - 4.1.1 Intersection operation (logical product) of fuzzy sets 112
  - 4.1.2 Union (logical sum) of fuzzy sets 126
  - 4.1.3 Compensatory operators 132
- 4.2 Fuzzy relations 136
- 4.3 Implication 149

5. Fuzzy Models 157
- 5.1 Structure, main elements and operations in fuzzy models 157
  - 5.1.1 Fuzzification 159
  - 5.1.2 Inference 160
    - 5.1.2.1 Premise evaluation 162

5.1.2.2 Determination of activated membership functions of conclusions in particular rules at given input values of a fuzzy model ... 167
5.1.2.3 Determination of the resulting membership function of the rule-base conclusion ... 172
5.1.3 Defuzzification of the resulting membership function of the rule-base conclusion ... 184
5.1.4 Example of fuzzy modeling ... 199
5.2 Significant features of rules, rule bases and fuzzy models ... 202
5.2.1 Local character of rules ... 202
5.2.2 Dependence of the number of rules on the number of inputs and fuzzy sets ... 204
5.2.3 Completeness of a fuzzy model ... 208
5.2.4 Consistency of the rule base ... 216
5.2.5 Continuity of the rule base ... 219
5.2.6 Redundancy of the rule base ... 222
5.3 Advice relating to rule base construction ... 224
5.4 Reduction of the rule base ... 229
5.5 Normalization (scaling) of the fuzzy model inputs and output 244
5.6 Extrapolation in fuzzy models ... 250
5.7 Types of fuzzy models ... 280
5.7.1 Mamdani models ... 281
5.7.2 Takagi-Sugeno models ... 301
5.7.3 Relational models ... 311
5.7.4 Global and local fuzzy models ... 316
5.7.5 Fuzzy multimodels ... 323
5.7.6 Neuro-fuzzy models ... 329
5.7.7 Alternative models ... 331
5.7.8 Similarity principles of the system and of the system model ... 338
5.7.9 Fuzzy classification ... 339

6. Methods of Fuzzy Modeling ... 363
6.1 Fuzzy modeling based on the system expert's knowledge ... 366
6.2 Creation of fuzzy, self-tuning models based on input/output measurement data of the system ... 373
6.2.1 Application of neuro-fuzzy networks for fuzzy model parameter tuning ... 374
6.2.1.1 Structuring and training of neural networks .. 374
6.2.1.2 Transformation of a Mamdani fuzzy model into a neuro-fuzzy network ... 388
6.2.1.3 Transformation of a Takagi-Sugeno fuzzy model into a neuro-fuzzy network ... 397
6.2.2 Tuning of fuzzy model parameters with the genetic algorithm method ... 400

6.3 Creation of self-organizing and self-tuning fuzzy models based on input/output measurement data of the system 405
6.3.1 Determination of significant and insignificant inputs of the model 406
6.3.2 Determining of fuzzy curves 410
6.3.3 Self-organization and self-tuning tuning of fuzzy model parameters 417
6.3.3.1 Self-organization and tuning of fuzzy models with the geometric method of the maximum absolute error 420
6.3.3.2 Self-organization and self-tuning of fuzzy models with clustering methods 452
6.3.3.3 Self-organization and self-tuning of fuzzy models with the searching method 487

7. Fuzzy Control 495
7.1 Static fuzzy controllers 495
7.2 Dynamic fuzzy controllers 500
7.3 The determination of structures and parameters for fuzzy controllers (organization and tuning) 511
7.3.1 The design of fuzzy controllers on the basis of expert knowledge concerning plant under control 512
7.3.2 The design of a fuzzy controller on the basis of a model of the expert controlling the plant 516
7.3.3 The design of a fuzzy controller on the basis of the model of controlled plant 522
7.3.3.1 Remarks concerning identification of models of dynamic plants 522
7.3.3.2 Some remarks concerning the identification of inverted models of dynamical plants 525
7.3.3.3 Tuning a fuzzy controller with an a priori chosen structure 554
7.3.3.4 Fuzzy control based on the Internal Model Control Structure (IMC structure) 561
7.3.3.5 Fuzzy control structure with an inverse of a plant model (InvMC structure) 583
7.3.3.6 Adaptive fuzzy control 599
7.3.3.7 Multivariable fuzzy control (MIMO) 601

8. The Stability of Fuzzy Control Systems 609
8.1 The stability of fuzzy control systems with unknown models of plants 614
8.2 The circle stability criterion 618
8.3 The application of hyperstability theory to analysis of fuzzy-system stability 625

8.3.1 The frequency domain representation of hyperstability conditions for control systems with a time invariant non-linear part 627
8.3.2 The time domain conditions for hyperstability of continuous, non-linear control systems containing a time-invariant non-linear part 655
8.3.3 The frequency domain conditions for hyperstability of discrete, non-linear control systems containing a time-invariant non-linear part 657

References 705

Index 725

1. Introduction

1.1 Essence of fuzzy set theory

Information accepted by methods based on conventional mathematics must be precise, for example, the speed of a car $v = 111$ (km/h). Such information can be represented graphically by means of the so-called singleton, Fig. 1.1.

Exact information can only be delivered by precision engineered measuring devices, whereas a man can directly estimate the speed of a car by applying such terms as low, medium and high. These imprecise evaluations can also be represented graphically, Fig. 1.2.

The functions "low", "medium" and "high", called **membership functions**, tell us if the given precise speed value is respectively low, medium or high. A man observing a car running at a speed of $v = 111$ (km/h) cannot evaluate its speed exactly, but he can estimate it roughly as a high speed, Fig. 1.2.

Such information can be defined as a **granule of information** (Zadeh 1979,1996). If 3 granules (low, medium, high) are not sufficient, one can increase the precision of evaluation applying, for example, 5 granules (very low, low, medium, high, very high), Fig. 1.3. A man can also decrease the precision of evaluation of the speed using only 2 granules (low, high). The granularity of information applied by a man is variable according to his requirements, mental powers or is otherwise dependent upon the context of information usage.

Information obtained from people is usually of less precision (large granularity), while information delivered by measuring devices is of higher precision

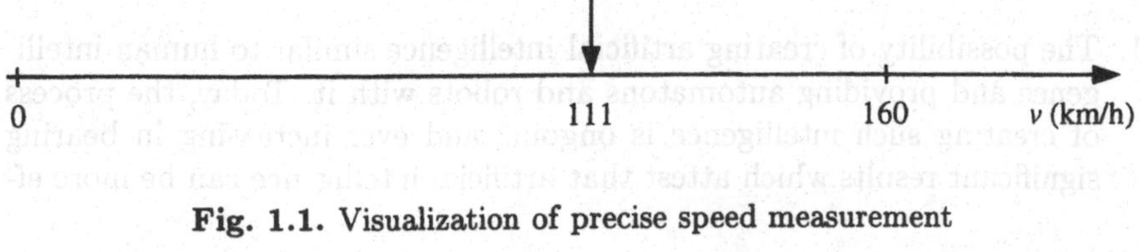

Fig. 1.1. Visualization of precise speed measurement

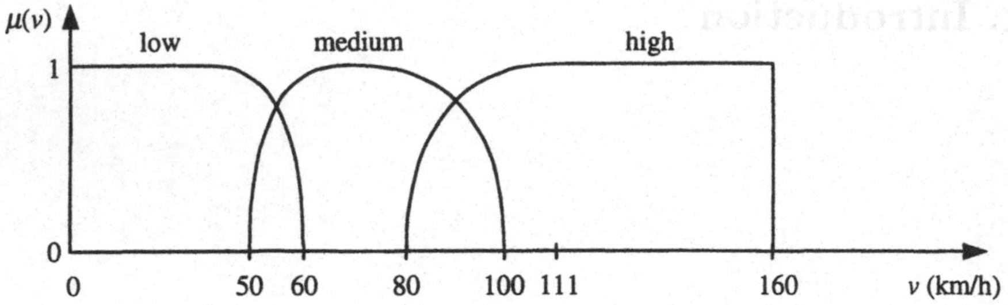

Fig. 1.2. Visualization of imprecise, rough speed evaluation

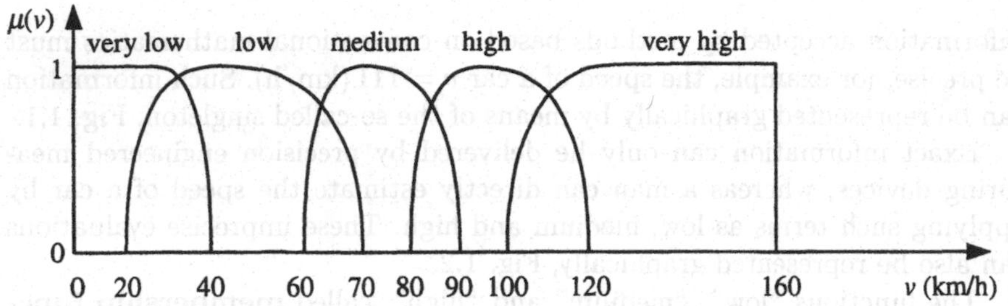

Fig. 1.3. Evaluation of speed using 5 granules of information

(small granularity). The **granularity of information** is defined by the width of a granule (membership function). And so, the granule "medium" can have various widths, according to the total number of granules of information used by a man, Fig. 1.4.

As you can see in Fig. 1.4, by decreasing the granularity of information we approach a limit: a granule of an infinitely small width called the **singleton**, which represents the precise information, i.e. such information which is employed by conventional mathematics.

The information represented by the granule of a finite (greater than zero) width has been called by Prof. Lofti Zadeh, the discoverer and creator of the concept of granularity, **fuzzy information**. The mathematics field using such information has been called **fuzzy set theory** (Zimmerman 1994). The most important element of this theory is **fuzzy logic**, applied for fuzzy modeling and control. For science and technology fuzzy set theory has opened new exploratory possibilities, which are described below.

1. The possibility of creating artificial intelligence similar to human intelligence and providing automatons and robots with it. Today, the process of creating such intelligence is ongoing and ever increasing in bearing significant results which attest that artificial intelligence can be more ef-

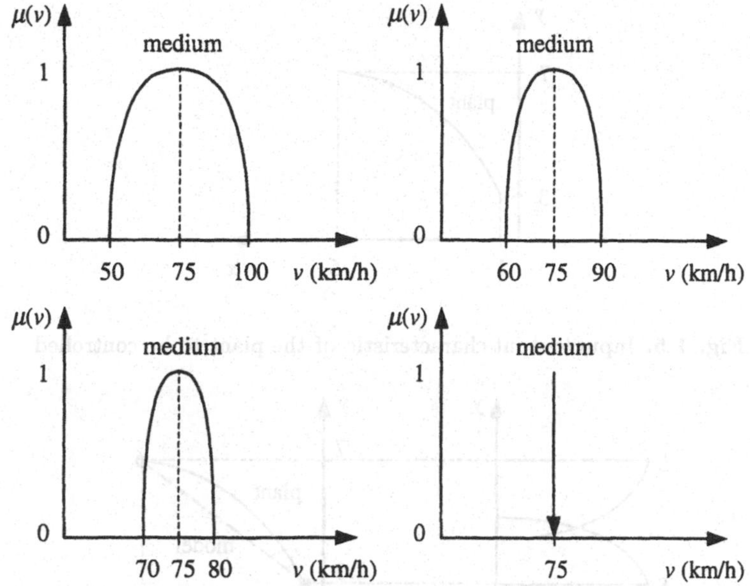

Fig. 1.4. Various width of the granule of information of the "medium" speed

fective than human intelligence in some well defined applications, e.g. in respect of quantity and speed of information processing.

2. The creation of computers programmed with words (Zadeh 1996). The application of such computers in robots and automatons makes it possible to control them and to "communicate with them" by means of human language using fuzzy notions. There presently exist devices for recognizing a limited number of words or word associations.
3. The application of information of any granularity for modeling, control, optimization and diagnostics of systems and objects. The use of greater granularity allows for reducing processed and stored information and for accelerating the operation of algorithms.
4. The possibility of adapting granularity of information according to the required accuracy of modeling, control, optimization, diagnostics, etc. Such adaptation is applied by man. The illustration of this statement is shown in Figs. 1.5 – 1.7.

Assume for the moment that someone controls a plant by realizing the input/output mapping shown in Fig. 1.5. At the outset they will remember the extreme states of the plant and generate in their own mind a model based on two rules given in Fig. 1.6. The model, for argument's sake, is a necessary approximation of the plant.

If the model represented in Fig. 1.6 is insufficiently exact, a man will try to increase its accuracy, bearing in mind the essential (Babuška 1995b), Fig. 1.7,

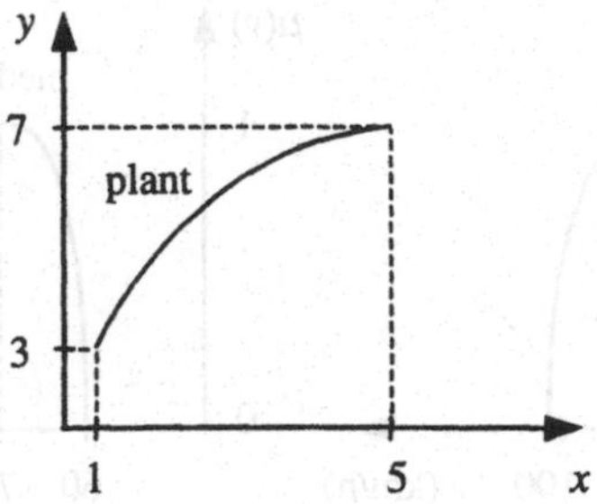

Fig. 1.5. Input/output characteristic of the plant to be controlled

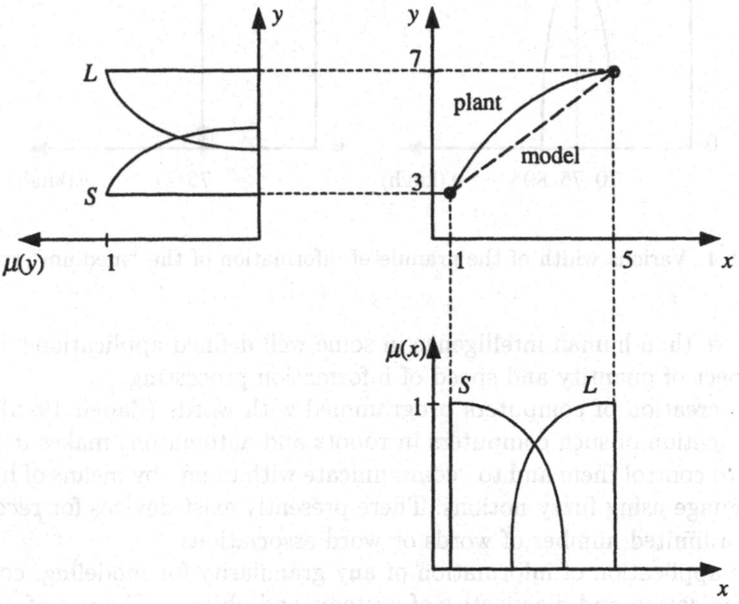

*R*1 : *IF* (x small) *THEN* (y small)
*R*2 : *IF* (x large) *THEN* (y large)

Fig. 1.6. Model of the plant based on two granules of information: small and large

medium state, thereby creating a new rule determining the operation of the plant and progressively introducing new, smaller granules of information. Moreover, if the model represented in Fig. 1.7 proves insufficient, a man can examine the next essential state of the plant, decrease the granularity of information, increase the number of verbal rules characterizing the operation of the plant and subsequently obtain a higher accuracy of modeling.

As psychological studies (Kruse 1994) have shown, an average capable man can remember only 5 to 9 characteristic states of a plant. Therefore, for

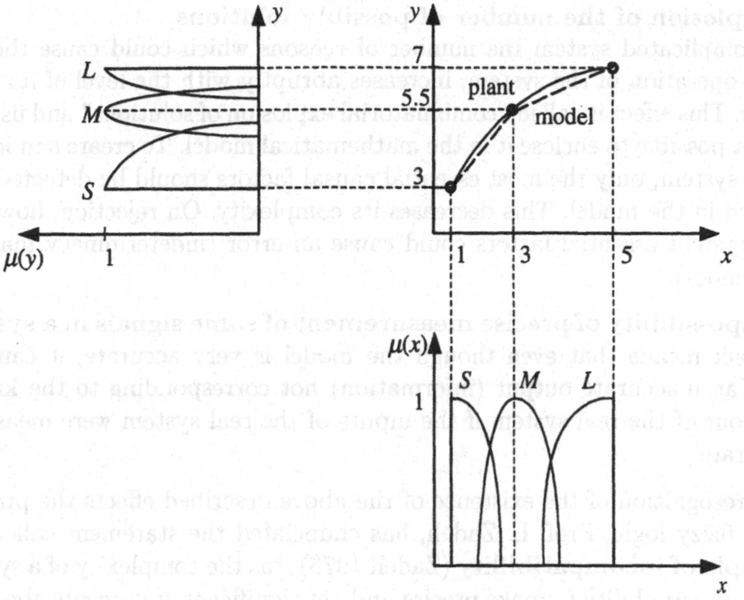

*R*1 : *IF* (*x* small) *THEN* (*y* small)
*R*1 : *IF* (*x* medium) *THEN* (*y* medium)
*R*3 : *IF* (*x* large) *THEN* (*y* large)

Fig. 1.7. Model of the plant based on three granules of information: small, medium, large

each variable, maximally 5 to 9 granules of information are applied. We note that in general, this kind of granularity is totally sufficient for controlling aircraft, vehicles and many other different objects and for solving thousands of everyday problems.

Since computer technology makes it possible to apply any granularity of information practically, we can get models of considerably higher accuracy. Retrospectively, experiments with modeling real world systems show that there are nearly always thresholds to limits of accuracy and exceeding such thresholds should not be rendered as costly objectives. Such situations arise when certain effects occur in complicated systems. They are described as follows.

1. **Existence of chaos**
Within the kernel of systems there are active disturbances which cannot be measured by us, or we don't know that they exist. Also, certain unexamined processes can occur in these systems. Their influences are dependent on magnitude and can render variable unforseeability of the system that can be attributed to chaotic events.

2. **Explosion of the number of possible solutions**
In a complicated system the number of reasons which could cause the observed operation of the system, increases abruptly with the level of its complexity. This effect is called "combinatorial explosion of solutions" and usually it is not possible to enclose it in the mathematical model. To create a model of such a system, only the most essential causal factors should be detected and involved in the model. This decreases its complexity. On rejection, however, less apparent essential factors could cause an error (indeterminacy margin) of the model.

3. **Impossibility of precise measurement of some signals in a system**
This fact means that even though the model is very accurate, it can calculate an inaccurate output (information) not corresponding to the known behaviour of the real system if the inputs of the real system were measured inaccurately.

In recognition of the existence of the above described effects the progenitor of fuzzy logic, Prof. L. Zadeh, has enunciated the statement called the **principle of incompatibillity** (Zadeh 1973): "as the complexity of a system increases, our ability to make precise and yet significant statements about its behaviour diminishes until a threshold is reached beyond which precision and significance (or relevance) become almost mutually exclusive characteristics".

Exact modeling using a very small granule of information is possible in the case of simple systems with a small number of inputs. In non-trivial systems, especially those having a greater number of inputs, we are enforced to apply information of a larger granule (fuzzy information).

1.2 Development of fuzzy set theory

Today fuzzy set theory attracts a high level of interest. In 1993 the number of related publications had been estimated at 15,000 to 16,000 (Altrock 1993). Now, in 2000, this number exceeds 27,000 and is increasing rapidly. As scientific conferences are organized there parallels an increase in the number of industrial applications. What causes are endemic to the great popularity of fuzzy set theory in the world of present-day science?

The development of fuzzy set theory was initiated by the American professor, Lofti Zadeh, who discovered the existence of fuzzy sets and in his seminal paper "Fuzzy Sets" (Zadeh 1965), presented the idea and first conception of the theory which has created possibilities for the fuzzy description of real systems. The main division of fuzzy set theory is **fuzzy logic** (Zimmermann 1994a) applied for prototype modeling and mainstream control of systems.

In the sixties the period of rapid development of computers and digital technology based on binary logic began. Generally, there was optimism that this logic would empower us to solve most technical and scientific problems. For that reason the rise of fuzzy logic went by almost unnoticed despite the fact that its conception can be acknowledged as a turning point. Nevertheless, some scientific circles understood its importance, developed it and brought it to fruition within the framework of industrial application. Some time later, interest in that school of thought on the part of binary technology advocates increased, because it has become evident that traditional models and mathematical methods were unable to solve many practical problems in spite of an enormous rate of calculation. A new methodology was needed whose requisite features are to be found in fuzzy logic.

As with robotics, fuzzy logic has met with greater interest outside its country of origin, the USA. And so, the first acknowledged application is the use of it in Europe, in the control of the steam generating system in a power plant, as realized by Assilian (Assilian 1974). The steam generator has been proved such a complicated non-linear system, that the application of various traditional, often very sophisticated methods have not given a solution for the problem of its control. Only fuzzy logic has allowed the synthesis of a controller satisfying all requirements. In 1976 it was applied in the automatic control system of a rotary furnace for cement production (Mamdani 1977). Even so, the first European and American applications did not result in any major change of interest in fuzzy logic. Similarly as in the case of robotics, the change has occurred in Japan, the first nation to become fully aware of the enormous potential of fuzzy logic and to introduce it widely (Bellon 1992).

The best known Japanese application of fuzzy logic was in the control system of the Sendai underground railway, utilized by the Hitachi company. The project was undertaken with the cooperation of an experienced driver whose knowledge provided parameters for its design. This system automatically decreased the speed of a train on entering a station, ensuring that the train stopped at a predetermined place. It also had the benefit of being a highly comfortable ride through mild acceleration and braking (Abel 1991). There were many other advantages in comparison with traditional control methods.

Tests and improvements of the system were carried out for two years, and their purpose was to check the new control method and to secure the greatest margin of safety for passengers. The success of the Sendai underground railway was such that 12 months later, 50 large Japanese companies worked on the development of their own fuzzy logic applications. In 1991 the contribution of Japan to the world production of articles based on this logic was estimated at billions of dollars. In absolute numbers this amounted to 80 per cent, according to the data of Market Intelligence Research. Starting from 1989, at least 5 scientific societies engaged in fuzzy logic have been established in Japan. They are:

1. Laboratory for International Fuzzy Engineering Research (LIFE),
2. Japan Society of Fuzzy Theory and Systems (SOFT),
3. Biomedical Fuzzy Systems Association (BMFSA),
4. Fuzzy Logic Systems Institute Iizuka (FLSI),
5. Center for Promotion of Fuzzy Logic.

The Japanese division of the international organization IFSA (International Fuzzy Systems Organization) has been carrying out its activity in Japan since 1986. Of these institutions, the most well known organization is LIFE.

The institute has been organized by the Japanese Ministry of International Trade and Industry and large industrial enterprises (49 in 1991), such as Honda, Kawasaki Steel, Tokyo Electric and others. Their purpose was to work out fuzzy methods for the needs of industry, trade, and all purpose decision making (for example, currency operations) etc. The best specialists in the field of fuzzy logic from Japanese universities and enterprises have been engaged at LIFE. In addition, large companies from outside Japan, for example Bosh, Zeiss, Siemens, Audi and Volkswagen have begun to sponsor the institute. The sponsors of LIFE send their engineers to the institute where they participate in training programs and carry out research under the guidance of specialists.

The fast development of fuzzy logic in Japan has resulted in practical applications, not only in industry but also in the manufacture of commonly used products. The Camcorder (video camera), for instance, was provided with a fuzzy picture stabilizer (Abel 1991). This was implemented in order to compensate for undesirable movements due to unskilled operator handling. The problem was difficult to solve with traditional methods because very often mobile objects to be filmed, such as people in motion, must be distinguished from induced disturbances. Another case is that of the automatic washing machine, controlled with only one push-button action (Zimmerman 1994). It's holistic nature aroused interest and met with approbation. Fuzzy control enabled the reduction of the number of push-buttons to one, resulting in considerable simplification of its operation and also in optimization of the washing process via recognition of type, quantity and grade of soiled clothing. Japanese companies have applied their fuzzy ingenuity to many other products such as the microwave cooker (Sanyo), anti-block system and automatic gear (Nissan), integrated control of running dynamics of a car (INVEC) and hard disk controllers in computers (reducing access time to the information to be searched).

Japanese engineers established an enormous number of patents in this field using their cutting edge position in research and applications of fuzzy logic. One company alone, Omron (Kyoto), was the owner of more than 700 patents in 1993.

Mass application of fuzzy logic in Japanese products became subject to worldwide scrutiny, especially in Europe, where mainly German industrialists and scientists resolved to oppose Japanese competitive dominance. The European foundation known as ELITE (European Laboratory for Intelligent Techniques Engineering Foundation) is based in Aachen, wherein development and promotion of artificial intelligence methods such as fuzzy logic and neural networks proceed with an emphasis on development of knowledge and research. The foundation organizes many international conferences, among them the yearly European conference EUFIT (European Congress on Intelligent Techniques and Soft Computing) which is dedicated to artificial intelligence.

Apart from the above-mentioned applications, the early nineties saw intensive fuzzy development in many miscellaneous secondary fields as well as non- technical ones. In the following list, a selection of applications should enable you to evaluate some of the possibilities offered by fuzzy logic. They are:

- artificial pacemaker control system (Akaiwa 1990; Kitamura 1991; Sugiura 1991),
- automotive vehicle control system (Altrock 1992),
- water boiler (Bien 1992),
- chemical reactors and devices (Altrock 1995; Bork 1993; Hanakuma 1989; Häck 1997; Höhmann 1993; Kolios 1994; Roffeld 1991),
- cooling systems (Becker 1994; Hakata 1990),
- conditioning and ventilation equipment (Tobi 1991; Watanabe 1990),
- refuse incinerating equipment (Altrock 1993; Fujiyoshi 1992; Ohnishi 1991),
- glass smelting furnace (Aoki 1990; Hishida 1992),
- blood pressure control system (Arita 1990),
- tumor diagnostics devices (Arita 1991),
- heart diseases warning system (Altrock 1993),
- crane or overhead crane control system (Altrock 1993; Watanabe 1991),
- pumping station (Chen 1992),
- picture processing (Fijiwara 1991; Franke 1994),
- fast battery charger (Altrock 1993),
- word recognition (Fujimoto 1989),
- diabetic therapy and blood glucose level control (Jacoby 1994; Kageyama 1990),
- power system (Hiyama 1991),
- metalworking equipment (Hsieh 1994),
- bioprocessor control (Hanss 1994),
- heating devices (Heider 1994),
- electric motor control (Kawai 1990; Lee 1992),
- welding equipment and processes (Murakami 1989; Reshuffled 1994),
- traffic control (traffic systems) (Sasaki 1988; Voit 1994),
- biomedicine (Takahashi 1990),
- room cleaning equipment (Yamashita 1992),
- desludging equipment (Yu 1990),
- water cleaning equipment (Altrock 1995).

Many books on the subject of fuzzy sets theory, e.g. (Altrock 1993,1995; Brown 1994; Bezdek 1981; Driankov 1993,1996; Gottwald 1993; Hung 1995; Kahlert 1994,1995; Knappe 1994; Kandel 1994; Kruse 1994; Kiendl 1997;

Kaufmann 1985; Koch 1996; Kacprzyk 1986,1992,1997; Nguyen 1995; Pedrycz 1993; Rutkowska 1997; Tilli 1991; Wang 1994a; Yager 1994,1995; Zimmermann 1994a,1994b), have already been published.

Several ready-for-use programs facilitating fuzzy modeling and control have been produced and are in distribution. Information about them can be found in (Ader 1996; Baldwin 1995a; Koch 1996; Kuhn 1994; Krieger 1994; Krone 1996c).

In Poland, initial research into fuzzy sets was carried out in the seventies (Kacprzyk 1977,1978). Amongst Polish scientists, Prof. E. Czogala, Prof. J. Kacprzyk and Prof. W. Pedrycz (listed in strict alphabetical order of surname) made significant contributions to worldwide development of the theory.

Although fuzzy set theory allows for the solving of problems which conventional methods could not cope with, it should not be considered as the only solution to the exclusion of all others. One would be mistaken were they to believe it could replace all known methods. Experience and practice have shown that fuzzy logic shall mainly be applied where hitherto known methods fail (Altrock 1993). If conventional methods yield good results they should be pursued.

2. Basic Notions of Fuzzy Set Theory

2.1 Fuzzy sets

Fuzzy sets are generally applied by people for qualitative evaluation of physical quantities, states of plants and systems and for comparison to each other. Each man can evaluate the height of temperature without using a thermometer on the basis of his own feelings, in the coarse scale, Fig. 2.1.

It should be noticed that quality evaluation is not quantity measure because the feature of these measures is their additivity.

Example:

1 (cm) + 1 (cm) = 2 (cm),

but: a small sum of money + a small sum of money = ?

The result of the operation doesn't always mean a large sum of money.

The notions of the "small" and "large" sum are fuzzy and subjective, depending on the person expressing them. Therefore quality evaluations can not be added up as quantity measures.

A man also applies a quality evaluation when he has accurate measuring instruments at his disposal. For example, the speedometers mounted in cars

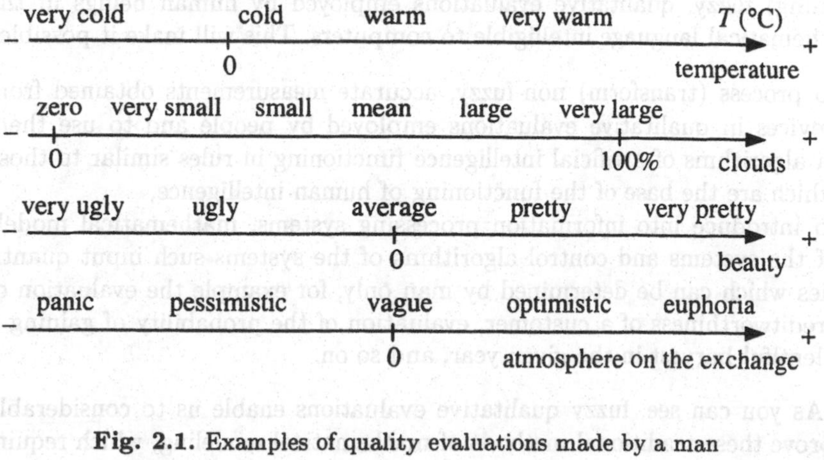

Fig. 2.1. Examples of quality evaluations made by a man

indicate the actual speed of the car. But drivers reporting their journeys speak mostly of:

- "I drove very fast",
- "I drove at a rate of approximately 100 (km/h)",
- "I drove at a rate of above 100 (km/h)".

The driver could try to remember the exact speed of his car per each successive second of his journey. However, this would be firstly: practically unfeasible, considering the limited capabilities of the human memory, and secondly: completely superfluous, since rough estimation of speed is sufficient for our needs, and allows us to reject the large amount of needless information and to limit our attention to the most essential information which can be processed quickly in order to come to a necessary decision.

In the world where we live there is a great number of such quantities which can not be evaluated with the use of measuring devices because no such devices have been constructed so far. These quantities are for example: beauty of a woman, order in a flat, danger of outbreak of war, the chance to meet with success in business, etc. But each person has his/her own inscrutable or only partly recognized "measuring instruments" which enable him/her to carry out qualitative measurements of such quantities and states which seem to be the most difficult, and present-day science has no means to cope with them. By employing such rough, fuzzy estimations a man can manage reality in his surroundings excellently, adapt to it, transform it, recognize (identify) systems existing in it, and then control them optimally or suboptimally.

Employing qualitative evaluations of reality man has developed in himself an excellently functioning logic and intelligence which robots do not possess, though intensive work is still carried on in this direction. For these reasons, scientists and engineers have conceived the idea of creating artificial intelligence which would imitate human intelligence and employ similar methods.

The first condition for creating such intelligence is formulating (transforming) fuzzy, qualitative evaluations employed by human beings in the mathematical language intelligible to computers. This will make it possible:

- to process (transform) non-fuzzy, accurate measurements obtained from devices in qualitative evaluations employed by people and to use them in algorithms of artificial intelligence functioning in rules similar to those which are the base of the functioning of human intelligence,
- to introduce into information processing systems, mathematical models of the systems and control algorithms of the systems such input quantities which can be determined by man only, for example the evaluation of creditworthiness of a customer, evaluation of the probability of gaining a plentiful harvest in the given year, and so on.

As you can see, fuzzy qualitative evaluations enable us to considerably improve these traditional methods of mathematical modeling, which require exact information about input values of a system. This should be effected through introducing information about those inputs of a system which could not be included because of lack of instruments enabling measurement of them (hybrid models consisting of non-fuzzy and fuzzy parts). Therefore, the fuzzy methods of the qualitative evaluation are not competitive against technical, exact measurements, but they are their complement enabling us to create a better picture (model) of reality.

The qualitative evaluation can be realized on the basis of the fuzzy set theory. In 1965 the concept of the fuzzy set was defined and introduced into scientific literature by the American scientist Lofti Zadeh (Zadeh 1965), who rendered extensive services in the development of the fuzzy set theory.

In the following, the basic concepts enabling us to understand and define a fuzzy set will be presented.

- **Linguistic variable**

The linguistic variable is an input, output or state variable, which we intend to evaluate applying linguistic evaluations called linguistic values.

Example: speed of a ship, voltage, and temperature.

In practical applications the linguistic variables are evaluated not only by means of linguistic values but also with the use of fuzzy numbers (Bertram 1994; Koch 1993), that is to say: in a mixed way.

- **Linguistic value**

The linguistic value is the verbal evaluation of a linguistic variable.

Example: very large negative, medium negative, medium positive, very large positive, old, young, pretty, average, nice, unpleasant, true, false.

Linguistic values appear in models together with linguistic variables to which they refer.

Example: high air pressure, strong water stream, young age (of a man), true information, and false information.

- **Fuzzy numbers**

The concept of fuzzy numbers will be explained in Chapter 3.

Examples of fuzzy numbers: approx. zero, approx. 5, more or less than 5, a little more than 9, approximately between 10 and 12.

If the variables of a system can be evaluated by a man using linguistic values with no information coming from technical measuring devices, on the basis of his own feelings only, then for the evaluation of variables employing fuzzy numbers technical measurements are necessary. Fuzzy numbers enable us to generalize the large amount of precise information observed by a man with measuring devices or coming from data bases, e.g. price of enterprise shares X_i, Fig. 2.2.

Information represented precisely (in a non-fuzzy way) in Fig. 2.2 can be generalized by means of a fuzzy number:

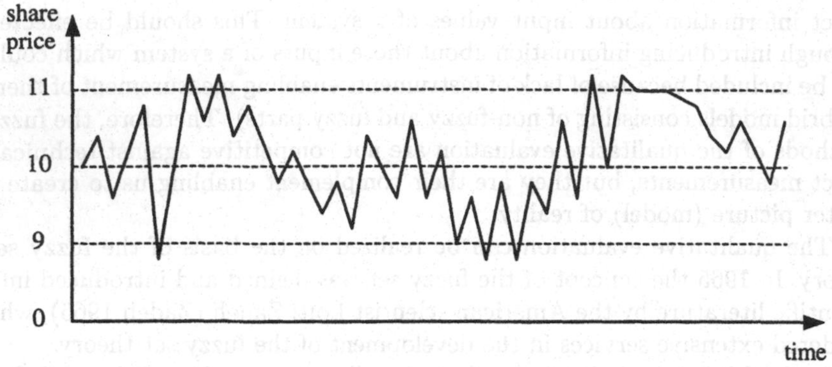

Fig. 2.2. Example of precise evaluation of a variable described by a great amount of information

- "approximately between 9 and 11" or
- "about 10".

In practice, for example (Abel 1991; Koch 1993), the mixed evaluation of the linguistic variables, e.g. with the use of the following scale, is applied:

- negative, **about zero**, positive,
- large negative, medium negative, small negative, **about zero**, small positive, medium positive, large positive.

- **Linguistic term-set of a variable**

The linguistic term-set of a variable is the set of all linguistic values applied for the evaluation of a given linguistic variable. The linguistic term-set is sometimes called the basic linguistic set (Bertram 1994), linguistic domain or linguistic universe (space) of discourse. It will be denoted with the capital Latin letters, e.g.

$$X_L = \{negative,\ positive\} = \{x_{L1},\ x_{L2}\}\ ,$$
$$Y_L = \{small,\ medium,\ large\} = \{y_{L1},\ y_{L2},\ y_{L3}\}\ .$$

The linguistic universe is a finite-dimensional set.

- **Universe of discourse of a variable**

The universe of discourse of a variable is the set of all numerical values which can actually be assumed by the variable in the system under examination or also such values which are essential for the problem to be solved (for the model of the system). The universe of discourse is also called :

- space of discourse (Bertram 1994),
- field of discourse (Abel 1991),
- space (Kacprzyk 1986; Yager 1994,1995),

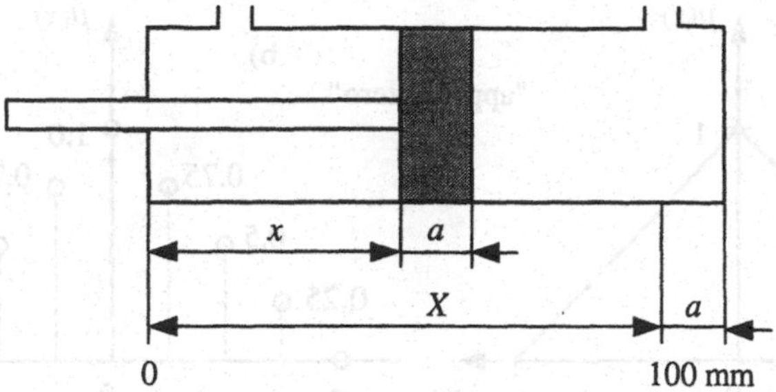

$$X = \{x:\ x \in R,\ 0 \le x \le 100\,(\text{mm})\}$$

Fig. 2.3. Continuous numerical range of positions of the piston x

- set (Kacprzyk 1986),
- domain of discourse,
- domain (Yager 1994,1995),
- basic range (Knappe 1994),
- reference set (Kruse 1994).

The adjective "numerical" enables us to distinguish it from the linguistic term-set. The universe of discourse of a variable will be denoted with the capital Latin letters, e.g.:

$X = \{x\}$ – infinite-dimensional (continuous) universe,
$X = \{x_1,\ x_2,\ \ldots,\ x_n\}$ – finite-dimensional, discrete universe.

Example, Fig. 2.3.
Example: the discrete universe of discourse.

$$X = \{x_1 = -1,\ x_2 = -0.75,\ \ldots,\ x_8 = 0.75,\ x_9 = 1\}$$

- **The power of the numerical universe of discourse**

The power of the numerical universe of discourse is the number of its elements.

$$\|X\| = n \tag{2.1}$$

- **Fuzzy set**

The fuzzy set A, in the certain numerical universe of discourse X, is a set of pairs:

$$A = \{(\mu_A^*(x), x)\}, \quad \forall x \in X, \tag{2.2}$$

where: μ_A is the membership function of the fuzzy set A, which assigns to each element $x \in X$ the grade of its membership μ_A^* in the fuzzy set A, considering that:

$$\mu_A(x) \in [0,1].$$

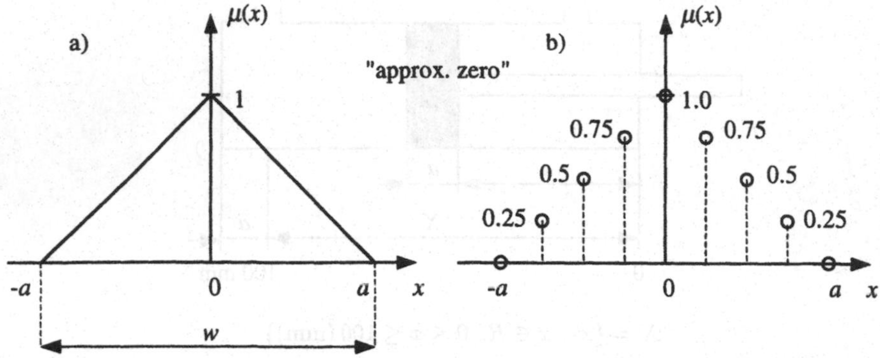

Fig. 2.4. Continuous (a) and discrete (b) graphical forms of the membership function of the fuzzy number "approx. zero"

The membership function maps the numerical universe X of a given variable in the interval [0,1]:

$$\mu_A : X \to [0,1] .$$

The concept of the fuzzy set enables us to formulate mathematically a notation of linguistic values and fuzzy numbers applied by people.

- **The power of a fuzzy set**

The power of a fuzzy set is defined as the number of pairs $(\mu_A^*(x), x)$ enclosed in the set.

$$\|A\| = n$$

The power of the fuzzy set A is equal to the power of the universe X of this set.

- **The membership function and grade of membership**

The **membership function** assigns to each element x of a given variable a certain value from the interval [0,1]:

$$\mu_A(x) : X \to [0,1], \quad \forall x \in X . \tag{2.3}$$

This value, called the **grade of membership**, informs us to which grade the element x belongs in the fuzzy set A. The membership function can be represented in the form of a :

- diagram (continuous or discrete diagram),
- mathematical formula,
- table,
- membership vector,
- sum or integral.

To represent the membership function in the form of a formula it is advantageous to introduce a logical variable w relating to the level of the variable x:

$$w = \begin{cases} 1 & \text{for } -a \leq x \leq a \\ 0 & \text{otherwise.} \end{cases} \tag{2.4}$$

Then the membership function represented in Fig. 2.4 can be written down in the following form:

$$\mu(x) = w\left(\frac{a-|x|}{a}\right) . \tag{2.5}$$

The discrete membership function can be represented in the form of Table 2.1.

Table 2.1. Example of the tabular membership function

$x \in X$	$x_1=a$	$x_2=-0.75a$	$x_3=-0.5a$	$x_4=-0.25a$	$x_5=0$	$x_6=0.25a$	$x_7=0.5a$	$x_8=0.75a$	$x_9=a$
$\mu_A(x)$	0.00	0.25	0.5	0.75	1.00	0.75	0.5	0.25	0.00

Note: Not only numbers, but also objects, persons and abstracts can be the elements x in the table. The table can inform, for example, about the membership of various companies in set A, which include prosperous firms, Table 2.2.

Table 2.2. Tabular membership function in the set of prosperous firms

$x \in X$	Firm 1	Firm 2	...	Firm $(n-1)$	Firm n
$\mu^*_A(x)$	0.4	0.5	...	1.00	1.00

If the order (sequence) of all n elements x_i in the universe of discourse X is fixed exactly, then the membership function can be given as the membership vector V_A (2.6):

$$V_A = \{\mu_A(x_1),\ \mu_A(x_2),\ \ldots,\ \mu_A(x_n)\} . \tag{2.6}$$

Example:

$$V_A = \{0.00,\ 0.25,\ 0.50,\ 0.75,\ 1.00,\ 0.75,\ 0.50,\ 0.25,\ 0.00\} .$$

Also used is the notation of the discrete fuzzy set A in the form of the sum (Zimmermann 1994a):

$$A = \frac{\mu_A(x_1)}{x_1} + \frac{\mu_A(x_2)}{x_2} + \ldots + \frac{\mu_A(x_n)}{x_n} = \sum_{i=1}^{n} \frac{\mu_A(x_i)}{x_i}. \tag{2.7}$$

The above notation means that the set A is the set-type sum (and not the arithmetic sum) of pairs $(\mu_A(x)/x)$.

Example:

$$A = 0.00/\text{-a} + 0.25/\text{-}0.75\text{a} + 0.50/\text{-}0.5\text{a} + 0.75/\text{-}0.25\text{a} + 1.00/0 + 0.75/0.25a + 0.50/0.5a + 0.25/0.75a + 0.00/a.$$

The continuous fuzzy set can be (Zimmermann 1994a) expressed in the form of the integral:

$$A = \int_X \frac{\mu_A(x)}{x}. \tag{2.8}$$

The above notation means that the fuzzy set A is the set-type integral of pairs $(\mu_A(x)/x)$.

Example: "Real numbers - approx. zero", Fig. 2.4.

$$A = \int_X w\left(\frac{a-|x|}{a}\right)/x \tag{2.9}$$

In practice, the elements x_i having the membership grade equal to zero are omitted in the notation of the membership function.

- **Empty fuzzy set**

The set A whose membership function $\mu_A(x)$ has the value of 0 for all elements of the universe of discourse X is called the empty set and denoted with the symbol $\emptyset$.

$$\emptyset : \mu_\emptyset(x) = 0, \quad \forall x \in X \tag{2.10}$$

- **Universal fuzzy set**

The fuzzy set where all elements of the universe of discourse have the grade of membership equal to 1 is called the universal set (Knappe 1994) and denoted with U.

$$U : \mu_U(x) = 1, \quad \forall x \in X \tag{2.11}$$

Both the empty set $\emptyset$ and universal set U are extreme sets. The relation:

$$\emptyset \le A \le U \tag{2.12}$$

applies to each fuzzy set A.

- **Normal fuzzy sets**

The range of the membership function need not be limited to the values between 0 and 1. It is possible to be within this range when only set-type

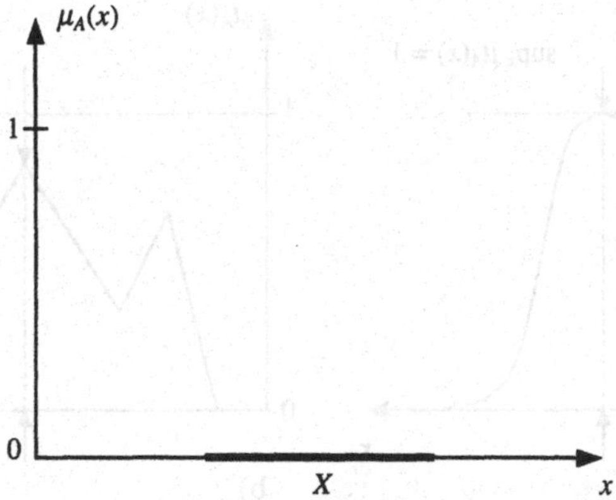

Fig. 2.5. Empty fuzzy set

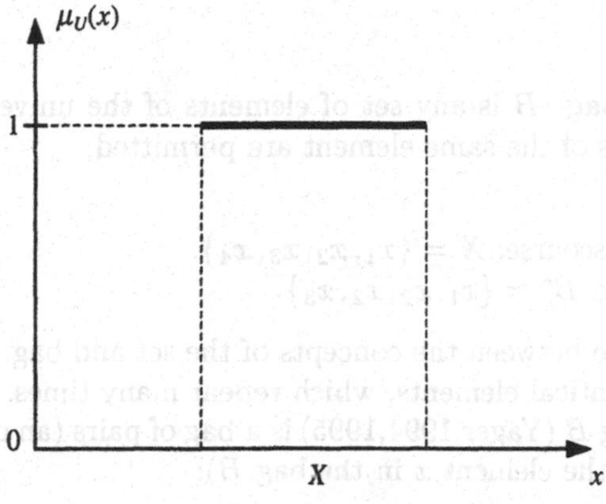

Fig. 2.6. Universal fuzzy set U

operations are carried out on fuzzy sets. If arithmetical operations are also carried out, then values greater than 1 can be obtained as a result of them.

If the maximal value of membership in a set will be denoted as $\sup_x \mu_A(x)$, then each non-empty fuzzy set A can be **normalized** (Knappe 1994; Zimmermann 1994a) to the set A_n in the interval of 0 to 1, by dividing the primary membership function by its maximal value.

$$\mu_{An} = \frac{\mu_A(x)}{\sup_x \mu_A(x)} \tag{2.13}$$

Normal (normalized) **fuzzy sets** are sets having a membership function which assumes a value between 0 and 1 (including 1).

Subnormal fuzzy sets are the sets whose maximal value of the membership function is less than 1. Subnormal sets arise as a result of various operations executed on normal sets.

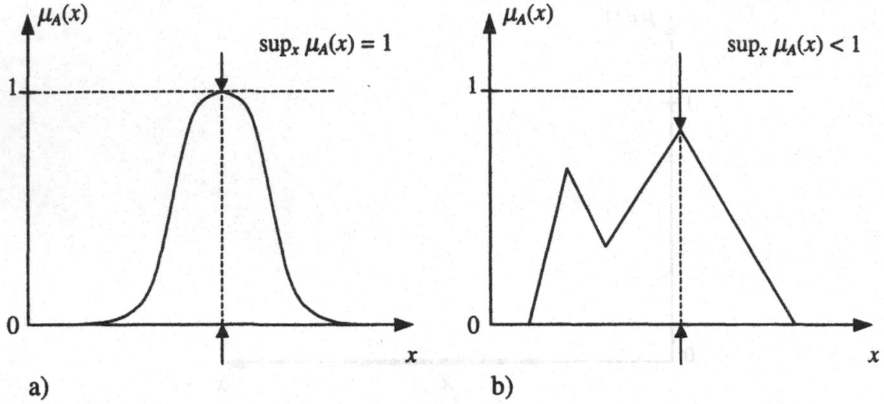

Fig. 2.7. Example of the normal set (a) and the subnormal set (b)

- **Bag**

The **non-fuzzy bag** B is any set of elements of the universe of discourse X. Multiple copies of the same element are permitted.

Example:
The universe of discourse: $X = \{x_1, x_2, x_3, x_4\}$.
The non-fuzzy bag: $B^* = \{x_1, x_2, x_2, x_3\}$.

The **difference** between the concepts of the set and bag: in the set there must not exist identical elements, which repeat many times.

The **fuzzy bag** B (Yager 1994,1995) is a bag of pairs (an element x, grade of membership of the element x in the bag B):

$$B = \{(x, \mu_B^*(x)), \quad \forall x \in X\} . \tag{2.14}$$

Example:

$$B = \left\{ \frac{0.7}{x_1}, \frac{0.9}{x_2}, \frac{0.6}{x_2}, \frac{0.5}{x_3} \right\} .$$

Fuzzy bags arise as a result of the execution of arithmetical (not set-type) operations on fuzzy sets, e.g. of the operation of a summation of several fuzzy sets (Yager 1994,1995). Because the same element can be repeated in the bag many times, the total grade of its membership (in the arithmetical and not set-type meaning) can be greater than 1.

Examples of fuzzy sets are shown in Figs. 2.8 – 2.10.

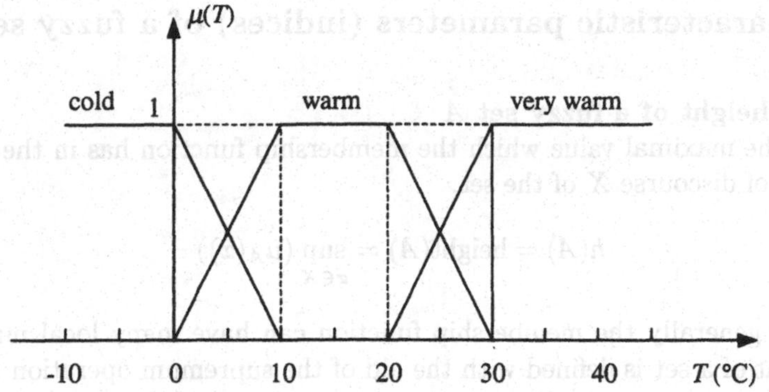

Fig. 2.8. A possible form of the fuzzy sets "cold", "warm" and "very warm" allowing for the qualitative evaluation of temperature

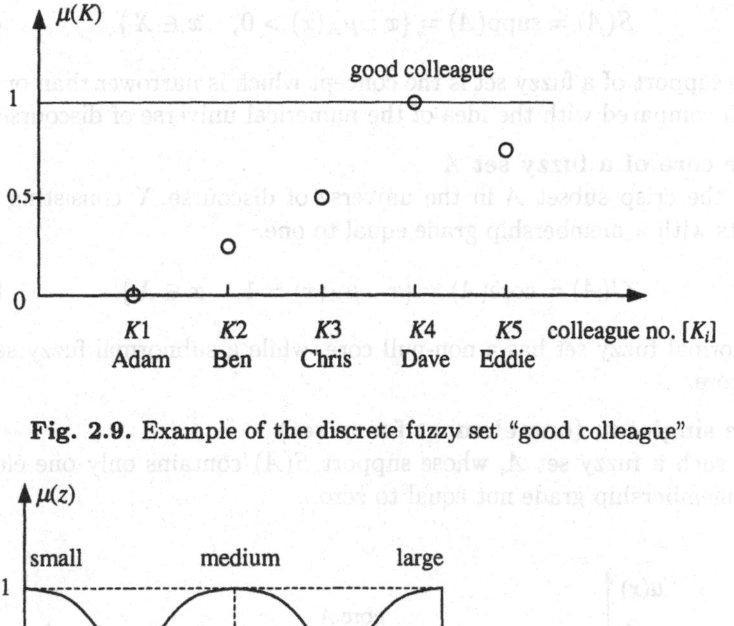

Fig. 2.9. Example of the discrete fuzzy set "good colleague"

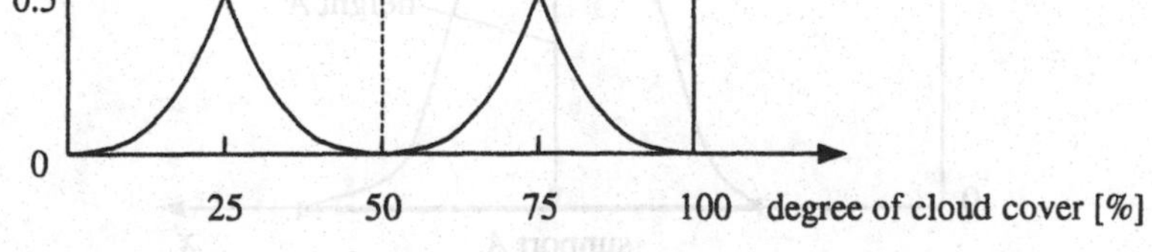

Fig. 2.10. Examples of fuzzy sets used for the rough evaluation of grades of cloud cover z

2.2 Characteristic parameters (indices) of a fuzzy set

- **The height of a fuzzy set A**

This is the maximal value which the membership function has in the whole universe of discourse X of the set.

$$h(A) = \text{height}(A) = \sup_{x \in X} (\mu_A(x)) \tag{2.15}$$

Since generally the membership function can have many local maxima, the height of a set is defined with the aid of the supremum operation (2.15).

- **The support of a fuzzy set A**

This is the crisp subset of the set A whose elements all have non-zero membership grades in the set A.

$$S(A) = \text{supp}(A) = \{x : \mu_A(x) > 0, \quad x \in X\} \tag{2.16}$$

The support of a fuzzy set is the concept which is narrower than or equal to when compared with the idea of the numerical universe of discourse.

- **The core of a fuzzy set A**

This is the crisp subset A in the universe of discourse X consisting of all elements with a membership grade equal to one.

$$C(A) = \text{core}(A) = \{x : \mu_A(x) = 1, \quad x \in X\} \tag{2.17}$$

A normal fuzzy set has a non-null core, while a subnormal fuzzy set has a null core.

- **The singleton (one-element fuzzy set)**

This is such a fuzzy set A, whose support $S(A)$ contains only one element with a membership grade not equal to zero.

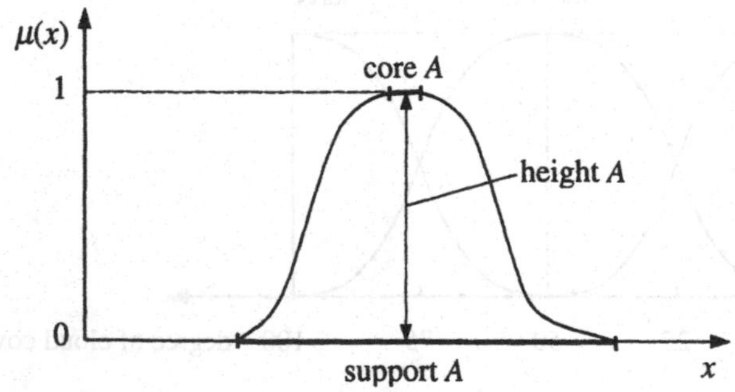

Fig. 2.11. Characteristic parameters of a fuzzy set

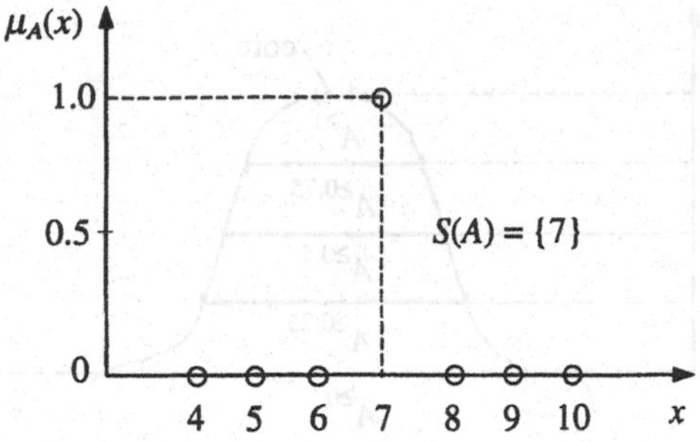

Fig. 2.12. Singleton – the one-element fuzzy set

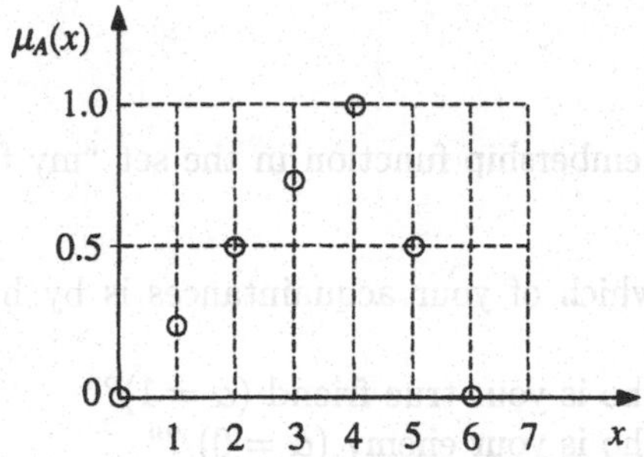

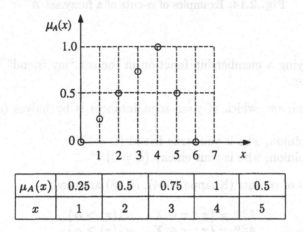

$\mu_A(x)$	0.25	0.5	0.75	1	0.5
x	1	2	3	4	5

Fig. 2.13. Examples of vertical representation of a fuzzy set (a discrete version)

- **Vertical representation of a fuzzy set**

The vertical representation of a fuzzy set A is the form of presentation of a fuzzy set as a set of pairs (an element x in the set A, membership grade of the element x in the set A). This form is most often used for the representation of fuzzy sets (Kruse 1994), Fig. 2.13.

- **Horizontal representation of a fuzzy set**

The horizontal representation of a fuzzy set A consists in presenting the fuzzy set with the aid of the so-called α-cuts A^α of this set, Fig. 2.14.

The concept of α-cuts is applied, because it makes it easier to identify the membership function with the method of knowledge acquisition from experts of the system (i.e. the expert method). If an expert of the system can not specify explicitly how big is the membership of individual elements of the universe of discourse in a fuzzy set, it is possible to ask a question that can be (usually) answered more easily, such as: "Which elements of the universe of discourse belong to a fuzzy set with at least a grade of α?"

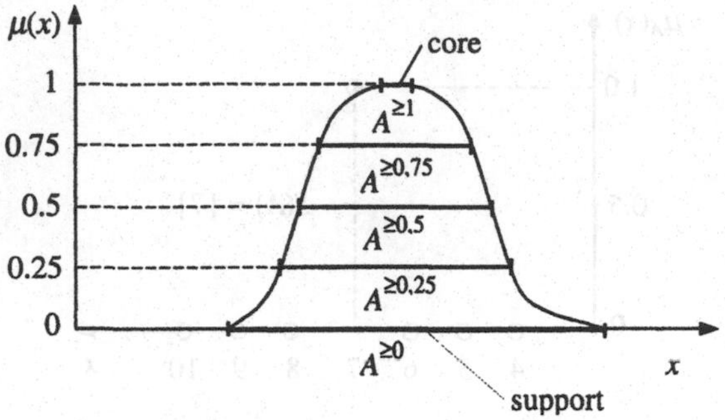

Fig. 2.14. Examples of α-cuts of a fuzzy set A

Example
When identifying a membership function in the set "my friend" we can ask a question like:

- "In your opinion, which of your acquaintances is by halves ($\alpha > 0.5$) a friend?"
- "In your opinion, who is your true friend ($\alpha = 1$)?"
- "In your opinion, who is your enemy ($\alpha = 0$) ?"

Two types of α- cuts (Knappe 1994), (2.18) are used:

$$\begin{aligned} A^{>\alpha} &= \{x : x \in X, \quad \mu_A(x) > \alpha\} , \\ A^{\geq\alpha} &= \{x : x \in X, \quad \mu_A(x) \geq \alpha\} . \end{aligned} \tag{2.18}$$

The α-cut, for $\alpha = 0$ corresponds with the support of the set $S(A)$, and for $\alpha = 1$ corresponds with the core of the set $C(A)$, Fig. 2.14.

On the basis of a sufficient number of α-cuts the membership function of a fuzzy set can be reproduced with required accuracy. In the case of discrete fuzzy sets the number of necessary cuts is finite, while in case of continuous fuzzy sets it is theoretically infinitely large, but practically finite.

When the membership of elements of the universe of discourse in particular α- cuts is known, we can define the approximate membership function $\mu_A^*(x)$ in the fuzzy set A:

$$\mu_A^*(x) = \sup_{\alpha\in[0,1]} (\alpha \cdot \mu_{A^{>\alpha}}(x)) ,$$

or

$$\mu_A^*(x) = \sup_{\alpha\in[0,1]} (\alpha \cdot \mu_{A^{\geq\alpha}}(x)) . \tag{2.19}$$

Example 2.2.1
Consider a given set $A = \{\frac{0.1}{1}, \frac{0.4}{2}, \frac{0.9}{3}, \frac{1.0}{4}, \frac{0.8}{5}, \frac{0.6}{6}, \frac{0.3}{7}, \frac{0.1}{8},\}$, Fig. 2.15.

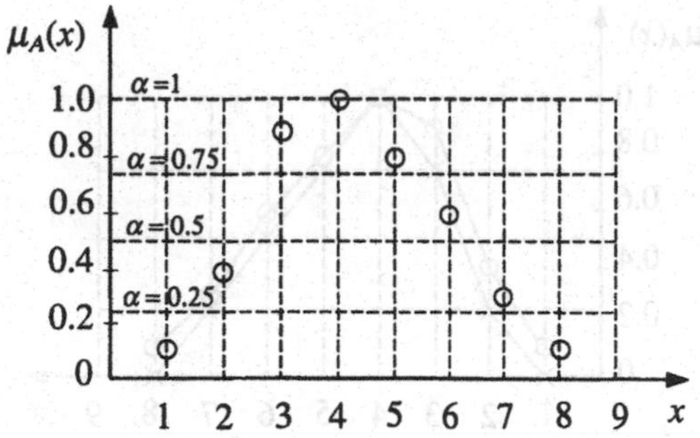

Fig. 2.15. Discrete membership function of the fuzzy set A

Cuts $A^{\geq\alpha}$:

$$A^{\geq 1} = \left\{\frac{1}{4}\right\},$$
$$A^{\geq 0.75} = \left\{\frac{1}{3}, \frac{1}{4}, \frac{1}{5}\right\},$$
$$A^{\geq 0.5} = \left\{\frac{1}{3}, \frac{1}{4}, \frac{1}{5}, \frac{1}{6}\right\},$$
$$A^{\geq 0.5} = \left\{\frac{1}{3}, \frac{1}{4}, \frac{1}{5}, \frac{1}{6}\right\},$$
$$A^{\geq 0.25} = \left\{\frac{1}{2}, \frac{1}{3}, \frac{1}{4}, \frac{1}{5}, \frac{1}{6}, \frac{1}{7}\right\},$$
$$A^{\geq 0} = \left\{\frac{1}{1}, \frac{1}{2}, \frac{1}{3}, \frac{1}{4}, \frac{1}{5}, \frac{1}{6}, \frac{1}{7}, \frac{1}{8}\right\}.$$

The membership grades in individual α-cuts can only be either 1 or 0. The reproduction of the set A on the basis of its α-cuts is:

$$\mu_A^*(x) = \sup_{\alpha\in[0,1]} \left(1 \cdot \left(\frac{1}{4}\right) + 0.75 \cdot \left(\frac{1}{3}, \frac{1}{4}, \frac{1}{5}\right) + 0.5 \cdot \left(\frac{1}{3}, \frac{1}{4}, \frac{1}{5}, \frac{1}{6}\right) + \right.$$
$$\left. +0.25 \cdot \left(\frac{1}{2}, \frac{1}{3}, \frac{1}{4}, \frac{1}{5}, \frac{1}{6}, \frac{1}{7}\right) + 0 \cdot \left(\frac{1}{1}, \frac{1}{2}, \frac{1}{3}, \frac{1}{4}, \frac{1}{5}, \frac{1}{6}, \frac{1}{7}, \frac{1}{8}\right)\right) =$$
$$= \left(\frac{0}{1}, \frac{0.25}{2}, \frac{0.75}{3}, \frac{1.0}{4}, \frac{0.75}{5}, \frac{0.5}{6}, \frac{0.25}{7}, \frac{0}{8}\right).$$

Comparing the primary membership function $\mu_A(x)$ with the function $\mu_A^*(x)$ reproduced on the basis of the α-cuts, Fig. 2.16, we can see that the reproduction is not exact. Its accuracy can be improved by the increase of the number of α-cuts or their better placement.

◊ ◊ ◊

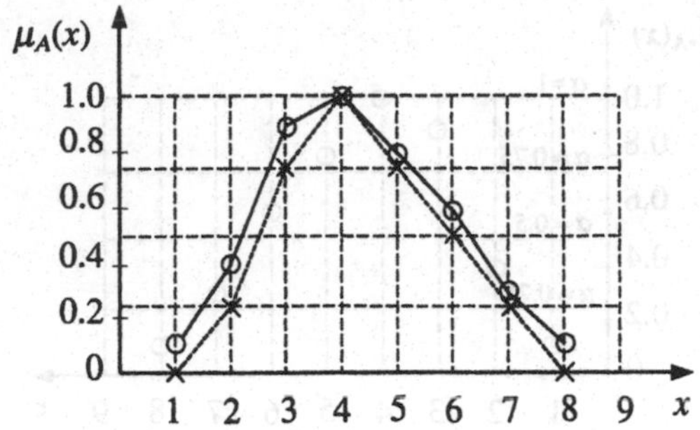

Fig. 2.16. Primary membership function $\mu_A(x)$ (continuous line) and the function $\mu_A^*(x)$ reproduced on the basis of α-cuts (dotted line)

- **The power (cardinality) of a fuzzy set**

The power $\|A\|$ or cardinal number, denoted by card(A), of a discrete fuzzy set, is the sum of the membership grades of all elements in the fuzzy set.

$$\|A\| = \text{card}(A) = \sum_{x \in S(A)} \mu_A(x)\,, \tag{2.20}$$

where: $S(A)$ is the support of the fuzzy set.

The power of a continuous fuzzy set is calculated by using the integral.

$$\|A\| = \text{card}(A) = \int_{x \in S(A)} \mu_A(x)dx \tag{2.21}$$

Integrating or summing up elements of the support of a set is sufficient because the membership of remaining elements of the universe of discourse in the fuzzy set is zero. The concept of the power of a fuzzy set enables us to compare various sets. The power of an empty fuzzy set is equal to zero.

- **The relative power of a fuzzy set**

In the case of a discrete fuzzy set the relative power is that part of the set power which falls to one element of the universe of discourse X of the set:

$$\|A\|_X = \frac{\sum_{x \in X} \mu_A(x)}{N}\,, \tag{2.22}$$

where: N – the number of elements in the universe of discourse of the set.

In the case of a continuous fuzzy set the relative power will be expressed with the formula:

$$\|A\|_X = \frac{\int_{x \in X} \mu_A(x)dx}{\int_{x \in X} dx}\,. \tag{2.23}$$

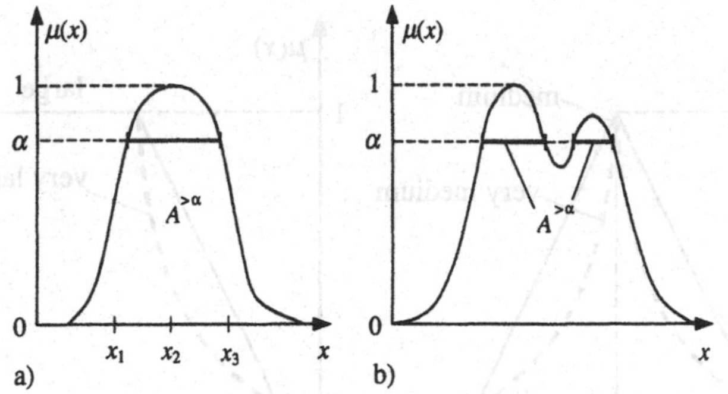

Fig. 2.17. Examples of a convex (a) and non-convex (b) fuzzy set

If the number N of elements in a discrete set or the length of the universe of discourse of a continuous set is infinitely large, then adding or integrating can be made on the support of the set $S(A)$.

- **Convex and non-convex fuzzy sets**

Examples of a convex and non-convex set are represented in Fig. 2.17.

A convex fuzzy set has the property that all its α-cuts are compact (closed), one-part intervals of the universe of discourse X. In the case of a non-convex fuzzy set there exist non-compact (unclosed) α-cuts consisting of many parts (Fig. 2.17b).

Non-convex sets come into existence as a result of performing set-type, algebraic and arithmetical operations on primary sets, which are convex as a rule. A convex fuzzy set satisfies the following conditions:

$$x_1 \leq x_2 \leq x_3 \Rightarrow \mu_A(x_2) \geq \min\left(\mu_A(x_1), \mu_A(x_3)\right), \quad \forall x_1, x_2, x_3 \in X, \quad (2.24)$$

or

$$\mu_A\left(\lambda x_1 + (1-\lambda)x_3\right) \geq \min\left(\mu_A(x_1), \mu_A(x_3)\right), \quad \forall \lambda \in [0,1] \text{ and } \forall x_1, x_3 \in X. \quad (2.25)$$

2.3 Linguistic modifiers of fuzzy sets

Linguistic modifiers enable us to create derivative fuzzy sets on the basis of primary sets. If there is (e.g.) the defined primary set "cold", then by the use of modifiers derivative sets "very cold" or "a little cold" can be created.

There are three main modifiers, also called operators:

- concentration operator,
- dilatation operator,
- operator of contrast intensification/decrease.

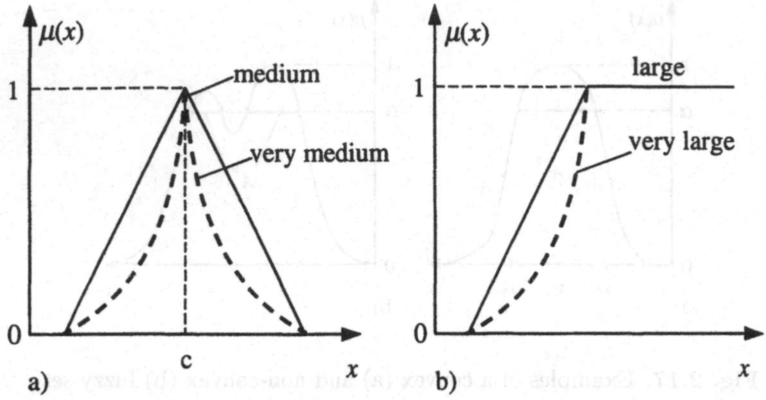

Fig. 2.18. An example of the operation of a concentration operator in the case of the middle fuzzy set (a) and utmost fuzzy set (b)

- **Modifier of concentration of a fuzzy set**

If A is the fuzzy set of the linguistic value l_i, then the concentration operator enable us to create a derivative value "very l_i". The operation of concentration is realized by means of the formula:

$$\mu_{\mathrm{CON}(A)}(x) = \mathrm{CON}\left(\mu_A(x)\right) = \mu_A(x)^2, \quad \forall x \in X. \tag{2.26}$$

In the case of the linguistic value "medium" with a triangular shape membership function, the operation of the concentration modifier is shown in Fig. 2.18a.

In the case of a triangular, internal membership function (Fig. 2.18a), the operation of the concentration operator "very" means that only those values x, which lie very near to the medium c of the support of a set, will be approved as "very medium" in a significant way. The concentration operator can also be used in the case of utmost fuzzy sets as for example "large" in Fig. 2.18b. However, in this case the modification can be abandoned. Instead of it a new utmost fuzzy set "very large" as in Fig. 2.19 can be introduced.

- **Dilatation operator of a fuzzy set**

This operator modifies the primary fuzzy set A in a linguistically defined way as "a little" or "more or less". Its operation is the opposite of the concentration operation.

$$\mu_{\mathrm{DIL}(A)}(x) = \mathrm{DIL}\left(\mu_A(x)\right) = \sqrt{\mu_A(x)}, \quad \forall x \in X. \tag{2.27}$$

The dilatation operator causes the extension of a membership function and increases its power. An example of the operation of this operator is represented in Fig. 2.20.

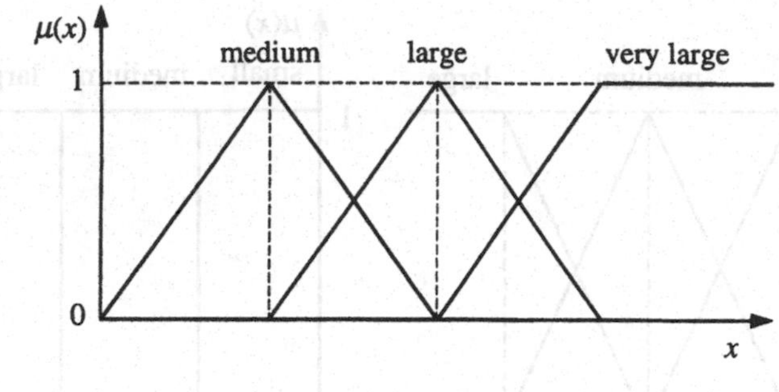

Fig. 2.19. Alternative for concentration of the set "large"

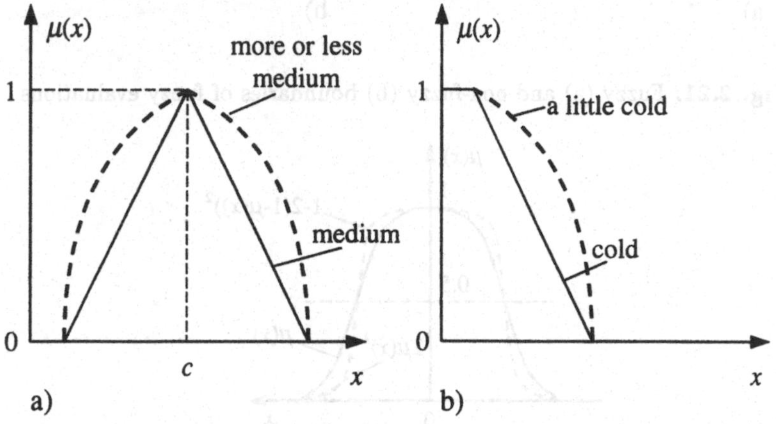

Fig. 2.20. Operation of a dilatation operator in the case of an internal (a) and external (b) fuzzy set

The dilatation operator acts in such a way that values x of the support of a set lying further from its middle c are counted in the set "more or less medium" to a higher degree than in the original set "medium".

- **Operator of contrast intensification of a fuzzy set**

The characteristic feature of evaluation using fuzzy concepts is the fuzziness of the boundary between the particular linguistic evaluations, Fig. 2.21.

In the case of the use of fuzzy linguistic evaluations, Fig. 2.21a, the boundary between the particular evaluations is not distinct and clear but fuzzy, in contrary to the non-fuzzy estimations, Fig. 2.21b, where the boundaries are distinct. The operator of contrast intensification enables us to liken the fuzzy set to the non-fuzzy one. Through the change of inclination angle of the wings of the membership function it enables the border of transition from one fuzzy set to another to appear more clearly, Fig. 2.22.

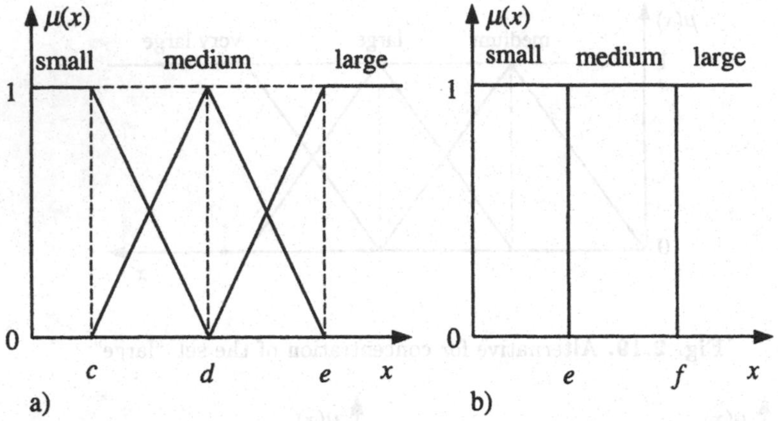

Fig. 2.21. Fuzzy (a) and non-fuzzy (b) boundaries of fuzzy evaluations

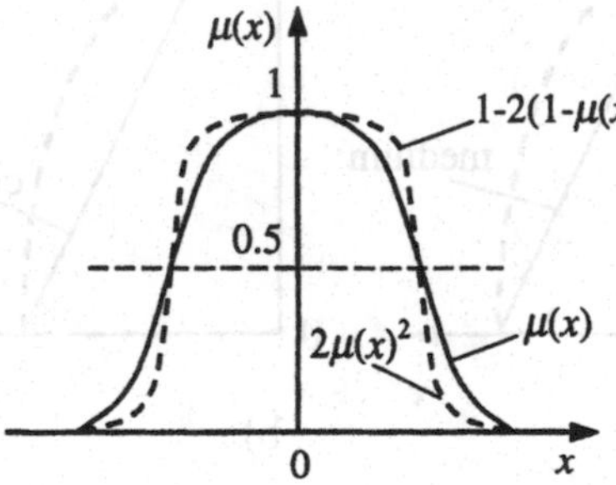

Fig. 2.22. Functioning of the operator of contrast intensification of a fuzzy set

Contrast intensification is carried out using two formulas – one for membership function with values less than 0.5 and another for membership functions with values more than or equal to 0.5, Fig. 2.23.

$$\mu_{\mathrm{INT}(A)}(x) = \mathrm{INT}\left(\mu_A(x)\right) = \begin{cases} 2(\mu_A(x))^2 & \text{for } \mu_A(x) < 0.5 \\ 1 - 2(1 - \mu_A(x))^2 & \text{otherwise} \end{cases} \tag{2.28}$$

Contrast intensification can be increased through the use of powers greater than 2. With an exponent approaching infinity the membership function $\mu_A(x)$ takes a rectangular shape and the set becomes a non-fuzzy one having distinct boundaries.

- **Operator of contrast decrease of a fuzzy set**

This is a contrary operator to the operator of contrast intensification. The decreasing of contrast intensification, denoted with the abbreviation BLR (blurring), can be carried out after the formula (2.29) (Kacprzyk 1986).

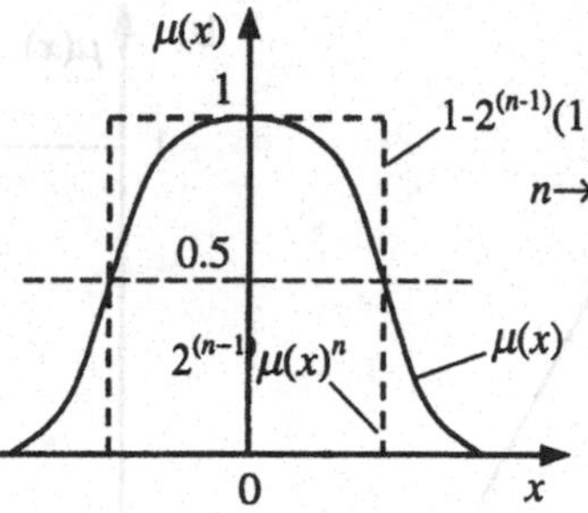

Fig. 2.23. The end form of contrast intensification of a fuzzy set – a non-fuzzy set

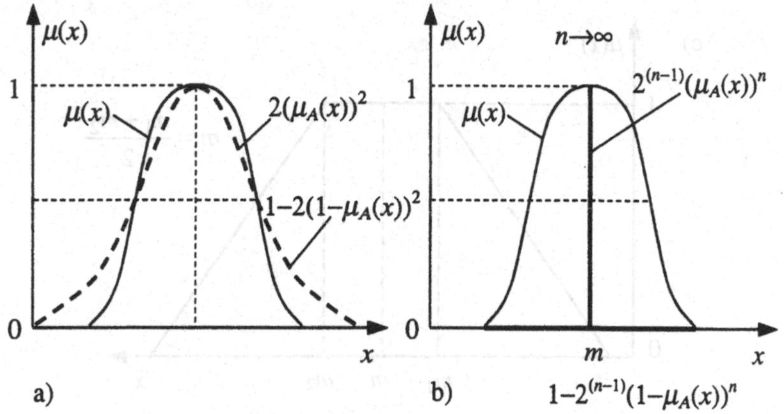

Fig. 2.24. Contrast decreasing of a fuzzy set at n=2 (a) and terminal form of a fuzzy set at $n \to \infty$ (b)

$$\mu_{\mathrm{BLR}(A)}(x) = \mathrm{BLR}\left(\mu_A(x)\right) = \begin{cases} 1 - 2(1-\mu_A(x))^2 & \text{for } \mu_A(x) < 0.5 \\ 2(\mu_A(x))^2 & \text{for } \mu_A(x) \geq 0.5 \end{cases} \tag{2.29}$$

To intensify the functioning of this operator one can use powers n greater than 1. In the case of very large powers a fuzzy set becomes converted into a singleton placed at the modal value m (Fig. 2.24b) of the set.

- **Modal value of a fuzzy set**

The modal value (Kahlert 1995) relates mainly to fuzzy sets with a core containing only one element of the universe of discourse X, Fig. 2.25a,b.

If the core of a fuzzy set contains more than one element, then the modal value can also be calculated as the mean value of the core, Fig. 2.25c.

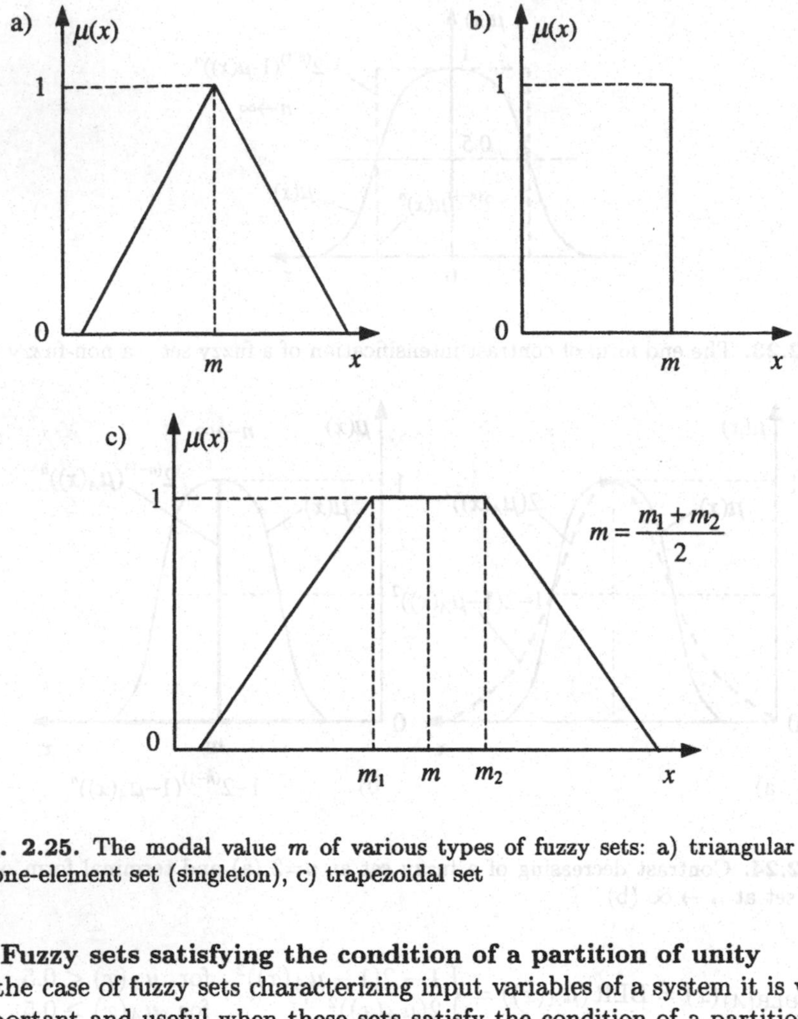

Fig. 2.25. The modal value m of various types of fuzzy sets: a) triangular set, b) one-element set (singleton), c) trapezoidal set

- **Fuzzy sets satisfying the condition of a partition of unity**

In the case of fuzzy sets characterizing input variables of a system it is very important and useful when these sets satisfy the condition of a partition of unity (2.30) (Brown 1994) (the sum of memberships of each element x from the universe of discourse is equal to 1):

$$\sum_h \mu_{A_h}(x) \equiv 1, \quad \forall x \in X, \tag{2.30}$$

where: h – number of the fuzzy set.

Fig. 2.26a represents the example of a set satisfying this condition, Figs. 2.26b and c represent sets not satisfying the above condition.

Satisfying the condition of a partition of unity usually gives the model a smoother surface in comparison with models employing sets of the type represented in Fig. 2.26b, while sets of the type represented in Fig. 2.26c cause the surface of a model to become flatter. Those fuzzy sets which do not satisfy this condition can be transformed so that the condition is satisfied.

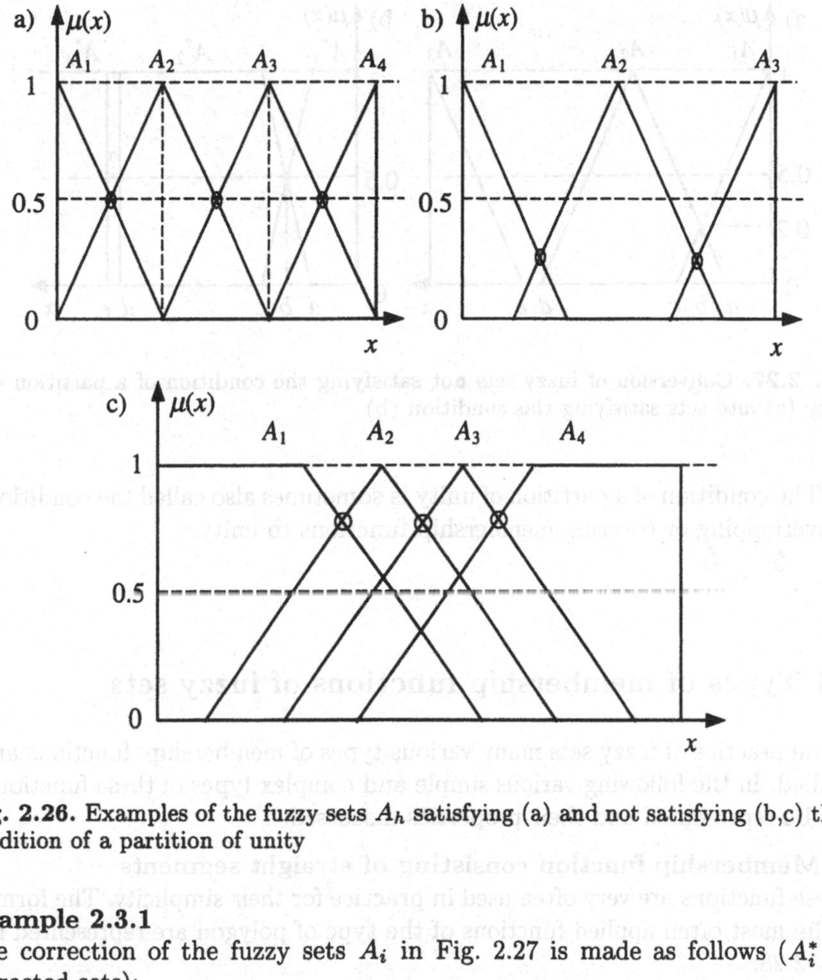

Fig. 2.26. Examples of the fuzzy sets A_h satisfying (a) and not satisfying (b,c) the condition of a partition of unity

Example 2.3.1
The correction of the fuzzy sets A_i in Fig. 2.27 is made as follows (A_i^* – corrected sets):

$$\mu_{A_1}^*(a) = \frac{\mu_{A_1}(a)}{\mu_{A_1}(a) + \mu_{A_2}(a)} = \frac{\mu_{A_1}(a)}{\mu_{A_1}(a)} = 1\,,$$

$$\mu_{A_1}^*(b) = \frac{\mu_{A_1}(b)}{\mu_{A_1}(b) + \mu_{A_2}(b)} = \frac{0.2}{0.4} = 0.5\,.$$

By carrying out the correction of the membership functions of the particular fuzzy sets A_h in Fig. 2.27a we obtain the corrected sets represented in Fig. 2.27b. However, it is remarkable that this operation can change the shape of the membership function and cause calculation problems in the case where the correction is made in ranges not covered by any fuzzy set, e.g. the range (d – e) in Fig. 2.27. Such ranges can come into existence during adjusting (adapting) fuzzy models based on input/output data (self-learning algorithms). The consequence is that zones arise which are insensitive to a change of the given input variable.

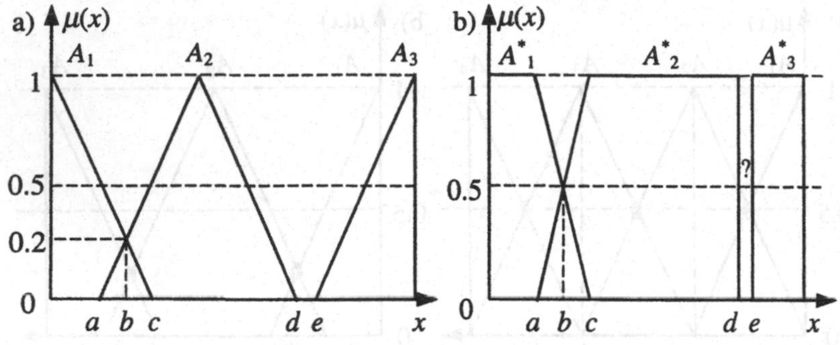

Fig. 2.27. Conversion of fuzzy sets not satisfying the condition of a partition of unity (a) into sets satisfying this condition (b)

The condition of a partition of unity is sometimes also called the condition of overlapping or covering membership functions to unity.

◊ ◊ ◊

2.4 Types of membership functions of fuzzy sets

In the practice of fuzzy sets many various types of membership functions are applied. In the following various simple and complex types of these functions will be represented and their properties discussed.

- **Membership function consisting of straight segments**

These functions are very often used in practice for their simplicity. The forms of the most often applied functions of the type of polygon are represented in Fig. 2.28.

The enormous advantage of the polygonal functions is that they can be defined using a minimal amount of information, in comparison with other membership functions. Data relating to the corner points of a function are hereby sufficient. It is of great importance in the case of modeling of systems on which we do not have much data. In practice, to define polygonal membership functions only data concerning their modal values are needed.

Example 2.4.1

Input/output values of real systems are usually limited to a certain range of variations. For example, the movement of the piston in a servomotor can vary in the range:

$$x_{min} \leq x \leq x_{max}\,.$$

One can assume that membership functions will have the form represented in Fig. 2.29.

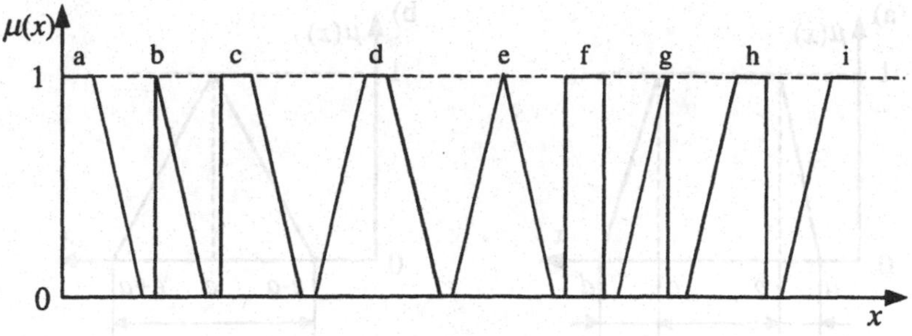

a – the left outside membership function
b,g – triangular asymmetrical membership function
c – trapezoidal asymmetrical membership function
d – trapezoidal symmetrical membership function
e – triangular symmetrical membership function
f – rectangular membership function
h – trapezoidal asymmetrical membership function
i – the right outside membership function

Fig. 2.28. Shapes of the most often applied segmentally-linear membership functions

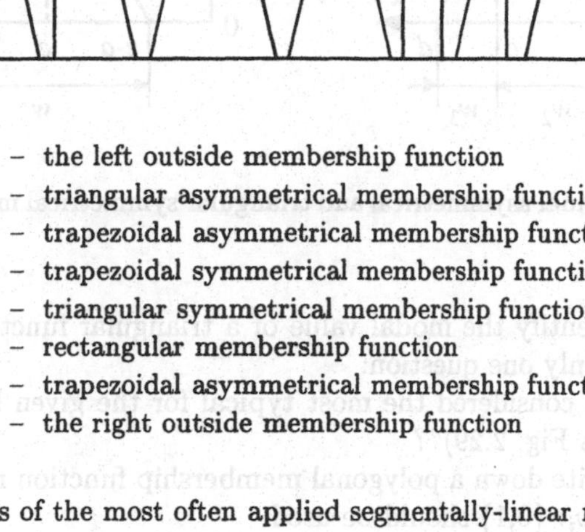

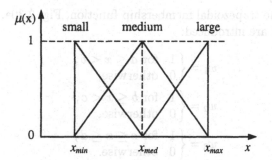

Fig. 2.29. Examples of membership functions when the range of variations of a variable is limited on both sides

In the case represented in Fig. 2.29 only 3 (instead of 9) parameters, i.e. the values $x_{min}, x_{med}, x_{max}$ are sufficient for the complete definition of 3 membership functions.

If an expert would place the modal value of the medium function in the centre of the interval, then only 2 parameters x_{min}, x_{max} are needed, because the values x_{med} can be calculated on the basis of them.

◊ ◊ ◊

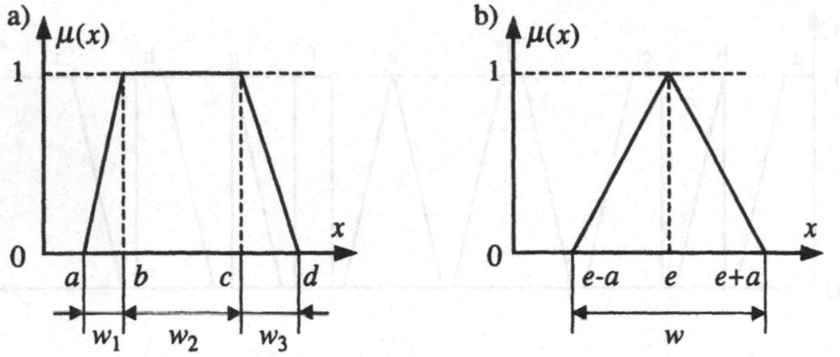

Fig. 2.30. Trapezoidal asymmetrical and triangular symmetrical membership functions

In order to identify the modal value of a triangular function (e.g. x_{med}) you need to ask only one question:
"What value x is considered the most typical for the given linguistic value (e.g. "medium" in Fig. 2.29)"?

In order to write down a polygonal membership function mathematically logical variables w_i: $\{0,1\}$ should be used.

Example 2.4.2
In the case of the trapezoidal membership function, Fig. 2.30a, the following logical variables are introduced:

$$w_1 = \begin{cases} 1 & \text{for } a \le x < b\,, \\ 0 & \text{otherwise,} \end{cases}$$
$$w_2 = \begin{cases} 1 & \text{for } b \le x < c\,, \\ 0 & \text{otherwise,} \end{cases} \tag{2.31}$$
$$w_3 = \begin{cases} 1 & \text{for } c \le x \le d\,, \\ 0 & \text{otherwise.} \end{cases}$$

The membership function of the type of asymmetrical trapezoid can be represented in the form:

$$\mu(x) = w_1\left(\frac{x-a}{b-a}\right) + w_2 + w_3\left(\frac{d-x}{d-c}\right). \tag{2.32}$$

In the case of the symmetrical triangular function, Fig. 2.30b, only one logical variable w must be introduced:

$$w = \begin{cases} 1 & \text{for } (e-a) \le x < (e+a)\,, \\ 0 & \text{otherwise.} \end{cases} \tag{2.33}$$

The membership function can be written down in the following form:

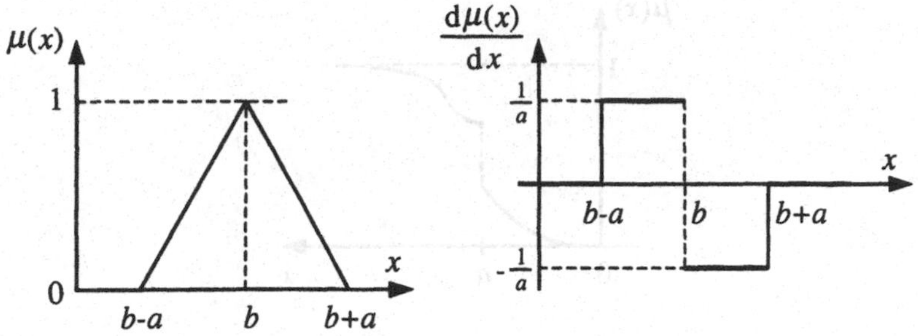

Fig. 2.31. Triangular membership function and its derivative

$$\mu(x) = w\left(\frac{a - |x - e|}{a}\right). \tag{2.34}$$

◊ ◊ ◊

Advantages of polygonal membership functions

1. A small amount of data is needed to define the membership function.
2. Ease of modification of parameters (modal values) of membership functions on the basis of measured values of the input→output of a system.
3. The possibility of obtaining input→output mapping of a model which is a hypersurface consisting of **linear** segments.
4. Polygonal membership functions mean the condition of a partition of unity (it means that the sum of membership grades for each value x amounts to 1) is easily satisfied.

Disadvantages of polygonal membership functions

1. Polygonal membership functions are not continuously differentiable.

Example 2.4.3

As can be seen in Fig. 2.31, the derivative of a membership function varies abruptly at points of discontinuity. The result is, that the model of a system containing such functions is also not continuously differentiable.

◊ ◊ ◊

In the opinion of some scientists (Preuss 1994a; Rummelhart 1986) not continuously differentiable membership functions make the process of adapting (teaching) fuzzy models difficult. However, experiments carried out by the author and his co-workers (Piegat 1996,1997a) have shown good adaptability of these membership functions.

- **Intuitive membership functions**

The membership functions which are used (often unconsciously) by a human can be termed intuitive membership functions. Research, mainly concerning

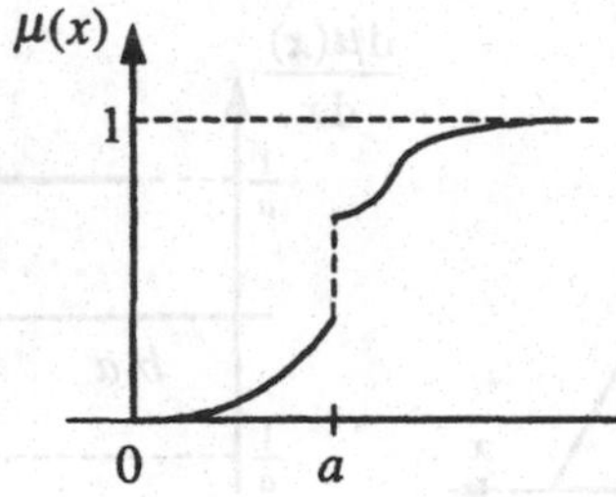

Fig. 2.32. Membership function with a discontinuity at point $x = a$

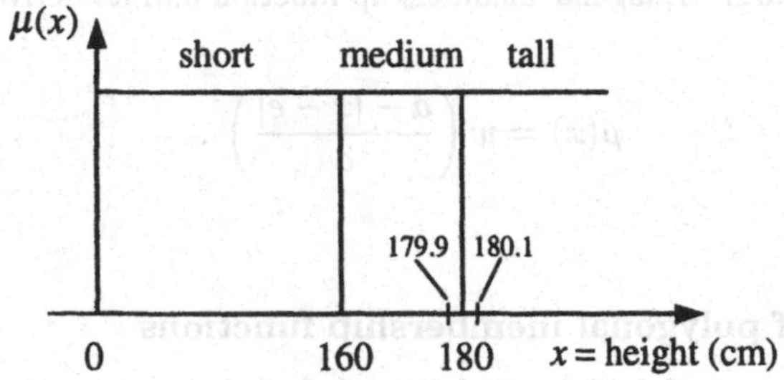

Fig. 2.33. Examples of discontinuous membership functions of the linguistic variable "height"

classification decisions (Altrock 1993), has shown that intuitive membership functions are characterized by some properties called Schwab axioms.

Axiom 1. The intuitive membership functions $\mu(x)$ are continuous in the whole range of the numerical universe of discourse X.

Any little variation of variable x causes no step variation of its qualitative evaluation made by a human. Such a shape of intuitive membership function as shown in Fig. 2.32 is very improbable.

Example 2.4.4

It is very improbable that people will employ rectangular membership functions for the qualitative evaluation of the height of a man, Fig. 2.33.

A man employing such membership functions would measure a person with a height of 179.9 (cm) as of medium height, and a person only 2 (mm) taller(180.1 (cm)) as tall height. Most people would admit that these evaluations are not right.

◊ ◊ ◊

Axiom 2. The first derivative of the intuitive membership function $\mu(x)$:

$$\dot{\mu}(x) = \frac{\mathrm{d}\mu(x)}{\mathrm{d}x},$$

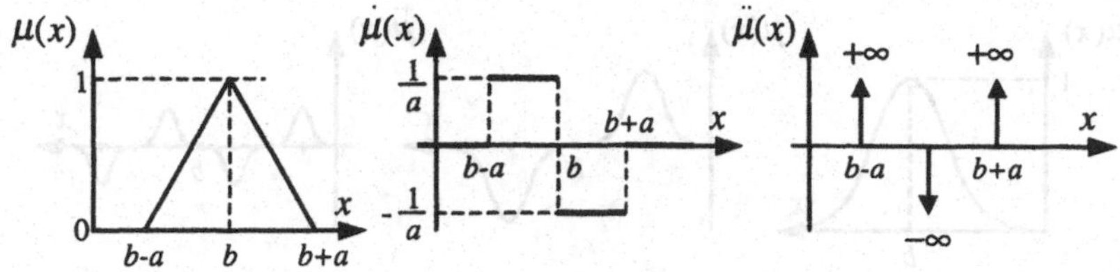

Fig. 2.34. Triangular membership functions and their derivatives

is continuous in the whole range of the numerical universe of discourse X.

This results from the observation that any small variation of variable x observed by a man causes no step variation in the speed of evaluation of this variable (i.e. the derivative of this evaluation).

Example 2.4.5
In the case of a triangular membership function, Fig. 2.34, for a minimal variation of variable x in the neighborhood of the point b an abrupt jump occurs not only in the value but also in the sign of the derivative $\dot{\mu}(x)$. The triangular membership function thus represents a very rough approximation of the human manner of evaluation.

However, this does not mean that they should not be used in fuzzy models since a model with triangular membership functions can be satisfactorily exact.

◊ ◊ ◊

Axiom 3. The second derivative of the intuitive membership function $\mu(x)$:

$$\ddot{\mu}(x) = \frac{\mathrm{d}^2\mu(x)}{\mathrm{d}x^2} ,$$

is continuous in the whole range of the universe of discourse.

Axiom 4. Curvatures of intuitive membership functions are minimal.

This means that a person selects from among various possible membership functions those which minimize the maximal value of the second derivative that can occur in these functions:

$$\mu(x) : \ \mu(x) = \min_{\mu} \left| \max_{X} \left(\frac{\mathrm{d}^2\mu(x)}{\mathrm{d}x^2} \right) \right| .$$

Example 2.4.6
In the case of the triangular membership function, Fig. 2.34, the curvatures at the points $(b-a)$, b, $(b+a)$ are so large that the second derivative $\ddot{\mu}(x)$ aims at infinitely large values. The function more suitable for the human manner of

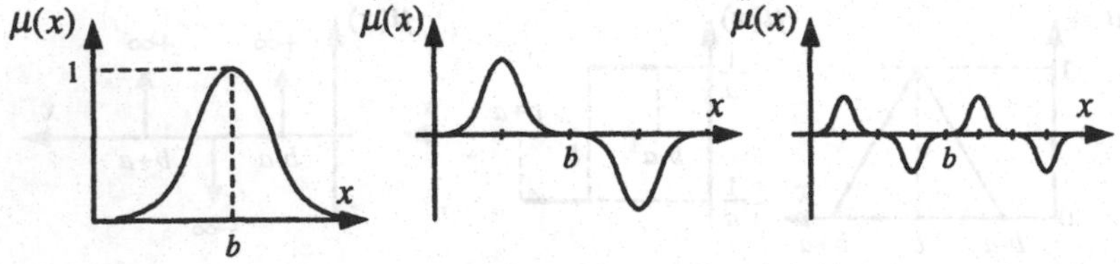

Fig. 2.35. Example of continuous membership function with little curvatures

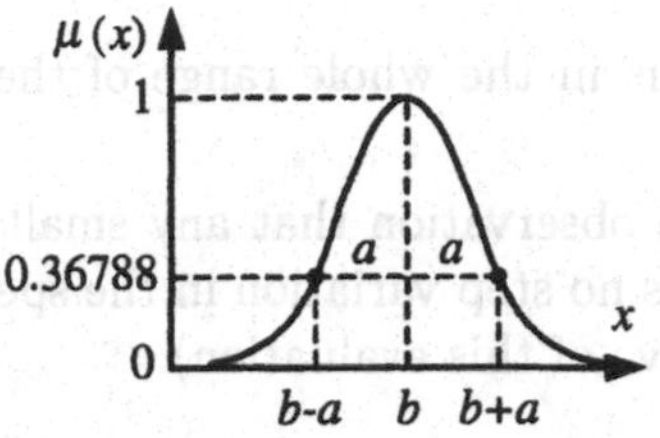

Fig. 2.36. Membership function of the Gaussian type

qualitative evaluation of variables is the one represented in Fig. 2.35 because it has little curvatures and a continuous first and second derivative.

◊ ◊ ◊

In what follows, various types of functions that can be applied for mathematical representation of intuitive membership functions will be discussed.

- **Gaussian symmetrical function**

The Gaussian function is expressed by the formula (2.35).

$$\mu(x) = \exp\left[-\left(\frac{x-b}{a}\right)^2\right] \tag{2.35}$$

Its shape is depicted in Fig. 2.36.

The shape of the Gaussian function, sometimes called the Gauss bell (Preuss 1994a), is determined by 2 parameters a and b, where b is the modal value of the function, and the parameter a determines its width. The Gaussian function has the width of $2a$ at the level of $\mu(x) = e^{-1} \cong 0.36788$. The identification of the modal value using the expert method can be made by putting a question about the most typical value x for the given fuzzy set.

Example 2.4.7

The value of $b = 170$ (cm) can be given as the most typical numerical value of the fuzzy set "medium height".

In order to determine the parameter a characterizing the width of a function using the expert method one can employ the concept of the **critical**

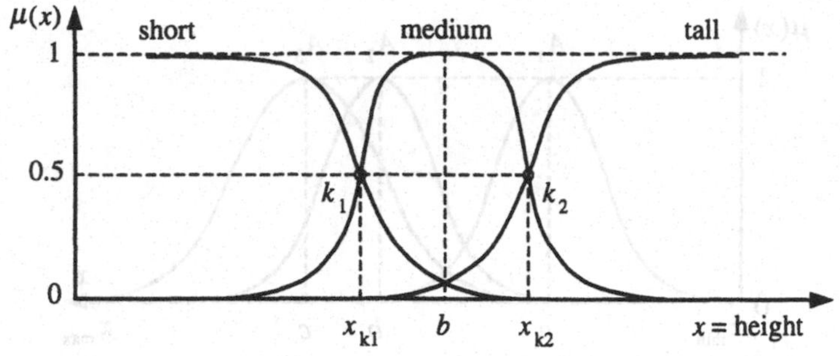

Fig. 2.37. Gaussian function as membership function for the fuzzy set "medium height"

point k of a membership function. This is the point of a membership function at which the grade of membership amounts to 0.5. Each Gaussian curve has 2 such points, Fig. 2.37.

If we assume that the adjoining membership functions intersect each other more or less at the height of $\mu(x_k) = 0.5$ (but this is not always met in fuzzy models), then the critical point k can be defined as the point having a coordinate x such that we are not able to decide whether it belongs more to the right or left fuzzy set (Altrock 1993).

If we are not able to decide if a man of height 165 (cm) should be counted in the group of the short-sized or medium-sized people, then we can assume that membership in both mentioned fuzzy sets is identical and amounts to 0.5. The height of 165 (cm) is thus the coordinate of the critical point of the membership function.

$$x_k = 165\,(\text{cm})$$

Knowing the modal value of the Gauss curve $b = 170$ (cm), one can calculate its second parameter a:

$$\mu(x_k) = \exp\left[-\left(\frac{x_k - 170}{a}\right)^2\right] = 0.5\,,$$

$$a = \frac{|x_k - b|}{\sqrt{\ln 2}} \cong 6\,(\text{cm})\,.$$

◊ ◊ ◊

The use of the concept of the critical point k of a membership function is particularly useful in identification of its parameters on the basis of an expert's opinion because a man most easily defines the characteristic limit values and other values distinguishing themselves in the basic set of the given variable. The expert is usually not able to determine exactly the membership grade of other elements, which do not distinguish themselves by anything in

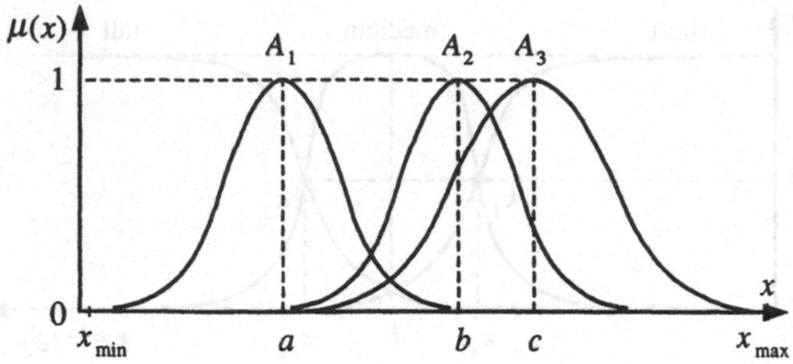

Fig. 2.38. Non-uniform spacing of the Gaussian membership functions of various widths in the universe of discourse X

the basic set. In the general case the surface of a fuzzy model with the use of the Gaussian function is globally and locally non-linear.

Advantages of the Gaussian membership function

1. The Gaussian functions facilitate obtaining smooth, continuously differentiable hypersurfaces of a fuzzy model.
2. The Gaussian functions facilitate theoretical analysis of fuzzy systems as they are continuously differentiable and infinitely differentiable (Brown 1994), i.e. they have derivatives of any grade.

Disadvantages of the Gaussian membership functions

1. The Gaussian function is symmetrical, and this means that the partition of unity condition is not satisfied, Fig. 2.38.
2. In the use of the Gaussian function it becomes necessary to identify a greater number of parameters (two parameters for each function) than in the case of triangular membership functions and this makes the adjustment of models more difficult.
3. The support of the Gaussian function is infinitely large. This means that each element x of the universe of discourse X belongs to each fuzzy set represented by this function, Fig. 2.38. This can not be in conformity with the intensions of the expert of the system to be modeled. However, in practice, membership of the elements x of the basic set which are further off the center of the Gaussian function is negligibly small, provided that the width of this function is not too large.
4. The use of the Gaussian function makes obtaining simple, locally linear surfaces of a fuzzy model difficult.

The support of the Gaussian function is infinitely large and therefore Gaussian membership functions with finite support (Werntges 1993) having the property of infinite differentiability have been introduced. The Gaussian function with the finite support (Brown 1994) is shown in Fig. 2.39.

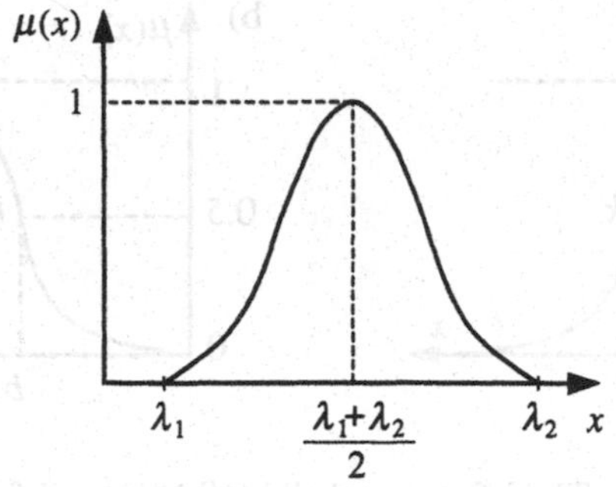

Fig. 2.39. Membership function of Gaussian type with finite support (λ_1, λ_2)

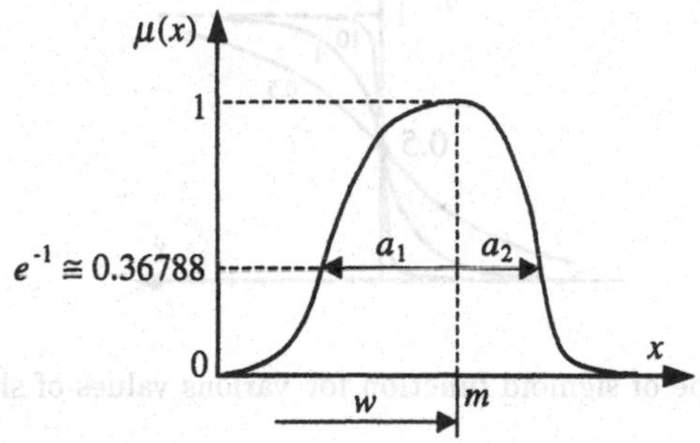

Fig. 2.40. Asymmetrical, infinitely differentiable Gaussian function

The shape of this function is described by the formula (2.36):

$$\mu(x) = \begin{cases} \exp\left[\frac{4(\lambda_2 - x)(x - \lambda_1) - (\lambda_2 - \lambda_1)^2}{4(\lambda_2 - x)(x - \lambda_1)}\right] & \text{for } \lambda_1 \le x \le \lambda_2, \\ 0 & \text{otherwise,} \end{cases} \tag{2.36}$$

where: λ_1, λ_2 are nodes (limits of the support) of the function. The Gaussian function (2.36) is symmetrical.

- **Asymmetrical Gaussian function**

Because the advantage of the Gaussian function is its infinite differentiability and its disadvantage is its symmetry, then an advantageous solution can be the use of the asymmetrical Gaussian function, Fig. 2.40.

The following logical variable, which tells us about the level of the variable x has been introduced:

$$w = \begin{cases} 1 & \text{for } -\infty < x \le m, \\ 0 & \text{otherwise.} \end{cases} \tag{2.37}$$

The asymmetrical Gaussian function is expressed by the formula:

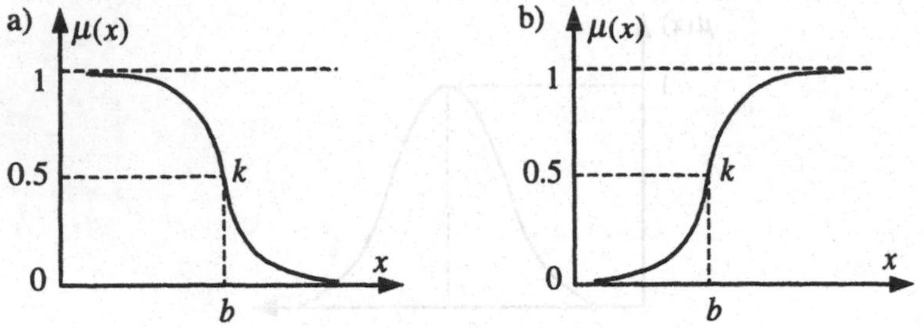

Fig. 2.41. The left (a) and right (b) sigmoid function

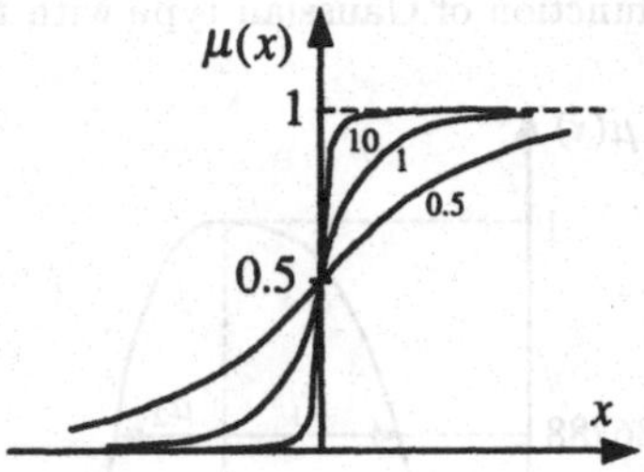

Fig. 2.42. Shape of sigmoid function for various values of slope factor a

$$\mu(x) = w \cdot \exp\left[-\left(\frac{x-m}{a_1}\right)^2\right] + (1-w) \cdot \exp\left[-\left(\frac{x-m}{a_2}\right)^2\right], \quad (2.38)$$

where: m – the modal value (center) of the function.

The use of the asymmetrical Gaussian function better enables us to realize the partition of unity condition (Brown 1994).

- **Sigmoid membership functions**

The Gauss functions are symmetrical functions and are suitable for representing internal fuzzy sets. For representing external sets the right and left sigmoid function can be used, Fig. 2.41.

The right sigmoid function is expressed by the formula:

$$\mu(x) = \frac{1}{1+\exp[-a \cdot (x-b)]} \cdot \quad (2.39)$$

The factor b determines the characteristic coordinate of the point k, at which the grade of membership in the fuzzy set amounts to 0.5. Therefore this parameter can be relatively easily identified by the expert. The factor a determines the slope of the function at the point of inflexion k. With the increase of this factor the slope of the function increases and at $a = 10$ the function is almost a step, Fig. 2.42.

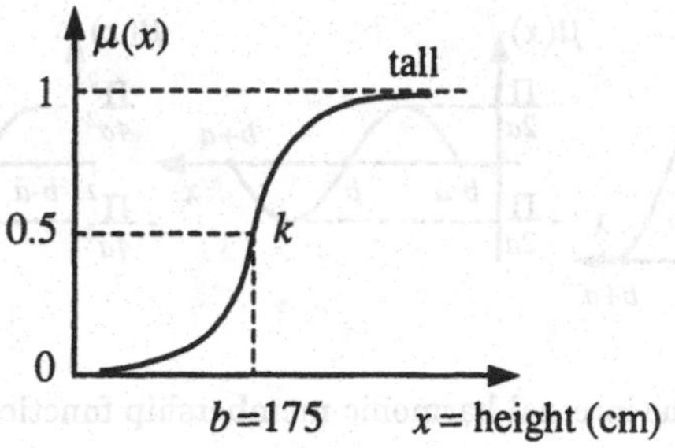

Fig. 2.43. Sigmoid membership function of the fuzzy set "tall height"

The slope factor a can be calculated by estimating (according to the expert's opinion) the value $x_{0.99}$, i.e the least value of the variable x which can be admitted with almost complete certainty to belong to a fuzzy set represented by the sigmoid function.

The sigmoid function achieves the value equal to 1.0 for $x \to \infty$:

$$\lim_{x\to\infty} \frac{1}{1+\exp[-a\cdot(x-b)]} = 1\,, \tag{2.40}$$

and therefore in practice we assume that, for example, the value 0.99 is accepted by us for complete membership of the variable x in the given fuzzy set.

Example 2.4.8
We assume the sigmoid function is one representing the fuzzy set "tall height", Fig. 2.43.

If we acknowledge that people of height 180 (cm) could be with quite absolute certainty ($\mu = 0.99$) acknowledged to be tall, then the slope factor a can be calculated from the formulas (2.41) and (2.42).

$$\mu(x_{0.99}) = \frac{1}{1+\exp[-a\cdot(x_{0.99}-b)]} \tag{2.41}$$

$$a = \frac{\ln(99)}{x_{0.99}-b} = \frac{\ln(99)}{180-175} \cong 0.919 \tag{2.42}$$

◊ ◊ ◊

The left sigmoid membership function, Fig. 2.41a, can be expressed by the formula:

$$\mu(x) = 1 - \frac{1}{1+\exp[-a\cdot(x-b)]} = \frac{\exp[-a\cdot(x-b)]}{1+\exp[-a\cdot(x-b)]}\,. \tag{2.43}$$

The left sigmoid function, as the right one, has its inflexion at the point $x = b$. The value of the function at this point is 0.5. The slope factor a of the left sigmoid function can be calculated from the formula:

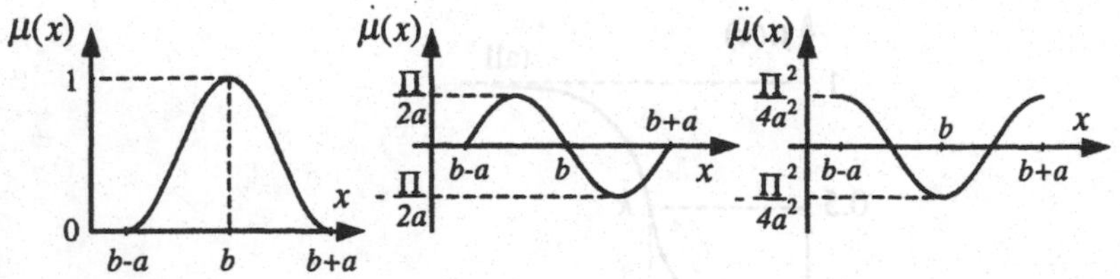

Fig. 2.44. Shape of the internal harmonic membership function and its derivatives

$$a = -\left(\frac{\ln(99)}{x_{0.99} - b}\right). \tag{2.44}$$

Advantages and disadvantages of the sigmoid function are the same as those of the Gaussian function.

- **Harmonic membership function**

The internal harmonic function is expressed with the formula:

$$\mu(x) = \begin{cases} 0.5 \cdot \left[1 + \cos\left(\pi \frac{x-b}{a}\right)\right] & \text{for } (b-a) \le x \le (b+a), \\ 0 & \text{otherwise.} \end{cases} \tag{2.45}$$

Its shape is depicted in Fig. 2.44.

Advantages of the harmonic membership function

1. The harmonic function has limited support $[(b-a), (b+a)]$ and this is what facilitates the identification of its parameters with the expert method.
2. The harmonic function makes obtaining smooth, differentiably continuous model surfaces easier, because it is infinitely differentiable.
3. The first derivative of the harmonic function has zero values at the ends of the support. Because of that, this function well satisfies the Schwab axioms (but not all).

Disadvantages of the harmonic membership function

1. The harmonic function is symmetrical, and at non-uniform distribution of membership functions in the space of discourse this makes the realization of the postulate of partition of unity difficult and negatively influences the quality of a model in regions indifferently covered by membership functions. To minimize this disadvantage one can use asymmetrical harmonic functions constructed in such a way as represented for the case of the asymmetrical Gaussian function (2.38).

The right external harmonic membership function is expressed by the formula:

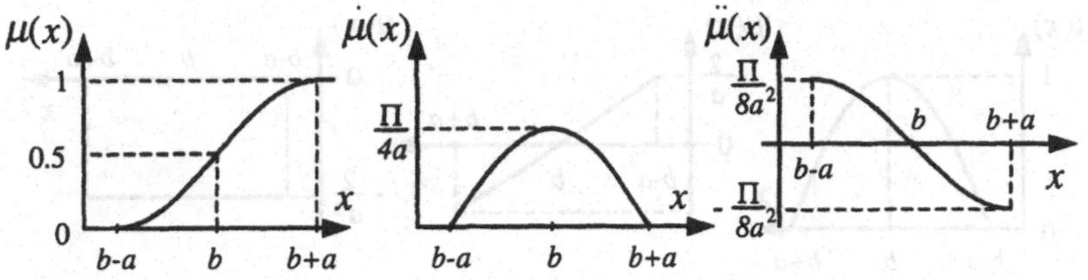

Fig. 2.45. Shape of the right external harmonic membership function and its derivatives

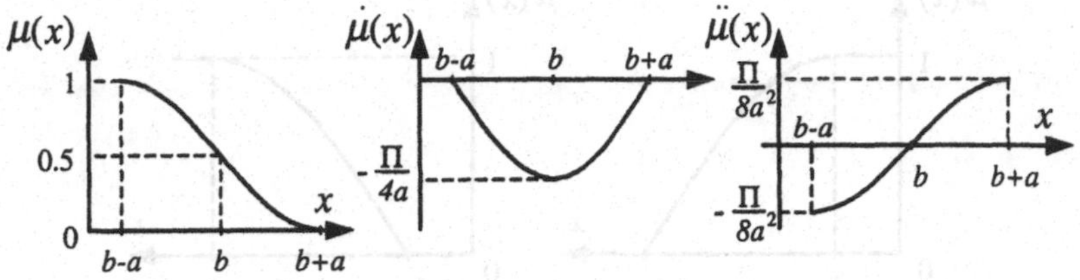

Fig. 2.46. Shape of the left external harmonic membership function and its derivatives

$$\mu(x) = \begin{cases} 1 & \text{for } x > (b+a), \\ 0.5 \cdot \left[1 + \sin\left(\pi \frac{x-b}{2a}\right)\right] & \text{for } (b-a) \le x \le (b+a), \\ 0 & \text{for } x < (b-a). \end{cases} \tag{2.46}$$

The shape of the right membership function is represented in Fig. 2.45. The first derivative of the right harmonic function, just as the internal function, has zero values at the ends of the support $[(b-a), b+a)]$. It has little curvature and because of that satisfies the Schwab axioms well (but not completely).

The left external harmonic membership function is expressed by the formula:

$$\mu(x) = \begin{cases} 0 & \text{for } x > (b+a), \\ 0.5 \cdot \left[1 - \sin\left(\pi \frac{x-b}{2a}\right)\right] & \text{for } (b-a) \le x \le (b+a), \\ 1 & \text{for } x < (b-a). \end{cases} \tag{2.47}$$

Its shape is shown in Fig. 2.46.

- **Polynomial membership functions**

The advantage of polynomial membership functions is the possibility to graduate their complexity depending on the number of the Schwab axioms to be satisfied by the membership functions.

The most simple polynomial internal membership function is the **second order function** expressed by the formula:

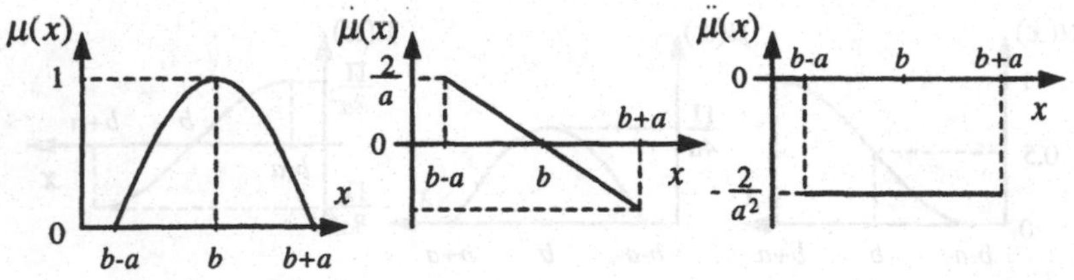

Fig. 2.47. Polynomial second order membership function and its derivatives

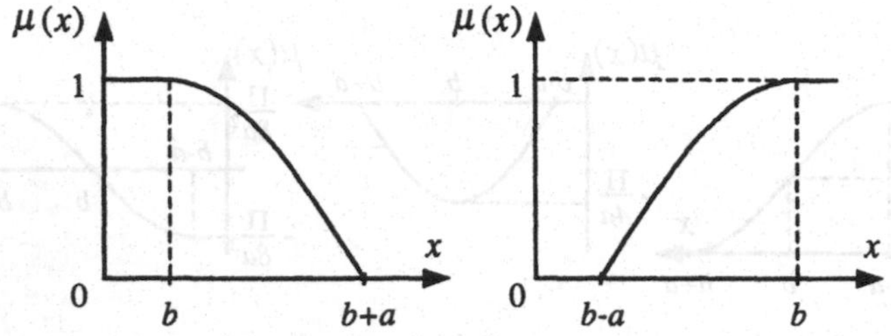

Fig. 2.48. Left and right external polynomial second order membership function

$$\mu(x) = \begin{cases} 1 - \left(\frac{x-b}{a}\right)^2 & \text{for } (b-a) \le x \le (b+a)\,, \\ 0 & \text{otherwise.} \end{cases} \tag{2.48}$$

The shape of this function is presented in Fig. 2.47.

The right external polynomial second order membership function is expressed by the formula:

$$\mu(x) = \begin{cases} 0 & \text{for } x < (b-a)\,, \\ 1 - \left(\frac{x-b}{a}\right)^2 & \text{for } (b-a) \le x \le b\,, \\ 1 & \text{for } x > b\,. \end{cases} \tag{2.49}$$

The left external polynomial second order membership function is expressed by the formula:

$$\mu(x) = \begin{cases} 1 & \text{for } x < b\,, \\ 1 - \left(\frac{x-b}{a}\right)^2 & \text{for } b \le x \le (b+a)\,, \\ 0 & \text{for } x > (b+a)\,. \end{cases} \tag{2.50}$$

The left and right second order membership functions are depicted in Fig. 2.48.

Advantages of the polynomial second order membership function

1. The function is continuously differentiable in the whole range of the support. It is thus smoother than a triangular function.

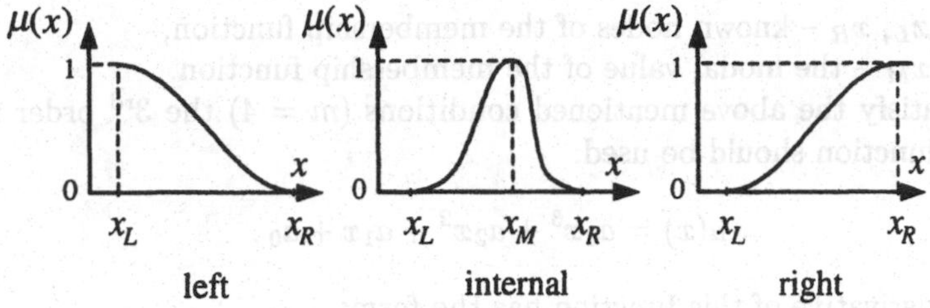

Fig. 2.49. Examples of polynomial membership functions

2. The parameters a, b are easy to identify by the expert.

Disadvantages of the polynomial second order membership function

1. The function satisfies Schwab axioms poorly. In particular, its derivative is not equal to 0 at the ends of the support (Fig. 2.47).
2. The function is symmetrical, which makes the realization of the postulate of partition of unity difficult.

To satisfy a greater number (and even all) of the Schwab axioms, we should use the ***n*-th order polynomial function**, where:

$$n = m - 1,$$

m – the number of conditions for the membership function, related to the Schwab axioms.

The polynomial n-th order membership function is described with the formula:

$$\mu(x) = \begin{cases} a_n x^n + a_{n-1} x^{n-1} + \ldots + a_1 x + a_0 & \text{for } x_L \leq x \leq x_R\,, \\ 0 \text{ or } 1 & \text{otherwise.} \end{cases} \tag{2.51}$$

Examples of shapes of external and internal membership functions are represented in Fig. 2.49.

The method of creating polynomial membership functions will be shown in the following example.

Example 2.4.9
Let us assume that in relation to the left polynomial membership function we lay down the following conditions connected with the Schwab axioms (the number of these conditions can be varied):

$$1.\ \mu(x_L) = 1\,,$$
$$2.\ \mu(x_R) = 0\,,$$
$$3.\ \dot{\mu}(x_L) = 0\,,$$
$$4.\ \dot{\mu}(x_R) = 0\,,$$

where: x_L, x_R – known nodes of the membership function,
x_M – the modal value of the membership function.

To satisfy the above mentioned conditions ($m = 4$) the 3rd order membership function should be used:

$$\mu(x) = a_3x^3 + a_2x^2 + a_1x + a_0 \,.$$

The derivative of this function has the form:

$$\mu(x) = 3a_3x^2 + 2a_2x + a_1 \,.$$

Assuming the values of the nodes of the function:

$$x_L = 0\,, \quad x_R = 1\,,$$

we obtain the set of 4 equations conforming to the conditions laid down in relation to the function:

$$\begin{aligned}
&1.\ a_0 = 1\,,\\
&2.\ a_1 = 0\,,\\
&3.\ a_3 + a_2 + 1 = 0\,,\\
&4.\ 3a_3 + 2a_2 = 0\,,
\end{aligned}$$

giving the solution:

$$\begin{aligned}
&1.\ a_0 = 1\,,\\
&2.\ a_1 = 0\,,\\
&3.\ a_2 = -3\,,\\
&4.\ a_3 = 2\,.
\end{aligned}$$

Finally, the left membership function is expressed by the formula:

$$\mu(x) = 2x^3 - 3x^2 + 1\,.$$

◊ ◊ ◊

Advantages of polynomial high order membership functions

1. Possibility of satisfying conditions laid down in relation to the membership function which are connected with the Schwab axioms, the position of critical points ($\mu(x_k) = 0.5$ and others.
2. Possibility of the considerable increase of accuracy of a fuzzy model and the adaptation of it to system under modeling owing to a large number of degrees of freedom of the polynomial membership functions.
3. The increase of possibilities of obtaining smooth, continuously differentiable surfaces of the input/output mapping of a fuzzy model.

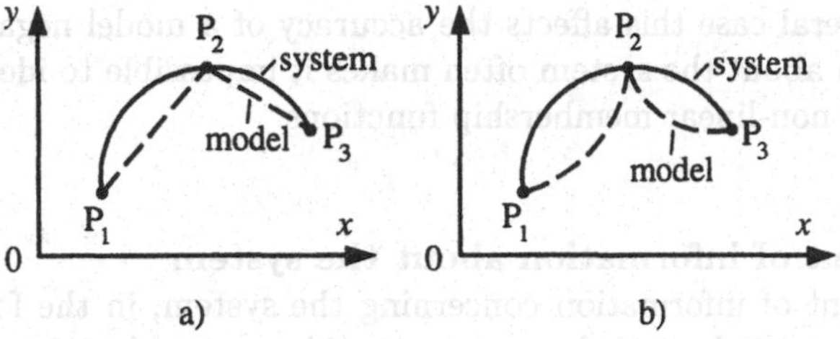

Fig. 2.50. Locally linear and locally non-linear model of a system

4. Ease of obtaining asymmetrical internal membership functions, which are conducive to increasing the accuracy of a model.

Disadvantages of polynomial high order membership functions

1. Difficulties with identification of a large number of parameters defining polynomial high order functions.
2. In the general case, the non-satisfying of the postulate of partition of unity.

- **Recommendations concerning selection of membership functions**

The selection of membership functions depends considerably on the amount of information about the system to be modeled, and also on the quality of the adjusting methods of the model which are at the modeler's disposal.

Small amount of information about the system

If the amount of information concerning the system is small, then the most simple membership functions consisting of straight segments should be used, because a much lesser – in comparison with other membership functions – amount of information is sufficient for the identification of their parameters. Segmentally-linear membership functions are conducive to obtaining locally linear surfaces of the model (if other elements of the fuzzy system have been properly selected). This advantageously affects the accuracy of the model for the small amount of data concerning the system.

Example 2.4.10

Assume that only 3 points of the surface of the system under modeling are known: $P_1 - P_3$, Fig. 2.50.

If no measured data testifying to the operation of the system between the points $P_1 - P_3$ are known, then the use of a model consisting of straight segments (Fig. 2.50a) is the safest. Applying membership functions consisting of not straight but curved segments, we also obtain the surface of a model consisting of curved segments (surface segments), but the direction and degree of curvature is difficult to foresee (Fig. 2.50b) because of the complexity of fuzzy models.

In the general case this affects the accuracy of a model negatively. Lack of information about the system often makes it impossible to identify all the parameters of non-linear membership functions.
◊ ◊ ◊

Large amount of information about the system
A large amount of information concerning the system, in the form of measured input/output data of the system, enables us to identify a large number of parameters of a fuzzy model. One can then apply more complicated membership functions defined by a larger number of parameters, e.g. the Gaussian or polynomial functions, and obtain a higher accuracy of the model than in the case of the use of simple straight segment membership functions. However, for the identification of a large number of parameters, highly effective adapting (adjusting) methods of a fuzzy model are needed. The person creating the model of the system has not always got them at his disposal. Additionally, more complicated membership functions are curvilinear and increase the degree of non-linearity of the model, in turn increasing the number of local extremes of the error function. This makes the process of identification very difficult. To overcome this difficulty, strenuous procedures of genetic algorithms (Davis 1991; Goldberg 1995; Kahlert 1995) must be applied. However, these procedures don't always lead to satisfactory results (Preuss 1995) considering the high degree of complexity of the model and the large number of its degrees of freedom. The experiences of the author (Piegat 1996,1997a) and other scientists (Hensel 1995) show that it is advantageous to apply the most simple straight segment membership functions which makes the process of adjusting (teaching) a fuzzy model easy and secures its high accuracy.

Some scientists (Altrock 1993) recommend applying the most simple membership functions at the first stage of modeling, and in the second stage carrying on a test of the modeling with more complicated membership functions in order to check if these functions allow for increasing the accuracy of the model. However, the opinion sometimes given (Altrock 1995; Zimmermann 1994) that the kind and form of membership function do not considerably affect the accuracy and quality of a fuzzy model is not true. This is testified by the results of research published, among others, in (Baglio 1994; Brown 1994).

2.5 Type 2 fuzzy sets

The type 1 fuzzy set is such a set A_1, in which each element x from the universe of discourse X is assigned a grade of membership $\mu^*_{A1}(x)$ in the set from the interval [0,1] determined on the basis of membership function $\mu_{A1}(x)$:

$$\mu_{A1}(x): \quad X \to [0,1]. \tag{2.52}$$

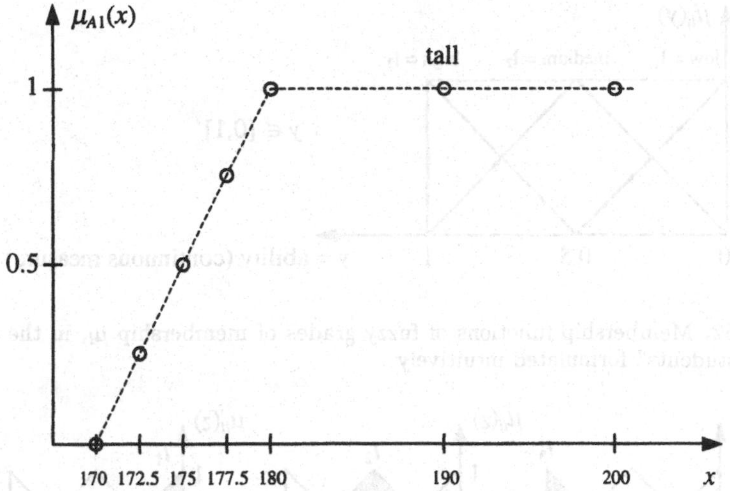

Fig. 2.51. Membership function of type 1 fuzzy set "tall height"

Example 2.5.1
The set A_1 = "tall height":

$$X = \{170\,\mathrm{cm}, 172.5\,\mathrm{cm}, 175\,\mathrm{cm}, 177.5\,\mathrm{cm}, 180\,\mathrm{cm}, 190\,\mathrm{cm}, 200\,\mathrm{cm}\},$$

$$A_1 = \left\{ \frac{0}{170\,\mathrm{cm}}, \frac{0.25}{172.5\,\mathrm{cm}}, \frac{0.5}{175\,\mathrm{cm}}, \frac{0.75}{177.5\,\mathrm{cm}}, \frac{1.0}{180\,\mathrm{cm}}, \frac{1.0}{190\,\mathrm{cm}}, \frac{1.0}{200\,\mathrm{cm}} \right\}.$$

The membership function in the set is represented in Fig. 2.51.

◊ ◊ ◊

However, often the grade of membership in the set can't be determined precisely, by means of a number, but only linguistically, by means of a fuzzy measure.

Example 2.5.2
X = {Andrew, Ben, Charlie} = $\{x_1, x_2, x_3\}$ – the set of students,
L = {low, medium, high} = $\{l_1, l_2, l_3\}$ – the set of fuzzy grades of membership in the set of "clever students",
A – the set of "clever students".

$$A_2 = \left\{ \frac{\text{high}}{\text{Andrew}}, \frac{\text{medium}}{\text{Ben}}, \frac{\text{low}}{\text{Charlie}} \right\} = \left\{ \frac{l_3}{x_1}, \frac{l_2}{x_2}, \frac{l_1}{x_3} \right\}$$

Fuzzy grades of membership in the set of "clever students" can be formulated intuitively, Fig. 2.52.

The set of "clever students" can be presented graphically as in Fig. 2.53.

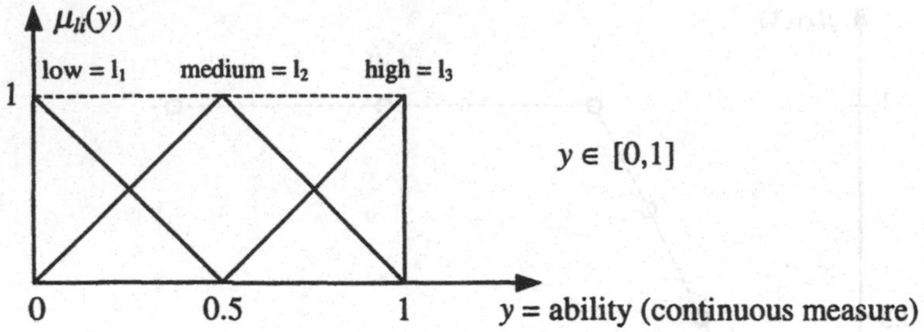

Fig. 2.52. Membership functions of fuzzy grades of membership μ_{li} in the set of "clever students" formulated intuitively

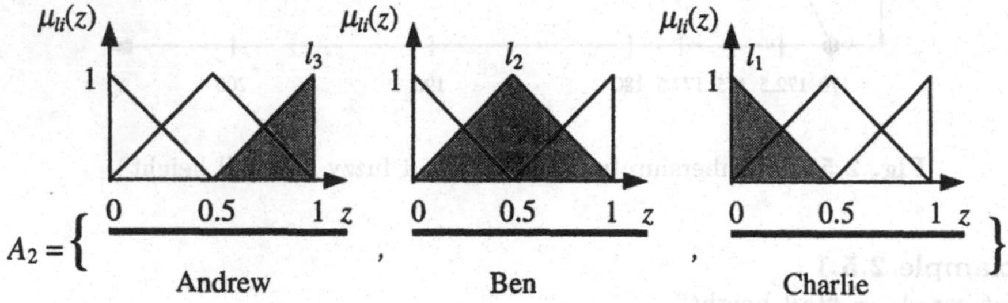

Fig. 2.53. Graphical form of representation of the set of "clever students"

In the evaluation of the ability of students, it is difficult to give precise grades of membership of a particular student in the set of "clever students" in the form of, for example:

$$\mu_{A_2}(\text{Andrew}) = 0.99\,.$$

The only reasonable measure of membership is a fuzzy measure of the type: high, medium, low membership.

◊ ◊ ◊

Other examples of the application of fuzzy membership grades:

- a set of charming girls,
- a set of the years of the high/low danger to the state,
- a set of creditworthy customers of a bank.

Fuzzy sets, in which grade of membership $\mu_A^*(x)$ in the set A is determined by means of a fuzzy measure (being also a fuzzy set) are called the type 2 sets. The type 2 fuzzy set, denoted with the index 2 to distinguish it from type 1 sets, is a set of pairs (2.53):

(fuzzy membership grade of the element x in the set A_2, an element x)

$$A_2 = \{(\mu^*_{A2}(x), x)\}, \quad \forall x \in X, \tag{2.53}$$

where: $\mu^*_{A2}(x)$ – grade of membership of the element x in the set A determined by membership function $\mu_{A2}(x)$, (2.54).

$$\mu_{A2}(x): \quad X \to L, \tag{2.54}$$

where: X – the universe of discourse of the set A_2,
$L = \{l_1, \ldots, l_m\}$ – the set of fuzzy membership grades in the set A_2,
$\{(\mu^*_{l_i}(y), y)\}$ – fuzzy grade of membership in the set A_2,
$\mu_{l_i}(y)$ – membership function of the fuzzy grade of membership l_i in the set A_2,
Y – the universe of discourse of fuzzy grades of membership,
$y: \ y \in Y$.

If the membership function of a type 1 set is the function of one variable $x \in X$, then the membership function of a type 2 set is the function of two variables (2.55):

$$\mu_{A2}(x, y): \quad X \to L, \ Y \to L. \tag{2.55}$$

Example 2.5.3
X – the set of passing cars designated with the number $c_i, i = 1 - 4$,
$X = \{c_1, c_2, c_3, c_4\}$,
A_2 – the set of cars exceeding the permissible speed $V = 60$ (km/h),

$$A_2 = \left\{\frac{N}{c_1}, \frac{Y}{c_2}, \frac{Y}{c_3}, \frac{N}{c_4}\right\},$$

where: $N =$ no – speed of a car accordant with the rules ($V \leq 60$ (km/h)),
$Y =$ yes – too high speed of a car ($V >$ (60 km/h)),
$L = \{N, Y\}$ – the set of fuzzy grades of membership.

The measurement of the speed of a car is carried out by a policeman evaluating the speed with the accuracy ±10 (km/h). In Fig. 2.54 membership function $\mu_L: \ V \to L$ is represented.

◊ ◊ ◊

Type 2 fuzzy sets can be applied to model systems in which the input and/or output quantities can be evaluated by means of fuzzy measures, Fig. 2.55.

Such evaluations are carried out, for example in economic and political systems where it is often necessary to employ intuition or feeling, owing to lack of information. Because fuzzy grades of membership (small, medium, large) can be represented in the form of fuzzy numbers, type 2 sets can be converted using the extension principle or simplified methods (representation L-R). More information on this subject can found in (Kacprzyk 1986; Zimmermann 1994a).

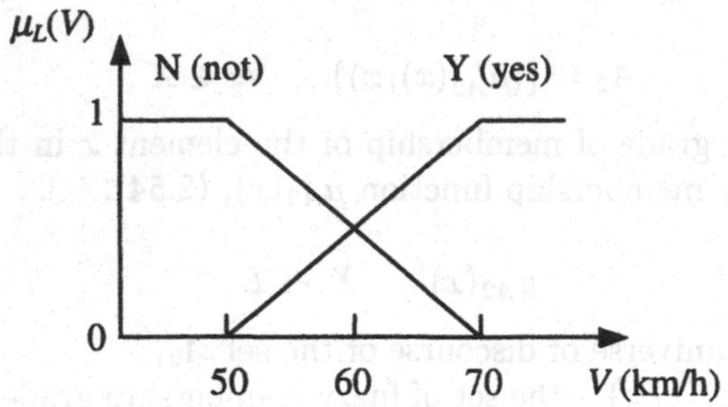

Fig. 2.54. Membership functions of fuzzy grades of membership in the set of cars exceeding the permissible speed

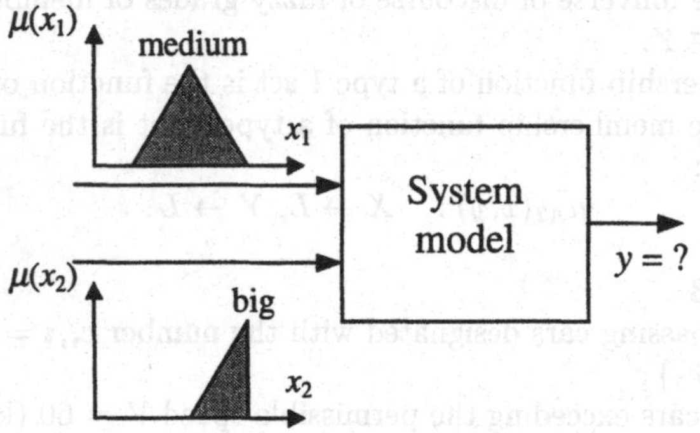

Fig. 2.55. Modeling of a system using fuzzy evaluation of the input quantities

2.6 Fuzziness and probability: two kinds of uncertainty

Fuzziness is often mistaken for probability. Therefore it is necessary to distinguish these concepts. In science (Altrock 1993) the two following types of uncertainty are distinguished (there are also other kinds):

1. stochastic uncertainty,
2. lexical uncertainty.

An example of stochastic uncertainty is the statement:

Probability of winning the first prize in a lottery in the system 6/49 amounts to 1/13983816.

The event "winning the first prize" is precisely determined here. This means 6 lucky hits have been made. The probability of obtaining the first prize can also be calculated precisely (Empacher 1970) owing to knowledge of all 6-element combinations from the set of 49 elements ($m = 6$, $n = 49$).

$$C_n^m = \frac{n!}{m!(n-m)!} \tag{2.56}$$

Stochastic uncertainty has been captured here quantitatively in the form of the probability of occurrence of a certain **exactly defined** event which is winning the first prize = 6 lucky hits.

Example of lexical uncertainty
In the statement: "***Probability of winning a great prize in a lottery in the system 6/49 is slight***" two designations occur:

1. great prize,
2. slight probability.

Both designations are fuzzy, imprecise and depend on the subjective feelings of involved persons. For persons of limited means the prize for 4 or 5 lucky hits can be "great", for very rich persons only 6 lucky hits will enter into account.

It is difficult to state precisely the concept "little probability" because an ordinary player at a lottery does not know the exact numerical value of probability of the first prize calculated from the formula (2.56). Usually he can not give even an approximate value of the probability, which is estimated by him intuitively and on the basis of his hope to be a winner. The man whose hope is at a high level plays a lottery, but when it is at a low level he declines to play.

The above example shows that lack of precise knowledge of reality is no drawback for a man in functioning in it and making decisions. For years scientists have tried to work out precise mathematical models of various fragments of reality. But till now only a small part of the phenomena in the universe has been modeled successfully. To create such models a great deal of knowledge about the phenomena to be modeled is necessary.

However, human beings can, irrespective of their education, create in their minds models of surrounding reality, models of machines to be operated, cars,airplanes etc, whose effectiveness is high. These models enclose such concepts as:

- big, small,
- pretty, ugly,
- reasonable, unreasonable,
- like X, unlike X, and so on.

They are imprecise lexical concepts and evaluations by means of which a man describes reality. The richer the vocabulary of a given man is, the more precisely he can formulate the description of the fragment of reality in which he is interested.

Summarizing:

- stochastic uncertainty means uncertainty of occurrence of an event, which is itself precisely defined,
- lexical uncertainty means uncertainty of the definition of this event.

Uncertainty of the definition means its fuzziness. Fuzzy systems theory is engaged in methods of creating models employing fuzzy concepts, which are used by people. It should be mentioned that people also employ, apart from lexical fuzzy concepts, intuitive concepts and pictures not connected at all with any vocabulary. There are people who know no language, there are also animals, which create intuitive, non-lexical information about reality enabling them to function and survive in it. The theory of **intuitive modeling** may probably be the continuation of the theory of fuzzy modeling in the future.

The difference between fuzziness and probability will be illustrated additionally by means of Example 2.6.1 based on (Bezdek 1993).

Example 2.6.1
Denote by X the set of all liquids, by A_D the set of drinkable liquids, by A_N – the set of non-drinkable liquids.

The grade of membership of spring water in the set of drinkable liquids amounts to 1, in the set of non-drinkable it amounts to 0. Assume that the grade of membership of river water derived at the mouth of the Vistula in the set of drinkable liquids amounts to 0.6 (this water is drunk by aquatic fowls) and in the set of non-drinkable ones 0.4. Let the bottle A be filled with just this river water.

The grade of membership of hydrochloric acid in the set of drinkable liquids amounts to 0 and in the set of non-drinkable ones to 1. Assume that we draw a bottle B from a basket in which there are 10 bottles – among them 6 bottles with spring water and 4 bottles with hydrochloric acid. The probability of drawing a drinkable liquid amounts to 0.6.

Assume that we must choose one of these two bottles.

Bottle A	Bottle B
$\mu_{A_D}(A) = 0.6$	$P_{A_D}(A) = 0.6$

where: $\mu_{A_D}(A)$ – grade of membership of liquid in the bottle A in the set of drinkable liquids A_D,
$P_{A_D}(A)$ – probability that liquid in the bottle B is a drinkable liquid.

What choice should be made in the situation when consumption of liquid from the bottle A is menacing for health and from the bottle B is menacing for life?

◊ ◊ ◊

3. Arithmetic of Fuzzy Sets

Fuzzy numbers may be applied, for example, in the modeling of a system of a known input/output mapping given in terms of a conventional mathematical model $y = f(\mathbf{X})$, where the input signals cannot be measured precisely but only approximately, e.g.:

$$x_1 = \text{"approx. 9"},$$
$$x_2 = \text{"approx. 10"},$$
$$y = x_1 + x_2 .$$

Thus, the system output value y can be calculated in a fuzzy number form, Fig. 3.1.

If a model of a system $y = f(\mathbf{X})$ has been given in a mathematical form that requires the application of such operations as addition, subtraction, multiplication or division, then a method of performing the operations for fuzzy numbers has to be defined. The operation methods are essential since they enable us to make use of fuzzy evaluations of input values made by a person with his/her senses or intuition and enter those evaluations into the conventional mathematical models of systems. Appropriate methods also allow us the construction of hybrid models consisting of fuzzy and conventional sections where the conventional sections can also be used to process the fuzzy information generated by the fuzzy fragments of the model.

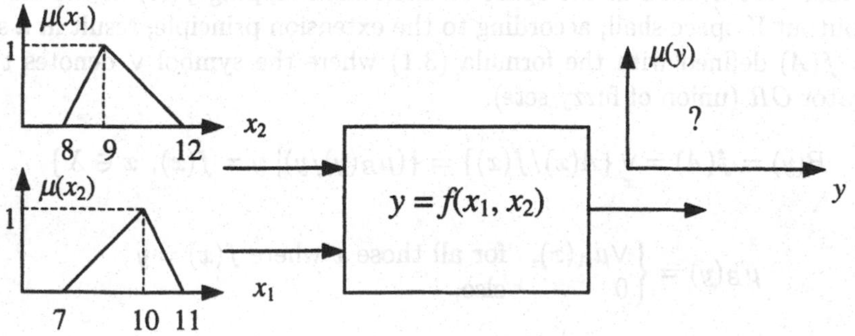

Fig. 3.1. Determination of a model's fuzzy output where the fuzzy information on input states is given

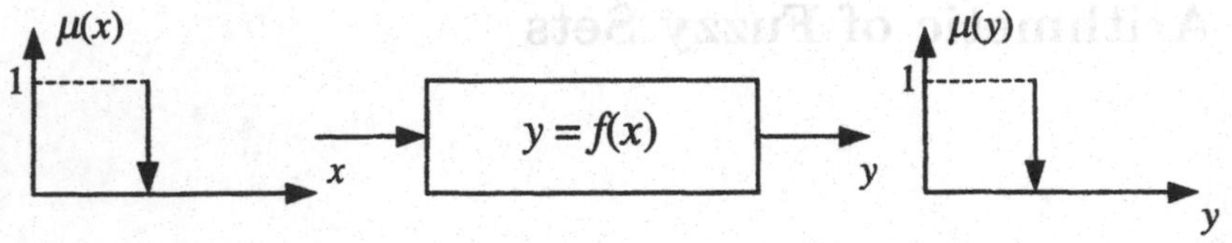

$$x \in X, \; y \in Y, \quad f: \; X \to Y$$

Fig. 3.2. Conventional SISO system (single input/single output) with a crisp input x and an output y

3.1 The extension principle

The conventional arithmetic provides the methods for performing the operations of addition, subtraction, multiplication or division on crisp numbers, e.g. 4, 5, 6. The fuzzy arithmetic defines the methods of performing the above operations on fuzzy numbers, e.g.:

about 4,
plus/minus 5,
approximately 6.

The fuzzy arithmetic defines basic mathematical operations on fuzzy numbers through an extension of the operations from the crisp numbers (Kacprzyk 1986; Kaufman 1985; Driankov 1993,1996). The method of the extension is given in Zadeh's extension principle. It is presented below in two versions, i.e. for SISO and MISO systems.

SISO System (single input/single output)

Let us have a conventional system of a single input / single output type performing a mapping f of an input X space onto an output Y space. If A is a fuzzy set defined in the space X then the mapping $f(A)$ of the set in the output Y space shall, according to the extension principle, result in a set $B = f(A)$ defined with the formula (3.1) where the symbol $\vee$ denotes the operator OR (union of fuzzy sets).

$$B(y) = f(A) = \bigvee_{x} \{A(x)/f(x)\} = \{(\mu_B(y)/y)| \; y = f(x), \; x \in X\}$$

$$\mu_B(y) = \begin{cases} \vee \mu_A(x), & \text{for all those } x \text{ where } f(x) = y\,, \\ 0 & \text{else,} \end{cases}$$

or in brief:

$$\mu_B(y) = \mu_{f(A)}(y) = \bigvee_{y=f(x)} \mu_A(x), \quad x \in X, \; y \in Y\,. \tag{3.1}$$

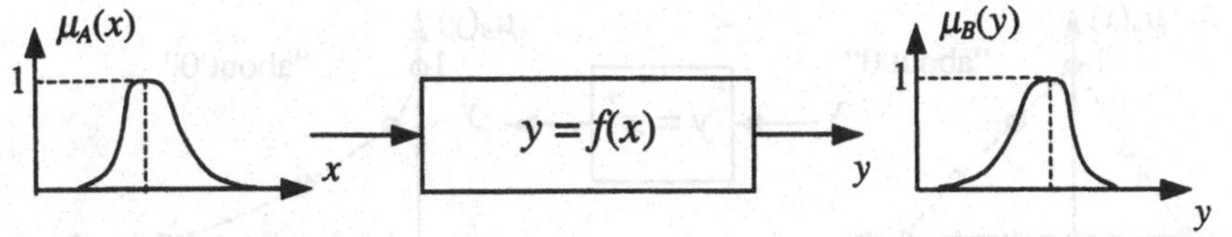

Fig. 3.3. Mapping of a fuzzy input x to a fuzzy output y by a conventional SISO system

In many cases of real systems the input and output values (e.g. current voltage or intensity) may be expressed by means of the real numbers, therefore in our further considerations the universes of discourse X, Y will be the real number spaces R. In general, the universes X, Y can be other spaces consisting of any arbitrary elements. If an input x and an output y meet the condition (3.2),

$$\forall x, y \in R, \tag{3.2}$$

then the single-argument (SISO system) extension principle is in fact the calculation of an output membership function $\mu_B(x)$ using the formula (3.3).

$$\mu_B(y) = \mu_{f(A)}(x) = \bigvee_{y=f(x)} \mu_A(x), \quad \forall x, y \in R. \tag{3.3}$$

If the operation $\vee$ is performed using the operator MAX then the formula (3.3) can be expressed in form (3.4) as follows:

$$\mu_B(y) = \underset{y=f(x)}{MAX}\, \mu_A(x), \quad \forall x, y \in R. \tag{3.4}$$

The concept of a single-argument extension operation is presented in Fig. 3.3.

Example 3.1.1
Let a SISO system perform a mapping $X \to Y$ where $y = x^2$. The state of the input x is given in the form of a fuzzy number $A(x)$ = "about 0" defined in Table 3.1.

Table 3.1. Fuzzy number $A(x)$, "about 0"

$\mu_A(x)$	0	0.5	1	0.66	0.33	0
x	−2	−1	0	1	2	3

A fuzzy output state $B(y)$ corresponding to the state of the input x has to be determined. The membership function is calculated according to the formula (3.5).

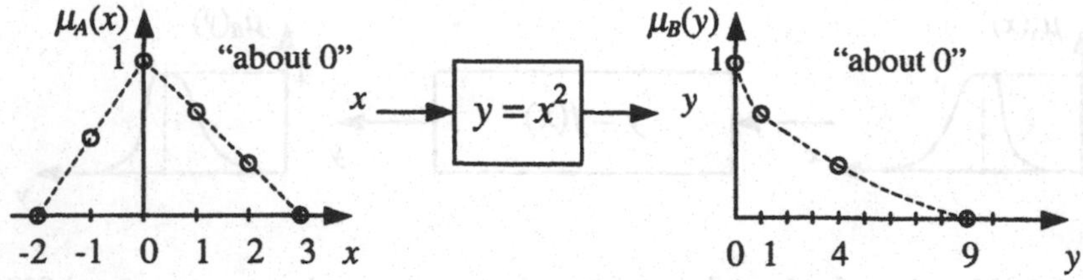

Fig. 3.4. Mapping of the fuzzy number from an input to an output of the conventional SISO system

$$\mu_B(y) = \underset{y=x^2}{MAX}\,\mu_A(x), \quad \forall x, y \in R \quad x \in X,\ y \in Y. \tag{3.5}$$

For example, the value $y = 1$ is obtained for $x = 1$ or $x = -1$. Thus, the formula to calculate the membership grade $\mu_B(y)$ for $y = 1$ can be expressed in form (3.6).

$$\mu_B(y) = \underset{y=1}{MAX}((0.5), (0.66)) = 0.66 \tag{3.6}$$

The fuzzy set B is presented in Table 3.2.

Table 3.2. Fuzzy number $B(y)$, "about 0"

$\mu_B(y)$	1	0.66	0.33	0
y	0	1	4	9

Fig. 3.4 presents the fuzzy number at the input to the system and its mapping to the system output.

◊ ◊ ◊

Below we will present the application of the single-argument extension principle to perform various mathematical operations involving fuzzy numbers.

- **Opposite fuzzy number**

An opposite fuzzy number for a fuzzy number A, i.e. $-A$, can be calculated using the extension principle where the applicable formula is given in the form (3.7).

$$\mu_{-A}(y) = \bigvee_{y=-x} \mu_A(x), \quad \forall x, y \in R \tag{3.7}$$

Example 3.1.2

Let us calculate an opposite number for a number $A =$ "about 2", Table 3.3. The calculation result is presented in Table 3.4 and in Fig. 3.5.

On the basis of the above we can see that the number A and its opposite number $-A$ are symmetrical about the ordinate axis.

◊ ◊ ◊

Table 3.3. Fuzzy number $A =$ "about 2"

$\mu_A(x)$	0	0.5	1	0.5	0
x	1	1.5	2	2.25	2.5

Table 3.4. Opposite fuzzy number $-A =$ "about -2"

$\mu_A(x)$	0	0.5	1	0.5	0
x	-1	-1.5	-2	-2.25	-2.5

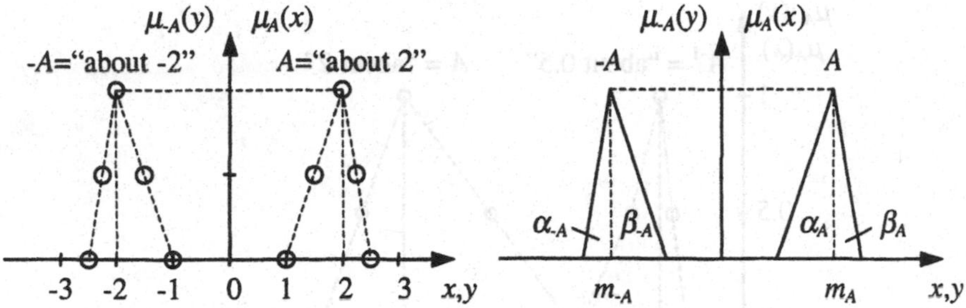

Fig. 3.5. Number $A =$ "about 2" and number $B = -A =$ "about -2"

The parameters of the opposite numbers are interrelated in the way presented in (3.8).

$$\begin{aligned} m_{-A} &= -m_A \\ \alpha_{-A} &= \beta_A \\ \beta_{-A} &= \alpha_A \end{aligned} \tag{3.8}$$

If the number A in its **L-R (left-right) representation** has a form (3.9):

$$A = (m_A, \alpha_A, \beta_A), \tag{3.9}$$

then the opposite number $-A$ has its L-R representation expressed with the formula (3.10):

$$-A = (-m_A, \beta_A, \alpha_A), \tag{3.10}$$

- **Inverse fuzzy number**

An inverse fuzzy number A^{-1} can be calculated using the extension principle formula (3.11).

$$\mu_{A^{-1}}(y) = \bigvee_{y=\frac{1}{x}} \mu_A(x), \quad \forall x, y \in R, \quad x \in X \tag{3.11}$$

Table 3.5. Fuzzy number $A =$ "about 2"

$\mu_A(x)$	0	0.5	1	0.5	0
x	1	1.5	2	2.25	2.5

Table 3.6. Inverse fuzzy number $A^{-1} =$ "about 0.5"

$\mu_A(x)$	0	0.5	1	0.5	0
x	1	0.66	0.5	0.44	0.4

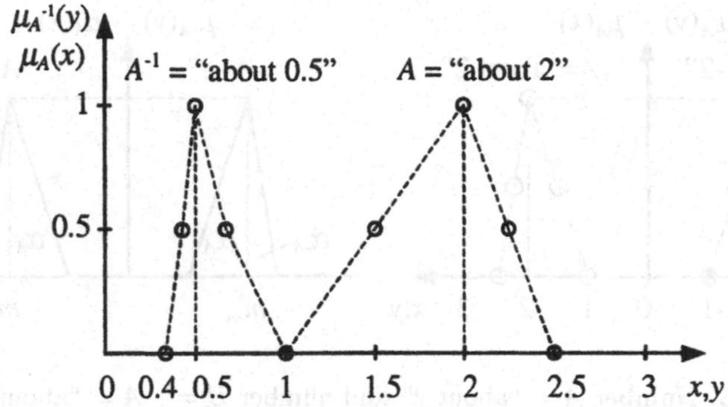

Fig. 3.6. Number $A =$ "about 2" and its inverse number A^{-1} calculated using the extension principle

Example 3.1.3

Let us calculate an inverse fuzzy number for a fuzzy number $A =$ "about 2", defined in Table 3.5. The results of calculations are shown in Table 3.6 and Fig. 3.6.

Fig. 3.6 illustrates the number A and its inverse number A^{-1}.

◊ ◊ ◊

The inversion of a fuzzy number can be defined in a simplified way using the L-R representation (Kacprzyk 1986; Kaufman 1985). Fig. 3.7 illustrates the notation used in L-R representation of A.

The characteristic points of a number A and its inverse A^{-1} are interrelated as shown in (3.12).

$$\begin{aligned} m_{A^{-1}} &= 1/m_A, \quad m_A \neq 0 \\ (m_{A^{-1}} - \alpha_{A^{-1}}) &= 1/(m_A + \beta_A) \\ (m_{A^{-1}} + \beta_{A^{-1}}) &= 1/(m_A - \alpha_A), \quad m_A \neq \alpha_A \end{aligned} \tag{3.12}$$

On the basis of the formulas (3.12) the left and right spreads (deviations) of the number A^{-1} (3.13) can be calculated.

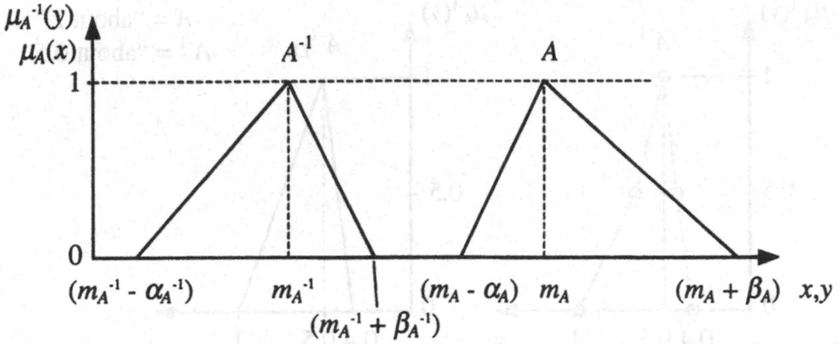

Fig. 3.7. Notation for the spreads (deviations) and nominal values of fuzzy numbers A and A^{-1}

$$\alpha_{A^{-1}} = \frac{\beta_A}{m_A(m_A + \beta_A)}, \quad m_A \neq 0$$

$$\beta_{A^{-1}} = \frac{\alpha_A}{m_A(m_A - \alpha_A)}, \quad m_A \neq \alpha_A \tag{3.13}$$

(3.14)

Thus, the inverse number A^{-1} has the L-R representation expressed by the formula (3.15).

$$A^{-1} \cong \left(\frac{1}{m_A}, \frac{\beta_A}{m_A(m_A + \beta_A)}, \frac{\alpha_A}{m_A(m_A - \alpha_A)}\right) \tag{3.15}$$

Example 3.1.4

Let us calculate an inverse number A^{-1} for $A = (2, 1, 0.5)$ using the L-R representation. The application of the formula (3.15) results in the following:

$$A^{-1} \cong (0.5, 0.1, 0.5)$$

Fig. 3.8 shows the comparison between the inverse number A^{-1}, obtained in the preceding example using the extension principle and the inverse number A^{-1}, obtained in this example using the L-R representation.

Fig. 3.8 proves that the formula (3.15) enables the calculation of a nominal value and the spreads of an inverse number precisely. However, the entire membership function described with the above mentioned formula is of an approximate character.

◊ ◊ ◊

When applying the extension principle the other single-argument operations on fuzzy numbers can be performed as well. Additional information on the above subject can be found in (Kacprzyk 1986; Kaufmann 1985).

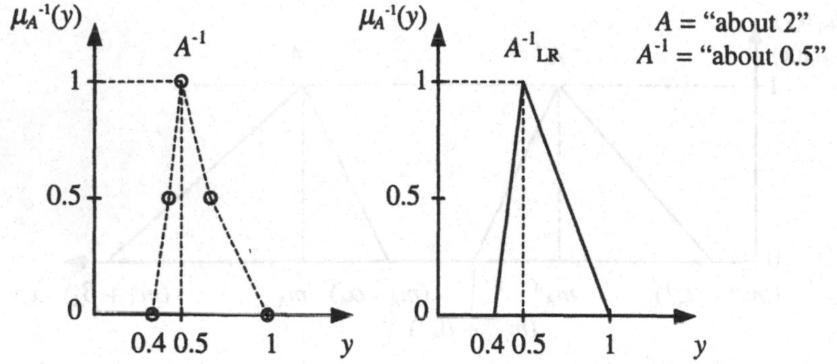

Fig. 3.8. Fuzzy number A^{-1} obtained through the extension principle formula and the simplified version of the same number A^{-1}_{LR} obtained with L-R representation formula

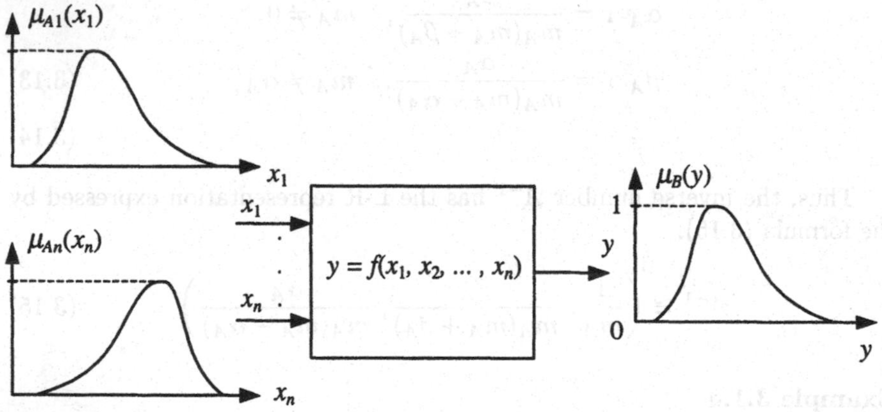

Fig. 3.9. The conventional MISO system with fuzzy inputs and a fuzzy output

MISO system (multiple input/single output)

Let us have a given conventional MISO system, Fig. 3.9, performing the mapping as follows:

$$y = (x_1, x_2, \ldots, x_n)$$

An input vector $\mathbf{X}$ is defined in the Cartesian product space of the universe of discourse of particular inputs $X_1 \times X_2 \times \cdots \times X_n$.

$$\mathbf{X} = \begin{bmatrix} x_1 \\ x_2 \\ \vdots \\ x_n \end{bmatrix}$$

The function f maps the set elements belonging to the universe of discourse of the input vector $\mathbf{X}$ onto the universe of discourse of the output Y, (3.16).

$$f: \quad X_1 \times X_2 \times \cdots \times X_n \to Y \tag{3.16}$$

If $A_1, \ldots, A_n$ are the fuzzy sets defined in the spaces $X_1, \ldots, X_n$ of particular inputs then the extension principle enables determination of a fuzzy set $B = f(A_1, \ldots, A_n)$ at the system output, where the output set is the result of the mapping of the input sets based on the formula (3.17).

$$\begin{aligned} B(y) &= \bigvee_{\substack{\text{for all those } X \in X_1 \times \cdots \times X_n, \\ \text{that } f(X)=y}} [A_1(x_1) \wedge A_2(x_2) \wedge \ldots \wedge A_n(x_n)] = \\ &= \{\mu_B(y)/y | \;\; y = f(\mathbf{X}), \; \mathbf{X} \in X_1 \times \cdots \times X_n\} \end{aligned} \tag{3.17}$$

The universes of discourse X_i and Y are usually the spaces of real numbers R.

In practice the extension operation in its multi-argument version is in fact the determination of a membership function of the system output using the formula (3.18).

$$\begin{aligned} \mu_B(y) = &\bigvee_{y=f(x_1,\ldots,x_n)} (\mu_{A1}(x_1) \wedge \mu_{A2}(x_2) \wedge \ldots \wedge \mu_{An}(x_n)) \, , \\ &\forall x_1, \ldots, x_n, y \in R \, , \end{aligned} \tag{3.18}$$

where: $\vee$ – denotes a set union operator of *MAX* type or an algebraic sum or other s-norms,

$\wedge$ – denotes a set intersection operator of *MIN* type or *PROD* type or other t-norms.

Below we shall present some applications of the extension principle used for performing the basic arithmetic operations in a system with two inputs.

3.2 Addition of fuzzy numbers

The addition of two fuzzy numbers is a mapping of an input vector:

$$\mathbf{X} = [x_1, x_2]^{\mathrm{T}} \, ,$$

defined in the Cartesian product space $R \times R$ onto an output y defined in the real space R, Fig. 3.10. If A_1 and A_2 are the fuzzy numbers then their sum is also a fuzzy number determined with the formula (3.19).

$$(A_1 + A_2)(y) = \bigvee_{y=x_1+x_2} [A_1(x_1) \wedge A_2(x_2)], \quad \forall x_1, x_2, y \in R \tag{3.19}$$

To calculate the sum of fuzzy numbers it is sufficient to determine the membership function $\mu_{A1+A2}(y)$ using the formula (3.20),

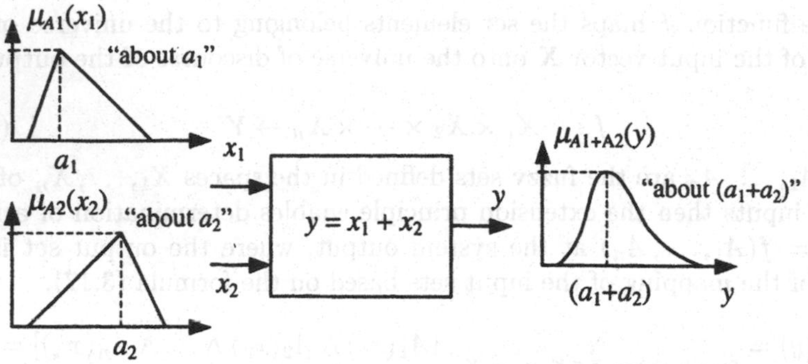

$$x_1 \in X_1 = R, \quad x_2 \in X_2 = R, \quad y \in Y = R$$

Fig. 3.10. Conventional system providing the addition of two fuzzy numbers

$$\mu_{A_1+A_2}(y) = \bigvee_{y=x_1+x_2} [\mu_{A1}(x_1) \wedge \mu_{A2}(x_2)], \quad \forall x_1, x_2, y \in R, \qquad (3.20)$$

where: $\vee$ – denotes a set union operator (e.g. s-norms),
$\wedge$ – denotes a set intersection operator (e.g. t-norms).

Example 3.2.1
Two fuzzy numbers A_1 = "about 5", Table 3.7, and A_2 = "about 7", Table 3.8, are given. Let us determine a fuzzy number $(A_1 + A_2)$.

Fig. 3.11 illustrates the method of calculation of the fuzzy number $(A_1 + A_2)$. On having completed the extension principle-based calculation we obtain the fuzzy number $(A_1 + A_2)$ = "about 12", presented in graphical form in Fig. 3.12 and Table 3.9.

Table 3.7. Fuzzy number A_1 = "about 5"

$\mu_{A1}(x1)$	0	0.333	0.666	1	0.5	0
x_1	2	3	4	5	6	7

Table 3.8. Fuzzy number A_2 = "about 7"

$\mu_{A2}(x2)$	0	0.5	1	0.666	0.333	0
x_2	5	6	7	8	9	10

Table 3.9. Fuzzy number $(A_1 + A_2)$ = "about 12"

$\mu_{A1+A2}(y)$	0	0	0.33	0.5	0.66	1	0.66	0.5	0.33	0	0
y	7	8	9	10	11	12	13	14	15	16	17

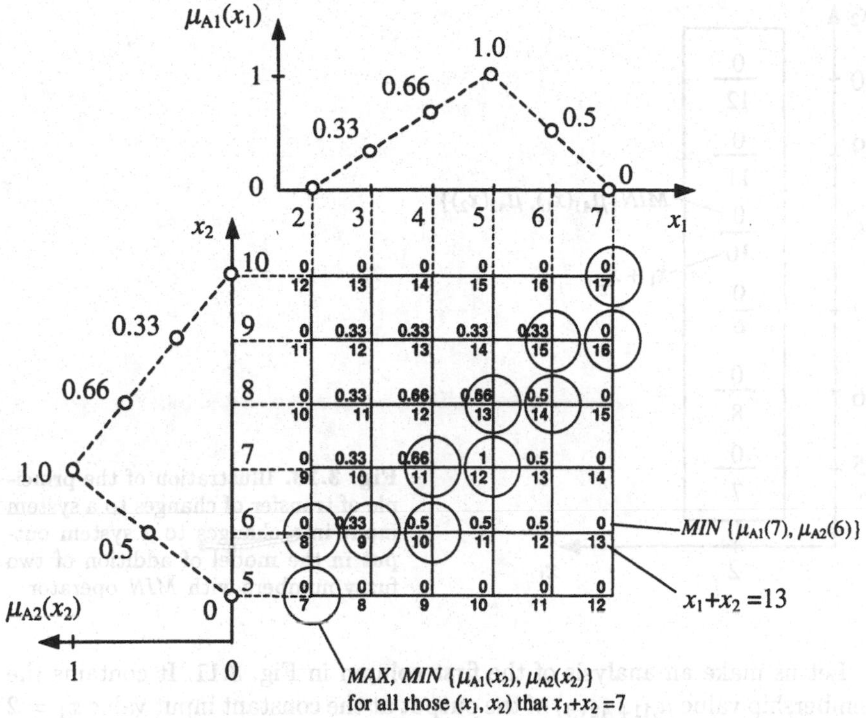

Fig. 3.11. The method of determination of the membership function for the fuzzy number (A_1+A_2), where A_1 = "about 5", A_2 = "about 7" and (A_1+A_2) = "about 12" obtained with the MAX (∨) and MIN (∧) operators

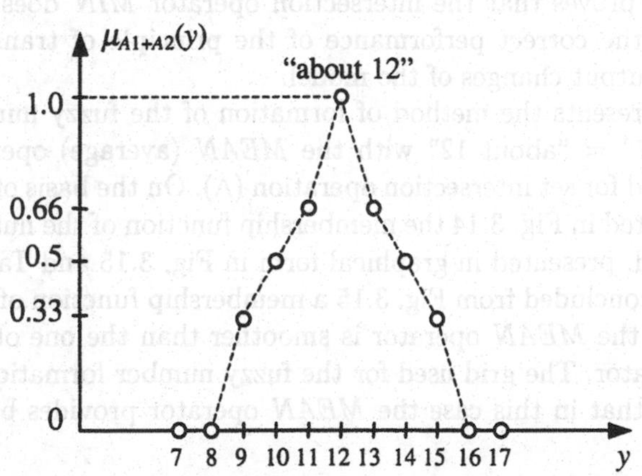

Fig. 3.12. The discrete membership function of the number "about 12" being the sum of the fuzzy numbers "about 5" + "about 7" obtained with the MAX (∨) and MIN (∧) operators

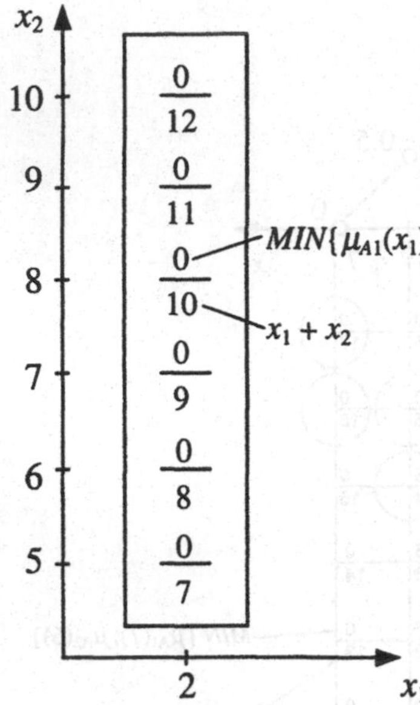

Fig. 3.13. Illustration of the principle of transfer of changes to a system input into changes to a system output in the model of addition of two fuzzy numbers with *MIN* operator

Let us make an analysis of the first column in Fig. 3.11. It contains the membership value $\mu_{A1+A2}(y)$ of the output at the constant input value $x_1 = 2$ and at the varying input value x_2 ranging from 5 up to 10. The above analysis proves that the membership value $\mu_{A1+A2}(y)$ remains constant (equal to 0), Fig. 3.13, although the output y of the modeled system varies from 7 up to 12.

The above proves that the intersection operator *MIN* does not provide (in this case) the correct performance of the principle of transfer of input changes into output changes of the model.

Fig. 3.14 presents the method of formation of the fuzzy number "about 5" + "about 7" = "about 12" with the *MEAN* (average) operator (Yager 1994,1995) used for set intersection operation ($\wedge$). On the basis of the calculation grid depicted in Fig. 3.14 the membership function of the number "about 12" is obtained, presented in graphical form in Fig. 3.15 and Table 3.10.

As can be concluded from Fig. 3.15 a membership function of a fuzzy sum obtained with the *MEAN* operator is smoother than the one obtained with the *MIN* operator. The grid used for the fuzzy number formation, Fig. 3.15, indicates also that in this case the *MEAN* operator provides better perfor-

Table 3.10. Fuzzy number $(A_1 + A_2)$ = "about 12"

$\mu_{A1+A2}(y)$	0	0.25	0.5	0.66	0.83	1	0.83	0.66	0.5	0.25	0
y	7	8	9	10	11	12	13	14	15	16	17

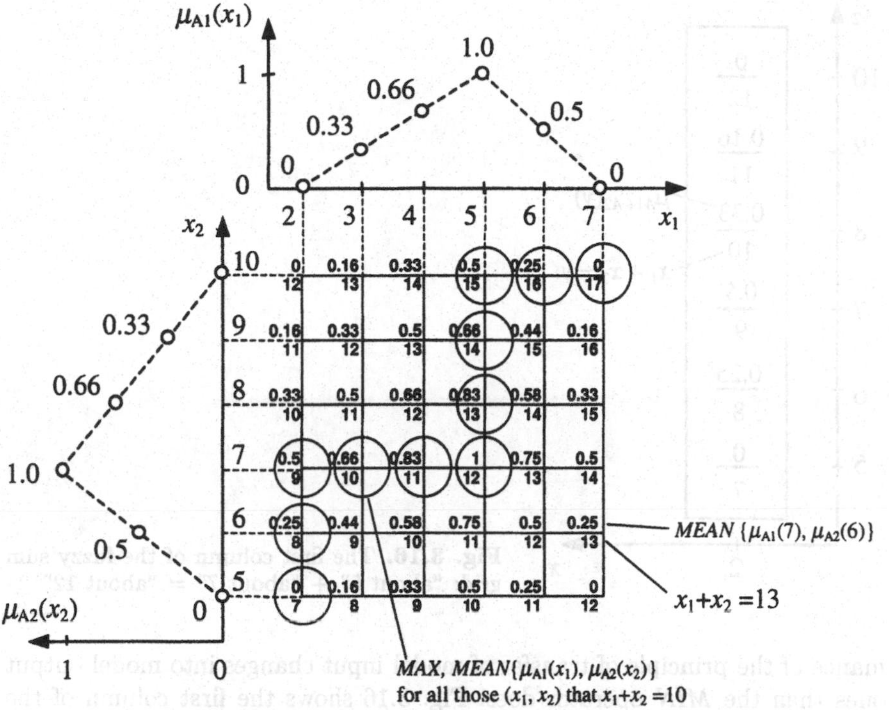

Fig. 3.14. Method of determination of the membership function for the fuzzy number $B = (A_1 + A_2)$, where A_1 = "about 5", A_2 = "about 7" and B = "about 12", with the use of the *MEAN* ($\wedge$) operator

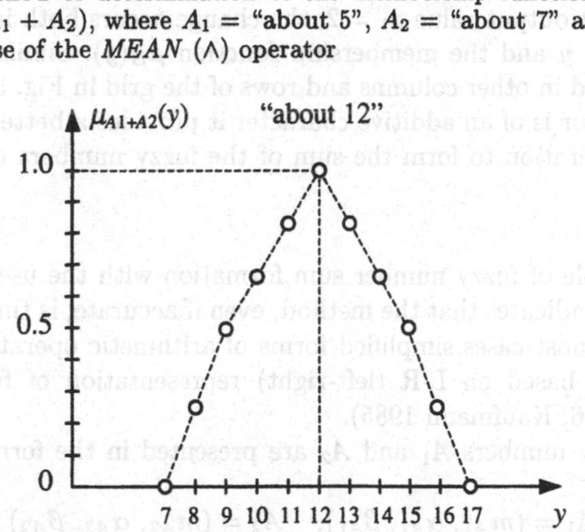

Fig. 3.15. Membership function of the fuzzy sum "about 5" + "about 7" = "about 12" obtained with the *MEAN* ($\wedge$) operator

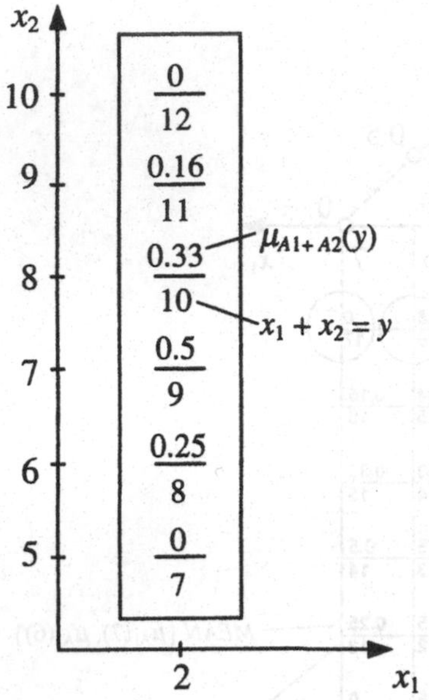

Fig. 3.16. The first column of the fuzzy sum grid: "about 5" + "about 7" = "about 12"

mance of the principle of transfer of model input changes into model output ones than the *MIN* operator does. Fig. 3.16 shows the first column of the fuzzy number grid presented in Fig. 3.15.

Fig. 3.16 allows us to presume that along with the change of the input x_2, at the constant output value $x_1 = 2$, the change occurs both in the modeled system output y and the membership function $\mu_B(y)$. Similar occurrences can be observed in other columns and rows of the grid in Fig. 3.14. Since the *MEAN* operator is of an additive character it provides a better performance of additive operation to form the sum of the fuzzy numbers than the *MIN* operator does.

◊ ◊ ◊

The example of fuzzy number sum formation with the use of the extension principle indicates that the method, even if accurate, is time-consuming. Therefore, in most cases simplified forms of arithmetic operations are used, i.e. the forms based on L-R (left-right) representation of fuzzy numbers (Kacprzyk 1986; Kaufmann 1985).

If the fuzzy numbers A_1 and A_2 are presented in the form of triples as follows:

$$A_1 = (m_{A1}, \alpha_{A1}, \beta_{A1}), \quad A_2 = (m_{A2}, \alpha_{A2}, \beta_{A2}), \tag{3.21}$$

and their sum in form of a triple (3.22):

$$(A_1 + A_2) = (m_{A1+A2}, \alpha_{A1+A2}, \beta_{A1+A2}), \tag{3.22}$$

then the parameters of the sum $(A_1 + A_2)$ and its components A_1 and A_2 are interrelated as shown in (3.23) and presented graphically in Fig. 3.17.

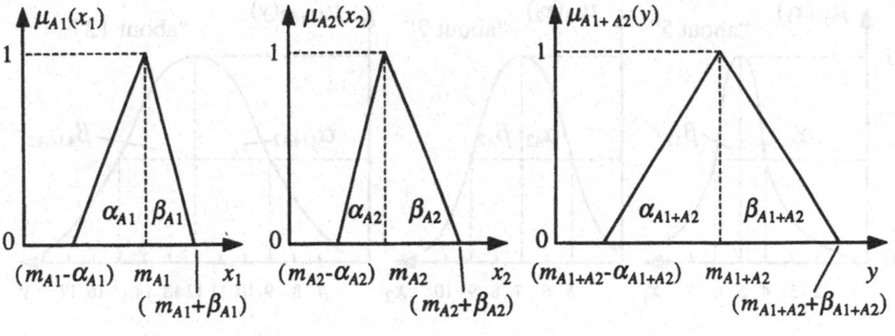

Fig. 3.17. Characteristic parameters of fuzzy numbers being added

$$\begin{aligned} m_{A1+A2} &= m_{A1} + m_{A2} \\ m_{A1+A2} - \alpha_{A1+A2} &= (m_{A1} - \alpha_{A1}) + (m_{A2} - \alpha_{A2}) \\ m_{A1+A2} + \beta_{A1+A2} &= (m_{A1} + \beta_{A1}) + (m_{A2} + \beta_{A2}) \end{aligned} \tag{3.23}$$

On the basis of relations presented in (3.23) the parameters of the sum $(A_1 + A_2)$ can be determined with the use of formulas (3.24).

$$\begin{aligned} \alpha_{A1+A2} &= \alpha_{A1} + \alpha_{A2} \\ \beta_{A1+A2} &= \beta_{A1} + \beta_{A2} \end{aligned} \tag{3.24}$$

Thus, the sum $(A_1 + A_2)$ in the form of the L-R representation can be expressed with the formula (3.25).

$$(A_1 + A_2) = (m_{A1} + m_{A2}\,, \alpha_{A1} + \alpha_{A2}\,, \beta_{A1} + \beta_{A2}) \tag{3.25}$$

Example 3.2.2

Let us calculate a sum of the two symmetrical fuzzy numbers $A_1(x_1)$ = "about 5" and $A_2(x_2)$ = "about 7", where their membership functions are determined using the formulas (3.26) and (3.27).

$$\mu_{A1}(x_1) = \frac{1}{1 + \left(\frac{x_1 - 5}{1}\right)^2}, \quad -\infty < x_1 < +\infty. \tag{3.26}$$

$$\mu_{A2}(x_2) = \frac{1}{1 + \left(\frac{x_2 - 7}{2}\right)^2}, \quad -\infty < x_2 < +\infty. \tag{3.27}$$

The L-R representation of the above numbers can be expressed in the form of (3.28) and (3.29).

$$A_1 = (m_{A1}, \alpha_{A1}, \beta_{A1}) = (5, 1, 1) \tag{3.28}$$

$$A_2 = (m_{A2}, \alpha_{A2}, \beta_{A2}) = (7, 2, 2) \tag{3.29}$$

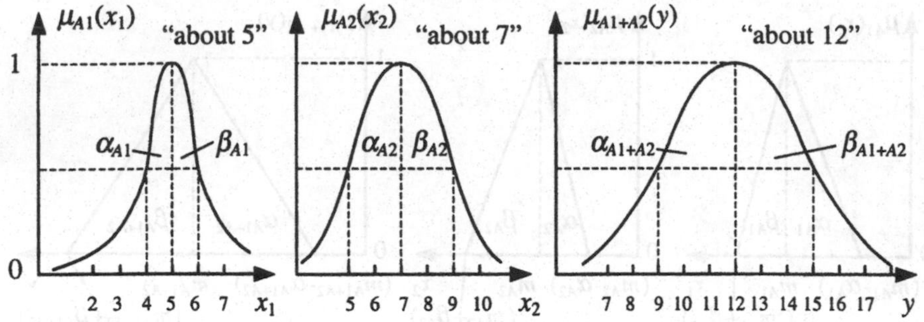

Fig. 3.18. Addition of two fuzzy numbers using the L-R representation method

The sum calculated on the basis of formula (3.25) is presented in (3.30),

$$(A_1 + A_2) = (5, 1, 1) + (7, 2, 2) = (12, 3, 3)\,, \tag{3.30}$$

and the sum membership function is presented in formula (3.31).

$$\mu_{A1+A2}(y) = \frac{1}{1 + \left(\frac{y-12}{3}\right)^2} \tag{3.31}$$

Fig. 3.18 depicts the fuzzy numbers A_1 and A_2 and the result of their addition. It can be seen that the spreads α, β of the sum $A_1 + A_2$ are bigger than the spreads of its particular components A_1 or A_2.

◊ ◊ ◊

Example 3.2.3
Let us calculate a sum of two triangular fuzzy numbers $A_1(x_1) =$ "about 5" and $A_2(x_2) =$ "about 7". The L-R representations of the above numbers are expressed with formulas (3.32) and (3.33).

$$\mu_{A1}(x_1) = \begin{cases} L\left(\frac{x-m}{\alpha}\right) = MAX\left(0, 1 + \frac{x_1-5}{3}\right), & \forall x_1 : x_1 < 5\,, \\ R\left(\frac{x-m}{\beta}\right) = MAX\left(0, 1 - \frac{x_1-5}{2}\right), & \forall x_1 : x_1 \geq 5\,. \end{cases} \tag{3.32}$$

$$\mu_{A2}(x_2) = \begin{cases} MAX\left(0, 1 + \frac{x_2-7}{2}\right), & \forall x_2 : x_2 < 7\,, \\ MAX\left(0, 1 - \frac{x_2-7}{3}\right), & \forall x_2 : x_2 \geq 7\,. \end{cases} \tag{3.33}$$

The L-R representation of the sum $(A_1 + A_2)$, calculated with the formula (3.25) is expressed with (3.34):

$$(A_1 + A_2) = (5, 3, 2) + (7, 2, 3) = (12, 5, 5)\,, \tag{3.34}$$

and the membership function has the form (3.35).

$$\mu_{A1+A2}(y) = \begin{cases} MAX\left(0, 1 + \frac{y-12}{5}\right), & \forall y : y < 12\,, \\ MAX\left(0, 1 - \frac{y-12}{5}\right), & \forall y : y \geq 12\,. \end{cases} \tag{3.35}$$

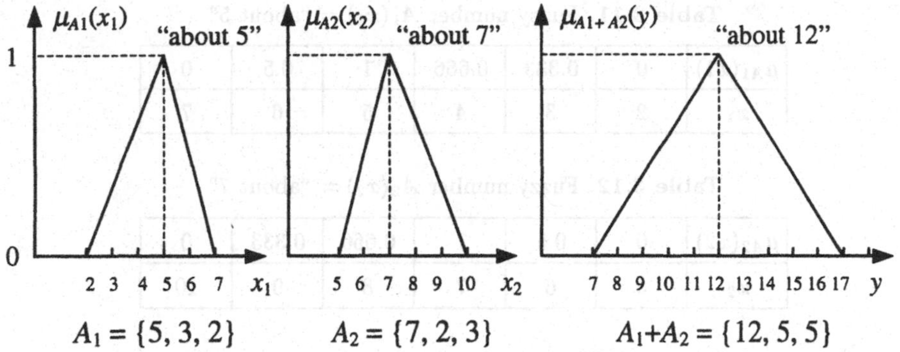

Fig. 3.19. Addition of two fuzzy numbers using the L-R representation method

The membership function $\mu_{A1+A2}(y)$ can be presented in a simplified form (3.36) since the fuzzy number $(A_1 + A_2)$ is symmetric ($\alpha = \beta = 5$).

$$\mu_{A1+A2}(y) = MAX\left(0, 1 - \left|\frac{y-12}{5}\right|\right), \quad \forall y: \ -\infty < y < +\infty . \tag{3.36}$$

The addition of fuzzy numbers is presented in Fig. 3.19.

◊ ◊ ◊

3.3 Subtraction of fuzzy numbers

Let $A_1(x_1)$ and $A_2(x_2)$ be the fuzzy numbers. Their difference can be calculated using the extension principle expressed with formula (3.37).

$$(A_1 - A_2)(y) = \bigvee_{y=x_1-x_2} [A_1(x_1) \wedge A_2(x_2)], \quad \forall x_1, x_2, y \in R \tag{3.37}$$

The calculation of a fuzzy number difference is in fact the calculation of its membership function using the formula (3.38),

$$\mu_{A_1-A_2}(y) = \bigvee_{y=x_1-x_2} [\mu_{A1}(x_1) \wedge \mu_{A2}(x_2)], \quad \forall x_1, x_2, y \in R, \tag{3.38}$$

where: $\vee$ – denotes a set union operator (e.g. *MAX* or other s-norms),
$\wedge$ – denotes a set intersection operator (e.g. *MIN*, *PROD* or other t-norms).

Example 3.3.1
Let us calculate the difference of two fuzzy numbers $(A_1 - A_2)$ given in the form of discrete samples, Table 3.11 and Table 3.12.

Table 3.11. Fuzzy number $A_1(x_1)$ = "about 5"

$\mu_{A1}(x1)$	0	0.333	0.666	1	0.5	0
x_1	2	3	4	5	6	7

Table 3.12. Fuzzy number $A_2(x_2)$ = "about 7"

$\mu_{A2}(x2)$	0	0.5	1	0.666	0.333	0
x_2	5	6	7	8	9	10

Table 3.13. Fuzzy number $(A_1 - A_2)$ = "about −2"

$\mu_{A1-A2}(y)$	0	0.16	0.33	0.5	0.66	0.83	1	0.75	0.5	0.25	0
y	−8	−7	−6	−5	−4	−3	−2	−1	0	1	2

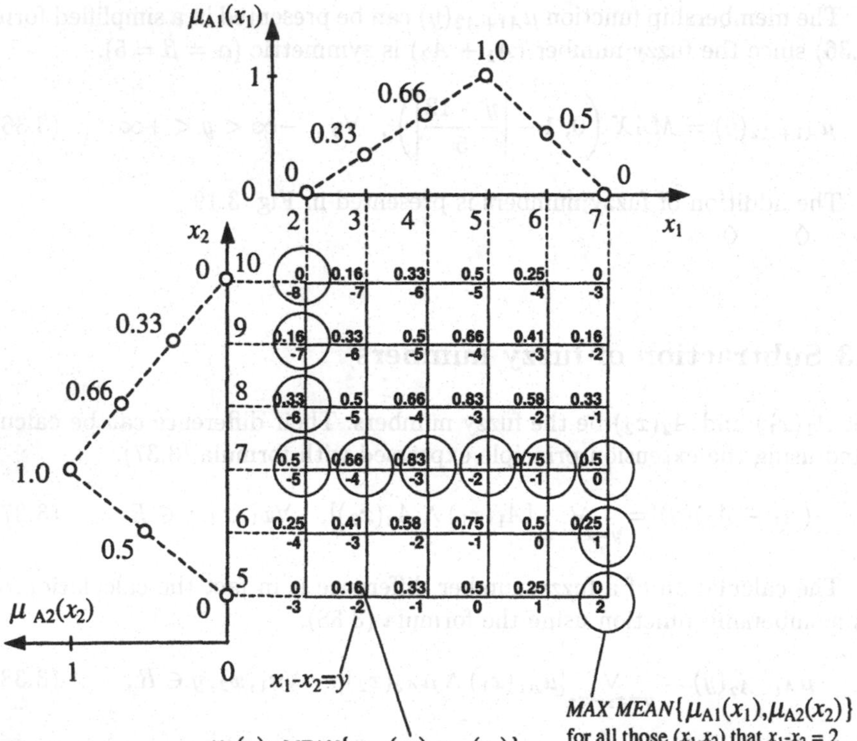

Fig. 3.20. Method of determination of the membership function of the fuzzy number $(A_1 - A_2)$ using the extension principle with the *MAX* (∨) and *MEAN* (∧) operators, (A_1 = "about 5", A_2 = "about 7", $(A_1 - A_2)$ = "about -2")

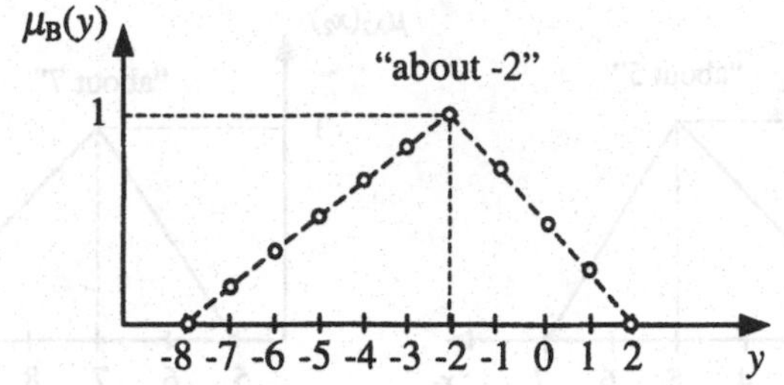

Fig. 3.21. Membership function of the fuzzy number "about -2" obtained as the result of subtraction of fuzzy numbers "about 5" and "about 7" using the extension principle

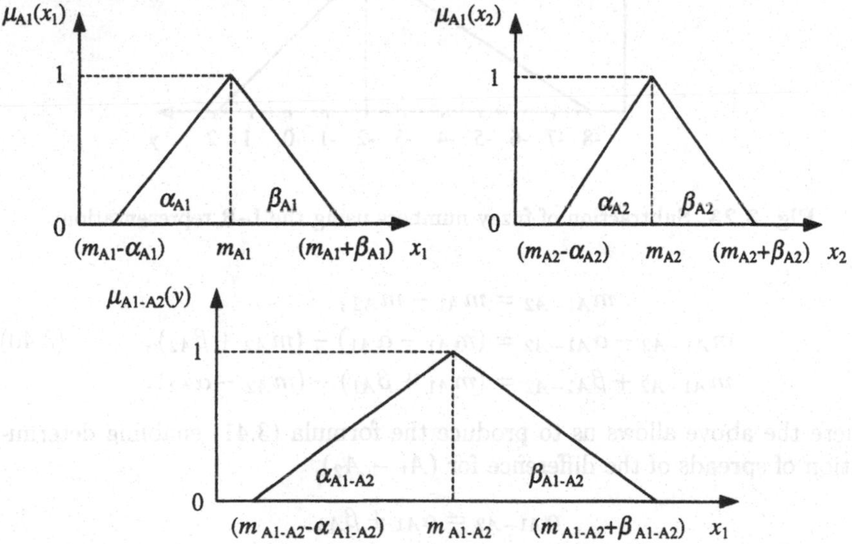

Fig. 3.22. Notation used in subtraction of L-R representation type fuzzy numbers $(A_1 - A_2)$

The method of subtracting the numbers A_1 and A_2 is shown in Fig. 3.20. On the basis of the calculation grid in Fig. 3.20 we obtain the fuzzy number $(A_1 - A_2) =$ "about -2" presented in Fig. 3.21 and Table 3.13.

◊ ◊ ◊

The use of L-R representation of fuzzy numbers, Fig. 3.22, enables the calculation of the difference in a simplified way. Let A_1 and A_2 have the form (3.39).

$$A_1 = (m_{A1}, \alpha_{A1}, \beta_{A1}), \quad A_2 = (m_{A2}, \alpha_{A2}, \beta_{A2}) \tag{3.39}$$

The parameters of the fuzzy numbers A_1, A_2 and the difference $(A_1 - A_2)$, Fig. 3.22, are interrelated as presented in (3.40),

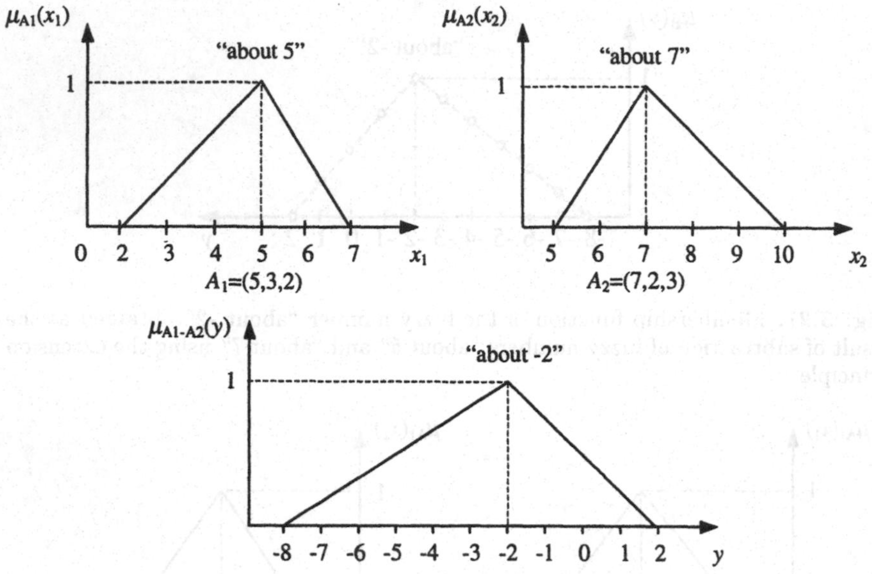

Fig. 3.23. Subtraction of fuzzy numbers using the L-R representation

$$
\begin{aligned}
m_{A1-A2} &= m_{A1} - m_{A2}\,,\\
m_{A1-A2} - \alpha_{A1-A2} &= (m_{A1} - \alpha_{A1}) - (m_{A2} + \beta_{A2})\,,\\
m_{A1-A2} + \beta_{A1-A2} &= (m_{A1} + \beta_{A1}) - (m_{A2} - \alpha_{A2})\,,
\end{aligned} \tag{3.40}
$$

where the above allows us to produce the formula (3.41) enabling determination of spreads of the difference for $(A_1 - A_2)$.

$$
\begin{aligned}
\alpha_{A1-A2} &= \alpha_{A1} + \beta_{A2}\\
\beta_{A1-A2} &= \alpha_{A2} + \beta_{A1}
\end{aligned} \tag{3.41}
$$

The difference $(A_1 - A_2)$ in the form of a L-R type triple can be expressed with the formula (3.42).

$$
\begin{aligned}
(A_1 - A_2) &= (m_{A1}, \alpha_{A1}, \beta_{A1}) - (m_{A2}, \alpha_{A2}, \beta_{A2}) =\\
&= (m_{A1} - m_{A2},\ \alpha_{A1} + \beta_{A2},\ \alpha_{A2} + \beta_{A1})
\end{aligned} \tag{3.42}
$$

Example 3.3.2

Let us calculate the difference between $A_1 = (5, 3, 2)$ and $A_2 = (7, 2, 3)$ using the L-R representation.

On the basis of the formula (3.42) we obtain the result (3.43).

$$
A_1 - A_2 = (5, 3, 2) - (7, 2, 3) = (-2, 6, 4) \tag{3.43}
$$

The subtraction result is presented in Fig. 3.23.

◊ ◊ ◊

It should be noted that the subtraction results for the triangular numbers "about 5" and "about 7" using the extension principle and the L-R representation are identical, Fig. 3.21 and Fig. 3.23.

3.4 Multiplication of fuzzy numbers

Let $A_1(x_1)$ and $A_2(x_2)$ be the fuzzy numbers. Their product $(A_1 \cdot A_2)$ can be calculated by using the extension principle with formula (3.44).

$$(A_1 \cdot A_2)(y) = \bigvee_{y=x_1 x_2} [A_1(x_1) \wedge A_2(x_2)], \quad \forall x_1, x_2, y \in R \tag{3.44}$$

The calculation of the product is in fact the calculation of the membership function of the product – formula (3.45),

$$\mu_{A_1 A_2}(y) = \bigvee_{y=x_1 x_2} [\mu_{A1}(x_1) \wedge \mu_{A2}(x_2)], \quad \forall x_1, x_2, y \in R, \tag{3.45}$$

where: $\vee$ – denotes a set union operator (e.g. *MAX* or other s-norms),
$\wedge$ – denotes a set intersection operator (e.g. *MIN*, *PROD* or other t-norms).

Example 3.4.1
Let us calculate a product $(A_1 \cdot A_2)$ of fuzzy numbers, where the membership functions are given in Table 3.14 and Table 3.15, respectively.

The procedure of product formation is depicted in Fig. 3.24, where on the basis of the calculation grid the fuzzy number $(A_1 \cdot A_2) =$ "about 35" is obtained, presented in Fig. 3.25.

From Fig. 3.25 it can be seen that the plane number "about 35", the product of two convex fuzzy numbers, **is not a convex number**. The latter results from the fact that the membership function of the number is presented in 2D- space $Y * M_{A1A2}$. The abscissa value y, shown in Fig. 3.25, does not contain any information on the elements (x_1, x_2) which have been involved in the product number formation. For example, the value $y = 42$ can be the result of $x_1 x_2 = 6 \cdot 7$ or $x_1 x_2 = 7 \cdot 6$ When the fuzzy product is presented in

Table 3.14. Fuzzy number $A_1 =$ "about 5"

$\mu_{A1}(x1)$	0	0.333	0.666	1	0.5	0
x_1	2	3	4	5	6	7

Table 3.15. Fuzzy number $A_2 =$ "about 7"

$\mu_{A2}(x2)$	0	0.5	1	0.666	0.333	0
x_2	5	6	7	8	9	10

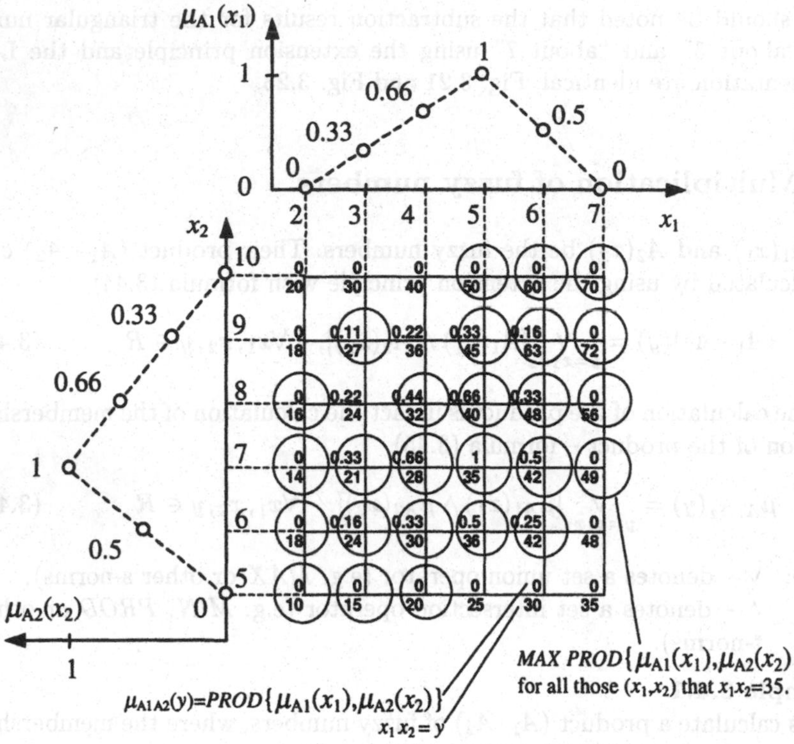

Fig. 3.24. The method of determination of the membership function of the fuzzy number $(A_1 \cdot A_2)$ using the extension principle with the *MAX* (∨) and *PROD* (∧) operators, (A_1 = "about 5", A_2 = "about 7")

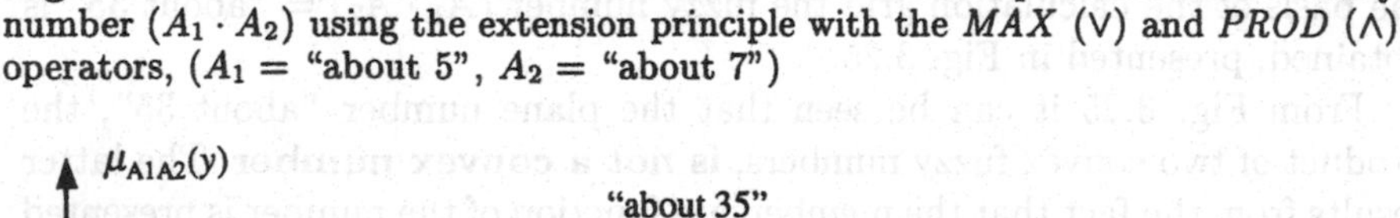

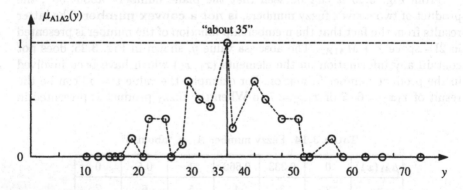

Fig. 3.25. The membership function of the fuzzy number "about 35" obtained as the result of multiplication of the numbers "about 5" and "about 7" using the *PROD* operator

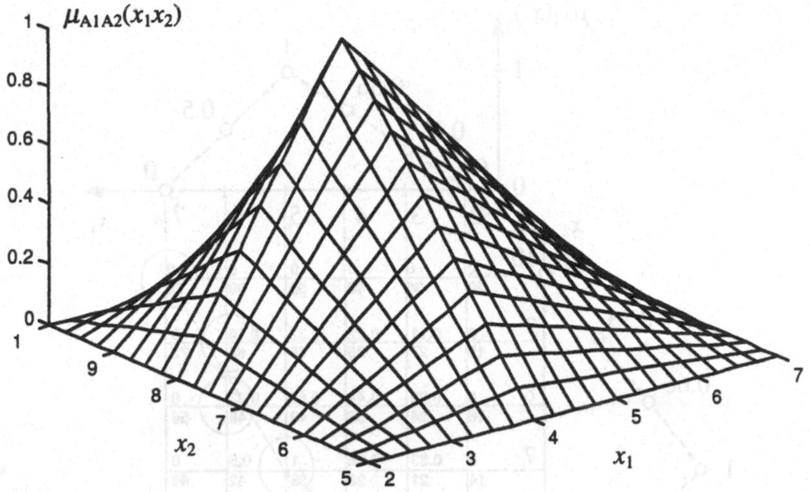

Fig. 3.26. The 3D-representation of the membership function $\mu_{A1A2}(x_1, x_2)$ of the product of fuzzy numbers "about 5" and "about 7" obtained with the *PROD* operator

the 3D-space $X_1 * X_2 * M_{A1A2}$, Fig. 3.26, then the number "about 35" is a convex number (where $M_{A1A2} : \mu_{A1A2}(y) \in M_{A1A2}$).

The characteristic of components (x_1, x_2) of the resulting number y are of great significance in practical applications. For example, the same electric power $L = u \cdot i$ can be obtained with different values of the voltage u and current intensity i; those are significant parameters in power supply systems, where the heat loss $Q = i^2 Rt$ strongly depends on the intensity i and not on the voltage u, (t – time, R – resistance).

Zadeh's extension principle defines a plane, 2D, simplified intersection of the product of two fuzzy numbers that may be presented in 4D (4-dimensional) space only. The product representation in 3D-space in Fig. 3.26 is already a certain simplification, containing reduced information (lack of coordinate y). Thus, the product representation in 2D-space is a very significant simplification which, however, might be used in cases where the numbers structure (x_1 and x_2 values) forming the product $y = x_1 \cdot x_2$ are not significant in the modeled system but only the resulting product value y is significant.

When the fuzzy numbers A_1 and A_2 are in the form of discrete samples then the extension principle (3.45) applied merely mechanically gives an incorrect image of the product $(A_1 \cdot A_2)$ that, in fact, is the convex number. The plane representation of a product is the intersection across the 3D product representation, joining the point of maximal membership value ($\mu_{A1A2} = 1$) with the point of the lowest product value using a straight line and with the point of the highest product value using another straight line. On having joined the above-mentioned points with a straight line we obtain an ap-

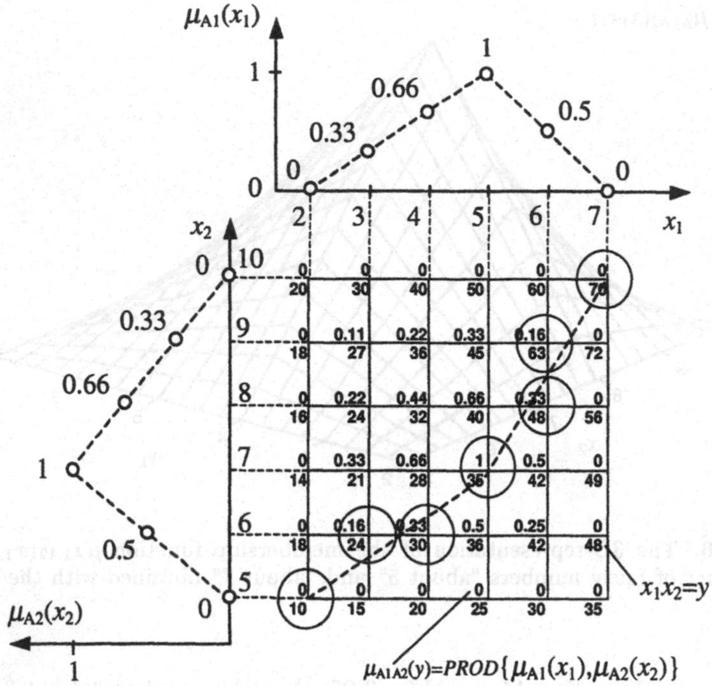

Fig. 3.27. The simplified intersection line across the 4D fuzzy number product $(A_1 \cdot A_2)$

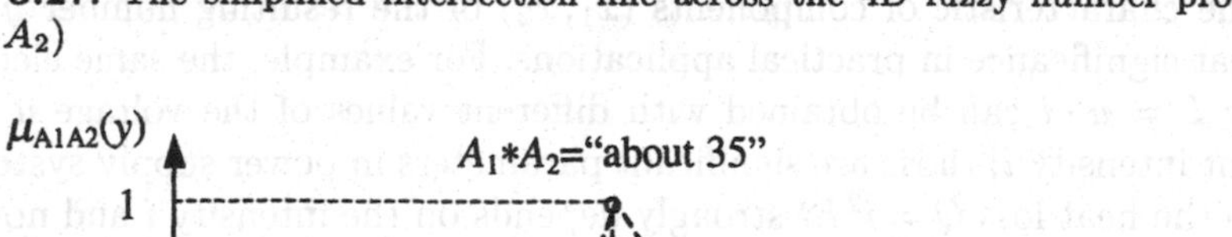

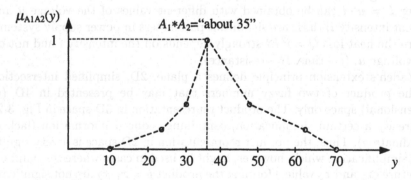

Fig. 3.28. The simplified 2D intersection across the fuzzy number product $(A_1 \cdot A_2)$

proximate plane representation of the fuzzy product that reflects the convex character of the fuzzy product in a more appropriate way, Fig. 3.27 and 3.28.

In this example the *PROD* (product) operator has been used to calculate the product of fuzzy numbers. Since the multiplication is of multiplicative character then that particular operator performs the principle of transfer of input changes into output changes in the best way, i.e. the changes of

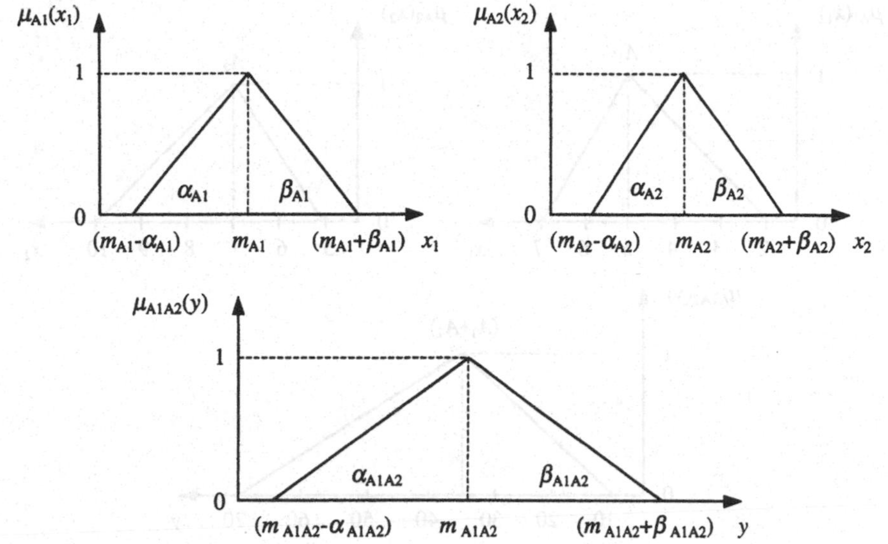

Fig. 3.29. Notation used when multiplying the two positive fuzzy numbers A_1 and A_2

the inputs (x_1, x_2) are transferred into the output y changes of the model similarly as in the real system. However, if e.g. $x_1 = 0$ then despite the changes in the input x_2 the changes are not transferred to the output of the fuzzy model $\mu_B(y)$ – just as in a real system, Fig. 3.27 (the first column of calculations). The *PROD* operator can be replaced with other t-norms, e.g. *MIN*. The multiplication results would be different, though.

◊ ◊ ◊

The application of L-R representation for fuzzy numbers, Fig. 3.29, enables us to calculate an approximate product $A_1 \cdot A_2$.

The parameters of the positive fuzzy numbers A_1, A_2 and their product $(A_1 \cdot A_2)$ are interrelated as shown in (3.46) below:

$$\begin{aligned} m_{A1A2} &= m_{A1} m_{A2} \\ m_{A1A2} - \alpha_{A1A2} &= (m_{A1} - \alpha_{A1})(m_{A2} - \alpha_{A2}) \\ m_{A1A2} + \beta_{A1A2} &= (m_{A1} + \beta_{A1})(m_{A2} + \beta_{A2}) \end{aligned} \tag{3.46}$$

On the basis of the above the characteristic parameters of the product $(A_1 \cdot A_2)$ can be determined; formula (3.47).

$$\begin{aligned} \alpha_{A1A2} &= m_{A1}\alpha_{A2} + m_{A2}\alpha_{A2} - \alpha_{A1}\alpha_{A2} \\ \beta_{A1A2} &= m_{A1}\beta_{A2} + m_{A2}\beta_{A1} + \beta_{A1}\beta_{A2} \end{aligned} \tag{3.47}$$

Thus, the product of two positive fuzzy numbers in the form of L-R representation can be expressed with the formula (3.48).

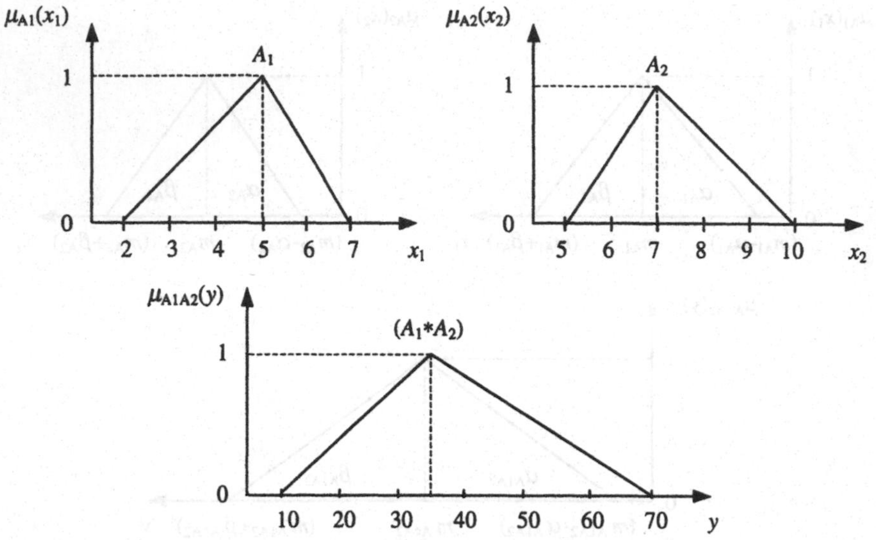

Fig. 3.30. Multiplication of two positive fuzzy numbers using L-R representation

$$\begin{aligned}(A_1 \cdot A_2) = (m_{A1}, \alpha_{A1}, \beta_{A1})(m_{A2}, \alpha_{A2}, \beta_{A2}) = (m_{A1}m_{A2}, \\ m_{A1}\alpha_{A2} + m_{A2}\alpha_{A1} - \alpha_{A1}\alpha_{A2}, m_{A1}\beta_{A2} + m_{A2}\beta_{A1} + \beta_{A1}\beta_{A2}),\end{aligned} \tag{3.48}$$

for $A_1 > 0,\ A_2 > 0$.

Example 3.4.2
Let us calculate a product of the following positive fuzzy numbers: $A_1 = (5, 3, 2)$ and $A_2 = (7, 2, 3)$, Fig. 3.30. On the basis of the formula (3.48) the following result is obtained:

$$(A_1 \cdot A_2) = (5, 3, 2) \cdot (7, 2, 3) = (35, 25, 35).$$

On having compared the Figures 3.28 and 3.30 it can be seen that the multiplication result obtained using the extension principle and using L-R representation is a fuzzy number in each case with identical spreads and identical nominal value even if the membership function shapes vary slightly. ◊ ◊ ◊

Multiplication of a positive number and a negative number
The multiplication of different sign fuzzy numbers with the use of the extension principle does not differ from the multiplication of positive fuzzy numbers at all.

Example 3.4.3
Let us calculate a product of the numbers A_1 and A_2 defined in Table 3.16 and Table 3.17, respectively.

Table 3.16. Fuzzy number A_1 = "about 5"

$\mu_{A1}(x1)$	0	0.333	0.666	1	0.5	0
x_1	2	3	4	5	6	7

Table 3.17. Fuzzy number A_2 = "about −7"

$\mu_{A2}(x2)$	0	0.5	1	0.666	0.333	0
x_2	−5	−6	−7	−8	−9	−10

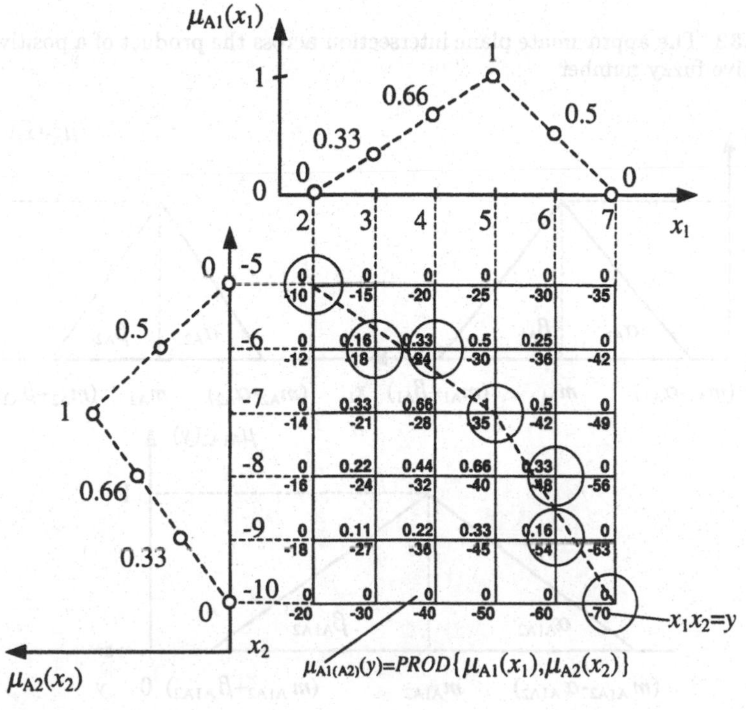

Fig. 3.31. The product of a positive and a negative fuzzy number and the simplified intersection line

The number A_2 = "about -7" is the opposite number to A_2 = "about 7", used beforehand in Example 3.4.2. The multiplication of A_1 and A_2 is presented in Fig. 3.31. The intersection in Fig. 3.31 in its 2D-representation is shown in Fig. 3.32.

◊ ◊ ◊

The approximate product of a positive and a negative fuzzy number can also be calculated using L-R representation, Fig. 3.33.

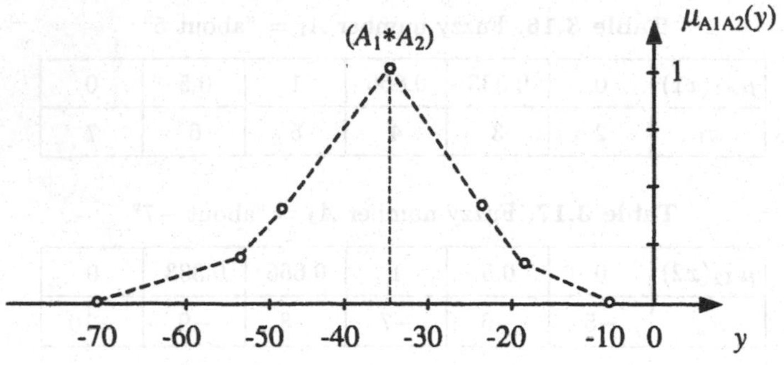

Fig. 3.32. The approximate plane intersection across the product of a positive and a negative fuzzy number

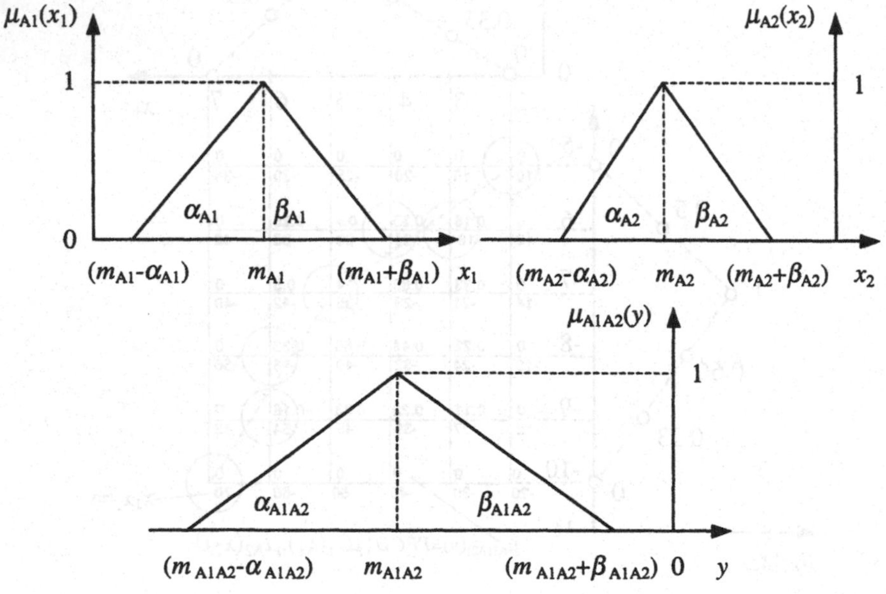

Fig. 3.33. Notation of parameters in the product of a positive fuzzy number A_1 and a negative fuzzy number A_2

The parameters of the numbers A_1 and A_2 presented in Fig. 3.33 are interrelated as shown in (3.49).

$$\begin{aligned} m_{A1A2} &= m_{A1}m_{A2} \\ m_{A1A2} - \alpha_{A1A2} &= (m_{A1} + \beta_{A1})(m_{A2} - \alpha_{A2}) \\ m_{A1A2} + \beta_{A1A2} &= (m_{A1} - \alpha_{A1})(m_{A2} + \beta_{A2}) \end{aligned} \tag{3.49}$$

On the basis of the relationships (3.49) the parameters of the product of a positive fuzzy number and a negative one can be determined with formula (3.50).

$$\begin{aligned}\alpha_{A1A2} &= m_{A1}\alpha_{A2} - m_{A2}\beta_{A1} + \alpha_{A2}\beta_{A1} \\ \beta_{A1A2} &= m_{A1}\beta_{A2} - m_{A2}\alpha_{A1} - \alpha_{A1}\beta_{A2}\end{aligned} \tag{3.50}$$

Thus, the product of a positive and a negative fuzzy number in L-R representation, where $A_1 > 0$ and $A_2 < 0$, has a form (3.51).

$$\begin{aligned}(A_1 \cdot A_2) \cong (m_{A1}, \alpha_{A1}, \beta_{A1})(m_{A2}, \alpha_{A2}, \beta_{A2}) = (m_{A1}m_{A2}, \\ m_{A1}\alpha_{A2} - m_{A2}\beta_{A1} + \alpha_{A2}\beta_{A1}, m_{A1}\beta_{A2} - m_{A2}\alpha_{A1} - \alpha_{A1}\beta_{A2})\end{aligned} \tag{3.51}$$

Example 3.4.4
Let us calculate the product of a positive fuzzy number A_1 and a negative fuzzy number A_2 using L-R representation.

$$\begin{aligned}A_1 &= \text{"about 5"} = (5, 3, 2) \\ A_2 &= \text{"about -7"} = (-7, 3, 2)\end{aligned}$$

According to the formula (3.51) the following result is obtained:

$$(A_1 \cdot A_2) = (5, 3, 2)(-7, 3, 2) = (-35, 35, 25)$$

The multiplication result is presented in Fig. 3.34. On having compared the result of multiplication of the positive fuzzy number A_1 and the negative fuzzy number A_2 (Fig. 3.32 and 3.34) using the extension principle and L-R representation, respectively, we can see the identical spreads α_{A1A2} and β_{A1A2} as well as the identical nominal value m_{A1A2} and the similarity in the membership function shape.
◊ ◊ ◊

Multiplication of a positive number and a negative number
For $A_1 > 0$ and $A_2 < 0$ the multiplication of L-R representations is performed with the formula (3.52).

$$\begin{aligned}(A_1 \cdot A_2) \cong (m_{A1}, \alpha_{A1}, \beta_{A1})(m_{A2}, \alpha_{A2}, \beta_{A2}) = (m_{A1}m_{A2}, \\ -m_{A1}\beta_{A2} + m_{A2}\alpha_{A1} + \alpha_{A1}\beta_{A2}, -m_{A1}\alpha_{A2} + m_{A2}\beta_{A1} - \alpha_{A2}\beta_{A1})\end{aligned} \tag{3.52}$$

Multiplication of negative fuzzy numbers
For $A_1 < 0$ and $A_2 < 0$ the multiplication is performed with the formula (3.53).

$$\begin{aligned}(A_1 \cdot A_2) \cong (m_{A1}, \alpha_{A1}, \beta_{A1})(m_{A2}, \alpha_{A2}, \beta_{A2}) = (m_{A1}m_{A2}, \\ -m_{A1}\beta_{A2} - m_{A2}\beta_{A1} - \beta_{A1}\beta_{A2}, -m_{A1}\alpha_{A2} - m_{A2}\alpha_{A1} + \alpha_{A1}\alpha_{A2})\end{aligned} \tag{3.53}$$

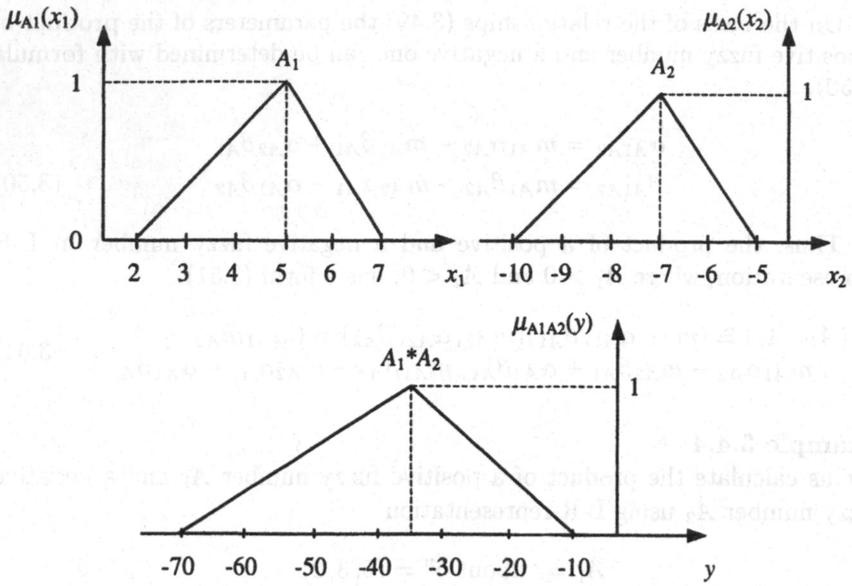

Fig. 3.34. The result of multiplying a positive fuzzy number A_1 and a negative fuzzy number A_2 using L-R representation

Multiplication of fuzzy zeroes

Let A_1 and A_2 be the fuzzy numbers "about zero" of different spreads. The multiplication of such numbers can be performed with the help of the extension principle, see Example 3.4.5.

Example 3.4.5

Let us calculate the product of the fuzzy numbers A_1 and A_2, representing the type "about zero", defined in Table 3.18 and Table 3.19, respectively.

Table 3.18. Fuzzy number A_1 = "about zero"

$\mu_{A1}(x1)$	0	0.333	0.666	1	0.5	0
x_1	−3	−2	−1	0	1	2

Table 3.19. Fuzzy number A_2 = "about zero"

$\mu_{A2}(x2)$	0	0.5	1	0.666	0.333	0
x_2	−2	−1	0	1	2	3

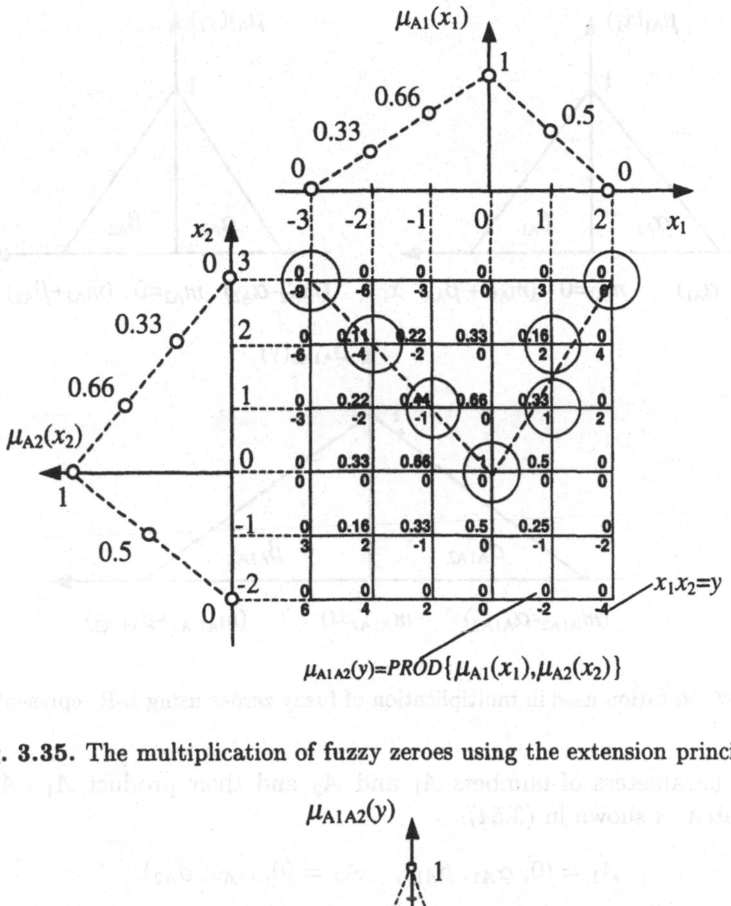

Fig. 3.35. The multiplication of fuzzy zeroes using the extension principle

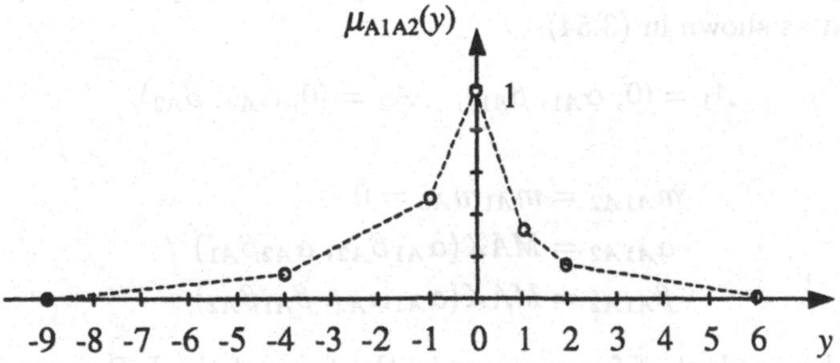

Fig. 3.36. The simplified plane intersection across the result of multiplying the fuzzy zeroes in reference to Fig. 3.35

The method of calculating the product $A_1 \cdot A_2$ using the extension principle is presented in Fig. 3.35. The product plane intersection of the above numbers is presented in Fig. 3.36.

◊ ◊ ◊

The multiplication of the fuzzy zeroes can also be performed using L-R representation, Fig. 3.37.

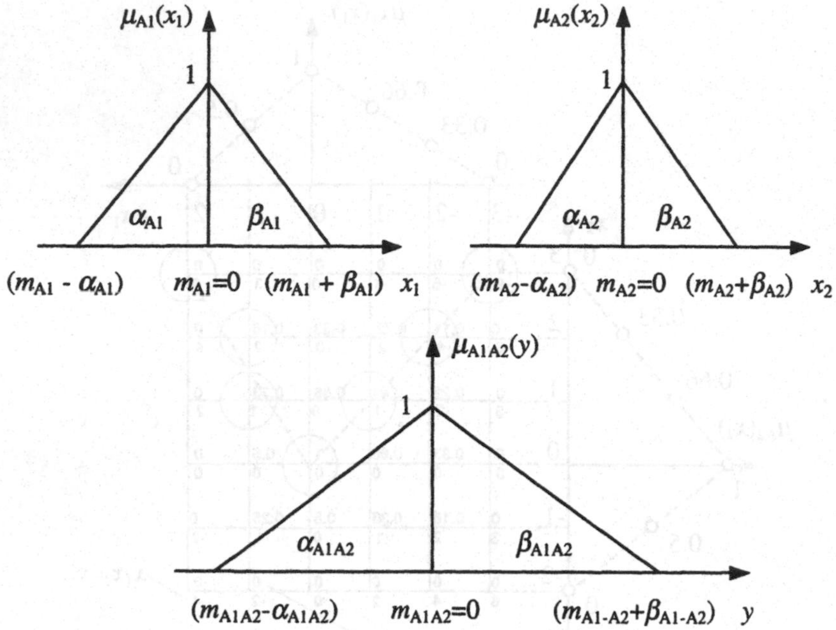

Fig. 3.37. Notation used in multiplication of fuzzy zeroes using L-R representation

The parameters of numbers A_1 and A_2 and their product $A_1 \cdot A_2$ are interrelated as shown in (3.54).

$$A_1 = (0,\, \alpha_{A1},\, \beta_{A1}), \quad A_2 = (0,\, \alpha_{A2},\, \beta_{A2})$$

$$\begin{aligned} m_{A1A2} &= m_{A1} m_{A2} = 0 \\ \alpha_{A1A2} &= MAX(\alpha_{A1}\beta_{A2},\, \alpha_{A2}\beta_{A1}) \\ \beta_{A1A2} &= MAX(\alpha_{A1}\alpha_{A2},\, \beta_{A1}\beta_{A2}) \end{aligned} \tag{3.54}$$

Thus, the product of fuzzy zeroes in the form of the L-R representation triple can be expressed as (3.55).

$$\begin{aligned} (A_1 \cdot A_2) &\cong (0, \alpha_{A1}, \beta_{A1})(0, \alpha_{A2}, \beta_{A2}) = \\ &= (0,\, MAX(\alpha_{A1}\beta_{A2},\, \alpha_{A2}\beta_{A1}),\, MAX(\alpha_{A1}\alpha_{A2},\, \beta_{A1}\beta_{A2})) \end{aligned} \tag{3.55}$$

Example 3.4.6
Let us calculate a product of the following fuzzy zeroes:

$$A_1 = (0, 3, 2), \quad A_2 = (0, 2, 3)\,.$$

On the basis of the formula (3.55) we determine the L-R representation of the product $A_1 \cdot A_2$ as follows:

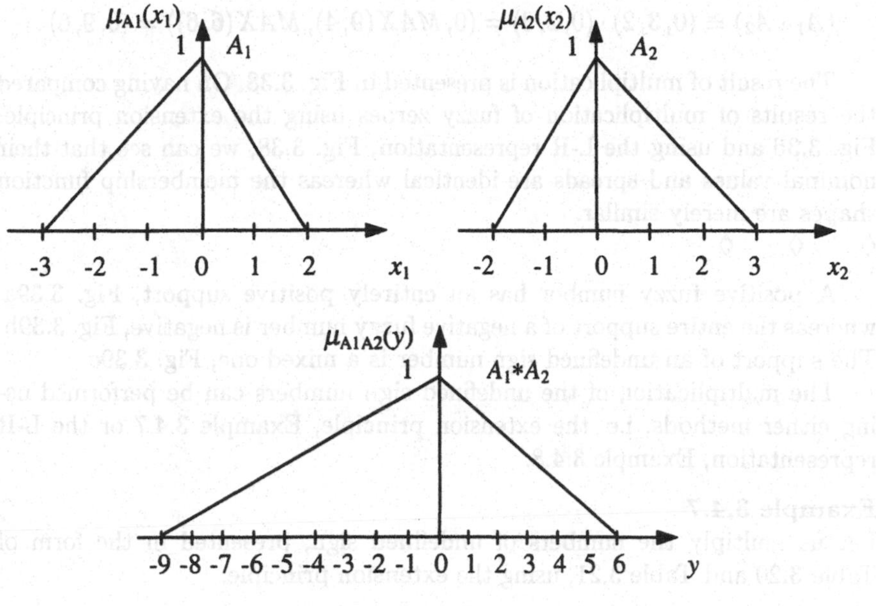

Fig. 3.38. The result of multiplication of fuzzy zeroes using L-R representation

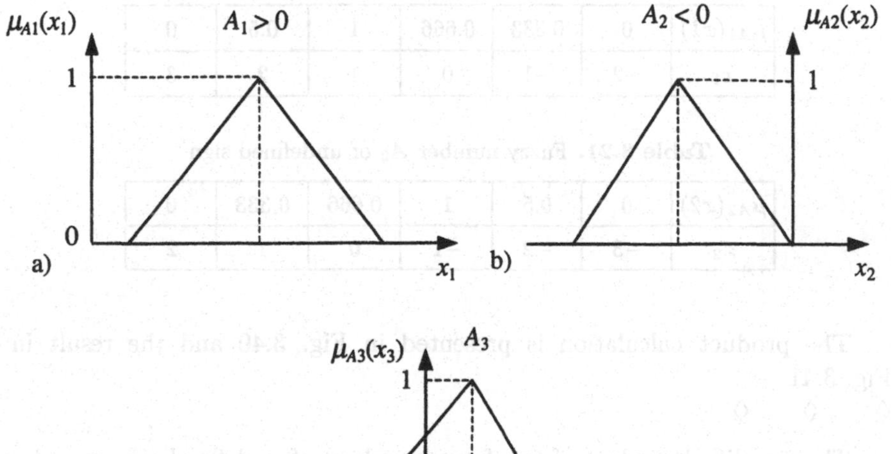

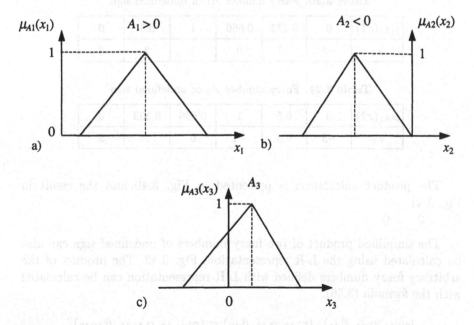

Fig. 3.39. A positive fuzzy number A_1 (a), a negative fuzzy number A_2 (b) and an undefined sign number A_3 (c)

$$(A_1 \cdot A_2) \cong (0,3,2) \cdot (0,2,3) = (0, MAX(9,4), MAX(6,6)) = (0,9,6)\,.$$

The result of multiplication is presented in Fig. 3.38. On having compared the results of multiplication of fuzzy zeroes using the extension principle, Fig. 3.36 and using the L-R representation, Fig. 3.38, we can see that their nominal values and spreads are identical whereas the membership function shapes are merely similar.

◊ ◊ ◊

A positive fuzzy number has an entirely positive support, Fig. 3.39a, whereas the entire support of a negative fuzzy number is negative, Fig. 3.39b. The support of an undefined sign number is a mixed one, Fig. 3.39c.

The multiplication of the undefined sign numbers can be performed using either methods, i.e. the extension principle, Example 3.4.7 or the L-R representation, Example 3.4.8.

Example 3.4.7

Let us multiply the numbers of undefined sign, presented in the form of Table 3.20 and Table 3.21, using the extension principle.

Table 3.20. Fuzzy number A_1 of undefined sign

$\mu_{A1}(x1)$	0	0.333	0.666	1	0.5	0
x_1	−2	−1	0	1	2	3

Table 3.21. Fuzzy number A_2 of undefined sign

$\mu_{A2}(x2)$	0	0.5	1	0.666	0.333	0
x_2	−3	−2	−1	0	1	2

The product calculation is presented in Fig. 3.40 and the result in Fig. 3.41.

◊ ◊ ◊

The simplified product of two fuzzy numbers of undefined sign can also be calculated using the L-R representation, Fig. 3.42. The product of the arbitrary fuzzy numbers defined with L-R representation can be calculated with the formula (3.56).

$$(m_{A1}, \alpha_{A1}, \beta_{A1}) \cdot (m_{A2}, \alpha_{A2}, \beta_{A2}) = (m_{A1A2}, \alpha_{A1A2}, \beta_{A1A2})\,,$$

where:

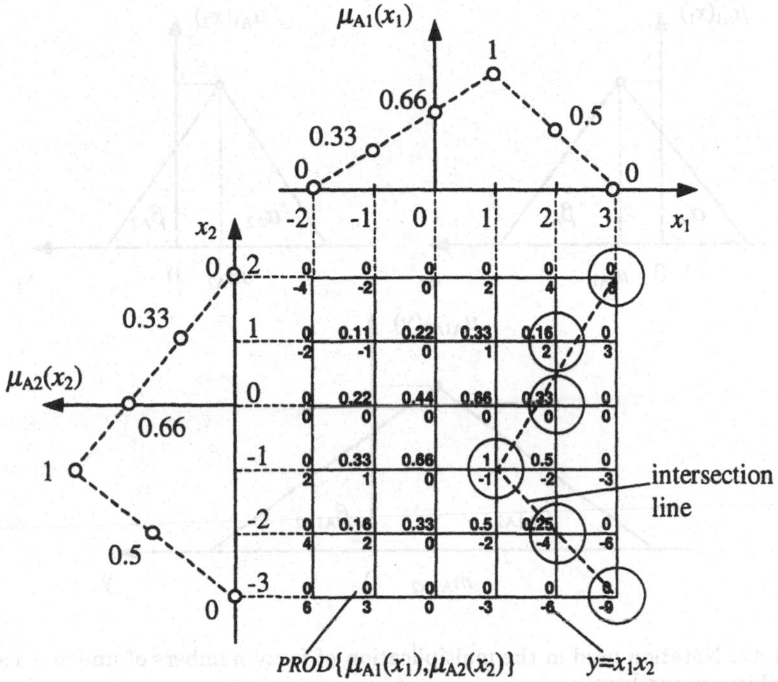

Fig. 3.40. The multiplication of fuzzy numbers of undefined sign using the extension principle

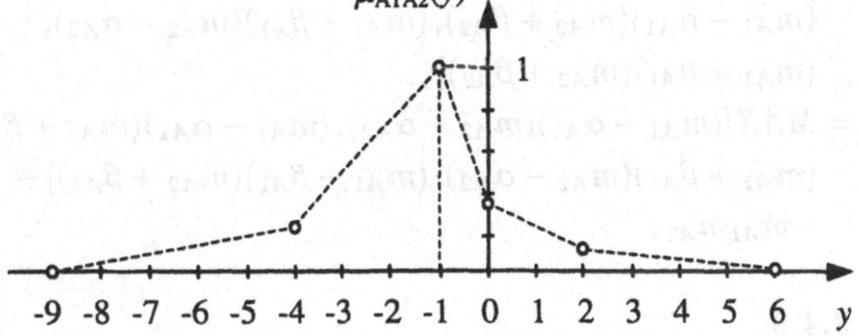

Fig. 3.41. The product of two fuzzy numbers of undefined sign: A_1 = "about 1", A_2 = "about -1", presented in the form of a plane intersection across the 4D-fuzzy number presented in Fig. 3.40

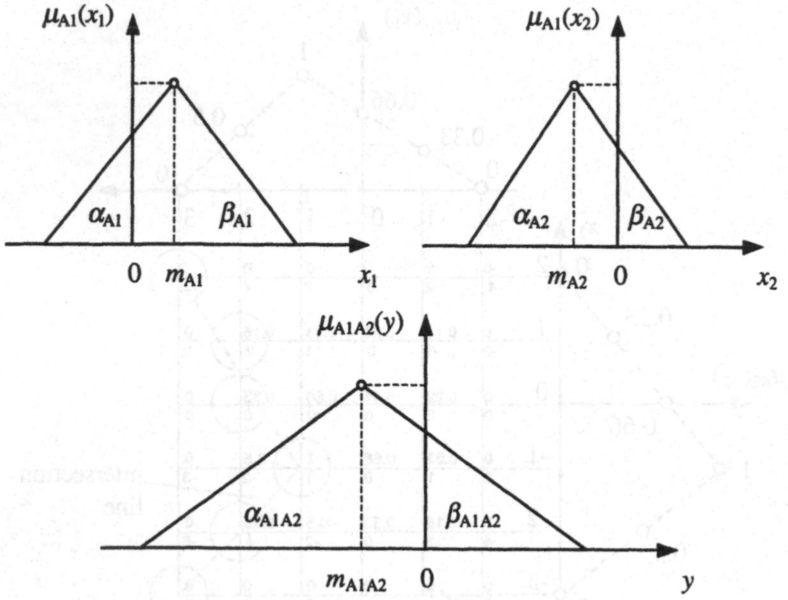

Fig. 3.42. Notation used in the multiplication of fuzzy numbers of undefined sign (any arbitrary numbers)

$$\begin{aligned} m_{A1A2} &= m_{A1}m_{A2}\,, \\ \alpha_{A1A2} &= m_{A1}m_{A2} - MIN[(m_{A1}-\alpha_{A1})(m_{A2}-\alpha_{A2}), \\ &\quad (m_{A1}-\alpha_{A1})(m_{A2}+\beta_{A2}), (m_{A1}+\beta_{A1})(m_{A2}-\alpha_{A2}), \\ &\quad (m_{A1}+\beta_{A1})(m_{A2}+\beta_{A2})]\,, \\ \beta_{A1A2} &= MAX[(m_{A1}-\alpha_{A1})(m_{A2}-\alpha_{A2}), (m_{A1}-\alpha_{A1})(m_{A2}+\beta_{A2}), \\ &\quad (m_{A1}+\beta_{A1})(m_{A2}-\alpha_{A2}), (m_{A1}+\beta_{A1})(m_{A2}+\beta_{A2})] - \\ &\quad - m_{A1}m_{A2}\,. \end{aligned} \tag{3.56}$$

Example 3.4.8

Let us calculate the product of two fuzzy numbers A_1 and A_2 of undefined sign:

$$A_1 = (1,3,2)\,,\quad A_2 = (-1,2,3)\,.$$

The multiplication with the formula (3.56) produces the result (3.57), presented also in Fig. 3.43.

$$(A_1 \cdot A_2) = (m_{A1A2}, \alpha_{A1A2}, \beta_{A1A2})\,,$$

where:

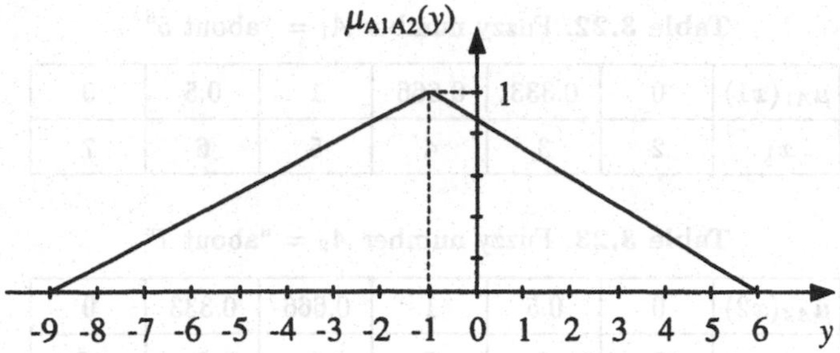

Fig. 3.43. The product of two fuzzy numbers of undefined sign: A_1 = "about 1" and A_2 = "about -1" obtained with the L-R representation method

$$\begin{aligned} m_{A1A2} &= -1, \\ \alpha_{A1A2} &= -1 - MIN(6, -4, -9, 6) = -1 + 9 = 8, \\ \beta_{A1A2} &= MAX(6, -4, -9, -6) - (-1) = 6 + 1 = 7. \end{aligned} \tag{3.57}$$

On having compared the Figures 3.41 and 3.43 we can see that the products of fuzzy numbers of undefined sign calculated using the extension principle or the L-R representation method have identical spreads and nominal values. The membership function shapes differ slightly, however.

◊ ◊ ◊

3.5 Division of fuzzy numbers

Let A_1 and A_2 be the fuzzy numbers. Their quotient (A_1/A_2) can be calculated using the extension principle regardless of the sign of the number, using the formula (3.58),

$$(A_1/A_2)(y) = \bigvee_{y=x_1/x_2} [A_1(x_1) \wedge A_2(x_2)], \quad \forall x_1, x_2, y \in R, \ x_2 \neq 0, \tag{3.58}$$

where: $\vee$ – denotes a set union operator (e.g. MAX, algebraic sum or other s-norms),
$\wedge$ – denotes a set intersection operator (e.g. MIN, $PROD$ or other t-norms).

The division of fuzzy numbers is in fact the calculation of the quotient membership function with the formula (3.59).

$$\mu_{A_1/A_2}(y) = \bigvee_{y=x_1/x_2} [\mu_{A1}(x_1) \wedge \mu_{A2}(x_2)], \quad \forall x_1, x_2, y \in R, \ x_2 \neq 0. \tag{3.59}$$

Example 3.5.1
Let us calculate the quotient of the fuzzy numbers A_1 and A_2, defined in Table 3.22 and Table 3.23, respectively.

Table 3.22. Fuzzy number A_1 = "about 5"

$\mu_{A1}(x1)$	0	0.333	0.666	1	0.5	0
x_1	2	3	4	5	6	7

Table 3.23. Fuzzy number A_2 = "about 7"

$\mu_{A2}(x2)$	0	0.5	1	0.666	0.333	0
x_2	5	6	7	8	9	10

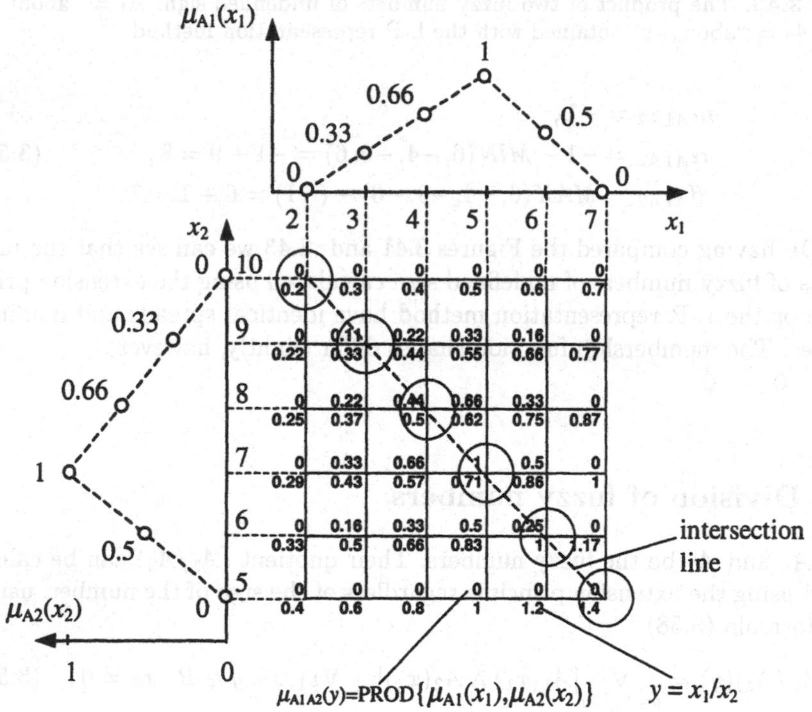

Fig. 3.44. The method of determination of the membership function of the fuzzy number quotient (A_1/A_2) using the extension principle

The method of quotient calculation using the extension principle is presented in Fig. 3.44. On having made the intersection across the fuzzy quotient, presented in Fig. 3.44, we obtain its plane 2D-representation as shown in Fig. 3.45.

The *PROD* ($\wedge$) operator has been used to calculate the quotient in this example. The application of that particular operator is advantageous in this case as the division operation A_1/A_2 can be presented as the multiplication of $A_1 \cdot A_2^{-1}$. Thus, the division is of multiplicative character and the operator

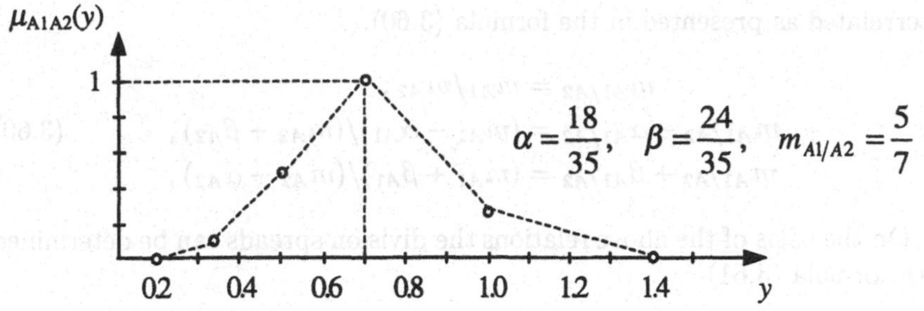

Fig. 3.45. The plane representation of the fuzzy quotient (A_1/A_2), where: $A_1 =$ "about 5", $A_2 =$ "about 7"

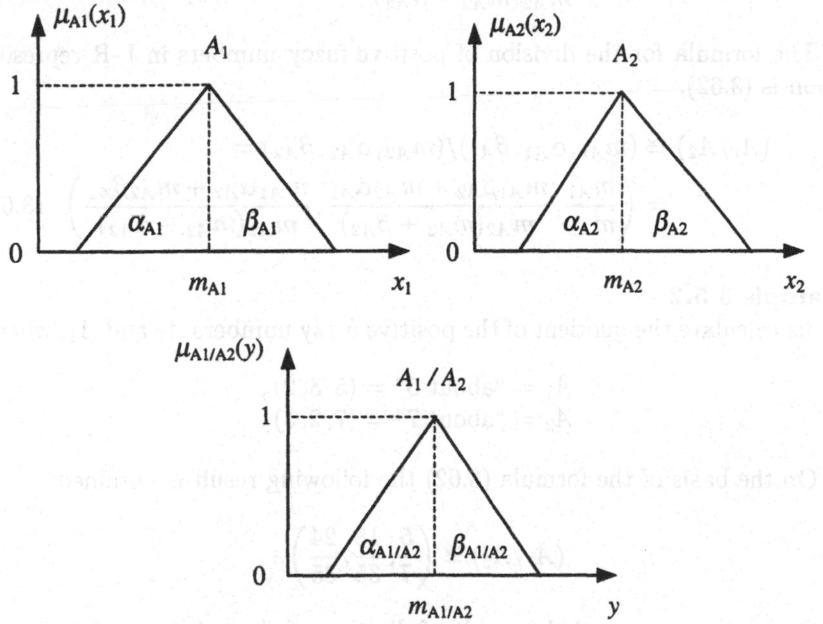

Fig. 3.46. Notation used in division of fuzzy numbers in L-R representation

PROD is the one maintaining the above feature well. Moreover, it enables us to keep the transfer principle of similar input and output changes of the system and of its model.

◊ ◊ ◊

The quotient of fuzzy numbers can be calculated approximately with the use of the L-R representation of fuzzy numbers, Fig. 3.46. If $A_1 = (m_{A1}, \alpha_{A1}, \beta_{A1})$ and $A_2 = (m_{A2}, \alpha_{A2}, \beta_{A2})$ and $A_1 > 0$ and $A_2 > 0$ then the parameters of A_1 and A_2 in L-R representation and the quotient (A_1/A_2) are interrelated as presented in the formula (3.60).

$$\begin{aligned} m_{A1/A2} &= m_{A1}/m_{A2}\,, \\ m_{A1/A2} - \alpha_{A1/A2} &= (m_{A1} - \alpha_{A1})/(m_{A2} + \beta_{A2})\,, \\ m_{A1/A2} + \beta_{A1/A2} &= (m_{A1} + \beta_{A1})/(m_{A2} - \alpha_{A2})\,, \end{aligned} \tag{3.60}$$

On the basis of the above relations the division spreads can be determined with formula (3.61).

$$\begin{aligned} \alpha_{A1/A2} &= \frac{m_{A1}\beta_{A2} + m_{A2}\alpha_{A1}}{m_{A2}(m_{A2} + \beta_{A2})}, \quad m_{A2} \neq 0\,, \\ \beta_{A1/A2} &= \frac{m_{A1}\alpha_{A2} + m_{A2}\beta_{A1}}{m_{A2}(m_{A2} - \alpha_{A2})}, \quad m_{A2} - \alpha_{A2} \neq 0\,. \end{aligned} \tag{3.61}$$

The formula for the division of positive fuzzy numbers in L-R representation is (3.62).

$$\begin{aligned} (A_1/A_2) &\cong (m_{A1}, \alpha_{A1}, \beta_{A1})/(m_{A2}, \alpha_{A2}, \beta_{A2}) = \\ &= \left(\frac{m_{A1}}{m_{A2}}, \frac{m_{A1}\beta_{A2} + m_{A2}\alpha_{A1}}{m_{A2}(m_{A2} + \beta_{A2})}, \frac{m_{A1}\alpha_{A2} + m_{A2}\beta_{A1}}{m_{A2}(m_{A2} - \alpha_{A2})}\right) \end{aligned} \tag{3.62}$$

Example 3.5.2
Let us calculate the quotient of the positive fuzzy numbers A_1 and A_2, where:

$$\begin{aligned} A_1 &= \text{“about 5”} = (5, 3, 2)\,, \\ A_2 &= \text{“about 7”} = (7, 2, 3)\,. \end{aligned}$$

On the basis of the formula (3.62) the following result is obtained:

$$(A_1/A_2) \cong \left(\frac{5}{7}, \frac{18}{35}, \frac{24}{35}\right).$$

On having compared the result of division of A_1 = “about 5” by A_2 = “about 7” either with the extension method, Fig. 3.45 or with L-R representation, Fig. 3.47, we can see that the quotient numbers are of identical spread and identical nominal value. Their membership function shapes are similar but not identical.

◊ ◊ ◊

Division of a positive number by a negative number
The division of this type can be performed using the extension method and using L-R representation, as in the case of other arithmetic operations on fuzzy numbers. For the division of L-R representations the quotient is obtained with the formula (3.63).

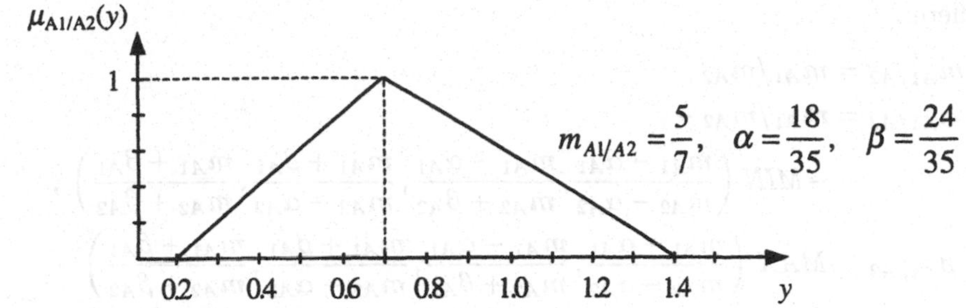

Fig. 3.47. Division of the fuzzy numbers "about 5" and "about 7" using L-R representation

$$(A_1/A_2) \cong (m_{A1}, \alpha_{A1}, \beta_{A1})/(m_{A2}, \alpha_{A2}, \beta_{A2}) = \\ = \left(\frac{m_{A1}}{m_{A2}}, \frac{m_{A1}\beta_{A2} - m_{A2}\beta_{A1}}{m_{A2}(m_{A2} + \beta_{A2})}, \frac{m_{A1}\alpha_{A2} - m_{A2}\alpha_{A1}}{m_{A2}(m_{A2} - \alpha_{A2})} \right) \tag{3.63}$$

Division of a negative number by a positive number

$A_1 < 0,\ A_2 > 0,\ A_2 \neq 0$

Division of this type can be performed using either the extension principle or L-R representation. In L-R representation the quotient is obtained with the formula (3.64).

$$(A_1/A_2) \cong (m_{A1}, \alpha_{A1}, \beta_{A1})/(m_{A2}, \alpha_{A2}, \beta_{A2}) = \\ = \left(\frac{m_{A1}}{m_{A2}}, \frac{-m_{A1}\alpha_{A2} + m_{A2}\alpha_{A1}}{m_{A2}(m_{A2} - \alpha_{A2})}, \frac{-m_{A1}\beta_{A2} + m_{A2}\beta_{A1}}{m_{A2}(m_{A2} + \beta_{A2})} \right) \tag{3.64}$$

Division of negative numbers

$A_1 < 0,\ A_2 < 0,\ A_2 \neq 0$

Division of this type can be performed using either the extension method or the L-R representation. In L-R representation the quotient is obtained with the formula (3.65).

$$(A_1/A_2) \cong (m_{A1}, \alpha_{A1}, \beta_{A1})/(m_{A2}, \alpha_{A2}, \beta_{A2}) = \\ = \left(\frac{m_{A1}}{m_{A2}}, \frac{-m_{A1}\alpha_{A2} - m_{A2}\beta_{A1}}{m_{A2}(m_{A2} - \alpha_{A2})}, \frac{-m_{A1}\beta_{A2} - m_{A2}\alpha_{A1}}{m_{A2}(m_{A2} + \beta_{A2})} \right) \tag{3.65}$$

Division of fuzzy numbers of undefined (arbitrary) sign

$A_1, A_2,\ A_2 \neq 0$

Division of this type can be performed using either the extension method or the L-R representation. In L-R representation the quotient is obtained with the formula (3.66).

$$(A_1/A_2) = (m_{A1}, \alpha_{A1}, \beta_{A1})/(m_{A2}, \alpha_{A2}, \beta_{A2}) = (m_{A1/A2}, \alpha_{A1/A2}, \beta_{A1/A2}),$$

where:

$$\begin{aligned}
m_{A1/A2} &= m_{A1}/m_{A2}\,, \\
\alpha_{A1/A2} &= m_{A1}/m_{A2} - \\
&\quad - MIN\left(\frac{m_{A1}-\alpha_{A1}}{m_{A2}-\alpha_{A2}}, \frac{m_{A1}-\alpha_{A1}}{m_{A2}+\beta_{A2}}, \frac{m_{A1}+\beta_{A1}}{m_{A2}-\alpha_{A2}}, \frac{m_{A1}+\beta_{A1}}{m_{A2}+\beta_{A2}}\right), \\
\beta_{A1/A2} &= MAX\left(\frac{m_{A1}-\alpha_{A1}}{m_{A2}-\alpha_{A2}}, \frac{m_{A1}-\alpha_{A1}}{m_{A2}+\beta_{A2}}, \frac{m_{A1}+\beta_{A1}}{m_{A2}-\alpha_{A2}}, \frac{m_{A1}+\beta_{A1}}{m_{A2}+\beta_{A2}}\right) - \\
&\quad - m_{A1}/m_{A2}\,.
\end{aligned} \tag{3.66}$$

3.6 Peculiarities of fuzzy numbers

The operations on fuzzy numbers can also involve crisp numbers. It is very important to make the appropriate distinction between the two types. Two numbers 0 and 1 are particularly significant. Fig. 3.48 presents the examples of the numbers in their fuzzy form (subscript $_f$) and crisp form (subscript $_{cr}$).

If A is a fuzzy number and $-A$ is its opposite number then the equation (3.67) is not correct.

$$A - A = 0_{cr} \tag{3.67}$$

However, the equation (3.68) is correct.

$$A - A = 0_f \tag{3.68}$$

The above is a consequence of the fact that the result of arithmetic operations on fuzzy numbers is always a fuzzy number but never a crisp number.

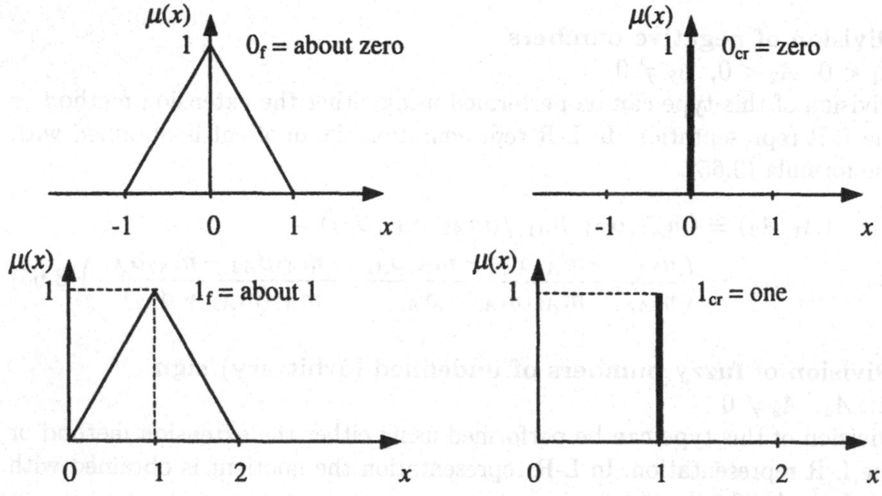

Fig. 3.48. Examples of zero and one in their fuzzy (f) and crisp (cr) forms

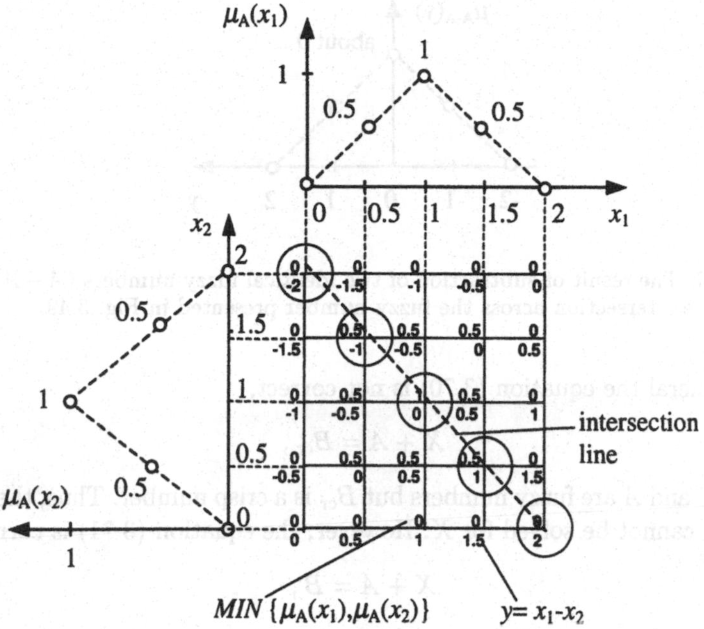

Fig. 3.49. Subtraction of two identical fuzzy numbers ($A - A$) using the extension principle

A fuzzy number cannot get reduced from an equation in the same way as a crisp number, i.e. through adding its opposite number ($-A$).

Example 3.6.1
Let us subtract two identical fuzzy numbers A, presented in Table 3.24.

Table 3.24. Fuzzy number A = "about 1"

$\mu_A(x)$	0	0.5	1	0.5	0
x	0	0.5	1	1.5	2

The method of forming the difference ($A - A$) using the extension principle (3.69) is presented in Fig. 3.49. The intersection of the difference ($A - A$) is presented in Fig. 3.50.

$$\mu_{A-A}(y) = \underset{y=x_1-x_2}{MAX}\ (MIN(\mu_A(x_1)\,,\mu_{-A}(x_2))), \quad \forall x_1, x_2, y \in R \qquad (3.69)$$

The example presented in Fig. 3.49 and 3.50 proves that the result of subtraction of two identical fuzzy numbers ($A - A$) is not a crisp zero but a fuzzy zero with support always greater than the support of the number A.
◊ ◊ ◊

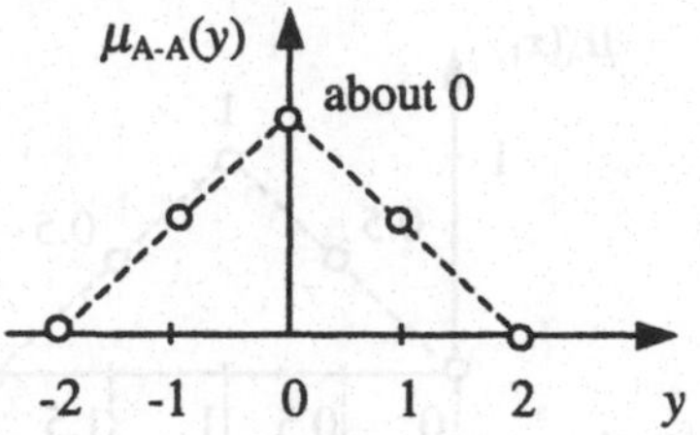

Fig. 3.50. The result of subtraction of two identical fuzzy numbers $(A - A)$ in the form of the intersection across the fuzzy number presented in Fig. 3.49

In general the equation (3.70) is not correct,

$$X + A = B_{cr}, \tag{3.70}$$

where: X and A are fuzzy numbers but B_{cr} is a crisp number. Thus, the above equation cannot be solved for X. However, the equation (3.71) is correct :

$$X + A = B, \tag{3.71}$$

where all component numbers are fuzzy.

The solution of the equation (3.71) is not in the form of equation (3.72) as it was in the crisp number case,

$$X = B - A, \tag{3.72}$$

since after having put the above number into the primary equation (3.71) we obtain the number C, i.e. the equation (3.73),

$$B - A + A = C \neq B, \tag{3.73}$$

which is **different** from B. The numbers C and B are of identical nominal value but of different spreads α and β respectively. The value X can be calculated with an approximate method of α-cuts (Knappe 1994) or with the L-R representation of fuzzy numbers; formulas (3.74), (3.75) and (3.76). If:

$$X = (m_X, \alpha_X, \beta_X), \quad A = (m_A, \alpha_A, \beta_A), \quad B = (m_B, \alpha_B, \beta_B), \tag{3.74}$$

then:

$$X + A = (m_X + m_A, \alpha_X + \alpha_A, \beta_X + \beta_A) = B = (m_B, \alpha_B, \beta_B). \tag{3.75}$$

Thus we obtain:

$$X = (m_B - m_A, \alpha_B - \alpha_A, \beta_B - \beta_A). \tag{3.76}$$

Example 3.6.2
Let us determine a fuzzy number X (unknown variable) in the equation (3.77):

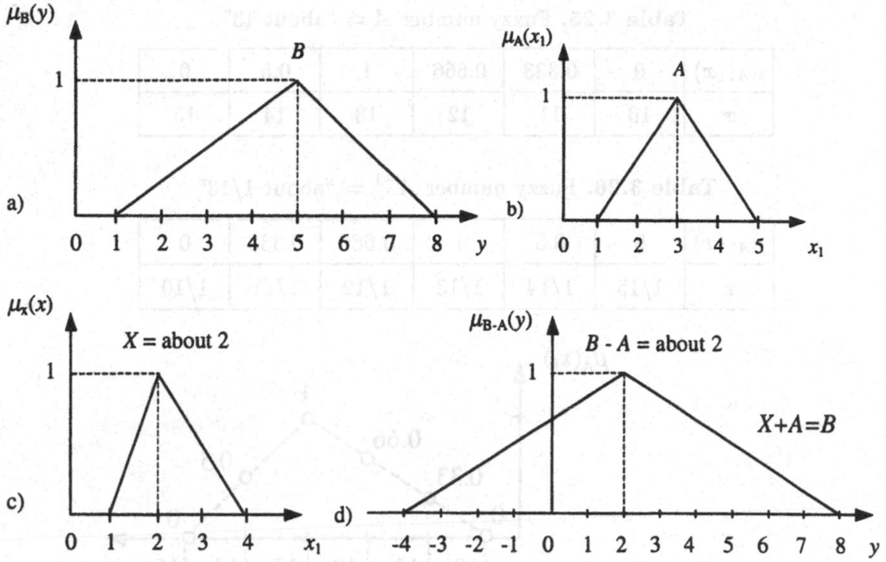

Fig. 3.51. The correct (c) solution X of the equation $X + A = B$ and the incorrect solution (d) equal to $B - A$, where the numbers A and B are depicted in (b) and (a) respectively

$$X + A = (5,4,3) = B\,, \tag{3.77}$$

where: $A = (3,3,1)$. Using equation (3.76) the following result is obtained:

$$X = (2,1,2)\,.$$

The above calculations are presented in Fig. 3.51.

◊ ◊ ◊

In general, the equation (3.78) is not correct.

$$A \cdot A^{-1} = 1_{\mathrm{cr}} \tag{3.78}$$

On the other hand, the equation (3.79) is correct.

$$A \cdot A^{-1} = 1_{\mathrm{f}} \tag{3.79}$$

The product of two fuzzy numbers, even if one of them is the opposite number to the other, is never crisp.

Example 3.6.3

Let us calculate a product of the numbers A and A^{-1}, defined in Table 3.25 and Table 3.26, using the extension principle.

The product calculation scheme is shown in Fig. 3.52. The plane intersection across the product $A \cdot A^{-1}$ is presented in Fig. 3.53.

◊ ◊ ◊

Table 3.25. Fuzzy number $A =$ "about 13"

$\mu_{A1}(x)$	0	0.333	0.666	1	0.5	0
x	10	11	12	13	14	15

Table 3.26. Fuzzy number $A^{-1} =$ "about 1/13"

$\mu_{A2}(x)$	0	0.5	1	0.666	0.333	0
x	1/15	1/14	1/13	1/12	1/11	1/10

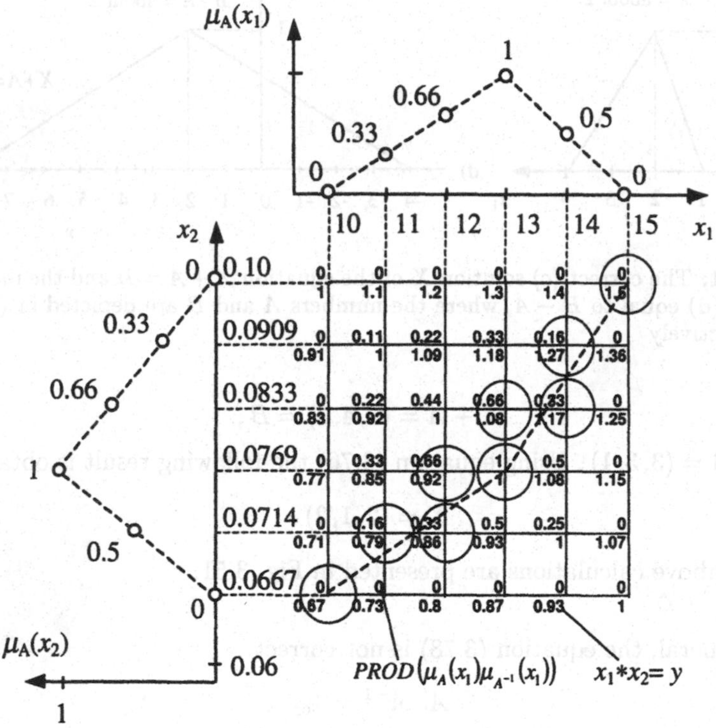

Fig. 3.52. The product of the fuzzy number A and its inverse number A^{-1} in 4D-form

Fig. 3.53 proves that the product of fuzzy number A and its opposite A^{-1} is a fuzzy "1" equal to "about 1". It is not possible to obtain a crisp number as a result of the multiplication of two fuzzy numbers. Therefore, the equation (3.80) is incorrect.

$$X \cdot A = B_{\mathrm{cr}} \tag{3.80}$$

is incorrect. However, the equation (3.81) is correct.

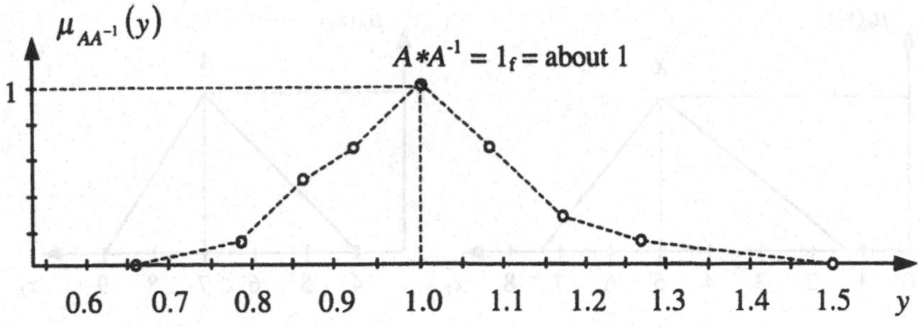

Fig. 3.53. The 2D-approximate intersection across the product of fuzzy numbers $A \cdot A^{-1}$, presented in Fig. 3.52

$$X \cdot A = B\,, \tag{3.81}$$

where: X, A, B are fuzzy numbers, B_{cr} is a crisp number.

The equation (3.81) cannot be solved for X by carrying out the multiplication of both sides of the equation by the inverse number A^{-1}, (3.82),

$$X \cdot A \cdot A^{-1} = B \cdot A^{-1}\,, \tag{3.82}$$

since $X \cdot A \cdot A^{-1} \neq X$. Nevertheless, the equation (3.81) can be solved for the fuzzy numbers in L-R representation form.

Example 3.6.4
Let us calculate a fuzzy number X in the equation (3.83),

$$X \cdot A = B\,, \tag{3.83}$$

where:

$$A = (7, 3, 2)\,,$$
$$B = (35, 29, 31)\,,$$
$$X = (m_X, \alpha_X, \beta_X)\,.$$

The equation (3.83) in L-R representation form can be expressed as (3.84).

$$(m_X, \alpha_X, \beta_X) \cdot (7, 3, 2) = (35, 29, 31) \tag{3.84}$$

On having performed the multiplication using the formula (3.48) the following is obtained:

$$(7m_X, 3m_X + 4\alpha_X, 2m_X + 9\beta_X) = (35, 29, 31)\,.$$

To determine the fuzzy number X the following system of 3 equations has to be solved:

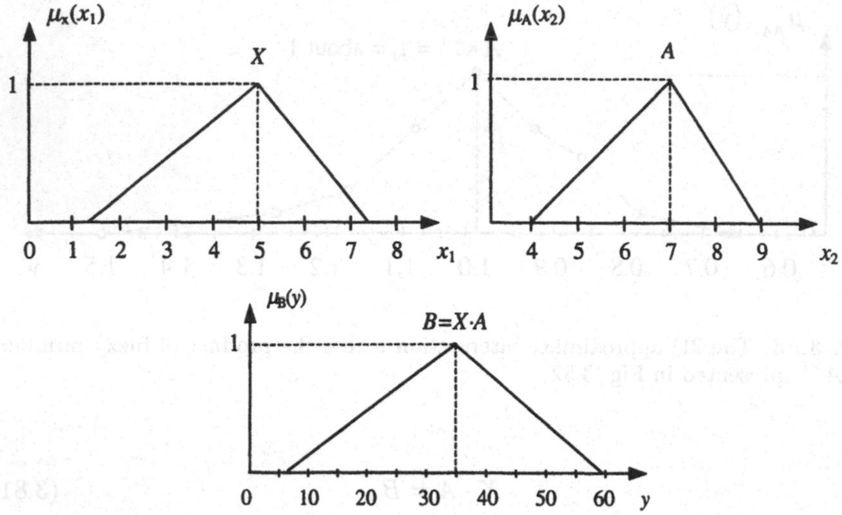

Fig. 3.54. Fuzzy numbers X and A and their multiplication result $B = X \cdot A$

$$7m_X = 35\,,$$
$$3m_X + 4\alpha_X = 29\,,$$
$$2m_X + 9\beta_X = 31\,.$$

The solution of the above system is the following:

$$m_X = 5\,,$$
$$\alpha_X = 3.5\,,$$
$$\beta_X = 2.33\,.$$

The fuzzy number X has the form: $X = (5, 3.5, 2.33)$ in L-R representation, Fig. 3.54. The wrong solution of the equation $X \cdot A = B$ is shown in Fig. 3.55

◊ ◊ ◊

The operations of addition and multiplication of fuzzy numbers refer to the following principles: commutation, association, distribution and neutral element, described with the formulas (3.85) – (3.88), respectively.

Commutation principle:

$$A + B = B + A\,,$$
$$A \cdot B = B \cdot A\,. \tag{3.85}$$

Association principle:

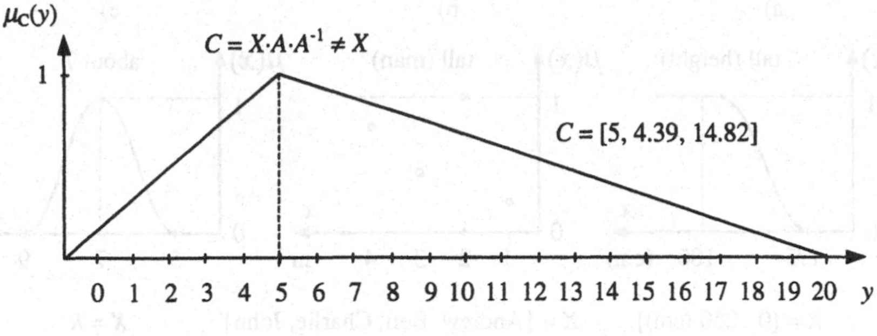

Fig. 3.55. The incorrect result $C = X \cdot A \cdot A^{-1}$ of the equation $X \cdot A = B$

$$\begin{aligned} A + (B + C) &= (A + B) + C\,, \\ A \cdot (B \cdot C) &= (A \cdot B) \cdot C\,. \end{aligned} \tag{3.86}$$

Distribution principle:

$$A \cdot (B + C) = (A \cdot B) + (A \cdot C)\,. \tag{3.87}$$

Neutral element principle:

$$\begin{aligned} A + 0_{cr} &= 0_{cr} + A = A\,, \\ A \cdot 1_{cr} &= 1_{cr} \cdot A = A\,. \end{aligned} \tag{3.88}$$

where: 0_{cr} – crisp zero,
1_{cr} – crisp one.

3.7 Differences between fuzzy numbers and linguistic values

Both quantities, i.e. fuzzy numbers (e.g. about 1) and linguistic values (e.g. low voltage), are in terms of mathematics represented by fuzzy sets. However, while a linguistic value can be defined either by the sets containing, as elements, the real numbers $(1, 2, 3 \ldots)$ or not numbers (John, Richard, Bill, ...), a fuzzy number is always defined only in a set of the real numbers R (e.g. about 7), Fig. 3.56.

Fuzzy numbers should according to the definition of (Knappe 1994; Zimmermann 1994a) be fuzzy, convex, normal sets with a single-element core x_0 (Fig. 3.56c) and a limited support. On the other hand, a linguistic value can be described both with a convex and non-convex membership function, and have a single- or multi-element or even empty core and an unlimited support. However, in practical applications fuzzy sets are used which according

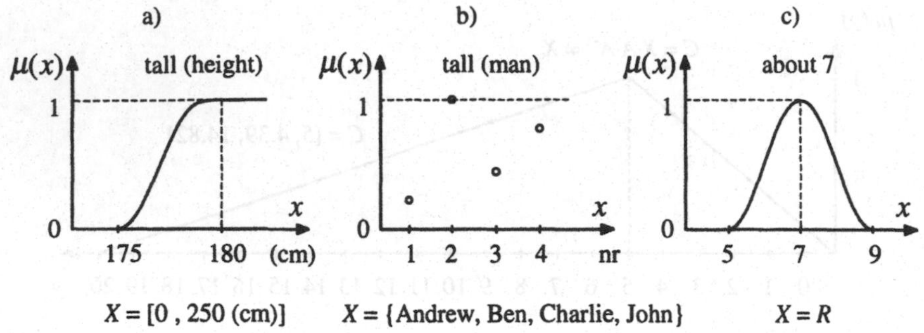

Fig. 3.56. Example membership functions of linguistic values "tall" (height), "tall" (man) and a fuzzy number "about 7"

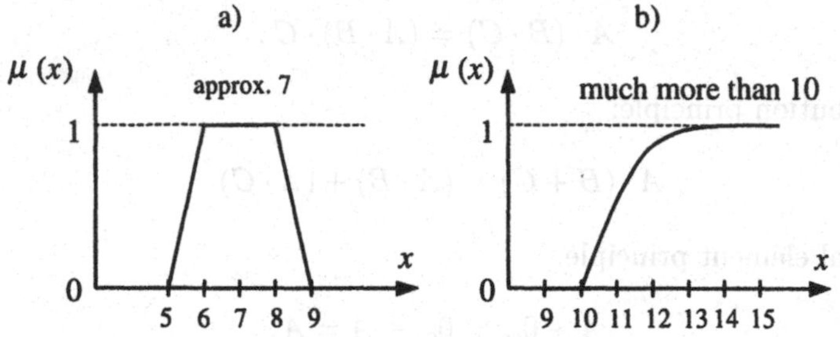

Fig. 3.57. Example of fuzzy sets containing numerical expressions in their labels

to the definition of (Knappe 1994; Zimmermann 1994a) are not fuzzy numbers but fuzzy intervals (Kacprzyk 1992; Knappe 1994; Zimmermann 1994a), Fig. 3.57a.

The trapezoid set (Fig. 3.57a) "approx. 7" is called a trapezoid "fuzzy number" (Zimmermann 1994a). The set "much more than 10", Fig. 3.57b, does not meet the requirements of a fuzzy number according to the definition: it has a multi-element core and one-side unlimited support. However, it contains a reference numerical value in its label. The labels of many linguistic values can be described with numerical expressions. For example, a linguistic value "very tall" (Fig. 3.56a) can be expressed as the height "much more than 175 cm". Similarly, the fuzzy number "about 7", (Fig 3.56c) when used as a linguistic variable describing the "dog's age" can be replaced with a linguistic value "medium" (age).

Regarding the above fuzzy modeling practicians often make use of mixed universes of discourse containing both linguistic values and fuzzy numbers (see Fig. 3.58), e.g.:

control error = {large negative, moderate negative, slight negative, **about zero**, slight positive, moderate positive, large positive}.

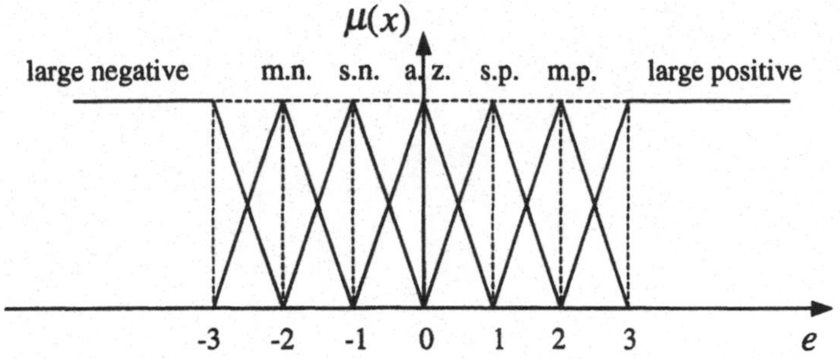

Fig. 3.58. Examples of membership functions of the linguistic variable "control error"

The above set can also be described with labels containing numerical expressions as follows:

control error = {much less than -2, about -2, about -1, about 0, about 1, about 2, much more than 2}.

The above example proves that linguistic values and fuzzy numbers may often be used alternatively. Therefore, several authors of publications on fuzzy systems, e.g. Kahlert (Kahlert 1995), Wang (Wang 1994) or Brown and Harris (Brown 1994), intentionally do not differentiate a linguistic value from a fuzzy number; they use the term **fuzzy set** as the most general one.

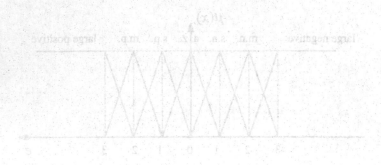

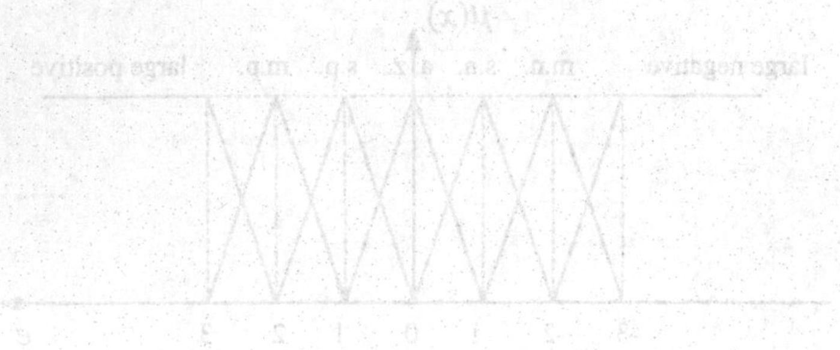

4. Mathematics of Fuzzy Sets

The main elements of fuzzy sets are logical rules of the type (4.1).

$$IF \ (x_1 \text{ is medium}) \ AND \ (x_2 \text{ is small}) \ THEN \ (y \text{ is large}) \tag{4.1}$$

To process information in such models many operations, mostly of a logical character, must be performed. A set of these operations and the ideas related to them can be labeled mathematics of fuzzy sets (Zimmermann 1994a). Its principles are presented below.

4.1 Basic operations on fuzzy sets

A fuzzy model of a real system contains logical rules reflecting the functioning of the system. In a system having two inputs (x_1, x_2) and one output y the rule can have the form (4.2):

$$\begin{aligned} IF \ & [(x_1 \text{ is small}) \ AND \ (x_2 \text{ is medium})] \ OR \\ & [(x_1 \text{ is medium}) \ AND \ (x_2 \text{ is small})] \ THEN \ (y \text{ is medium}), \end{aligned} \tag{4.2}$$

where: "small", "medium" are fuzzy sets (fuzzy evaluations of the states of the variables in the system),
IF-THEN, *AND*, *OR* – logical connectives (aggregation operators of fuzzy sets).

If the fuzzy sets (small, medium) are used for the evaluation of the input and output states of a system, then the logical connectives determine the qualitative relationship between these states by joining fragments of a rule to get a whole. The accuracy of a fuzzy model depends on both the manner of defining the fuzzy sets used (their number, shape and parameters of membership function) and the kind of logical connectives used.

The main kinds of connectives, called logical operators, are:

- *AND*, $\cap$, $\wedge$ – the intersection (logical product) operator of sets,
- *OR*, $\cup$, $\vee$ – the union (logical sum) operator of sets,
- *NOT*, $^{-}$, $\neg$ – the negation (logical complement) operator of sets.

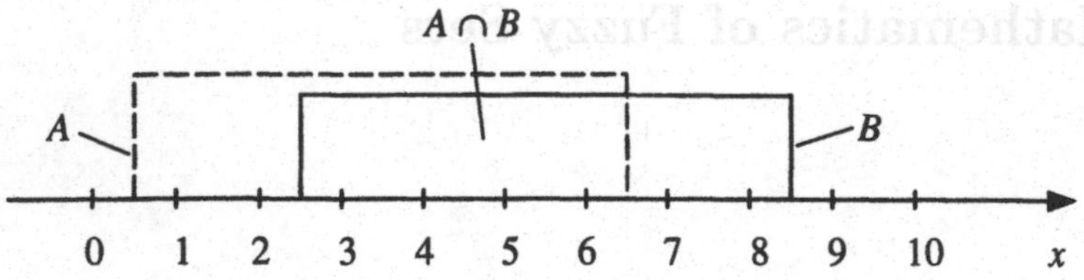

$A = \{1,2,3,4,5,6\}$, $B = \{3,4,5,6,7,8\}$, $A \cap B = \{3,4,5,6\}$

Fig. 4.1. Example of the logical product of the non-fuzzy sets $A \cap B$

Many various mathematical forms can represent the logical operators and selecting them properly is a problem. The condition for proper selection is, at minimum, knowledge of the basic forms of these operators.

4.1.1 Intersection operation (logical product) of fuzzy sets

Fuzzy logic has developed on the basis of classical, non-fuzzy, bivalued logic. Its initiator, Lofti Zadeh, noticed imperfections of classical logic in the modeling of reality. By introducing the idea of fuzzy sets (Zadeh 1965) he created possibilities for the improvement of models containing logical connectives. He defined the intersection operation of fuzzy sets to be the expansion of that operation on non-fuzzy sets. This means that operations on non-fuzzy sets should be a particular case of operations on fuzzy sets. This postulate can often be found in the literature, e.g. in (Yager 1994,1995), although practical reasons (attempting to improve the accuracy of fuzzy models) mean that one also introduces operators not satisfying it (Yager 1994,1995).

In non-fuzzy logic the logical product of sets A and B is defined without the use of membership functions (Driankov 1993,1996; Poradnik 1971), (4.3).

$$A \cap B = \{x : \ x \in A \text{ and } x \in B\} \tag{4.3}$$

Example 4.1.1.1
An example of the calculation of the logical product of the non-fuzzy sets $A \cap B$ is shown in Fig. 4.1.
◊ ◊ ◊

The main properties of the intersection operation of the non-fuzzy sets $A \cap B$ defined in the universe of discourse X, to be expected also in the case of fuzzy sets, are the features (4.4) – (4.9) given below.

Commutativity:

$$A \cap B = B \cap A . \tag{4.4}$$

This property means that the ordering of the sets involved in the operation is unimportant for the final result.

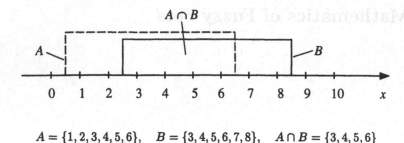

Associativity:

$$(A \cap B) \cap C = A \cap (B \cap C). \tag{4.5}$$

This property means that when creating the logical product of many sets we can calculate it stepwise, using the products of pairs of the sets. The order in which the sets are paired is unimportant for the final result.

Idempotency:

$$A \cap A = A. \tag{4.6}$$

Absorption by the empty set $\emptyset$:

$$A \cap \emptyset = \emptyset. \tag{4.7}$$

Identity:

$$A \cap X = A, \tag{4.8}$$

where: X – the universe of discourse.

Exclusive contradiction:

$$A \cap \bar{A} = \emptyset. \tag{4.9}$$

Below it will be shown that not all these properties can be transferred to the intersection operation of fuzzy sets, e.g. the property (4.9) is such a one. In the environment of fuzzy sets this operation can be performed with the use of various methods and therefore its sense is non-unique. The non-uniqueness will be presented in Example 4.1.1.2.

Example 4.1.1.2
Two fuzzy sets A and B defined by (4.10) and (4.11) are given. A – the set of cheap cars, x_i – the notation of a car:

$$A = \left\{ \frac{1}{x_1}, \frac{0.8}{x_2}, \frac{0.6}{x_3}, \frac{0.4}{x_4}, \frac{0.2}{x_5}, \frac{0}{x_6} \right\}, \tag{4.10}$$

B – the set of luxury cars:

$$B = \left\{ \frac{0}{x_1}, \frac{0.2}{x_2}, \frac{0.4}{x_3}, \frac{0.6}{x_4}, \frac{0.8}{x_5}, \frac{1}{x_6} \right\}. \tag{4.11}$$

Simultaneously there is $A = \bar{B}$, and this is represented in Fig. 4.2.

A set $C = A \cap B$ containing cheap and at the same time luxury cars has to be determined. Since $A = \bar{B}$, then in the case of non-fuzzy sets we would obtain the empty set, according to the property (4.9) ($A \cap \bar{A} = \emptyset$). What will the result be in the case of the fuzzy sets to be considered?

A car x_4 is cheap to the degree $\mu_A(x_4) = 0.4$ and luxury to the degree $\mu_B(x_4) = 0.6$. To which degree $\mu_{A \cap B}(x_4)$ is it cheap and luxury at the same time? How can this degree be determined using the membership grades $\mu_A(x_4)$ and $\mu_B(x_4)$ in the respective sets?

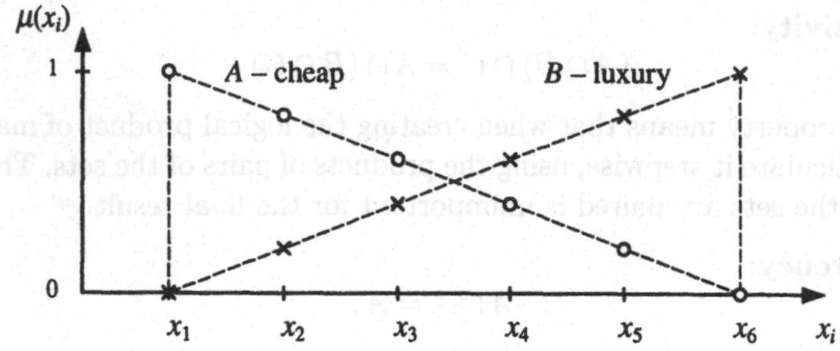

Fig. 4.2. Discrete membership functions of cars x_i in the set of cheap cars (A) and luxury cars (B)

In (Zadeh 1965) L. Zadeh proposed to calculate the membership function of the product of the sets by means of the *MIN* operator, the formula (4.12).

$$\mu_{A\cap B}(x) = MIN(\mu_A(x),\, \mu_B(x)), \quad \forall x \in X\,. \tag{4.12}$$

This operator was the first operator extending the intersection operation of non-fuzzy sets ∩ to fuzzy sets. By applying the formula (4.12) to calculate the set $A \cap B$ of the cars that are both cheap and luxurious, we obtain the formula (4.13). The fuzzy set resulting from this operation is illustrated in Fig. 4.3.

$$B = \left\{ \frac{0}{x_1}, \frac{0.2}{x_2}, \frac{0.4}{x_3}, \frac{0.4}{x_4}, \frac{0.2}{x_5}, \frac{0}{x_6} \right\}, \tag{4.13}$$

◇ ◇ ◇

The *MIN* operator can be represented in algebraic form:

$$MIN(x_1, x_2) = \frac{x_1 + x_2 - |x_1 - x_2|}{2} = \frac{x_1 + x_2 - (x_1 - x_2)\cdot sgn(x_1 - x_2)}{2}, \tag{4.14}$$

where:

$$sgn(x_1 - x_2) = \begin{cases} -1 & \text{for } x_1 - x_2 < 0\,, \\ 0 & \text{for } x_1 - x_2 = 0\,, \\ 1 & \text{for } x_1 - x_2 > 0\,. \end{cases}$$

The *MIN* operator in the form (4.14) is called "*hardMIN*", because the reversal of sign of the difference $(x_1 - x_2)$ causes a drastic change in the value $sgn(x_1 - x_2)$ and the output of the operator, Fig. 4.4.

To make the operation of the $MIN(x_1 - x_2)$ operator softer, one can apply a special form (4.15) of the two-input *sgn* operator that can be found in the literature on this subject.

$$sgn_\delta(x_1 - x_2) = \frac{x_1 - x_2}{\sqrt{(x_1 - x_2)^2 + \delta^2}}, \tag{4.15}$$

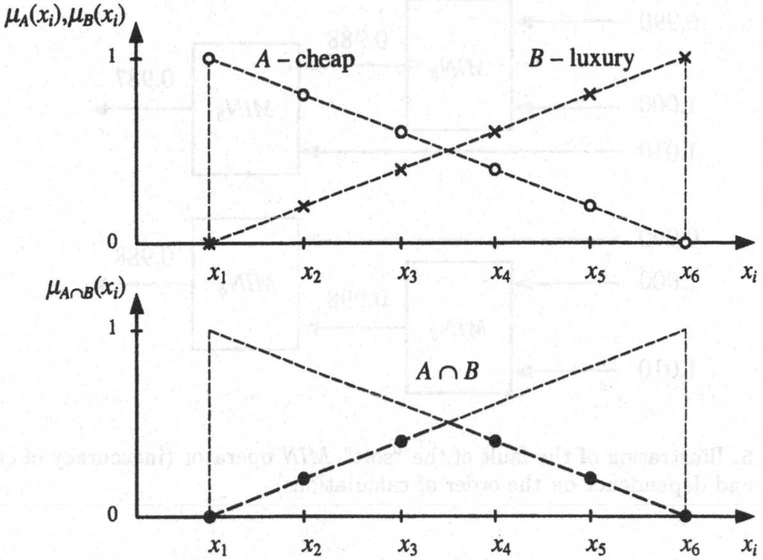

Fig. 4.3. The product of the fuzzy sets $A \cap B$ of cheap (A) and luxury (B) cars obtained by means of the *MIN* operator

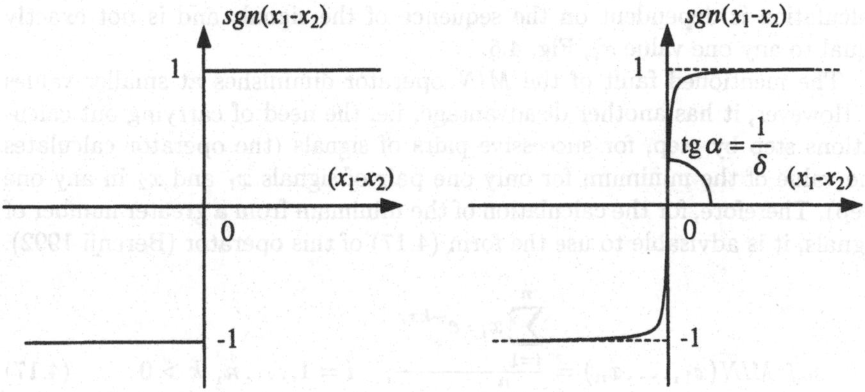

Fig. 4.4. "Hard" (a) and "soft" (b) $sgn(x_1 - x_2)$

where: δ – a small number, e.g. 0.05. Increasing the value δ makes the operation of $sgn(x_1 - x_2)$ "softer", Fig. 4.4b.

By using the "soft" *sgn*, we obtain as a result the "soft" $MIN_\delta(x_1 - x_2)$ operator determined by the formula (4.16).

$$MIN_\delta(x_1, x_2) = \frac{x_1 + x_2 + \delta^2 + \sqrt{(x_1 - x_2)^2 + \delta^2}}{2} \tag{4.16}$$

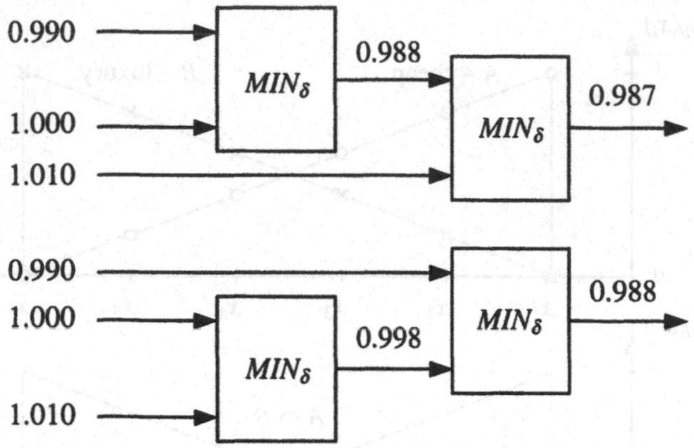

Fig. 4.5. Illustration of the fault of the "soft" *MIN* operator (inaccuracy of calculations and dependence on the order of calculations)

The "soft" *sgn* means that the operation of controllers and fuzzy models becomes smoother. It smoothes out sharp edges of the surface of the input → output mapping. However, it also has a disadvantage. When the soft *sgn* is calculated from a greater (than 2) number of signals x_i, the result of the calculation is dependent on the sequence of the signals and is not exactly equal to any one value x_i, Fig. 4.5.

The mentioned fault of the *MIN* operator diminishes at smaller values δ. However, it has another disadvantage, i.e. the need of carrying out calculations step by step, for successive pairs of signals (the operator calculates the value of the minimum for only one pair of signals x_1 and x_2 in any one step). Therefore, for the calculation of the minimum from a greater number of signals, it is advisable to use the form (4.17) of this operator (Berenji 1992).

$$softMIN(x_1, \ldots, x_n) = \frac{\sum_{i=1}^{n} x_i \cdot e^{-kx_i}}{\sum_{i=1}^{n} e^{-kx_i}}, \quad i = 1, \ldots, n, \; k > 0 . \tag{4.17}$$

The operation of the *softMIN* operator is illustrated in Fig. 4.6. Together with the increase in the value of the factor k ($k \to \infty$), the operation of the *softMIN* operator becomes more and more similar to the operation of the "hard" *MIN*. Practically, for the value $k > 100$ great accuracy of calculation is already obtained. However, it should be remembered that the operation of the operator becomes "harder" as the value k increases.

Summarizing, the "hard" *MIN* operator has the following advantages and disadvantages.

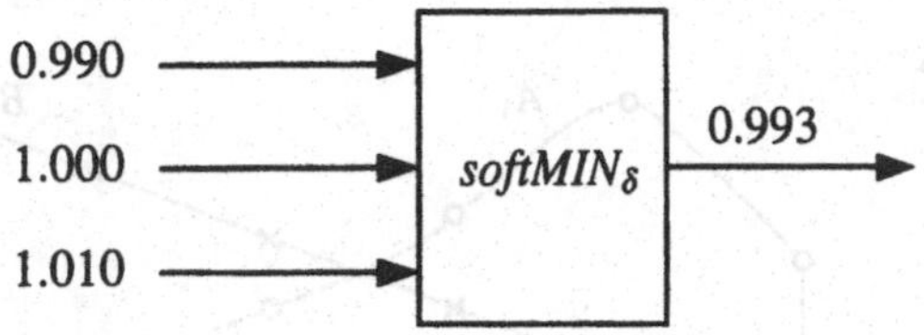

Fig. 4.6. Example of the calculation of the value *softMIN*(0.990; 1.000; 1.010) from the formula (4.17)

Advantages:

1. The calculations are carried out quickly and in a simple way, resulting in a decrease of the load on computers and microprocessors. It makes it possible to use cheap microprocessors as fuzzy controllers.
2. The possibility of "softening" the action of the *MIN* operator. However, this results in the increase of the number of calculations associated with decreasing accuracy.

Disadvantages:

1. In general, the accuracy of a model is worse than when other operators are used.
2. The smoothness of the surface of a model is worse than in the case of using other operators.
3. The occurrence of insensitivity and drastic changes at the output of a model and the output of a fuzzy controller containing the *MIN* operators.

The analysis of the result of the intersection operation of sets $A \cap B$ with the use of the *MIN* operator represented in Fig. 4.3 attests to disadvantage no. 3. Only membership in the set of luxury cars decides whether the cars x_1, x_2, x_3 belong to the set $A \cap B$. The fact, whether the cars are cheaper or more expensive, has no influence on this membership. In the case of the car x_3 we obtain the result:

$$\mu_{A\cap B}(x_3) = \mu_B(x_3) = 0.4\,.$$

According to this model, irrespective of how cheap car x_3 would be (even when the price of that car would amount to zero), the grade of membership $\mu_{A\cap B}(x_3)$ would be the same (0.4). The discussed fault is illustrated in Fig. 4.7.

The example represented in this drawing shows that the use of the *MIN* operator to realize the intersection operation of sets causes part of the information to be lost. The reason is that this operator takes into account only the fact that one membership grade is less than another, whereas the value of the difference of the membership grades is not taken into account by it.

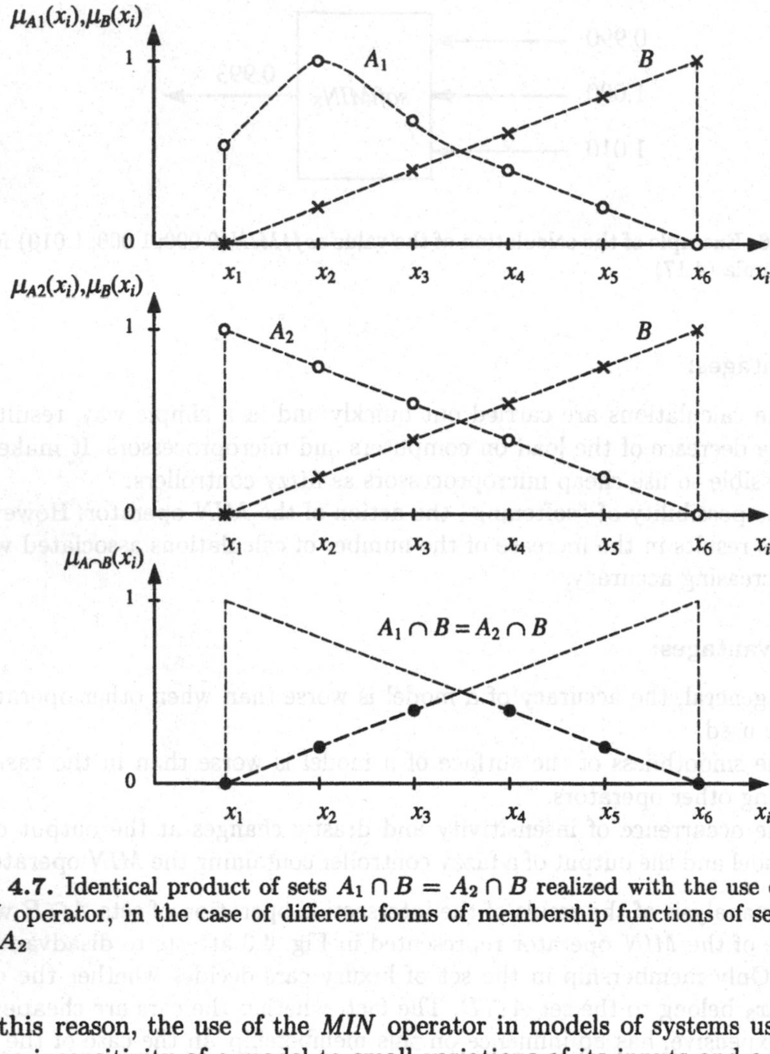

Fig. 4.7. Identical product of sets $A_1 \cap B = A_2 \cap B$ realized with the use of the *MIN* operator, in the case of different forms of membership functions of sets A_1 and A_2

For this reason, the use of the *MIN* operator in models of systems usually causes insensitivity of a model to small variations of its inputs and a drastic switching over of the output state after a certain level of the inputs has been exceeded (Piegat 1995a). This feature is very inconvenient in modeling systems having a smooth surface of the input → output mapping.

In some systems, where the method of processing information is similar to the logical one (the majority of dependencies between inputs and output of the system are of a logical character), the use of these operators can have advantages. The above-mentioned disadvantages of the *MIN* operator mean that the range of its use is decreasing. In 1995 the participants in the 5th Workshop "Fuzzy Control" (October 16–17, 1995, Witten, Germany) were asked for their opinion regarding the use of the *MIN* operator. It showed that most specialists participating in that conference preferred the *PROD* operator (logical product) rather than the *MIN* operator (Pfeiffer 1996),

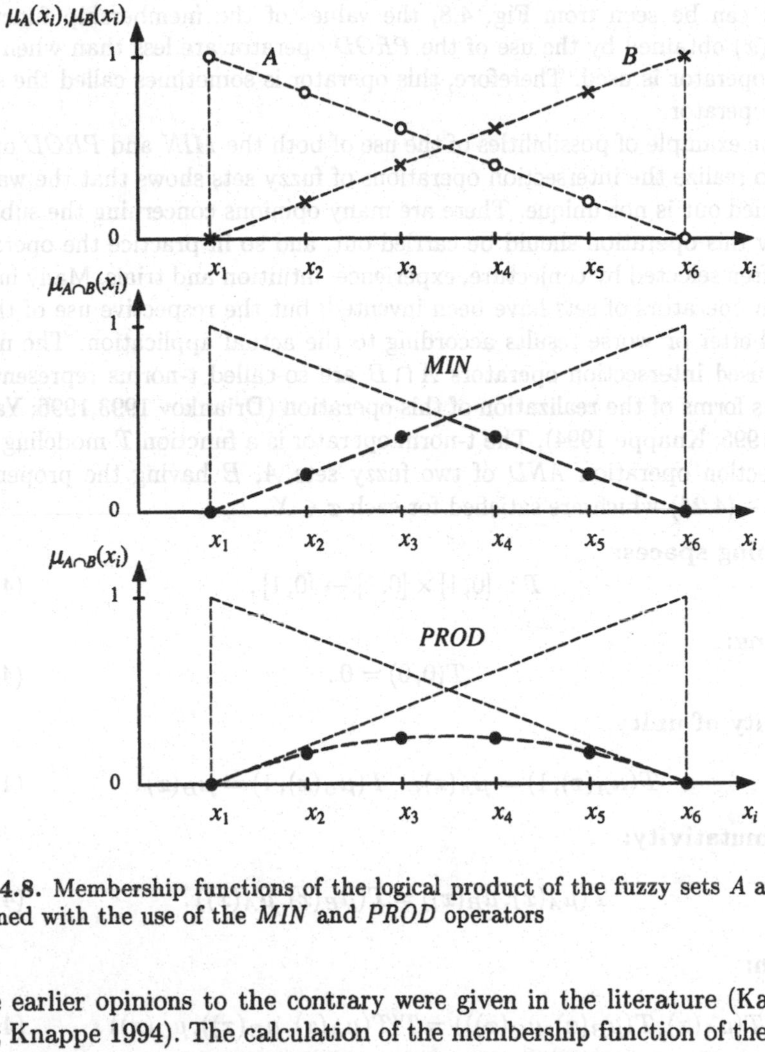

Fig. 4.8. Membership functions of the logical product of the fuzzy sets A and B obtained with the use of the *MIN* and *PROD* operators

while earlier opinions to the contrary were given in the literature (Kahlert 1995; Knappe 1994). The calculation of the membership function of the product of fuzzy sets using the *PROD* operator is carried out according to the formula (4.18).

$$\mu_{A\cap B}(x) = \mu_A(x) \cdot \mu_B(x), \quad \forall x \in X . \tag{4.18}$$

The advantage of the *PROD* operator is that the product value $\mu_{A\cap B}(x)$ is quantitatively dependent on the actual values of both the components of membership function $\mu_A(x)$ and $\mu_B(x)$ (apart from when the value of one function is zero). You can see that here the loss of information is not so considerable as in the case of the *MIN* operator where the value $\mu_{A\cap B}(x)$ is dependent only on the lesser value (in the given range of x) of the components $\mu_A(x)$ or $\mu_B(x)$. The comparison of the results of logical product calculation using the *MIN* and *PROD* operators is represented in Fig. 4.8.

As can be seen from Fig. 4.8, the values of the membership function $\mu_{A\cap B}(x)$ obtained by the use of the *PROD* operator are less than when the *MIN* operator is used. Therefore, this operator is sometimes called the sub-*MIN* operator.

The example of possibilities of the use of both the *MIN* and *PROD* operator to realize the intersection operations of fuzzy sets shows that the way it is carried out is not unique. There are many opinions concerning the subject of how this operation should be carried out, and so in practice the operator $\cap$ is often selected by conjecture, experience, intuition and trials. Many intersection operators of sets have been invented; but the respective use of them gives better or worse results according to the actual application. The most often used intersection operators $A \cap B$ are so-called t-norms representing various forms of the realization of this operation (Driankov 1993,1996; Yager 1994,1995; Knappe 1994). The t-norm operator is a function T modeling the intersection operation *AND* of two fuzzy sets A, B having the properties (4.19) – (4.24) which are satisfied for each $x \in X$.

Mapping spaces:

$$T: \ [0,1] \times [0,1] \rightarrow [0,1]\,. \tag{4.19}$$

Zeroing:

$$T(0,0) = 0\,. \tag{4.20}$$

Identity of unity:

$$T(\mu_A(x), 1) = \mu_A(x), \quad T(\mu_B(x), 1) = \mu_B(x)\,. \tag{4.21}$$

Commutativity:

$$T(\mu_A(x), \mu_B(x)) = T(\mu_B(x), \mu_A(x))\,. \tag{4.22}$$

Union:

$$T(\mu_A(x), T(\mu_B(x), \mu_C(x))) = T(T(\mu_A(x), \mu_B(x)), \mu_C(x))\,. \tag{4.23}$$

Monotonicity:

$$\mu_A(x) \le \mu_C(x), \ \mu_B(x) \le \mu_D(x) \Rightarrow T(\mu_A(x), \mu_B(x)) \le T(\mu_C(x), \mu_D(x))\,. \tag{4.24}$$

The property of commutativity says that the order of sets is unimportant for carrying out the operation. The property of union tells us that the intersection operation of a greater (than 2) number of sets should be carried out successively, but the order of creating pairs of sets has no effect on the final result. The property of monotonicity means that the result of the operation does not diminish when the values of its arguments become larger.

These operators are divided into parameterized and non-parameterized t-norms. The effect of the action of non-parameterized t-norms is constant. In

Table 4.1. Some non-parameterized t-norm operators

Operator name	Formula
minimum (*MIN*)	$\mu_{A\cap B}(x) = MIN(\mu_A(x), \mu_B(x))$
product (*PROD*)	$\mu_{A\cap B}(x) = \mu_A(x) \cdot \mu_B(x)$
Hamacher product	$\mu_{A\cap B}(x) = \dfrac{\mu_A(x) \cdot \mu_B(x)}{\mu_A(x) + \mu_B(x) - \mu_A(x) \cdot \mu_B(x)}$
Einstein product	$\mu_{A\cap B}(x) = \dfrac{\mu_A(x) \cdot \mu_B(x)}{2 - (\mu_A(x) + \mu_B(x) - \mu_A(x) \cdot \mu_B(x))}$
drastic product	$\mu_{A\cap B}(x) = \begin{cases} MIN(\mu_A(x), \mu_B(x)) & \text{for } MAX(\mu_A, \mu_B) = 1 \\ 0 & \text{otherwise} \end{cases}$
bounded difference	$\mu_{A\cap B}(x) = MAX(0, \mu_A(x) + \mu_B(x) - 1)$

Table 4.2. Some parameterized intersection operators of fuzzy sets

Operator name	Formula
Dubois-intersection operator	$\mu_{A\cap B}(x,\alpha) = \dfrac{\mu_A(x) \cdot \mu_B(x)}{MAX[\mu_A(x), \mu_B(x), \alpha]}, \ \alpha \in [0,1]$ $\alpha = 0$: $\mu_{A\cap B}(x,\alpha) = MIN(\mu_A(x), \mu_B(x))$ $\alpha = 1$: $\mu_{A\cap B}(x,\alpha) = PROD(\mu_A(x), \mu_B(x))$
Hamacher-intersection operator	$\mu_{A\cap B}(x,\gamma) = \dfrac{\mu_A(x) \cdot \mu_B(x)}{\gamma + (1-\gamma)(\mu_A(x) + \mu_B(x) - \mu_A(x) \cdot \mu_B(x))}, \ \gamma \geq 0$ $\gamma = 0$: $\mu_{A\cap B}(x,\gamma) = HamacherPROD$ $\gamma = 1$: $\mu_{A\cap B}(x,\gamma) = PROD(\mu_A(x), \mu_B(x))$ $\gamma = 2$: $\mu_{A\cap B}(x,\gamma) = EinsteinPROD$ $\gamma \to \infty$: $\mu_{A\cap B}(x,\gamma) = drasticPROD$
Yager-intersection operator	$\mu_{A\cap B}(x,p) = 1 - MIN\Big(1, ((1-\mu_A(x))^p + (1-\mu_B(x))^p)^{\frac{1}{p}}\Big), \ p \geq 1$ $p = 1$: $\mu_{A\cap B}(x,p) =$ bounded difference $p \to \infty$: $\mu_{A\cap B}(x,p) = MIN(\mu_A(x), \mu_B(x))$

contrast, the effect of parameterized t-norms varies qualitatively and quantitatively when any parameter which is the degree of freedom of the operator has been changed. The most well known non-parameterized t-norms are listed in Table 4.1. The most often used parameterized t-norms are specified in Table 4.2 where the dependence of the operators upon their parameters is also given.

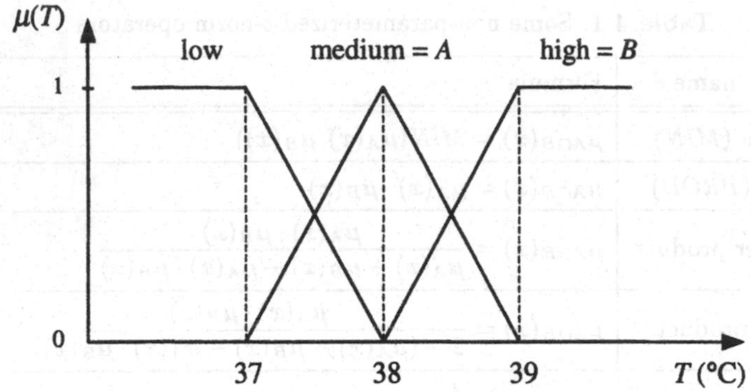

Fig. 4.9. Membership functions of the linguistic value "fever"

Example 4.1.1.3
The example illustrates the operation of fuzzy t-norms. In Fig. 4.9 the membership functions of the linguistic variable "fever" are given.

The task is to determine the membership function of fever T in the fuzzy set $C =$ "medium *AND* high temperature" ($C = A \cap B$) using various non-parameterized operators. The results are shown in Fig. 4.10.

According to binary non-fuzzy logic the fever can not be medium *AND* high at the same time. Fuzzy logic shows that such a set can exist. The membership function of this set is not strictly defined and is dependent upon which t-norm operator has been used.
◊ ◊ ◊

As can be seen in Fig. 4.10, the *MIN* operator allows us to obtain the highest values of membership functions. Therefore, other t-norm operators are sometimes called sub-*MIN*-operators or sub-*MIN*-norms (Knappe 1994), Fig. 4.11.

By using the sub-*MIN*-operators (t-norms) smaller values of the membership function $\mu_{A \cap B}(x)$ of the product of sets are obtained than when the *MIN*-operator is applied. This means that the sub-*MIN*-operators are stricter and require the conditions A and B of the fuzzy product to be satisfied to a higher degree. Therefore, the *MIN*-operator is considered (Driankov 1993) to be the most optimistic among t-norms.

According to the degree of optimism, t-norms can be ordered in the following sequence:

minimum > Hamacher product > algebraic product > Einstein product > bounded difference > drastic product.

To realize the intersection operation operators which are not t-norms (i.e. operators not satisfying the conditions of a t-norm) are also used. An

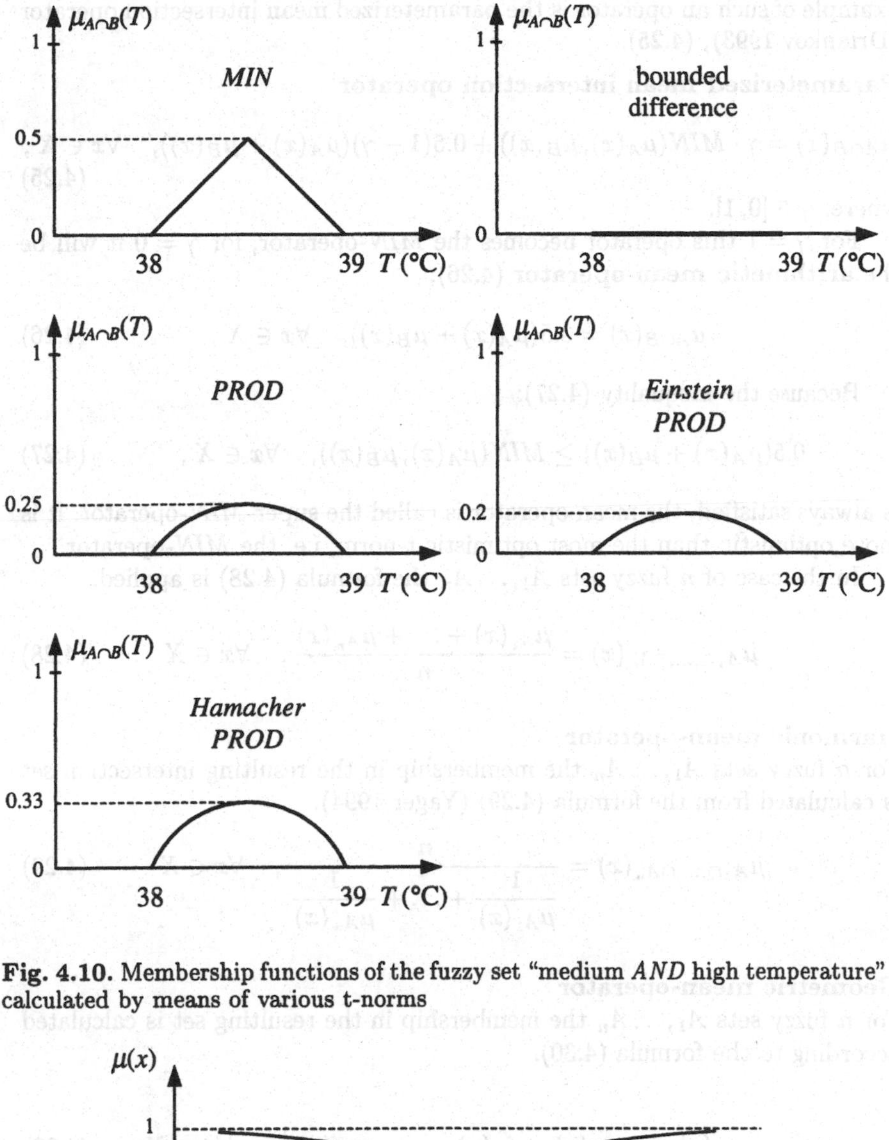

Fig. 4.10. Membership functions of the fuzzy set "medium *AND* high temperature" calculated by means of various t-norms

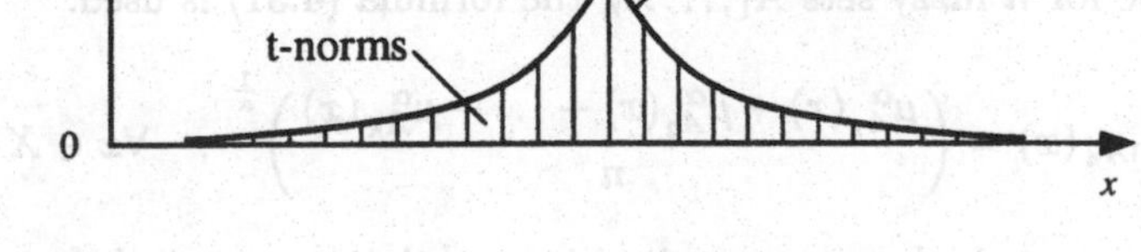

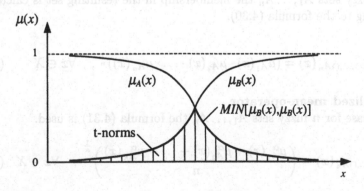

Fig. 4.11. The relationship of the *MIN*-operator to other t-norms

example of such an operator is the parameterized mean intersection operator (Driankov 1993), (4.25).

Parameterized mean intersection operator

$$\mu_{A\cap B}(x) = \gamma \cdot MIN(\mu_A(x), \mu_B(x)) + 0.5(1-\gamma)(\mu_A(x) + \mu_B(x)), \quad \forall x \in X, \tag{4.25}$$

where: $\gamma \in [0, 1]$.

For $\gamma = 1$ this operator becomes the *MIN*-operator, for $\gamma = 0$ it will be the **arithmetic mean-operator** (4.26):

$$\mu_{A\cap B}(x) = 0.5(\mu_A(x) + \mu_B(x)), \quad \forall x \in X, \tag{4.26}$$

Because the inequality (4.27):

$$0.5(\mu_A(x) + \mu_B(x)) \geq MIN(\mu_A(x), \mu_B(x)), \quad \forall x \in X, \tag{4.27}$$

is always satisfied, the mean operator is called the super-*MIN*-operator. It is more optimistic than the most optimistic t-norm, i.e. the *MIN*-operator.

In the case of n fuzzy sets $A_1, \ldots A_n$ the formula (4.28) is applied.

$$\mu_{A_1\cap,\ldots,\cap A_n}(x) = \frac{\mu_{A_1}(x) + \ldots + \mu_{A_n}(x)}{n}, \quad \forall x \in X \tag{4.28}$$

Harmonic mean-operator

For n fuzzy sets $A_1, \ldots A_n$ the membership in the resulting intersection set is calculated from the formula (4.29) (Yager 1994).

$$\mu_{A_1\cap,\ldots,\cap A_n}(x) = \frac{n}{\dfrac{1}{\mu_{A_1}(x)} + \ldots + \dfrac{1}{\mu_{A_n}(x)}}, \quad \forall x \in X \tag{4.29}$$

Geometric mean-operator

For n fuzzy sets $A_1, \ldots A_n$ the membership in the resulting set is calculated according to the formula (4.30).

$$\mu_{A_1\cap,\ldots,\cap A_n}(x) = (\mu_{A_1}(x) \cdot \mu_{A_2}(x) \cdot \ldots \cdot \mu_{A_n}(x))^{\frac{1}{n}}, \quad \forall x \in X \tag{4.30}$$

Generalized mean-operator

In this case for n fuzzy sets $A_1, \ldots A_n$ the formula (4.31) is used.

$$\mu_{A_1\cap,\ldots,\cap A_n}(x) = \left(\frac{\mu^{\alpha}_{A_1}(x) + \mu^{\alpha}_{A_2}(x) + \ldots + \mu^{\alpha}_{A_n}(x)}{n}\right)^{\frac{1}{\alpha}}, \quad \forall x \in X \tag{4.31}$$

This operator is the parameterized intersection operator: the parameterized quantity is the parameter α. For:

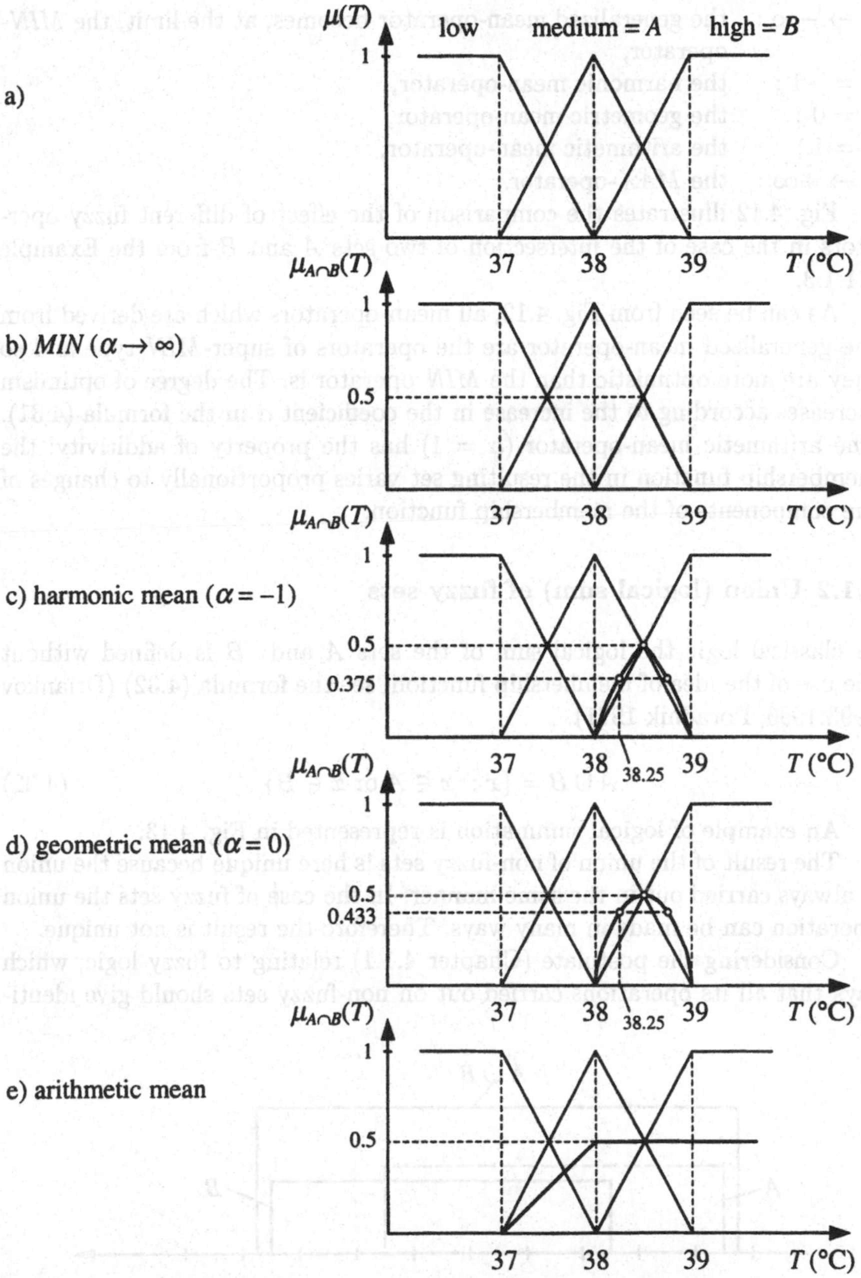

Fig. 4.12. The comparison of the results of the realization of the intersection of the fuzzy sets A and B carried out with the use of the MIN-operator and mean-operators

$\alpha \to -\infty$: the generalized mean-operator becomes, at the limit, the *MIN*-operator,
$\alpha = -1$: the harmonic mean-operator,
$\alpha = 0$: the geometric mean-operator,
$\alpha = 1$: the arithmetic mean-operator,
$\alpha \to +\infty$: the *MAX*-operator.

Fig. 4.12 illustrates the comparison of the effect of different fuzzy operators in the case of the intersection of two sets A and B from the Example 4.1.1.3.

As can be seen from Fig. 4.12, all mean-operators which are derived from the generalized mean-operator are the operators of super-*MIN* type and so they are more optimistic than the *MIN* operator is. The degree of optimism increases according to the increase in the coefficient α in the formula (4.31). The arithmetic mean-operator ($\alpha = 1$) has the property of additivity: the membership function in the resulting set varies proportionally to changes of the components of the membership function.

4.1.2 Union (logical sum) of fuzzy sets

In classical logic the logical sum of the sets A and B is defined without the use of the idea of membership function, see the formula (4.32) (Driankov 1993,1996; Poradnik 1971).

$$A \cup B = \{x : x \in A \text{ or } x \in B\} \tag{4.32}$$

An example of logical summation is represented in Fig. 4.13.

The result of the union of non-fuzzy sets is here unique because the union is always carried out in the same manner. In the case of fuzzy sets the union operation can be made in many ways. Therefore the result is not unique.

Considering the postulate (Chapter 4.1.1) relating to fuzzy logic, which says that all its operations carried out on non-fuzzy sets should give identi-

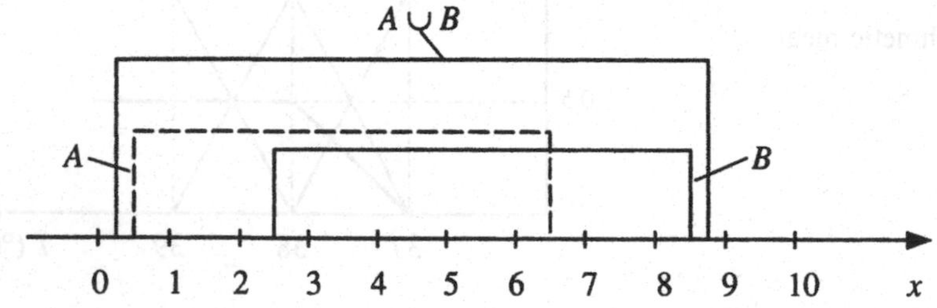

$A = \{1,2,3,4,5,6\}$, $B = \{3,4,5,6,7,8\}$, $A \cup B = \{1,2,3,4,5,6,7,8\}$

Fig. 4.13. Example of the logical sum of non-fuzzy sets

cal results to the operations of classical logic, it is expected that the union operation of fuzzy sets would have the properties (4.33) – (4.38).

Commutativity:

$$A \cup B = B \cup A. \tag{4.33}$$

This property means that the order of sets involved in the union operation is unimportant for the final result.

Associativity:

$$A \cup (B \cup C) = (A \cup B) \cup C = A \cup B \cup C. \tag{4.34}$$

When creating the union of many sets we can do it successively for pairs of sets. The order of succession of creating pairs is unimportant.

Idempotency:

$$A \cup A = A. \tag{4.35}$$

Union with the empty set $\emptyset$:

$$A \cup \emptyset = A. \tag{4.36}$$

Absorption by the universe of discourse X:

$$A \cup X = X, \tag{4.37}$$

Union with the complementary set $\bar{A}$:

$$A \cup \bar{A} = X. \tag{4.38}$$

In the following example it will be shown that not all properties of the union operation of non-fuzzy sets can be transferred into fuzzy sets.

Example 4.1.2.1
Assume there is the set A of cheap cars (4.39) and the set $B = \bar{A}$ of luxury cars (4.40), x_i – a car number.

$$A = \left\{ \frac{1}{x_1}, \frac{0.8}{x_2}, \frac{0.6}{x_3}, \frac{0.4}{x_4}, \frac{0.2}{x_5}, \frac{0}{x_6} \right\} \tag{4.39}$$

$$B = \left\{ \frac{0}{x_1}, \frac{0.2}{x_2}, \frac{0.4}{x_3}, \frac{0.6}{x_4}, \frac{0.8}{x_5}, \frac{1}{x_6} \right\} \tag{4.40}$$

The universe of discourse:

$$X = \left\{ \frac{1}{x_1}, \frac{1}{x_2}, \frac{1}{x_3}, \frac{1}{x_4}, \frac{1}{x_5}, \frac{1}{x_6} \right\}.$$

A set $C = A \cup B = A \cup \bar{A}$ of cheap or luxury cars has to be determined.

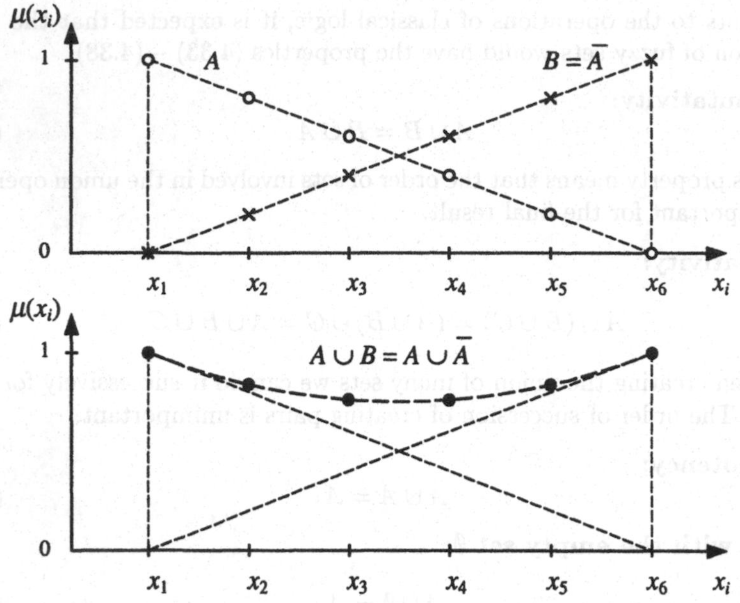

Fig. 4.14. The membership functions of the sets A and B and their logical sum

If the operator of the type of algebraic sum (4.41) were used to realize the union of the sets $A \cup B$, then we would obtain the set (4.42), as a result.

$$\mu_C(x) = \mu_A(x) + \mu_B(x) - \mu_A(x) \cdot \mu_B(x) \tag{4.41}$$

$$C = A \cup B = \left\{ \frac{1}{x_1}, \frac{0.84}{x_2}, \frac{0.76}{x_3}, \frac{0.76}{x_4}, \frac{0.84}{x_5}, \frac{1}{x_6} \right\} \tag{4.42}$$

The membership functions of the sets A, B, C are shown in Fig. 4.14.

◊ ◊ ◊

As can be seen from the above represented example, if the operator of algebraic sum has been used to carry out the union of fuzzy sets $A \cup \bar{A}$, then the result is not conformable to the sixth property ($A \cup \bar{A} = X$) of this operation that is always true in the case of non-fuzzy sets.

The first operators which were proposed to realize the union of fuzzy sets were (Zadeh 1965) the MAX-operator and algebraic sum. As fuzzy logic has been developing, the number of those operators has been increasing. At present, the most popular union operators of sets are the t-conorms also called the s-norms.

The s-norm or t-conorm operator is the function S realizing the union operation OR of two fuzzy sets A and B having the properties (4.43) – (4.48) for each $x \in X$.

Table 4.3. Non-parameterized s-norms

Operator name	Formula
maximum (MAX)	$\mu_{A\cup B}(x) = MAX(\mu_A(x), \mu_B(x))$
algebraic sum	$\mu_{A\cup B}(x) = \mu_A(x) + \mu_B(x) - \mu_A(x)\cdot\mu_B(x)$
Hamacher sum	$\mu_{A\cup B}(x) = \dfrac{\mu_A(x) + \mu_B(x) - 2\cdot\mu_A(x)\cdot\mu_B(x)}{1 - \mu_A(x)\cdot\mu_B(x)}$
Einstein sum	$\mu_{A\cup B}(x) = \dfrac{\mu_A(x) + \mu_B(x)}{1 + \mu_A(x)\cdot\mu_B(x)}$
drastic sum	$\mu_{A\cup B}(x) = \begin{cases} MAX(\mu_A(x), \mu_B(x)) & \text{for } MIN(\mu_A, \mu_B) = 0 \\ 1 & \text{otherwise} \end{cases}$
bounded sum	$\mu_{A\cup B}(x) = MIN(1,\, \mu_A(x) + \mu_B(x))$

Input/output mapping space:

$$S:\ [0,1]\times[0,1]\rightarrow[0,1]\,. \tag{4.43}$$

The property of zeroing:

$$S(0,0) = 0\,. \tag{4.44}$$

The action when a pair contains a natural element $\mu_B(x) = 0$**:**

$$S(\mu_A(x), 0) = S(0, \mu_A(x)) = \mu_A(x)\,. \tag{4.45}$$

The property of commutativity:

$$S(\mu_A(x), \mu_B(x)) = S(\mu_B(x), \mu_A(x))\,. \tag{4.46}$$

The property of associativity (association of any pairs):

$$S(\mu_A(x), S(\mu_B(x), \mu_C(x))) = S(S(\mu_A(x), \mu_B(x)), \mu_C(x))\,. \tag{4.47}$$

The feature of monotonicity:

$$\mu_A(x) \le \mu_C(x),\ \mu_B(x) \le \mu_D(x) \Rightarrow S(\mu_A(x), \mu_B(x)) \le S(\mu_C(x), \mu_D(x))\,. \tag{4.48}$$

The s-norm operators are divided into the parameterized and non-parameterized ones. The non-parameterized operators act in a constant way. The most often used operators of this type are listed in Table 4.3. The most often used parameterized s-norms are specified in Table 4.4.

The particular s-norms differ one from another in the degree of optimism. The biggest result of calculations is given by the operator *drastic sum*, the

Table 4.4. Parameterized s-norms

Operator name	Formula
Hamacher-union operator	$\mu_{A\cup B}(x,\gamma)=\dfrac{\mu_A(x)+\mu_B(x)+(\gamma-1)\cdot\mu_A(x)\cdot\mu_B(x)}{1+\gamma\cdot\mu_A(x)\cdot\mu_B(x)}$, $\gamma\geq -1$ $\gamma=-1$: $\mu_{A\cup B}(x,\gamma)=$ Hamacher sum $\gamma=0$: $\mu_{A\cup B}(x,\gamma)=$ algebraic sum $\gamma=1$: $\mu_{A\cup B}(x,\gamma)=$ Einstein sum $\gamma\to\infty$: $\mu_{A\cup B}(x,\gamma)=$ drastic sum
Yager-union operator	$\mu_{A\cup B}(x,p)=MIN\left\{1,\ [(\mu_A(x))^p+(\mu_B(x))^p]^{\frac{1}{p}}\right\}$, $p\geq 1$ $p=1$: $\mu_{A\cup B}(x,p)=$ bounded sum $p\to\infty$: $\mu_{A\cup B}(x,p)=MAX(\mu_A(x),\mu_B(x))$

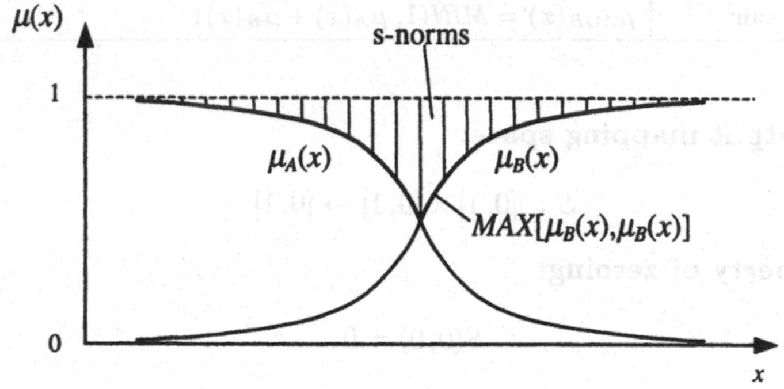

Fig. 4.15. The relationship between the MAX-operator and other s-norm operators

smallest one by the MAX-operator. The sequence of the s-norms according to the degree of optimism is as follows:

drastic sum > bounded sum > Einstein sum > algebraic sum > Hamacher sum > MAX

Because the calculation of the membership in the set $A\cup B$ with the use of the MAX-operator gives the smallest result, all the s-norms operators except for the MAX-operator are called **super-MAX-operators**, Fig. 4.15.

The t-norm and s-norm operators create complementary pairs satisfying the condition (4.49).

$$T[\mu_A(x),\mu_B(x)]=1-S[1-\mu_A(x),1-\mu_B(x)] \tag{4.49}$$

If the t-norm is given, the complementary (in relation to it) s-norm can be calculated. In Table 4.5 the **complementary pairs of the t-norms and s-norms** are given.

Table 4.5. The complementary pairs of the t-norms and s-norms

t-norm	complementary s-norm
MIN	*MAX*
algebraic product	algebraic sum
Hamacher product	Hamacher sum
Einstein product	Einstein sum
drastic product	drastic sum
bounded product	bounded sum
parameterized Hamacher-intersection operator	parameterized Hamacher-union operator
parameterized Yager-intersection operator	parameterized Yager-union operator

For the realization of the union operation of sets, the *OR*-operators which are not s-norms are also applied (they do not satisfy the conditions of a s-norm). An example of such an operator is the **parameterized union mean-operator of sets** (4.50) (Driankov 1993).

$$\mu_{A\cup B}(x) = \gamma \cdot MAX[\mu_A(x), \mu_B(x)] + 0.5 \cdot (1-\gamma) \cdot [\mu_A(x) + \mu_B(x)],$$
$$\gamma \in [0,1], \quad \forall x \in X. \tag{4.50}$$

For $\gamma = 1$ the operator (4.50) becomes the *MAX*-operator, for $\gamma = 0$ it becomes the arithmetic mean-operator.

For the union operation of fuzzy sets the **algebraic sum operator** (4.51) can also be used.

$$\mu_{A_1\cup,\ldots,\cup A_n}(x) = \mu_{A_1}(x)+,\ldots,+\mu_{A_n}(x), \quad \forall x \in X. \tag{4.51}$$

This operator is the most optimistic of all union operators of sets and has the property of additivity. The resulting membership function increases proportionally to the increase in the component functions in the formula (4.51). Therefore, this operator as well as the arithmetic mean-operator can be called a linear operator. The use of them in fuzzy models is conducive to obtaining linear surfaces of the sectors in input/output mapping of these models. The linear operators transform fuzzy models involved in the operation in the so-called fuzzy bags (Yager 1994,1995). They can be used for performing operations on fuzzy bags described in Chapter 2.

Example 4.1.2.3

Let A be the set of fast cars and B the set of comfortable cars, x_i – the designation of a car,

$$A = \left\{ \frac{1}{x_1}, \frac{1}{x_2}, \frac{1}{x_3}, \frac{1}{x_4} \right\},$$

$$B = \left\{ \frac{0.6}{x_1}, \frac{0.7}{x_2}, \frac{0.9}{x_3}, \frac{1}{x_4} \right\}.$$

We want to purchase a fast *OR* comfortable car. By applying the *MAX* s-norm we obtain the result (4.52).

$$C = A \cup B = \left\{ \frac{1}{x_1}, \frac{1}{x_2}, \frac{1}{x_3}, \frac{1}{x_4} \right\} \cup \left\{ \frac{0.6}{x_1}, \frac{0.7}{x_2}, \frac{0.9}{x_3}, \frac{1}{x_4} \right\} = \left\{ \frac{1}{x_1}, \frac{1}{x_2}, \frac{1}{x_3}, \frac{1}{x_4} \right\} \tag{4.52}$$

As a result of the use of the *MAX*-operator we obtain the information that each car can be purchased because their membership in the set "fast *OR* comfortable" is the same and equal to 1. By using the arithmetic sum-operator the sets A and B are transformed in a fuzzy bag (4.53) in the first step and then in a fuzzy set (4.54) in the second step.

Step 1

$$\begin{aligned} A \cup B &= \left\{ \frac{1}{x_1}, \frac{1}{x_2}, \frac{1}{x_3}, \frac{1}{x_4} \right\} \cup \left\{ \frac{0.6}{x_1}, \frac{0.7}{x_2}, \frac{0.9}{x_3}, \frac{1}{x_4} \right\} \\ &= \left\{ \frac{1}{x_1}, \frac{0.6}{x_1}, \frac{1}{x_2}, \frac{0.7}{x_2}, \frac{1}{x_3}, \frac{0.9}{x_3}, \frac{1}{x_4}, \frac{1}{x_4} \right\} \end{aligned} \tag{4.53}$$

Step 2

$$A \cup B = \left\{ \frac{1.6}{x_1}, \frac{1.7}{x_2}, \frac{1.9}{x_3}, \frac{2}{x_4} \right\} \tag{4.54}$$

As a result of the use of the *arithmetic sum*-operator we obtain the information that the car which should be purchased (car x_4) is the one whose membership in the bag "fast or comfortable" is the highest (the membership in (4.54) can be normalized to the interval [0,1]). It seems that in many situations people making decisions apply such an operator because a man is still inclined to take into account many circumstances when he makes a decision.

◊ ◊ ◊

4.1.3 Compensatory operators

Both the t-norms and s-norms are the so-called supposed operators, i.e. they are based on suppositions concerning the way intersection and union operations of sets are performed in the human mind. Up to now the problem has not been completely explained, perhaps because various people use miscellaneous ways to realize the above-mentioned operations, depending on their character, mood and the actual situation.

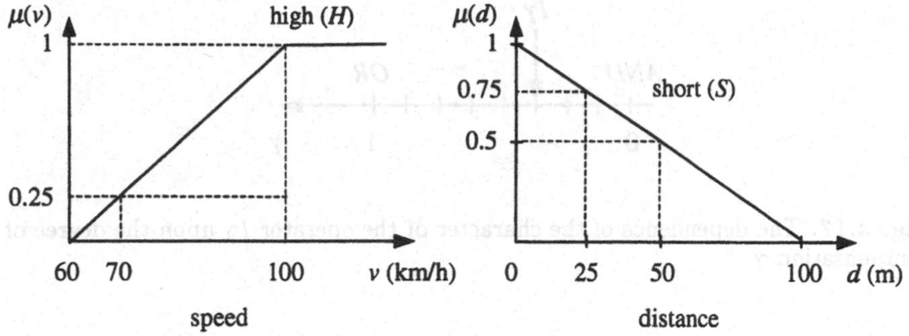

Fig. 4.16. The membership functions of the fuzzy sets "high" speed and "short" distance

Research concerning the operators applied by a man, which was carried out by Zimmermann (Altrock 1993; Zimmermann 1979,1987), led to the formulation of the idea of the compensatory operators. The relevance of the compensation will be explained in the example of the reasoning of a driver who approaches an obstacle in the street at a fast speed – the rule(4.55).

$$\begin{aligned}&\mathit{IF}\ (\text{the speed is high})\ \mathit{AND}\ (\text{an obstacle is near})\\&\mathit{THEN}\ (\text{brake very strongly})\end{aligned} \tag{4.55}$$

By denoting the speed of the car by v (km/h) and the distance to the obstacle by d (m), the premise A can be given in the form (4.56).

$$A \;=\; A_1\ \mathit{AND}\ A_2 \;=\; (v = H)\ \mathit{AND}\ (d = S) \tag{4.56}$$

The greater the truth-value of the premise is, the stronger the braking must be.

Let us assume that the membership functions of the linguistic values "high" speed and "short" distance have, respectively, the forms represented in Fig. 4.16.

Now, let us investigate 3 possible situations to which the car can be exposed and how the driver using the PROD-operator evaluates them.

Situation 1 – the car approaches the obstacle

$$v = 70\,(\text{km/h}) \qquad \mu_H(v) = 0.25$$

$$d = 50\,(\text{m}) \qquad \mu_S(d) = 0.5$$

The degree of satisfying the premise A and in turn the degree of the braking force (the degree of the activation of the conclusion resulting from the rule (4.55)) is equal to:

$$\mu_A(v,d) = \mu_H(v) \cdot \mu_S(d) = 0.25 \cdot 0.5 = 0.125\,.$$

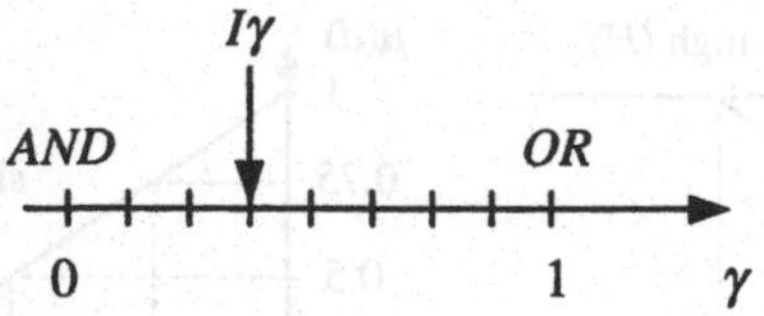

Fig. 4.17. The dependence of the character of the operator $I\gamma$ upon the degree of compensation γ

Situation 2 – the car is very close to the obstacle

$$v = 70\,(\text{km/h}) \qquad \mu_H(v) = 0.25$$
$$d = 25\,(\text{m}) \qquad \mu_S(d) = 0.75$$

The degree of satisfying the premise:

$$\mu_A(v,d) = \mu_H(v) \cdot \mu_S(d) = 0.25 \cdot 0.75 = 0.1875\,.$$

Situation 3 – the car crashes into the obstacle

$$v = 70\,(\text{km/h}) \qquad \mu_H(v) = 0.25$$
$$d = 0\,(\text{m}) \qquad \mu_S(d) = 1$$

The degree of satisfying the premise A:

$$\mu_A(v,d) = \mu_H(v) \cdot \mu_S(d) = 0.25 \cdot 1 = 0.25\,.$$

The analysis of situations 1 – 3 shows that when the *PROD*-operator is used to realize the operation *AND* in the premise, the degree to which it is satisfied $\mu_A(v,d)$, and thus also the degree of the braking force, does not vary appropriately fast in spite of the fast approaching danger. The driver keeping the rule strictly (4.55) with the *PROD*-operator would crash into the obstacle. Considering the above, is the rule (4.55) false? No, it is not.

The analysis of people's behavior shows that people apply the so-called **compensatory principle** modifying the *AND*-operation by combining it to a certain degree with the *OR*-operation. The measure of compensation is the degree of compensation γ, Fig. 4.17.

On the basis of the experimental results of investigations of people's decisions Zimmermann proposed the $I\gamma$-intersection operator in the form of (4.57).

$$\mu_A = \left(\prod_{i=1}^{m} \mu_{Ai}\right)^{(1-\gamma)} \left[1 - \prod_{i=1}^{m}(1-\mu_{Ai})\right]^{\gamma}, \tag{4.57}$$

where: γ – the degree of compensation, $0 \leq \gamma \leq 1$,
μ_A – the degree of satisfying of the whole premise $A = A_1 \cap \ldots \cap A_n$,
μ_{Ai} – the degrees of satisfying of the component premises.

If $\gamma = 0$, then the whole premise is evaluated on the basis of the *AND*-intersection operation only – by means of the *PROD* operator, the formula (4.58).

$$\mu_A = \prod_{i=1}^{m} \mu_{Ai} \,. \tag{4.58}$$

If $\gamma = 1$, then the whole premise is evaluated on the basis of the formula (4.59) realizing only the *OR*-operation:

$$\mu_A = 1 - \prod_{i=1}^{m} (1 - \mu_{Ai}) \,. \tag{4.59}$$

The functioning of the operator (4.59) is approximate to that of the *MAX* operator, although, as is easily noticeable, its functioning is better because the $I\gamma$-operator takes into account all component premises and not only the premise satisfied to the highest degree.

A driver most likely changes the value γ depending on the situation: for a distant obstacle he applies a small value and close to an obstacle a large value γ. By using the $I\gamma$-operator, when $\gamma = 1$, in situation 3, we obtain a completely different result from that when the *PROD*-operator is used.

Situation 3 – the car crashes into the obstacle

$$v = 70\,(\text{km/h}) \qquad \mu_H(v) = 0.25$$
$$d = 0\,(\text{m}) \qquad \mu_S(d) = 1$$

The degree of satisfying of the premise A calculated with the use of the operator $I\gamma$, $\gamma = 1$, is equal to:

$$\mu_A(v, d) = 1 - (1 - 0.25)(1 - 1) = 1 \,.$$

The full degree of satisfying the premise imposes, according to the rule (4.55), to press the brake as strongly as possible, which is, of course, the most natural reaction in such a situation. Naturally, the rule (4.55) with the $I\gamma$ -operator would suggest stronger and stronger braking earlier, as the distance to the obstacle is decreasing.

Because the compensation coefficient γ can vary in the range of $0 \leq \gamma \leq 1$, there exists a problem connected with the selection of its optimal value. In (Altrock 1993] it is recommended to select γ for technical applications from the range:

$$0.1 \leq \gamma \leq 0.4 \,.$$

A pragmatic method is the choice of the mean value from this range, $\gamma = 0.25$, at the beginning, and then the examination of the accuracy of a

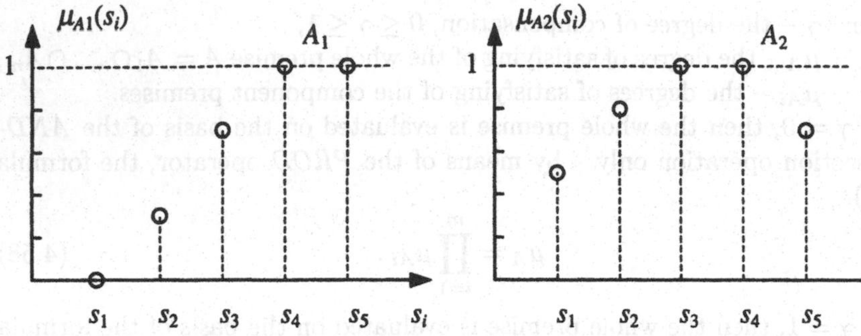

Fig. 4.18. The membership function in the subsets A_1 (good students) and A_2 (highly proficient students)

fuzzy model based on this value. If the accuracy is not satisfactory, it is recommended to correct the coefficient stepwise, assuming a step of $\Delta\gamma = 0.01$, and to examine the model accuracy.

4.2 Fuzzy relations

In Section 4.1 operations on fuzzy sets defined over one-dimensional universe of discourse X have been considered. It is illustrated in Example 4.2.1.

Example 4.2.1

In a set of students two subsets have been separated: the subset A_1 of good students and the subset A_2 of highly proficient students. The set of good and highly proficient students: $A_1 \wedge A_2$ should be determined.
X – the set of students:

$$X = \{s_1, s_2, \ldots, s_5\}.$$

A_1 – the subset of good students:

$$A_1 = \{(s_1, 0), (s_2, 0.3), (s_3, 0.7), (s_4, 1), (s_5, 1)\}.$$

A_2 – the subset of highly proficient students:

$$A_2 = \{(s_1, 0.5), (s_2, 0.8), (s_3, 1), (s_4, 1), (s_5, 0.7)\}.$$

By using the *MIN*-operator for the realization of the intersection operation of the subsets $\wedge$ we obtain:

$$\begin{aligned} A_1 \wedge A_2 &= \{(s_i, MIN(\mu_{A1}(s_i), \mu_{A2}(s_i))\} \\ &= \{(s_1, 0), (s_2, 0.3), (s_3, 0.7), (s_4, 1), (s_5, 0.7)\}. \end{aligned}$$

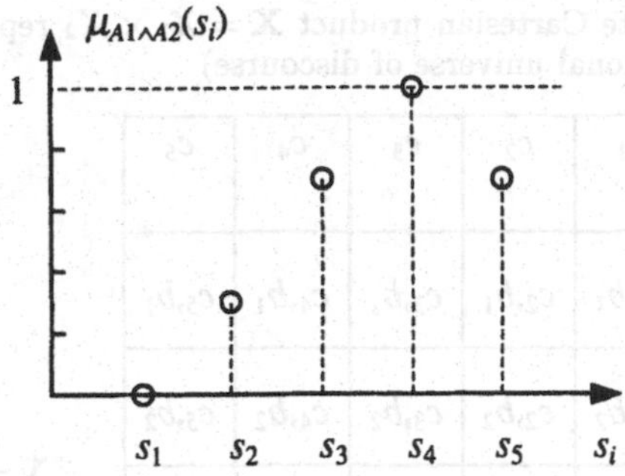

Fig. 4.19. The membership function in the set $A_1 \wedge A_2$ (good and highly proficient students)

The set $A_1 \wedge A_2$ is represented in Fig. 4.19.

Owing to the fact that both the subsets have been defined over the same one-dimensional universe of discourse X, the result of the intersection operation can be represented on the surface, in two-dimensional space. Both the subset A_1 and the subset A_2 can be considered to be simple sets because the grades of membership $\mu_{Ai}(s_i)$ are assigned to the single elements s_i of the universe of discourse X.

◊ ◊ ◊

Apart from one-dimensional universes of discourse there are multidimensional domains being the Cartesian product $\mathbf{X}$ of the component universes of discourse $X_1, \ldots, X_n$ of various quantities. This is illustrated by Example 4.2.2.

Example 4.2.2

X_1 – the set of citizens,

$$X_1 = \{c_1, c_2, \ldots, c_5\},$$

X_2 – the set of banks,

$$X_2 = \{b_1, b_2, \ldots, b_5\}.$$

The Cartesian product $\mathbf{X} = X_1 \times X_2$ is the set of all possible pairs (c_i, b_j), $i = 1, \ldots, 5$, $j = 1, \ldots, 5$, Table 4.6.

◊ ◊ ◊

Before the idea of fuzzy relations is explained, it will be advantageous to acquaint oneself with the concept of classical relation (Empacher 1970).

The **classical relation** (bivariant, binary) – one of the prime ideas of mathematical logic – is a property of pairs of objects and describes a certain interrelation existing between the objects. The idea of the classical relation is illustrated in Example 4.2.3.

Table 4.6. The discrete Cartesian product $\mathbf{X} = X_1 \times X_2$ represented in the form of a table (two-dimensional universe of discourse)

b_j \ c_i	c_1	c_2	c_3	c_4	c_5
b_1	c_1,b_1	c_2,b_1	c_3,b_1	c_4,b_1	c_5,b_1
b_2	c_1,b_2	c_2,b_2	c_3,b_2	c_4,b_2	c_5,b_2
b_3	c_1,b_3	c_2,b_3	c_3,b_3	c_4,b_3	c_5,b_3
b_4	c_1,b_4	c_2,b_4	c_3,b_4	c_4,b_4	c_5,b_4
b_5	c_1,b_5	c_2,b_5	c_3,b_5	c_4,b_5	c_5,b_5

$\mathbf{X} = X_1 \times X_2$

Example 4.2.3

The one-dimensional component sets X_1 and X_2 are given. X_1 – the set of citizens,

$$X_1 = \{c_1, c_2, \ldots, c_5\},$$

X_2 – the set of banks,

$$X_2 = \{b_1, b_2, \ldots, b_5\}.$$

An example of a classical relation over the set $\mathbf{X} = X_1 \times X_2$ is the relation "one having an account in", Table 4.6. The set $\mathbf{X}$ is here the universe of discourse. The relation can have the following form:

$$\mathbf{R} = \{(c_1, b_2), (c_3, b_4), (c_4, b_1), (c_5, b_3)\}.$$

The relation $\mathbf{R}$ consists of the pairs (c_i, b_j) and so it is the binary relation assigning the citizens c_i to the banks b_j in which they have opened their accounts. This relation can be described by using the membership function $\mu(c_i, b_j)$ which can be represented in three-dimensional space, Fig. 4.20.

The relation $\mathbf{R}$ can be also represented in the form of the matrix $\mathbf{R}$.

$$\mathbf{R} = \begin{bmatrix} 0&0&0&1&0 \\ 1&0&0&0&0 \\ 0&0&0&0&1 \\ 0&0&1&0&0 \\ 0&0&0&0&0 \end{bmatrix}$$

Because citizen c_2 has not opened an account anywhere, there are only zeros in the second column of the matrix $\mathbf{R}$.

◊ ◊ ◊

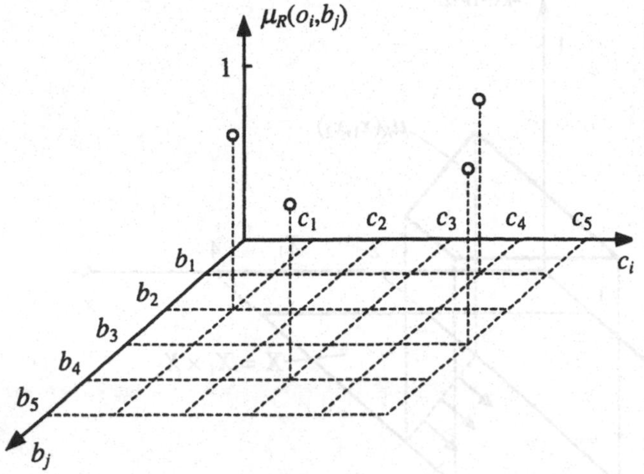

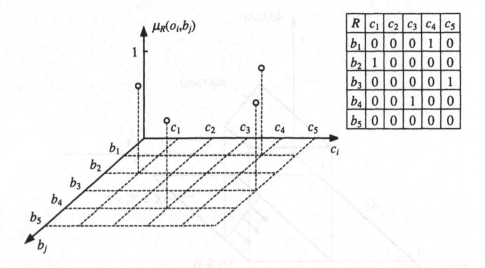

R	c_1	c_2	c_3	c_4	c_5
b_1	0	0	0	1	0
b_2	1	0	0	0	0
b_3	0	0	0	0	1
b_4	0	0	1	0	0
b_5	0	0	0	0	0

Fig. 4.20. The representation of the relation **R** in the form of the three-dimensional membership function$\mu(c_i, b_j)$ and in the form of the relation matrix

The matrix **R** need not be quadratic. It depends on the number of elements in the component universes of discourse X_i. The relation from Example 4.2.3 has a discrete character. In Example 4.2.4 a continuous relation is represented.

Example 4.2.4
Assume that two sets of real numbers X_1, X_2 are given:

$$X_1 = \{x_1 : \ 2 \leq x_1 \leq 4\},$$
$$X_2 = \{x_2 : \ 1 \leq x_2 \leq 5\}.$$

Let us determine the relation "smaller, equal to" or "$\leq$" defined over the Cartesian product $\mathbf{X} = X_1 \times X_2$:

$$\mathbf{R} = \{(x_1, x_2) : \ x_1 \leq x_2\}.$$

The relation is of a continuous character. Its membership function is represented in Fig. 4.21.
◊ ◊ ◊

The Definition 4.2.1 determines a classical n-variant relation **R** defined over the universe of discourse $\mathbf{X} = X_1 \times \ldots \times X_n$.

Definition 4.2.1
The classical n-variant relation **R** determined over the universe of discourse:

$$\mathbf{X} = X_1 \times \ldots \times X_n,$$

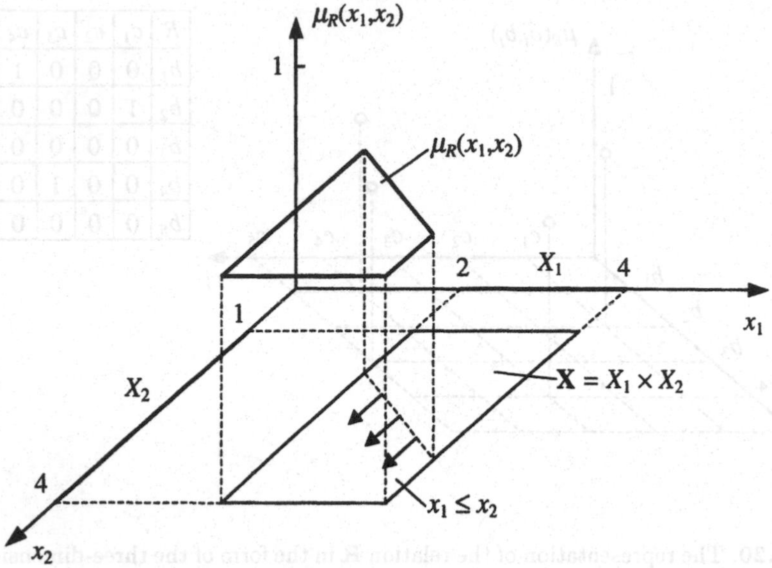

Fig. 4.21. The membership function $\mu_R(x_1, x_2)$ in the form of a continuous surface spread over the continuous universe of discourse $\mathbf{X} = X_1 \times X_2$

is an ordered set of n-tuples having the form:

$$\mathbf{R} = \{((x_1, \ldots, x_n), \mu_R(x_1, \ldots, x_n)) \mid (x_1, \ldots, x_n) \in X_1 \times \ldots \times X_n\},$$

where:

$$\mu_R(x_1, \ldots, x_n) = \begin{cases} 1 & \text{if } (x_1, \ldots, x_n) \in \mathbf{R}, \\ 0 & \text{otherwise,} \end{cases}$$

is the membership function of the relation $\mathbf{R}$.

As we know, the membership function of a classic relation realizes the mapping of the universe of discourse $\mathbf{X}$ onto the discrete set $\{0, 1\}$:

$$\mu_R : X_1 \times \ldots \times X_n \to \{0, 1\}.$$

The difference between the fuzzy relation and the classical relation is that for the membership function the continuous interval $[0, 1]$ has been introduced instead of the discrete set consisting of two elements $\{0, 1\}$.

Definition 4.2.2
The fuzzy n-variant relation $\mathbf{R}$ determined on the universe of discourse $\mathbf{X} = X_1 \times \ldots \times X_n$ is the ordered set of n-tuples having the form of:

$$\mathbf{R} = \{((x_1, \ldots, x_n), \mu_R(x_1, \ldots, x_n)) \mid (x_1, \ldots, x_n) \in X\},$$

where:

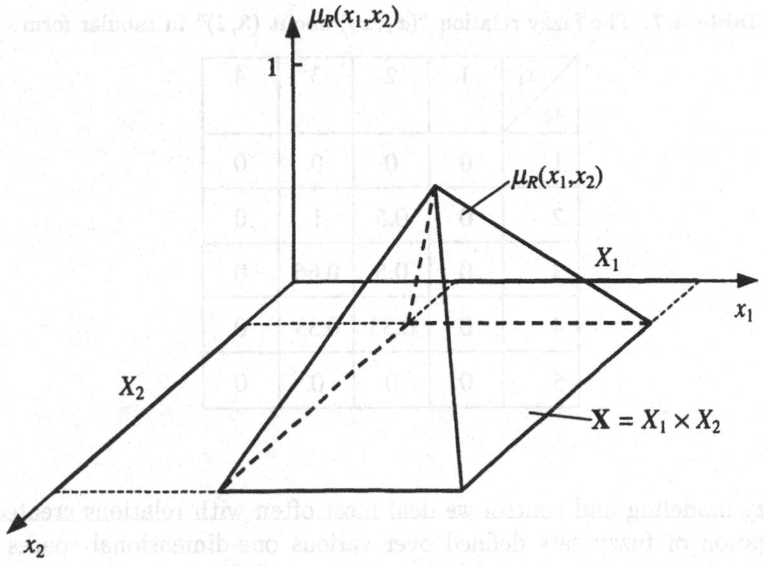

Fig. 4.22. Example of the continuous membership function of the fuzzy relation

$$\mu_R(x_1, \ldots, x_n) : X_1 \times \ldots \times X_n \rightarrow [0, 1] ,$$

is the membership function of the relation $\mathbf{R}$ realizing the mapping of the universe of discourse $\mathbf{X}$ onto the continuous interval $[0, 1]$.

In the general case the membership function μ_R of the relation is a hypersurface in $(n+1)$-dimensional space. An example of the membership function for $n = 2$ is shown in Fig. 4.22.

The membership functions of fuzzy relations for discontinuous universes of discourse can be represented in the form of relation tables giving the grade of membership $\mu_R(x_1, \ldots, x_n)$ for each discrete n-tuple. This is illustrated by Example 4.2.5.

Example 4.2.5
Two discrete component universes of discourse X_1 and X_2 are given:

$$X_1 = \{1, 2, 3, 4\} , \quad X_2 = \{1, 2, 3, 4, 5\} .$$

Table 4.7 exemplifies grades of membership for the relation "(x_1, x_2) about $(3, 2)$" defined over the set $\mathbf{X} = X_1 \times X_2$.

In Table 4.7 we can see, for example, that the pair $(x_1, x_2) = (2, 3)$ belongs to the relation $\mathbf{R}$ to the degree 0.5 or, otherwise, the pair $(2, 3)$ is similar, in the sense of relation, to the pair (3, 2) to the degree 0.5.
◊ ◊ ◊

Fuzzy relations can be defined directly over the n-tuples of a multidimensional universe of discourse $X_1 \times \ldots \times X_n$, as in Example 4.2.5. However,

Table 4.7. The fuzzy relation "(x_1, x_2) about $(3, 2)$" in tabular form

x_2 \ x_1	1	2	3	4
1	0	0	0	0
2	0	0.5	1	0
3	0	0.5	0.66	0
4	0	0.33	0.33	0
5	0	0	0	0

in fuzzy modeling and control we deal most often with relations created by aggregation of fuzzy sets defined over various one-dimensional spaces. An example of this is the rule of the type of *IF* ... *THEN* ..., (4.60)

$$IF\ (x_1 = \text{small})\ AND\ (x_2 = \text{large})\ THEN\ (y = \text{medium}), \tag{4.60}$$

where the component premises (x_1 = small), (x_2 = large) are combined by the logical connectives *AND*, *OR*. They create the binary fuzzy relation **R** having the membership function $\mu_R(x_1, x_2)$ which determines the degree to which the premise is satisfied for the actual numerical values of the arguments x_1, x_2. The aggregation of the fuzzy sets "small" (S) and "large" (L) can be accomplished using operators of the t-norm type in the case of the connective *AND*, (4.61), or the s-norm type for the connective *OR*.

$$\mu_R(x_1, x_2) = T(\mu_S(x_1), \mu_L(x_2)) \tag{4.61}$$

Assume that the membership functions $\mu_S(x_1)$ and $\mu_L(x_2)$ have the form shown in Fig. 4.23.

Both the fuzzy sets S and L are defined over various universes of discourse X_1 and X_2 which can represent, in general cases, different physical quantities (e.g. current voltage and intensity). Therefore, these sets can not be aggregated directly, as in the case of fuzzy sets defined on one universe of discourse. Firstly, such sets should be transformed in certain special fuzzy relations defined on the Cartesian product $X_1 \times X_2$ and called the cylindrical extensions. Only when this has been done can the aggregation be carried out. The **cylindrical extension** is determined by the Definition 4.2.3.

Definition 4.2.3

If X_1 and X_2 are non-fuzzy sets and a fuzzy set A is defined on X_1, then the cylindrical extension $\mathbf{A}^*$ of the set A onto the universe of discourse $X_1 \times X_2$ is the relation determined as the Cartesian product of the sets A and X_2, i.e. $A \times X_2$:

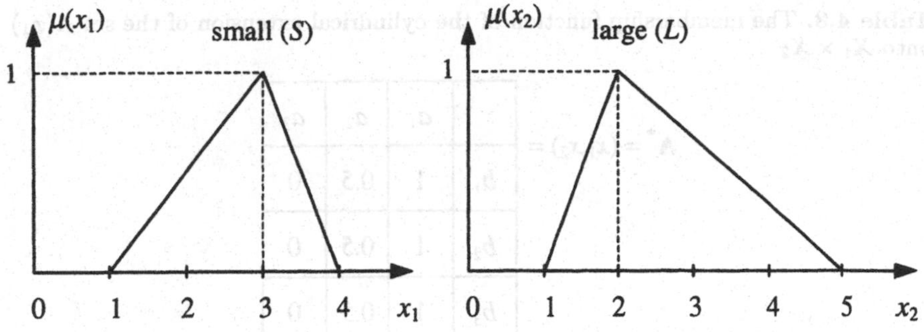

Fig. 4.23. The membership functions of the fuzzy sets "small" and "large" used in the relation (4.61)

$$A^*(x_1, x_2) = A(x_1) \wedge X_2(x_2) = A(x_1) \wedge 1 = A(x_1)\,,$$

for all the pairs $(x_1, x_2) \in X_1 \times X_2$.

In the case of the cylindrical extension of a set $A(x_1)$ onto a n-dimensional universe of discourse $X_1 \times \ldots \times X_n$ the extension operation is accomplished according to the formula:

$$A^*(x_1, \ldots, x_n) = A(x_1) \wedge X_2 \wedge \cdots \wedge X_n = A(x_1)\,,$$

for all n-tuples $(x_1, \ldots, x_n) \in X_1 \times \ldots \times X_n$. The cylindrical extension is illustrated by Example 4.2.6.

Example 4.2.6
Let us determine the cylindrical extension of the set $A(x_1)$ onto the discrete set $X_1 \times X_2$. The universes of discourse are:

$$X_1 = \{a_1, a_2, a_3\}\,, \quad X_2 = \{b_1, b_2, b_3\}\,,$$

the fuzzy set is:

$$A = \{1/a_1, 0.5/a_2, 0/a_3\}\,.$$

The cylindrical extension of the set $A(x_1)$ onto $X_1 \times X_2$ is given in Table 4.8.

◊ ◊ ◊

The next example illustrates the continuous cylindrical extension.

Example 4.2.7
Assume that there are two fuzzy sets "small' (S) and "large" (L), Fig. 4.23, defined over the sets X_1 and X_2. The universes of discourse are:

$$X_1 = [1, 4], \quad X_2 = [1, 5]\,.$$

Table 4.8. The membership function of the cylindrical extension of the set $A(x_1)$ onto $X_1 \times X_2$

$\mathbf{A}^*=(x_1,x_2)=$

	a_1	a_2	a_3
b_1	1	0.5	0
b_2	1	0.5	0
b_3	1	0.5	0

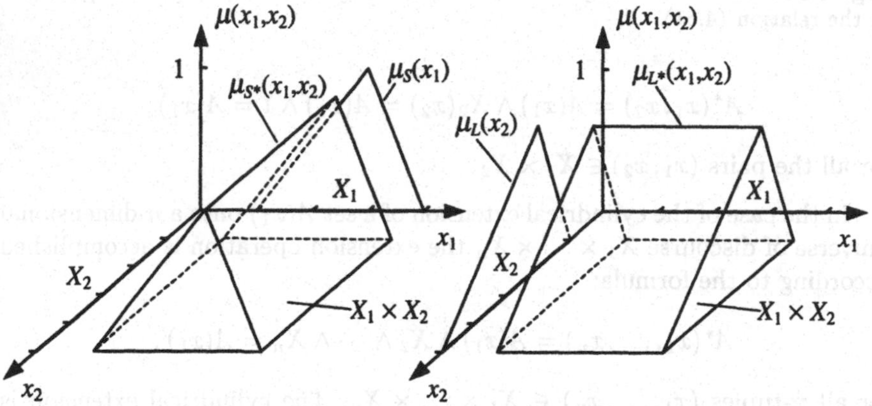

Fig. 4.24. The cylindrical extensions $\mathbf{S}^*(x_1,x_2)$ and $\mathbf{L}^*(x_1,x_2)$ of the fuzzy sets $S(x_1)$ and $L(x_2)$ onto the two-dimensional universe of discourse $X_1 \times X_2$

The membership functions $\mu_S(x_1)$ and $\mu_L(x_2)$ are given in Fig. 4.23. Fig. 4.24 represents the cylindrical extensions $\mathbf{S}^*(x_1,x_2)$ and $\mathbf{L}^*(x_1,x_2)$ of the fuzzy sets $S(x_1)$ and $L(x_2)$ onto $X_1 \times X_2$.

◊ ◊ ◊

The membership function of the relation $\mathbf{R}(x_1,x_2)$ determined by the formula (4.61) can be calculated using, for example, the *MIN*-operator as a t-norm.

$$\mu_R(x_1,x_2) = MIN(\mu_{S^*}(x_1,x_2), \mu_{L^*}(x_1,x_2)), \quad (x_1,x_2) \in X_1 \times X_2 \,. \tag{4.62}$$

The result of this operation is represented in Fig. 4.25.

If the rule (4.60) had, in the premise, the logical connective of the type *OR*:

$$IF\ (x_1 = S)\ OR\ (x_2 = L)\ THEN\ (y = M)\,,$$

then for calculating the membership function of the premise, one of the s-norms should be used, for instance the *MAX*:

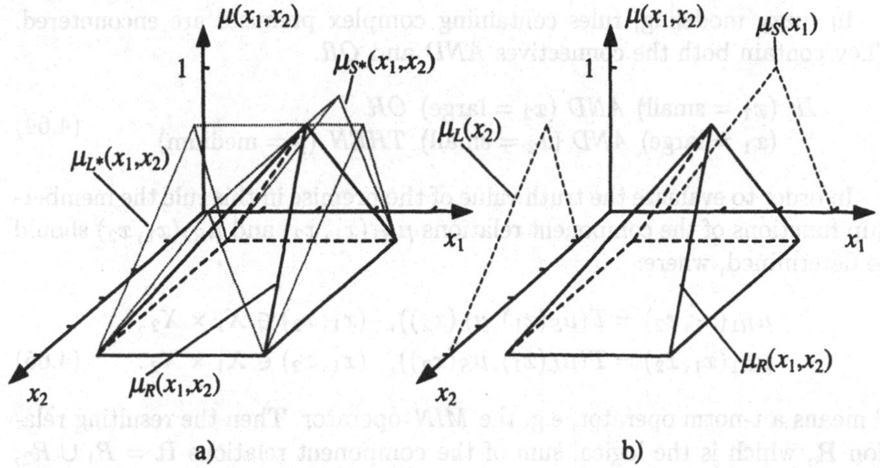

Fig. 4.25. The formation of the membership function of the relation $\mu_R(x_1, x_2)$ using the cylindrical extension of the component sets S and L (a) and the result of the operation (b) obtained with the use of the *MIN*-operator

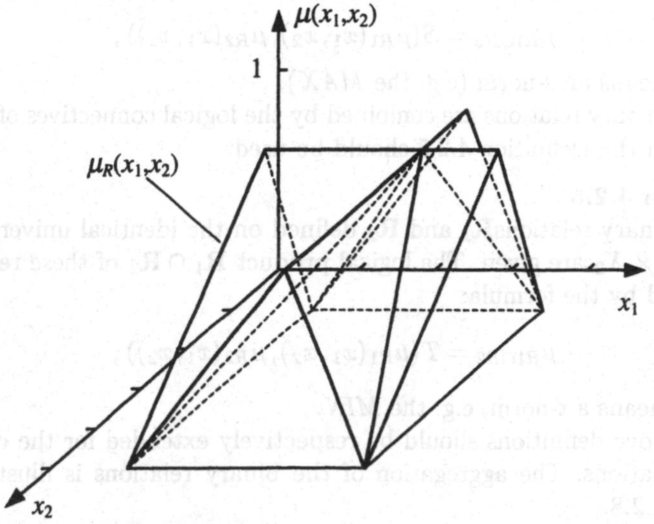

Fig. 4.26. The membership function of the relation $\mu_R(x_1, x_2)$ according to the formula (4.63) created by means of the *MAX*-operator

$$\mu_R(x_1, x_2) = MAX(\mu_{S^*}(x_1, x_2), \mu_{L^*}(x_1, x_2)), \quad (x_1, x_2) \in X_1 \times X_2 \,. \quad (4.63)$$

Then we would obtain the membership function $\mu_R(x_1, x_2)$ shown in Fig. 4.26.

In fuzzy modeling, rules containing complex premises are encountered. They contain both the connectives *AND* and *OR*.

$$\begin{aligned} &IF\ (x_1 = \text{small})\ AND\ (x_2 = \text{large})\ OR \\ &(x_1 = \text{large})\ AND\ (x_2 = \text{small})\ THEN\ (y = \text{medium}) \end{aligned} \tag{4.64}$$

In order to evaluate the truth value of the premise in this rule the membership functions of the component relations $\mu_{R1}(x_1, x_2)$ and $\mu_{R2}(x_1, x_2)$ should be determined, where:

$$\begin{aligned} \mu_{R1}(x_1, x_2) &= T(\mu_S(x_1), \mu_L(x_2)), \quad (x_1, x_2) \in X_1 \times X_2\,, \\ \mu_{R1}(x_1, x_2) &= T(\mu_L(x_1), \mu_S(x_2)), \quad (x_1, x_2) \in X_1 \times X_2\,. \end{aligned} \tag{4.65}$$

T means a t-norm operator, e.g. the *MIN*-operator. Then the resulting relation $\mathbf{R}$, which is the logical sum of the component relations $\mathbf{R} = R_1 \cup R_2$, should be calculated on the basis of Definition 4.2.4.

Definition 4.2.4
Two binary relations $\mathbf{R}_1$ and $\mathbf{R}_2$ defined on the same universe of discourse $X_1 \times X_2$ are given. Then the sum of these relations $\mathbf{R}_1 \cup \mathbf{R}_2$ is given by the formula:

$$\mu_{R1 \cup R2} = S(\mu_{R1}(x_1, x_2), \mu_{R2}(x_1, x_2))\,,$$

where S means an s-norm (e.g. the *MAX*).

If two binary relations are combined by the logical connectives of the type *AND*, then the Definition 4.2.5 should be used.

Definition 4.2.5
Two binary relations$\mathbf{R}_1$ and $\mathbf{R}_2$ defined on the identical universe of discourse $X_1 \times X_2$ are given. The logical product $\mathbf{R}_1 \cap \mathbf{R}_2$ of these relations is determined by the formula:

$$\mu_{R1 \cap R2} = T(\mu_{R1}(x_1, x_2), \mu_{R2}(x_1, x_2))\,,$$

where T means a t-norm, e.g. the *MIN*.

The above definitions should be respectively extended for the case of n-variant relations. The aggregation of the binary relations is illustrated by Example 4.2.8.

Example 4.2.8
Let us determine the membership function $\mu_R(x_1, x_2)$ of the premise composed of two component premises (4.66).

$$IF\ (x_1 = S)\ AND\ (x_2 = L)\ OR\ (x_1 = L)\ AND\ (x_2 = S) \tag{4.66}$$

The membership functions of individual fuzzy sets are given in Fig. 4.27.

The membership function of the first component premise can be calculated using the *MIN*-operator. In the case of the second component premise the membership function can be calculated similarly.

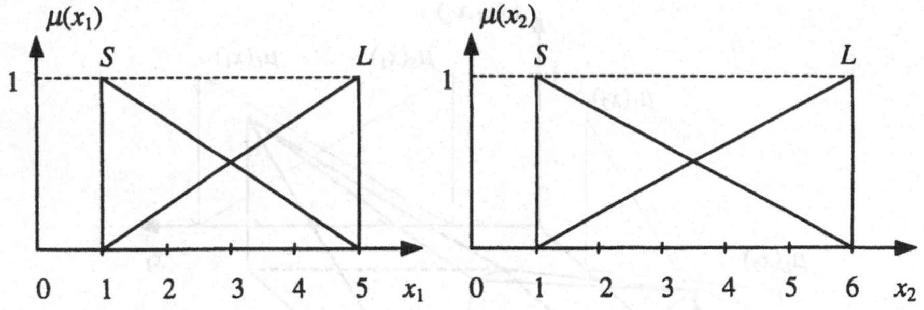

Fig. 4.27. The membership functions of the fuzzy sets in the premise of the rule (4.66)

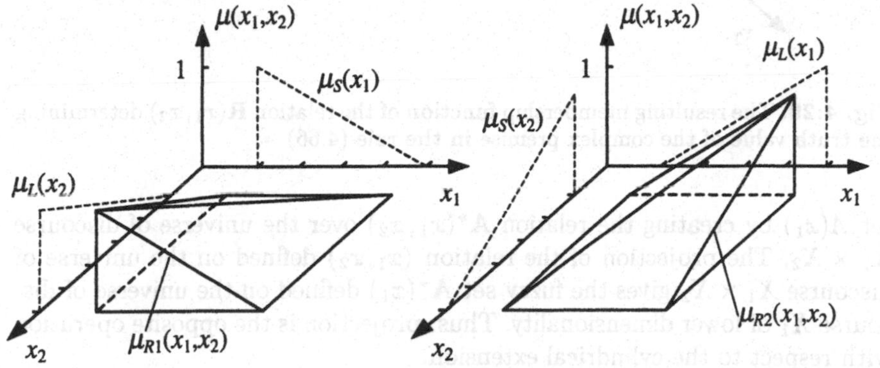

Fig. 4.28. The membership functions of the component relations $\mathbf{R}_1$ and $\mathbf{R}_2$ of the premise of the rule (4.67)

$$\mu_{R1}(x_1, x_2) = MIN(\mu_S(x_1), \mu_L(x_2))$$
$$\mu_{R1}(x_1, x_2) = MIN(\mu_L(x_1), \mu_S(x_2)) \tag{4.67}$$

The membership functions of the component relations are represented in Fig. 4.28.

To carry out the *OR*-operation in the rule (4.66) the *MAX*-operator can be selected as an s-norm. Then the membership function of the resulting relation $\mathbf{R} = \mathbf{R}_1 \cup \mathbf{R}_2$ is calculated from the formula (4.68).

$$\mu_R(x_1, x_2) = MAX(\mu_{R1}(x_1, x_2), \mu_{R2}(x_1, x_2)) \tag{4.68}$$

This function is represented graphically in Fig. 4.29.

◊ ◊ ◊

In fuzzy models an opposite operation with respect to the cylindrical extension is also used. It is called **projection**. The cylindrical extension increases the dimensionality of the universe of discourse X_1 of the given fuzzy

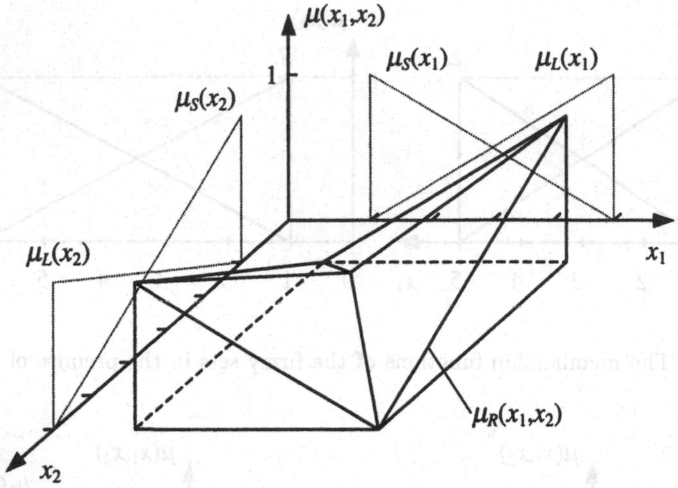

Fig. 4.29. The resulting membership function of the relation $\mathbf{R}(x_1, x_2)$ determining the truth value of the complex premise in the rule (4.66)

set $A(x_1)$ by creating the relation $\mathbf{A}^*(x_1, x_2)$ over the universe of discourse $X_1 \times X_2$. The projection of the relation (x_1, x_2) defined on the universe of discourse $X_1 \times X_2$ gives the fuzzy set $\mathbf{A}^*(x_1)$ defined on the universe of discourse X_1 of lower dimensionality. Thus, projection is the opposite operation with respect to the cylindrical extension.

Definition 4.2.6
If $\mathbf{A}$ is a fuzzy relation defined over the universe of discourse $X_1 \times X_2$, then the **projection** of this relation onto the universe of discourse X_1 is the fuzzy set A^* determined as follows:

$$A^*(x_1) = \underset{x_1}{Proj}\,\mathbf{A}(x_1, x_2) = \underset{x_2}{MAX}[\mathbf{A}(x_1, x_2)]$$

The projection is illustrated by Example 4.2.9.

Example 4.2.9
The relation $\mathbf{A}$, Table 4.9, defined on the universe of discourse $\mathbf{X} = X_1 \times X_2$, is given. Let us determine its projection onto the universe of discourse X_1.

$$X_1 = \{a_1, a_2, a_3\}\,, \quad X_2 = \{b_1, b_2, b_3\}\,,$$

$$Proj\,\mathbf{A} = \underset{x_2}{MAX}[\mathbf{A}(x_1, x_2)] = A^*(x_1) = \left(\frac{1}{a_1}, \frac{0.5}{a_2}, \frac{0}{a_3}\right)$$

The accomplished projection is represented in Fig. 4.30.

◊ ◊ ◊

Table 4.9. The discrete membership function of the relation $A(x_1, x_2)$

x_2 \ x_1	a_1	a_2	a_3
b_1	1	0.5	0
b_2	05	0.5	0
b_3	0	0	0

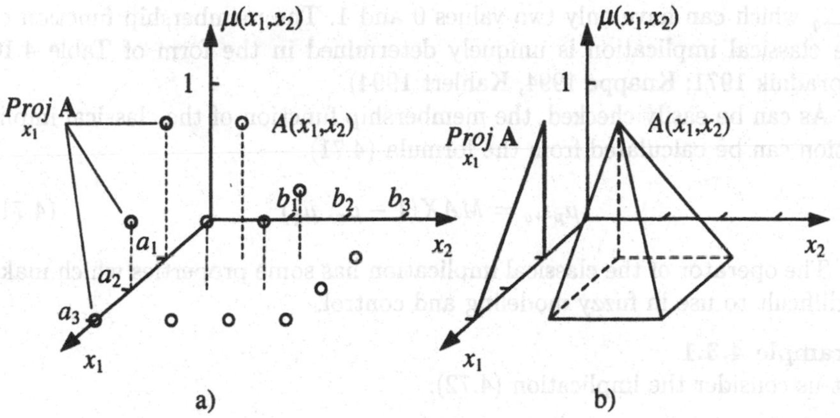

Fig. 4.30. Graphical illustration of fuzzy projection of a discrete (a) and continuous (b) relation

4.3 Implication

Implication is a kind of relation having the form of a rule used in the process of reasoning. A distinction can be made between classical and fuzzy implication.

Classical implication has the form (4.69) (Poradnik 1971).

$$IF \quad p \quad THEN \quad q \tag{4.69}$$

Its abbreviated form is expressed by (4.70).

$$p \rightarrow q\,, \tag{4.70}$$

where: p – the statement called the antecedent (premise),
q – the statement called the consequent (conclusion, result).

In classical logic, statements can be absolutely true ($\mu_p = 1$, $\mu_q = 1$) or untrue ($\mu_p = 0$, $\mu_q = 0$). The implication can be true or untrue according to the actual value of μ_p and μ_q (the truth of the antecedent and consequent).

Table 4.10. The membership function of the classical implication $\mu_{p\to q}$

μ_p	μ_q	$\mu_{p\to q}$
1	1	1
1	0	0
0	1	1
0	0	1

The truth-value of the implication is determined by its membership function $\mu_{p\to q}$ which can have only two values 0 and 1. The membership function of the classical implication is uniquely determined in the form of Table 4.10 (Poradnik 1971; Knappe 1994, Kahlert 1994).

As can be easily checked, the membership function of the classical implication can be calculated from the formula (4.71).

$$\mu_{p\to q} = MAX(1 - \mu_p, \mu_q) \tag{4.71}$$

The operator of the classical implication has some properties which make it difficult to use in fuzzy modeling and control.

Example 4.3.1
Let us consider the implication (4.72).

$$IF \text{ (the age of a car } x = \text{new) } THEN \text{ (fuel consumption } y = \text{small)} \tag{4.72}$$

The universe of discourse X of the variable "the age of a car" will be represented in binary form (new: $x = 1$, old: $x = 0$). Similarly, the universe of discourse Y of the variable "fuel consumption" (small: $y = 1$, large: $y = 0$).

The statement: (the age of a car = new) = p = the antecedent,
The statement: (fuel consumption = low) = q = consequent.

The question can be put: when is the implication (4.72) true ($\mu_{p\to q} = 1$) and when is it untrue ($\mu_{p\to q} = 0$)?

By substituting possible linguistic values (new, old) for x and (small, large) for y we obtain four states S_i of the implication.

S1: *IF* (the age of a car = new) *THEN* (fuel consumption = small),

$$\mu_p = 1, \quad \mu_q = 1, \quad \mu_{p\to q} = 1\,.$$

For x = new, y = small, the implication is true.

S2: *IF* (the age of a car = new) *THEN* (fuel consumption = large),

$$\mu_p = 1, \quad \mu_q = 0, \quad \mu_{p\to q} = 0\,.$$

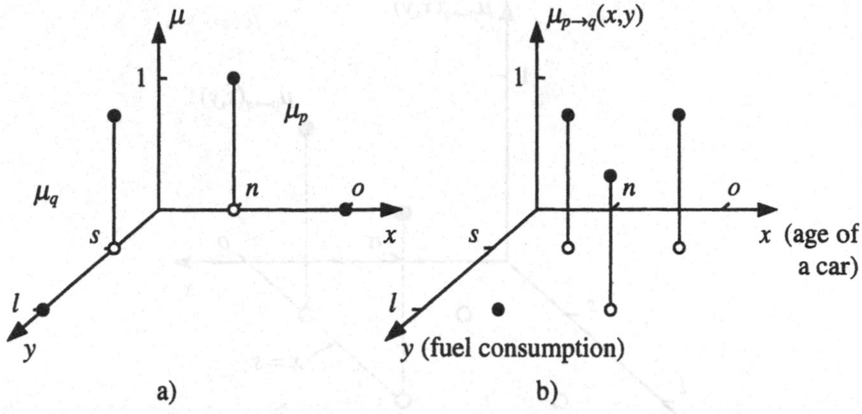

Fig. 4.31. The discrete membership functions of the premise μ_p and conclusion μ_q (a) and the membership function of the implication $\mu_{p\to q}$ defined on the Cartesian product $X \times Y$ (n – new, o – old, s – small, l – large) (b)

For $x =$ new and $y =$ large the implication (4.72) is untrue. It is intelligible because the premise (the age of a car = new) has not changed. And so, the changed conclusion (fuel consumption = large) can not be true.

S3: *IF* (the age of a car = old) *THEN* (fuel consumption = small),

$$\mu_p = 0, \quad \mu_q = 1, \quad \mu_{p\to q} = 1\,.$$

For $x =$ old and $y =$ small the implication (4.72) is true. This results from the fact that the considered implication (4.72) concerns only the state: (the age of a car = new), whereas it says nothing about the opposite state (the age of a car = old). According to classical logic, both conclusions (fuel consumption = small) and (fuel consumption = large) – given in S4 – can be true here.

S4: *IF* (the age of a car = old) *THEN* (fuel consumption = large),

$$\mu_p = 0, \quad \mu_q = 0, \quad \mu_{p\to q} = 1\,.$$

The implication (4.72) is true for these x,y values. Explanation as in S3. The membership function of the discrete implication $\mu_{p\to q}$ considered in the example is shown in Fig. 4.31.

The fault of the operator of the classical implication is that if its premise p is not satisfied even to the lowest degree ($\mu_p = 0$), then the implication is true and this activates the mutually exclusive conclusions (fuel consumption is small) and (fuel consumption is large), Fig. 4.32.

◊ ◊ ◊

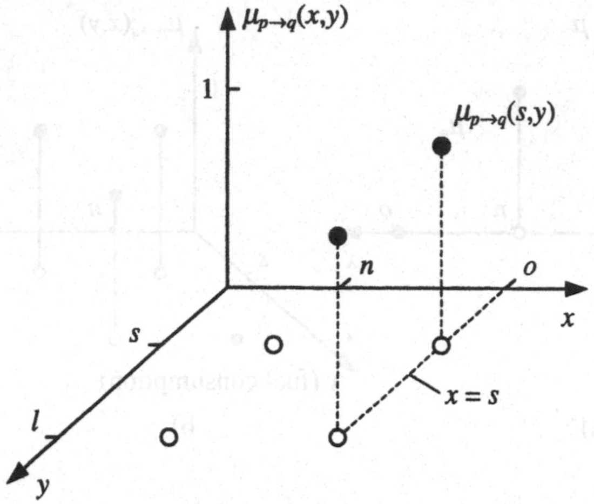

Fig. 4.32. The membership functions of the implication $\mu_{p\to q}(x,y)$ for $x = o$ (old)

In the case of fuzzy control systems, when many fuzzy rules are activated at the same time, the use of the operator of classical logic would disadvantageously affect the control process (Kahlert 1994). In this case implication operators of more unique operation are needed. Therefore, in fuzzy control and in many problems of fuzzy modeling another implication, namely the Mamdani implication, is most often used. The Mamdani implication will be described in the following.

Fuzzy implication

The fuzzy implication is the rule R whose simplest form is expressed by (4.73),

$$IF\ (x = A)\ THEN\ (y = B)\,, \tag{4.73}$$

where: $(x = A)$ is a premise (antecedent) and $(y = B)$ is a conclusion (consequent).

A and B are the fuzzy sets defined by their membership functions $\mu_A(x)$ and $\mu_B(y)$ and the universes of discourse X and Y. The fuzzy implication is determined by the symbol (4.74).

$$A \to B \tag{4.74}$$

The difference between the classical and fuzzy implication is that in the case of the classical implication the premise and conclusion is either absolutely true or absolutely untrue, whereas in the case of the fuzzy implication their partial truth, contained in the continuous interval [0,1], is allowed. Such an approach is very advantageous because in practice we rarely deal with situations where the premises of rules are completely satisfied. For this reason it can not be assumed that the conclusion is absolutely true.

Table 4.11. The operators of fuzzy implication

Łukasiewicz implication	$MIN(1, 1 - \mu_A(x) + \mu_B(y))$
Kleene-Dienes implication	$MAX(1 - \mu_A(x), \mu_B(y))$
Kleene-Dienes-Łukasiewicz implication	$1 - \mu_A(x) + \mu_A(x) \cdot \mu_B(y)$
Gödel implication	$\begin{cases} 1 & \text{for } \mu_A(x) \leq \mu_B(y) \\ \mu_B(y) & \text{otherwise} \end{cases}$
Yager implication	$(\mu_A(x))^{\mu_B(y)}$
Zadeh implication	$MAX(1 - \mu_A(x), MIN(\mu_A(x), \mu_B(y)))$

The fuzzy implication is determined, as every other fuzzy relation, by the membership function $\mu_{A\to B}(x, y)$ defined in the universe of discourse which is the Cartesian product $X \times Y$ of the domains of the premise and conclusion.

The membership function of the implication $\mu_{A\to B}(x, y)$ is the base for the so-called fuzzy reasoning (Chapter 5.1.2) which makes it possible to calculate the output of a fuzzy model (controller) for a given state of its inputs. To determine this function on the basis of the membership function of the premise $\mu_A(x)$ and conclusion $\mu_B(y)$ an appropriate operator of the implication should be used. The **Mamdani implication operator** is based on the assumption that the truth-value of the conclusion $\mu_B(y)$ can not be higher than the degree to which the premise $\mu_A(x)$ is satisfied, formula (4.75).

$$\mu_{A\to B}(x, y) = MIN(\mu_A(x), \mu_B(y)) \tag{4.75}$$

Such an assumption is intuitively intelligible. For example, in the case of the rule (4.76):

IF (the age of a car = new) *THEN* (fuel consumption = small), (4.76)

it is intelligible that if a car is not completely new, then its fuel consumption can not be so small as in the case of a completely new car. Apart from the Mamdani operator in fuzzy control the algebraic product operator *PROD* (4.77) is also applied:

$$\mu_{A\to B}(x, y) = \mu_A(x) \cdot \mu_B(y) \tag{4.77}$$

Besides the operators of fuzzy implication represented so far, many other operators have also been elaborated. Their use can have various results according to the actual problem. The operators are represented in Table 4.11.

According to the research results published in (Knappe 1994) the operator which has the best properties, considering a certain set of criteria assumed in this publication, is the Łukasiewicz operator. The remaining operators given

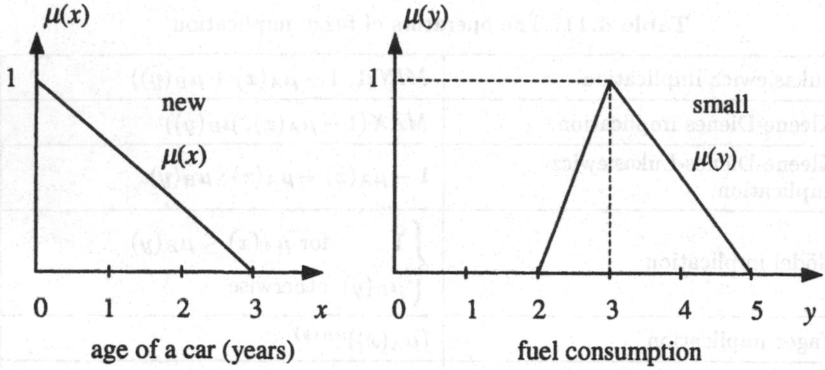

Fig. 4.33. The membership functions of the fuzzy sets "new" and "small" used in the premise and conclusion

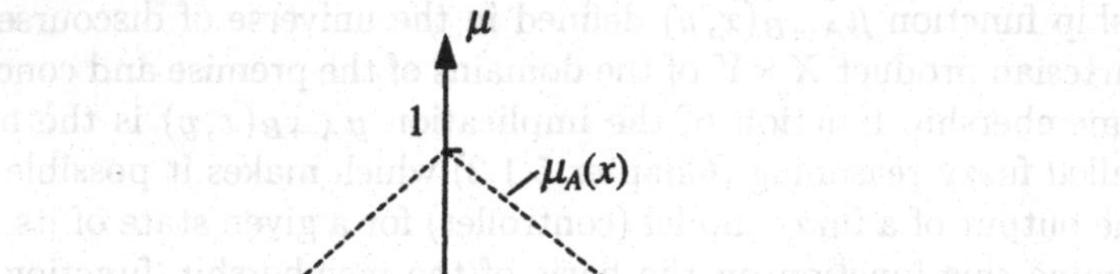

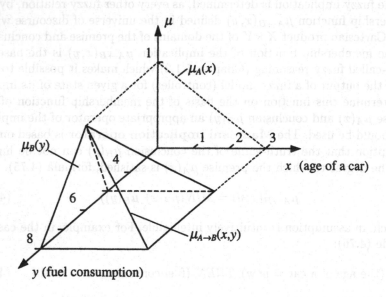

Fig. 4.34. The membership function $\mu_{A \to B}(x, y)$ of the implication (4.78) obtained with the use of the Mamdani implication operator

in Table 4.11 are ordered according to the decreasing degree to which these criteria are satisfied. Example 4.3.2 shows the method of creating the membership functions of the implication $\mu_{A \to B}(x, y)$ with the use of the Mamdani operator.

Example 4.3.2

Let us consider the fuzzy implication (4.78),

$$\textit{IF} \text{ (the age of a car } x = \text{new) } \textit{THEN} \text{ (fuel consumption } y = \text{small)} \quad (4.78)$$

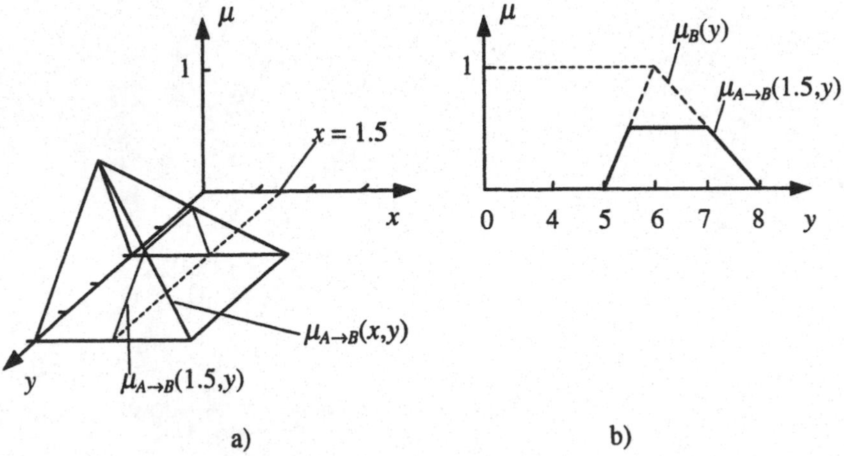

Fig. 4.35. The membership function of the implication $\mu_{A\to B}(x,y)$ for the given value of the variable $x = x_0 = 1.5$ (a) and its projection onto the plane $\{\mu, y\}$

where the fuzzy sets "new" and "small" are defined by the membership functions $\mu_n(x)$ and $\mu_s(y)$ given in Fig. 4.33.

The membership function of the implication (4.78) is represented in Fig. 4.34. By using an appropriate method of inference and having the given value x_0 of the variable x in the premise the membership function of the conclusion $\mu_{A\to B}(x_0, y)$ can be determined, which can then be used for calculating the crisp value of the output y_0 of a fuzzy model, Fig. 4.35. This problem is described in Sections 5.1.2. and 5.1.3.

◊ ◊ ◊

5. Fuzzy Models

5.1 Structure, main elements and operations in fuzzy models

The structure of a typical fuzzy model for a 2-inputs/1-output system is depicted in Fig. 5.1.

On the fuzzy model inputs, two crisp, numeric values x_1^*, x_2^* have been introduced. The block **FUZZIFICATION** calculates their membership grades in particular fuzzy sets A_i, B_j of the inputs. To carry out this operation the FUZZIFICATION-block must be provided with precisely defined membership functions $\mu_{Ai}(x_1)$, $\mu_{Bj}(x_2)$ of the inputs. Examples of membership functions of the inputs are depicted in Fig. 5.2.

The membership grades $\mu_{Ai}(x_1^*)$, $\mu_{Bj}(x_2^*)$ calculated and given in the FUZZIFICATION-block output inform us how high membership is of the

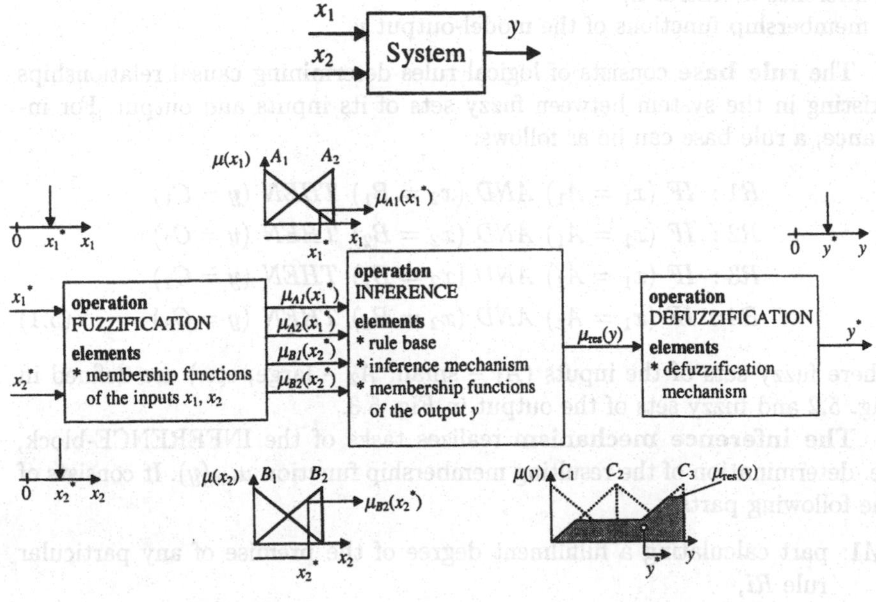

Fig. 5.1. Structure of a fuzzy model of a 2-inputs/1-output system

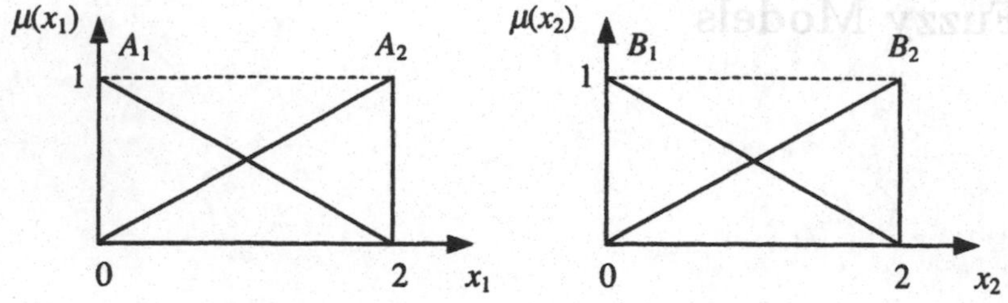

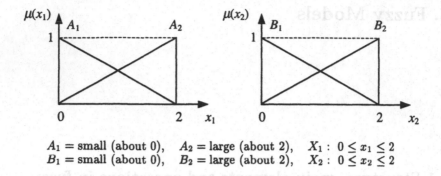

A_1 = small (about 0), A_2 = large (about 2), $X_1: 0 \le x_1 \le 2$
B_1 = small (about 0), B_2 = large (about 2), $X_2: 0 \le x_2 \le 2$

Fig. 5.2. Examples of membership functions of fuzzy sets and definition of the universe of discourse

numerical values x_1^*, x_2^* in particular fuzzy sets, i.e. how small (A_1, B_1) or large (A_2, B_2) these values are.

The **INFERENCE**-block gets as inputs the membership degrees $\mu_{Ai}(x_1^*)$, $\mu_{Bj}(x_2^*)$ and calculates as an output a so-called resulting membership function $\mu_{\text{res}}(y)$ of the model output, Fig. 5.1. This function usually has a complicated shape and its determination is made by means of inference, which can be carried out in many ways. To accomplish calculations the INFERENCE-block has to include the following, strictly defined elements:

- rule base,
- inference mechanism,
- membership functions of the model-output y.

The **rule base** consists of logical rules determining causal relationships existing in the system between fuzzy sets of its inputs and output. For instance, a rule base can be as follows:

$$\begin{aligned}
R1 &: \ IF\ (x_1 = A_1)\ AND\ (x_2 = B_1)\ THEN\ (y = C_1) \\
R2 &: \ IF\ (x_1 = A_1)\ AND\ (x_2 = B_2)\ THEN\ (y = C_2) \\
R3 &: \ IF\ (x_1 = A_2)\ AND\ (x_2 = B_1)\ THEN\ (y = C_2) \\
R4 &: \ IF\ (x_1 = A_2)\ AND\ (x_2 = B_2)\ THEN\ (y = C_3)
\end{aligned} \tag{5.1}$$

where fuzzy sets of the inputs (A_1 – small, A_2 – large, etc.) are defined in Fig. 5.2 and fuzzy sets of the output in Fig. 5.3.

The inference mechanism realizes tasks of the INFERENCE-block, i.e. determination of the resulting membership function $\mu_{\text{res}}(y)$. It consists of the following parts:

IM1: part calculating a fulfillment degree of the premise of any particular rule *Ri*,

IM2: part calculating an activated membership function of the conclusion of any particular rule *Ri*,

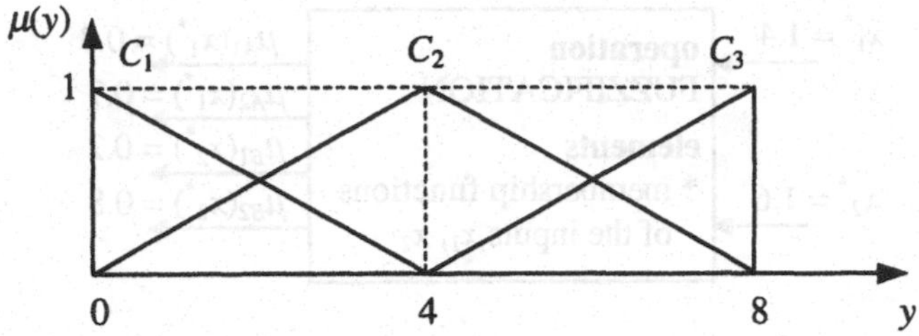

C_1 = small (about 0), C_2 = mean (about 4), C_3 = large (about 8)

Fig. 5.3. Examples of membership functions of fuzzy sets of the model output and definition of the output universe of discourse

IM3: part determining the resulting membership function $\mu_{res}(y)$ of the output based on activated conclusions of individual rules.

An example of an inference mechanism of a 2-input system:

IM1: premise aggregation in rules with the *PROD*-operator for the set intersection *AND* along with the *MAX*-operator for the set union *OR*,
IM2: determination of activated membership functions of the rule conclusions with the Mamdani-implication operator,
IM3: determination of the resulting output membership function $\mu_{res}(y)$ (accumulation) with the *MAX*-operator.

The **DEFUZZIFICATION**-block calculates from the resulting membership function $\mu_{res}(y)$ a crisp, numeric output value y^*, being an effect of the numeric inputs' values x_1, x_2. This operation is accomplished by a **defuzzification mechanism** determining the calculation method. An example of a defuzzification mechanism: the Center of Gravity method.

In what follows, particular blocks of a fuzzy model and various possibilities of the selection of their elements will be presented.

5.1.1 Fuzzification

In the fuzzification-block shown in Fig. 5.4 membership grades of the numeric values of the model inputs in fuzzy sets of the inputs are calculated.

The membership grade of the input value $x_1^* = 1.4$ in the fuzzy set A_1 (small) equal to 0.3 means that this input value is similar to the most typical small value (0) at the grade equal to 0.3. On the other hand, the truth that $x_1^* = 1.4$ is large equals 0.7. So this value of x_1 is more similar to the typical large value (2) than to the typical small one (0).

To calculate membership grades in particular fuzzy sets their membership functions must be precisely defined in respect of quality (function type) and of quantity (function parameters). Both the parameters and the shape

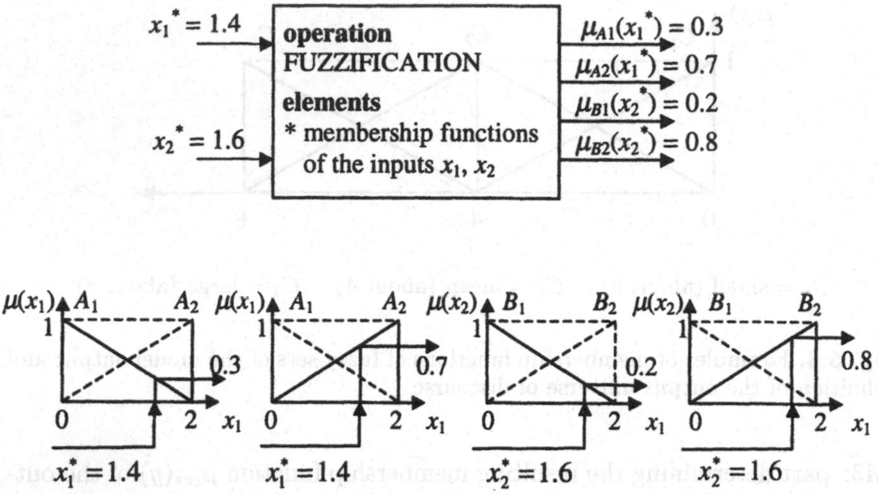

A_1 = small (ab. 0), A_2 = large (ab. 2), B_1 = small (ab. 0), B_2 = large (ab. 2)

Fig. 5.4. FUZZIFICATION-block and the operation example

of the membership functions strongly influence the model accuracy (Baglio 1994). An example of a mathematical description of membership functions (see Fig. 5.4) is given by (5.2).

$$\mu_{A1}(x_1) = 0.5(2 - x_1), \quad \mu_{A2}(x_1) = 0.5x_1,$$
$$\mu_{B1}(x_2) = 0.5(2 - x_2), \quad \mu_{B2}(x_2) = 0.5x_2. \tag{5.2}$$

In the fuzzification process a crisp input vector $\mathbf{X}^*$ is transformed into vector $\mathbf{M}$ of membership grades which are simultaneously inputs of the INFERENCE-block.

$$\mathbf{X}^* = \begin{bmatrix} x_1^* \\ x_2^* \end{bmatrix} \xrightarrow{\text{fuzzification}} \mathbf{M} = \begin{bmatrix} \mu_{A1}(x_1^*) \\ \mu_{A2}(x_1^*) \\ \mu_{B1}(x_2^*) \\ \mu_{B2}(x_2^*) \end{bmatrix}$$

5.1.2 Inference

The INFERENCE-block determines, from the inputs' membership grades $\mu_{Ai}(x_1)$, $\mu_{Bj}(x_2)$, the resulting membership function $\mu_{\text{res}}(y)$ of the model output, Fig. 5.5. The inference operation involves the following steps:

1. evaluation of fulfillment degrees of particular rules (to be precise, premises of the rules),

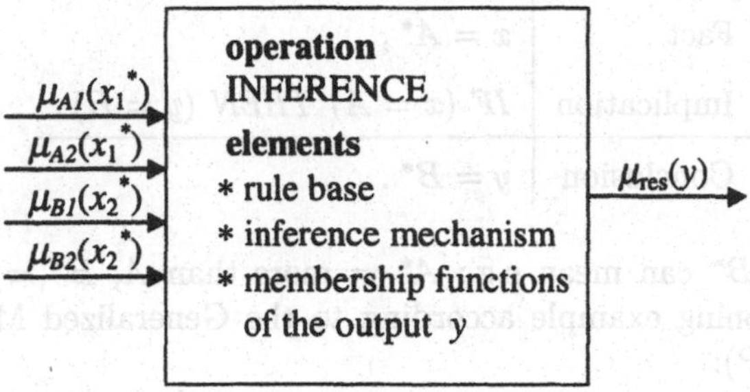

Fig. 5.5. INFERENCE-block of a fuzzy model

2. determination of activated membership functions of conclusions of particular rules,
3. determination of the resulting membership function of the conclusion of all rules in the rule base.

Classical logic elaborated many reasoning methods called tautologies. One of the best known is **Modus Ponens**. The reasoning process in Modus Ponens is as follows:

Fact	$x = A$,
Implication	*IF* $(x = A)$ *THEN* $(y = B)$,
Conclusion	$y = B$.

In the classical Modus Ponens tautology the truth value of the premise $(x = A)$ and conclusion $(y = B)$ is allowed to assume only two discrete values 1 or 0 and the fact considered $(x = \ldots)$ must fully agree with the implication premise:

$$IF\ (x = A)\ THEN\ (y = B).$$

Only then may the implication be used in the reasoning process. Both the premise and the rule conclusion must be formulated in a strict, deterministic way. Statements with non-precise fuzzy formulations as:

$$(x = \text{about } A),$$
$$(x = \text{more than } A),$$
$$(y = \text{more or less } B),$$

are not accepted. In fuzzy modeling and control approximate reasoning has been applied. It enables the use of fuzzy formulations in premises and conclusions. The approximate reasoning based on the **Generalized Modus Ponens** tautology is:

Fact	$x = A^*$,
Implication	*IF* $(x = A)$ *THEN* $(y = B)$,
Conclusion	$y = B^*$.

Where A^*, B^* can mean e.g.: A^* = more than A, B^* = more or less B, etc. A reasoning example according to the Generalized Modus Ponens tautology (GMP):

Fact	route of the trip is very long,
Implication	*IF* (route of the trip is long) *THEN* (travelling time is long),
Conclusion	travelling time is very long.

Generalized Modus Ponens reasoning doesn't always give good results. It can be exemplified as follows:

Fact	time of lying in the sun is very long,
Implication	*IF* (time of lying in the sun is long) *THEN* (the skin gets bronze),
Conclusion	the skin gets very bronze.

The approximate reasoning based on GMP gives the conclusion (the skin gets bronze). However, after a too long time of lying in the sun our skin frequently gets, not bronze, but red as a consequence of sunburn.

The GMP-tautology may be used if the rule has extrapolative features (Knappe 1994) enabling application of this rule also in cases when x is more or less equal to A (the fact doesn't accurately agree with the rule premise). The conclusions drawn (y is more or less equal to B) also have to "more or less" agree with reality. In practice, in fuzzy control and modeling the inference scheme based on GMP proves to be correct and is universally used. An example is given in Fig. 5.6.

5.1.2.1 Premise evaluation

To carry out fuzzy inference, initially, an evaluation of the premise fulfillment degree (truth value) of any particular rule has to be made. This degree, unlike classical logic, can take not only values 0 or 1 but also fractional values in the interval $[0, 1]$. If the premise fulfillment degree of a given rule equals 0 then this rule will not participate in the inference process. The higher this degree, the more will the rule influence the inference result.

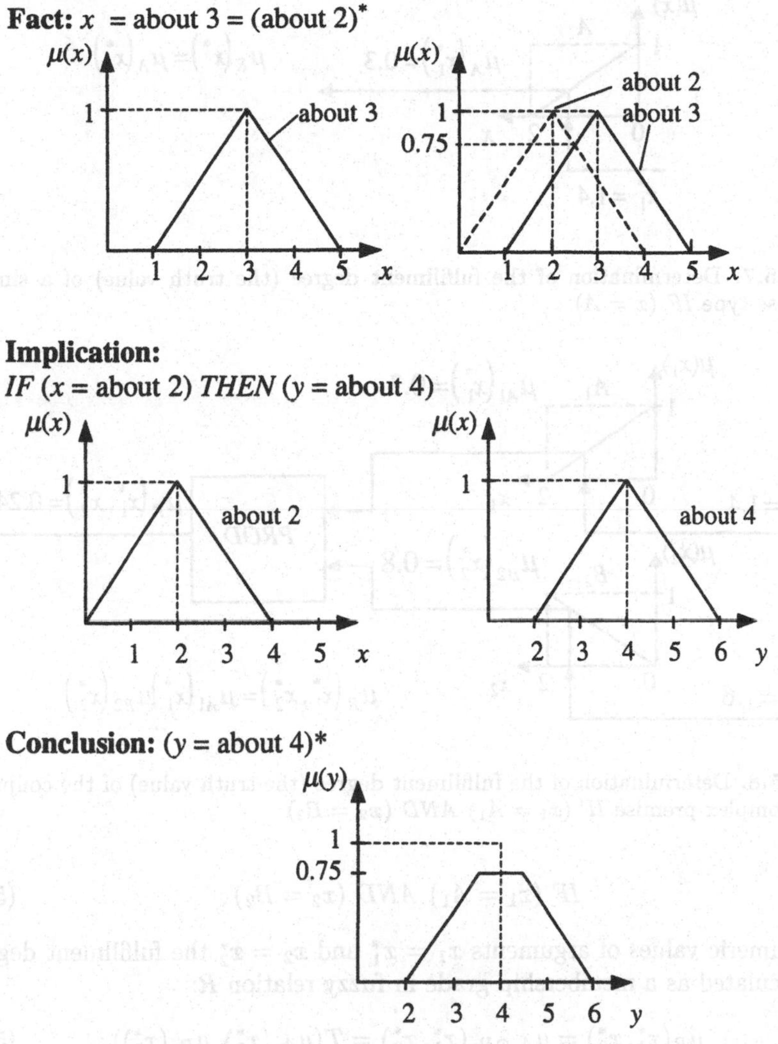

Fig. 5.6. Inference example with the use of the Generalized Modus Ponens tautology (method of conclusion determination has been explained in Section 5.1.2.2)

The method of calculation of the fulfillment degree of a premise depends on its form. In the case of the simplest premise, see (5.3):

$$IF\ (x = A)\,, \tag{5.3}$$

for $x = x^*$, the fulfillment degree $\mu_R(x^*)$ is equal to the membership grade of the value x^* in the set A, Fig. 5.7.

In the case of a complex premise, composed of two simple premises connected by logical conjunction *AND* (a conjunctive premise) as given by (5.4):

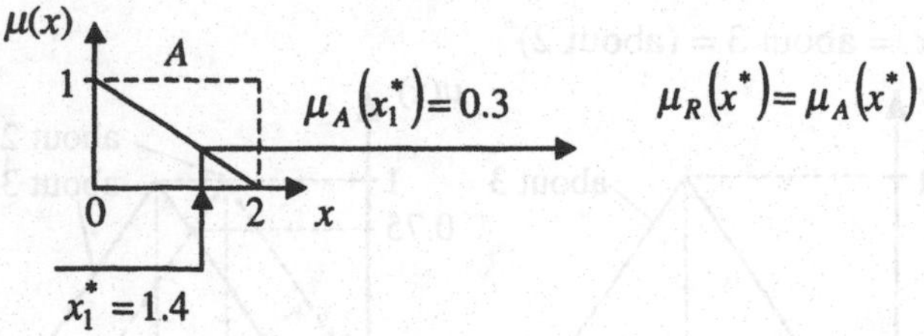

Fig. 5.7. Determination of the fulfillment degree (the truth value) of a simple premise type *IF* $(x = A)$

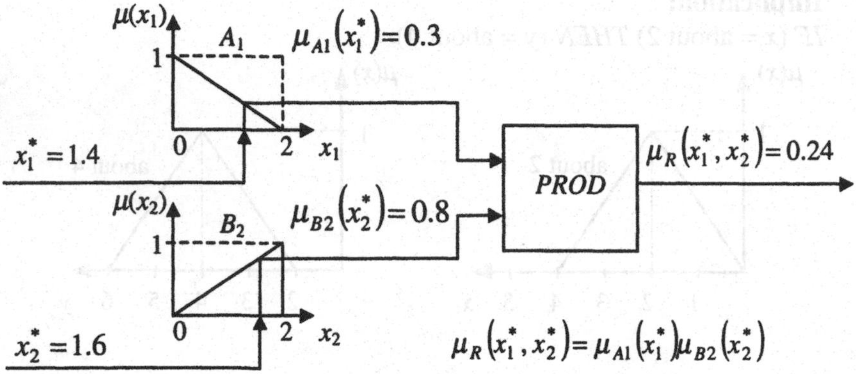

Fig. 5.8. Determination of the fulfillment degree (the truth value) of the conjunctive complex premise *IF* $(x_1 = A_1)$ *AND* $(x_2 = B_2)$

$$IF\ (x_1 = A_1)\ AND\ (x_2 = B_2)\,, \tag{5.4}$$

for numeric values of arguments $x_1 = x_1^*$ and $x_2 = x_2^*$ the fulfillment degree is calculated as a membership grade in fuzzy relation R:

$$\mu_R(x_1^*, x_2^*) = \mu_{A_1 \cap B_2}(x_1^*, x_2^*) = T(\mu_{A_1}(x_1^*), \mu_{B_2}(x_2^*))\,, \tag{5.5}$$

where A_1, B_2 are fuzzy sets and T is one of the t-norm operators, for instance *PROD*, Fig. 5.8.

If a complex premise consists of two simple premises connected by logical conjunction *OR* (alternative premise) as (5.6):

$$IF\ (x_1 = A_1)\ OR\ (x_2 = B_2)\,, \tag{5.6}$$

then for given values of arguments $x_1 = x_1^*$ and $x_2 = x_2^*$ the fulfillment degree of the premise is calculated as a membership grade in the relation R, (5.7),

$$\mu_R(x_1^*, x_2^*) = \mu_{A_1 \cup B_2}(x_1^*, x_2^*) = S(\mu_{A_1}(x_1^*), \mu_{B_2}(x_2^*))\,, \tag{5.7}$$

where S means one of the s-norm operators, e.g. *MAX*, as in Fig. 5.9.

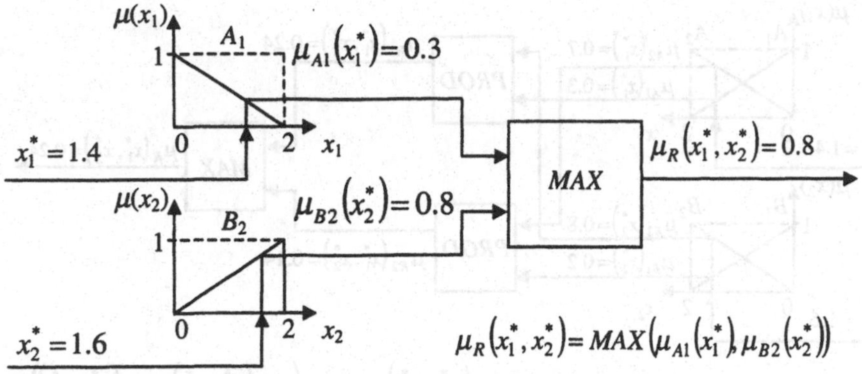

Fig. 5.9. Determination of the fulfillment degree of the alternative complex premise *IF* $(x_1 = A_1)$ *OR* $(x_2 = B_2)$

Premises can be more complex than the premise given by (5.4) or by (5.6). They may consist of many parts connected by conjunctions *AND*, *OR*, as exemplified by (5.8):

$$IF\ (x_1 = A_1)\ AND\ (x_2 = B_2)\ OR\ (x_1 = A_2)\ AND\ (x_2 = B_1)\,. \qquad (5.8)$$

In calculating the fulfillment degree of complex premises first the set intersection operation *AND* has to be made and next the set union operation *OR*. Definitions 4.2.4 and 4.2.5 can be employed here. The calculation of the fulfillment degree of the premise (5.8) for the argument values $x_1^* = 1.4$, $x_2^* = 1.6$ is depicted in Fig. 5.10.

Calculation of the fulfillment degree of complex premises consisting of simple ones is sometimes called **aggregation** (Knappe 1994). In examples given in Figs. 5.7 – 5.10 the premise arguments $x_1^* = 1.4$, $x_2^* = 1.6$ were crisp numeric values. However, premise arguments (inputs' values of a fuzzy model) can also take the form of fuzzy sets A_i^*, B_j^* different from fuzzy sets A_i, B_j which occur in these premises.

Let us consider a simple premise:

$$IF\ (x = A)\,,$$

in a situation where the model input x^* has the form of a fuzzy number A^*. The similarity degree, or in brief, similarity of two fuzzy sets A and A^*, being simultaneously a fulfillment degree $\mu_R(x^*)$ of the rule premise, can be calculated according to (5.9).

$$h = \underset{x \in X}{MAX}\ MIN(\mu_A(x), \mu_{A^*}(x)) \qquad (5.9)$$

In an operation defined by (5.9) a common part $A \cap A^*$ of the fuzzy sets has to be found in the first step and the maximum of it in the second step, see Fig. 5.11a.

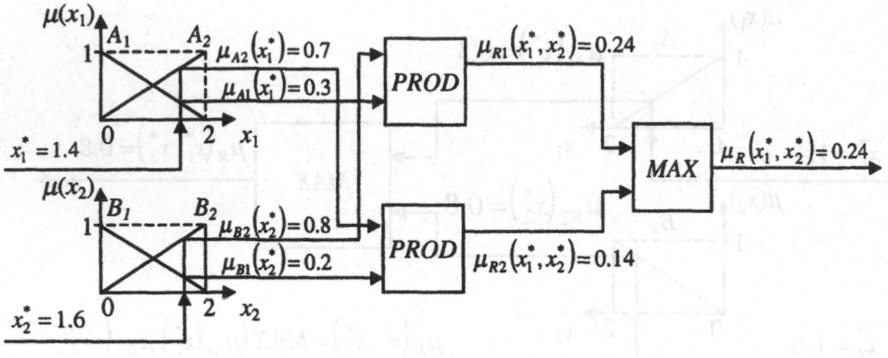

$$\mu_R(x_1^*, x_2^*) = MAX(\mu_{R1}(x_1^*, x_2^*), \mu_{R2}(x_1^*, x_2^*))$$

$$\mu_{R1}(x_1^*, x_2^*) = \mu_{A1}(x_1^*)\mu_{B2}(x_2^*)$$

$$\mu_{R2}(x_1^*, x_2^*) = \mu_{A2}(x_1^*)\mu_{B1}(x_2^*)$$

Fig. 5.10. Determination of the fulfillment degree of the complex premise: *IF* $(x_1 = A_1)$ *AND* $(x_2 = B_2)$ *OR* $(x_1 = A_2)$ *AND* $(x_2 = B_1)$

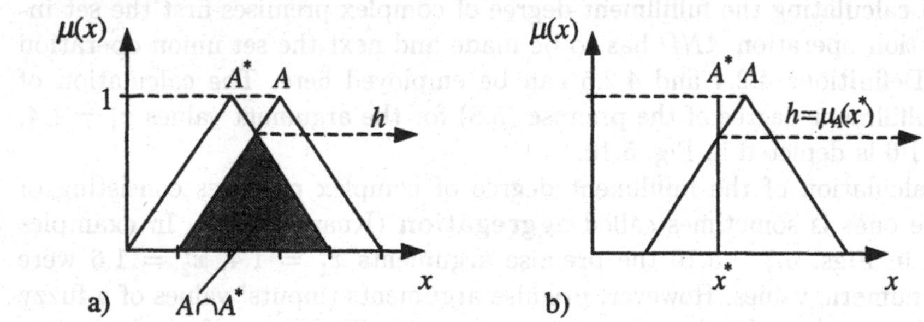

Fig. 5.11. Determination of the simplest similarity measure h of two fuzzy sets A and A^* (a) and a special case of a fuzzy set $A^* = x^*$ (singleton), (b)

If the fuzzy set A^* is a one-element set as in Fig. 5.11b, then the similarity h is equal to the membership grade of x^* in the set A: $\mu_A(x^*)$. Determination of the premise fulfillment degree of particular rules gives information on which rules and to what extent they should participate in the process of inference. It also enables determination of activated membership functions of conclusions in particular rules at given inputs' values x_i^* of a fuzzy model.

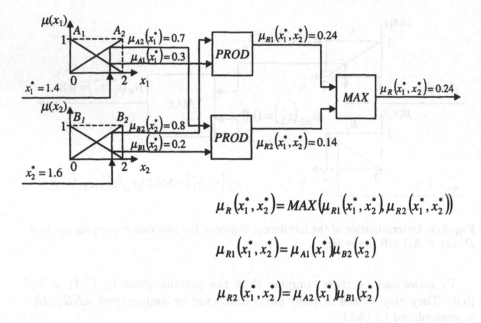

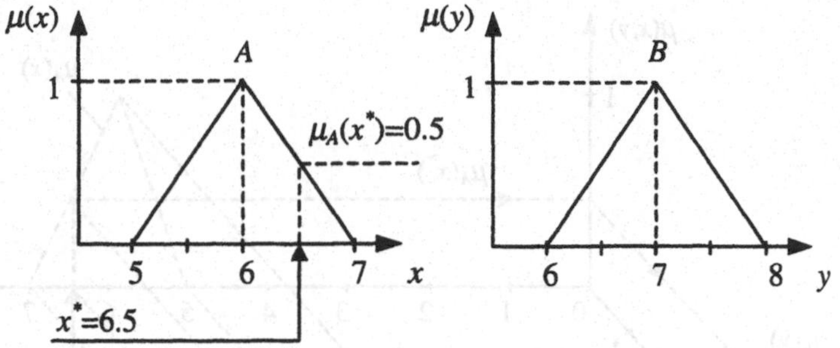

Fig. 5.12. Membership functions of fuzzy sets A and B in the rule (5.10)

5.1.2.2 Determination of activated membership functions of conclusions in particular rules at given input values of a fuzzy model

Determination of activated membership functions of conclusions in particular rules is based on the fulfillment degree of their premises. This operation can be called **inference in rules** and is accomplished with the use of operators of fuzzy implication described in Section 4.3. Here the operation will be explained in more detail and will also be exemplified. Let us assume an inference should be made in the rule (5.10):

$$IF\ (x = A)\ THEN\ (y = B), \tag{5.10}$$

where membership functions $\mu_A(x)$, $\mu_B(y)$ are depicted in Fig. 5.12 and the input value $x^* = 6.5$.

As can be seen in the figure, the fulfillment degree of the rule premise is equal to 0.5. Using the Mamdani-implication an activated membership function of the implication $A \to B$, being a certain fuzzy relation R, can be determined. The surface of this relation is portrayed in Fig. 5.13.

$$\begin{aligned} &\mu_R(x,y) = MIN(\mu_A(x), \mu_B(y)) \\ &R : A \to B \end{aligned} \tag{5.11}$$

The surface of the membership function of the implication $\mu_R(x,y)$ was composed of cylindrical extensions of the sets $A(x)$ and $B(y)$ on the Cartesian-product space $X \times Y$. For a given input value $x = x^*$ a two-dimensional membership function $\mu_R(x^*,y)$ of the implication is obtained. It is a special cut of the full three-dimensional function $\mu_R(x,y)$. Projection of the function $\mu_R(x^*,y)$ on the plane $\{\mu, y\}$ denoted $\mu_{B^*}(y)$ is the result of the inference carried out in the rule.

It can easily be noted that according to an actual input value x^* the inference result $\mu_{B^*}(y)$ differs more or less from the original membership function $\mu_B(y)$. In a special case, when the fulfillment degree of the premise

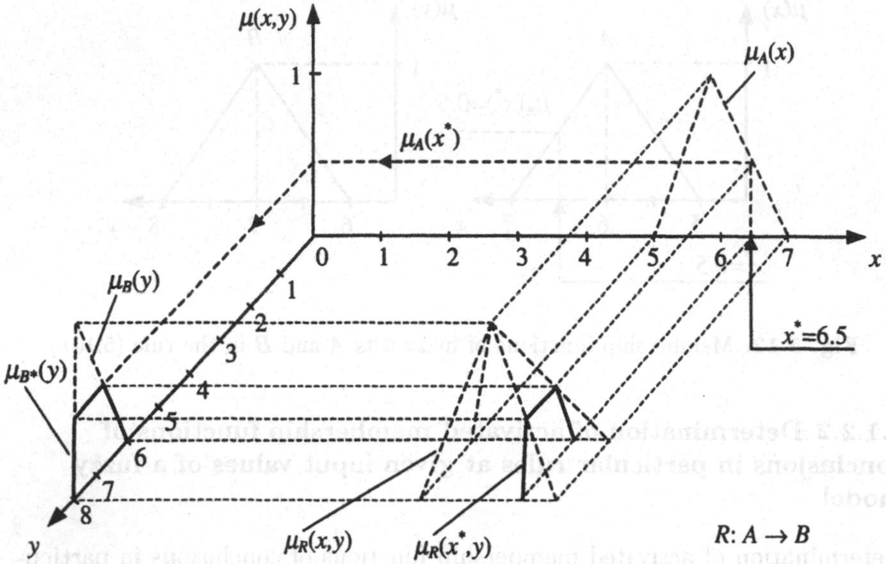

Fig. 5.13. Illustration of a fuzzy inference in the rule *IF* $(x = A)$ *THEN* $(y = B)$ at the input value $x = x^*$

$\mu_A(x^*) = 1$, the functions $\mu_{B^*}(y)$ and $\mu_B(y)$ are identical. The membership function $\mu_{B^*}(y)$ is sometimes called a **modified membership function of the rule conclusion** and the fuzzy set B^* a modified fuzzy set B of the rule conclusion (Knappe 1994).

To determine the modified (activated) membership function $\mu_{B^*}(y)$ of the conclusion we do not have to determine the three-dimensional membership function $\mu_R(x, y)$ of the implication. In practice it is done more simply. As can be noticed in Fig. 5.13, the functions $\mu_R(x^*, y)$ and $\mu_{B^*}(y)$ are identical and can be achieved in the case of the Mamdani-implication (5.11) by simple limitation of the conclusion membership function $\mu_B(y)$ to the level of the fulfillment degree $\mu_A(x^*)$ of the rule premise. In practice, the inference operation with the Mamdani-implication operator is made as in Fig. 5.14.

Other implication operators cause in effect other modifications of the conclusion membership function. Fig. 5.15 depicts an inference with the *PROD*-operator.

In a case when the input x^* is a fuzzy set A^* differing from the set A in the rule premise:

$$IF\ (x = A)\ THEN\ (y = B),$$

the modified (activated) membership function $\mu_{B^*}(y)$ of the conclusion can be determined with the aid of Zadeh's *Compositional Rule of Inference* (Kahlert 1994,1995) given by Definition 5.1.2.2.1.

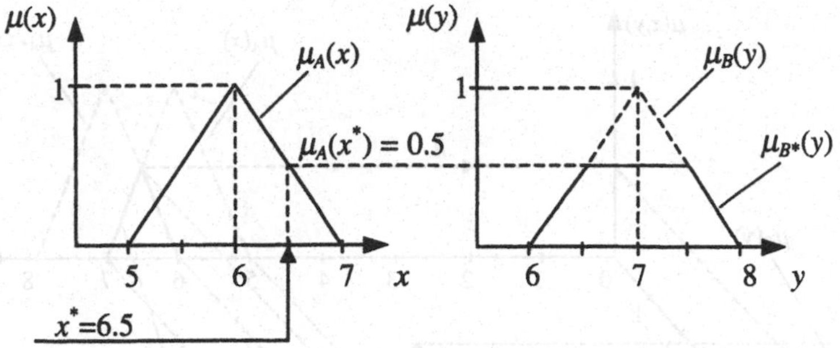

Fig. 5.14. Simplified method of the inference in a rule with the use of the Mamdani-implication operator

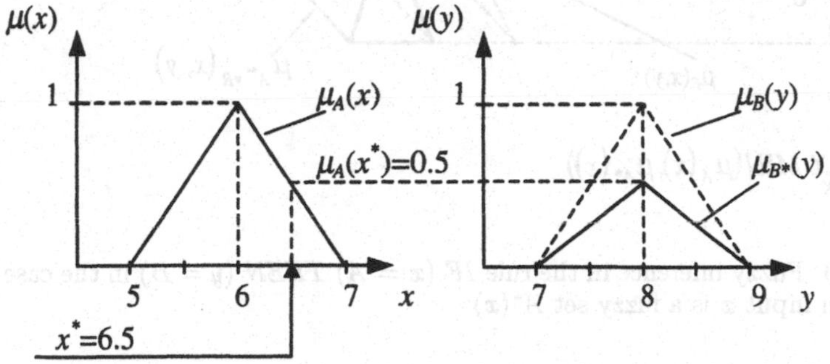

Fig. 5.15. Inference with the implication operator *PROD*

Definition 5.1.2.2.1

Let A^* be a fuzzy set of the universe of discourse X, and R a two-argument fuzzy relation defined in the universe of discourse $X \times Y$. Composition of A^* and R (denoted $A^* \circ R$) results in a fuzzy set B^* defined in the universe of discourse Y by a membership function $\mu_{B^*}(y)$ of the following form:

$$\mu_{B^*}(y) = \underset{x \in X}{MAX}\, MIN(\mu_{A^{**}}(x,y), \mu_R(x,y)),$$

where $A^{**}(x,y)$ is the cylindrical extension of the set $A^*(x)$ to the universe of discourse $X \times Y$.

The way the modified conclusion membership function $\mu_{B^*}(y)$ is constructed is depicted in Fig. 5.16.

As results from Fig. 5.16, composition of the relation $R(x,y)$ and the fuzzy set $A^*(x)$, or in practice its cylindrical extension $A^{**}(x,y)$, gives a fuzzy set of the composition $A^* \circ R(x,y)$, whose membership function is a three-dimensional surface.

Projection of the composition $A^* \circ R$ on the universe of discourse Y results in the modified fuzzy set of the conclusion B^*, where membership function

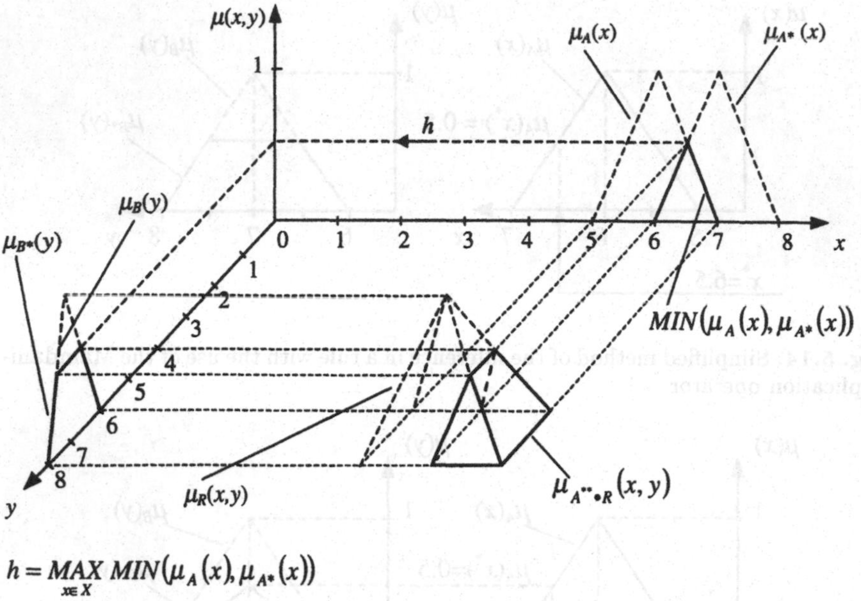

Fig. 5.16. Fuzzy inference in the rule *IF* $(x = A)$ *THEN* $(y = B)$ in the case when the model input x is a fuzzy set $A^*(x)$

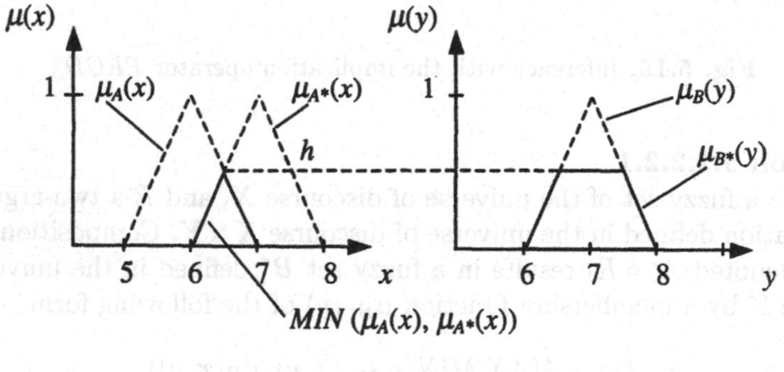

Fig. 5.17. Simplified inference method in the case when the model input x^* is a fuzzy set $A^*(x)$

$\mu_{B^*}(y)$ is the function $\mu_B(y)$ limited to the height h, where h denotes the fulfillment degree of the rule premise defined by (5.12).

$$h = \underset{x \in X}{MAX}\ MIN(\mu_A(x), \mu_{A^*}(x)) \tag{5.12}$$

So it appears in practice that if the fuzzy model input is also a fuzzy set A^* the simplified inference method presented in Fig. 5.17 can be used.

Table 5.1. Discrete membership function of the premise set A

x	5	5.5	6	6.5	7
$\mu_A(x)$	0	0.5	1	0.5	0

Table 5.2. Discrete membership function of the conclusion set B

y	6	6.5	7	7.5	8
$\mu_B(y)$	0	0.5	1	0.5	0

Table 5.3. Membership function of the relation $R = A \to B$

$x \backslash y$	6	6.5	7	7.5	8	
5	0	0	0	0	0	
5.5	0	0.5	0.5	0.5	0	
6	0	0.5	1	0.5	0	
6.5	0	0.5	0.5	0.5	0	$\leftarrow \mu_R(x^*, y) = \mu_{B^*}(y)$
7	0	0	0	0	0	

In the case when modified membership functions $\mu_A(x)$, $\mu_B(y)$ are of discrete forms (Table 5.1 and 5.2), the modified conclusion membership function $\mu_{B^*}(y)$ can be determined using tables or relational matrices. An illustration of this is Example 5.1.2.2.1.

Example 5.1.2.2.1

Let us determine the modified (activated) conclusion membership function $\mu_{B^*}(y)$ of the rule *IF* $(x = A)$ *THEN* $(y = B)$ in the case when input x is a crisp number: $x = x^* = 6.5$, and membership functions $\mu_A(x)$ and $\mu_B(x)$ are of a discrete form, Table 5.1 and 5.2. We create the relational Table 5.3 for the relation $R = A \to B$ according to (5.13). The table contains values $\mu_R(x, y)$.

$$\mu_R(x, y) = MIN(\mu_A(x), \mu_B(x)) \tag{5.13}$$

The membership function $\mu_{B^*}(y)$ of the rule conclusion for input $x^* = 6.5$ is a row in the relational table corresponding to the actual input value $x^* = 6.5$. The same result will be achieved restricting membership degrees in Table 5.2 to the level 0.5 which corresponds to the input value $x^* = 6.5$. This simplified procedure of the discrete conclusion membership function $\mu_{B^*}(y)$ determination is shown in Fig. 5.18.

◊ ◊ ◊

If rule premises consist of many simple rules connected with logical connectives then the resulting truth value of the complete premise should be calculated (the premise aggregation) following principles given in Section 5.1.2.1

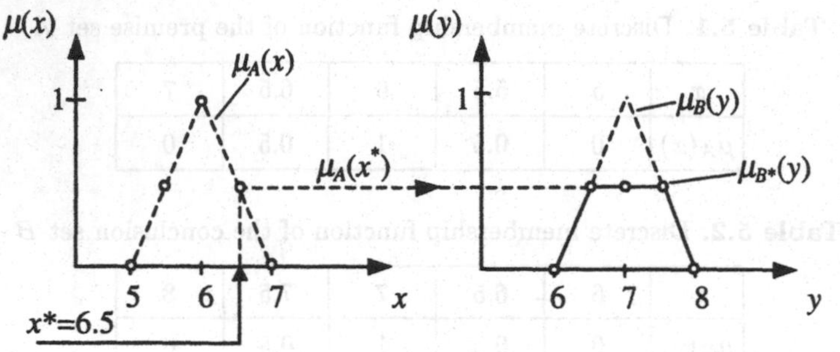

Fig. 5.18. Simplified inference in the case of discrete membership functions $\mu_A(x)$ and $\mu_B(x)$

and as a second step an inference has to be carried out in a way explained in this section.

5.1.2.3 Determination of the resulting membership function of the rule-base conclusion

As an inference result in particular rules (Ri) of the rule base, for m rules, m modified membership functions of conclusions are obtained. On their bases one resulting membership function of the conclusion of the whole rule base is to be determined. The process of the determination of the general conclusion is sometimes referred to as **accumulation** (Knappe 1994). The accumulation can be accomplished in many ways since many operators can be applied here. Below, the most frequently used ways of accumulation are presented. Let us consider Example 5.1.2.3.1.

Example 5.1.2.3.1
A fuzzy model with rule base (5.14) is given.

$$\begin{aligned} R1 &: \ IF\ (x = A_1)\ THEN\ (y = B_1) \\ R2 &: \ IF\ (x = A_2)\ THEN\ (y = B_2) \end{aligned} \tag{5.14}$$

Membership functions of fuzzy sets used in rules are presented in Fig. 5.19. A resulting membership function $\mu_{\text{res}}(y)$ of the complete rule base for the input $x = x^* = 1.4$ is to be determined.

All component rules can be combined in one complex rule (5.15).

$$R\colon\ IF\ (x = A_1)\ THEN\ (y = B_1)\ \ OR\ \ IF\ (x = A_2)\ THEN\ (y = B_2) \tag{5.15}$$

The rule R consists of two simple rules $R1$ and $R2$ connected with the logical connective OR, which can be written as:

$$R = R1 \cup R2\,. \tag{5.16}$$

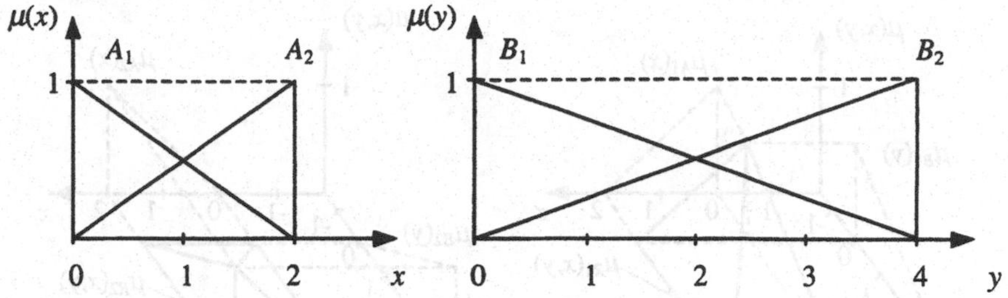

Fig. 5.19. Membership functions of fuzzy sets used in the rule base (5.14)

Each of the rules is a two-argument fuzzy relation (implication). The resulting relation can be determined according to Def. 4.2.4 using one of the s-norms, e.g. MAX-operator. The membership function $\mu_R(x, y)$ of the resulting relation R can be determined on the basis of the membership functions of the component relations (implications), formula (5.17).

$$\mu_R(x, y) = MAX(\mu_{R1}(x, y), \mu_{R2}(x, y)) \tag{5.17}$$

The resulting membership function $\mu_{\text{res}}(y)$ of the complete rule base conclusion for the given input value $x = x^*$ can be determined with formula (5.18) describing a cut through the relation surface for $x = x^*$.

$$\mu_{\text{res}}(y) = \mu_R(x^*, y) \tag{5.18}$$

This method will be referred to as **method 1**. Its successive steps are presented in Fig. 5.20a, b, c, d. The resulting membership function $\mu_{\text{res}}(y)$ can be shown more precisely in the 2-dimensional coordinate system as in Fig. 5.21.

Method 2 for determination of the resulting membership function $\mu_{\text{res}}(y)$ of the rule base conclusion consists of the following steps: first, modified membership functions $\mu_{B_i^*}(y)$ of particular rule conclusions are determined, and next, using one of the s-norms, e.g. the MAX-operator, the resulting function $\mu_{\text{res}}(y)$ according to formula (5.19) is found.

$$\mu_{\text{res}}(y) = MAX(\mu_{B_1^*}(y), \mu_{B_2^*}(y)) \tag{5.19}$$

Modified membership functions $\mu_{B_i^*}(y)$ of the particular rule conclusions can be found with methods described in Section 5.1.2.2. In practice the simplified method 2 is mostly applied. The manner the function $\mu_{\text{res}}(y)$ can be determined with this method is depicted in Figs. 5.22 and 5.23.

◊ ◊ ◊

If determination of modified membership functions $\mu_{B_i^*}(y)$ of particular rule conclusions is carried out with the MIN-operator (Fig. 5.22) and their

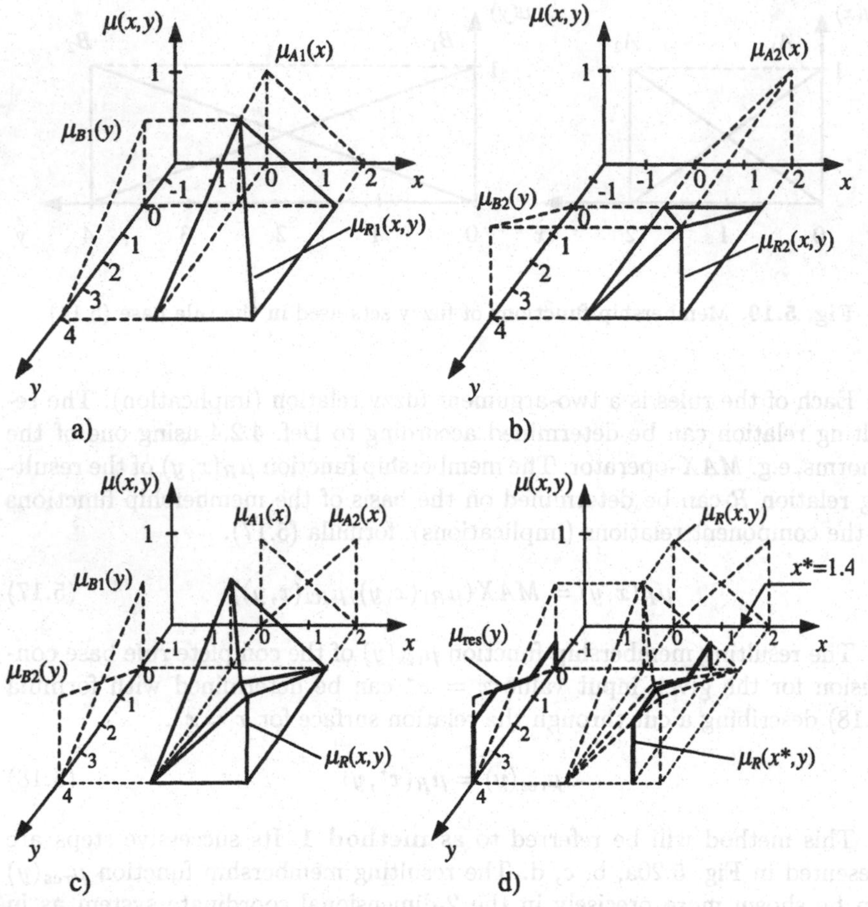

Fig. 5.20. Determination of the resulting membership function $\mu_{\text{res}}(y)$ of the rule base conclusion from the cut $\mu_R(x^*, y)$ of the resulting membership function $\mu_R(x, y)$ of the relation $R = R1 \cup R2$

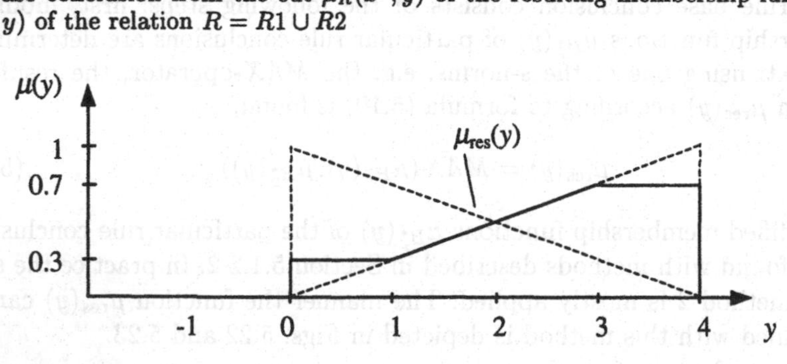

Fig. 5.21. Resulting membership function $\mu_{\text{res}}(y)$ of the rule base conclusion achieved with method 1

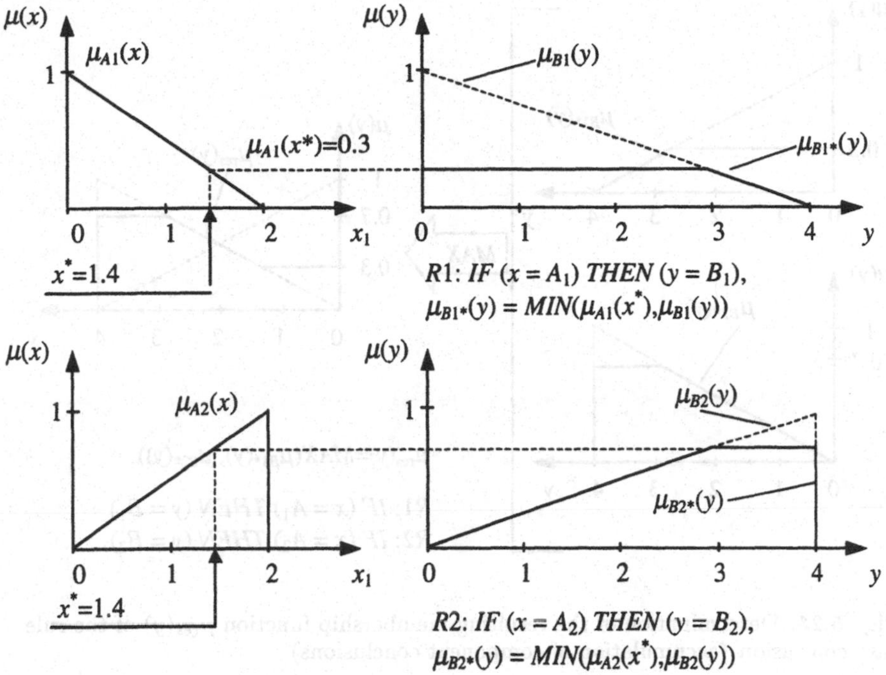

Fig. 5.22. Determination of modified membership functions $\mu_{res}(y)$ of particular rule conclusions with the simplified method (inference in rules)

accumulation to the resulting membership function $\mu_{res}(y)$ with the *MAX*-operator, then the entire operation is called ***MAX-MIN* inference**.

If function $\mu_{B_i^*}(y)$ is determined with the *PROD*-operator and function $\mu_{res}(y)$ with the *MAX*-operator, then the operation is denominated ***MAX-PROD* inference**. Using other t-norms and s-norms we get the next kinds of inference. Below, an example of inference in a fuzzy model with discrete membership functions is presented.

Example 5.1.2.3.2

A fuzzy model with rule base (5.20) is given.

$$\begin{aligned} R1: \; & IF\ (x = A_1)\ THEN\ (y = B_1) \\ R2: \; & IF\ (x = A_2)\ THEN\ (y = B_2) \end{aligned} \tag{5.20}$$

Membership functions of fuzzy sets used in the rules are depicted in Fig. 5.24 and in Tables 5.4 – 5.7.

The resulting membership function of the rule base conclusion $\mu_{res}(y)$ should be determined with the application of *MAX-MIN* inference for the input signal $x = x^* = 1.4$.

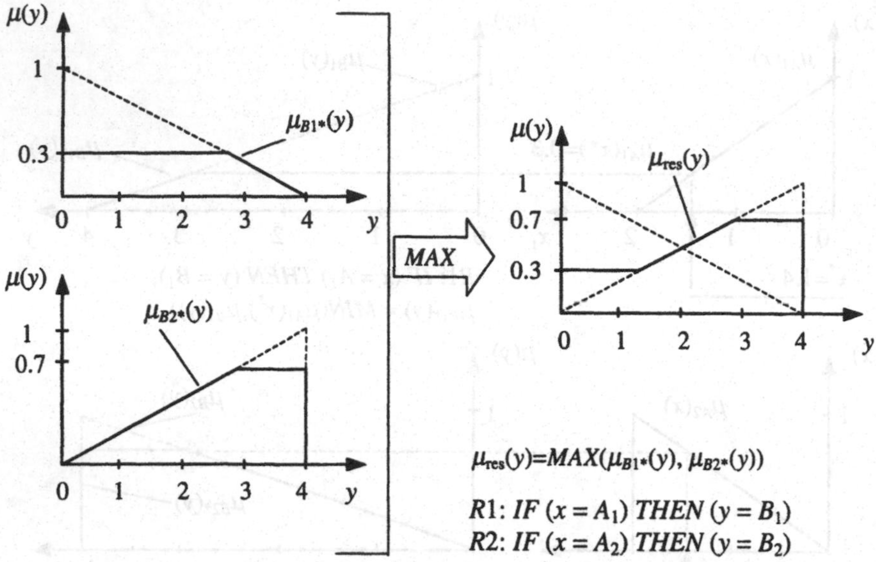

Fig. 5.23. Determination of the resulting membership function $\mu_{\text{res}}(y)$ of the rule base conclusion (accumulation of component conclusions)

Table 5.4. Discrete fuzzy set A_1

x	0	0.4	0.8	1.0	1.4	1.8	2.0
$\mu_{A_1}(x)$	1.0	0.8	0.6	0.5	0.3	0.1	0

Table 5.5. Discrete fuzzy set A_2

x	0	0.4	0.8	1.0	1.4	1.8	2.0
$\mu_{A_1}(x)$	0	0.2	0.4	0.5	0.7	0.9	1.0

Table 5.6. Discrete fuzzy set B_1

y	0	0.8	1.6	2.0	2.8	3.6	4.0
$\mu_{B_1}(y)$	1.0	0.8	0.6	0.5	0.3	0.1	0

Table 5.7. Discrete fuzzy set B_2

y	0	0.8	1.6	2.0	2.8	3.6	4.0
$\mu_{B_2}(y)$	0	0.2	0.4	0.5	0.7	0.9	1.0

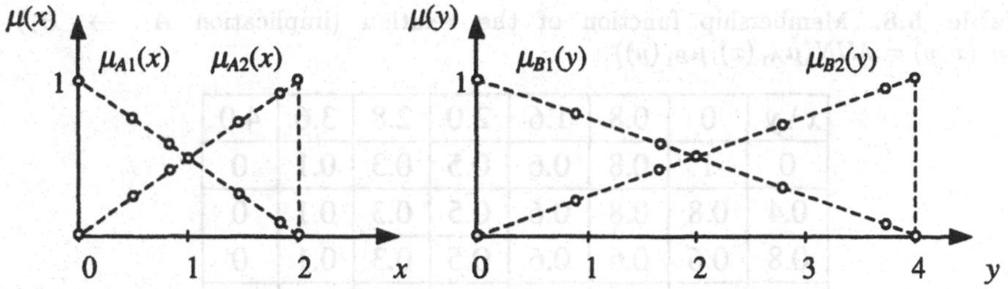

Fig. 5.24. Discrete membership functions of fuzzy sets used in rules (5.20)

Method 1

A cut of the membership function of the relation $R = R1 \cup R2$, given by formula (5.21), is to be found.

$$R = IF\ (x = A_1)\ THEN\ (y = B_1)\ \ OR\ \ IF\ (x = A_2)\ THEN\ (y = B_2) \tag{5.21}$$

First, the membership function of the implication relations $R1$ and $R2$ representing respective rules should be determined from formulas (5.22) and (5.23) (inference in rules).

$$\mu_{R1}(x,y) = \mu_{A_1 \to B_1} = MIN(\mu_{A_1}(x), \mu_{B_1}(y)) \tag{5.22}$$

$$\mu_{R2}(x,y) = \mu_{A_2 \to B_2} = MIN(\mu_{A_2}(x), \mu_{B_2}(y)) \tag{5.23}$$

The obtained relations $R1$ and $R2$ are given in Table 5.8 and 5.9.

From the component relations $R1$ and $R2$ the relation $R = R1 \cup R2$ should be calculated according to formula (5.24) and next a cut of this relation at $x^* = 1.4$, being the resulting membership function $\mu_{\text{res}}(y)$ of the rule base conclusion, is to be determined.

$$\mu_R(x,y) = MAX(\mu_{R1}(x,y), \mu_{R2}(x,y)) \tag{5.24}$$

The relation achieved is shown in Table 5.10. The obtained function $\mu_{\text{res}}(y)$ is depicted in Fig. 5.25.

Method 2

In this method, first, in a simplified way, modified membership functions of particular rules are determined and next their accumulation is made. The modified membership function $\mu_{B_1^*}(y)$ of the rule $R1$ conclusion is determined according to formula (5.25) and function $\mu_{B_2^*}(y)$ for the rule $R2$ according to (5.26).

$$\mu_{B_1^*}(y) = \mu_{R1}(x^*, y) = MIN(\mu_{A_1}(x^*), \mu_{B_1}(y)) = MIN(0.3, \mu_{B_1}(y)) \tag{5.25}$$

$$\mu_{B_2^*}(y) = \mu_{R2}(x^*, y) = MIN(\mu_{A_2}(x^*), \mu_{B_2}(y)) = MIN(0.7, \mu_{B_2}(y)) \tag{5.26}$$

From Tables 5.4 – 5.7 and formulas (5.25) – (5.26) functions $\mu_{B_1^*}(y)$, $\mu_{B_2^*}(y)$ in Table 5.11 and 5.12 are achieved.

Table 5.8. Membership function of the relation (implication $A_1 \rightarrow B_1$) $\mu_{R1}(x, y) = MIN(\mu_{A_1}(x), \mu_{B_1}(y))$

$x \backslash y$	0	0.8	1.6	2.0	2.8	3.6	4.0
0	1	0.8	0.6	0.5	0.3	0.1	0
0.4	0.8	0.8	0.6	0.5	0.3	0.1	0
0.8	0.6	0.6	0.6	0.5	0.3	0.1	0
1.0	0.5	0.5	0.5	0.5	0.3	0.1	0
1.4	0.3	0.3	0.3	0.3	0.3	0.1	0
1.8	0.1	0.1	0.1	0.1	0.1	0.1	0
2.0	0	0	0	0	0	0	0

Table 5.9. Membership function of the relation (implication $A_2 \rightarrow B_2$) $\mu_{R2}(x, y) = MIN(\mu_{A_2}(x), \mu_{B_2}(y))$

$x \backslash y$	0	0.8	1.6	2.0	2.8	3.6	4.0
0	0	0	0	0	0	0	0
0.4	0	0.2	0.2	0.2	0.2	0.2	0.2
0.8	0	0.2	0.4	0.4	0.4	0.4	0.4
1.0	0	0.2	0.4	0.5	0.5	0.5	0.5
1.4	0	0.2	0.4	0.5	0.7	0.7	0.7
1.8	0	0.2	0.4	0.5	0.7	0.9	0.9
2.0	0	0.2	0.4	0.5	0.7	0.9	1.0

Table 5.10. Membership function of the relation $R = R1 \cup R2$ and resulting membership function of the rule base conclusion $\mu_{\text{res}}(y) = \mu_R(1.4, y)$

$x \backslash y$	0	0.8	1.6	2.0	2.8	3.6	4.0
0	1	0.8	0.6	0.5	0.3	0.1	0
0.4	0.8	0.8	0.6	0.5	0.3	0.1	0
0.8	0.6	0.6	0.6	0.5	0.3	0.1	0
1.0	0.5	0.5	0.5	0.5	0.3	0.1	0
1.4	0.3	0.3	0.3	0.3	0.3	0.1	0
1.8	0.1	0.1	0.1	0.1	0.1	0.1	0
2.0	0	0	0	0	0	0	0

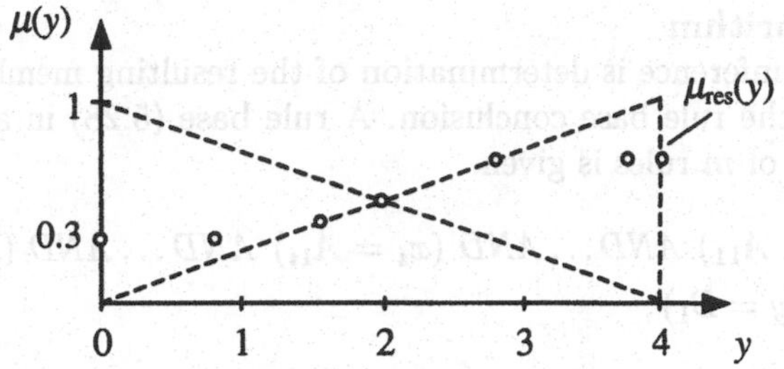

Fig. 5.25. Graphical form of the resulting membership function $\mu_{\mathrm{res}}(y)$ of the rule base conclusion constructed with data from Table 5.10

Table 5.11. Modified membership function $\mu_{B_1^*}(y)$ of the rule $R1$

y	0	0.8	1.6	2.0	2.8	3.6	4.0
$\mu_{B_1^*}(y)$	0.3	0.3	0.3	0.3	0.3	0.1	0

Table 5.12. Modified membership function $\mu_{B_2^*}(y)$ of the rule $R2$

y	0	0.8	1.6	2.0	2.8	3.6	4.0
$\mu_{B_2^*}(y)$	0	0.2	0.4	0.5	0.7	0.7	0.7

Table 5.13. Resulting membership function $\mu_{\mathrm{res}}(y)$ of the rule base conclusion

y	0	0.8	1.6	2.0	2.8	3.6	4.0
$\mu_{\mathrm{res}}(y)$	0.3	0.3	0.4	0.5	0.7	0.7	0.7

Accumulation of modified membership functions $\mu_{B_1^*}(y)$ and $\mu_{B_2^*}(y)$ is realized using formula (5.27). Its results are given in Table 5.13.

$$\mu_{\mathrm{res}}(y) = MAX(\mu_{B_1^*}(y), \mu_{B_2^*}(y)) \tag{5.27}$$

Comparing Tables 5.10 and 5.13 one can easily note that inferences realized both with method 1 and method 2 yielded identical results. Such a situation occurs when *MAX-MIN* inference is used. When, instead of *MAX* and *MIN* operators other s-norms and t-norms are applied, results of inference with method 1 and 2 can be different!

◊ ◊ ◊

In practice the most often used method, because of its lesser labour-consumption, is method 2. The general inference algorithm using this method is presented below.

Inference algorithm
The aim of the inference is determination of the resulting membership function $\mu_{\text{res}}(y)$ of the rule base conclusion. A rule base (5.28) in a conjunctive form consisting of m rules is given.

$$
\begin{aligned}
R1:\ & \mathit{IF}\ (x_1 = A_{11})\ \mathit{AND} \ldots \mathit{AND}\ (x_i = A_{1i})\ \mathit{AND} \ldots \mathit{AND}\ (x_n = A_{1n}) \\
& \mathit{THEN}\ (y = B_1), \\
& \vdots \\
Rj:\ & \mathit{IF}\ (x_1 = A_{j1})\ \mathit{AND} \ldots \mathit{AND}\ (x_i = A_{ji})\ \mathit{AND} \ldots \mathit{AND}\ (x_n = A_{jn}) \\
& \mathit{THEN}\ (y = B_j), \\
& \vdots \\
Rm:\ & \mathit{IF}\ (x_1 = A_{m1})\ \mathit{AND} \ldots \mathit{AND}\ (x_i = A_{mi})\ \mathit{AND} \ldots \mathit{AND}\ (x_n = A_{mn}) \\
& \mathit{THEN}\ (y = B_m), \qquad (5.28)
\end{aligned}
$$

where: $A_{11}, \ldots, A_{ji}, \ldots, A_{mn}$ – fuzzy sets of premises,
$B_1, \ldots, B_m$ – fuzzy sets of conclusions,
$x_1, \ldots, x_n$ – inputs of the fuzzy model,
$x_1^*, \ldots, x_n^*$ – values of the inputs of the fuzzy model,
y – output of the model.

Step 1
Determination of the fulfillment degree h of particular rule conclusions (premise aggregation) is accomplished using the formula (5.29),

$$
\begin{aligned}
h_1 &= T(\mu_{A_{11}}(x_1^*), \ldots, \mu_{A_{1n}}(x_n^*)), \\
&\vdots \\
h_j &= T(\mu_{A_{j1}}(x_1^*), \ldots, \mu_{A_{jn}}(x_n^*)), \\
&\vdots \\
h_m &= T(\mu_{A_{m1}}(x_1^*), \ldots, \mu_{A_{mn}}(x_n^*)), \qquad (5.29)
\end{aligned}
$$

where T marks one of the t-norm operators. According to inquiries carried out among fuzzy logic specialists (Pfeifer 1996) the most frequently used t-norm is the *PROD*-operator. Other operators, which are not t-norms, can also be applied.

Step 2
The aim of this step is determination of modified membership functions $\mu_{B_j^*}(y)$ of particular rule conclusions (inference in rules), formula (5.30). This operation is carried out only in rules with premises fulfilled to degree $h > 0$. Such rules can be referred to as "activated". Non-activated rules ($h = 0$) don't take part in the inference.

$$\mu_{B_1^*}(y) = T(h_1, \mu_{B_1}(y)),$$
$$\vdots$$
$$\mu_{B_j^*}(y) = T(h_j, \mu_{B_j}(y)),$$
$$\vdots$$
$$\mu_{B_m^*}(y) = T(h_m, \mu_{B_m}(y)), \tag{5.30}$$

Step 3
In this step the resulting membership function $\mu_{\text{res}}(y)$ is determined by means of the accumulation of modified membership functions of particular rule conclusions using formula (5.31),

$$\mu_{\text{res}}(y) = \mu_{B^*}(y) = S(\mu_{B_1^*}(y), \dots, \mu_{B_m^*}(y)), \tag{5.31}$$

where S denotes one of the s-norms, e.g. MAX, and $B^* = B_1^* \cup \dots \cup B_m^*$ a fuzzy set of the resulting conclusion of the rule base. Apart from s-norms other operators realizing operation OR may be applied.

The presented algorithm uses the conjunctive form (5.28) of the rule base. It is not a restriction because disjunctive or mixed, conjunctive-disjunctive forms can be transformed in the conjunctive ones. E.g. a mixed form of the rule (5.32):

$$\begin{aligned} &IF\ (x_1 = A_{11})\ AND\ (x_2 = A_{12})\ \ OR\ \ (x_1 = A_{21})\ AND\ (x_2 = A_{22}) \\ &THEN\ (y = B_1), \end{aligned} \tag{5.32}$$

can be replaced by two conjunctive ones (5.33):

$$\begin{aligned} &IF\ (x_1 = A_{11})\ AND\ (x_2 = A_{12})\ THEN\ (y = B_1), \\ &IF\ (x_1 = A_{21})\ AND\ (x_2 = A_{22})\ THEN\ (y = B_1). \end{aligned} \tag{5.33}$$

The rule (5.34):

$$IF\ (x_1 = A_1)\ THEN\ (y = B_1)\ OR\ (y = B_2), \tag{5.34}$$

can be replaced by two simple rules (5.35):

$$\begin{aligned} &IF\ (x_1 = A_1)\ THEN\ (y = B_1), \\ &IF\ (x_1 = A_1)\ THEN\ (y = B_2). \end{aligned} \tag{5.35}$$

In the inference not only different s-norm and t-norm operators but also other operators realizing the intersection and union of fuzzy sets may be used. What influence has the operator type on the inference result?

Fig. 5.26 shows inference in the model with 2 rules (5.36) from Example 5.1.2.3.1 carried out with different operators at input value $x^* = 1.4$.

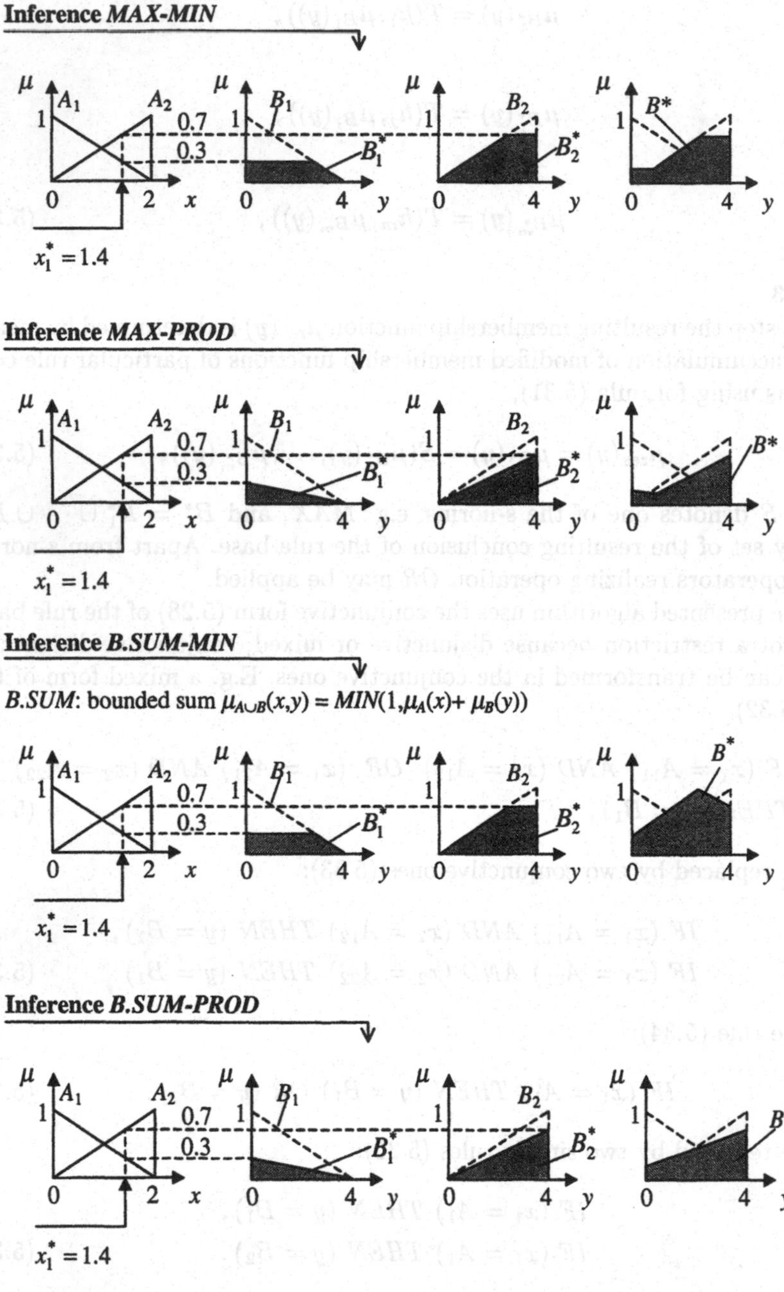

Fig. 5.26. Inference results with different t- and s-norm operators

$$R1: \ IF \ (x_1 = A_1) \ THEN \ (y = B_1)$$
$$R2: \ IF \ (x_1 = A_2) \ THEN \ (y = B_2) \qquad (5.36)$$

Analyzing Fig. 5.26 one can easily note that different logical operators used in the inference process result in very different fuzzy sets B^* of the rule base conclusion (different conclusion membership functions $\mu_{\text{res}}(y)$). It gives rise to the question, which operators should first of all be used in fuzzy models or **which operators give the highest accuracy of fuzzy modeling and control?**

In the case of self-learning (self-tuning) models and controllers, the types of operators used are of lesser importance, since in the course of the model (controller) learning, its degrees of freedom (parameters of membership functions) are changed so as to increase its accuracy as much as possible. Therefore choice of less appropriate operators is here at least partially compensated by the tuning process.

In the case of non-tunable fuzzy models and controllers, influence of the operators used is much stronger because the choice of inappropriate operators can not be compensated by anything. Then a trial and error method should be used and the model (controller) performance with different operator combinations should be investigated. As a result, the best operators can be determined.

A certain indication as to which operators are better or worse can be the frequency of their application by fuzzy modeling and control specialists (their popularity with the specialists). Analysis of literature on the subject seems to give evidence that premises aggregation is mostly accomplished with the *PROD*-operator, which, contrary to the *MIN*-operator, reacts on all input changes of the model (operator *MIN* reacts only on changes of the input whose membership is the smallest one). In the inference process the most often used combination is the *MAX-MIN* one. Of advantage is also the application of the *MEAN*-operator in premises aggregation and *SUM-MIN* combination in inference. The *SUM*-operator (unbounded sum), contrary to the *MAX*-operator, in the calculation process of the function $\mu_{\text{res}}(y)$, takes into account all component functions $\mu_{B_i^*}(y)$ of particular rules, whereas the *MAX*-operator takes into account only the function with the greatest membership for a given output y value. Inference with the *SUM*-operator is therefore more "democratic" than inference with the *MAX*-operator, which might be called "dictatorship" of the most activated rule.

As a result of inference, we obtain the resulting membership function $\mu_{\text{res}}(y)$ representing fuzzy set B^* of the complete rule base. If we are going to calculate crisp output y^* of the model (controller) a defuzzification of the fuzzy reference result has to be carried out. Defuzzification methods will be discussed in the next section.

5.1.3 Defuzzification of the resulting membership function of the rule-base conclusion

By the defuzzification of a fuzzy set $B^*(y)$, obtained as a result of inference, we are to understand the operation of determining the crisp value y^* representing this set in the most "reasonable" way. Of course, different criteria of the evaluation of "reasonability" of the representative y^* of the fuzzy set B^* can exist. The number of these criteria is reflected in the number of defuzzification methods, among which the most well-known are:

- the Middle of Maxima method (MM),
- the First of Maxima method (FM),
- the Last of Maxima method (LM),
- the Center of Gravity method (CG),
- the Center of Sums method (CS),
- the Height method (H).

These methods will be discussed in more detail below.

The Middle of Maxima method

The membership function in a fuzzy set can be understood as a function that informs us about the similarity between particular elements of the set and the most typical element in this set. An example is given in Fig. 5.27.

According to the membership function of "medium" height, a man of 170 (cm) in height is the typical representative of this height category (the membership = 1), while a man of 175 (cm) in height is "medium" to the degree of 0.5 and "tall" to the degree of 0.5. Otherwise speaking, he is partly similar to the medium-height man and to the tall man. Similarly, we can state that the most typical representative of the fuzzy set B^*, resulting from the inference characterized by the membership function $\mu_{B^*}(y) = \mu_{\text{res}}(y)$, is this value y^* whose membership function is the highest.

However, the set of such values can often contain more than one element and even an infinite number of elements as is shown in Fig. 5.28. The resolution of such a situation is to admit the mean value from the formula (5.37):

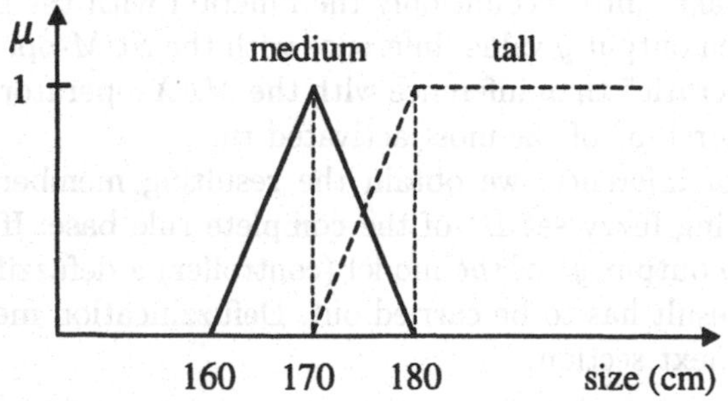

Fig. 5.27. The fuzzy set of "medium" size

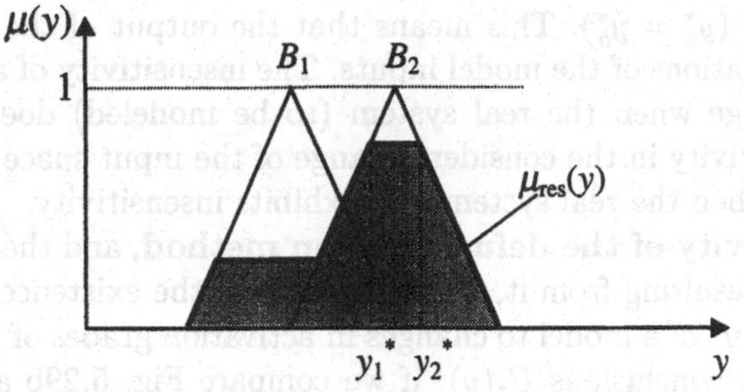

Fig. 5.28. The resulting membership function with the infinite number of elements y of the highest membership ($y_1^* \le y \le y_2^*$)

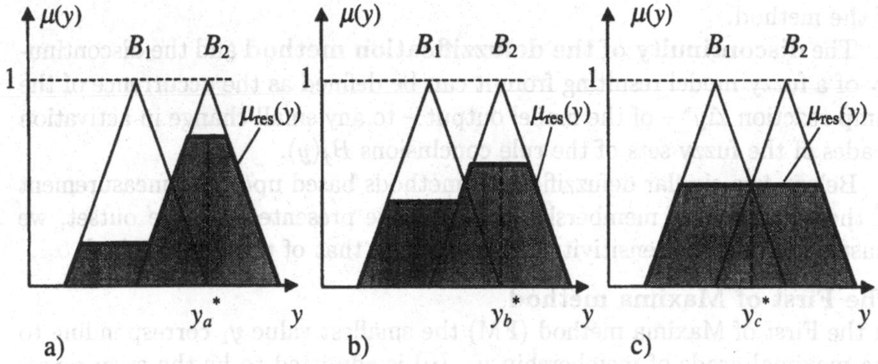

Fig. 5.29. The illustration of disadvantages of the Middle of Maxima method (MM)

$$y^* = 0.5(y_1^* + y_2^*) \tag{5.37}$$

to represent the resulting set of the conclusion. That is why the method is called: the Middle of Maxima method (MM).

The **advantage** of this method is the simplicity of calculations making the use of cheaper microprocessors in control systems easier. But the simplicity of calculations is obtained at the price of certain disadvantages.

The **disadvantage** of the method is that the result of defuzzification is affected only by the fuzzy set B_j that is most activated. The less activated sets have no effect. This also means that the result y^* is influenced only by those rules of the rule base which have this set in their conclusion (it is often only one rule). This way the defuzzification becomes "undemocratic" because not all rules take part in the "voting". The result of this is shown in Fig. 5.29.

In Fig. 5.29b the grade of activation of the set B_1 has increased in relation to Fig. 5.29a. Meanwhile, the grade of activation of the set B_2 has decreased. This is the result of variations in the input values x_i of the model. But the result of the defuzzification – the output of the model y^* – is identical for both cases a) and b) ($y_a^* = y_b^*$). This means that the output of the model is not sensitive to variations of the model inputs. The insensitivity of a fuzzy model is a disadvantage when the real system (to be modeled) does not exhibit such an insensitivity in the considered range of the input space. But, it is no disadvantage when the real system also exhibits insensitivity.

The **sensitivity of the defuzzification method**, and the sensitivity of a fuzzy model resulting from it, can be defined as the existence of a reaction of the output Δy^* of a model to changes in activation grades of the fuzzy sets of the rule base conclusions $B_j(y)$. If we compare Fig. 5.29b and c, then it can easily be noticed that a drastic step change of the defuzzification result y^* has occurred there, since y_c differs significantly from y_b. This means that a small change in the activation grade of the sets B_1 and B_2 has caused a big jump of the model output Δy^*. This property is called the discontinuity of the method.

The **discontinuity of the defuzzification method** and the discontinuity of a fuzzy model resulting from it can be defined as the occurrence of the jump reaction Δy^* – of the model output – to any small change in activation grades of the fuzzy sets of the rule conclusions $B_j(y)$.

Below, two similar defuzzification methods based upon the measurement of the maximum of membership function are presented. At the outset, we must say that their sensitivity is greater than that of the MM method.

The First of Maxima method

In the First of Maxima method (FM) the smallest value y_1 corresponding to the maximal grade of membership $\mu_{\text{res}}(y)$ is admitted to be the crisp representative y^* of the resulting conclusion fuzzy set. As is shown in Fig. 5.30a, with the increase in activation grade of the most activated set (B_2), its representative $y^* = y_1$ moves to the side of the modal value y_{m2} of this set. If the activation grade of the set B_2 decreases, the representative $y^* = y_1$ moves back from the modal value of the set B_2 to the side of its modal value y_{m1}.

The **advantages** of the FM method:

- low cost of calculation,
- greater sensitivity (in comparison with the MM method) to variations in the activation grade of the rule base conclusions.

The **disadvantages** of the FM method:

- discontinuity,
- the fact that in the defuzzification process only the most activated set B_j is taken into account.

The Last of Maxima method

The Last of Maxima method (LM) assumes the greatest value y_2 corresponding to the maximal grade of membership $\mu_{\text{res}}(y)$ to be the crisp representative y^* of the resulting conclusion fuzzy set, Fig. 5.31.

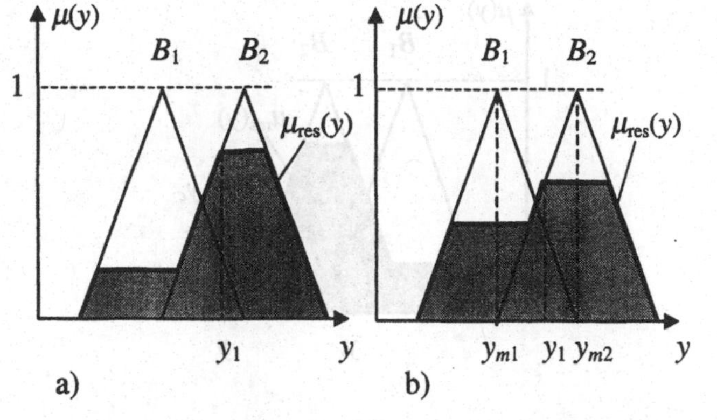

Fig. 5.30. Defuzzification using the First of Maxima method (FM), $y^* = y_1$

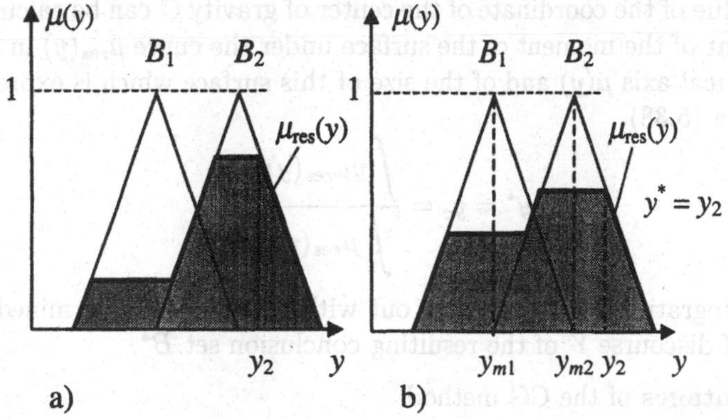

Fig. 5.31. Defuzzification using the Last of Maxima method (LM), $y^* = y_2$

The LM method has the same advantages and disadvantages as the FM method plus additionally one further disadvantage that is presented below. In the case when the activation of the set B_2 (deciding the choice of the representative y^*) decreases and that of the set B_1 increases (i.e. the significance of the set B_1 in the reasoning process increases), Fig. 5.31b, then the value $y^* = y_2$ should approach the modal value y_{m1} of the set B_1. Instead of this we observe the opposite occurring: the value y_2 moves away from the modal value.

The Center of Gravity method

The Center of Gravity method (CG) assumes that the crisp representative y^* of the resulting conclusion fuzzy set B^* defined by the membership function $\mu_{\text{res}}(y) = \mu_{B^*}(y)$ is the coordinate y_c of the center of gravity of the surface under the curve determined by this function, Fig. 5.32.

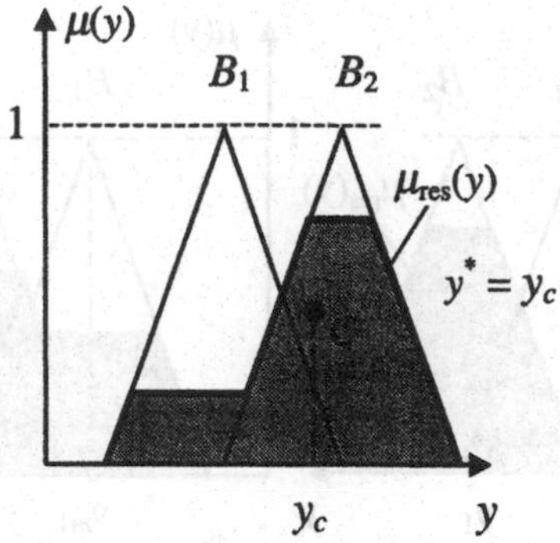

Fig. 5.32. Defuzzification using the Center of Gravity method (CG)

The value of the coordinate of the center of gravity C can be calculated as the quotient of the moment of the surface under the curve $\mu_{\mathrm{res}}(y)$ in relation to the vertical axis $\mu(y)$ and of the size of this surface which is expressed by the formula (5.38).

$$y^* = y_c = \frac{\int y\mu_{\mathrm{res}}(y)\mathrm{d}y}{\int \mu_{\mathrm{res}}(y)\mathrm{d}y} \tag{5.38}$$

The integration is to be carried out within the limits determined by the universe of discourse Y of the resulting conclusion set B^*.

The **advantages** of the CG method

- All activated membership functions of the conclusions (all active rules) take part in the defuzzification process. The CG method is "democratic". It guarantees that the sensitivity of a fuzzy model to changes in its inputs is greater than in the case of the FM, LM, and MM methods.

The **disadvantages** of the CG method

- **Large cost of calculation** connected with the integration of a surface of irregular shape, particularly when membership functions not consisting of straight segments (e.g. the Gaussian function) are used. To realize the integration it is necessary to determine the intersection points of the particular component membership functions $\mu_{B_j}(y)$, to divide the surface into sectors, and integrate in the range of these sectors, Fig. 5.33.

By using rectangular membership functions the calculations can be made easier, Fig. 5.34. The calculations can be further simplified by the use of the rectangular membership functions of the fuzzy sets B_j of the same width l, Fig. 5.35.

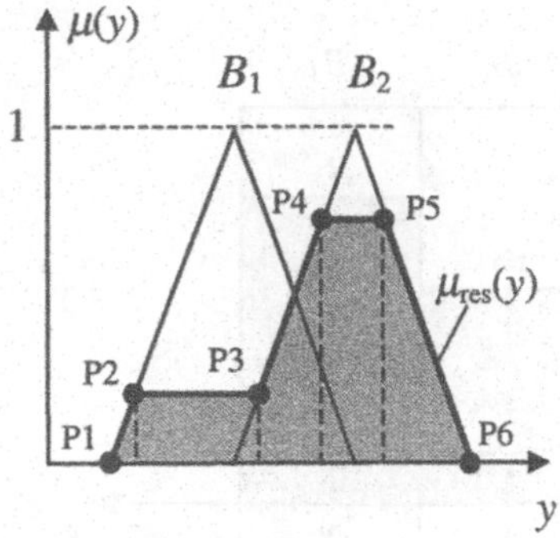

Fig. 5.33. The division of the surface to be integrated into sectors

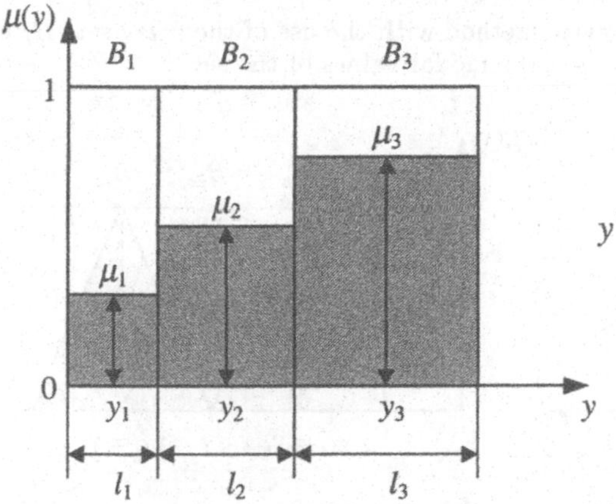

$$y^* = y_c = \frac{\sum_{i=1}^{3} \mu_i l_i y_i}{\sum_{i=1}^{3} \mu_i l_i}$$

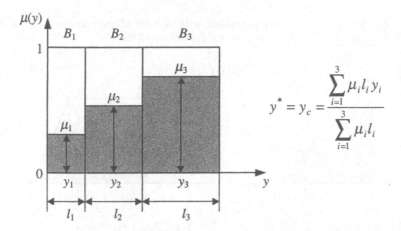

Fig. 5.34. The Center of Gravity method with the use of rectangular membership functions of the sets B_j of the rule base conclusion, y_i – the modal values of the sets

A disadvantage of simplifying the defuzzification by the use of rectangular membership functions $\mu_{B_j}(y)$ is that one has to limit oneself to only one shape of the function, whereas other shapes of the membership function can be more suited to the modeled system and can provide a higher accuracy of modeling.

- **The narrowing of the range of defuzzification** is the next disadvantage of the Center of Gravity method, Fig. 5.36.

In the case of the original version of the CG method the output y^* of a fuzzy model (controller) can not achieve the minimal (maximal) value $(y_{\min}, y_{\max})$ of the possible setting range even when the boundary fuzzy sets of the rule base conclusion B_1 or B_3 have been maximally activated. Therefore the fuzzy model could not work the same way as the modeled system and

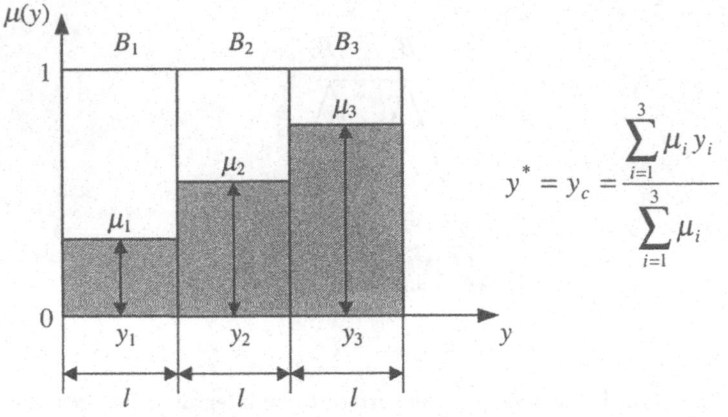

$$y^* = y_c = \frac{\sum_{i=1}^{3} \mu_i y_i}{\sum_{i=1}^{3} \mu_i}$$

Fig. 5.35. The Center of Gravity method with the use of the fuzzy sets B_j having identical supports (widths) l, y_i – the modal values of the sets

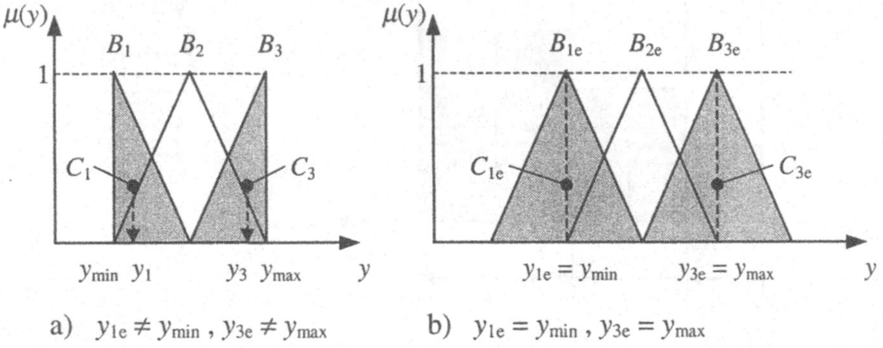

Fig. 5.36. The narrowing of the range of defuzzification in the original version of the Center of Gravity method (a) and elimination of this disadvantage in the extended version of the CG method

the fuzzy controller could not generate larger control signals. That would decrease the quality of control (e.g. the limitation of turning the ship rudder). This disadvantage can be removed through extending the boundary fuzzy sets, Fig. 5.36b, causing the coordinates of the centers of gravity of these sets to coincide with the boundaries of the range of operation (y_{min}, y_{max}). This method is called the **Extended Center of Gravity method** (ECG).

- **The insensitivity of the method** when only one membership function $\mu_{B_j}(y)$ of the output is activated is the next fault of this method. If several rules have the identical conclusion (the set B_2 in Fig. 5.37) or only one rule is activated, then the coordinate of the center of gravity y_c does not change although the grade of activation (Fig. 5.37a and b) of the resulting set has

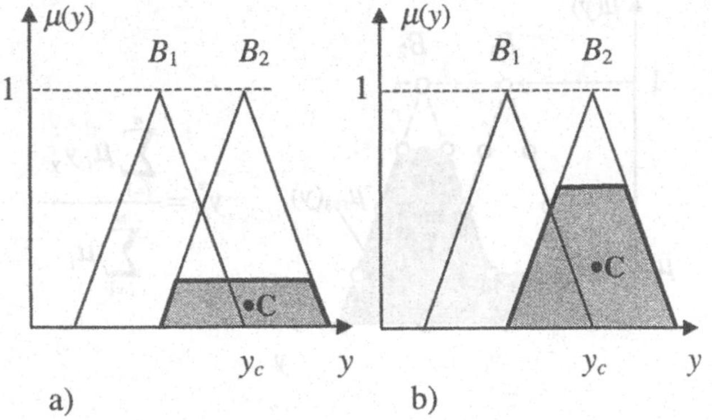

Fig. 5.37. The Center of Gravity method when only one fuzzy set $B_j(y)$ of the output of a model is activated

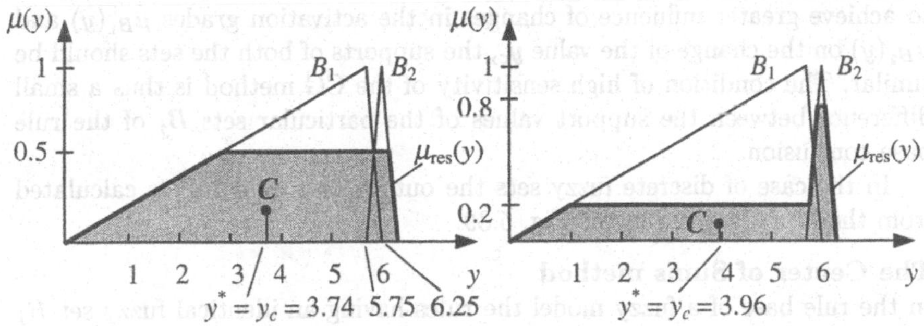

Fig. 5.38. Illustration of the small influence on the result of defuzzification caused by a change in the activation grade of the conclusion sets B_1, B_2

been changed. This means that the model is insensitive to changes at the input. This disadvantage can be limited by assigning different, not the same fuzzy sets B_j to particular rules. Also, such defuzzification methods as the MM, CS, and singleton method (if several rules are assigned to one singleton) have the above fault.

- **The decrease in the sensitivity** of the CG method when the support values of the sets $B_j(y)$ of the output of a fuzzy model differ greatly is also a disadvantage of this method. The problem is represented in Fig. 5.38.

The example represented in Fig. 5.38 shows that a large change in the activation grade of the component sets (μ_{B_1}: 0.5 – 0.2, μ_{B_2}: 0.5 – 0.8) causes minimal displacement of the coordinate of the center of gravity ($y^* = y_c$: 3.74 – 3.96). The reason for such a situation is the great difference between the surfaces of the component sets B_1 and B_2 which results from the great difference between the supports of both the sets (B_1: 6, B_2: 0.5). In order

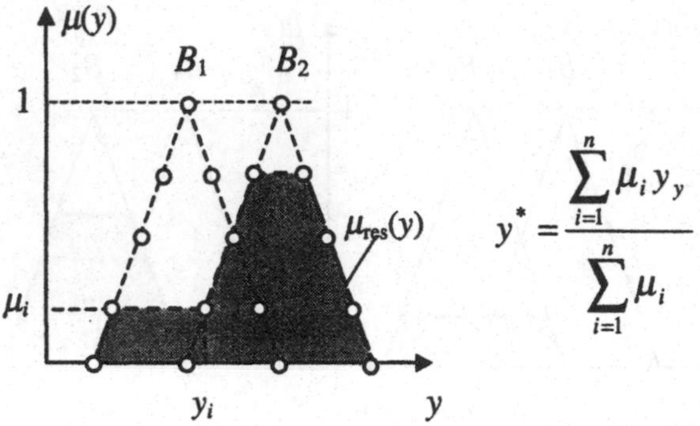

Fig. 5.39. Discrete variant of defuzzification using the Center of Gravity method (CG)

to achieve greater influence of changes in the activation grades $\mu_{B_1}(y)$ and $\mu_{B_2}(y)$ on the change of the value y_c, the supports of both the sets should be similar. The condition of high sensitivity of the CG method is thus a small difference between the support values of the particular sets B_j of the rule base conclusion.

In the case of discrete fuzzy sets the output of a model y^* is calculated from the formula given in the Fig. 5.39.

The Center of Sums method

In the rule base of a fuzzy model the rules having an identical fuzzy set B_j in the conclusions occur frequently. An example is the rule base (5.39).

$$\begin{aligned} R1: \ & IF\ (x_1 = A_{11})\ AND\ (x_2 = A_{21})\ THEN\ (y = B_1) \\ R2: \ & IF\ (x_1 = A_{11})\ AND\ (x_2 = A_{22})\ THEN\ (y = B_2) \\ R3: \ & IF\ (x_1 = A_{12})\ AND\ (x_2 = A_{21})\ THEN\ (y = B_2) \\ R4: \ & IF\ (x_1 = A_{12})\ AND\ (x_2 = A_{22})\ THEN\ (y = B_3) \end{aligned} \tag{5.39}$$

The rules $R2$ and $R3$ have, in the conclusion, the identical set B_2. Let us assume that the input values x_1^* and x_2^* are the same and the sets B_j in the conclusions are activated to the degree shown in Fig. 5.40.

The fuzzy set B_2 is activated by two rules $R2$ and $R3$. If we use the MAX operator to calculate the membership function $\mu_{\mathrm{res}}(y) = \mu_{B^*}(y)$, then we obtain the resulting set $B^* = B_1^* \cup B_{22}^* \cup B_{23}^* \cup B_3^*$ represented in Fig 5.41.

As results from Fig. 5.41 show, the greatest influence on the position of the center of gravity C, and so on the result of defuzzification, comes from the set B_3 (the rule $R3$), which is activated to the highest degree (0.8). But the set B_2 is activated by two rules $R2$ and $R3$ and the total degree of its activation ($0.4 + 0.6 = 1.0$) is higher than the grade of activation of the set B_3 by the rule $R3$.

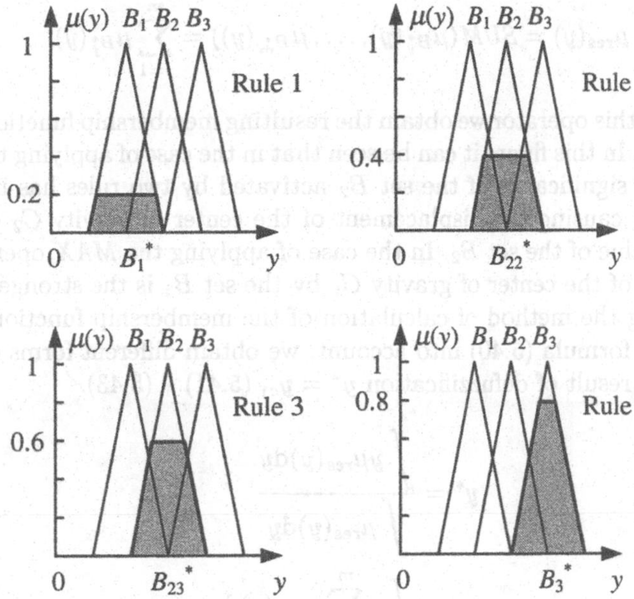

Fig. 5.40. An example of the activation of the sets B_j by the particular rules Ri (5.39)

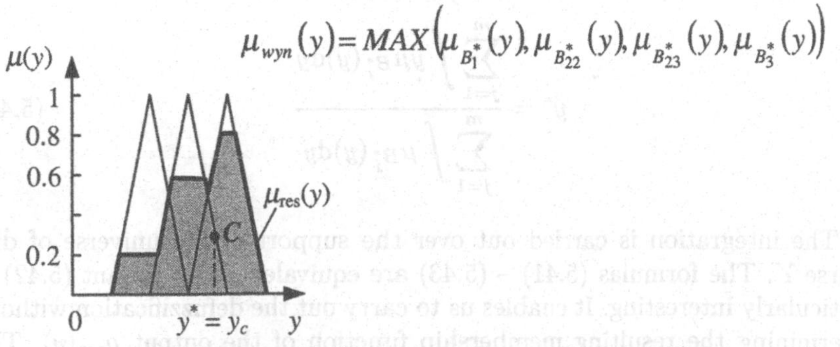

same fuzzy sets B_j to particular rules. Also, such defuzzification methods as the MM, CS, and singleton method (if several rules are assigned to one singleton) have the above fault.

- **The decrease in the sensitivity** of the CG method when the support values of the sets $B_j(y)$ of the output of a fuzzy model differ greatly is also a disadvantage of this method. The problem is represented in Fig. 5.38.

The example represented in Fig. 5.38 shows that a large change in the activation grade of the component sets (μ_{B_1}: 0.5 – 0.2, μ_{B_2}: 0.5 – 0.8) causes minimal displacement of the coordinate of the center of gravity ($y^* = y_c$: 3.74 – 3.96). The reason for such a situation is the great difference between the surfaces of the component sets B_1 and B_2 which results from the great difference between the supports of both the sets (B_1: 6, B_2: 0.5). In order

$$\mu_{\text{res}}(y) = SUM(\mu_{B_1^*}(y), \ldots, \mu_{B_m^*}(y)) = \sum_{j=1}^{m} \mu_{B_j^*}(y) \tag{5.40}$$

By using this operator we obtain the resulting membership function shown in Fig. 5.42b. In this figure it can be seen that in the case of applying the SUM operator the significance of the set B_2 activated by two rules has increased considerably, causing the displacement of the center of gravity C_2 closer to the modal value of the set B_2. In the case of applying the MAX operator the "attraction" of the center of gravity C_1 by the set B_3 is the strongest.

By taking the method of calculation of the membership function $\mu_{\text{res}}(y)$ according to formula (5.40) into account, we obtain different forms of calculation of the result of defuzzification $y^* = y_c$, (5.41) – (5.43).

$$y^* = \frac{\int y\mu_{\text{res}}(y)\mathrm{d}y}{\int \mu_{\text{res}}(y)\mathrm{d}y} \tag{5.41}$$

$$y^* = \frac{\int y \sum_{j=1}^{m} \mu_{B_j^*}(y)\mathrm{d}y}{\int \sum_{j=1}^{m} \mu_{B_j^*}(y)\mathrm{d}y} \tag{5.42}$$

$$y^* = \frac{\sum_{j=1}^{m} \int y\mu_{B_j^*}(y)\mathrm{d}y}{\sum_{j=1}^{m} \int \mu_{B_j^*}(y)\mathrm{d}y} \tag{5.43}$$

The integration is carried out over the support of the universe of discourse Y. The formulas (5.41) – (5.43) are equivalent. The variant (5.42) is particularly interesting. It enables us to carry out the defuzzification without determining the resulting membership function of the output $\mu_{\text{res}}(y)$. The defuzzification can be performed on the basis of knowledge of the inference results in the particular rules $\mu_{B_j^*}(y)$.

If the expression (5.44) is called the moment of the set $B_j^*(y)$:

$$\int y\mu_{B_j^*}(y)\mathrm{d}y = M_j\,, \tag{5.44}$$

in relation to the vertical axis $\mu(y)$, Fig. 5.43, and the expression (5.45) the area S_j of the set $B_j^*(y)$,

$$\int \mu_{B_j^*}(y)\mathrm{d}y = S_j\,, \tag{5.45}$$

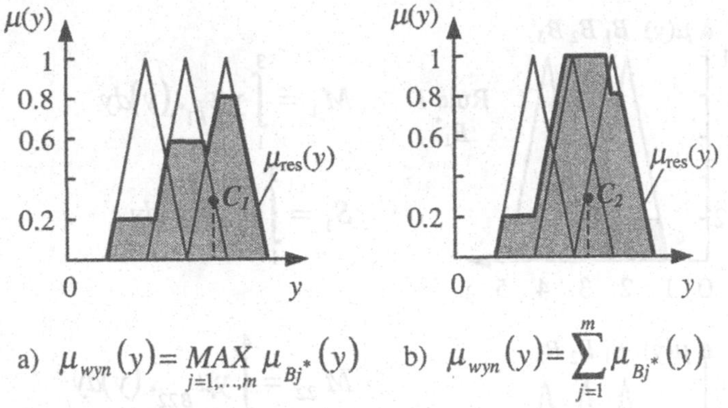

Fig. 5.42. Comparison of the results of accumulation of the component sets B_j^* by the use of the *MAX* and *SUM* operators

then the formula (5.43) can be written in the form of the quotient of the sum of the moments M_j and the sum of the areas S_j, (5.46).

$$y^* = \frac{\sum_{j=1}^{m} M_j}{\sum_{j=1}^{m} S_j} \tag{5.46}$$

This method of defuzzification is illustrated in Fig. 5.43. The method of calculating the moments M_j and areas S_j of fuzzy sets is shown in Fig. 5.44.

The **advantages** of the CS method

- Decreased cost of calculation in comparison with the CG method.
- The participation of all rules in the reasoning process, which has an advantageous influence in many fuzzy models and controllers.

The remaining advantages and disadvantages are the same as in the case of the CG method.

When the formulas (5.41) and (5.42) are used, the Center of Sums method (CS) becomes in reality the Center of Gravity method (CG) combined with the inference of the *SUM-MIN* type, where *SUM* is the unlimited (arithmetic) sum operator.

In the case of discrete membership functions the result of defuzzification y^* is calculated from the formula (5.47).

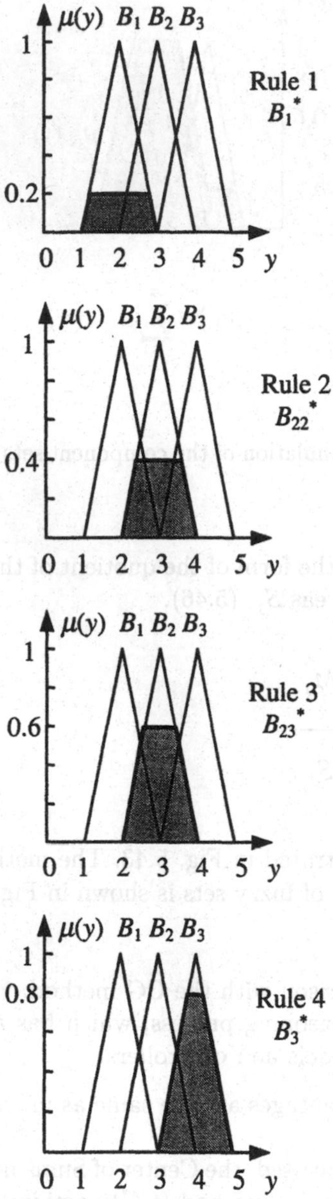

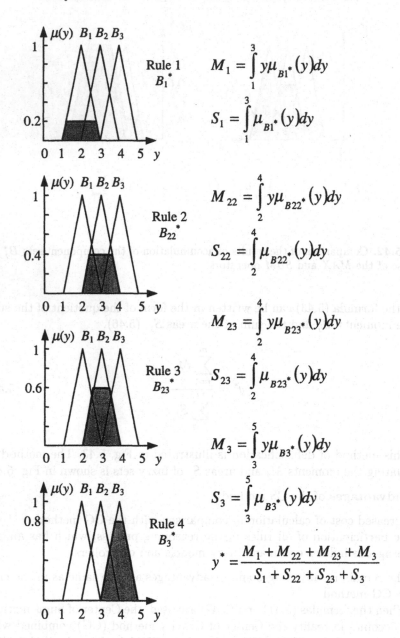

$$M_1 = \int_1^3 y\mu_{B1^*}(y)dy$$

$$S_1 = \int_1^3 \mu_{B1^*}(y)dy$$

$$M_{22} = \int_2^4 y\mu_{B22^*}(y)dy$$

$$S_{22} = \int_2^4 \mu_{B22^*}(y)dy$$

$$M_{23} = \int_2^4 y\mu_{B23^*}(y)dy$$

$$S_{23} = \int_2^4 \mu_{B23^*}(y)dy$$

$$M_3 = \int_3^5 y\mu_{B3^*}(y)dy$$

$$S_3 = \int_3^5 \mu_{B3^*}(y)dy$$

$$y^* = \frac{M_1 + M_{22} + M_{23} + M_3}{S_1 + S_{22} + S_{23} + S_3}$$

Fig. 5.43. The illustration of defuzzification using the Center of Sums method (CS)

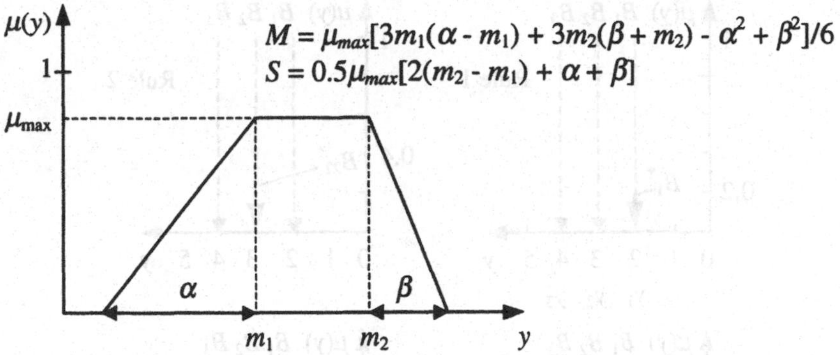

Fig. 5.44. The method of calculating the moment M and area S of a fuzzy set in the form of a trapezoid, triangle ($m_1 = m_2$) and rectangle ($\beta = \alpha = 0$)

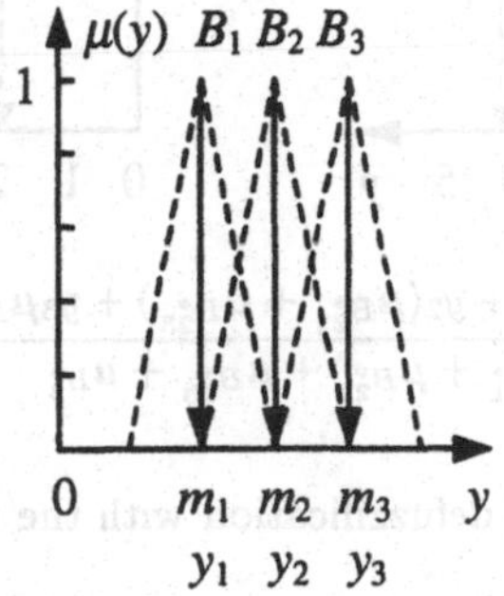

Fig. 5.45. The replacement of the fuzzy sets B_j by one-element sets (singletons)

$$y^* = \frac{\sum_{i=1}^{l} y_i \sum_{j=1}^{m} \mu_{B_j^*}(y_i)}{\sum_{i=1}^{l} \sum_{j=1}^{m} \mu_{B_j^*}(y_i)}, \tag{5.47}$$

where: l – the number of elements of the discrete universe of discourse Y,
m – the number of rules of a fuzzy model.

The Height method

The Height method (H) is the simplified discrete version of the Center of Sums method (CS). Each fuzzy set $B_j(y)$ of the model output is replaced here by a singleton (one-element set) placed at the modal value $y_j = m_j$ of this set, Fig. 5.45. Therefore this method is also called **Singleton method**.

As a result of inference in the particular rules the singletons are activated identically to all other types of fuzzy sets. To calculate the output of a model

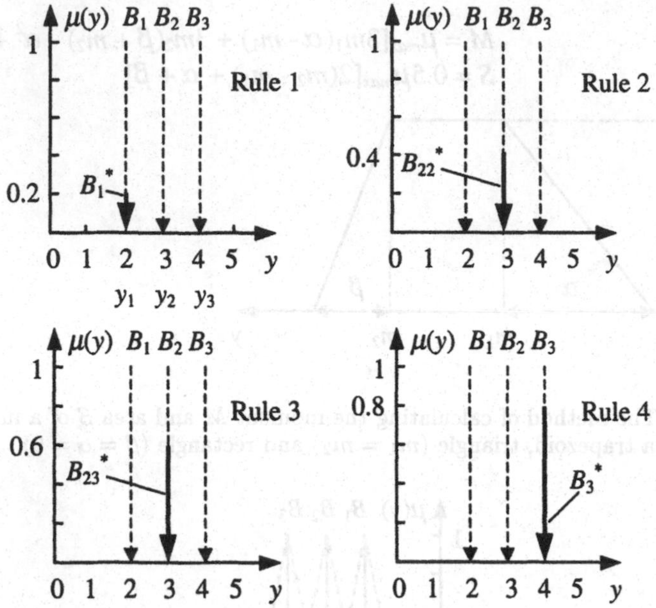

$$y^* = \frac{y_1\mu_{B_1^*} + y_2(\mu_{B_{22}^*} + \mu_{B_{23}^*}) + y_3\mu_{B_3^*}}{\mu_{B_1^*} + \mu_{B_{22}^*} + \mu_{B_{23}^*} + \mu_{B_3^*}} = 3.667$$

Fig. 5.46. The example of defuzzification with the use of the Height method

y^* (the result of defuzzification) we use the CS method. Fig. 5.46 shows the Height method applied to the example of defuzzification from Fig. 5.43, (the rule base (5.39)).

The result of defuzzification with the use of the Height method is calculated according to the formula (5.48):

$$y^* = \frac{\sum_{j=1}^{m} y_j \mu_{B_j^*}(y)}{\sum_{j=1}^{m} \mu_{B_j^*}(y)}, \tag{5.48}$$

where: m – the number of rules.

The **advantages** of the Height method

- Considerable decrease in calculation costs in comparison with the CG and CS methods.
- The length of the supports of the output sets B_j has no influence on the result of defuzzification y^*.

- The shape of the membership function $\mu_{B_j}(y)$ has no influence on defuzzification. (This feature can be a disadvantage in several problems).
- Continuity.
- Sensitivity.

The Height method (the Singleton method) is often used in fuzzy modeling and control, first of all because of its computational simplicity but also the remaining advantages. If the sets A_{ij} of the input quantities x_i are fuzzy sets (they are not singletons as the output sets are), then the model (controller) retains its fuzzy character.

5.1.4 Example of fuzzy modeling

In order to illustrate all operations involved in the processing of information in a fuzzy model we will present the fuzzy model of a system having two inputs and one output, which realizes the known mapping (5.49).

$$\begin{aligned} y &= x_1 + x_2 \\ X_1 &= [0, 10] \\ X_2 &= [0, 10] \\ Y &= [0, 20] \end{aligned} \tag{5.49}$$

Knowledge of the real mapping realized by the system will make it possible to evaluate the accuracy of the fuzzy model. However, we have to bear in mind that in most problems of modeling, the input/output mapping realized by a real system is not known in mathematical form but in the form of numerical measurement information of the system inputs and output or from knowledge of the system gained by the operator/expert of the system in the course of observation of its behavior.

For the inputs and output of the system the membership functions represented in Fig. 5.47 and the rules (5.50) are assumed.

$$\begin{aligned} r1 &: \ IF\ (x_1 = S)\ AND\ (x_2 = S)\ THEN\ (y = S) \\ r2 &: \ IF\ (x_1 = S)\ AND\ (x_2 = L)\ THEN\ (y = M) \\ r3 &: \ IF\ (x_1 = L)\ AND\ (x_2 = S)\ THEN\ (y = M) \\ r4 &: \ IF\ (x_1 = L)\ AND\ (x_2 = L)\ THEN\ (y = L) \end{aligned} \tag{5.50}$$

Among the rules (5.50) there are two rules ($r2$ and $r3$) having the identical conclusion ($y = M$). These rules can be combined to create one rule $R2$, which allows the reduction of rules to three. As a result we obtain the rule base (5.51).

$$\begin{aligned} R1 &: IF\ (x_1 = S)\ AND\ (x_2 = S)\ THEN\ (y = S) \\ R2 &: IF\ (x_1 = S)\ AND\ (x_2 = L)\ OR \\ &\qquad (x_1 = L)\ AND\ (x_2 = S)\ THEN\ (y = M) \\ R3 &: IF\ (x_1 = L)\ AND\ (x_2 = L)\ THEN\ (y = L) \end{aligned} \tag{5.51}$$

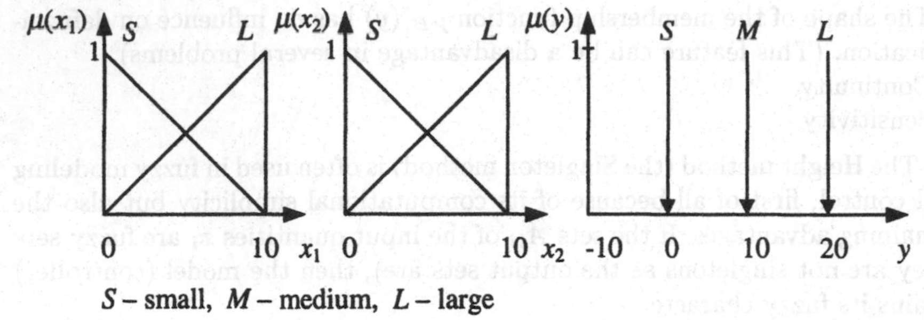

Fig. 5.47. The membership functions assumed in the fuzzy model of the system (5.49)

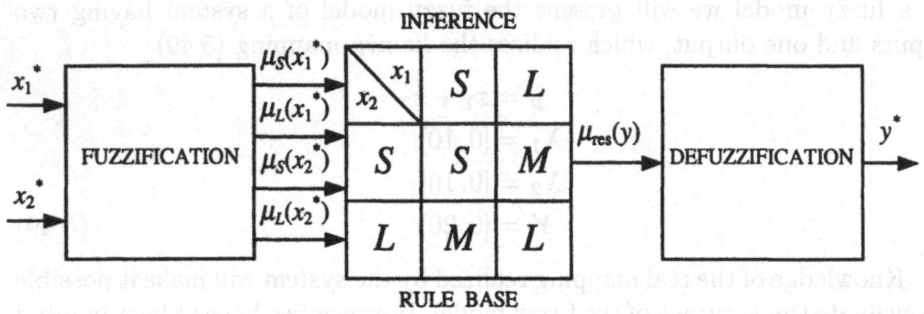

Fig. 5.48. The general scheme of the fuzzy model to be considered

The union of the rules is not absolutely necessary. Also the model (5.50) with the base consisting of 4 rules can be used. The general scheme of a fuzzy model is represented in Fig. 5.48.

The elements of the model:

- inference mechanism: *MAX-MIN*,
- aggregation of premises: *MIN* and *MAX* operators.

In order to compare various situations, the output y^* of the fuzzy model and the output y of the modeled system for the state of the inputs: $x_1^* = 2.5$, $x_2^* = 7.5$, will now be calculated.

The course of the calculations is illustrated in Fig. 5.49. The output y of the modeled system, which realizes the mapping $y = x_1 + x_2$ is equal to 10 for the input values $x_1^* = 2.5$, $x_2^* = 7.5$. The output y^* of the fuzzy model is in this case identical to the output of the modeled system. It must be noted that this case is special. Generally, fuzzy models only approximate the modeled systems and the output values are not identical but more or less similar, depending on the accuracy of the model.

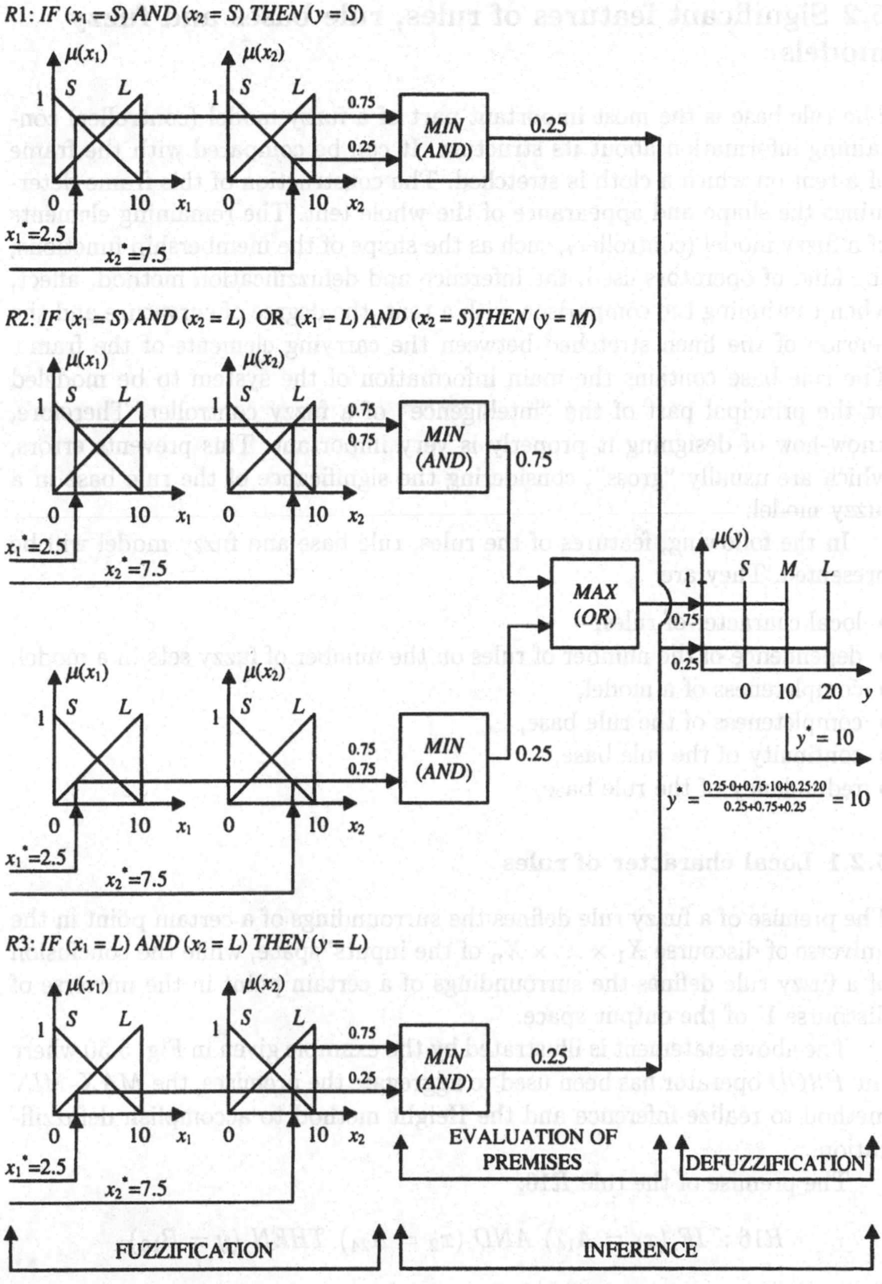

Fig. 5.49. The scheme of calculating the output y^* of the fuzzy model for the input values $x_1^* = 2.5$, $x_2^* = 7.5$

5.2 Significant features of rules, rule bases and fuzzy models

The rule base is the most important part of a fuzzy model (controller) containing information about its structure. It can be compared with the frame of a tent on which a cloth is stretched. The construction of this frame determines the shape and appearance of the whole tent. The remaining elements of a fuzzy model (controller), such as the shape of the membership functions, the kind of operators used, the inference and defuzzification method, affect, when continuing the comparison with a tent, the degree of curvature and the tension of the linen stretched between the carrying elements of the frame. The rule base contains the main information of the system to be modeled or the principal part of the "intelligence" of a fuzzy controller. Therefore, know-how of designing it properly is very important. This prevents errors, which are usually "gross", considering the significance of the rule base in a fuzzy model.

In the following, features of the rules, rule base and fuzzy model will be presented. They are:

- local character of rules,
- dependence of the number of rules on the number of fuzzy sets in a model,
- completeness of a model,
- completeness of the rule base,
- continuity of the rule base,
- redundancy of the rule base.

5.2.1 Local character of rules

The premise of a fuzzy rule defines the surroundings of a certain point in the universe of discourse $X_1 \times \ldots \times X_n$ of the inputs' space, while the conclusion of a fuzzy rule defines the surroundings of a certain point in the universe of discourse Y of the output space.

The above statement is illustrated by the example given in Fig. 5.50 where the *PROD* operator has been used to aggregate the premises, the *MAX-MIN* method to realize inference and the Height method to accomplish defuzzification.

The premise of the rule $R16$:

$$R16: \; IF \; (x_1 = A_{14}) \; AND \; (x_2 = A_{24}) \; THEN \; (y = B_{16}),$$

determines the surroundings of the point having the coordinates (a_{14}, a_{24}) and the conclusion of this rule assigns the surroundings of the point $y = b_{16}$ to the point determined by the premise. If the state of the inputs (x_1, x_2) exactly agrees with the point (a_{14}, a_{24}), then the output y of the model is exactly equal to the value b_{16}. The surface of the model in Fig. 5.50 consists of 9 segments whose corners (nodes) are determined by the individual rules.

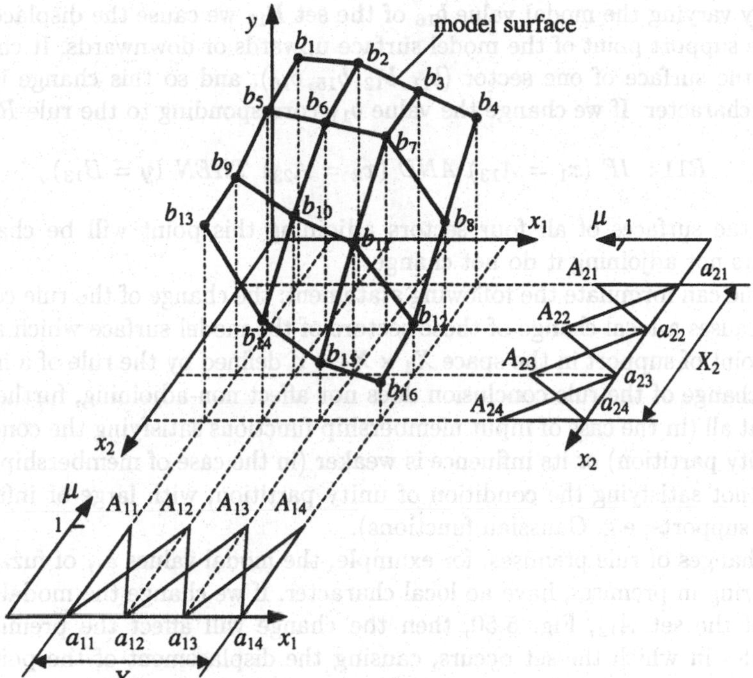

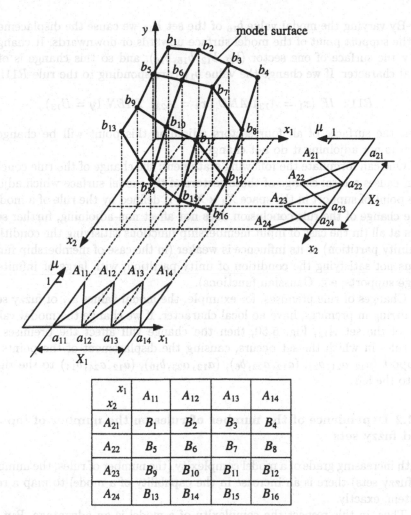

x_1 / x_2	A_{11}	A_{12}	A_{13}	A_{14}
A_{21}	B_1	B_2	B_3	B_4
A_{22}	B_5	B_6	B_7	B_8
A_{23}	B_9	B_{10}	B_{11}	B_{12}
A_{24}	B_{13}	B_{14}	B_{15}	B_{16}

Rule base

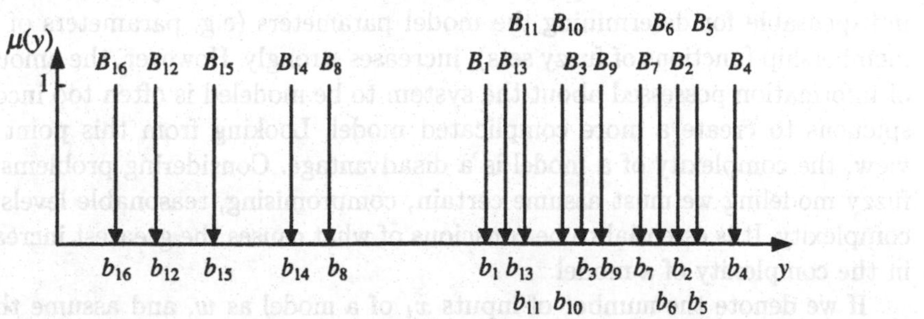

Fig. 5.50. The rule base, membership functions and surface of the mapping $\mathbf{X} = X_1 \times X_2 \to Y$

By varying the modal value b_{16} of the set B_{16} we cause the displacement of the support point of the model surface upwards or downwards. It changes only the surface of one sector $(b_{11}, b_{12}, b_{15}, b_{16})$, and so this change is of a local character. If we change the value b_{11} corresponding to the rule $R11$:

$$R11: \ IF\ (x_1 = A_{13})\ AND\ (x_2 = A_{23})\ THEN\ (y = B_{13}),$$

then the surfaces of all four sectors adjoining this point will be changed. Sectors not adjoining it do not change.

One can formulate the following statement: the change of the rule conclusion causes a local change of those sectors of the model surface which adjoin the point of support in the space $X_1 \times X_2 \times Y$ defined by the rule of a model. The change of the rule conclusion does not affect non-adjoining, further sectors at all (in the case of input membership functions satisfying the condition of unity partition) or its influence is weaker (in the case of membership functions not satisfying the condition of unity partition, with large or infinitely large supports, e.g. Gaussian functions).

Changes of rule premises, for example, the modal values a_{ij} of fuzzy sets occurring in premises, have no local character. If we change the modal value a_{12} of the set A_{12}, Fig. 5.50, then the change will affect the premises of all rules in which the set occurs, causing the displacement of the points of support (a_{12}, a_{21}, b_2), (a_{12}, a_{22}, b_6), (a_{12}, a_{23}, b_{10}), (a_{12}, a_{24}, b_{14}) to the right or to the left.

5.2.2 Dependence of the number of rules on the number of inputs and fuzzy sets

With increasing grade of a model complexity (the number of rules, the number of fuzzy sets) there is an increase in the capability of a model to map a real system, exactly.

Thus, in this respect the complexity of a model is an advantage. But at the same time the amount of information about a modeled system, which is indispensable for determining the model parameters (e.g. parameters of all membership functions of fuzzy sets), increases strongly. However, the amount of information possessed about the system to be modeled is often too inconspicuous to create a more complicated model. Looking from this point of view, the complexity of a model is a disadvantage. Considering problems of fuzzy modeling we must assume certain, compromising, reasonable levels of complexity. It is essential to be conscious of what causes the greatest increase in the complexity of a model.

If we denote the number of inputs x_i of a model as w, and assume that each input variable is characterized by the identical number of fuzzy sets z, then the number of rules r containing simple premises is determined (Kahlert 1995) by formula (5.52).

$$r = z^w \tag{5.52}$$

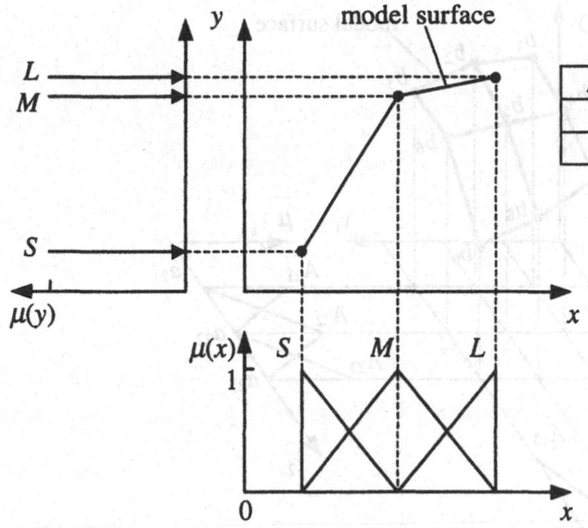

Ri	R1	R2	R3
x	S	M	L
y	S	M	L

Rule base

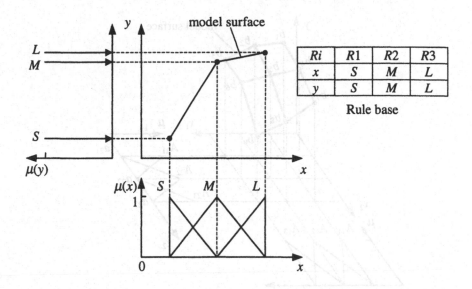

Fig. 5.51. The number of rules in a model with one input characterized by three fuzzy sets S, M, L

From this formula it results that the number of rules r depends exponentially on the number of inputs w and the number of fuzzy sets z of a model. In order to show the reader how strong this dependence is, the comparison of a model with one input and with two inputs, Figs. 5.51 – 5.53, will be presented.

When comparing Figs. 5.51 and 5.52 one can see that the increase in the number of inputs from one to two has caused an increase in the number of rules from 3 to 9. To totally define a model with one input, 6 parameters of membership functions must be defined, and a model with two inputs – 15 parameters. In Fig. 5.53 a model with two inputs is presented. Each of the inputs is characterized by four fuzzy sets.

Comparison of Figs. 5.52 and 5.53 shows that the increase in the number of fuzzy sets from 3 to 4 has caused an increase in the number of rules from 9 to 16 and the number of parameters of membership functions from 15 to 24. Assuming the same number of fuzzy sets A_{ij} for each input and assigning different fuzzy sets B_j to each rule, the number of parameters p characterizing membership functions can be calculated from the formula (5.53).

$$p = r + w \cdot z = z^w + w \cdot z \tag{5.53}$$

Table 5.14 shows the rapid increase in the number of rules and parameters of membership functions which must be determined with the increase in the number of inputs w.

The increase in the number of rules and parameters of membership functions with the increase of the number of inputs of a model is so rapid that

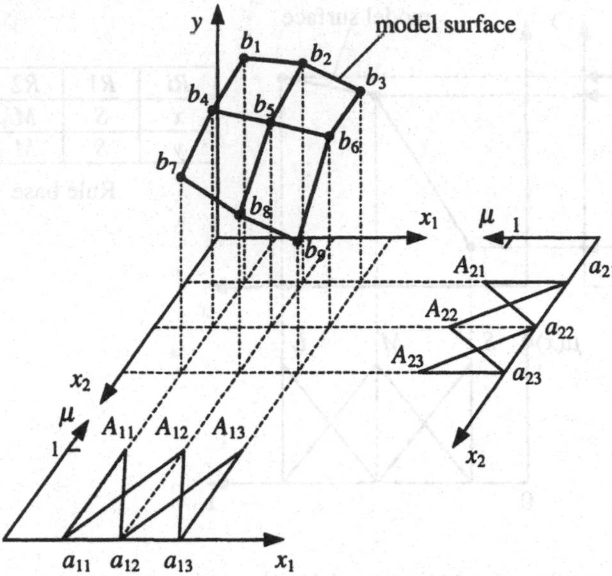

x_2 / x_1	A_{21}	A_{22}	A_{23}
A_{11}	B_1	B_2	B_3
A_{12}	B_4	B_5	B_6
A_{13}	B_7	B_8	B_9

Rule base

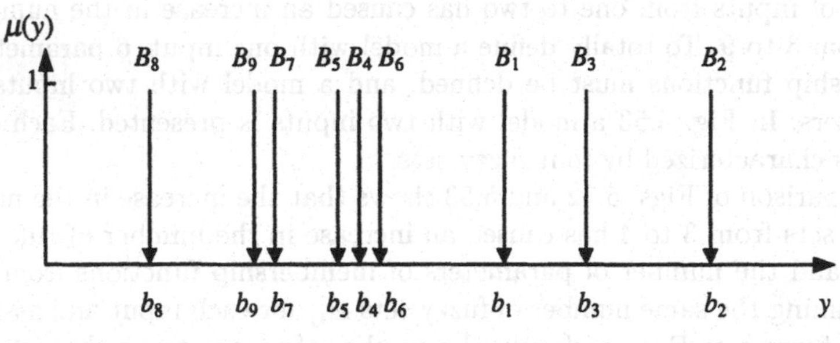

Fig. 5.52. **The number of rules in a model with two inputs characterized by three fuzzy sets**

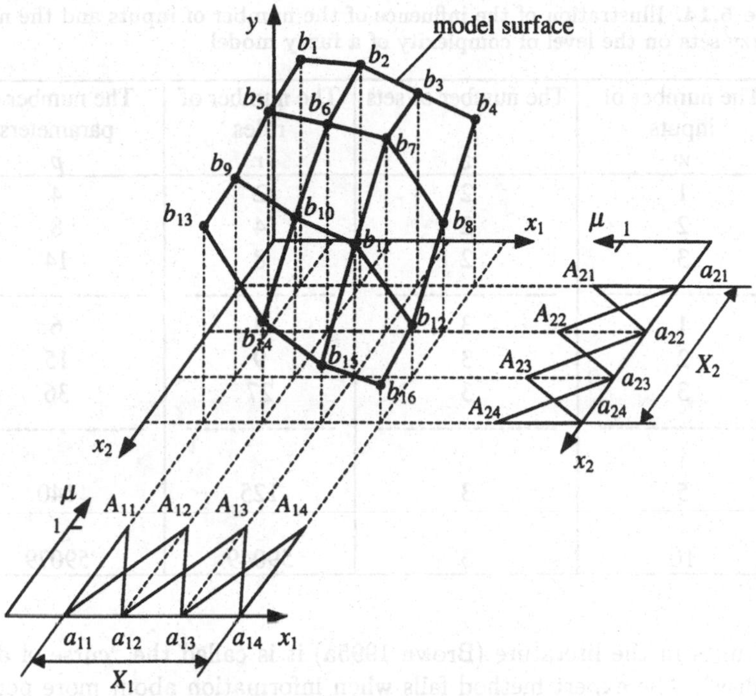

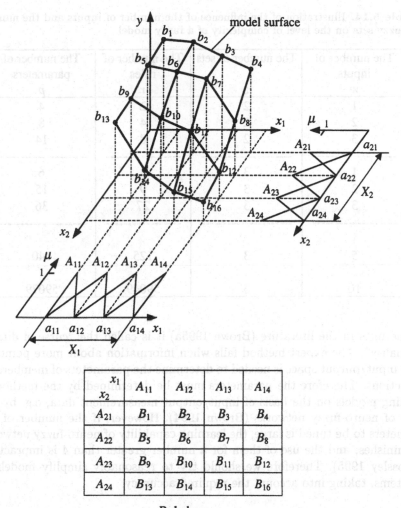

x_1 / x_2	A_{11}	A_{12}	A_{13}	A_{14}
A_{21}	B_1	B_2	B_3	B_4
A_{22}	B_5	B_6	B_7	B_8
A_{23}	B_9	B_{10}	B_{11}	B_{12}
A_{24}	B_{13}	B_{14}	B_{15}	B_{16}

Rule base

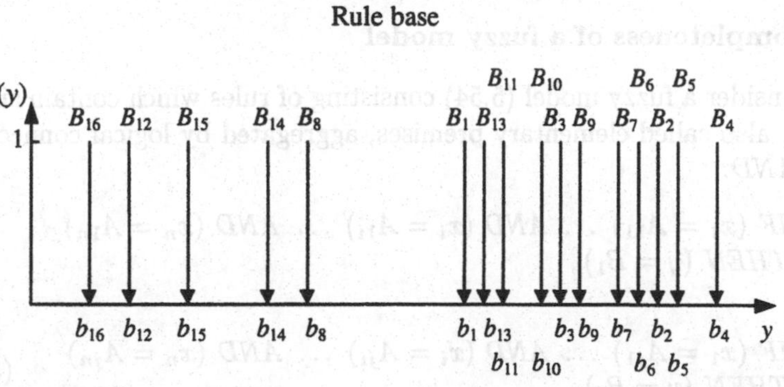

Fig. 5.53. The number of rules in a model with two inputs characterized by four fuzzy sets

Table 5.14. Illustration of the influence of the number of inputs and the number of fuzzy sets on the level of complexity of a fuzzy model

The number of inputs w	The number of sets z	The number of rules r	The number of parameters p
1	2	2	4
2	2	4	8
3	2	8	14
1	3	3	6
2	3	9	15
3	3	27	36
⋮			
5	3	125	140
⋮			
10	3	59049	59079

sometimes in the literature (Brown 1995a) it is called the "curse of dimensionality". The expert method fails when information about more points in the input/output space is needed to determine the parameters of membership functions. Therefore the parameters must be determined by the method of tuning models on the basis of input/output measurement data, e.g. by the use of neuro-fuzzy networks (Brown 1994). However, if the number of parameters to be tuned is large, the learning capability of neuro-fuzzy networks diminishes, and the use of them for a number greater than 4 is impractical (Bossley 1995). Therefore we should try to reasonably simplify models of systems, taking into account the required accuracy.

5.2.3 Completeness of a fuzzy model

Let us consider a fuzzy model (5.54) consisting of rules which contain simple premises, also called elementary premises, aggregated by logical connectives of type *AND*.

$$\begin{aligned}
R1:\ & IF\ (x_1 = A_{11})\ \dots\ AND\ (x_i = A_{1i})\ \dots\ AND\ (x_n = A_{1n}) \\
& THEN\ (y = B_1)\,, \\
& \vdots \\
Rj:\ & IF\ (x_1 = A_{j1})\ \dots\ AND\ (x_i = A_{ji})\ \dots\ AND\ (x_n = A_{jn}) \\
& THEN\ (y = B_j)\,, \\
& \vdots \\
Rm:\ & IF\ (x_1 = A_{m1})\ \dots\ AND\ (x_i = A_{mi})\ \dots\ AND\ (x_n = A_{mn}) \\
& THEN\ (y = B_m)\,.
\end{aligned} \tag{5.54}$$

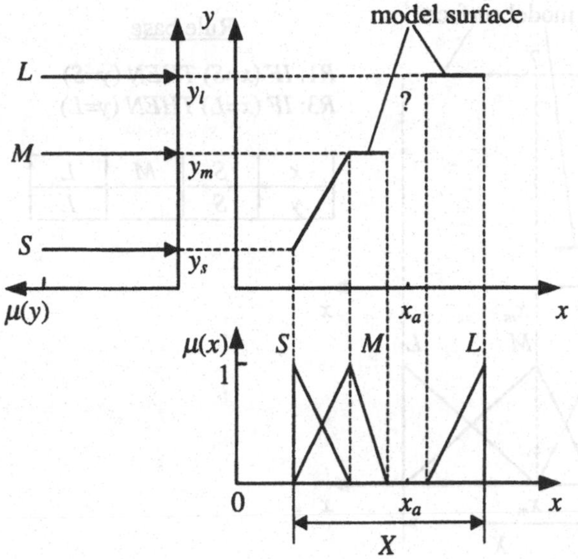

Rule base

R1: *IF (x=S) THEN (y=S)*
R2: *IF (x=M) THEN (y=M)*
R3: *IF (x=L) THEN (y=L)*

x	S	M	L
y	S	M	L

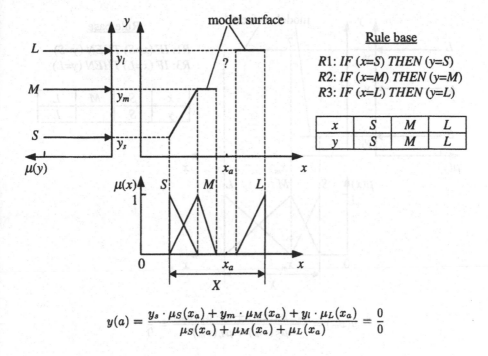

$$y(a) = \frac{y_s \cdot \mu_S(x_a) + y_m \cdot \mu_M(x_a) + y_l \cdot \mu_L(x_a)}{\mu_S(x_a) + \mu_M(x_a) + \mu_L(x_a)} = \frac{0}{0}$$

Fig. 5.54. Incompleteness of a fuzzy model in the case of incomplete fuzzy partition of the universe of discourse X of the inputs' space

The universe of discourse $\mathbf{X}$ of the numerical space of the inputs is determined by the Cartesian product of the universes of discourse X_i, $i = 1, \ldots, n$, of the particular inputs: $\mathbf{X} = X_1 \times X_2 \times \ldots \times X_n$. The universe of discourse of the numerical output space is denoted by the symbol Y.

Definition 5.2.3.1
A fuzzy model is complete, if it can assign a certain state of the output y^* to any and every state of the inputs $\mathbf{x}^* = (x_1^*, \ldots, x_n^*)$ of the universe of discourse $\mathbf{X}$. A **fuzzy model is incomplete** if there are such states of the inputs $\mathbf{x}^*$ to which it can not assign any state of the output y^*.

The possibility (capability) of creating incomplete fuzzy models is evidenced by the models represented in Fig. 5.54 and 5.55. The completeness of a set should not be mistaken for its accuracy. A complete model can be inexact, but the condition for achieving high accuracy of a model is its completeness.

As shown in Fig. 5.54 the reason for the incompleteness of a model can be incomplete fuzzy partition of the universe of discourse X of the input space.

Definition 5.2.3.2
The fuzzy **partition** of the universe of discourse X_i of the variable x_i is **complete** if the following relationship is satisfied:

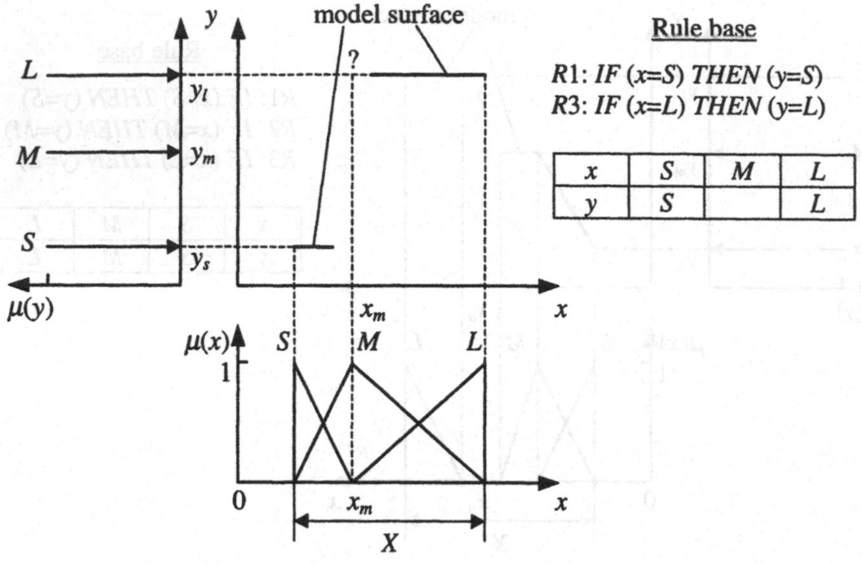

x	S	M	L
y	S		L

$$y(x_m) = \frac{y_s \cdot \mu_S(x_m) + y_l \cdot \mu_L(x_m)}{\mu_S(x_m) + \mu_L(x_m)} = \frac{0}{0}$$

Fig. 5.55. Incompleteness of a fuzzy model when the rule base is incomplete

$$\sum_{j=1}^{m} \mu_{A_{ji}}(x_i^*) > 0, \quad x_i^* \in X_i,$$

where: m – the number of fuzzy sets A_{ji} assumed for the variable x_i.

The Definition 5.2.3.2 says that each of the values x_i^* of the variable x_i contained in the universe of discourse X_i belongs to at least one fuzzy set A_{ji}. The incomplete fuzzy partition of the universes of discourse, such as in Fig. 5.54, is not an abstract thing and can be encountered in scientific publications. It occurs in incorrectly designed self-learning fuzzy models. In the course of learning, parameters of membership functions are changed, the membership functions are displaced and narrowed or widened. If no appropriate preventative measures have been introduced, then an interval can arise which is not covered by any fuzzy set A_{ji}.

The model represented in Fig. 5.55 is incomplete because it can not determine the output y for the input state $x = x_m$. It is caused by the incompleteness of the rule base. The rules contain in their premises the linguistic state of the input ($x = S$, $x = L$). The conclusions inform us which linguistic state of the output is assigned to the linguistic input state mentioned in the premises. The rule base from Fig. 5.55 does not contain any rule informing us about the output state y of the model for the input $x = M$. If the input

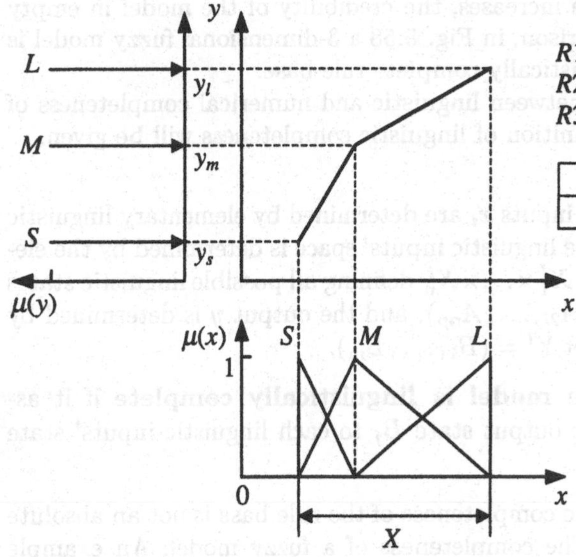

Rule base

*R*1: *IF* (*x*=*S*) *THEN* (*y*=*S*)
*R*2: *IF* (*x*=*M*) *THEN* (*y*=*M*)
*R*3: *IF* (*x*=*L*) *THEN* (*y*=*L*)

x	*S*	*M*	*L*
y	*S*	*M*	*L*

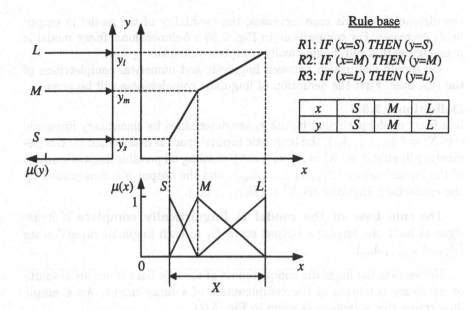

Fig. 5.56. The fuzzy model with complete partition of the universe of discourse X of the input space and with a complete linguistic rule base

state x is exactly medium ($\mu_M(x) = 1$), which is the case when $x = x_m$, then none of the rules $R1$, $R2$, is activated and the output of the model can not be calculated because the result of defuzzification is the indeterminate symbol 0/0.

For comparison, in Fig. 5.56 the surface of the fuzzy model from Fig. 5.55 is shown, whose rule base has been completed and both the rule base and the partition of the universe of discourse of the input set X are complete.

The complete fuzzy model from Fig. 5.56 is a more exact model of a real system than the incomplete models from Fig. 5.54 and 5.55. It can easily be seen that incomplete models differ significantly from a complete model. In Fig. 5.57 a 3-dimensional model with the linguistically incomplete rule base is represented. The sectors of high accuracy are marked with continuous lines, whereas the sectors of low accuracy or low reliability of calculations with broken lines. The accuracy of a model decreases as the distance from the points of support b_i of a model determined by the well-known rules given in the rule base becomes larger and larger. This model is not capable of calculating the output value y for the inputs' states (a_{1i}, a_{2j}) given in Fig. 5.57 and for intermediate states lying at the boundaries of the universe of discourse $\mathbf{X} = X_1 \times X_2$.

Some missing rules, e.g. missing information about which fuzzy set of the output B_k corresponds to the linguistic inputs' state determined by the logical product $A_{1i} \wedge A_{2j}$ (empty fields of the rule base) means that the model is credible only in the zone defined by the given rules, Fig. 5.57, whereas as the distance from this zone increases, the credibility of the model in empty fields decreases. For comparison, in Fig. 5.58 a 3-dimensional fuzzy model is presented with a full, linguistically complete rule base.

We should distinguish between linguistic and numerical completeness of the rule base. First the definition of linguistic completeness will be given.

Definition 5.2.3.3

In a fuzzy model particular inputs x_i are determined by elementary linguistic sets $X_i^l = (A_{i1}, \ldots, A_{ir})$, the linguistic inputs' space is determined by the elementary linguistic set $\mathbf{X}^l = X_1^l \times \ldots \times X_n^l$ defining all possible linguistic states of the inputs' vector $(A_{1k}, A_{2l}, \ldots, A_{np})$, and the output y is determined by the elementary linguistic set $Y^l = (B_1, \ldots, B_m)$.

The **rule base of the model is linguistically complete** if it assigns at least one linguistic output state B_j to each linguistic inputs' state $(A_{1k}, A_{2l}, \ldots, A_{np})$.

Notice that the linguistic completeness of the rule base is not an absolute or necessary condition of the completeness of a fuzzy model. An example illustrating this statement is given in Fig. 5.59.

In the case shown in Fig. 5.59 the completeness of the fuzzy model has been achieved in spite of the incompleteness of the rule base owing to the appropriate shape of the membership functions $\mu_S(x)$, $\mu_L(x)$. Therefore this model can calculate the output value y for each input value x, also for x_m, which is impossible in the example from Fig. 5.55.

In the literature there are different definitions of completeness of rule bases, but they relate to the completeness that can be determined as a numerical completeness (Kahlert 1995; Driankov 1993,1996). In what follows the definition of numerical completeness based on Kahlert's definition of completeness (Kahlert 1995) will be given (without specifying whether numerical or linguistic completeness).

Definition 5.2.3.4

A numerically complete rule base is the base in which each of the crisp numerical inputs' states $(x_1^*, \ldots, x_N^*)$ activates at least one rule (the conclusion of this rule).

Because the activation of "one rule at least" allows for calculating the output of a model, this definition agrees in practice with the Definition 5.2.3.1 of a complete fuzzy model. According to this definition the linguistically complete rule base can be numerically incomplete even if the fuzzy partition of the universes of discourse of the inputs' space is incomplete, Fig. 5.54. But linguistically incomplete rule bases can be numerically complete if membership functions with adequately wide supports have been assumed, Fig. 5.59.

In the literature on fuzzy systems, the concept of complete rule bases is understood to mean linguistically complete rule bases. Summarizing the subject, one can put the following questions:

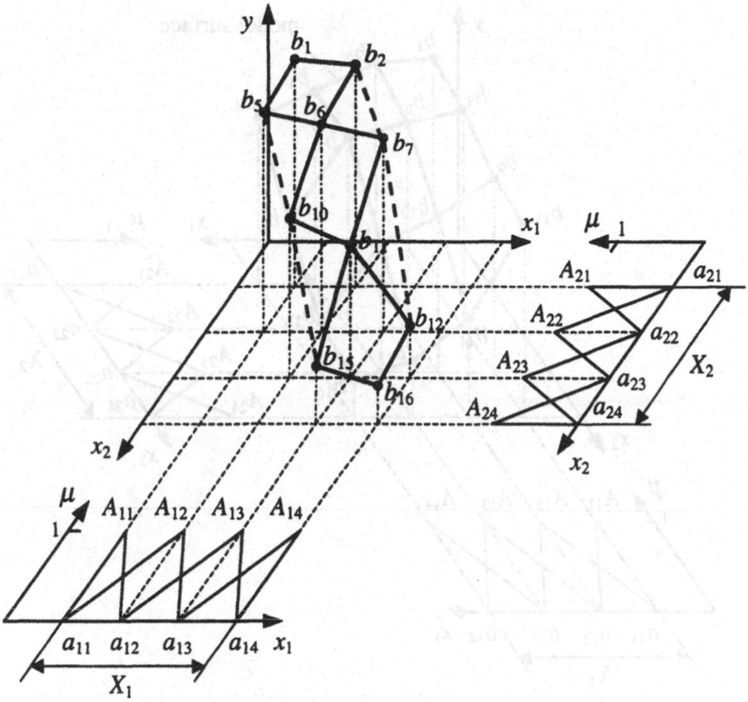

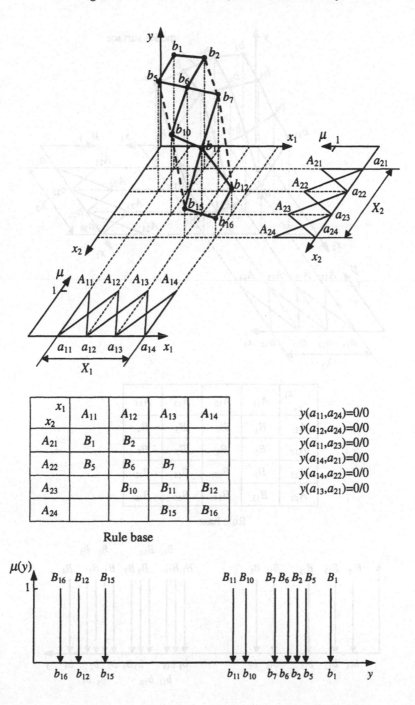

x_1 / x_2	A_{11}	A_{12}	A_{13}	A_{14}
A_{21}	B_1	B_2		
A_{22}	B_5	B_6	B_7	
A_{23}		B_{10}	B_{11}	B_{12}
A_{24}			B_{15}	B_{16}

Rule base

$y(a_{11},a_{24})=0/0$
$y(a_{12},a_{24})=0/0$
$y(a_{11},a_{23})=0/0$
$y(a_{14},a_{21})=0/0$
$y(a_{14},a_{22})=0/0$
$y(a_{13},a_{21})=0/0$

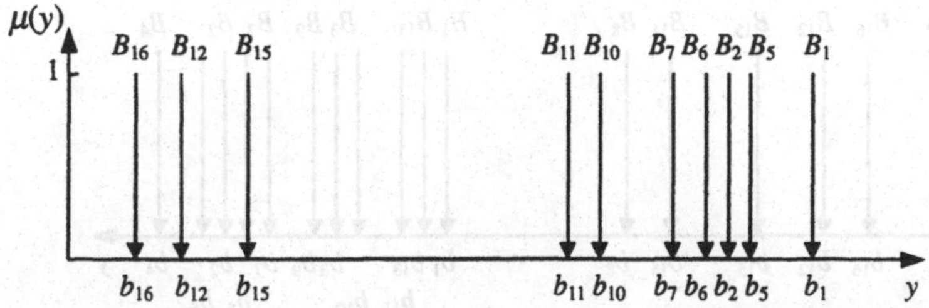

Fig. 5.57. The fuzzy model with the linguistically incomplete rule base and complete fuzzy partition of the universe of discourse $\mathbf{X} = X_1 \times X_2$ of the inputs' space

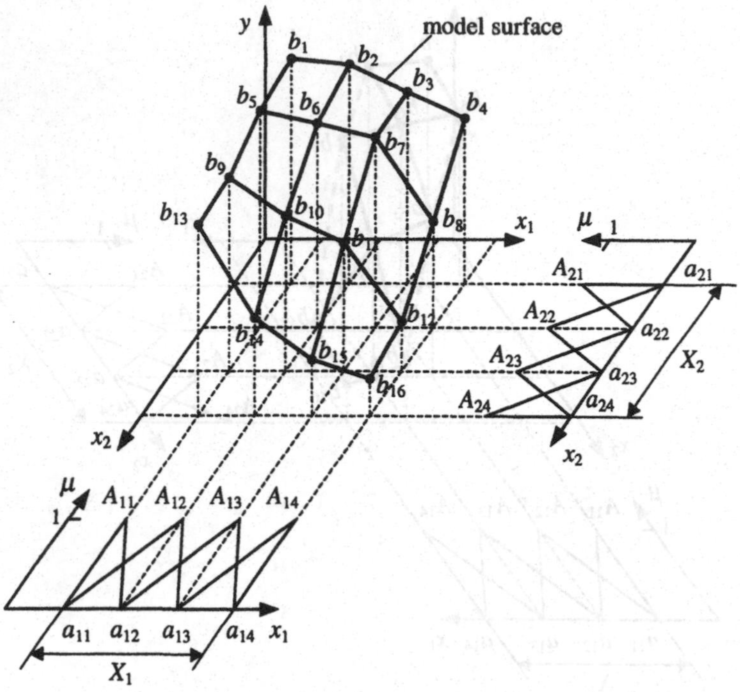

x_2 \ x_1	A_{11}	A_{12}	A_{13}	A_{14}
A_{21}	B_1	B_2	B_3	B_4
A_{22}	B_5	B_6	B_7	B_8
A_{23}	B_9	B_{10}	B_{11}	B_{12}
A_{24}	B_{13}	B_{14}	B_{15}	B_{16}

Rule base

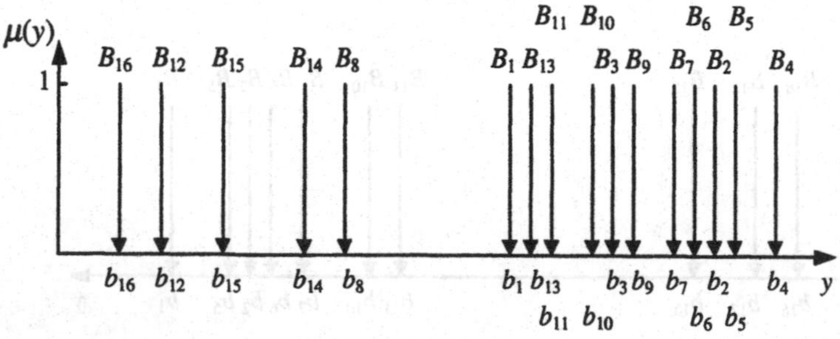

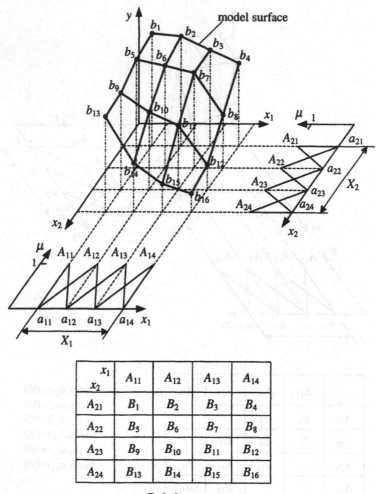

Fig. 5.58. The 3-dimensional fuzzy model with the linguistically complete rule base and with complete partition of the universes of discourse of the inputs' space X_1 and X_2

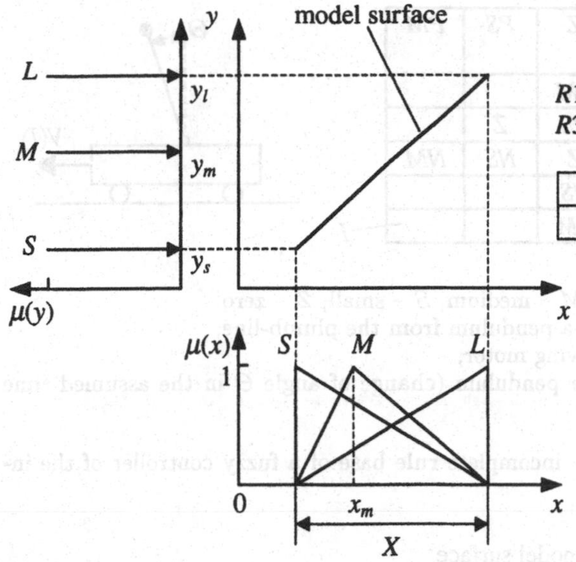

Rule base

R1: IF (x=S) THEN (y=S)
R3: IF (x=L) THEN (y=L)

x	*S*	*M*	*L*
y	*S*		*L*

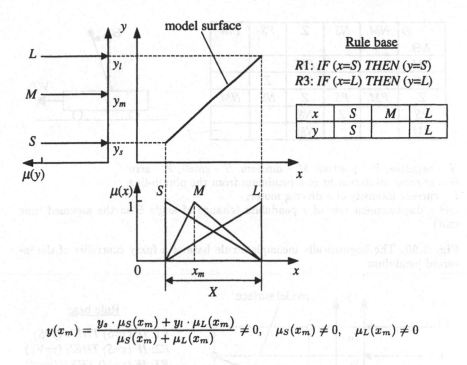

$$y(x_m) = \frac{y_s \cdot \mu_S(x_m) + y_l \cdot \mu_L(x_m)}{\mu_S(x_m) + \mu_L(x_m)} \neq 0, \quad \mu_S(x_m) \neq 0, \quad \mu_L(x_m) \neq 0$$

Fig. 5.59. The complete fuzzy model with a linguistically incomplete rule base

1. Must a fuzzy model be complete?
2. Must the rule base be linguistically complete?
3. Must the base rule be numerically complete?

The answers to questions 1 and 3 are the same because the substance of the completeness of a model agrees with the substance of the numerical completeness of the rule base. A fuzzy model can be incomplete when the places of its discontinuity lie in the zones of the input/output space where the fuzzy model (controller) does not work. Similarly the rule base need not be linguistically complete if the input/output states of a model not mentioned in the rules do not occur in practice in its operation. For instance, the fuzzy controller (Kosko 1992) with a linguistically incomplete base of knowledge, Fig. 5.60, stabilizes the inverted pendulum in the vertical position.

However, the use of incomplete models is dangerous. Therefore, before the implementation of such a model (controller) it has to be examined carefully whether it will come into non-defined states under real conditions.

ΔΘ \ Θ	NM	NS	Z	PS	PM
NM			PM		
NS			PS	Z	
Z	PM	PS	Z	NS	NM
PS		Z	NS		
PM			NM		

I

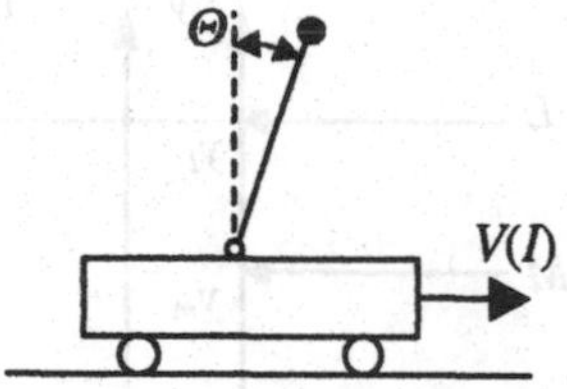

N – negative, P – positive, M – medium, S – small, Z – zero,
Θ – angular displacement of a pendulum from the plumb-line,
I – current intensity of a driving motor,
$\Delta\Theta$ – displacement rate of a pendulum (change of angle Θ in the assumed time unit)

Fig. 5.60. The linguistically incomplete rule base of a fuzzy controller of the inverted pendulum

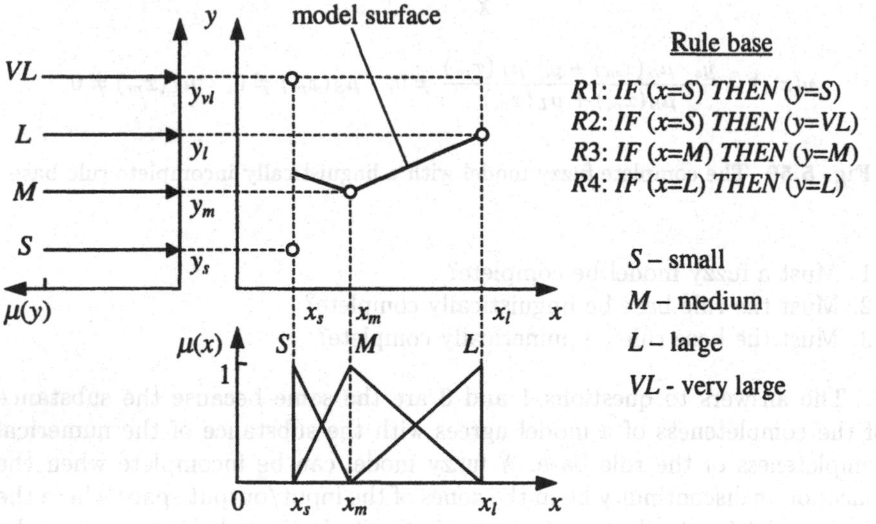

Fig. 5.61. The rule-base with "very" inconsistent rules $R1$ and $R2$ as well as the obtained surface of the $X \to Y$ mapping of the model

5.2.4 Consistency of the rule base

Definition 5.2.4.1
The rule base is consistent (conformable) if it does not contain incompatible rules, i.e. the rules having the same premises but different conclusions.

The example of a model with inconsistent rules is represented in Fig. 5.61.

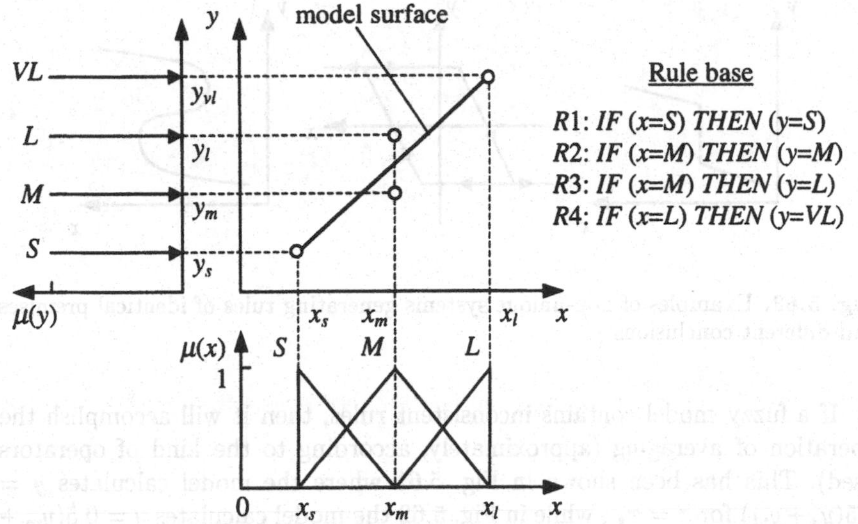

Fig. 5.62. The rule base containing the "weakly" inconsistent rules $R2$ and $R3$ as well as the obtained surface of the mapping $X \to Y$ of the model

In the model in Fig. 5.61 the rules $R1$ and $R2$ contain the identical premises (x = Small) but different conclusions (y = Small) and (y = Very Large). The conclusions of these rules are diametrically different (Small and Very Large). Thus one can state that the inconsistency of the rules is "strong".

In Fig. 5.62 a model with inconsistent rules is also given. But in this case one can state "weak" inconsistency of the rules because the modal values of the inconsistent conclusions M and L lie near each other.

As can be seen, inconsistency of rules can be more or less according to the relative positions of the modal values of the conclusions of inconsistent rules. What is the reason for the existence of inconsistent rules in the base?

The first reason can be a mistake made while generating rules, in particular when there is a large number of them. The second reason can be the non-uniqueness of the system to be modeled. Such a system generates non-unique input/output measurement information. In this connection one input state x^* can correspond to different output states. Examples are given in Fig. 5.63.

Then the "inconsistent" rules are not really inconsistent because they give true information about the system. In order to avoid the occurrence of inconsistent rules in the case of a non-unique system (e.g. hysteresis) we must eliminate its non-uniqueness arising when the model has too few inputs. For example, the model of hysteresis shown in Fig. 5.63 becomes unique if it is represented in the 3-dimensional coordinate system with the inputs $x(k)$, $y(k-1)$ and output $y(k)$ (Piegat 1995c).

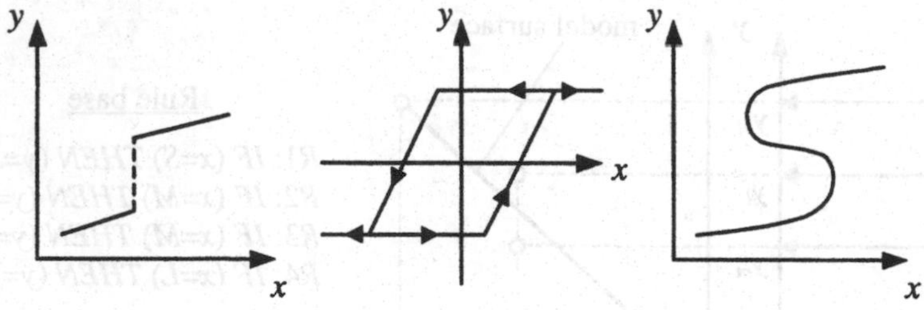

Fig. 5.63. Examples of non-unique systems generating rules of identical premises and different conclusions

If a fuzzy model contains inconsistent rules, then it will accomplish the operation of averaging (approximately, according to the kind of operators used). This has been shown in Fig. 5.61 where the model calculates $y = 0.5(y_s + y_{vl})$ for $x = x_s$, while in Fig. 5.62 the model calculates $y = 0.5(y_m + y_l)$ for $x = x_m$.

As has been mentioned earlier, the inconsistency of rules can be more or less. An interesting definition enabling us to evaluate the level of this inconsistency is given in (Leichtfried 1995). In this definition a fuzzy model containing the rules of the following type is considered:

$$R1: \; IF\ (x_1 = A_{j1})\ \ldots\ AND\ (x_i = A_{ji})\ \ldots AND\ (x_n = A_{jn})$$
$$THEN\ (y = B_1),$$

where: $x_1, \ldots, x_n$ – input values,
$A_{1i}, \ldots, A_{mi}$ – fuzzy sets of the inputs x_i,
$B_1, \ldots, B_m$ – fuzzy sets of the output y,
X_i – the universe of discourse of the variable x_i,
$\mathbf{X} = X_1 \times \ldots \times X_n$ – the universe of discourse of the inputs' space of the model.

If the *PROD* operator is used for the *AND* operation, then the degree of satisfying the premise of the rule *Rj* can be calculated for the inputs' state $\mathbf{x}^* = (x_1^*, \ldots, x_n^*)$ from the formula (5.55).

$$\mu^j(\mathbf{x}^*) = \mu_{A_{j1}}(x_1^*) \cdot \ldots \cdot \mu_{A_{jn}}(x_n^*), \quad \forall \mathbf{x}^* \in \mathbf{X}. \tag{5.55}$$

Definition 5.2.4.2
The base of the m rules Rj, $j = 1, \ldots, m$, is complete and consistent when the relationship (5.56) is satisfied:

$$\sum_{j=1}^{m} \mu^j(\mathbf{x}^*) = 1, \quad \forall \mathbf{x}^* \in \mathbf{X}. \tag{5.56}$$

The above relationship means that the sum of the degrees of satisfying the premises of all rules for any inputs' state $\mathbf{x}^* \in \mathbf{X}$ is equal to 1. If the sum of the degrees of satisfying the premises is less than 1, then for the inputs' state $\mathbf{x}^*$ the rule-base is incomplete. If this sum is greater than 1 then there is inconsistency of the rules for the inputs' state $\mathbf{x}^*$.

The condition (5.56) of summing up the premise values to unity can be understood as the basis of the decision-making process. The rule base determines the decision $y^*(\mathbf{x}^*)$ on the basis of the degrees of satisfying the rule premises $\mu^j(\mathbf{x}^*)$. If the sum of the degrees is less than 1, then the decision $y^*(\mathbf{x}^*)$ is insufficiently grounded, and this means that knowledge of the system contained in the rule base is incomplete. If the sum of the degrees of satisfying the premises is greater than 1, then this fact testifies to the existence of an inconsistency of the rules. The situation occurring in this case is such that if the degree of satisfying of one rule is total and amounts to 1, then in the rule base there are also other rules whose premises are partly satisfied.

As a result, the conclusion (output of the model) $y^*(\mathbf{x}^*)$ is not determined only by the totally activated rule, and the information contained in this rule becomes untrue (the rule stops "telling" the truth). The example is given in Fig. 5.64.

In the example in Fig. 5.64 for $x = x_m = 20$ the premise of the rule $R2$ is satisfied totally, $\mu_S(20) = 1$. In this situation, according to the information given by this rule:

$$IF\ (x \text{ about } 20)\ THEN\ (y \text{ about } 14)\,,$$

the output y should achieve the value $y^* = y_m = 14$. However, because at $x = x_m$ the remaining rules $R1$ and $R2$ are also activated, they take part in the decision-making y^* and change the result to $y^* = 15$. The model surface does not cross the point $P(20, 14)$ indicated by the rule $R2$ but crosses another point, namely $P_1(20, 15)$.

Models in which the sum of the degrees of satisfying the rule premises is not equal to 1 can be used and are often used because this condition is difficult to realize in each case. Examples of such models can be easily found in the literature, for example (Knappe 1995; Altrock 1993; Zimmermann 1994a) and others. The models also achieve high accuracy when their parameters are adequately tuned. Their disadvantage is missing transparency. The model calculates other output values than those given in the rules.

5.2.5 Continuity of the rule base

The continuity of the rule base is determined by the Definition 5.2.5.1 (Driankov 1993,1996).

Definition 5.2.5.1

The rule base is continuous if there are in it no adjoining rules Rj, Rk with the fuzzy sets B_j, B_k of the conclusion whose product $B_j \cap B_k = \emptyset$ (i.e.

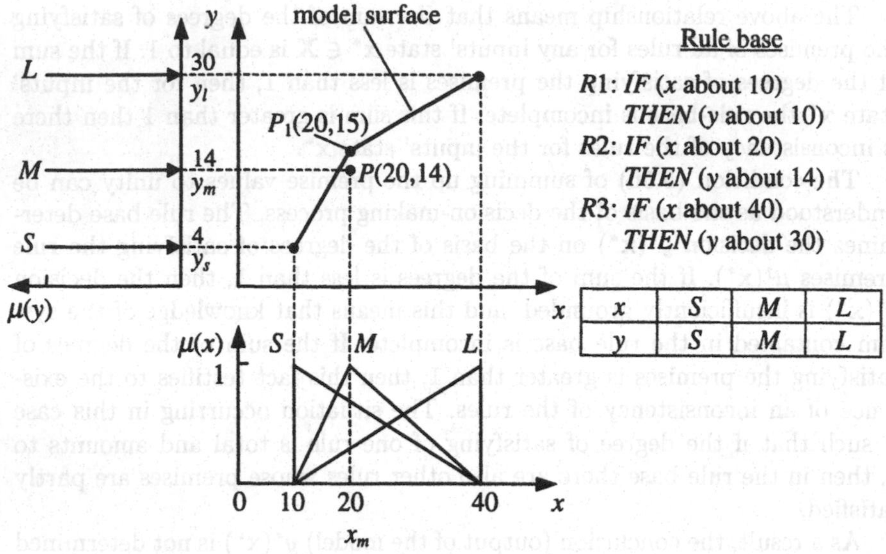

$$\mu_S = (40 - x)/30\,, \quad \mu_L = (x - 10)/30$$

$$\mu_S(x_m) = 2/3\,, \quad \mu_M(x_m) = 1\,, \quad \mu_L(x_m) = 1/3\,, \quad \sum_{j=1}^{3} \mu^j(x_m) = 2$$

$$y^*(x_m) = \frac{y_s \cdot \mu_S(x_m) + y_m \cdot \mu_M(x_m) + y_l \cdot \mu_L(x_m)}{\mu_S(x_m) + \mu_M(x_m) + \mu_L(x_m)} = \frac{4 \cdot \frac{2}{3} + 14 \cdot 1 + 40 \cdot \frac{1}{3}}{2} = 15$$

Fig. 5.64. The example of a fuzzy model in which the sum of the degrees of satisfying the premises is greater than 1 for some inputs' states

empty). For any y belonging to the universe of discourse Y of the output, the relationship (5.57) is satisfied.

$$\mu_{B_j}(y) \cdot \mu_{B_k}(y) \neq 0\,, \quad \forall y : y \in Y\,. \tag{5.57}$$

By the term "adjoining rules" the reader should understand the rules determined in the adjoining squares of the rule table, Fig. 5.65. In Fig. 5.66 a fuzzy model having a continuous rule-base is shown, and in Fig. 5.65 another model with a discontinuous rule base. In the model from Fig. 5.65 there are many adjoining sets which do not satisfy the condition of continuity (5.57). Such sets are, for example B_1, B_5, from the adjoining rules $R1$, $R2$. The comparison of the model surface with the continuous, Fig. 5.66, and discontinuous, Fig. 5.65, rule base shows that the continuity of the rule base increases the smoothing out of the model surface, while the discontinuity of the rule-base contributes to the rise of great heights (steep slopes) in the model surface. Naturally, at the continuous rule bases such regular surfaces,

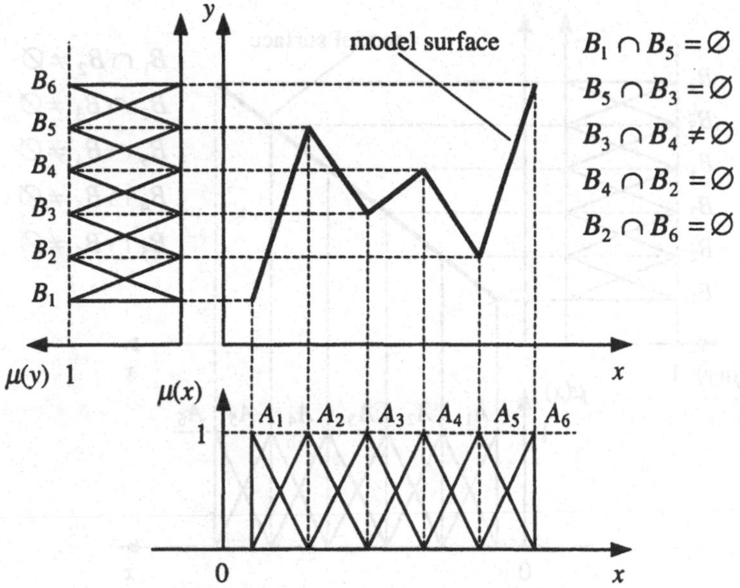

Ri	R1	R2	R3	R4	R5	R6
x	A_1	A_2	A_3	A_4	A_5	A_6
y	B_1	B_5	B_3	B_4	B_2	B_6

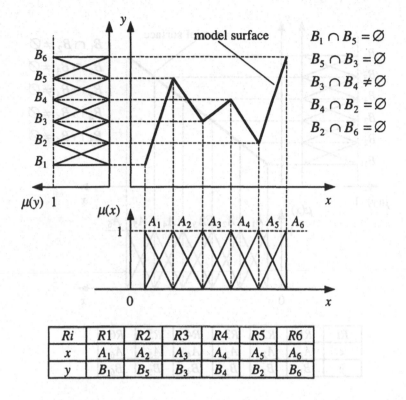

Fig. 5.65. The fuzzy model with the discontinuous rule base. Defuzzification by the use of the H (singletons) method

as in Fig. 5.66 are not always obtained. It depends on the arrangement of the modal values of the output sets B_j.

It raises the question: should the rule bases in fuzzy models (controllers) be continuous? It depends on the surface of the mapping $X \rightarrow Y$ of the system being modeled. If the surface contains big steep slopes, then it can not be exactly mapped by means of a model with the continuous rule base. If the surface of a model is smooth, then the continuous rule base will model it exactly. In the case of fuzzy controllers, most often a smooth surface of the inputs/output mapping is desired. Big steep slopes of this surface mean strong and drastic changes of the control variable of the plant. In such a case the continuity of the rule base is recommended. However, it need not concern each case because processes (plants, systems) requiring drastic changes of the controlling values, and thus the discontinuous rule base of the controller, also exist.

In the case of a model having one input x, one rule can have two adjoining rules at most, whereas in a model with two inputs x_1 and x_2 there are up to 8 adjoining rules. The number of rules increases steeply with the increase of the

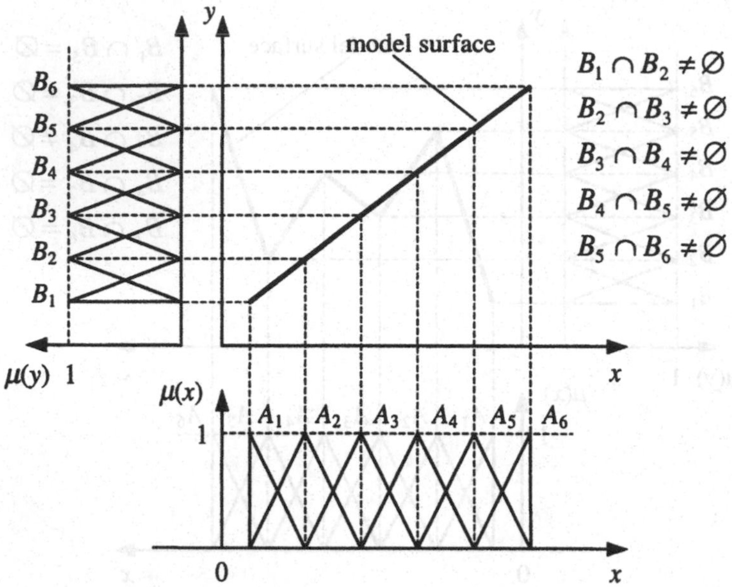

Ri	R1	R2	R3	R4	R5	R6
x	A_1	A_2	A_3	A_4	A_5	A_6
y	B_1	B_2	B_3	B_4	B_5	B_6

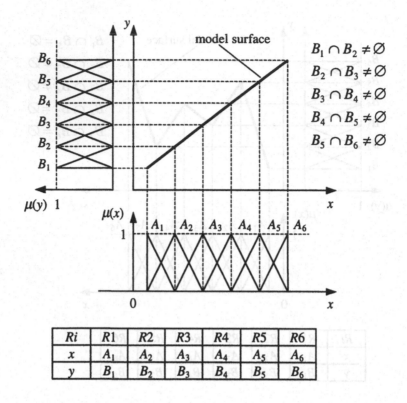

Fig. 5.66. The fuzzy model with the continuous rule base. Defuzzification by the use of the H (singleton) method

number of model inputs. In this connection the evaluation of the continuity of the rule base becomes more and more difficult because it can not be carried out visually by an inspection of the rule table.

5.2.6 Redundancy of the rule base

In fuzzy models one can meet two or more identical rules (identical premise, identical conclusion). The reasons for such a situation can be as follows:

1. the mistake of the constructor of the rule base (when there is a larger number of rules);
2. the generation, in the case of a self-organizing fuzzy model, of additional, identical rules in order to intensify their conclusions.

Case 1 is obvious, the redundant rules should be eliminated. Case 2 must be explained, see Fig. 5.67.

In the case of the model $M1$ the system surface at the point $x = x_m$ differs considerably from the surface of the real system (the points P_1 and P). Such

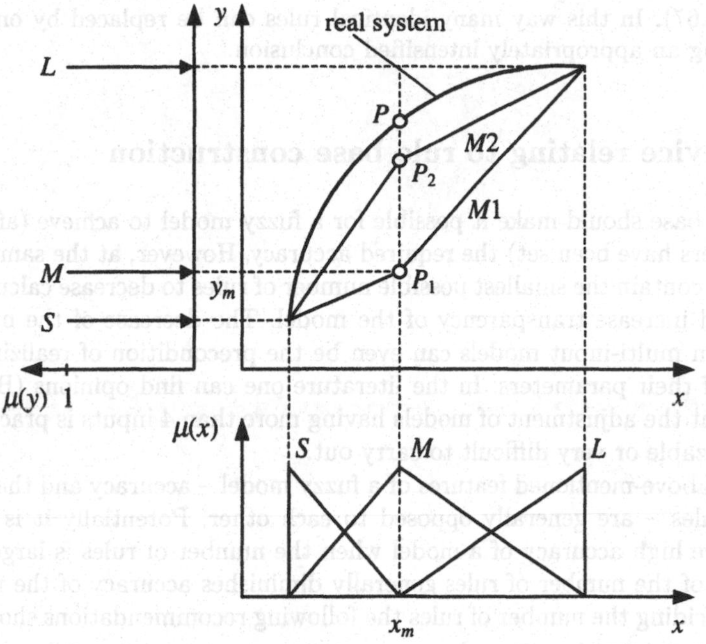

Rule base of the model $M1$			
Ri	$R1$	$R2$	$R3$
x	S	M	L
y	S	M	L

Rule base of the model $M2$				
Ri	$R1$	$R2$	$R2$	$R3$
x	S	M	M	L
y	S	M	M	L

Fig. 5.67. Comparison of the fuzzy models $M1$ and $M2$, without redundancy and with redundancy of the rule base

a big error has arisen in consequence of improperly adjusting the parameter y_m of the membership function of the output $\mu_M(y)$. In this situation the self-learning model can additionally generate the identical rule $R2$. Two identical rules $R2$ can be replaced by one rule $R2^*$ containing the logically summed conclusion according to the formula (5.58).

$$(IF\ (x = M)\ THEN\ (y = M)) \cup (IF\ (x = M)\ THEN\ (y = M)) = \\ = IF\ (x = M)\ THEN\ (y = M \cup M) \tag{5.58}$$

If the logical summing is realized by means of the MAX operator, then the resulting set $M^* = M \cup M = M$. If other operators, e.g. the SUM operator, are used, (the membership functions are summed) the resulting set M^* is equal to $M \cup M \neq M$, which causes the intensification of the resulting conclusion of the model and can diminish its error (the points P and P_2

in Fig. 5.67). In this way many identical rules can be replaced by one rule containing an appropriately intensified conclusion.

5.3 Advice relating to rule base construction

The rule base should make it possible for a fuzzy model to achieve (after its parameters have been set) the required accuracy. However, at the same time it should contain the smallest possible number of rules to decrease calculation costs and increase transparency of the model. The decrease of the number of rules in multi-input models can even be the precondition of realizing the tuning of their parameters. In the literature one can find opinions (Bossley 1995) that the adjustment of models having more than 4 inputs is practically non-realizable or very difficult to carry out.

Two above-mentioned features of a fuzzy model – accuracy and the number of rules – are generally opposed to each other. Potentially it is easier to achieve high accuracy of a model when the number of rules is large. The decrease of the number of rules generally diminishes accuracy of the model. When deciding the number of rules the following recommendations should be taken into account.

1. The number of rules increases as the partition grid of the universe of discourse $\mathbf{X}$ of the inputs' space of a model becomes denser.
2. The density of the partition grid should increase at greater creasing of the mapping surface $\mathbf{X} \rightarrow Y$ of the modeled system.
3. When the density of the partition grid (the number of rules) is constant the accuracy of a model can be increased by adequately arranging the points of support of the model surface which are defined by its rules.

In the following, explanations concerning the remarks 1 – 3 are presented. If the arrangement of the measurement points (x_1, x_2) of the system inputs is given, then the universe of discourse $\mathbf{X}$ of the inputs' space of a model can also be determined, Fig. 5.68.

After the universe of discourse $\mathbf{X}$ of the inputs' space of a model has been fixed, the density of its partition network should be chosen. If we think or know that the surface of the mapping $\mathbf{X} \rightarrow Y$ of a system is strongly non-linear and creased, Fig. 5.69, we should apply the denser partition grid. When the surface is plane (approximately linear), the partition is not necessary, Fig. 5.70.

After the density partition grid has been chosen one can start to create the rules defining the support points of the model surface. There are two fundamental methods of locating the points of support.

A. Locating in the corners of the rectangular sectors of the partition grid.
B. Locating in the center of the network sectors.

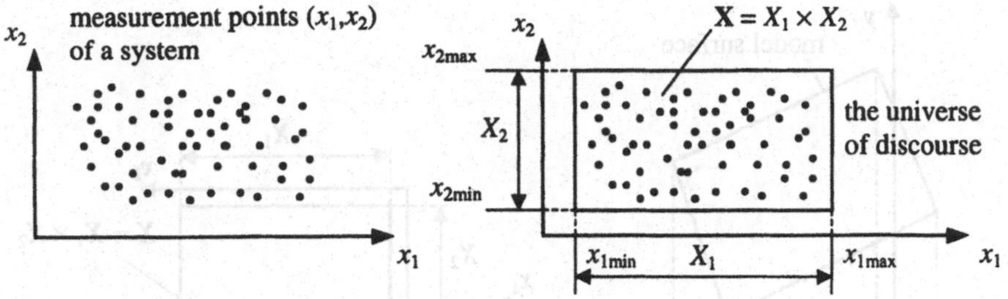

Fig. 5.68. Determination of the universe of discourse $\mathbf{X}$ of the model on the basis of the arrangement of the input measurement points (x_1, x_2) of the system to be modeled

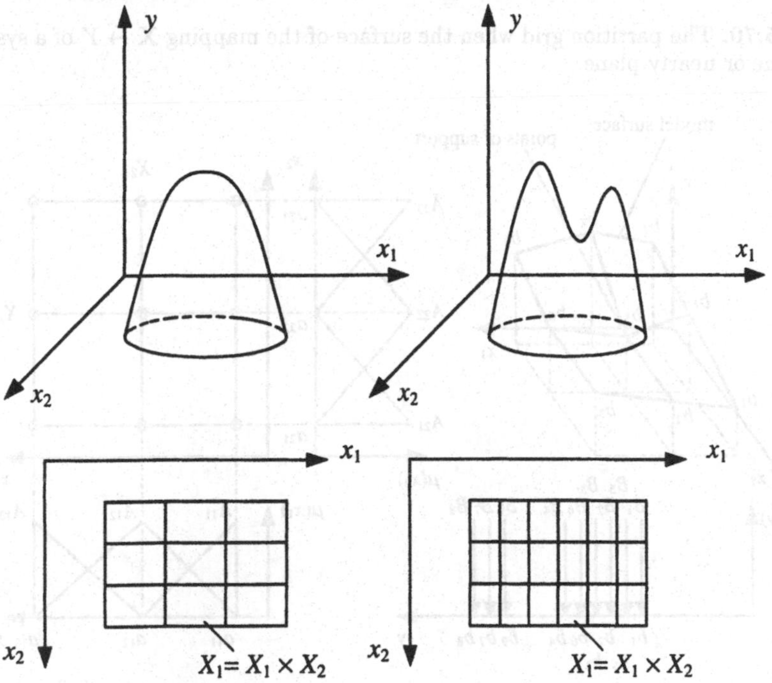

Fig. 5.69. The illustration of the situation when the density of the partition grid of the inputs' space must be increased as the surface of the mapping $\mathbf{X} \rightarrow Y$ of the modeled system becomes more creased

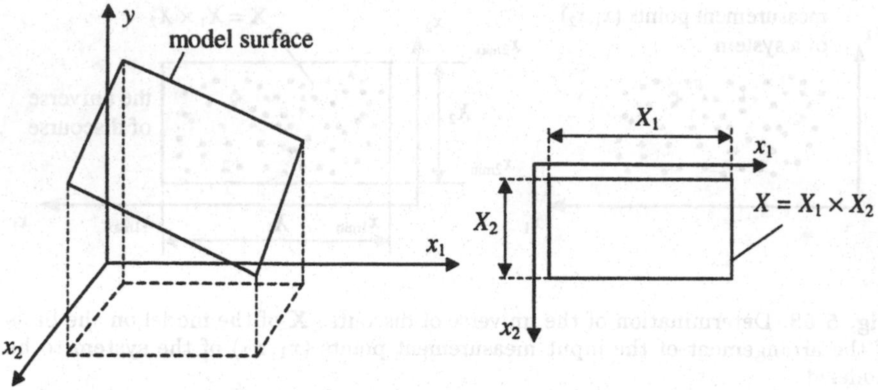

Fig. 5.70. The partition grid when the surface of the mapping $\mathbf{X} \rightarrow Y$ of a system is plane or nearly plane

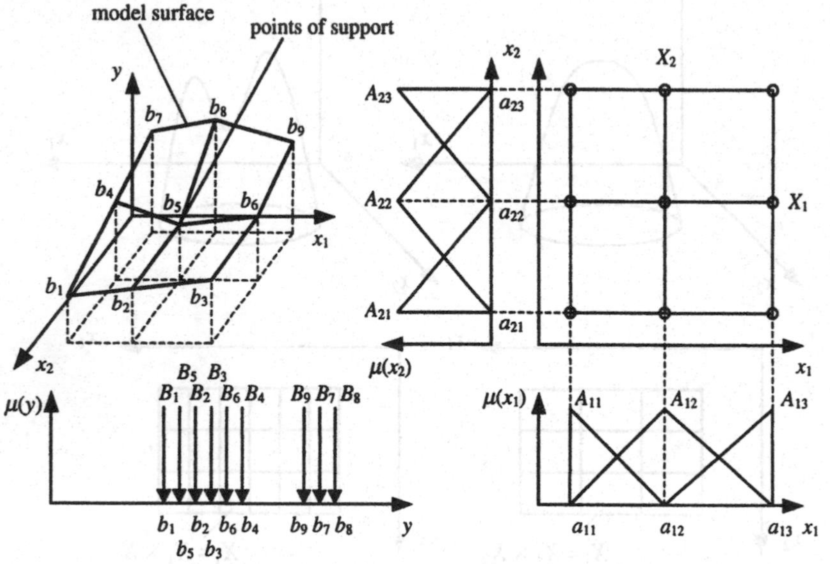

Fig. 5.71. The example of a fuzzy model with the points of support located in the corners of the partition sectors of the universe of discourse $\mathbf{X} = X_1 \times X_2$ of the inputs' space

Fig. 5.71 shows the example of the use of method A. When defining the rules for the corners of the rectangular sectors of the inputs' space the model surface under each sector is defined by four support points corresponding to its corners. In the case of the model from Fig. 5.71 the surface under the whole universe of discourse $\mathbf{X} = X_1 \times X_2$ is defined by nine rules:

$$\begin{aligned}
R1 &: IF\ (x_1 = A_{11})\ AND\ (x_2 = A_{21})\ THEN\ (y = B_1)\\
R2 &: IF\ (x_1 = A_{12})\ AND\ (x_2 = A_{21})\ THEN\ (y = B_2)\\
R3 &: IF\ (x_1 = A_{13})\ AND\ (x_2 = A_{21})\ THEN\ (y = B_3)\\
R4 &: IF\ (x_1 = A_{11})\ AND\ (x_2 = A_{22})\ THEN\ (y = B_4)\\
R5 &: IF\ (x_1 = A_{12})\ AND\ (x_2 = A_{22})\ THEN\ (y = B_5)\\
R6 &: IF\ (x_1 = A_{13})\ AND\ (x_2 = A_{22})\ THEN\ (y = B_6)\\
R7 &: IF\ (x_1 = A_{11})\ AND\ (x_2 = A_{23})\ THEN\ (y = B_7)\\
R8 &: IF\ (x_1 = A_{12})\ AND\ (x_2 = A_{23})\ THEN\ (y = B_8)\\
R9 &: IF\ (x_1 = A_{13})\ AND\ (x_2 = A_{23})\ THEN\ (y = B_9)
\end{aligned} \tag{5.59}$$

These rules inform us about the output value of the model for the input state (x_1, x_2) which corresponds exactly to the corners of the sectors. For example, for $x_1 = a_{12}$, $x_2 = a_{22}$ the output state is equal to $y = b_5$. Between the sector corners the fuzzy model carries out interpolation whose nature is dependent on the method of inference, defuzzification, and the type of membership functions.

Method B of locating the rules (points of support) in the center of the sectors is illustrated in Fig. 5.72.

By using method B of defining the rules in the centers of the partition sectors of the inputs' space, for the identical universe of discourse $\mathbf{X}$ as in Fig. 5.72, we obtain the rule base (5.60) containing only 4 rules.

$$\begin{aligned}
R1 &: IF\ (x_1 = A_{11})\ AND\ (x_2 = A_{21})\ THEN\ (y = B_3)\\
R2 &: IF\ (x_1 = A_{11})\ AND\ (x_2 = A_{22})\ THEN\ (y = B_1)\\
R3 &: IF\ (x_1 = A_{12})\ AND\ (x_2 = A_{21})\ THEN\ (y = B_4)\\
R4 &: IF\ (x_1 = A_{12})\ AND\ (x_2 = A_{22})\ THEN\ (y = B_2)
\end{aligned} \tag{5.60}$$

As shown in Figs. 5.71 and 5.72, method B allows us to create a fuzzy model with a smaller number of rules than is necessary when method A is used. The smaller number of rules means that less measurement information is needed to model the system.

The rules (5.60) inform us exactly about the output states of a model at the points corresponding (more or less) to the sector centers. Between these points the model calculates the output realizing fuzzy interpolation. Outside of the points of support the model exhibits saturation, Fig. 5.72, having the values b_i determined by the nearest support points. Its accuracy in this range with respect to the surface of the modeled system is usually

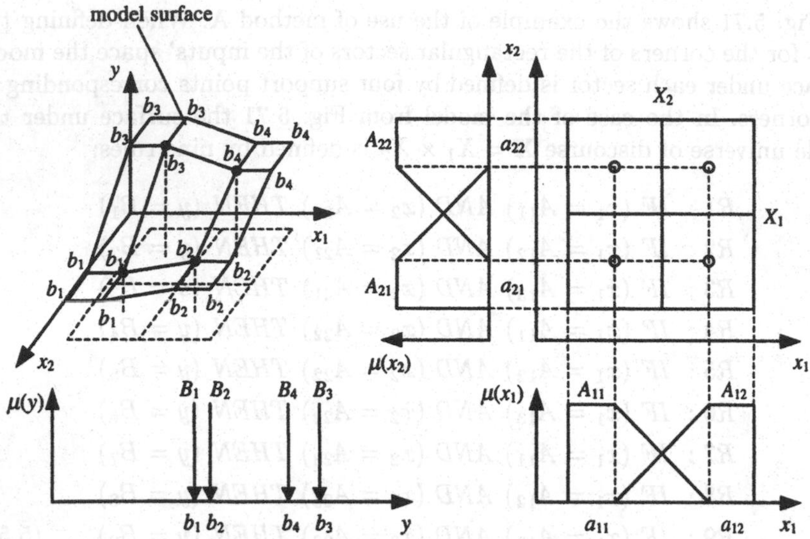

Fig. 5.72. The example of a fuzzy model with the support points (rules) defined in the centers of the partition sectors of the universe of discourse $\mathbf{X} = X_1 \times X_2$

low. Summarizing, one can state the following advantages and disadvantages or faults of both methods of defining the rules.

Method A (support points in the sector corners)

- Advantages:
  - increased accuracy of a model, also at the boundaries of the universe of discourse $\mathbf{X}$ of the inputs' space.
- Disadvantages:
  - a greater number of rules, which leads to minor transparency of a model,
  - more information is needed to determine the rules.

Method B (support points at the sector centers)

- Disadvantages:
  - minor accuracy of a model in comparison with method A, in particular at the boundaries of the universe of discourse.
- Advantages:
  - a minor number of rules in comparison with method A, which leads to greater transparency of a model,
  - less measurement information is needed to determine the rules.

At the beginning of the tuning process of a model, the parameters of the points of support a_{ij}, b_k can be arranged, for example, uniformly. In the course of tuning, the support points (parameters of the fuzzy sets in the rules) shift to the positions securing greater and greater accuracy of the model. The greater number of support points (rules) makes it possible (potentially) to achieve greater accuracy of a fuzzy model, provided that the tuning process has been carried out effectively. However, with a greater number of rules it becomes more and more difficult to tune the model.

5.4 Reduction of the rule base

The major difficulty of the tuning process of multidimensional self-learning fuzzy models, (e.g. neuro-fuzzy networks based on the regular hyper-rectangular partition of the inputs' space) is the great number of parameters which must be tuned. The number increases by leaps and bounds as the number of inputs and the number of fuzzy sets used for evaluating the individual inputs of a model increases. This problem, discussed in Section 5.2.2, called in the literature "curse of dimensionality" (Bossley 1995) is the subject of research carried out by specialists. One of the ways for reducing this problem is to abandon the regular partition of the inputs' space and to use the irregular grid (Su 1995; Kwon 1994) where the grid lines are not straight segments. The second way is to abandon the grid partition and use non-grid strategies of the inputs' space partition (Brown 1995a), such as:

- the rectangular partition (k-d tree partition),
- the quadratic partition (quad tree partition).

The essence of these partitions is explained in Fig. 5.73.

The idea of introducing the non-grid partitions consists of decreasing the number of fuzzy sectors. The partition of the inputs' space is denser where the surface of the mapping $\mathbf{X} \to Y$ of the modeled system changes more drastically (large steep slopes, irregularities), while it is sparser where the surface of the system is smoother.

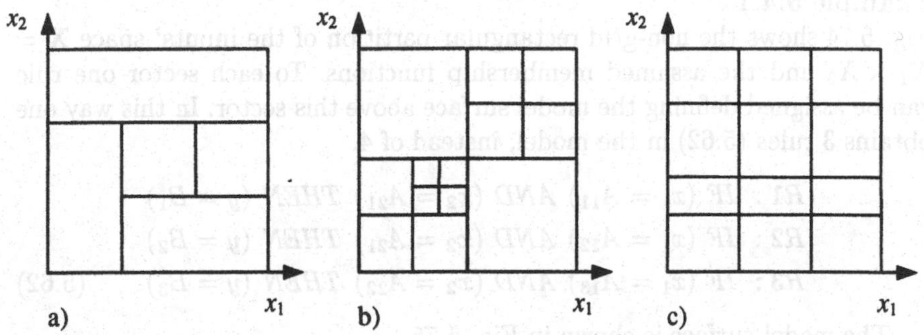

Fig. 5.73. The non-grid partitions of the inputs' space: rectangular partition (a), quadratic partition (b) as well as the grid partition (c)

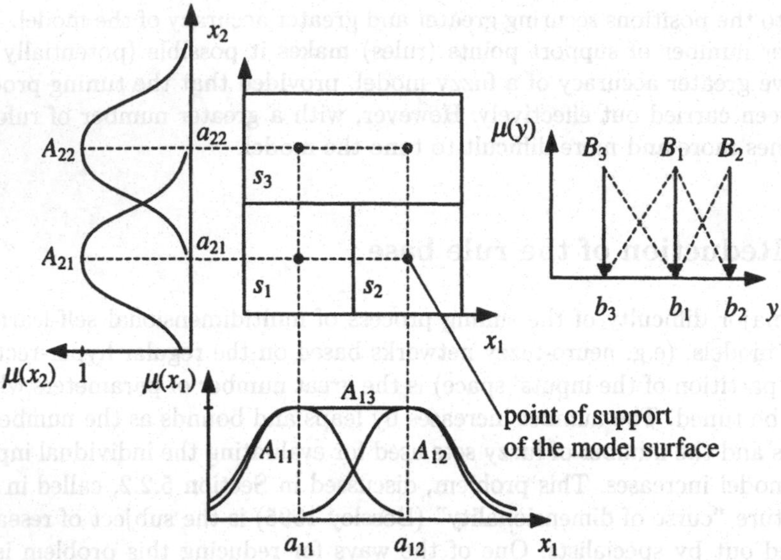

Fig. 5.74. The non-grid partition of the inputs' space in 3 sectors $S_1 - S_3$ and the assumed membership functions of fuzzy sets

The surface over each partition sector can be defined by the use of one rule only. Here it is advantageous to use the Takagi-Sugeno models described in Section 5.7.7 in which the rule conclusion is a function (usually a linear one) and not a fuzzy set. An example of such a rule is given by the formula (5.61).

$$IF\ (x_1 = A_{11})\ AND\ (x_2 = A_{21})\ THEN\ (y = a_{11}x_1 + a_{21}x_2 + a_{01}) \qquad (5.61)$$

Nevertheless the Mamdani models, as in Example 5.4.1, can also be used here.

Example 5.4.1

Fig. 5.74 shows the non-grid rectangular partition of the inputs' space $\mathbf{X} = X_1 \times X_2$ and the assumed membership functions. To each sector one rule can be assigned defining the model surface above this sector. In this way one obtains 3 rules (5.62) in the model, instead of 4.

$$\begin{aligned} R1 &: IF\ (x_1 = A_{11})\ AND\ (x_2 = A_{21})\ THEN\ (y = B_1) \\ R2 &: IF\ (x_1 = A_{12})\ AND\ (x_2 = A_{21})\ THEN\ (y = B_2) \\ R3 &: IF\ (x_1 = A_{13})\ AND\ (x_2 = A_{22})\ THEN\ (y = B_3) \end{aligned} \qquad (5.62)$$

The model surface is shown in Fig. 5.75.

Defining the large sector S_3 of the inputs' space $\mathbf{X}$, Fig. 5.74, by means of only one rule was possible owing to the use of the membership function A_{13} having a long core extended along nearly the whole width of this sector.

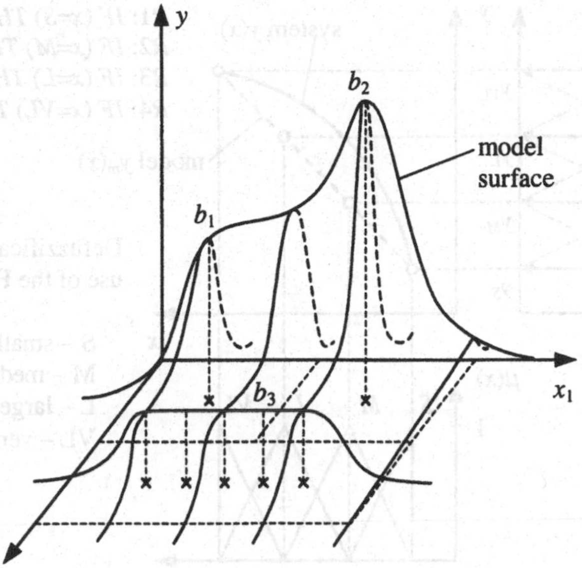

Fig. 5.75. The surface of the model (5.62) based on the non-grid partition of the inputs' space

In order to use the non-grid partition of the inputs' space, initially it is essential to know the degree of variability of the surface of the modeled system above different zones of this space. Only then can we reasonably decide if the partition should be denser or sparser. The required information can be obtained from, for example, the clustering analysis of the input/output measurement samples of the modeled system (Babuška 1996), Section 6.3.3.2.

It should be noticed that the rule base (5.62) is linguistically incomplete because not all possible combinations of the fuzzy sets of the inputs A_{1i}, A_{2j} occur in it. But it is numerically complete owing to the use of the Gaussian membership functions having infinitely large supports. At any state of the inputs $(x_1^*, x_2^*) \in X_1 \times X_2$ at least one rule is always activated. Therefore the output of a model can be calculated independently of the parameters of these functions (right and left spread).

Defining the large sector S_3 was possible owing to the use of the fuzzy set A_{13}, Fig. 5.74, having an adequately long core. The use of such membership functions is one of the methods enabling us to decrease the number of rules. The next method involves reducing the number of fuzzy sets of a model enabling us to decrease the number of rules and/or to simplify their form (the decrease in the number of component premises in the rules). Its essence will be explained by means of examples (Piegat 1997c).

◊ ◊ ◊

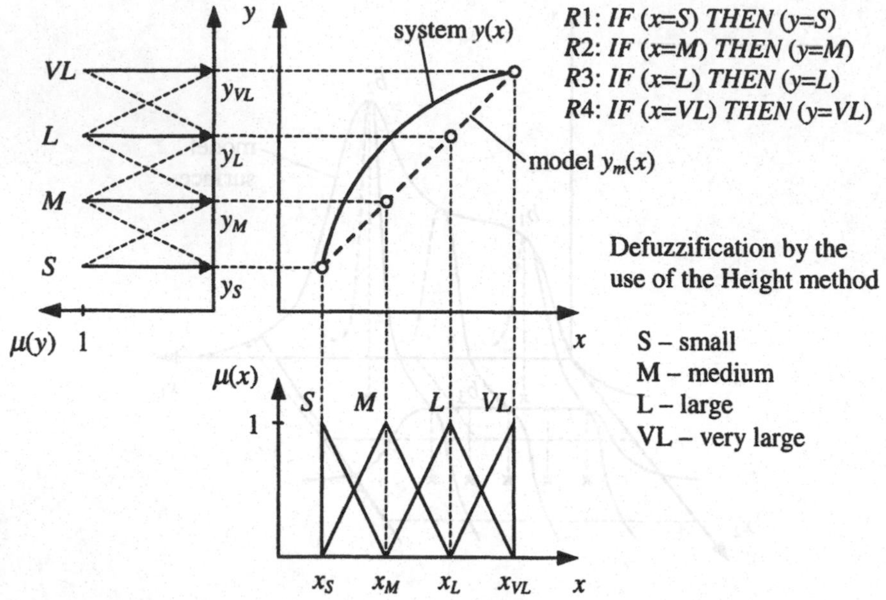

Fig. 5.76. The surface of the modeled system and of the model at the beginning of the adaptation process

Example 5.4.2
Let us consider an adaptive fuzzy model, which can tune its own parameters on the basis of input/output measurement data of the modeled system. Let us assume that at the beginning of the adaptation process a uniform arrangement of membership functions has been chosen, Fig. 5.76. This figure also shows the surface of the modeled system and the initial surface of the fuzzy model.

Assume that as a result of tuning the model, one has obtained the parameters of membership functions and the model surface as shown in Fig. 5.77. The modal values x_m and x_l of the fuzzy sets "medium" and "large" approximate each other. In this connection there exists the possibility that the union of these sets into one set $M^* = M \cup L$ will not cause an excessive decrease in the accuracy of the model measured, for example, by the sum of the absolute error values – see formula (5.63).

$$I = \sum_{i=1}^{n} |y(x_i) - y_m(x_i)| , \tag{5.63}$$

where: n – the number of the inputs/output measurement samples of the system.

Simultaneously with the union of the sets $M \cup L$ of the variable x into the resulting set M^* we should also unite the corresponding sets M and L of

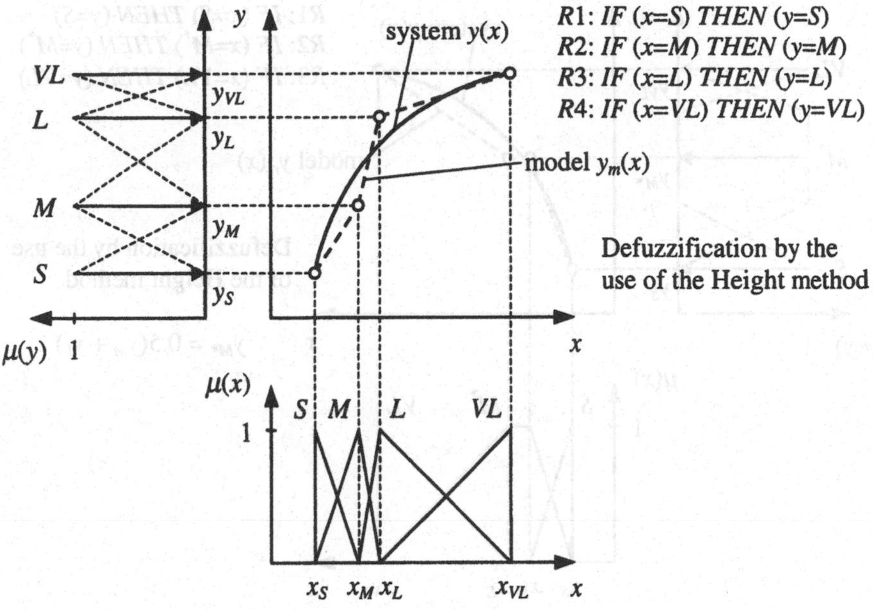

Fig. 5.77. The surface of the modeled system and the model after the parameters have been adjusted

the variable y in the rule conclusions. The union of the sets can be realized, for example, by the use of the *SUM* operator, formula (5.64).

$$\mu_{M^*}(x) = SUM(\mu_M(x), \mu_L(x))$$
$$\mu_{M^*}(y) = SUM(\mu_M(y), \mu_L(y)) \tag{5.64}$$

The result of this method of uniting the sets and the obtained model surface are represented in Fig. 5.78.

As can be seen in Fig. 5.78 the reduction of the number of sets by uniting them has caused (in this case) no essential change of the model accuracy. But the number of rules has decreased from 4 to 3. In the case of models with many inputs the decrease of the number of rules is considerably greater (Section 5.2.2).

Instead of calculating the resulting set M^* according to formula (5.64) we can also determine the resulting set M^* by using the simplified method consisting in locating the modal values x_{M*} and y_{M*} of these sets in the middle, between the modal values of the sets M and L, according to formula (5.65). In this case we obtain a slightly different model surface represented in Fig. 5.79.

$$x_{M*} = 0.5(x_M + x_L)$$
$$y_{M*} = 0.5(y_M + y_L) \tag{5.65}$$

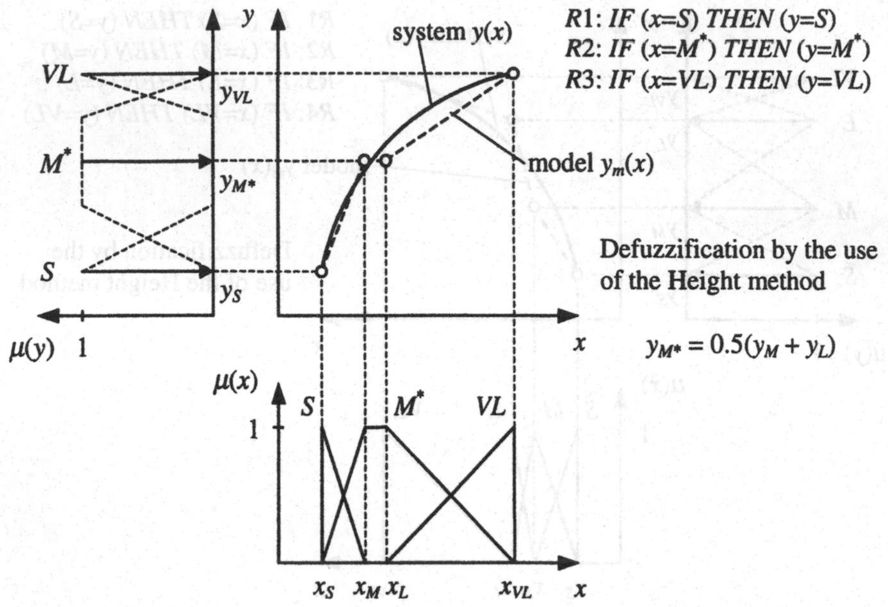

Fig. 5.78. The surface of the system to be modeled and the model after the fuzzy sets have been reduced ($M \cup L = M^*$)

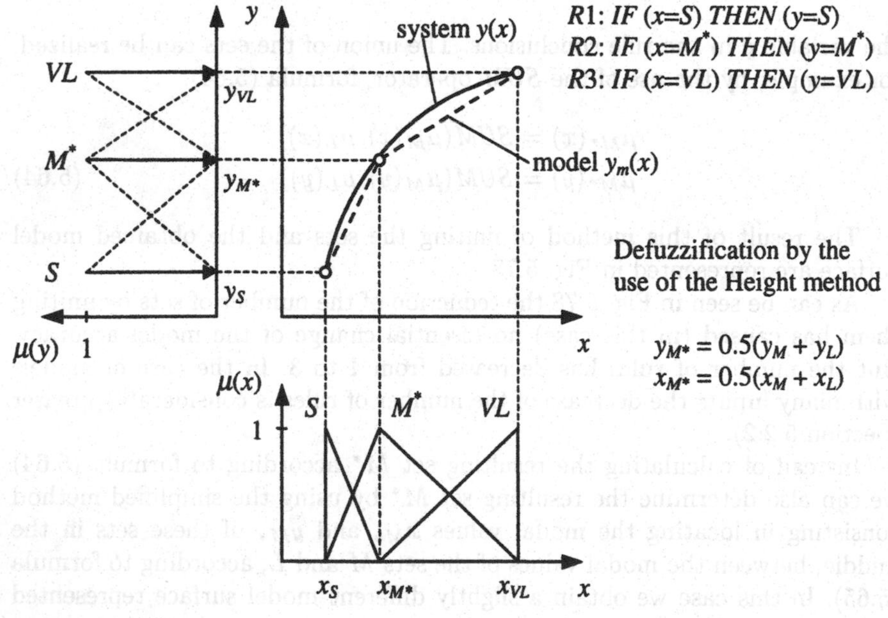

Fig. 5.79. The surface of the modeled system and the model after the fuzzy sets M and L have been united using the simplified method

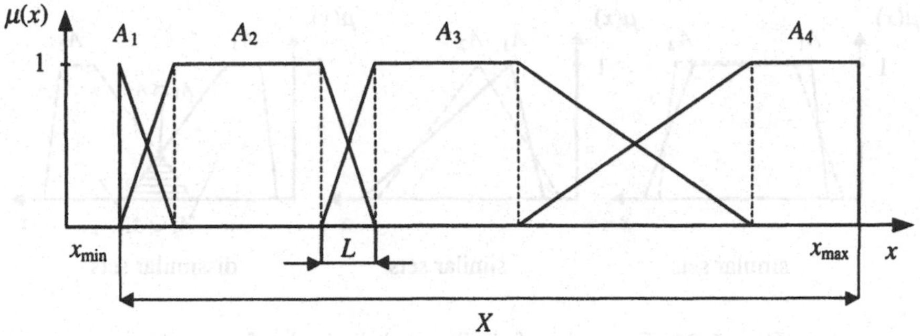

$$B(A_2, A_3) = \frac{L}{x_{max} - x_{min}} \leq \lambda$$

Fig. 5.80. The example of adjoining trapezoidal fuzzy sets lying near each other

The union of two fuzzy sets can be realized using the non-simplified method (5.64) or the simplified method (5.65). The factor determining which of them should be selected is the decrease in the accuracy of a model resulting from the use of the given method. In some cases the accuracy of a model can be even higher after the union of the sets.

◊ ◊ ◊

The fuzzy sets used in Example 5.4.2 have membership functions satisfying the condition of unity partition. This case refers to two adjoining sets in the inputs' space having similar modal values. Trapezoidal sets can be united if they are adjoining and their cores are located near each other.

The concept of the "nearness" of sets is not connected only with the distance L of the cores of the sets, Fig. 5.80, but it must also take into account the length of the set support $(x_{max} - x_{min})$. Therefore only the adjoining sets A_i, A_{i+1} whose relative nearness (proximity) B expressed by the formula (5.66) is less than a certain limit value can be considered to be "near sets".

$$B(A_i, A_{i+1}) = \frac{L_{A_i, A_{i+1}}}{x_{max} - x_{min}} \leq \lambda \qquad (5.66)$$

Defining the value λ depends on the designer of the model. At greater values of λ a larger decrease in the accuracy of a simplified model should be expected.

If the membership functions of a model do not satisfy the condition of unity partition, then in the course of tuning, free changes of spread, core length and modal values of these functions occur. In such a case we can obtain, as a result of tuning, membership functions more or less overlapping each other, Fig. 5.81.

In order to select sets suitable for unification the idea of set similarity S (Babuška 1996) can be used. Various measures according to the assumed cri-

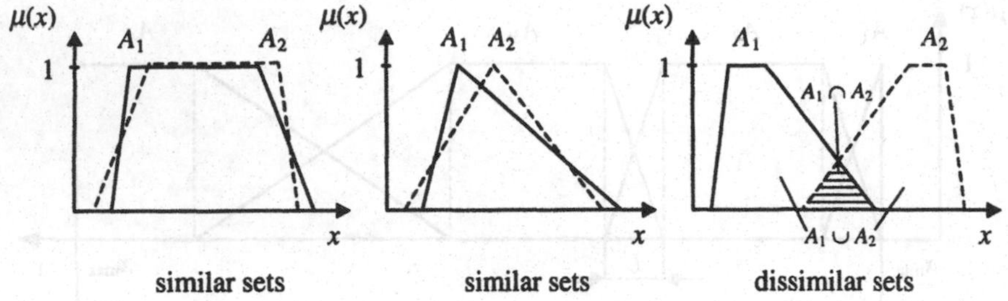

Fig. 5.81. Examples of similar and dissimilar fuzzy sets

terion have been elaborated. The easily intelligible measure of the similarity of two sets A_1 and A_2 is the measure expressed by the formula (5.67):

$$S(A_1, A_2) = \frac{|A_1 \cap A_2|}{|A_1 \cup A_2|} = \frac{\int_{x_{\min}}^{x_{\max}} MIN[\mu_{A1}(x), \mu_{A2}(x)]\mathrm{d}x}{\int_{x_{\min}}^{x_{\max}} MAX[\mu_{A1}(x), \mu_{A2}(x)]\mathrm{d}x}, \tag{5.67}$$

where: $x_{\min}, x_{\max}$ – the limits of the universe of discourse X, Fig. 5.79,
$|\cdot|$ – the cardinality (power) of a fuzzy set.

In the case of discrete membership functions the similarity of sets is calculated from the formula (5.68):

$$S(A_1, A_2) = \frac{|A_1 \cap A_2|}{|A_1 \cup A_2|} = \frac{\sum_{j=1}^{m} MIN[\mu_{A1}(x_j), \mu_{A2}(x_j)]}{\sum_{j=1}^{m} MAX[\mu_{A1}(x_j), \mu_{A2}(x_j)]}, \tag{5.68}$$

where: $x_j \in X_D$,
X_D – the discrete universe of discourse of the variable x.

The formulas (5.67) and (5.68) mean that the similarity of two sets A_1 and A_2 is the greater the more their common part $A_1 \cap A_2$, Fig. 5.80, equals their sum $A_1 \cup A_2$. If the sets A_1 and A_2 agree totally then $A_1 \cap A_2 = A_1 \cup A_2 = 1$ and their similarity $S(A_1, A_2) = 1$.

If two sets A_1 and A_2 defined in the inputs' space X are sufficiently similar, the expression (5.69) is true:

$$S(A_1, A_2) \geq \delta\,, \tag{5.69}$$

where: $\delta : \delta \in [0, 1]$ – the minimal limiting similarity of sets, then the union of them can be carried out according to formula (5.70).

$$A = A_1 \cup A_2$$
$$\mu_A(x) = MAX[\mu_{A1}(x), \mu_{A1}(x)] \tag{5.70}$$

With the aid of the sum A we substitute in the rules both the set A_1 and A_2, and this allows us to simplify the rule base. For example, if the rule base of a model with one input x and one output y has the form (5.71):

$$\begin{aligned} R1 &: \; IF\ (x = A_1)\ THEN\ (y = B_1), \\ R2 &: \; IF\ (x = A_2)\ THEN\ (y = B_2), \\ R3 &: \; IF\ (x = A_3)\ THEN\ (y = B_3), \\ R4 &: \; IF\ (x = A_4)\ THEN\ (y = B_4), \end{aligned} \tag{5.71}$$

and the fuzzy sets A_3 and A_4 are acknowledged to be similar and replaced with the set $A_3^* = A_3 \cup A_4$, then by inserting in the rules the set A_3^* instead of A_3 and A_4 we obtain the rule base (5.72).

$$\begin{aligned} R1 &: \; IF\ (x = A_1)\ THEN\ (y = B_1), \\ R2 &: \; IF\ (x = A_2)\ THEN\ (y = B_2), \\ R3 &: \; IF\ (x = A_3^*)\ THEN\ (y = B_3), \\ R4 &: \; IF\ (x = A_3^*)\ THEN\ (y = B_4). \end{aligned} \tag{5.72}$$

Because the rules $R3$ and $R4$ have identical premises, they can be united according to the formula (5.73):

$$\begin{aligned} &[IF\ (x = A_3^*)\ THEN\ (y = B_3)] \cup [IF\ (x = A_3^*)\ THEN\ (y = B_4)] = \\ &IF\ (x = A_3^*)\ THEN\ (y = B_3 \cup B_4) = \\ &IF\ (x = A_3^*)\ THEN\ (y = B_3^*), \end{aligned} \tag{5.73}$$

where: $B_3^* = B_3 \cup B_4$.

Then the rule base (5.72) reduces itself to the base (5.74) containing the smaller number of rules.

$$\begin{aligned} R1 &: \; IF\ (x = A_1)\ THEN\ (y = B_1), \\ R2 &: \; IF\ (x = A_2)\ THEN\ (y = B_2), \\ R3 &: \; IF\ (x = A_3^*)\ THEN\ (y = B_3^*). \end{aligned} \tag{5.74}$$

If the decrease in model accuracy caused by the reduced rule base (5.74) is not too excessive, then this base can be accepted. The example given below shows how a large reduction in the number of rules can be obtained in a two-input model.

Example 5.4.3

Let us consider the model of a system having 2 inputs and one output and the linguistically complete rule base (5.75), where:

A_1, A_2, A_3 – fuzzy sets of the input x_1,
B_1, B_2, B_3 – fuzzy sets of the input x_2,
$C_1 \dots C_9$ – fuzzy sets of the output y.

$$\begin{aligned}
R1:\ & IF\ (x_1 = A_1)\ AND\ (x_2 = B_1)\ THEN\ (y = C_1)\\
R2:\ & IF\ (x_1 = A_1)\ AND\ (x_2 = B_2)\ THEN\ (y = C_2)\\
R3:\ & IF\ (x_1 = A_1)\ AND\ (x_2 = B_3)\ THEN\ (y = C_3)\\
R4:\ & IF\ (x_1 = A_2)\ AND\ (x_2 = B_1)\ THEN\ (y = C_4)\\
R5:\ & IF\ (x_1 = A_2)\ AND\ (x_2 = B_2)\ THEN\ (y = C_5)\\
R6:\ & IF\ (x_1 = A_2)\ AND\ (x_2 = B_3)\ THEN\ (y = C_6)\\
R7:\ & IF\ (x_1 = A_3)\ AND\ (x_2 = B_1)\ THEN\ (y = C_7)\\
R8:\ & IF\ (x_1 = A_3)\ AND\ (x_2 = B_2)\ THEN\ (y = C_8)\\
R9:\ & IF\ (x_1 = A_3)\ AND\ (x_2 = B_3)\ THEN\ (y = C_9)
\end{aligned} \qquad (5.75)$$

Let us assume that the sets $A_2(x_1)$ and $A_3(x_1)$ are similar and can be united into the set $A_2^*(x_1)$. This set $A_2^*(x_1)$ can then replace $A_2(x_1)$ and $A_3(x_1)$ in the rules ($A_2 \cong A_2^*$, $A_3 \cong A_2^*$). Then the pairs of rules ($R4, R7$), ($R5, R8$), ($R6, R9$) will have the identical premises and can be united by the use of the OR operation obtaining the rule-base (5.76).

$$\begin{aligned}
R1:\ & IF\ (x_1 = A_1)\ AND\ (x_2 = B_1)\ THEN\ (y = C_1)\\
R2:\ & IF\ (x_1 = A_1)\ AND\ (x_2 = B_2)\ THEN\ (y = C_2)\\
R3:\ & IF\ (x_1 = A_1)\ AND\ (x_2 = B_3)\ THEN\ (y = C_3)\\
R4:\ & IF\ (x_1 = A_2^*)\ AND\ (x_2 = B_1)\ THEN\ (y = C_4 \cup C_7)\\
R5:\ & IF\ (x_1 = A_2^*)\ AND\ (x_2 = B_2)\ THEN\ (y = C_5 \cup C_8)\\
R6:\ & IF\ (x_1 = A_2^*)\ AND\ (x_2 = B_3)\ THEN\ (y = C_6 \cup C_9)
\end{aligned} \qquad (5.76)$$

The decrease in the number of fuzzy sets by 1 makes it possible in this case to diminish the number of rules by 3. We can accept the simplified model (5.76) only when the decrease in accuracy in relation to the original model (5.75) is not considerable. Example 5.4.4 shows that such a situation does not always take place.

◊ ◊ ◊

Example 5.4.4

The original fuzzy model is given in Fig. 5.82. Because the modal values of the sets A_2 and A_3 are near each other we can try to unite them into the set A_2^*. The fuzzy set obtained in this way is represented in Fig. 5.83.

As can be easily concluded from Fig. 5.83, the union of the sets A_2 and A_3 as well as the reduction of the number of rules has caused a significant decrease in the model accuracy. This decrease could be partly foreseen on the basis of the large distance of the singletons $(y_3 - y_2)$ representing the sets B_2 and B_3, Fig. 5.82.

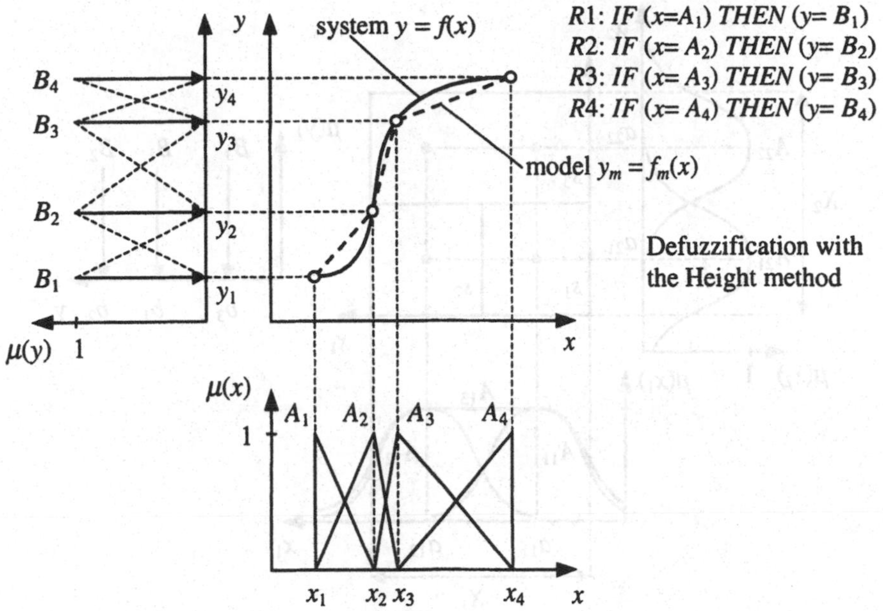

Fig. 5.82. The surface of the modeled system and the surface of the fuzzy model with 4 rules

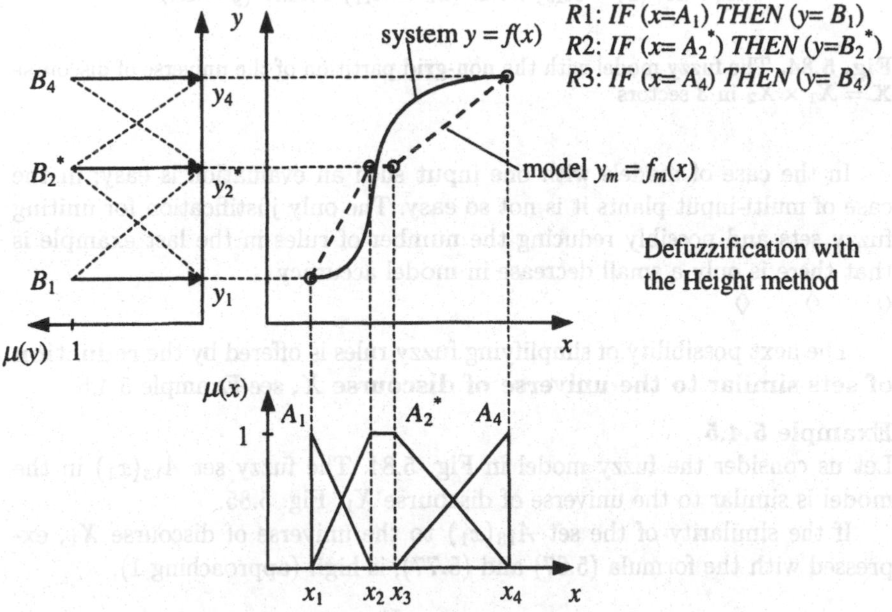

Fig. 5.83. The surface of the modeled system from Fig. 5.82 and the surface of the model with the decreased number of rules

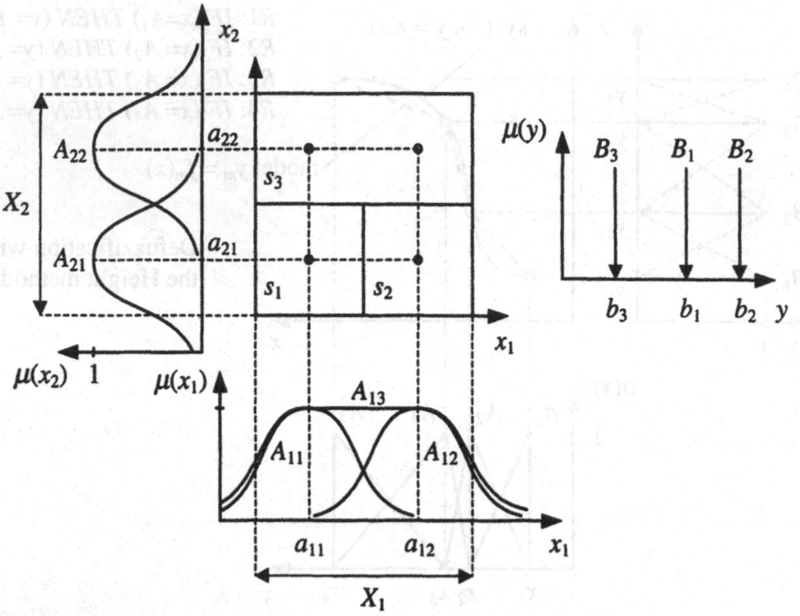

$$
\begin{aligned}
R1:\ & IF\ (x_1 = A_{11})\ AND\ (x_2 = A_{21})\ THEN\ (y = B_1)\\
R2:\ & IF\ (x_1 = A_{12})\ AND\ (x_2 = A_{21})\ THEN\ (y = B_2)\\
R3:\ & IF\ (x_1 = A_{13})\ AND\ (x_2 = A_{22})\ THEN\ (y = B_3)
\end{aligned}
$$

Fig. 5.84. The fuzzy model with the non-grid partition of the universe of discourse $\mathbf{X} = X_1 \times X_2$ in 3 sectors

In the case of models with one input such an evaluation is easy; in the case of multi-input plants it is not so easy. The only justification for uniting fuzzy sets and possibly reducing the number of rules in the last example is that there is only a small decrease in model accuracy.

◊ ◊ ◊

The next possibility of simplifying fuzzy rules is offered by the **reduction of sets similar to the universe of discourse** X, see Example 5.4.5.

Example 5.4.5

Let us consider the fuzzy model in Fig. 5.84. The fuzzy set $A_{13}(x_1)$ in the model is similar to the universe of discourse X_1, Fig. 5.85.

If the similarity of the set $A_{13}(x_1)$ to the universe of discourse X_1, expressed with the formula (5.67) and (5.77), is high (approaching 1),

$$
S(A_{13}(x_1), X_1) = \frac{|A_{13}(x_1) \cap X_1|}{|A_{13}(x_1) \cup X_1|} = \frac{\int\limits_{x_{1\min}}^{x_{1\max}} \mu_{A_{13}}(x_1)\mathrm{d}x_1}{x_{1\max} - x_{1\min}} \geq \delta\,, \tag{5.77}
$$

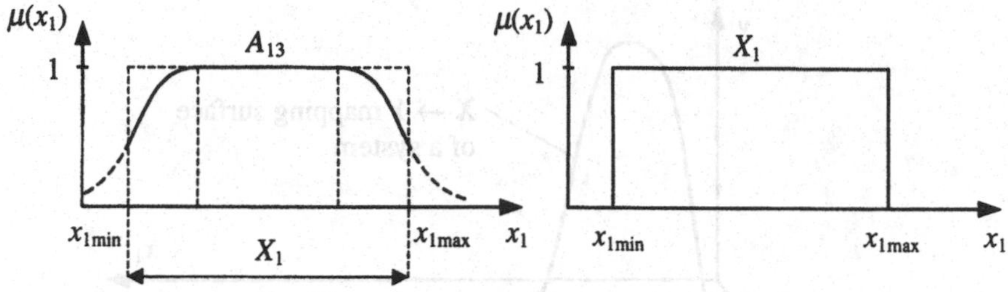

Fig. 5.85. The membership functions of the set $A_{13}(x_1)$ and universe of discourse X_1

then this set can be substituted in the rules with the set X_1. Because the membership in the universe of discourse is equal to 1 in the whole range of the support of this set, the component premises containing the set $A_{13} \cong X_1$ can be eliminated from the rules, thereby simplifying their form. In the case of the model from Fig. 5.84 we obtain the simplified rule base in the form (5.78).

$$\begin{aligned} &R1: \ IF \ (x_1 = A_{11}) \ AND \ (x_2 = A_{21}) \ THEN \ (y = B_1) \\ &R2: \ IF \ (x_1 = A_{12}) \ AND \ (x_2 = A_{21}) \ THEN \ (y = B_2) \\ &R3: \ IF \ \qquad\qquad\qquad\qquad\ (x_2 = A_{22}) \ THEN \ (y = B_3) \end{aligned} \tag{5.78}$$

If the fuzzy model with the simplified rule base (5.78) has only slightly less accuracy in comparison with the original model, then the new rule base can be accepted.

The **reduction of a fuzzy model** can be achieved **by the use of the method of local models** (Bossley 1995; Babuška 1995c). This method consists in the use of the set of local models instead of one global model defined on the whole universe of discourse $\mathbf{X}$ of the inputs' space. The local models are characterized by the differentiated density of the partition grid defined for the individual separate sectors of the inputs' space. The method of local models can be used in the case of systems whose surface of the $\mathbf{X} \to Y$ mapping consists both of areas of little slope ("plateaus"), and those of large slope ("mountains"), Fig. 5.86.

The system surface above the sector S_1 of the universe of discourse $\mathbf{X}$, Fig. 5.86, is characterized by strong differentiation in height and slope. For its exact modeling a greater number of support points (defined by the model rules) is needed than in the case of the nearly even surface in the sector S_2. For this reason the grid of fuzzy partition in the sector S_1 should be considerably denser than in the sector S_2, Fig. 5.87.

If we use the partition grid of the same density in the whole range of the universe of discourse $\mathbf{X} = X_1 \times X_2$, then the number of rules defining each of the support points would amount to 98. However, we can use differentiated grids. If we use the denser grid for the sector S_1, then the number of rules

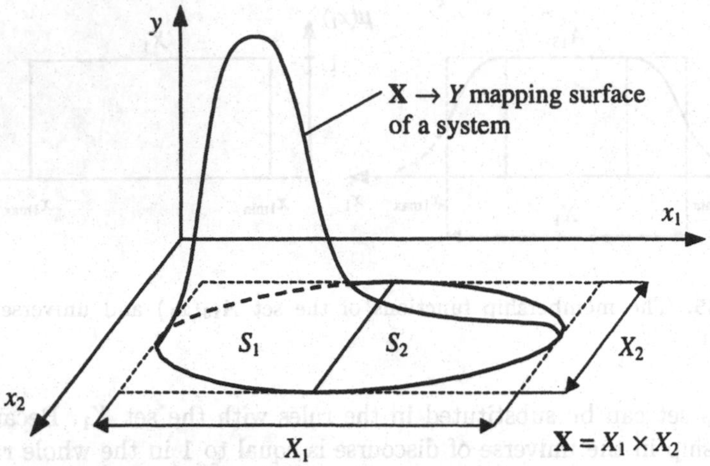

Fig. 5.86. System with strongly differentiated slope of the input/output surface

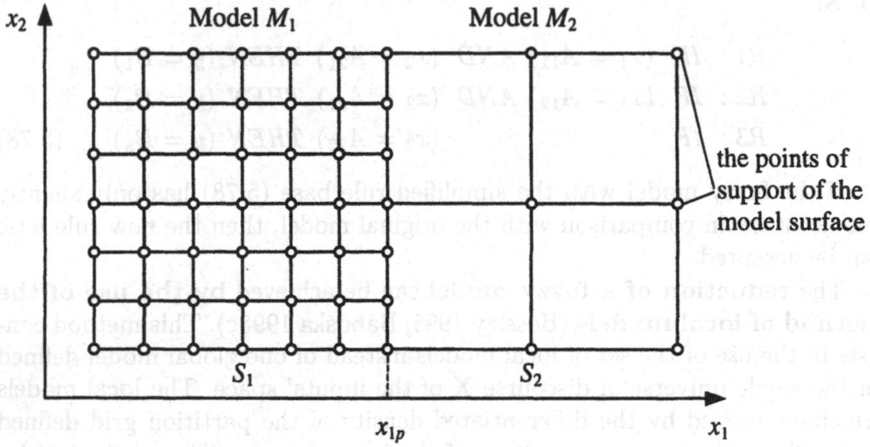

Fig. 5.87. The differentiated density of the fuzzy partition of the universe of discourse for different local models M_1 and M_2

in this sector will amount to 49 and in the sector S_2 covered with a sparser grid to 9 (total 58 rules). So, the reduction of the number of rules owing to the use of two local models is considerable.

The condition for the good operation of a model consisting of many local models is the continuity of the surface in the region where the local models join. This condition will be satisfied when the outputs of a model y calculated for the coordinates of the joint edge on the basis of different adjacent local models are the same. In the case of two local models M_1 and M_2 from Fig. 5.87

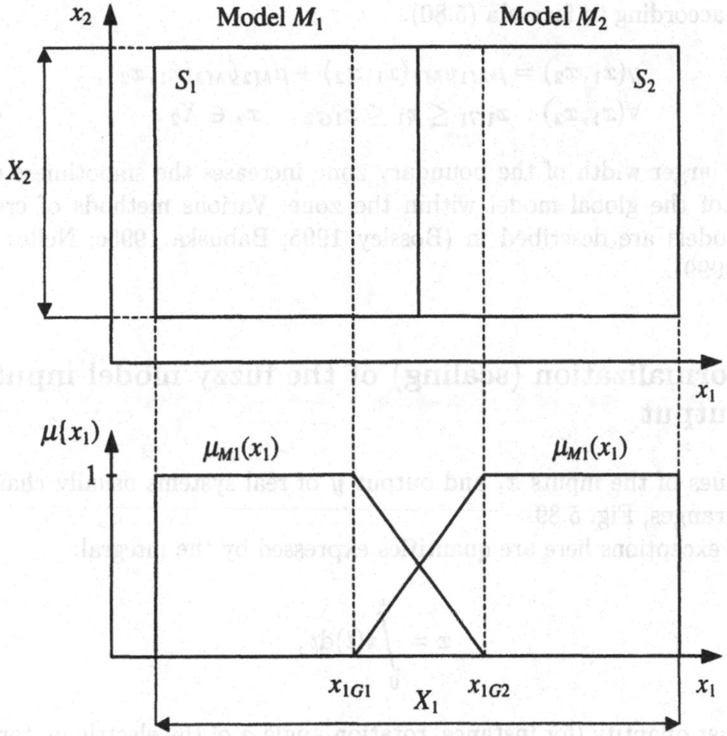

Fig. 5.88. Membership functions for adjoining local models with the determination of the boundary zone $x_{1G1} \leq x_1 \leq x_{1G2}$

this condition is expressed with the formula (5.79).

$$y_{M1}(x_{1P}, x_2) = y_{M2}(x_{1P}, x_2) \tag{5.79}$$

The condition (5.79) imposes a special construction on models adjoining each other, which consists in the interdependence of their parameters. This method is difficult to realize particularly when the number of inputs is great. But also in such a case the continuity of the surface of the global model can be achieved by fuzzily uniting the local models. In order to do this we must determine the boundary zones in the adjoining models and calculate the output of the global model y on the basis of all outputs y_{Mi} of the adjoining models multiplied by the grades of membership in the boundary zone. The method is explained in Fig. 5.88.

In the area of the sector S_1 not lying in the boundary zone the output of the global model y is equal to the output of the local model y_{M1}. Similarly, in the area of the sector S_2 also not belonging to the boundary zone the output of the model $y = y_{M2}$. However the output of the global model for inputs in the boundary zone is calculated on the basis of the outputs of both the local models according to formula (5.80).

$$\begin{aligned} y(x_1,x_2) &= \mu_{M1}y_{M1}(x_1,x_2) + \mu_{M2}y_{M2}(x_1,x_2), \\ \forall(x_1,x_2) &: \ x_{1G1} \le x_1 \le x_{1G2}, \quad x_2 \in X_2 . \end{aligned} \tag{5.80}$$

The larger width of the boundary zone increases the smoothness of the surface of the global model within the zone. Various methods of creating local models are described in (Bossley 1995; Babuška 1995c; Nelles 1998; Nelles 1999).

5.5 Normalization (scaling) of the fuzzy model inputs and output

The values of the inputs x_i and output y of real systems usually change in limited ranges, Fig. 5.89.

The exceptions here are quantities expressed by the integral:

$$x = \int_0^t z(t)\mathrm{d}t ,$$

of another quantity (for instance, rotation angle α of the electric motor shaft increases so long as the power supply of the motor is switched on) or quantities being the derivative ($x = \partial z/\partial t$) of another quantity (e.g. the derivative of the error signal in a controller). Integrals and derivatives can theoretically grow to infinitely big values. However, in real systems their values are often also limited (although they can be very high) considering the limited

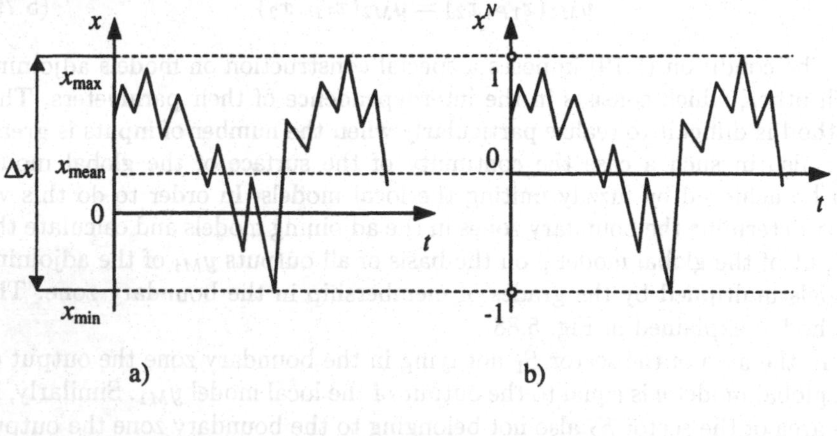

Fig. 5.89. The graph of a real quantity x during the system operation (a) and the graph of the quantity x^N after normalization using one of methods (b)

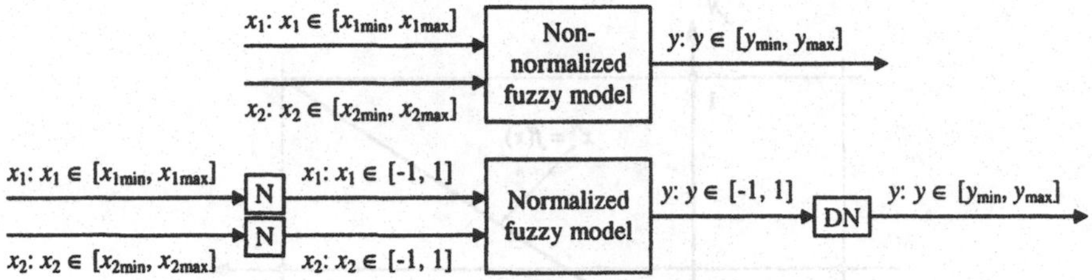

Fig. 5.90. Normalization of the input and output of a model (N – operation of normalization, x^N – normalized quantity, DN – operation of denormalization)

power and operation rate of actuators, operation time of a system etc. The occurrence of signal limitations in a system can be confirmed by observation and measurements. If the limitations $x_{\max}$ and $x_{\min}$ are known, then their normalization, also called scaling (Driankov 1996; Yager 1995; Kahlert 1995), can be carried out. The normalization of a value x, which varies in the interval $[x_{\min}, x_{\max}]$ consists in reducing it by appropriate scaling to the normalized interval $[-1, 1]$. Also the interval $[0, 1]$ is used. The essence of the normalization of signals of a fuzzy model is represented in Fig. 5.90.

What advantages are provided by normalization?

1. In the case of qualitatively similar real systems we obtain similar normalized fuzzy models. In the case of the control of qualitatively similar processes we obtain similar normalized fuzzy controllers. It allows a designer to gain the feel and knowledge of designing models and controllers.
2. Experience and feel gained at designing models and controllers for qualitatively similar systems allows the designer to design a model or controller quickly and roughly that thereafter only needs to be tuned (often only slightly).

The normalization for the interval $[-1, 1]$ is illustrated in Fig. 5.91. The advantage of the normalization from Fig. 5.91 is that the range $[-1, 1]$ is fully used. The disadvantage is that zeros of the quantity x and x^N, do not cover (overlap) what can sometimes be of some importance.

Simplified normalization is also used (Kahlert 1995) which consists in dividing the quantity x only by a certain constant coefficient, Fig. 5.92. The advantage of simplified normalization is lower calculation cost and the fact that the normalized quantity and non-normalized one have a common origin (zero point). The disadvantage is that the interval $[-1, 1]$ is not fully used.

As is shown in Fig. 5.92 the normalized quantity does not achieve $x^N = -1$ for $x = x_{\min}$. Therefore the simplified normalization should be used mainly for symmetrical variability intervals of the signals $|x_{\min}| = |x_{\max}|$. In Fig. 5.93 the normalization in the interval $[0, 1]$ has been shown.

After the normalized output y^N has been calculated by a normalized fuzzy model we must carry out the denormalization of it, Fig. 5.89. Denormaliza-

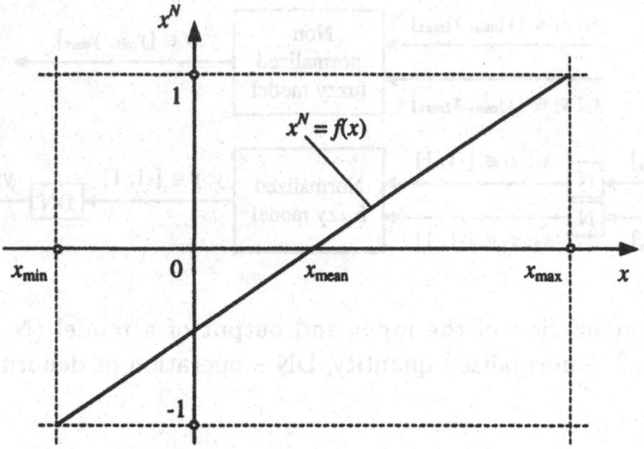

$$x^N = \frac{x - x_{mean}}{x_{max} - x_{min}} \qquad x_{mean} = 0.5(x_{max} + x_{min})$$

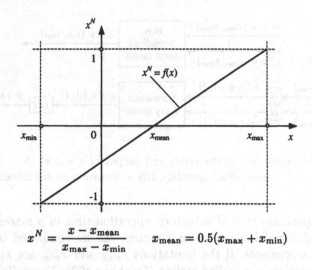

Fig. 5.91. Normalization of the quantity x in the interval $[-1, 1]$

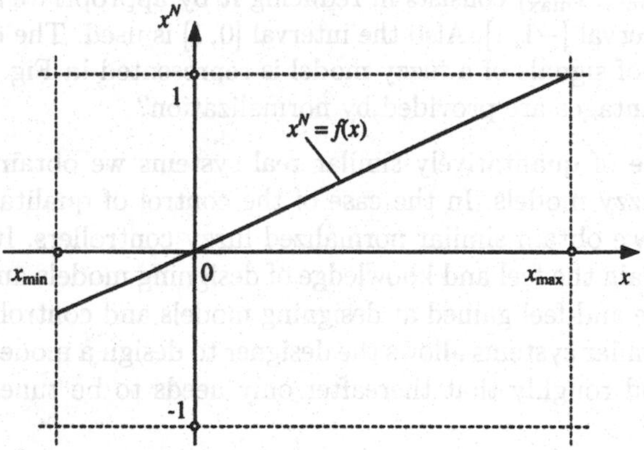

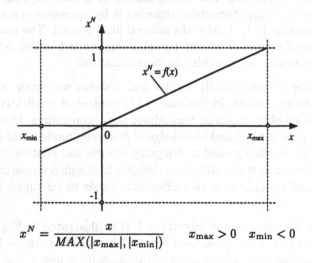

$$x^N = \frac{x}{MAX(|x_{max}|, |x_{min}|)} \qquad x_{max} > 0 \quad x_{min} < 0$$

Fig. 5.92. Simplified normalization of the quantity x in the interval $[-1, 1]$

tion is the inverse of normalization. To perform it, the maximal (y_{max}) and minimal (y_{min}) output values of the system to be modeled or eventually the maximal and minimal values of the quantity generated by the actuator in the control system must be known.

Denormalization formulas result directly from normalization formulas, Figs. 5.91 – 5.93. We must remember that the normalized output y^N of a normalized model changes in the symmetrical interval $[-1, 1]$, whereas the denormalized output y often changes in the asymmetrical interval $[y_{min}, y_{max}]$. Thus, it is usually necessary here, contrary to normalization, to map a sym-

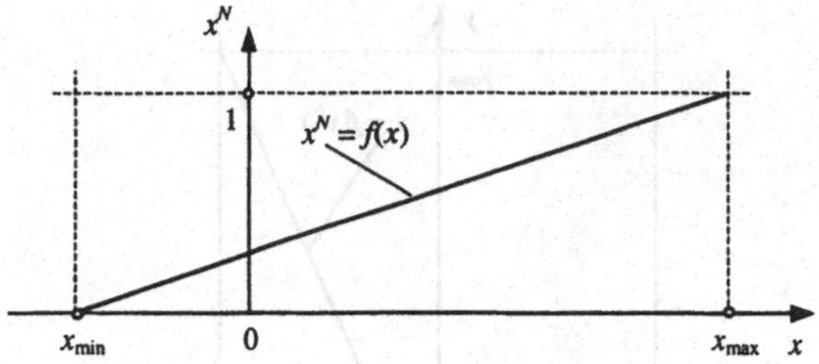

$$x^N = \frac{x - x_{min}}{x_{max} - x_{min}}$$

Fig. 5.93. Normalization of the quantity x in the interval $[0, 1]$

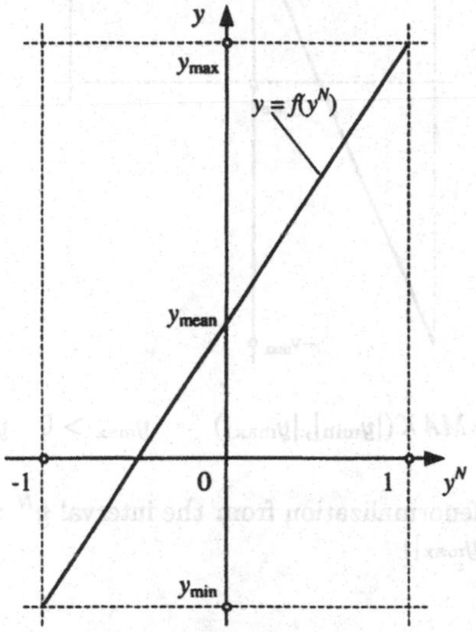

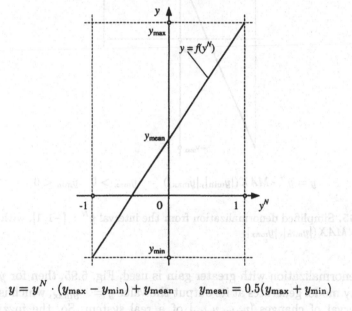

$$y = y^N \cdot (y_{max} - y_{min}) + y_{mean} \qquad y_{mean} = 0.5(y_{max} + y_{min})$$

Fig. 5.94. Denormalization $y^N \rightarrow y$ from the interval $y^N : [-1, 1]$ with the full use of the interval: $y : \ [y_{min}, y_{max}]$

metrical interval into an asymmetrical one. It slightly complicates a simplified denormalization but has no influence on full denormalization, Fig. 5.94.

The advantage of the denormalization illustrated in Fig. 5.94 is the full utilization of the model output $[y_{min}, y_{max}]$ Its disadvantage is that the zeros on the scale y^N and on the scale y do not overlap (the zero on the scale y^N corresponds with the value $y = y_{mean}$.

Simplified denormalization can be carried out using two methods illustrated in Figs. 5.95 and 5.96.

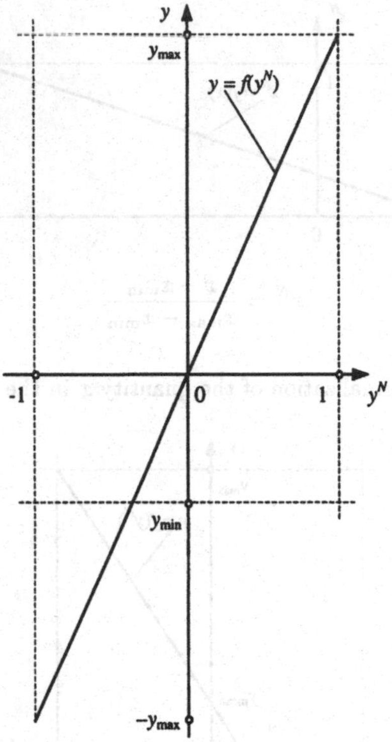

$$y = y^N \cdot MAX(|y_{min}|, |y_{max}|) \qquad y_{max} > 0 \quad y_{min} < 0$$

Fig. 5.95. Simplified denormalization from the interval y^N : $[-1, 1]$, with the gain equal to $MAX(|y_{min}|, |y_{max}|)$

If denormalization with greater gain is used, Fig. 5.95, then for $y^N = -1$ the fuzzy model generates at its output the value $y = -y_{max}$, which is outside the interval of changes $[y_{min}, y_{max}]$ of a real system. So, the fuzzy model will calculate unreal outputs. In the case of fuzzy models such a situation is usually not acceptable. In the case of fuzzy controllers it can be accepted because then the actuator will set $y = y_{min}$ (saturation effect). However, such a denormalization introduces additional non-linearity into a control system in the form of limitation (saturation) of the signal. Fig. 5.96 shows another variant of simplified denormalization: denormalization with less gain.

In the case of simplified denormalization with less gain, Fig. 5.96, the fuzzy model will not generate the maximal output value for $y^N = 1$, as in a real system. The model will not map this system exactly. The presented problems connected with simplified denormalization vanish if the interval of change $[y_{min}, y_{max}]$ is symmetrical, i.e. $y_{max} = -y_{min}$. Fig. 5.97 represents the denormalization from the interval y^N : $[0, 1]$.

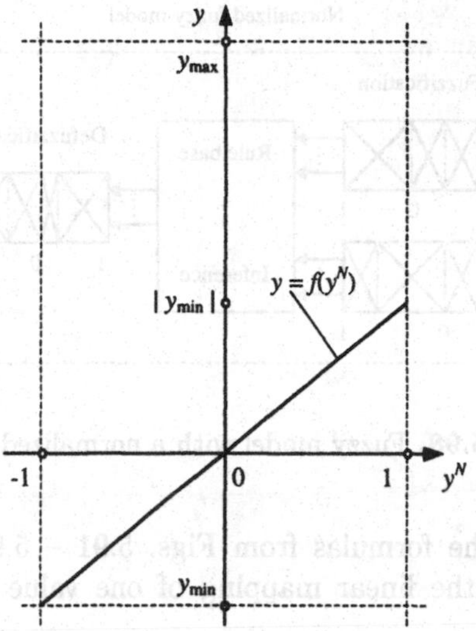

$$y = y^N \cdot MIN(|y_{min}|, |y_{max}|) \qquad y_{max} > 0 \quad y_{min} < 0$$

Fig. 5.96. Simplified denormalization from the interval y^N : $[-1, 1]$, with the gain equal to $MIN(|y_{min}|, |y_{max}|)$

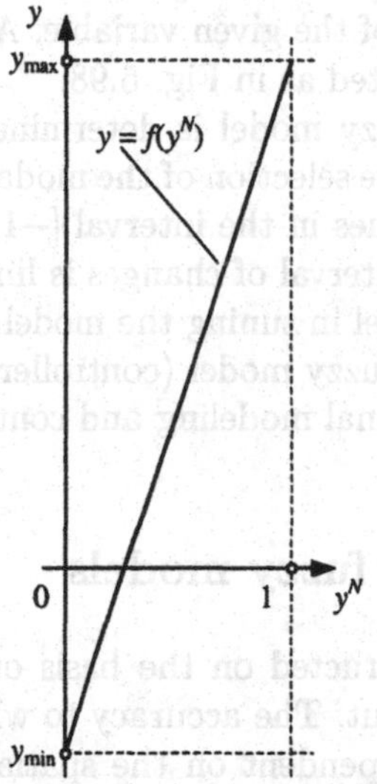

$$y = y^N \cdot (y_{max} - y_{min}) + y_{min}$$

Fig. 5.97. The denormalization from the interval y^N : $[0, 1]$ into the interval $y : [y_{min}, y_{max}]$

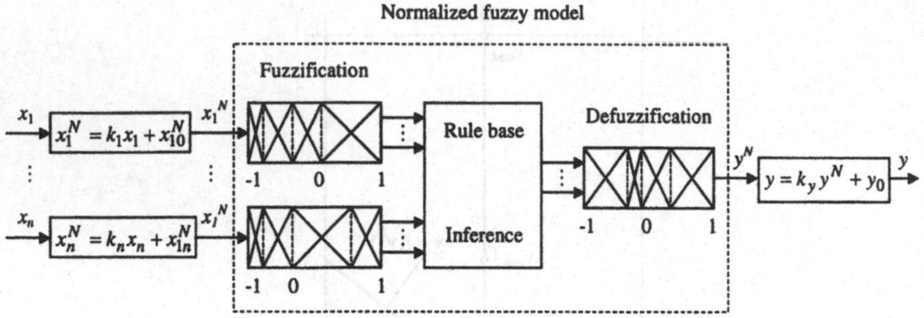

Fig. 5.98. Fuzzy model with a normalized part

The analysis of the formulas from Figs. 5.91 – 5.97 shows that these formulas accomplish the linear mapping of one value into another of the type:

$$x^N = k_x \cdot x + x_0^N \quad \text{or} \quad x^N = k_x \cdot x \,,$$
$$y = k_y \cdot y^N + y_0 \quad \text{or} \quad y = k_y \cdot y^N \,,$$

depending on the kind of chosen normalization (denormalization). The coefficients k_x, k_y are constant gains and x_0^N, y_0 are constants dependent on the limit values of the interval of the given variable. As a result the fuzzy model (controller) can be represented as in Fig. 5.98.

If the rule base of a fuzzy model is determined, then the tuning of the model consists in appropriate selection of the modal values of individual fuzzy sets. By changing these values in the interval $[-1, 1]$ we change gains of the fuzzy model. Because the interval of changes is limited to $[-1, 1]$, it is easier to get experience and the feel in tuning the model. It is unnecessary to carry out the normalization of a fuzzy model (controller) in each case. However, it is applied in many professional modeling and control systems.

5.6 Extrapolation in fuzzy models

Fuzzy models can be constructed on the basis of measurement samples of the system inputs and output. The accuracy to which such models map real systems is very strongly dependent on the spatial distribution of the measurement samples used for the modeling. The most advantageous situation is when the measurement samples are distributed uniformly in the inputs' space, Fig. 5.99.

In modeling practice it very often occurs that the universe of discourse is not fully covered by measurement samples. It relates particularly to larger systems, e.g. economic, biological and ecological ones where we can not ourselves carry out measurements actively, by adjusting various values of the

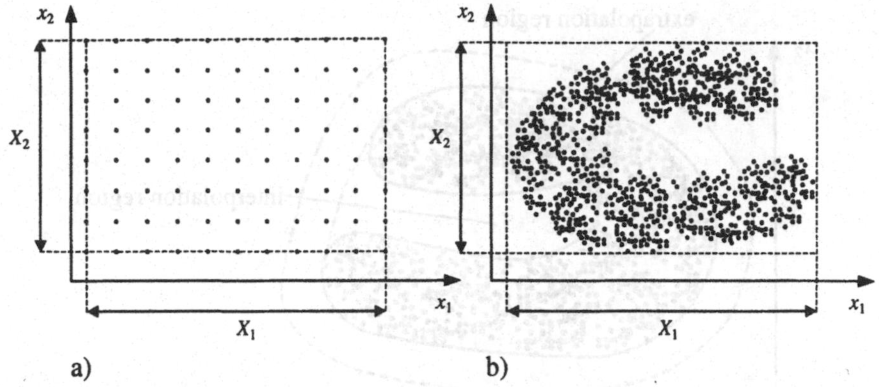

Fig. 5.99. The uniform regular distribution of measurement samples of the system to be modeled in the universe of discourse $X_1 \times X_2$ (a) and a non-uniform distribution containing empty regions (b)

system inputs and measuring its output, as we can, for instance, in the measurement of the speed of a ship in relation to the rotational speed of the propeller as well as the angle of its blades. In the case of many systems we can only passively observe the state of the inputs and output, e.g. the number of unemployed people in relation to the number of employment offers, the number of people leaving school and the level of unemployment benefit. In the universe of discourse of the system to be modeled, internal or external regions not covered by samples often occur, as is shown in Fig. 5.99b. Although measurement data on the behavior of the system in those regions are lacking, at least approximate knowledge about how the system will behave in those regions is often required. The knowledge can often be created on the basis of measurement samples already determined. Sometimes the qualitative expert's knowledge about the system to be modeled can additionally be used to complete the information resulting from the samples.

The extension of the model surface onto internal regions not covered by measurement samples is called **interpolation** and onto external ones **extrapolation**, Fig. 5.100.

Investigations on modeling these regions of the input space which are not covered with measurement samples are fairly intensive at present. They are referred to as **incomplete information research** (Liao 1999). The analysis of the subject literature shows that interpolation problems are investigated more often, for instance (Kóczy 1993; Ullrich 1998; Dubois 1992), whereas fuzzy extrapolation problems rarely are.

Meanwhile, extrapolation is much required in practice because we often have to predict the behavior of systems outside the regions where they have been operating up to the present. Some examples are:

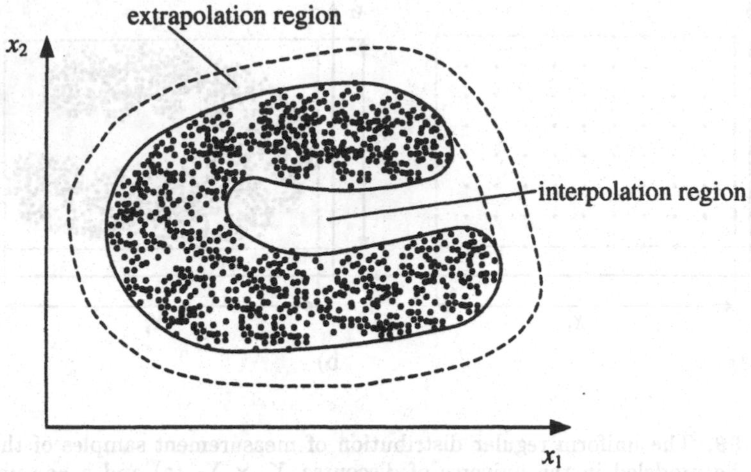

Fig. 5.100. Interpolation and extrapolation regions of a model

- modeling and selling prognoses of products with a relatively short "lifetime" caused by "moral" aging, as in the case of computers. We have at our disposal only scarce data from a short selling period. The data doesn't inform us about all aspects of the selling process. But we have to take decisions about how many inputs and which inputs should be changed to increase the sell;
- predictive control: the controller should predict the next state of the plant on the basis of the present and previous states and calculate the appropriate control signal for the next step;
- predicting future values of stock prices from their previous values (modeling of time series);
- predictive coding of images (Tian-Hu 1998): "predicted values for different patterns are defined using linear extrapolation from available neighborhood pixel values".

As the above examples show, good extrapolation offers great practical advantages and is essential. Every day each of us uses it to predict future occurrences more or less successfully. Some scientists are opponents of model extrapolation (Niederliński 1997), especially of static ones. Therefore, it would be of advantage to present the general meaning and idea of extrapolation.

What does extrapolation of a model from its validated universe of discourse onto a new, extended and not validated region mean?

Let us assume we want to predict an output y – value of the system for the input value $x = c$ placed outside the validated region used hitherto. An example here can be a bridge that according to previously made calculations and experiments can be loaded maximally with a weight of 35 tons. But in war conditions we must quickly transport to the other side of the river a

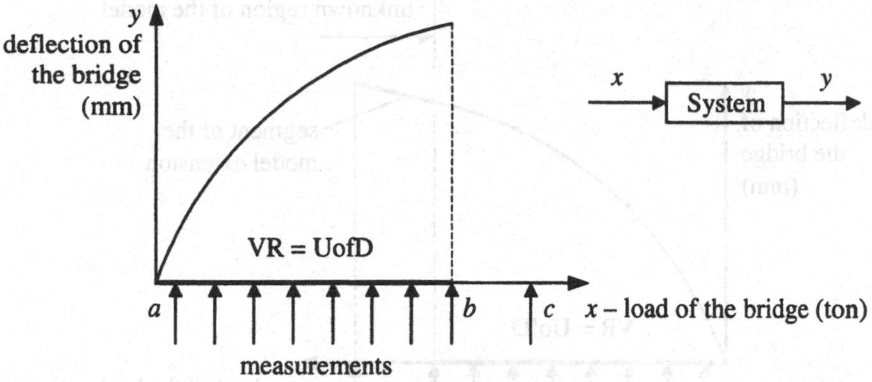

Fig. 5.101. Example of a characteristic (model) $y = f(x)$ of a SISO system determined on the basis of the knowledge possessed hitherto about the modeled system, $a \leq x \leq b$ – universe of discourse (UofD) of the model equal to the region validated by hitherto possessed input/output measurement samples

weight of 37 tons. Can we risk it? Under the excessive load, will the bridge merely deflect more than allowed or will it break? Bridges are constructed with a certain margin of safety and any bridge should be able to carry a load a little higher than the permissible limit, but our bridge is fairly old and partly rusty. What decision should be taken? Should we decide to transport a weight which is a little greater than the permissible one or shouldn't we?

Taking the decision "to transport the load" means the assumption of a continuous extrapolation of the bridge characteristic in the extended universe of discourse as for example shown in Fig. 5.102.

Taking the decision "not to transport the load" means the assumption of a hypothesis that under the increased load $x = c$ the bridge will break, Fig. 5.103.

The example of the bridge explains that **extrapolation of a system model (characteristic) into an unknown region is of hypothetical character**, which cannot be proved since we have no (measurement) information about the system in the new region in the moment of decision taking. These hypotheses can be confirmed, or not, only experimentally on the basis of future data about the system. It depends only on us how we want to extrapolate the model in the Unknown – it is our risk and our responsibility. But after we have assumed the kind of extrapolation we have a right to know quantitatively what we can hope for in the new extended model region. To this aim we must know the possible extrapolation kinds. Basic extrapolation methods offered us by conventional mathematics will be presented below.

Extrapolation methods used in "conventional" mathematical models $y = f(x_1, x_2, \ldots, x_n)$ of a system (Bronsztejn 1996) are well known. In the following we limit ourselves to a system with one input x whose model has a form

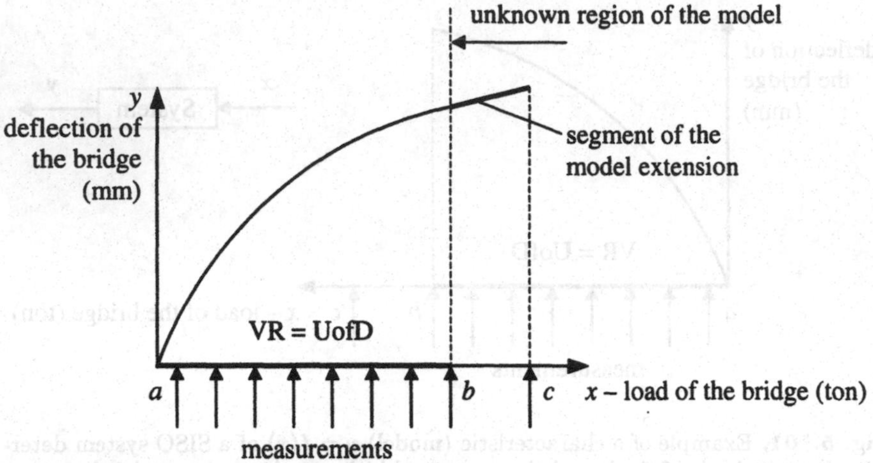

Fig. 5.102. Example of a continuous extrapolation of the bridge characteristic in the extended universe of discourse

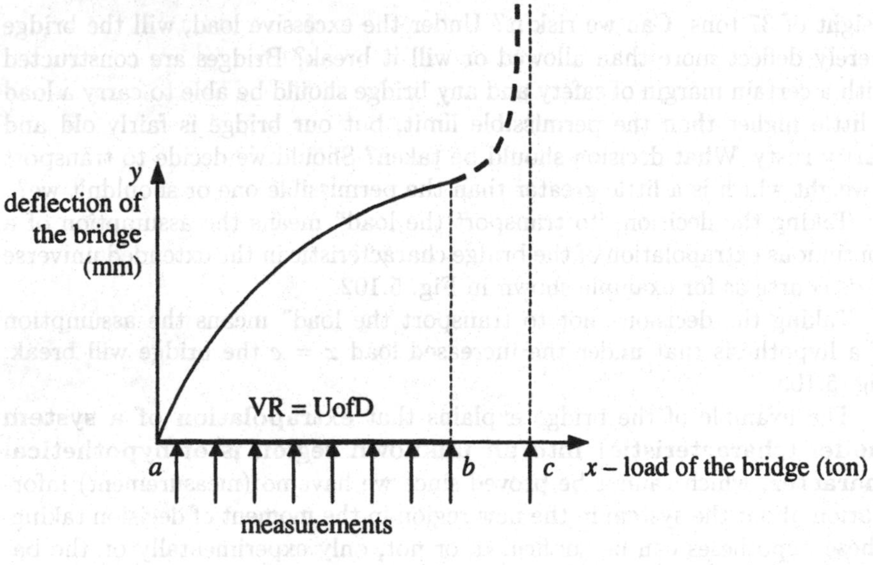

Fig. 5.103. Extrapolation of the known bridge characteristic in the region of increased load $x > b$ assuming breaking of the bridge under the load $x = c$

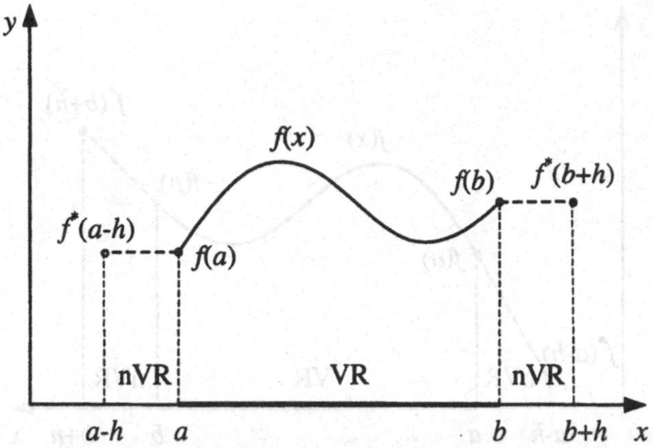

Fig. 5.104. Zero-order extrapolation of the function $f(x)$, $[a, b]$ is the validated region (VR), $x < a$ and $x > b$ are not validated regions (nVR) of the function

$y = f(x)$. If the model $y = f(x)$ is continuous and has continuous derivatives at the border points of its universe of discourse $X = [a, b]$, then by applying Taylor's expansion an approximate value $f^*(x)$ at the point $x = b + h$ placed in the close neighborhood of the region validated by measurement samples (VR) can be calculated according to the formula (5.81).

$$f(b+h) \cong f(b) + \frac{h}{1!}f'(b) + \frac{h^2}{2!}f''(b) + \ldots + \frac{h^n}{n!}f^{(n)}(b) + \ldots \qquad (5.81)$$

where: n – the extrapolation order.

The simplest kind of extrapolation is the **zero-order extrapolation** expressed by (5.82) and illustrated in Fig. 5.104.

$$\begin{aligned} f^*(b+h) &= f(b) \\ f^*(a-h) &= f(a) \end{aligned} \qquad (5.82)$$

The zero-order extrapolation is the simplest one. It uses only one piece of information from the VR: the border function value $f(a)$ or $f(b)$.

The **first-order extrapolation** is expressed by the formula (5.83) and depicted in Fig. 5.105.

$$\begin{aligned} f^*(b+h) &= f(b) + h\dot{f}(b) \\ f^*(a-h) &= f(a) - h\dot{f}(a) \end{aligned} \qquad (5.83)$$

The first-order extrapolation uses not only information about the border function value $f(a)$ or $f(b)$ but also about the derivative values $\dot{f}(a)$ or $\dot{f}(b)$ on the VR borderlines. Therefore, the probability of achieving better results, more comformable with the system behavior in an unknown external

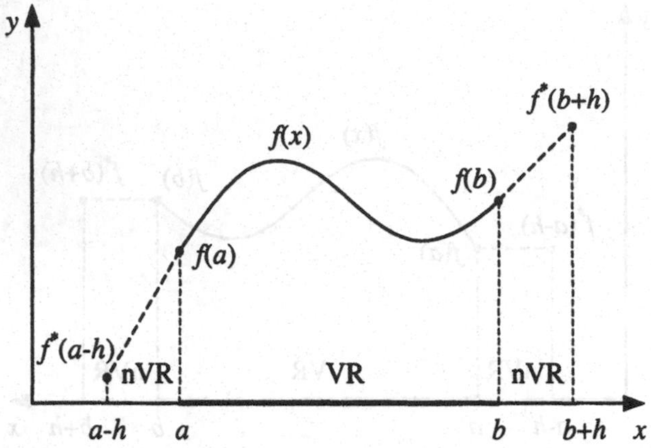

Fig. 5.105. First-order extrapolation: VR – region validated by measurements, nVR – not validated model region

neighborhood, seems higher here than in the case of using the zero-order extrapolation. Applying the second-order extrapolation (5.84) or a still higher order, one can further increase this probability. However, we must keep in mind that it is always only the probability, and not the certainty, which can be scientifically proved. Which kind of extrapolation would be better in the case of a particular system can not be stated a priori, until measurements in the unknown region have been made. But **in many situations we have to make decisions based on our extrapolated knowledge** contained in limited system models and also in fuzzy models.

$$\begin{aligned} f^*(b+h) &= f(b) + h\dot{f}(b) + \frac{h^2}{2}\ddot{f}(b) \\ f^*(a-h) &= f(a) - h\dot{f}(a) + \frac{h^2}{2}\ddot{f}(a) \end{aligned} \tag{5.84}$$

How can extrapolation of a fuzzy model be accomplished? Let us consider the simplified **problem of the increase in earnings**.

Example 5.6.1

A supermarket concern invested different amounts of money ($ millions) in the development of its network in particular years. As a result, it gained different earnings increases ($ millions) each year. The data are presented in Table 5.15.

The management of the concern considers the possibility of investing an amount of $230 millions in the building of new supermarkets in 2000. What earnings increase ΔE can be expected?

The numerical data given in Table 5.15 is the only quantitative information which can be used to forecast the investments' effects in 2000. On its basis a simple fuzzy model including 3 rules (5.85) can be created. The model is represented in Fig. 5.106.

Table 5.15. Capital expenditures of the concern and their financial effects

Year	1997	1998	1999	2000
Capital expenditure CE [$ millions]	100	150	210	230
Earnings increase ΔE [$ millions]	220	270	300	?

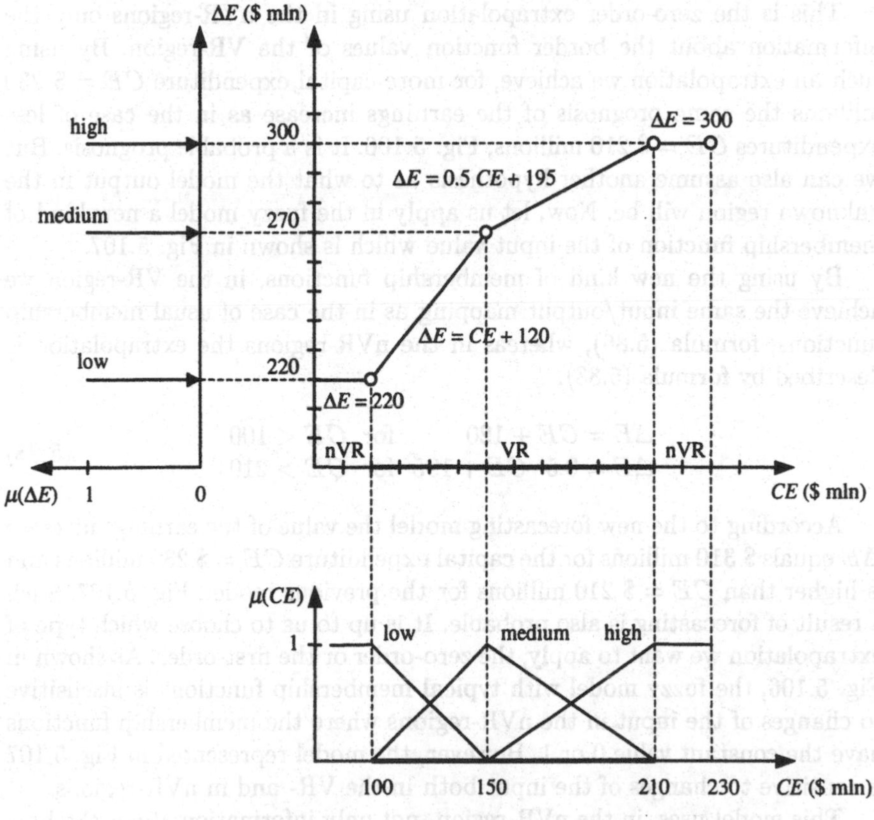

Fig. 5.106. The input/output mapping of a fuzzy model determining interdependence between capital expenditures CE and earnings increase ΔE of the supermarket concern with membership functions limited to the interval [0, 1]

$$\begin{aligned}
&IF\ (\text{capital expenditure is low})\ THEN\ (\text{earnings increase is low})\\
&IF\ (\text{capital expenditure is medium})\ THEN\ (\text{earnings increase is medium})\\
&IF\ (\text{capital expenditure is high})\ THEN\ (\text{earnings increase is high})
\end{aligned} \tag{5.85}$$

The mapping function $\Delta E = f(CE)$ (5.86) of the fuzzy model in the range VR can be readily derived for particular regions of the VR, Fig. 5.106.

$$\begin{aligned}
\Delta E &= CE + 120 && \text{for } 100 \le CE \le 150\\
\Delta E &= 0.5 \cdot CE + 195 && \text{for } 150 < CE \le 210
\end{aligned} \tag{5.86}$$

If we use the membership functions: low, medium and high limited to the interval [0, 1] in the regions nVR, Fig. 5.106, then the model surface will be saturated and described by the formula (5.87).

$$\begin{aligned} \Delta E &= 220 \;\text{ for }\; CE < 100 \\ \Delta E &= 300 \;\text{ for }\; CE > 210 \end{aligned} \tag{5.87}$$

This is the zero-order extrapolation using in the nVR-regions only the information about the border function values of the VR-region. By using such an extrapolation we achieve, for more capital expenditure $CE =$ \$ 230 millions the same prognosis of the earnings increase as in the case of less expenditures $CE =$ \$ 210 millions, Fig. 5.106. It is a probable prognosis. But we can also assume another hypothesis as to what the model output in the unknown region will be. Now, let us apply in the fuzzy model a new kind of membership function of the input value which is shown in Fig. 5.107.

By using the new kind of membership functions, in the VR-region we achieve the same input/output mapping as in the case of usual membership functions, formula (5.86), whereas in the nVR-regions the extrapolation is described by formula (5.88).

$$\begin{aligned} \Delta E &= CE + 120 && \text{for }\; CE < 100 \\ \Delta E &= 0.5 \cdot CE + 195 && \text{for }\; CE > 210 \end{aligned} \tag{5.88}$$

According to the new forecasting model the value of the earnings increase ΔE equals \$ 310 millions for the capital expenditure $CE =$ \$ 230 millions and is higher than $CE =$ \$ 210 millions for the previous model, Fig. 5.107. Such a result of forecasting is also probable. It is up to us to choose which type of extrapolation we want to apply, the zero-order or the first-order. As shown in Fig. 5.106, the fuzzy model with typical membership functions is insensitive to changes of the input in the nVR-regions where the membership functions have the constant value 0 or 1. However, the model represented in Fig. 5.107 is sensitive to changes of the input both in the VR- and in nVR-regions.

This model uses, in the nVR-region, not only information about the border output value $\Delta E =$ \$ 300 millions, but also information about the inclination of the model surface in the adjacent border part of the VR. Thus it is the first-order extrapolation.

◊ ◊ ◊

Can the extension of the membership interval above 1 and below 0 (negative membership) be explained using sound logic? Let us consider it with the example of estimating the height of people.

Example 5.6.2

Let us assume that the membership function of the linguistic variable "height" has been determined as in Fig. 5.108.

Those elements x whose membership $\mu(x)$ in the fuzzy set equals 1 can be called the "typical elements" of the given set. According to the assumed

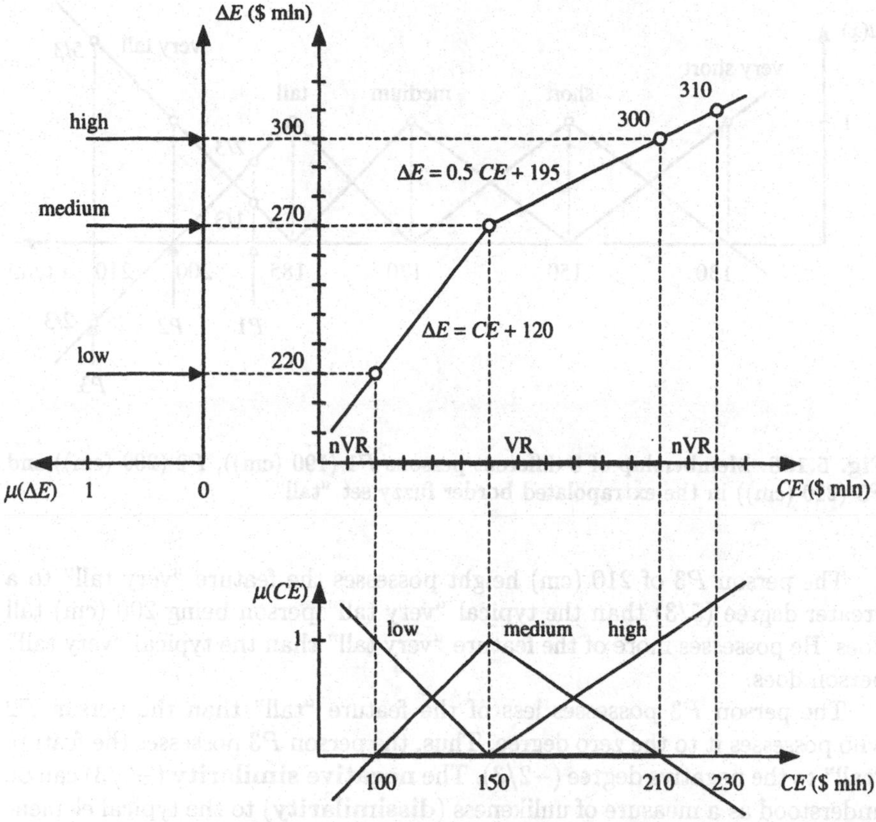

Fig. 5.107. The input/output mapping of the fuzzy model determining interdependence between capital expenditures CE and earnings increase ΔE of the supermarket concern with border membership functions not limited to the interval [0, 1]

membership functions (Fig. 5.108) the typical medium-sized man is a man being 170 (cm) tall. He entirely possesses the feature "medium", i.e. to the degree (1). The typical tall man is a man of 185 (cm) height. He entirely possesses (1) the feature "tall". The typical "very tall" height is equal to 200 (cm). **The membership $\mu(x)$ of an element x in a fuzzy set can be understood as the similarity of the element to the typical element of the given set.**

The person $P1$ of 190 (cm) height is similar to the degree of 2/3 to the typical "tall" man and to the degree of 1/3 to the "very tall" one. He possesses the feature "tall" to the degree of 2/3 and "very tall" to the degree of 1/3.

The person $P2$ of 200 (cm) height is entirely, to the degree of 1, similar to the typical "very tall" person and to the degree of 0 similar to the typical "tall" person. He entirely possesses the feature "very tall" but does not possess the feature "tall".

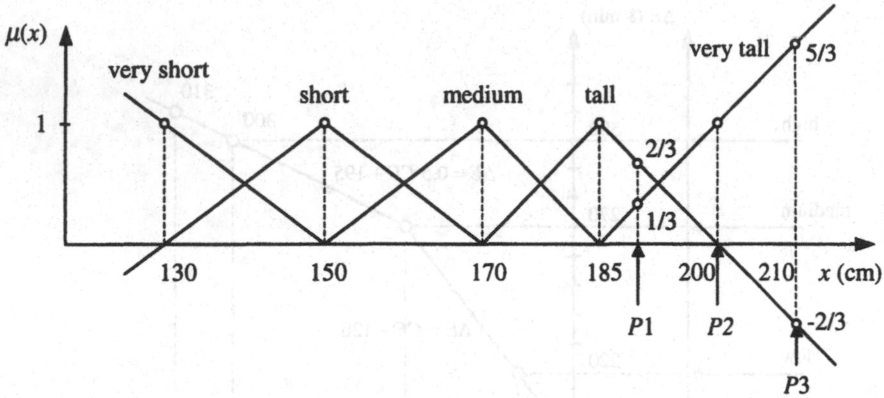

Fig. 5.108. Membership of 3 different persons $P1$ (190 (cm)), $P2$ (200 (cm)) and $P3$ (210 (cm)) in the extrapolated border fuzzy set "tall"

The person $P3$ of 210 (cm) height possesses the feature "very tall" to a greater degree (5/3) than the typical "very tall" person being 200 (cm) tall does. He possesses more of the feature "very tall" than the typical "very tall" person does.

The person $P3$ possesses less of the feature "tall" than the person $P2$ who possesses it to the zero degree. Thus, the person $P3$ possesses the feature "tall" to the negative degree $(-2/3)$. The **negative similarity** $(-2/3)$ can be understood as a measure of unlikeness (**dissimilarity**) to the typical element of the given set.

If we assume that similarity is a relative measure of the amount of a feature possessed by the given element in relation to the amount of this feature possessed by the most typical element of the given fuzzy set, then the possibility of the existence of the membership in the set which is more than 1 as well as negative becomes comprehensible. In this situation, **why should membership functions restricted to the interval $[0,1]$ always be used for internal sets and why can functions falling outside this interval only be used for external sets?**

Let us consider the fuzzy set medium height in Fig. 5.109.

Each of the sets means a certain height class. Individual classes differ one from another and therefore they can be noticed and distinguished. A typical "medium"-sized man must differ from a typical man of "short" or "tall" height, i.e. he must not be very similar to typical people of adjacent classes. This means that the function of membership in the internal height class "medium" should decrease as it nears the adjacent height classes "short" and "tall" and have a small or zero value at their typical elements (150 and 185 (cm)).

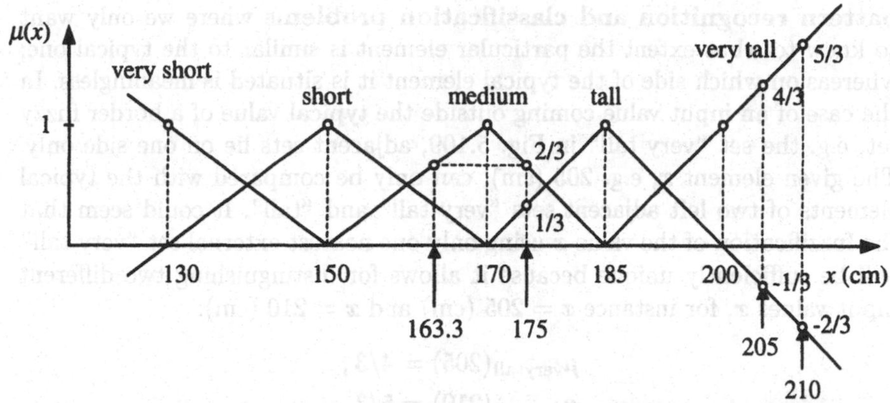

Fig. 5.109. Fuzzy sets very short, short, medium, tall, very tall and their typical elements 130 (cm) (very short), 150 (cm) (short), 170 (cm) (medium), 185 (cm) (tall), 200 (cm) (very tall)

If the function "medium" increases as it nears the center of the adjacent function "tall", then it will possess in this center a value greater than 1. It would mean that the typical "tall" man of 185 (cm) height is more "medium" than the typical "medium"-sized man (170 (cm)). **In the case of the internal sets each element x can be compared with two typical elements of the nearest adjacent classes.** For instance, the height of 175 (cm), Fig. 5.109, can be compared with the typical height of 170 (cm) of the "medium" class and 185 (cm) of the "tall" class. Only in this way can the height be coded uniquely by means of two grades of membership:

$$\mu_{\text{medium}}(175) = 2/3\,,$$
$$\mu_{\text{tall}}(175) = 1/3\,.$$

If we code this height using membership in only one class, e.g. "medium", then the coding would be non-unique because the height of 163.3 (cm) also has the same membership in this class:

$$\mu_{\text{medium}}(175) = 2/3\,,$$
$$\mu_{\text{medium}}(163.3) = 2/3\,.$$

This results from the fact that the internal membership functions are two-armed or even symmetrical. Therefore, **to secure the uniqueness of coded input values (fuzzification) it is necessary to use membership in two adjacent fuzzy sets. Uniqueness is significant in models of input/output causalities of systems** where we need to know not only the elements' membership in the set but also if the element lies to the right or left of the typical element of the set. **Uniqueness is not significant in**

pattern recognition and classification problems where we only want to know to what extent the particular element is similar to the typical one; whereas on which side of the typical element it is situated is meaningless. In the case of an input value coming outside the typical value of a border fuzzy set, e.g. the set "very tall" in Fig. 5.109, adjacent sets lie on one side only. The given element x, e.g. 205 (cm), can only be compared with the typical elements of two left adjacent sets "very tall" and "tall". It could seem that the fuzzification of the value x using only one nearest external set "very tall" will be sufficiently unique because it allows for distinguishing two different input values x, for instance $x = 205$ (cm) and $x = 210$ (cm):

$$\mu_{\text{verytall}}(205) = 4/3\,,$$
$$\mu_{\text{verytall}}(210) = 5/3\,,$$

on the basis of membership grades. But a fuzzy model coding external values x with the use of only one membership function would accomplish the zero-order and not the first-order extrapolation. It results from the fact that in the extrapolation region only one rule would always be activated, and according to the general defuzzification formula (5.38) or (5.48) a model with only one rule will always calculate a constant output value equal to the position y_B of a singleton representing a fuzzy conclusion of the rule $y = \mu_A * y_B/\mu_A$, where μ_A means the truth value of the rule premise.

To achieve the first-order extrapolation it is necessary to have information about the inclination of the fuzzy model surface in its border region. This information is given by two sets: the last border set and the last but one border set. In Fig. 5.109 they are the "very tall" and "tall" sets.

If border membership functions are restricted to the interval $[0, 1]$ as in Fig. 5.110, then the input values outside the typical values of external sets, e.g. 205 (cm) and 210 (cm), are not coded (fuzzified) the unique way, since:

$$\mu_{\text{verytall}}(205) = 1\,,$$
$$\mu_{\text{verytall}}(210) = 1\,,$$
$$\mu_{\text{tall}}(205) = 0\,,$$
$$\mu_{\text{tall}}(210) = 0\,.$$

This means that after they have been coded (after fuzzification) the fuzzy model will not distinguish them. The model output will be constant for each input value $x > 200$ (cm). Such a fuzzy model will accomplish the zero-order extrapolation.

◊ ◊ ◊

Membership in a fuzzy set can be understood in terms of truth (as the truth about membership in a fuzzy set or possessing the feature of the given set). Classical logic has applied two values of truth 0 and 1, the set {0,1}. These values can be called "crisp truth".

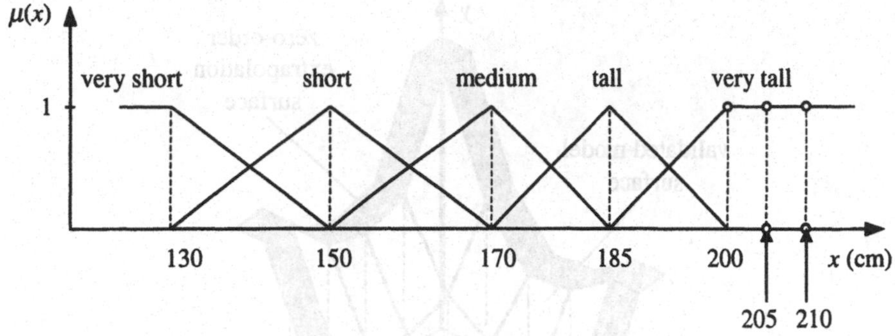

Fig. 5.110. Classification of a man's height using external fuzzy sets restricted to the interval [0,1]

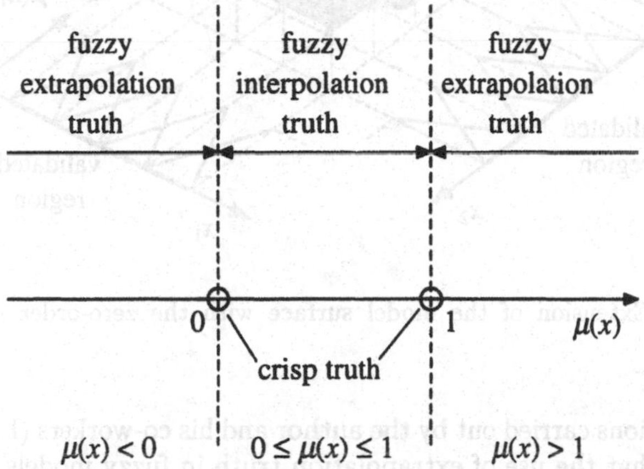

Fig. 5.111. Kinds of truth about membership in fuzzy sets

However fuzzy logic also uses fractional truth-values from the interval $[0,1]$. Because a fuzzy model based on such a truth interval generates the interpolation surface spread between the points of the input/output space which are determined by logical rules (example: the rules (5.85), Fig. 5.106), the author proposes to call the **truth from the interval** $[0,1]$ **"the interpolation truth"**. The **truth coming from outside the interval** $[0,1]$ allows for the first-order fuzzy extrapolation. Therefore it can be called **"the extrapolation truth"**. The intervals of the particular kinds of truth are represented in Fig. 5.111.

Up to the present the concepts of the fuzzy zero-order extrapolation and the first-order one have been illustrated by examples of systems with 1 (one) input. In Figs. 5.112 and 5.113 the difference between the zero-order extrapolation surface and the first-order one for a system with 2 inputs is shown.

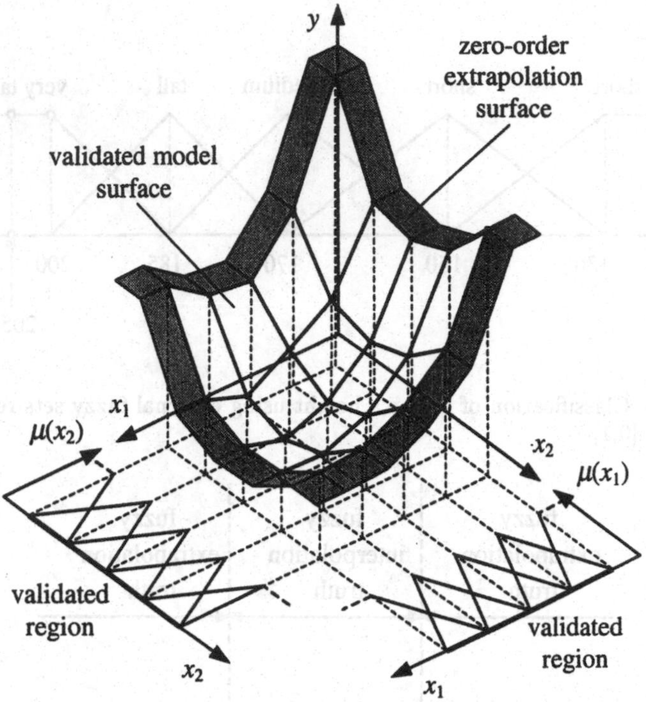

Fig. 5.112. Extension of the model surface with the zero-order extrapolation method

Investigations carried out by the author and his co-workers (Piegat 1997c) have shown that the use of extrapolation truth in fuzzy models has a good, practical point.

Fuzzy models in the form of neuro-fuzzy networks can be tuned using the error back propagation method (Brown 1994). The inputs of neuro-fuzzy networks are stimulated with the measurement samples of the system to be modeled. The network calculates its output and the error of the output which is used to correct the parameters of membership functions, e.g. a_1, a_2, a_3, a_4 in Fig. 5.114.

The correction of parameters is made until their (sub-)optimal values minimizing the mean error of the network have been found. The tuning usually starts with random values of parameters. If these values are positioned close to each other, as in Fig. 5.114, then in the case of using limited border membership functions, as in Fig. 5.114a, part of the measurement samples placed in the model insensitivity zone (derivatives equal to 0) does not cause any correction of the parameters to be tuned and the tuning process slows down.

If unlimited border membership functions, as in Fig. 5.114b, are used, then insensitivity zones do not occur and all measurement samples cause the correction of the parameters. The learning process proceeds faster. The in-

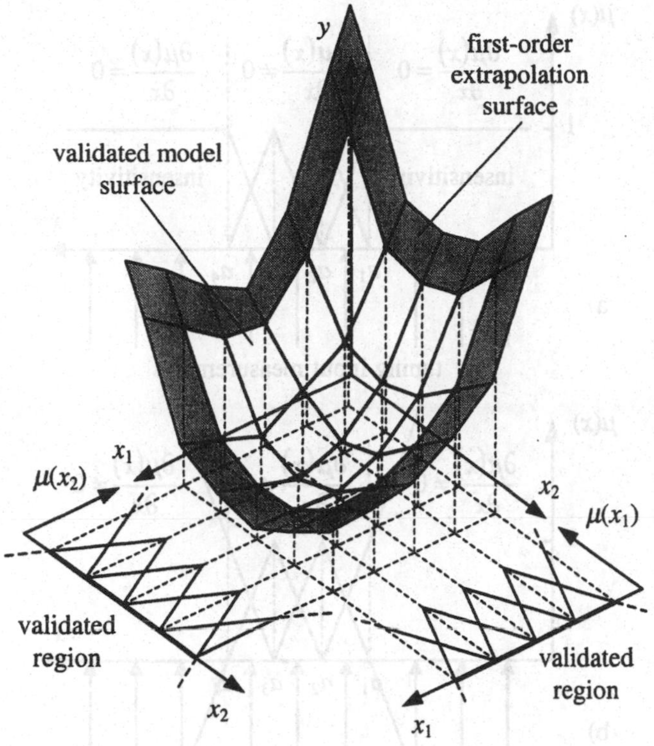

Fig. 5.113. Extension of the model surface with the first-order extrapolation method

fluence of extrapolation truth on the tuning process of a neuro-fuzzy network was investigated experimentally (Piegat 1997c).

Example 5.6.3

In the experiment the influence of extrapolation truth on the speed and accuracy of the tuning process of a neuro-fuzzy controller was investigated. The task consisted in tuning the controller to operate as a classical PID controller, Fig. 5.115. The structure of the neuro-fuzzy controller is represented in Fig. 5.116.

A triangular signal $e(k)$ of the amplitude equal to 1 and duration of 10 (s) was used as the learning signal. Responding this signal, represented in Fig. 5.117a, the model controller generated the output signal represented in Fig. 5.117b.

Two neuro-fuzzy controllers: the controller C_I – applying interpolation truth only, and the controller C_{IE} – using both interpolation and extrapolation truth were tuned with the signals represented in Fig. 5.117. Each of the controllers was tuned through a period of 500 epochs. Five tuning ex-

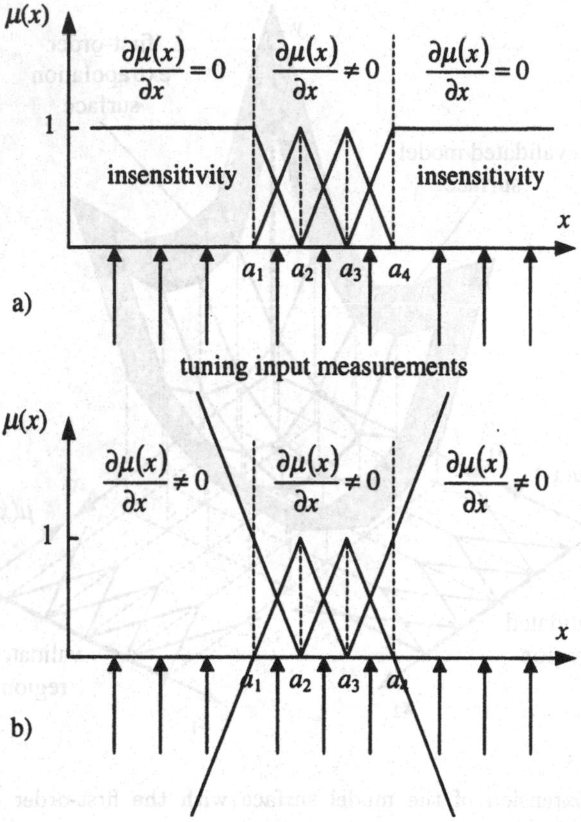

Fig. 5.114. Different border membership functions and their influence on the creation of insensitivity zones slowing down the tuning process of neuro-fuzzy networks; a) limited border membership functions, b) unlimited border functions

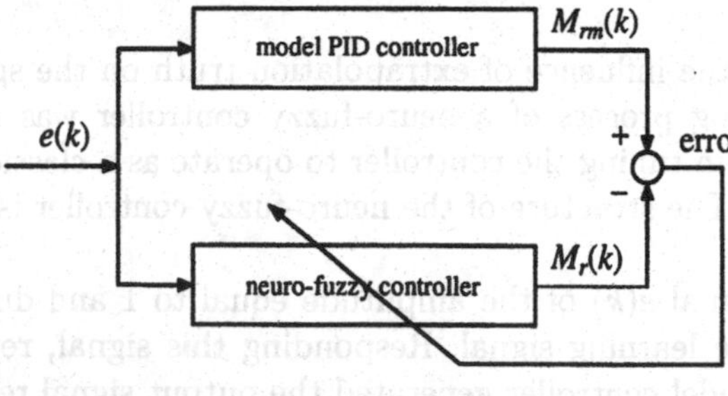

Fig. 5.115. The scheme of a neuro-fuzzy controller tuning

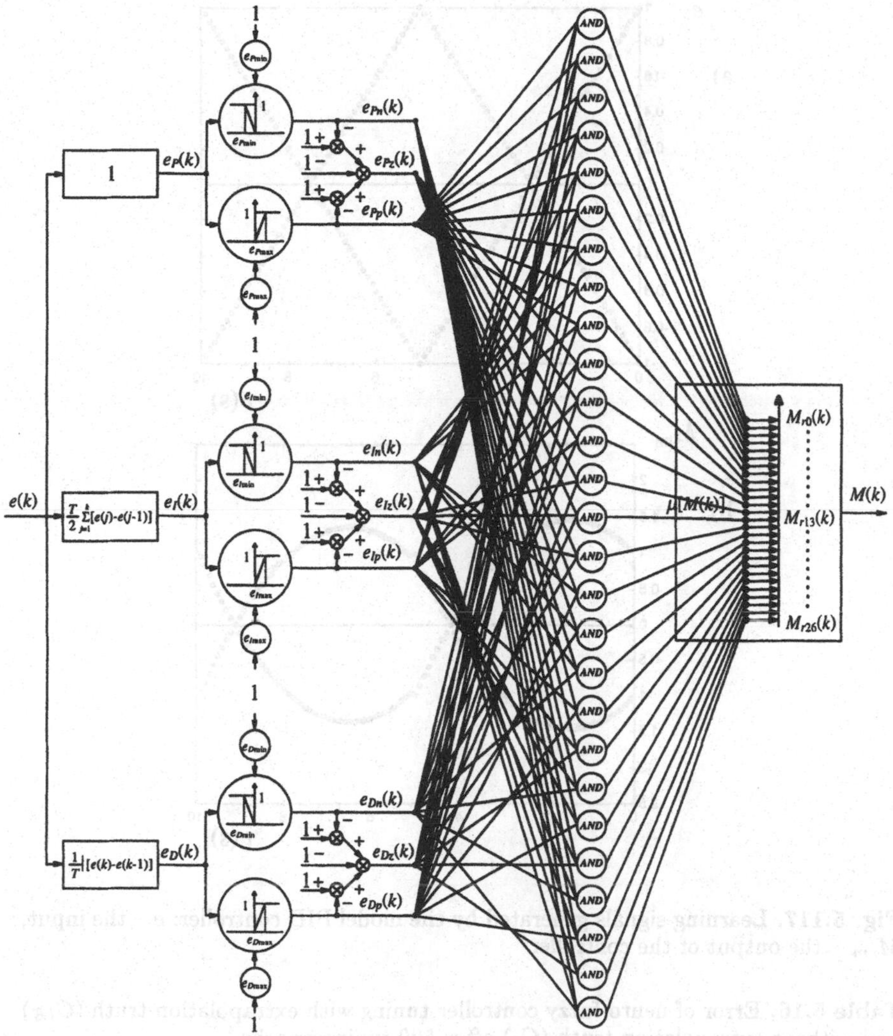

Fig. 5.116. The neuro-fuzzy controller used in the experiment

periments were carried out starting from five different initial states $S_1 - S_5$. Table 5.16 illustrates the results.

Table 5.16 shows clear superiority of the controller C_{IE} based on extrapolation truth in respect of the tuning speed. This controller achieved higher accuracy after the chosen time period. Other experiments have shown that the superiority of the controller C_{IE} to C_I grows considerably as the amplitude of the input signal $e(k)$ is increased, Fig. 5.117. If the amplitude is decreased, then at an appropriately small value the tuning speeds of both the controllers become the same. The experiments thus show the disadvan-

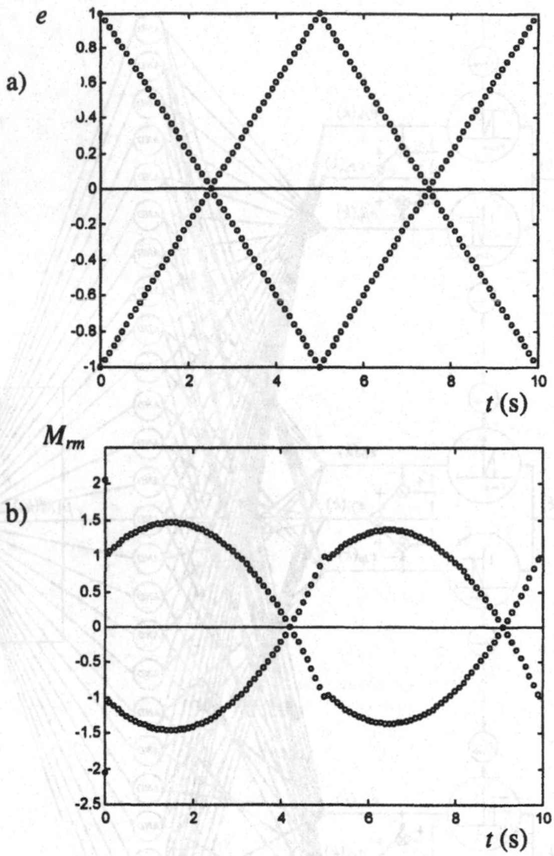

Fig. 5.117. Learning signals generated by the model PID controller: e – the input, M_{rm} – the output of the controller

Table 5.16. Error of neuro-fuzzy controller tuning with extrapolation truth (C_{IE}) and without extrapolation truth (C_I) after 500 tuning epochs

Starting point	Absolute mean error	
	C_I	C_{IE}
S_1	0,331	0,168
S_2	0,330	0,108
S_3	0,466	0,442
S_4	0,296	0,074
S_5	0,369	0,009

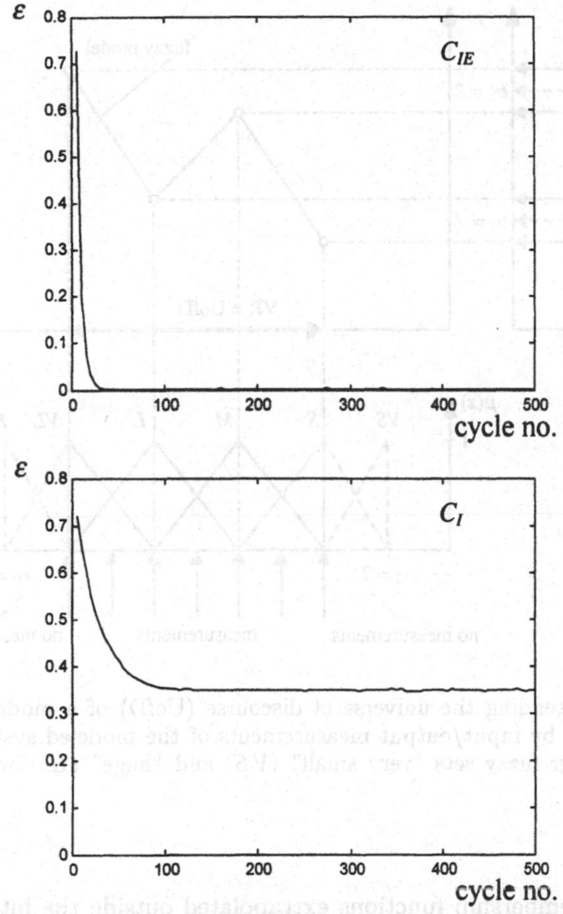

Fig. 5.118. **Comparison of the speed and accuracy of the controller tuning with extrapolation truth (C_{IE}) and without extrapolation truth (C_I)**

tageous influence of the insensitivity zones of membership functions on the tuning speed of neuro-fuzzy networks. Fig. 5.118 shows the typical differences between the tuning speed and accuracy of two controllers C_I and C_{IE}.

◊ ◊ ◊

The above-mentioned experiments prove that practical advantages result from the use of extrapolation truth in fuzzy models. Additionally, this truth enables us to better reconstruct numerical input signals on the basis of their fuzzy codes: the problem formulated by Bartolan and Pedrycz in (Bartolan 1997).

Below, **explanations to** a few, **mostly repeated questions and doubts concerning the first-order extrapolation** are given. They allow for better and deeper understanding of the idea.

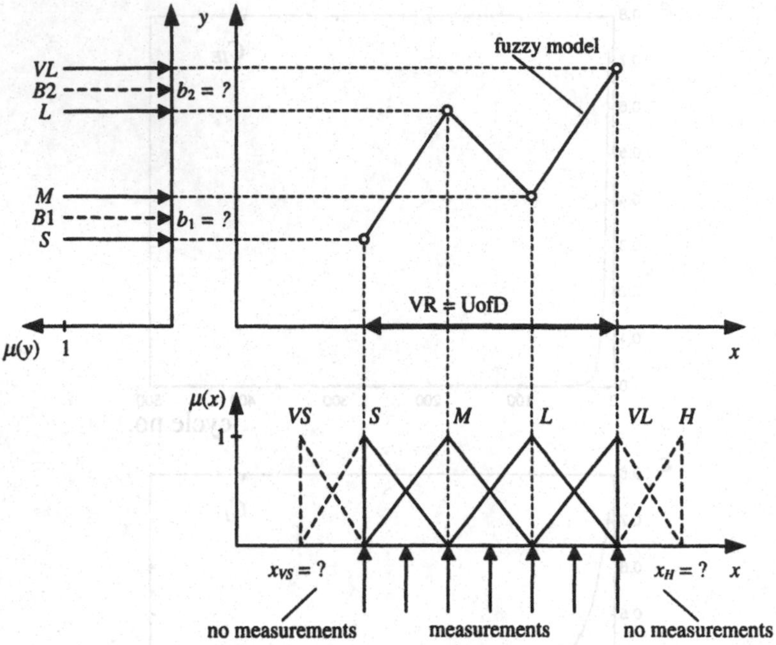

Fig. 5.119. Extending the universe of discourse (UofD) of a model outside the region validated by input/output measurements of the modeled system by introducing new outer fuzzy sets "very small" (VS) and "huge" (H) for input x and output y

Question 1
The idea of membership functions extrapolated outside the interval [0,1] is not really necessary. Membership functions can be as usual, but new outer terms can be added, and so the same effect is achieved without violating the principles of fuzzy set theory.

Answer 1
Let us consider an example of a simple fuzzy model with rule base (5.89) and membership functions depicted in Fig. 5.119.

$$\begin{aligned} &R1: \; IF\; (x = \text{small})\; THEN\; (y = \text{small}) \\ &R2: \; IF\; (x = \text{mean})\; THEN\; (y = \text{large}) \\ &R3: \; IF\; (x = \text{large})\; THEN\; (y = \text{mean}) \\ &R4: \; IF\; (x = \text{very large})\; THEN\; (y = \text{very large}) \end{aligned} \tag{5.89}$$

Extension of the model by introducing new outer fuzzy sets "very small" (VS) and "huge" (H) requires:

a) determining modal values x_{VS} and x_H of these sets, Fig. 5.119,

b) introducing new rules into the model, which assign appropriate output sets $y = B1$ and $y = B2$ to the new input sets $x = VS$ and $x = H$,

$$R0: \ IF\ (x = \text{very small})\ THEN\ (y = B1)$$

$$R5: \ IF\ (x = \text{huge})\ THEN\ (y = B2)$$

c) determining modal values of the new outer sets $B1$ and $B2$ of the output y, Fig. 5.119.

How can we accomplish requirements a) and c) if no measurements of the input x and output y of the modeled system in the new extended model domain are given?

Determination of the modal values on the basis of the fuzzy model in the validated region would mean extrapolation. In this case certain kinds of extrapolation (zero-, first- or higher order must be chosen). Determining the new modal values without using information from the validated region (thus without any information) would only be blind divination, based on nothing.

Question 2
Assuming that the proposed extension of the membership interval above 1 and below 0 is correct, it can be said in the case of height evaluation that only a set "very tall" , Fig. 5.120b, is sufficient to represent the heights for all persons. Then the concept of membership becomes useless since we can denote (code) the heights for all persons by only using this set.

Answer 2
An example of fuzzy classification of a man's height is represented in Fig. 5.120a. Really, heights for all persons can be denoted (coded) using only the set "very tall" extended above 1 and below 0. However this coding yields information only on how a particular height is similar to the typical height "very tall" equal to 200 (cm), while it does not yield information on to what extent the height is "very short", "short", "medium" or "tall". Using only one membership function (one class) for the height evaluation we can not make any reasonable classification. Therefore extending outer fuzzy sets does not mean that other sets (classes) are unnecessary.

Question 3
If the membership $\mu_A(x) = 1$ means "x in the set A", then there is no meaning for $\mu_A(x) > 1$. If the membership $\mu_A(x) = 0$ means "x is not in the set A", then there is no meaning for $\mu_A(x) < 0$.

Answer 3
Let us consider the example of a fuzzy model presented in Fig. 5.121. The fuzzy model has rules of the type: *IF* $(x = A_i)$ *THEN* $(y = B_j)$, e.g. *IF* (x = very large) *THEN* (y = very large). Which 2 significant pieces of information does a membership function representing a fuzzy set in a rule premise deliver?

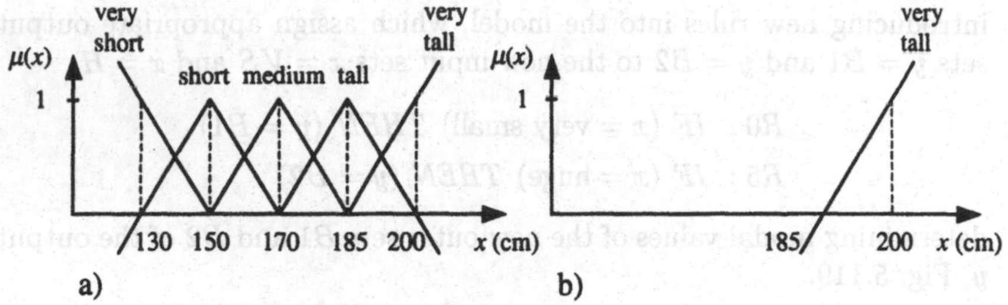

Fig. 5.120. Example of fuzzy classification of a man's height (a) and the fuzzy set "very tall" extended above 1 and below 0 (b)

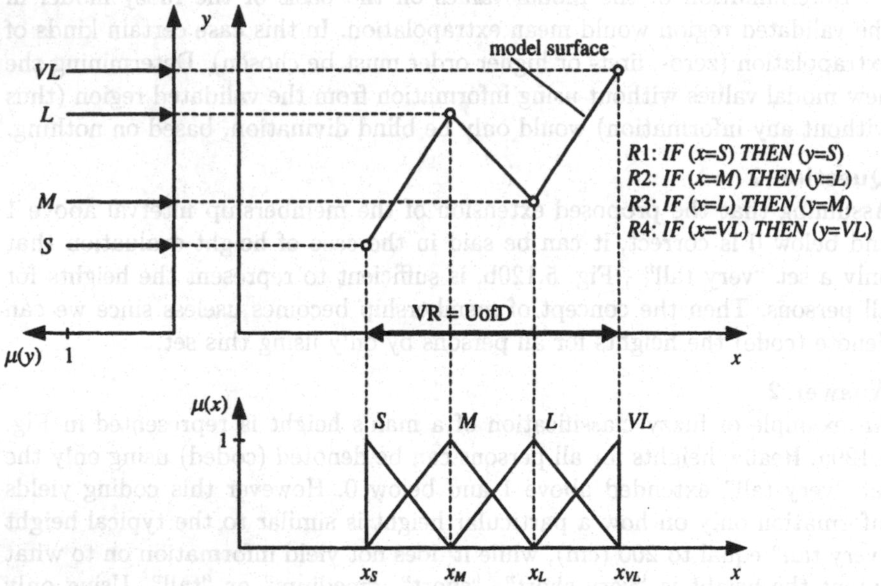

Fig. 5.121. Example of a fuzzy model of a system SISO, where: S – small, M – medium, L – large, VL – very large

1. The membership function gives an interval of the input x in which the rule is valid and is activated. E.g. rule *R4*: *IF* $(x = VL)$ *THEN* $(y = VL)$ is valid only in the interval $x_L < x < x_{VL}$. If input x is outside this interval (support of membership function) then the rule conclusion is not activated (fired) at all and does not take part in inference. The above also refers to multidimensional membership functions of composed premises in MISO-system rules.
2. The membership function of a rule premise informs us how much the conclusion of the rule has to be activated at a given input x-value (input vector $\mathbf{X}$ in the case of MISO-systems). At x equal to the

modal value of the premise set ($x = x_{VL}$), the conclusion of rule $R4$: *IF* ($x = VL$) *THEN* ($y = VL$) is activated to a degree equal to 1. At $x = x_L$ (Fig. 5.121) the rule conclusion is activated to a degree equal to 0 and at x in the interval $x_L < x < x_{VL}$ to a fractional degree. Activation strength of a rule conclusion depends on the distance of the input value x from the modal value of the membership function of the rule premise and on the shape of that function. This shape can not be just any shape (in the general case). It has to be chosen or determined so that a fuzzy model has as high an accuracy as possible (Baglio 1994).

If we want to extend a fuzzy model to the right then we must extend the validity region of its right-hand border rules. In the case of the model in Fig. 5.121 these are rules $R4$ and $R3$. Extension of a rule validity region means extending a validity region of a membership function of the rule premise. If validity regions of the border membership functions "large" and "very large" are extended in a way shown in Fig. 5.122a we get the zero-order extrapolation of the model. If extension is carried out in a way shown in Fig. 5.122b we get the first-order extrapolation. We have to make a decision – which kind of extrapolation should be used in a given system.

In the model from Fig. 5.122a, at $x > x_{VL}$ conclusion of the border rule $R4$: *IF* ($x = VL$) *THEN* ($y = VL$) is activated always to degree 1. In the model from Fig. 5.122b always to degree greater than 1. If we want to realize the first-order extrapolation we must use activation $\mu(x) > 1$. So, the first answer to Question 3, whether $\mu_A(x) > 1$ has a meaning, is: $\mu_A(x) > 1$ **can** in the case of input/output models of systems **mean that the condition (premise) of a rule is more than fulfilled, which results in increased activation of the rule conclusion** (the above refers only to border sets but not to internal input fuzzy sets – see the explanation in Example 5.6.2). The statement "the condition is more than fulfilled" is frequently used by people in everyday language. Appropriately **membership** $\mu_A(x) < 0$ **can mean that condition of a rule is less than not fulfilled which results in negative activation of the rule conclusions.**

$\mu_A(x)$ is mostly understood as a membership grade of element x in fuzzy set A. The grade of membership in a (normalized) fuzzy set is positive and its supremum equals 1 (Zimmermann 1991). Membership grade $\mu_A(x)$ can also be comprehended as similarity of a given element x to the typical element x_A of set A which fully, to degree 1, belongs to the set (typical warm temperature – univariable membership function, typical creditworthy bank customer – multivariable membership function). The number of typical elements can also be infinite in the case of trapezoidal membership functions.

Membership functions used until now, such as for example high in Fig. 5.123a, inform us only about to what degree a value x belongs to a given set. **If** $\mu_A(x)$ **can be considered as a similarity degree then a question arises why elements** x **greater than the typical element of the right border fuzzy set** (e.g. $x > x_H$ in Fig. 5.123c) **must always**

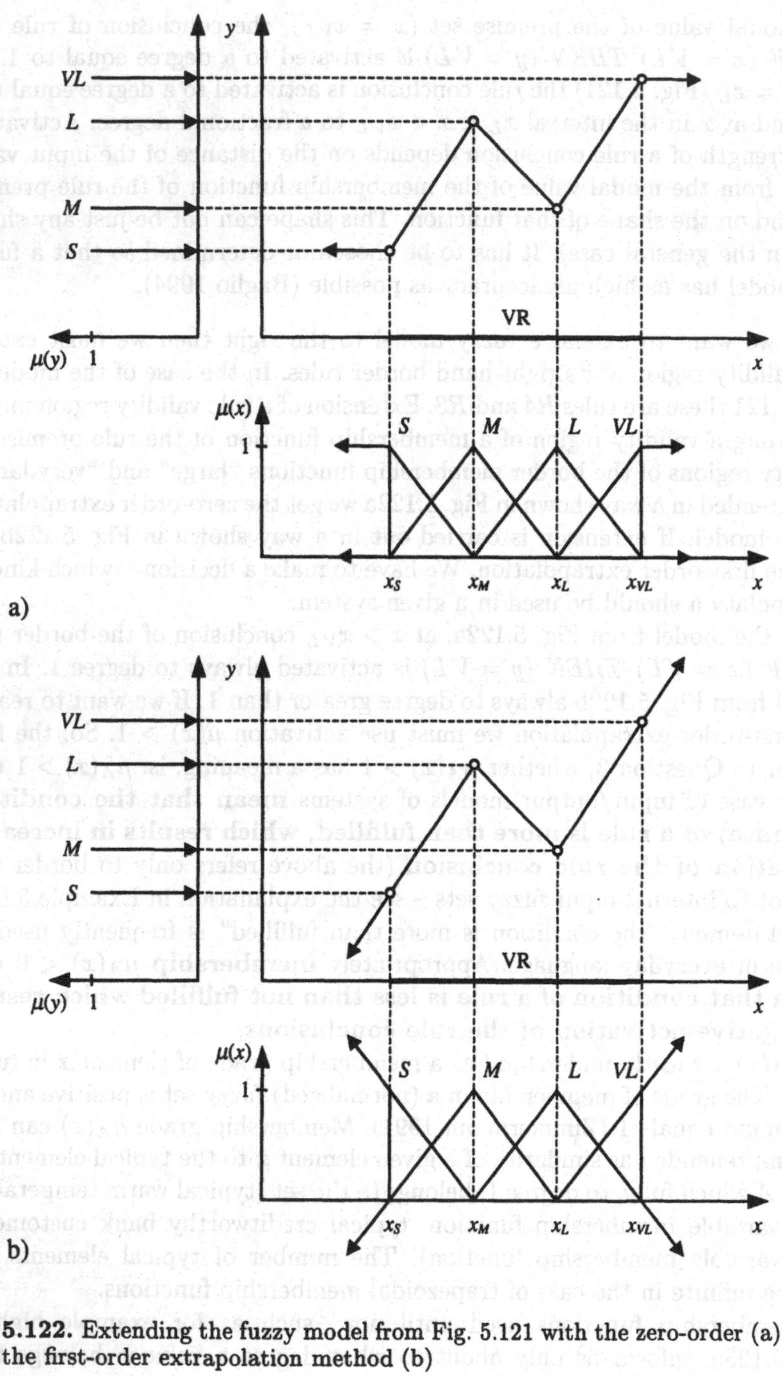

Fig. 5.122. Extending the fuzzy model from Fig. 5.121 with the zero-order (a) and with the first-order extrapolation method (b)

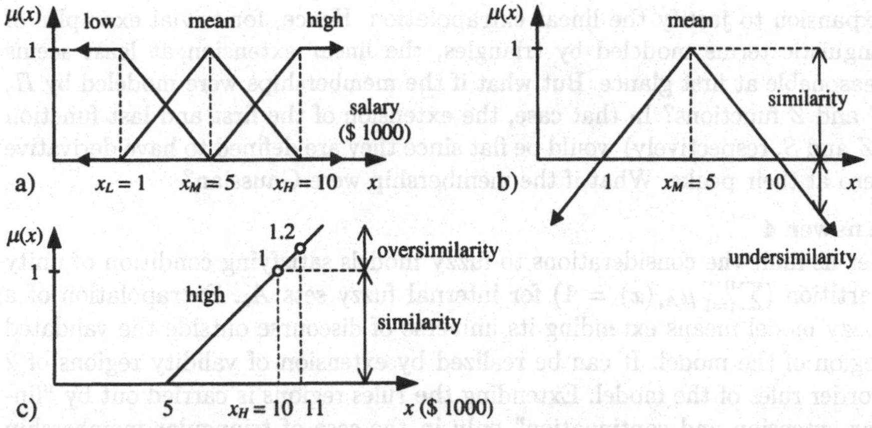

Fig. 5.123. Membership functions of the linguistic variable x = salary, x_L = \$1000, x_M = \$5000, x_H = \$10000 – typical elements of sets: low, mean and high

be treated, independently of their values, as equally similar to the typical element? Let us note that elements to the left of the typical element x_H ($x < x_H$) have differentiated similarity but elements to the right do not.

Why could we not assume that a fuzzy set A is a set of elements x with a distinguished typical (characteristic) element x_A (or subset of typical elements) of similarity $\mu_A(x) = 1$ and of other elements of similarity varying in such limits, which are advantageous to the solution of the problem under consideration? According to the above interpretation, fuzzy set A could be defined as follows:

$$A = \{(x, \mu_A(x))| \; x_A, \; \mu_{A\min} \leq \mu_A(x) \leq \mu_{A\max}, \; x \in X\}. \qquad (5.90)$$

For instance, fuzzy set A_3 = high salary in Fig. 5.123c would be defined as below:

$$A_3 = \{(x, \mu_{A_3}(x) = \frac{x - 5000}{5000}| \; x_{A_3} = 10000, \; 0 \leq \mu_{A_3}(x) \leq 199\,999, \; x \in [0, 1000\,000]\}.$$

A normalized fuzzy set could be meant to be a set which assigns to its typical element (or to a subset of typical elements) a similarity degree equal to 1. Interval $0 < \mu_A(x) \leq 1$ could be referred to as a **similarity interval**, $-\infty < \mu_A(x) \leq 0$ as a **dissimilarity** or unlikeness **interval**, and $1 < \mu_A(x) < \infty$ as an **oversimilarity interval**, Fig. 5.123b and c. In a particular specific problem different membership intervals, not necessarily limited to [0, 1], could be used.

Question 4

The idea of extrapolation reflects a linear extension, continuation of standard triangular and trapezoidal membership functions. It relies on the Taylor series expansion to justify the linear extrapolation. Hence, for trivial examples of linguistic terms modeled by triangles, the linear extension at least seems reasonable at first glance. But what if the memberships were modeled by Π, S and Z functions? In that case, the extension of the first and last function (Z and S, respectively) would be flat since they are defined to have derivative zero at their peaks. What if the membership were Gaussian?

Answer 4

Let us limit the considerations to fuzzy models satisfying condition of unity partition ($\sum_{i=1}^{n} \mu_{A_i}(x) = 1$) for internal fuzzy sets A_i. Extrapolation of a fuzzy model means extending its universe of discourse outside the validated region of the model. It can be realized by extension of validity regions of 2 border rules of the model. Extending the rules regions is carried out by "linear extension and continuation" only in the case of triangular membership functions but not in the case of trapezoidal functions as will be shown below. **First-order extrapolation means, according to the formal mathematical definition** (5.83), **a linear extension,** continuation of the inclined model surface from its inner border region into its outer neighborhood. The surface inclination is influenced mainly by support points of the surface. Positions of these points are determined by conclusions of particular rules at inputs x_i equal to modal values (typical elements) of fuzzy sets in the rule premises (see Section 5.2.1). Therefore, **extension of border membership functions should be made so that they pass through points of the border functions corresponding with their modal (typical) values.** An example of extrapolation of a fuzzy model with trapezoidal membership function is presented in Fig. 5.124.

Only **in the case where triangular membership functions are used in a model, can the influence of border rules be extended by linear continuation of border membership functions**, Fig. 5.125.

A way in which first-order extrapolation is made in the case of models with Gaussian membership functions, slightly violating the partition of unity condition ($\sum_{i=1}^{n} \mu_{A_i}(x) = 1$), is depicted in Fig. 5.126.

Segments extending a model with the Gaussian functions are straight since first-order extrapolation is, according to its definition, linear. Straight segments extending border membership functions pass through intersection points of the border functions with perpendiculars positioned at their modal (typical) values. In the case of the membership function VL in Fig. 5.126 these are points a_2 and a_3, and of the function L points a_1 and a_4. Owing to this, at the end of the validated region (VR) the model output does not vary stepwise. First-order extrapolation at Gaussian input membership functions strongly violating the condition of unity partition is also possible but more complicated because more than 2 border rules must be taken into account. For purposes of comparison, Fig. 5.127 presents a way of realizing the zero-order extrapolation for Gaussian membership functions and Fig. 5.128 for trapezoidal membership functions.

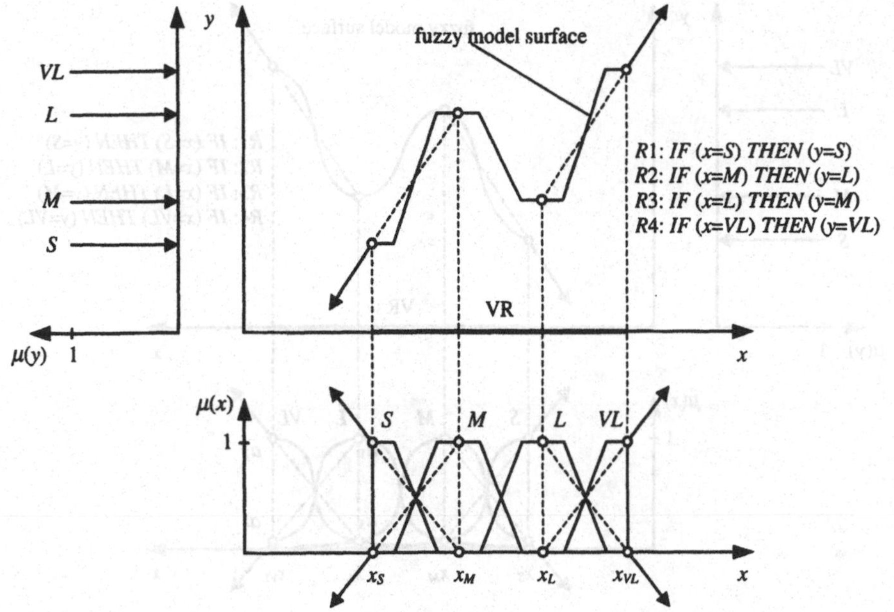

Fig. 5.124. First-order extrapolation of a fuzzy model with trapezoidal membership functions of input x

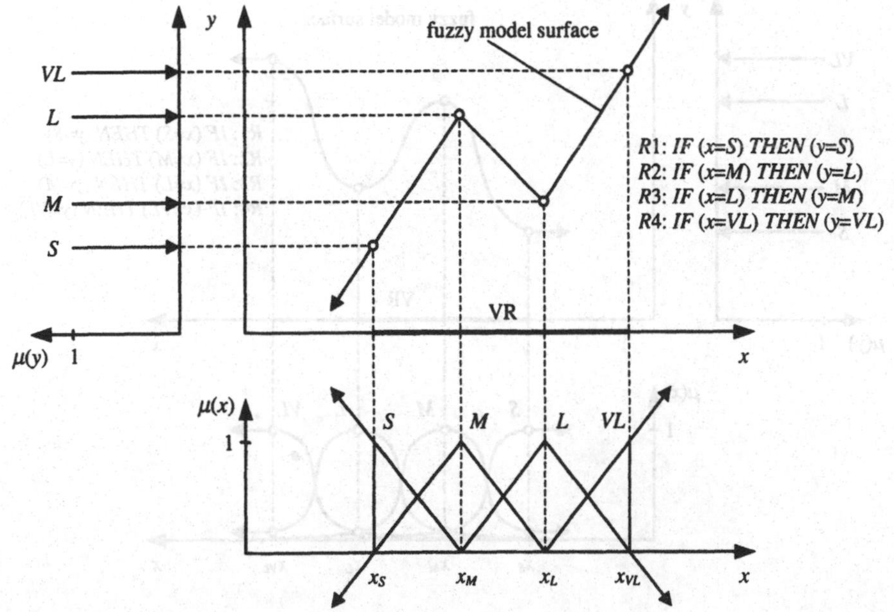

Fig. 5.125. First-order extrapolation of a fuzzy model with triangular membership functions of the model input

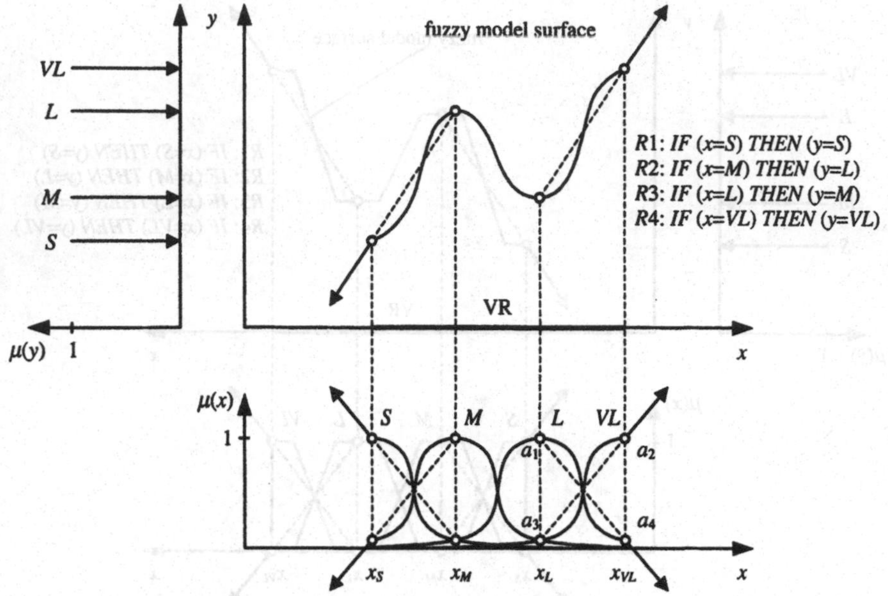

Fig. 5.126. First-order extrapolation of a fuzzy model with Gaussian membership functions of the model input

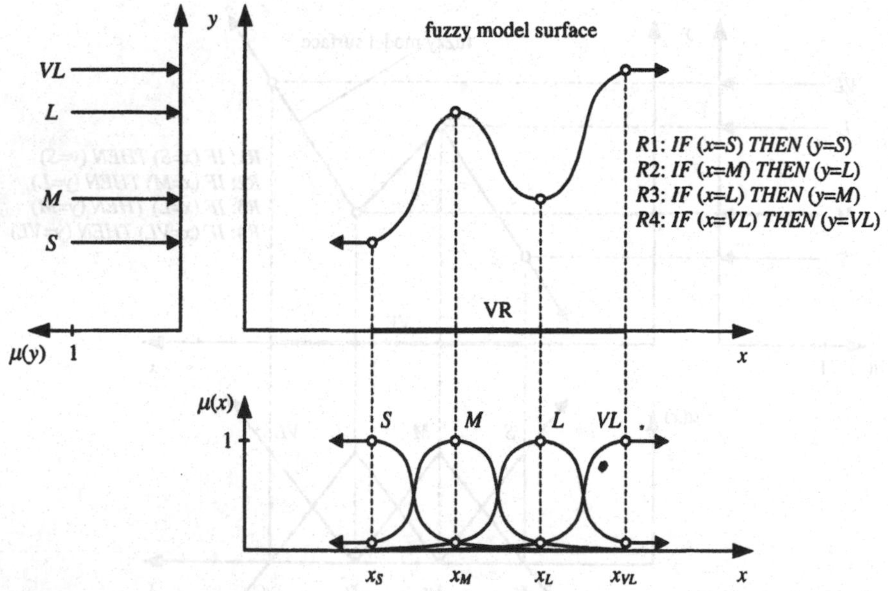

Fig. 5.127. Zero-order extrapolation of a fuzzy model with Gaussian membership functions, VR – validated region of the model

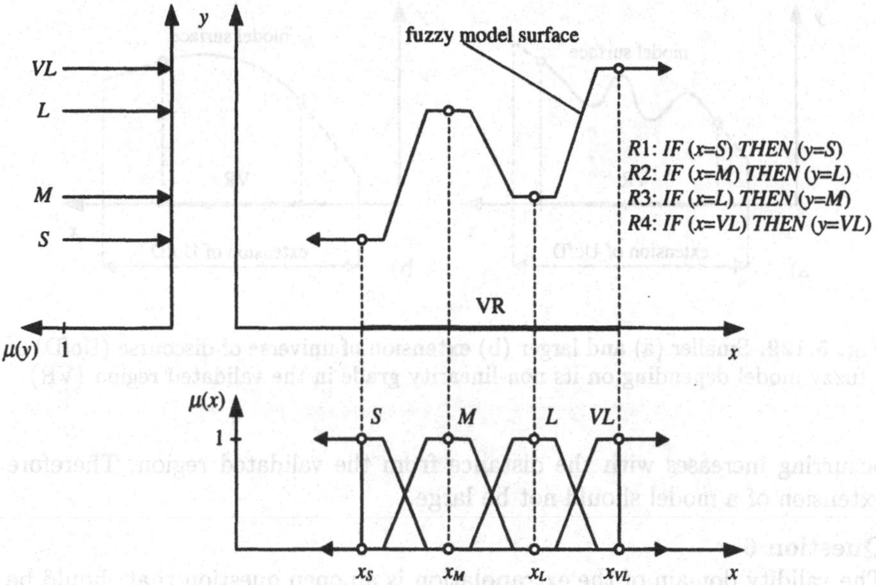

Fig. 5.128. Zero-order extrapolation of a fuzzy model with trapezoidal membership functions of the model input

Question 5

Why should we assume that the definition of membership functions should be extendable outside the range where data exists? The connection between the input/output mapping of the modeled system and the particular choice of membership functions to be used is not unique. So, why should linear extension of models really be valid? There is no theoretical explanation. Blindly extending the boundaries can give outliers in the data very high memberships, a non-desirable characteristic.

Answer 5

Extension of membership functions outside the range where data exists is necessary if we want to extrapolate a fuzzy model. The way we realize the extrapolation (zero-order, first-order, higher orders) depends on our choice and is of a hypothetical character. Whether the extrapolation method chosen is correct can not be proved at the moment of making the choice because no data about the system behavior in the unknown, extended region exists. Justification of the choice can be confirmed only later, after appropriate data from the extended region has been obtained. The extrapolation method chosen by us determines the way border membership functions are extended. **Linear extension of fuzzy models is not compulsory**. We may choose other extrapolation methods, which in our opinion would probably give better results. Large errors (outliers) of the extended fuzzy model occur when the wrong extrapolation method has been chosen. The probability of the outliers

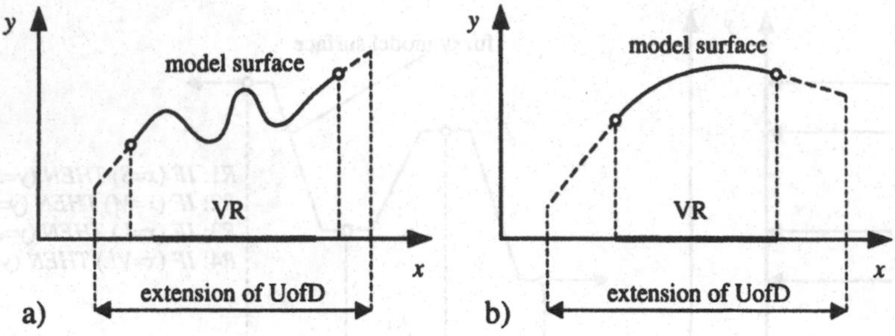

Fig. 5.129. Smaller (a) and larger (b) extension of universe of discourse (UofD) of a fuzzy model depending on its non-linearity grade in the validated region (VR)

occurring increases with the distance from the validated region. Therefore extension of a model should not be large.

Question 6
The validity domain of the extrapolation is an open question that should be definitely addressed.

Answer 6
Extension of the model domain shouldn't be large, preferably 105% of the original domain, than 150%. If variability (non-linearity) of the model surface is strong, a smaller extension should be chosen, Fig. 5.129a. If the model curvature is small as in Fig. 5.129b, a greater extension range can be chosen.

Question 7
There is no advantage demonstrated by the out-of-range membership values.

Answer 7
Owing to knowledge of extrapolation methods of fuzzy models we can extend their domains adding spaces which are in the close vicinities of their validated regions (Example 5.6.1) to original domains. The advantage of the application of the first-order extrapolation in neuro-fuzzy models (networks) is an increase in the tuning speed demonstrated in Example 5.6.3.

5.7 Types of fuzzy models

As fuzzy logic develops, new types of fuzzy models are elaborated (Babuška 1995b,c; Pedrycz 1994a; Yager 1994; Brown 1994). The aim of creating new models is to achieve greater accuracy, dimensionality and also the desire to simplify the structure. The necessity of creating new models is also caused by the great variety of real systems, different grades of accessibility of information about these systems and various forms of information.

The main advantage of fuzzy models in comparison with conventional mathematical models is the possibility of elaborating them on the basis of far lesser amounts of information about a system. The information can be of an inexact, fuzzy character. In the following, two fundamental types of fuzzy models will be presented and relationships between them will be discussed.

The most important and most often used type of a fuzzy model is the Mamdani model. This model and other derivative models will be presented in the present section.

5.7.1 Mamdani models

The idea of linguistic fuzzy models imitating the human way of thinking was elaborated by Zadeh in his pioneer works. Mamdani (Mamdani 1974,1977) applied this idea to fuzzy control of dynamic plants, showing at the same time how a model of a man-controller operating a plant can be created. The way of modeling given by Mamdani met with great interest and approbation owing to its simplicity and comprehensibility. At present it is the most often used method of modeling although other types of models are elaborated too. Among them the most important are the Takagi-Sugeno models (presented in the next section). In the Mamdani models the system to be modeled is considered to be a black box characterized by lack of information about physical phenomena occurring inside (Babuška 1995a,b).

The aim is to elaborate a model which will accomplish the mapping of its inputs (the vector $\mathbf{X}$) on the output Y (in the following we will limit ourselves to considering the MISO system with one output) in a way which approximates a real system as closely as possible (e.g. in the sense of mean absolute error). This mapping means a certain geometric surface, in the following called the mapping surface, existing in the space of the Cartesian product $\mathbf{X} \times Y$.

The Mamdani model is a set of rules where each rule defines one fuzzy point in this space. The set of fuzzy points creates a fuzzy graph in which the interpolation between the points depends on which elements are used in the fuzzy logic apparatus.

Example 5.7.1.1
A system SISO to be modeled accomplishes the mapping $y = (x-2)^2 + 1$, Fig. 5.130. The Mamdani fuzzy model of the system can have the form of a set of rules (5.91) as well as membership functions represented in Fig. 5.130a.

$$\begin{aligned} R1:\ & IF\ (x = A_1)\ THEN\ (y = B_1), \\ R2:\ & IF\ (x = A_2)\ THEN\ (y = B_2), \\ R3:\ & IF\ (x = A_3)\ THEN\ (y = B_3), \end{aligned} \tag{5.91}$$

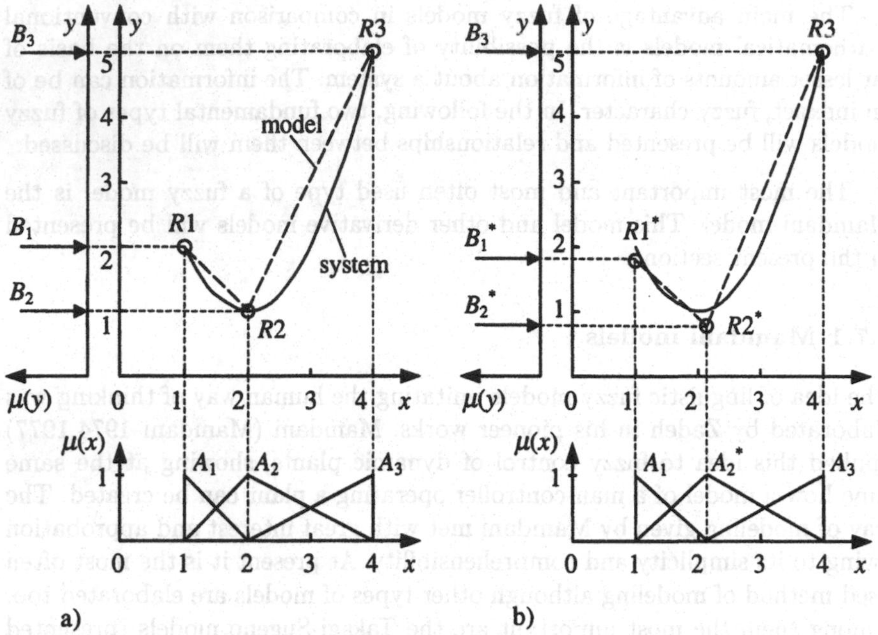

Fig. 5.130. Illustration showing how the placement of "important" points influences the fuzzy model's accuracy

where: A_1 = about 1, A_2 = about 2, A_3 = about 4,
B_1 = about 2, B_2 = about 1, B_3 = about 5,
$x: \quad 1 \leq x \leq 4.$

Each rule determines an important typical feature of the system's behavior which geometrically corresponds with a point in the space $\mathbf{X} \times Y$. The "important" points of the model can be situated directly on the characteristic of the real system, as in Fig. 5.130a. Then the characteristic of the model coincides at these points with the characteristic of the system to be modeled. Then the model rules "tell" the truth about the system, for instance the rule $R1$:

$$IF\ (x \text{ is about } 1)\ THEN\ (y \text{ is about } 2)\,,$$

determines the point $R1$, which is an important point of the system and of the model simultaneously.

However, the important point of a model need not always lie on the characteristic (surface) of a real system. As Fig. 5.130b shows, if these points are situated at other places, then higher accuracy of the model can be achieved. In this case the parameters of membership functions change (new fuzzy sets: A_2^*, B_1^*, B_2^*), and the rules have the form (5.92):

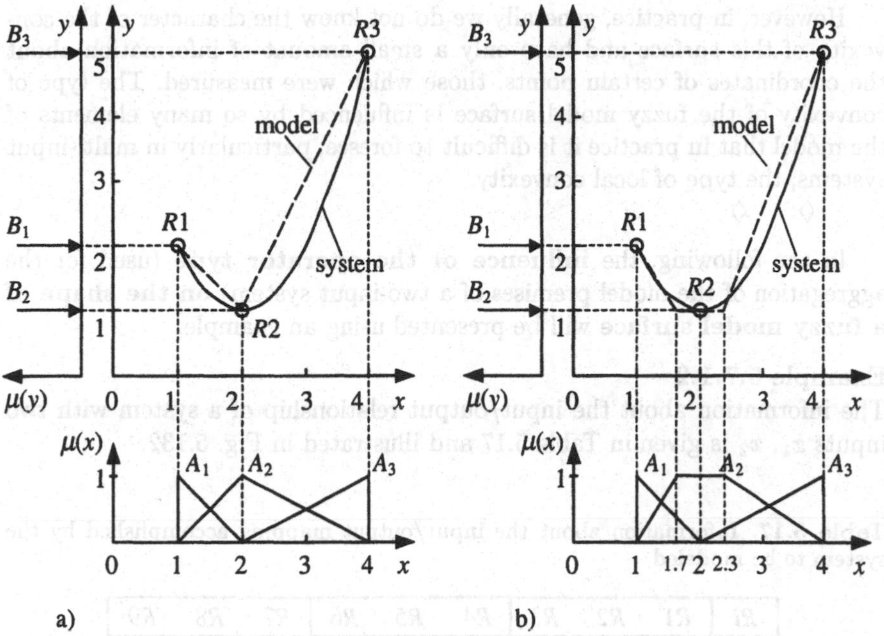

Fig. 5.131. The influence of the change of the functions of membership in the fuzzy set A_2 on the shape of the model characteristic (5.91)

$$\begin{aligned} R1^* &: \ IF\ (x = A_1)\ THEN\ (y = B_1^*), \\ R2^* &: \ IF\ (x = A_2^*)\ THEN\ (y = B_2^*), \\ R3 &: \ IF\ (x = A_3)\ THEN\ (y = B_3), \end{aligned} \tag{5.92}$$

The rules $R1^*$ and $R2^*$ in (5.92) do not tell the truth about the system because the points determined by them do not lie on the characteristic of the real system. But mean accuracy can be higher here than in the case of the model from Fig. 5.130a.

How the characteristic of a fuzzy model runs between "important" points determined by particular rules depends on the fuzzy logic apparatus used (the way of accomplishing fuzzification, defuzzification, etc.). If in the example represented in Fig. 5.131a another membership function is introduced for the set A_2, it will change the shape of the model characteristic, Fig. 5.131b.

As results from Fig. 5.131, the introduction of a trapezoidal membership function for the set A_2 causes the change of the interpolation type realized by the model between the "important" points $R1$, $R2$, $R3$ of the model. The interpolation is of a non-linear character though it is locally linear. The non-linear interpolation between "important" points of the Mamdani model can increase its accuracy, provided that the camber of the model surface between the given points will be similar to that of the system surface.

However, in practice, generally we do not know the character of the convexity of this surface and have only a small amount of information about the coordinates of certain points, those which were measured. The type of convexity of the fuzzy model surface is influenced by so many elements of the model that in practice it is difficult to foresee, particularly in multi-input systems, the type of local convexity.

◊ ◊ ◊

In the following, the **influence of the operator type** (used for the aggregation of the model premises of a two-input system) **on the shape of a fuzzy model surface** will be presented using an example.

Example 5.7.1.2
The information about the input/output relationship of a system with two inputs x_1, x_2 is given in Table 5.17 and illustrated in Fig. 5.132.

Table 5.17. Information about the input/output mapping accomplished by the system to be modeled

Ri	*R1*	*R2*	*R3*	*R4*	*R5*	*R6*	*R7*	*R8*	*R9*
x_1	1	1	1	3	3	3	5	5	5
x_2	2	4	6	2	4	6	2	4	6
y	-3	1	-3	1	5	1	-3	1	-3

As the projections of the points *Ri* lie at the nodes of the regular rectangular grid in the inputs' space $X_1 \times X_2$ of a plant, Figs. 5.132 and 5.133, these points can be used directly for creating the rule base (5.93) and inputs' and output membership functions.

$$
\begin{aligned}
R1:&\ \mathit{IF}\ (x_1 \text{ about } 1)\ \mathit{AND}\ (x_2 \text{ about } 2)\ \mathit{THEN}\ (y \text{ about } -3),\\
R2:&\ \mathit{IF}\ (x_1 \text{ about } 1)\ \mathit{AND}\ (x_2 \text{ about } 4)\ \mathit{THEN}\ (y \text{ about } 1),\\
R3:&\ \mathit{IF}\ (x_1 \text{ about } 1)\ \mathit{AND}\ (x_2 \text{ about } 6)\ \mathit{THEN}\ (y \text{ about } -3),\\
R4:&\ \mathit{IF}\ (x_1 \text{ about } 3)\ \mathit{AND}\ (x_2 \text{ about } 2)\ \mathit{THEN}\ (y \text{ about } 1),\\
R5:&\ \mathit{IF}\ (x_1 \text{ about } 3)\ \mathit{AND}\ (x_2 \text{ about } 4)\ \mathit{THEN}\ (y \text{ about } 5),\\
R6:&\ \mathit{IF}\ (x_1 \text{ about } 3)\ \mathit{AND}\ (x_2 \text{ about } 6)\ \mathit{THEN}\ (y \text{ about } 1),\\
R7:&\ \mathit{IF}\ (x_1 \text{ about } 5)\ \mathit{AND}\ (x_2 \text{ about } 2)\ \mathit{THEN}\ (y \text{ about } -3),\\
R8:&\ \mathit{IF}\ (x_1 \text{ about } 5)\ \mathit{AND}\ (x_2 \text{ about } 4)\ \mathit{THEN}\ (y \text{ about } 1),\\
R9:&\ \mathit{IF}\ (x_1 \text{ about } 5)\ \mathit{AND}\ (x_2 \text{ about } 6)\ \mathit{THEN}\ (y \text{ about } -3),
\end{aligned}
\tag{5.93}
$$

or in the general form of:

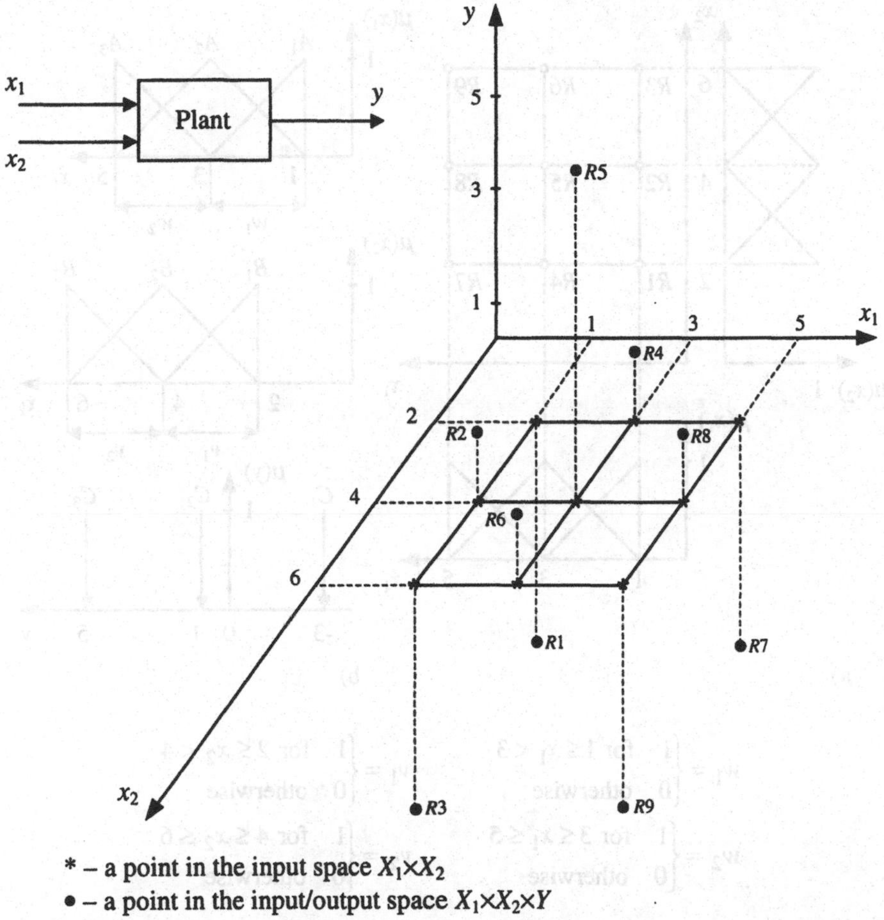

Fig. 5.132. The position of the measurements points (information about a system) in the input space $X_1 \times X_2$ and in the input/output space $X_1 \times X_2 \times Y$

$$Rl: \ IF\ (x_1 = A_i)\ AND\ (x_2 = B_j)\ THEN\ (y = C_k) \text{ where: } i, j, k = 1, 2, 3\,. \tag{5.94}$$

The membership functions of the inputs can be described by using logical sector variables w_i, v_j defined in Fig. 5.133b.

$$\begin{aligned}
\mu_{A1}(x_1) &= 0.5(3 - x_1)w_1 \\
\mu_{A2}(x_1) &= 0.5(x_1 - 1)w_1 + 0.5(5 - x_1)w_2 \\
\mu_{A3}(x_1) &= 0.5(x_1 - 3)w_2 \\
\mu_{B1}(x_2) &= 0.5(4 - x_2)v_1 \\
\mu_{B2}(x_2) &= 0.5(x_2 - 2)v_1 + 0.5(6 - x_2)v_2 \\
\mu_{B3}(x_2) &= 0.5(x_2 - 4)v_2
\end{aligned} \tag{5.95}$$

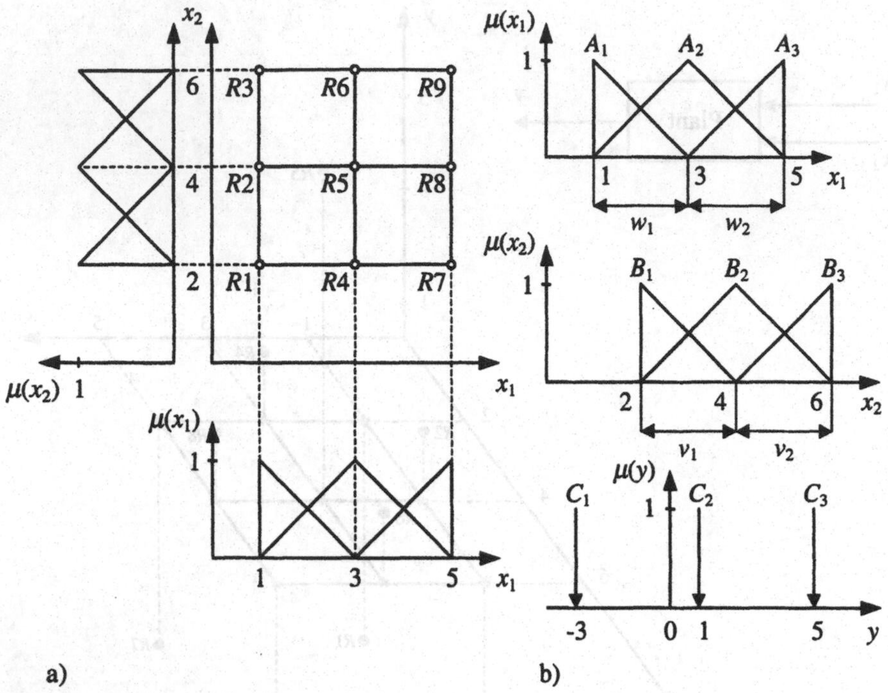

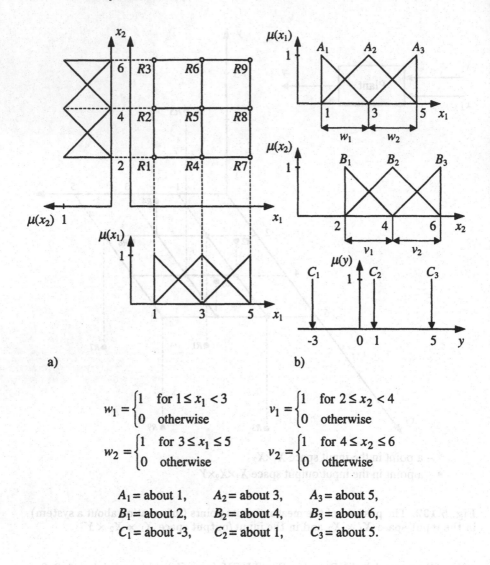

$$w_1 = \begin{cases} 1 & \text{for } 1 \le x_1 < 3 \\ 0 & \text{otherwise} \end{cases} \qquad v_1 = \begin{cases} 1 & \text{for } 2 \le x_2 < 4 \\ 0 & \text{otherwise} \end{cases}$$

$$w_2 = \begin{cases} 1 & \text{for } 3 \le x_1 \le 5 \\ 0 & \text{otherwise} \end{cases} \qquad v_2 = \begin{cases} 1 & \text{for } 4 \le x_2 \le 6 \\ 0 & \text{otherwise} \end{cases}$$

A_1 = about 1, A_2 = about 3, A_3 = about 5,
B_1 = about 2, B_2 = about 4, B_3 = about 6,
C_1 = about -3, C_2 = about 1, C_3 = about 5.

Fig. 5.133. Rectangular network in the inputs' space of a system (a), the assumed membership functions of the inputs and output (b), as well as the definition of logical sector variables w_i, v_j

Illustration of the influence of the operator type used for the intersection operation of sets (*AND*)
By using the *PROD*-operator (algebraic product) for the *AND* operation occurring in the rule premises (5.93) we achieve the formula (5.96) determining the output of the fuzzy model:

$$y^*(x_1,x_2) = \frac{\mu_{A1}(x_1)\mu_{B1}(x_2)(-3)+\mu_{A1}(x_1)\mu_{B2}(x_2)(1)+\mu_{A1}(x_1)\mu_{B3}(x_2)(-3)}{M^*}+$$
$$+\frac{\mu_{A2}(x_1)\mu_{B1}(x_2)(1)+\mu_{A2}(x_1)\mu_{B2}(x_2)(5)+\mu_{A2}(x_1)\mu_{B3}(x_2)(1)}{M^*}+$$
$$+\frac{\mu_{A3}(x_1)\mu_{B1}(x_2)(-3)+\mu_{A3}(x_1)\mu_{B2}(x_2)(1)+\mu_{A3}(x_1)\mu_{B3}(x_2)(-3)}{M^*},$$

$$\text{where:} \quad M^* = \sum_{i=1}^{3}\sum_{j=1}^{3}\mu_{Ai}(x_1)\mu_{Bj}(x_2). \tag{5.96}$$

By using the *MIN*-operator for the realization of the intersection operation of sets (*AND*) in the rules (5.93) we get the formula (5.97) describing the model output y^{**}.

$$y^*(x_1,x_2) = \frac{-3MIN[\mu_{A1}(x_1),\mu_{B1}(x_2)]+1MIN[\mu_{A1}(x_1),\mu_{B2}(x_2)]-3MIN[\mu_{A1}(x_1),\mu_{B3}(x_2)]}{M^{**}}+$$
$$+\frac{1MIN[\mu_{A2}(x_1),\mu_{B1}(x_2)]+5MIN[\mu_{A2}(x_1),\mu_{B2}(x_2)]+1MIN[\mu_{A2}(x_1),\mu_{B3}(x_2)]}{M^{**}}+$$
$$+\frac{-3MIN[\mu_{A3}(x_1),\mu_{B1}(x_2)]+1MIN[\mu_{A3}(x_1),\mu_{B2}(x_2)]-3MIN[\mu_{A3}(x_1),\mu_{B3}(x_2)]}{M^{**}},$$

$$\text{where:} \quad M^{**} = \sum_{i=1}^{3}\sum_{j=1}^{3} MIN[\mu_{Ai}(x_1),\mu_{Bj}(x_2)]. \tag{5.97}$$

Because the formulas (5.96) and (5.97) are different, then in the general case the outputs $y^*(x_1,x_2)$ and $y^{**}(x_1,x_2)$ will be different for the same input vector $[x_1,x_2]^{\mathrm{T}}$, although in certain points they can be identical, e.g. at the points indicated by the rule premises (5.93), i.e. at the "important" model points. This means that the interpolation realized by the model (5.93) using the *PROD*-operator and *MIN*-operator is different. Fig. 5.134 represents the surfaces of fuzzy models achieved by the use of the *PROD*- and *MIN*-operator.

Table 5.18 gives example values of the inputs y^* and y^{**} for such input vectors $[x_1,x_2]^{\mathrm{T}}$ which were not used to construct the model.

As results from Fig. 5.134 show, the surface of the fuzzy model with the *PROD*-operator is smoother than that with the *MIN*-operator. In order to state which model is more accurate it would be necessary to have a test set of input/output measurement vectors of the system to be modeled and to investigate the absolute (or square) mean error at these points. It would be interesting to investigate what type of interpolation is achieved in the Mamdani model between the well-known "important" points indicated by the rules.

Table 5.18. Example output values of the fuzzy model (5.93) calculated using the *PROD*- and *MIN*-operators

x_1	1.5	2.5	3.5	4.33
x_2	2.5	2.5	3.5	5.33
PROD y^*	−1	1	3	−0.33
MIN y^{**}	−0.33	1	2.66	0.2

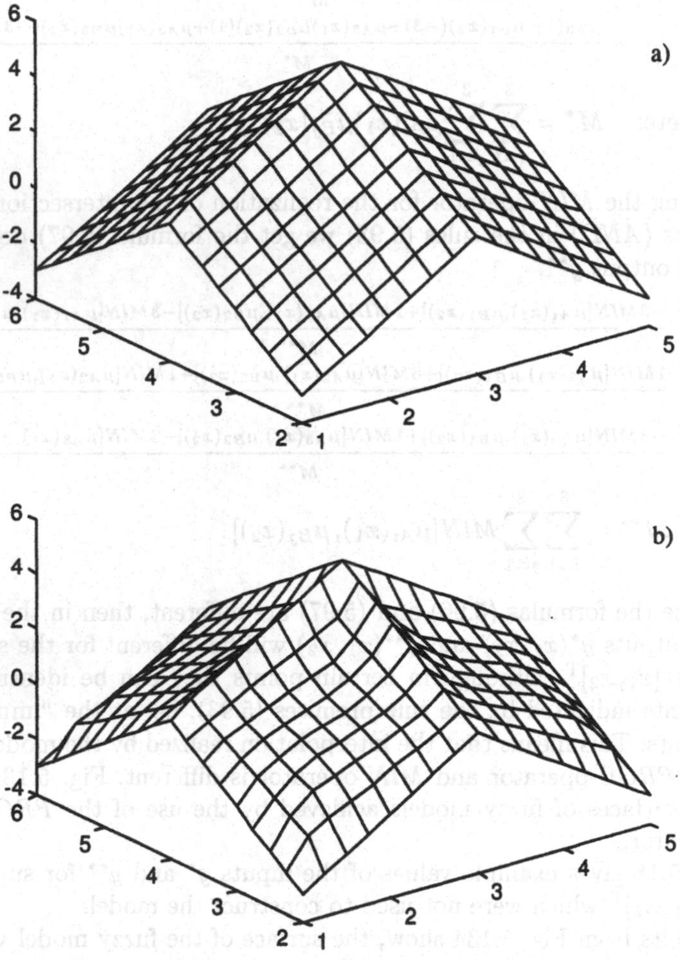

Fig. 5.134. The surfaces of the input/output mapping of the model (5.93) achieved by the use of the *PROD*-operator (a) and *MIN*-operator (b)

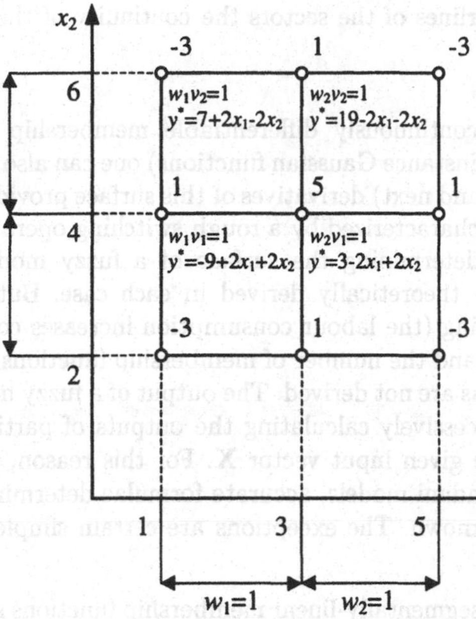

Fig. 5.135. The formulas for calculating the interpolation surfaces of the model (5.93) (with *PROD*-operator) in the particular sectors of the input space $X_1 \times X_2$

In the case of the model considered in the example, the description of the interpolation surface can be derived by the use of the formulas (5.96) and (5.97). This surface is different in each sector of the inputs' space $X_1 \times X_2$. In Fig. 5.135 the interpolation surfaces in the case of using the *PROD*-operator are given.

In the discussed example the interpolation surfaces are linear though the non-linear *PROD*-operator has been used in the model. This case is special for the reason that the support points of the model shown in Fig. 5.132 and in Table 5.17 corresponding with the nodes of the grid from Fig. 5.135 are situated so that four linear segments can be passed through them. However, if the *PROD*-operator and segmentally-linear membership functions for the inputs and singleton functions for the output are used, then non-linear interpolation functions of multi-linear type (multi-linear functions) containing a product element are usually achieved.

In the case of three inputs such a function has the form (5.98):

$$y = a_0 + a_1 x_1 + a_2 x_2 + a_3 x_3 + a_4 x_1 x_2 + a_5 x_1 x_3 + a_6 x_2 x_3 + a_7 x_1 x_2 x_3 \,. \quad (5.98)$$

The interpolation functions of the Mamdani model have, at the sector borderlines, common vertices (nodes) and common edges (identical values of the function). It can be checked in the example given in Fig. 5.135. Therefore on the borderlines of the sectors the continuity of the model surface is maintained.

◊ ◊ ◊

By applying continuously differentiable membership functions of the model inputs (for instance Gaussian functions) one can also achieve the continuity of the first (and next) derivatives of this surface provided that operators of the *MIN*-type characterized by a rough switching operation are not used.

The formula determining the surface of a fuzzy model in open form, $y = f(\mathbf{X})$, can be theoretically derived in each case. But in practice it is so labour-consuming (the labour consumption increases considerably as the number of inputs and the number of membership functions becomes greater) that these formulas are not derived. The output of a fuzzy model is calculated as a result of successively calculating the outputs of particular elements of the model for the given input vector $\mathbf{X}$. For this reason, in practice, when applying the Mamdani models, accurate formulae determining interpolation surfaces are not known. The exceptions are certain simpler cases described below:

- when applying segmentally-linear membership functions satisfying the condition of unity partition for the inputs and singleton functions for the output, the *MEAN* (mean value) and *SUM* (unlimited sum) operator, we achieve a globally non-linear model surface composed of locally linear segments accomplishing the functions of type (5.99):

$$y = a_0 + a_1x_1 + a_2x_2 + \ldots + a_nx_n\,, \tag{5.99}$$

where: n – the number of inputs,

- when applying segmentally-linear membership functions satisfying the condition of unity partition for the inputs and singleton functions for the output, the *PROD*-operator for intersection operation and the *SUM*-operator for union operation of sets, we achieve a model surface composed of multilinear segments realizing the function of type (5.98).

In remaining cases we achieve, in the Mamdani models, different and difficult-to-determine non-linear interpolation surfaces.

Linguistic and non-linguistic Mamdani models

At the beginning, nearly only the linguistic labels of the type "small", "large" were used in the Mamdani models. The fuzzy models containing such labels of fuzzy sets are called linguistic models. However, in practice, it was noticed that the assigning of linguistic labels to fuzzy sets often has little sense or meaning. Examples are given in Fig. 5.136.

In the example in Fig. 5.136a the fuzzy set "about 9" is so close to the set "about 10" that it would be difficult to call it the "mean" one. In the example in Fig. 5.136b the great number of fuzzy sets makes it difficult to use linguistic labels. In this case a great number of labels would be necessary.

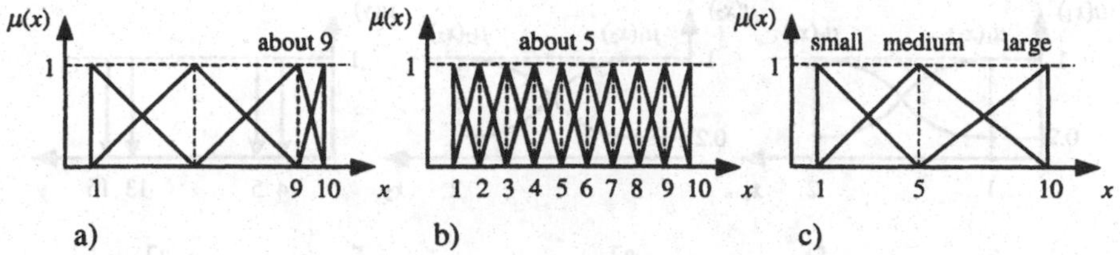

Fig. 5.136. Examples illustrating the rationale of assigning linguistic labels to fuzzy sets (c) or presenting the sets in the form of fuzzy numbers (a,b)

It would cause difficulties in distinguishing them, e.g.: small, nearly small, medium-small, nearly medium, medium and so on. It is more practical here to use labels in form of fuzzy numbers, e.g.: about 1, about 2, about 3 etc. They give better orientation as regards the position of each set.

Fig. 5.136c gives the example where the use of linguistic labels of the type small, medium, large is sensible. The distances between the cores of particular sets are appropriately large and the number of sets is small. Fuzzy models containing labels of sets in the form of fuzzy numbers are called non-linguistic models.

When do the rules in the Mamdani model not "tell" the truth?
The truth of the information contained in the rule premises and conclusions of the Mamdani model is of great importance for understanding the input/output relationship of a model and a modeled system. The best situation is when the rules "tell" the truth, i.e. if one of the model rules has the form:

$$IF\ (x_1 \text{ about } 2)\ AND\ (x_2 \text{ about } 1)\ THEN\ (y \text{ about } 5)\,,$$

then after applying the vector $[x_1, x_2]^{\mathrm{T}} = [2, 1]$ to the model input the model calculates the output $y = 5$, i.e. the output given by the conclusion of the above rule. However, the next example shows that such a situation does not always occur.

Example 5.7.1.3
Let us consider the fuzzy model of a two-input plant having the rule base (5.100) and membership functions represented in Fig. 5.137.

$$\begin{aligned}
R1:&\ IF\ (x_1 \text{ about } 1)\ AND\ (x_2 \text{ about } 1)\ THEN\ (y \text{ about } 4)\,,\\
R2:&\ IF\ (x_1 \text{ about } 1)\ AND\ (x_2 \text{ about } 2)\ THEN\ (y \text{ about } 13)\,,\\
R3:&\ IF\ (x_1 \text{ about } 2)\ AND\ (x_2 \text{ about } 1)\ THEN\ (y \text{ about } 5)\,,\\
R4:&\ IF\ (x_1 \text{ about } 2)\ AND\ (x_2 \text{ about } 2)\ THEN\ (y \text{ about } 16)\,, \qquad (5.100)
\end{aligned}$$

In the course of calculating the model output y for the input vectors given in the rule premises, i.e.:

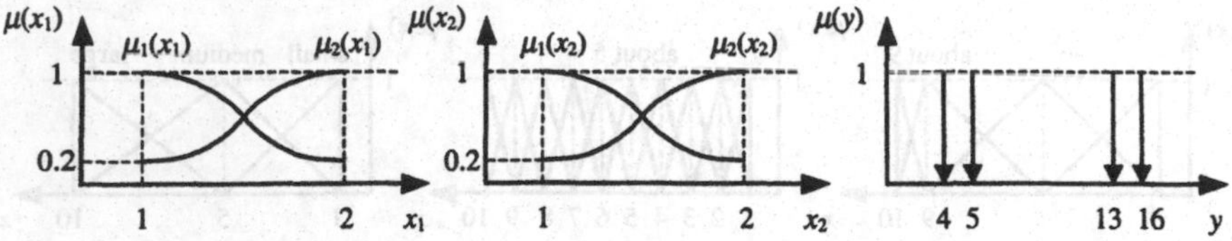

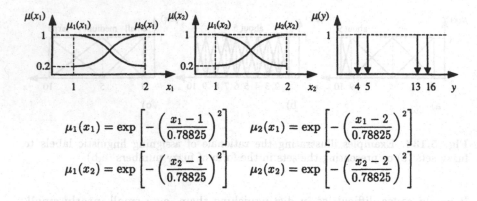

$$\mu_1(x_1) = \exp\left[-\left(\frac{x_1-1}{0.78825}\right)^2\right] \qquad \mu_2(x_1) = \exp\left[-\left(\frac{x_1-2}{0.78825}\right)^2\right]$$

$$\mu_1(x_2) = \exp\left[-\left(\frac{x_2-1}{0.78825}\right)^2\right] \qquad \mu_2(x_2) = \exp\left[-\left(\frac{x_2-2}{0.78825}\right)^2\right]$$

Fig. 5.137. Gaussian functions of membership in fuzzy sets of the inputs and output of an exemplified two-input system

$$\mathbf{X}(R1) = [1,1]^{\mathrm{T}}, \ \mathbf{X}(R2) = [1,2]^{\mathrm{T}}, \ \mathbf{X}(R3) = [2,1]^{\mathrm{T}}, \ \mathbf{X}(R4) = [2,2]^{\mathrm{T}}, \tag{5.101}$$

all rules are activated each time, not only the rule that has the given vector $\mathbf{X}(Ri)$ in its premise.

Because all rules always participate in calculating the output, it differs from the value y_i given in the conclusion of the rule Ri. For example, when calculating the model output for the input vector $\mathbf{X}(R4) = [2,2]^{\mathrm{T}}$, the grades of satisfying the premises of the particular rules (using the *PROD*-operator for the *AND* operation) are as follows:

$$\mu_{R1}(2,2) = \mu_1(x_1)\mu_1(x_2) = 0.04\,,$$
$$\mu_{R2}(2,2) = \mu_1(x_1)\mu_2(x_2) = 0.2\,,$$
$$\mu_{R3}(2,2) = \mu_2(x_1)\mu_1(x_2) = 0.2\,,$$
$$\mu_{R4}(2,2) = \mu_2(x_1)\mu_2(x_2) = 1\,.$$

The grades of truth of the premises μ_{Ri} activate the conclusions of all rules. The model output is calculated according to the following formula:

$$y_m = \frac{\sum_{i=1}^{4} y_i\mu_{Ri}}{\sum_{i=1}^{4}\mu_{Ri}} = \frac{4\cdot 0.04 + 13\cdot 0.2 + 5\cdot 0.2 + 16\cdot 1}{0.04 + 0.2 + 0.2 + 1} = 13.72\,.$$

The output $y_m(2,2)$ of the model is other than the value $y_4 = 16$ given in the conclusion of the rule $R4$. Table 5.19 presents the comparison of the model outputs y_{mi} for the input vectors (5.101) occurring in the rule premises (5.100) with the values y_i occurring in the conclusions of the rules Ri. As results from Table 5.19 show, the outputs y_{mi} calculated by the model are

Table 5.19. The comparison of the outputs y_{mi} calculated by the model (5.100) with the rule conclusions y_i in this model

Rule Ri		$R1$	$R2$	$R3$	$R4$
Input vector	x_1	1	1	2	2
	x_2	1	2	1	2
Conclusion	y_i	4	13	5	16
Model output	y_{mi}	5.72	11.94	6.61	13.72

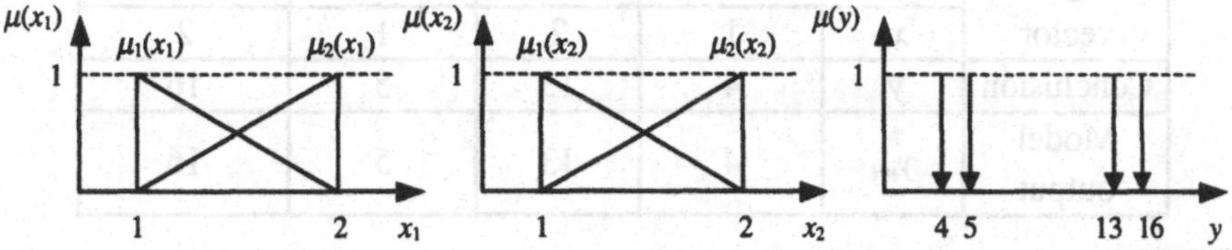

Fig. 5.138. Segmentally-linear membership functions satisfying the condition of unity partition of the inputs, and singleton functions of the output of the exemplified two-input system

other than the values y_i in the rule conclusions. Thus the rules do not "tell" the truth. Now, we shall apply in the model under consideration, different types of membership functions, namely the functions satisfying the condition of unity partition, Fig. 5.138.

For the realization of the *AND* operation in the rule premises (5.100) the *PROD*-operator will be used. When calculating the output y_{m4} of the model for the input vector $\mathbf{X}(R4) = [2, 2]^T$ the grade of truth of the premises of the particular rules is as follows:

$$\mu_{R1}(2,2) = \mu_1(x_1)\mu_1(x_2) = 0\,,$$
$$\mu_{R2}(2,2) = \mu_1(x_1)\mu_2(x_2) = 0\,,$$
$$\mu_{R3}(2,2) = \mu_2(x_1)\mu_1(x_2) = 0\,,$$
$$\mu_{R4}(2,2) = \mu_2(x_1)\mu_2(x_2) = 1\,.$$

Because only one premise has the grade of truth different from zero, only one rule ($R4$) will be activated. The output of the fuzzy model is equal to the value y_4 given in the conclusion of this rule.

$$y_m = \frac{\sum_{i=1}^{4} y_i \mu_{Ri}}{\sum_{i=1}^{4} \mu_{Ri}} = \frac{4 \cdot 0 + 13 \cdot 0 + 5 \cdot 0 + 16 \cdot 1}{0 + 0 + 0 + 1} = 16$$

The rule $R4$ is now consistent with the model calculations. Similar situations exist in the case of the remaining rules. The model output y_{mi} for the input vectors contained in the premises of the particular rules are given in Table 5.20.

Table 5.20. The comparison of the outputs y_{mi} calculated by the model (5.100) having triangular membership functions with the rule conclusions y_i in this model

Rule Ri		$R1$	$R2$	$R3$	$R4$
Input vector	x_1	1	1	2	2
	x_2	1	2	1	2
Conclusion	y_i	4	13	5	16
Model output	y_{mi}	4	13	5	16

As results from Table 5.20 show, in the model (5.100) with triangular membership functions (satisfying the condition of unity partition) for the inputs, all rules give the information consistent with the model calculation results.
◊ ◊ ◊

The rules will "tell" the truth when:

- for the realization of the *AND* operation we use t-norms and of the *OR* operation s-norms, i.e. the operators satisfying the conditions $\mu_{A\cap\emptyset}(x) = 0$ and $\mu_{A\cap B}(x) \leq 1$ (these conditions are not satisfied, for example, by the *MEAN*-operator or *SUM* (arithmetic sum)-operator),
- as the membership functions of the fuzzy sets of the model inputs we use such functions whose support is finite (Gaussian functions do not satisfy this condition) and which have the value equal to zero at the modal value point of the adjacent function where $\mu(x) = 1$). Examples are shown in Fig. 5.139.

In Fig. 5.140 examples of the functions which do not allow for achieving the effect of the rules "telling" the truth are represented.

The influence of remote (not adjacent) support points of the Mamdani model on local interpolation

A fuzzy model can be constructed in two ways. Either so that the shape of the interpolation surface in the given sector of the input space will be affected only by the model support points adjacent to this sector, or so that also some further (sometimes even all), not adjacent support points will affect it. In the second case the tuning of the parameters of membership functions (adaptation of the model on the basis of measurement data of a real system)

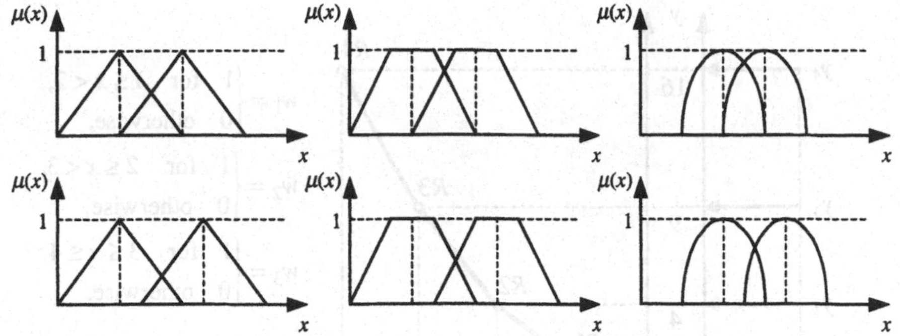

Fig. 5.139. Examples of membership functions of the model inputs enabling us to achieve the effect of the rules "telling" the truth

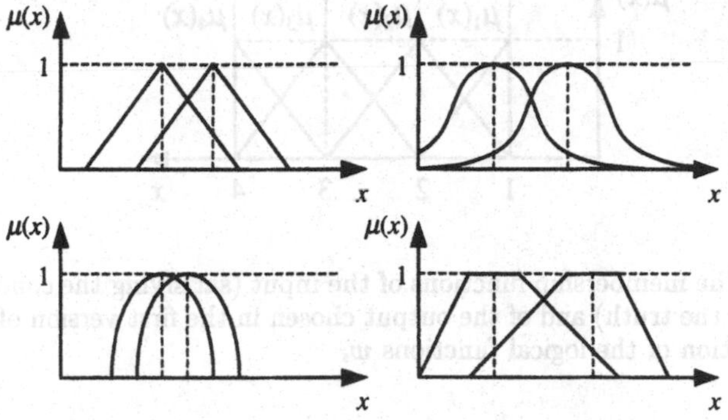

Fig. 5.140. Examples of membership functions of the model inputs which do not allow for achieving the effect of the rules of the Mamdani model "telling" the truth

becomes significantly more difficult. The problem will be explained through an example.

Example 5.7.1.4

Let us consider a one-input/one-output system. Measurement information about the system is given in Table 5.21. This data can be used to construct a fuzzy model of the system.

Table 5.21. The measurement data of a SISO system

x	1	2	3	4
y	1	4	9	16

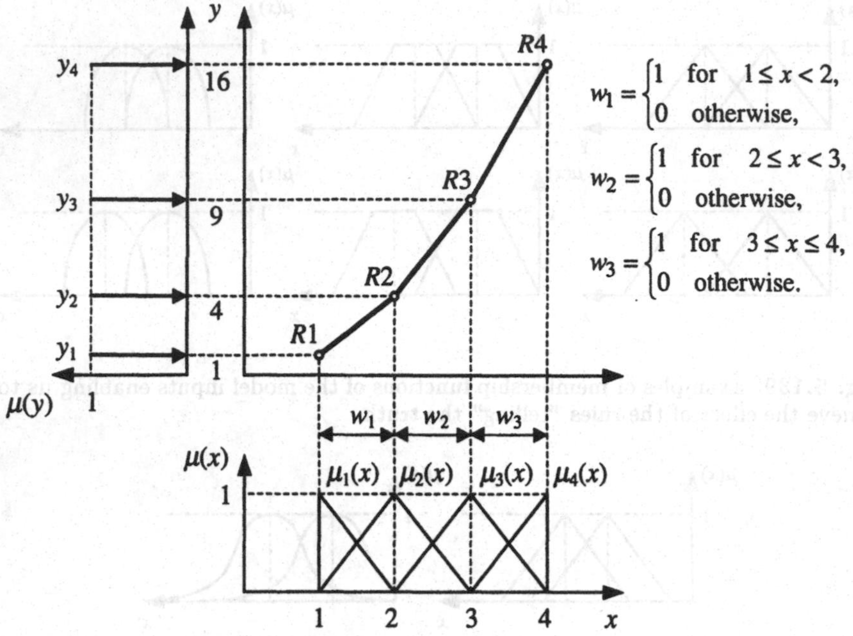

Fig. 5.141. The membership functions of the input (satisfying the condition of the rules "telling" the truth) and of the output chosen in the first version of the model, and the definition of the logical functions w_i

In the first version of the model we assume triangular membership functions satisfying the condition of unity partition for the inputs, and singleton functions for the output, Fig. 5.141. The rules of the model have the form (5.102):

$$\begin{aligned} R1: &\ IF\ (x \text{ about } 1)\ THEN\ (y \text{ about } y_1), \\ R2: &\ IF\ (x \text{ about } 2)\ THEN\ (y \text{ about } y_2), \\ R3: &\ IF\ (x \text{ about } 3)\ THEN\ (y \text{ about } y_3), \\ R4: &\ IF\ (x \text{ about } 4)\ THEN\ (y \text{ about } y_4), \end{aligned} \tag{5.102}$$

where: $y_1 = 1$, $y_2 = 4$, $y_3 = 9$, $y_4 = 16$.

By using the logical variables w_i defined in Fig. 5.141, we can determine the formulae (5.103) for the particular membership functions of the input x.

$$\begin{aligned} \mu_1(x) &= (2 - x)w_1 , \\ \mu_2(x) &= (x - 1)w_1 + (3 - x)w_2 , \\ \mu_3(x) &= (x - 2)w_2 + (4 - x)w_3 , \\ \mu_4(x) &= (x - 3)w_3 . \end{aligned} \tag{5.103}$$

The formula for calculating the model output in the whole universe of discourse $1 \leq x \leq 4$ has the form (5.104).

$$y = y_1(2-x)w_1+y_2[(x-1)w_1+(3-x)w_2]+y_3[(x-2)w_2+(4-x)w_3]+y_4(x-3)w_3 \tag{5.104}$$

The interpolation formulae for the sectors of the space between the particular "important" points of the model, Fig. 5.141, can be derived on the basis of (5.104).

$$\begin{aligned} R1 \; - \; R2: \; w_1 = 1\,, \quad & y = y_1(2-x) + y_2(x-1) \\ R2 \; - \; R3: \; w_2 = 1\,, \quad & y = y_2(3-x) + y_3(x-2) \\ R3 \; - \; R4: \; w_3 = 1\,, \quad & y = y_3(4-x) + y_4(x-3) \end{aligned} \tag{5.105}$$

Analyzing the formulae (5.105) it can easily be noticed that the interpolation functions of the model between its support points $Ri \; - \; R(i+1)$ are affected only by the parameters y_i, y_{i+1} of the points lying at the border of the sector to be considered. Local interpolation surfaces can be determined here on the basis of the local measurement information about the system. Such an effect has been achieved owing to the use of the input membership functions, which have a zero value at the modal value point of the adjacent membership function.

If we use functions of such a type as well as t-norms and s-norms (for the realization of the intersection and union operations of sets in a fuzzy model), in the case of a two-input system, we make the interpolation surface dependent only on the coordinates of the nearest support points corresponding with the corners of the rectangular sector of the input space $[x_1, x_2]$, Fig. 5.142, or with the corners of the (hyper-)rectangular sector, if the number of inputs is greater.

Then the interpolation surface passes exactly through the support points and its shape depends on different elements of the model, such as the type of assumed inference, defuzzification, and the type of operators used. The rules of the fuzzy model "tell" the truth about its outputs in this case. Now let us consider another variant of the input membership functions represented in Fig. 5.143.

The membership functions of the input from Fig. 5.143 can be described by the formulae (5.106) containing the logical variables w_i.

$$\begin{aligned} \mu_1(x) &= 0.25[(5-x)(w_1+w_2+w_3)]\,, \\ \mu_2(x) &= 0.25[(x+2)w_1+(6-x)(w_2+w_3)]\,, \\ \mu_3(x) &= 0.25[(x+1)(w_1+w_2)+(7-x)w_3]\,, \\ \mu_4(x) &= 0.25[x(w_1+w_2+w_3)]\,. \end{aligned} \tag{5.106}$$

The rule base of the model is given by the formula (5.107).

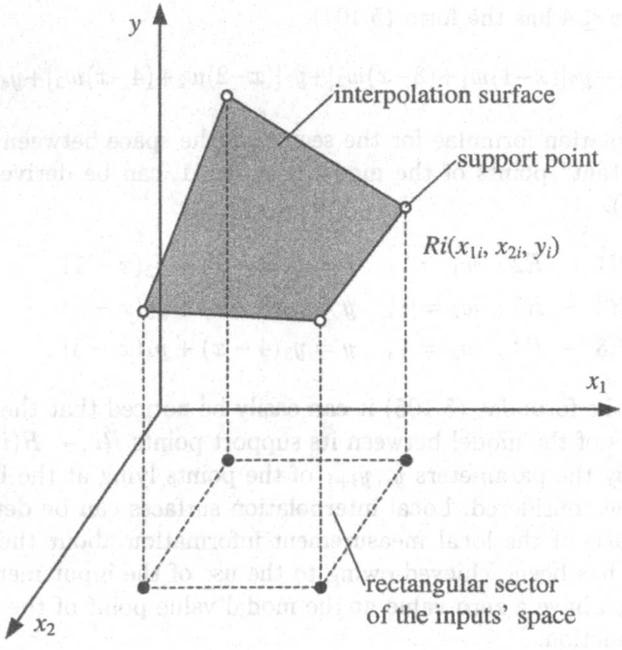

Fig. 5.142. Illustration of the dependence of the local interpolation surface only on the nearest adjacent support points

$$\begin{aligned}
R1^*&: IF\ (x \text{ about } 1)\ THEN\ (y \text{ about } y_1^*)\,,\\
R2^*&: IF\ (x \text{ about } 2)\ THEN\ (y \text{ about } y_2^*)\,,\\
R3^*&: IF\ (x \text{ about } 3)\ THEN\ (y \text{ about } y_3^*)\,,\\
R4^*&: IF\ (x \text{ about } 4)\ THEN\ (y \text{ about } y_4^*)\,,
\end{aligned} \tag{5.107}$$

On the basis of (5.106) and (5.107) one can derive the formula (5.108) determining the model surface.

$$\begin{aligned}
y = {} & 0.1y_1^*[(5-x)(w_1+w_2+w_3)] + 0.1y_2^*[(x+2)w_1 + (6-x)(w_2+w_3)] + \\
& + 0.1y_3^*[(x+1)(w_1+w_2) + (7-x)w_3] + 0.1y_4^*[x(w_1+w_2+w_3)]
\end{aligned} \tag{5.108}$$

The interpolation surface between the particular support points $Ri - R(i+1)$ is expressed by the formulae (5.109).

$$\begin{aligned}
R1 - R2:&\ w_1 = 1\,, \quad y = 0.1[y_1^*(5-x) + y_2^*(x+2) + y_3^*(x+1) + y_4^*x]\\
R2 - R3:&\ w_2 = 1\,, \quad y = 0.1[y_1^*(5-x) + y_2^*(6-x) + y_3^*(x+1) + y_4^*x]\\
R3 - R4:&\ w_3 = 1\,, \quad y = 0.1[y_1^*(5-x) + y_2^*(6-x) + y_3^*(7-x) + y_4^*x]
\end{aligned} \tag{5.109}$$

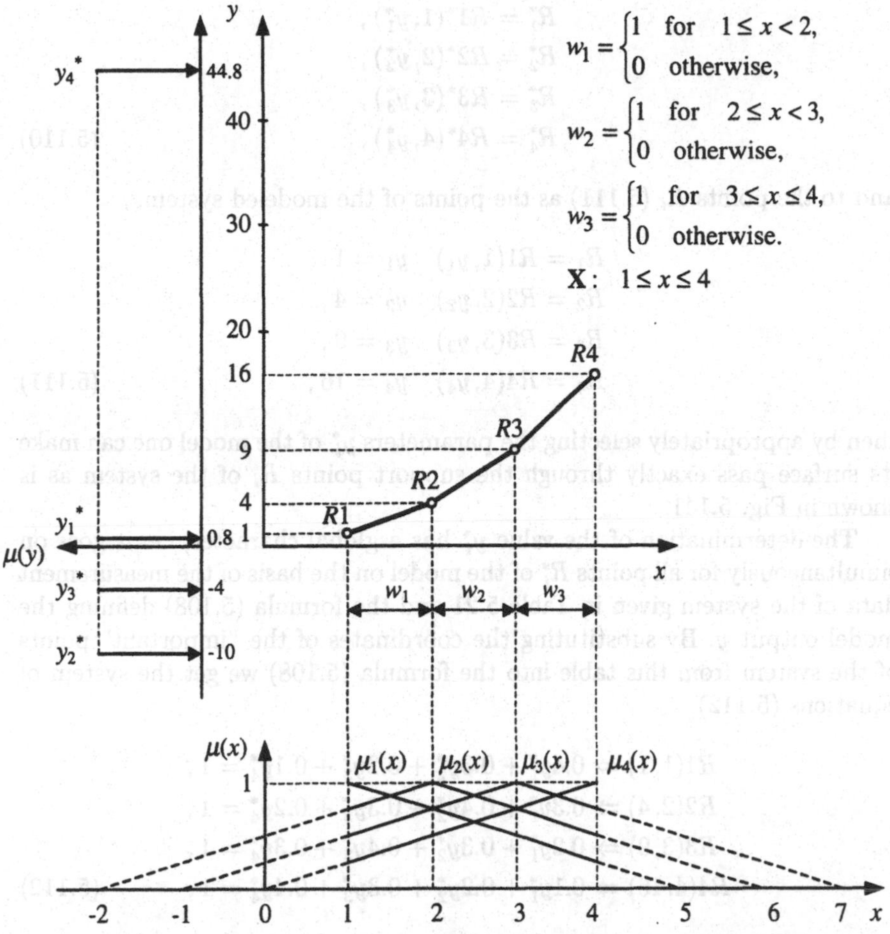

Fig. 5.143. The membership functions of the input (not satisfying the condition of the rules "telling" the truth) and of the output chosen in the second version of the model and the definitions of the logical functions w_i

Analyzing the formulae (5.109) one can easily notice that in this version of the model the local interpolation surfaces which are spread between each pair of the adjacent support points depend not only on those points but also on all further support points. The reason for this is that the coordinates y_1^*, y_2^*, y_3^*, y_4^* given in the conclusions of all rules (5.107) (which do not "tell" the truth in this version) occur in each of the interpolation formulae (5.109).

If we refer to points R_i^* (5.110) as the support points of the model,

$$\begin{aligned}
R_1^* &= R1^*(1, y_1^*)\,,\\
R_2^* &= R2^*(2, y_2^*)\,,\\
R_3^* &= R3^*(3, y_3^*)\,,\\
R_4^* &= R4^*(4, y_4^*)\,,
\end{aligned} \tag{5.110}$$

and to the points R_i (5.111) as the points of the modeled system,

$$\begin{aligned}
R_1 &= R1(1, y_1) \quad y_1 = 1\,,\\
R_2 &= R2(2, y_2) \quad y_2 = 4\,,\\
R_3 &= R3(3, y_3) \quad y_3 = 9\,,\\
R_4 &= R4(4, y_4) \quad y_4 = 16\,,
\end{aligned} \tag{5.111}$$

then by appropriately selecting the parameters y_i^* of the model one can make its surface pass exactly through the support points R_i of the system as is shown in Fig. 5.141.

The determination of the value y_i^* has a global character, i.e. it goes on simultaneously for all points R_i^* of the model on the basis of the measurement data of the system given in Table 5.21 and the formula (5.108) defining the model output y. By substituting the coordinates of the "important" points of the system from this table into the formula (5.108) we get the system of equations (5.112),

$$\begin{aligned}
R1(1,1) &\Rightarrow 0.4y_1^* + 0.3y_2^* + 0.2y_3^* + 0.1y_4^* = 1\,,\\
R2(2,4) &\Rightarrow 0.3y_1^* + 0.4y_2^* + 0.3y_3^* + 0.2y_4^* = 1\,,\\
R3(3,9) &\Rightarrow 0.2y_1^* + 0.3y_2^* + 0.4y_3^* + 0.3y_4^* = 1\,,\\
R4(4,16) &\Rightarrow 0.1y_1^* + 0.2y_2^* + 0.3y_3^* + 0.4y_4^* = 1\,,
\end{aligned} \tag{5.112}$$

whose solution gives the result (5.113),

$$y_1^* = 0.8\,, \quad y_2^* = -10\,, \quad y_3^* = -4\,, \quad y_4^* = 44.8\,. \tag{5.113}$$

After taking into account (5.113) the rule base (5.107) assumes the form (5.114):

$$\begin{aligned}
&R1^*: \ \mathit{IF}\ (x \text{ about } 1)\ \mathit{THEN}\ (y \text{ about } 0.8)\,,\\
&R2^*: \ \mathit{IF}\ (x \text{ about } 2)\ \mathit{THEN}\ (y \text{ about } -10)\,,\\
&R3^*: \ \mathit{IF}\ (x \text{ about } 3)\ \mathit{THEN}\ (y \text{ about } -4)\,,\\
&R4^*: \ \mathit{IF}\ (x \text{ about } 4)\ \mathit{THEN}\ (y \text{ about } 44.8)\,,
\end{aligned} \tag{5.114}$$

Analysis of the results of using the membership functions from Fig. 5.143 with the supports extended beyond the modal values of the adjacent membership functions gives the following conclusions.

- The interpolation surface of a model is qualitatively of the same type as in the case of membership functions not extended beyond the modal values of adjacent membership functions (in the example under consideration – the linear one). Thus the extension of the supports of membership functions does not change the quality of interpolation.
- The tuning of model parameters must go on globally here with the use of all support points of a system and not locally with the exclusive use of adjacent points only. The global tuning is significantly more difficult than the local one, considering the phenomenon of "the curse of dimensionality".
- The model achieved after the tuning has been completed is correct (the model surface passes through the support points R_i of the system), but the rules achieved Ri^* "do not tell the truth" making the understanding of the model difficult and decreasing its transparency.

The use of membership functions of the support not going out beyond the range of adjacent membership functions (in particular the functions satisfying the condition of unity partition) in the Mamdani models is thus advantageous for many reasons presented in the example considered.

◊ ◊ ◊

5.7.2 Takagi-Sugeno models

The Takagi-Sugeno (TS) models were described in (Takagi 1985) for the first time. These models are sometimes also called Takagi-Sugeno-Kang models (Nquyen 1995; Yager 1994,1995), quasi-linear models or fuzzy linear models (Babuška 1995a,b). Takagi-Sugeno models differ from Mamdani models in the form of their rules. If in the case of the Mamdani model of a system having one input/one output the rule has the form (5.115),

$$IF\ (x \text{ is } A)\ THEN\ (y \text{ is } B)\,, \tag{5.115}$$

(where: A,B – fuzzy sets of the type "small", "near 5") then in the case of the TS model the rules have the form (5.116):

$$IF\ (x \text{ is } A)\ THEN\ (y = f(x))\,. \tag{5.116}$$

Their conclusion contains the function $f(x)$ and not a fuzzy set. This function can be non-linear, although usually linear functions are applied. Then the TS rules have the form (5.117),

$$IF\ (x \text{ is } A)\ THEN\ (y = ax + b))\,. \tag{5.117}$$

If the rule base of the model of a SISO system has the form (5.118):

$$\begin{aligned} R1:\ & IF\ (x \text{ is } A_1)\ THEN\ (y = f_1(x))\,, \\ & \vdots \\ Rm:\ & IF\ (x \text{ is } A_m)\ THEN\ (y = f_m(x))\,, \end{aligned} \tag{5.118}$$

then the model output is calculated on the basis of the grade of activation of the particular conclusions f_i, $i = 1, \ldots, m$, which is determined by formula (5.119).

$$y = \frac{\sum_{i=1}^{m} \mu_{Ai}(x) f_i(x)}{\sum_{i=1}^{m} \mu_{Ai}(x)} \tag{5.119}$$

Certain features of the TS models will be presented in Example 5.7.2.1.

Example 5.7.2.1
Let us consider the TS model of a SISO system with the rule base (5.120) and membership functions represented in Fig. 5.144.

$$\begin{aligned} R1: &\ IF\ (x \text{ is } A_1)\ THEN\ (y = -x + 3)\,, \\ R2: &\ IF\ (x \text{ is } A_2)\ THEN\ (y = (4x - 10)/3)\,, \\ R3: &\ IF\ (x \text{ is } A_3)\ THEN\ (y = (-x + 24)/3)\,, \end{aligned} \tag{5.120}$$

If the logical variables are determined by the formulae (5.121),

$$\begin{aligned} w_1 &= \begin{cases} 1 & \text{for } 0 \le x < 2\,, \\ 0 & \text{otherwise,} \end{cases} \\ w_2 &= \begin{cases} 1 & \text{for } 2 \le x < 4\,, \\ 0 & \text{otherwise,} \end{cases} \\ w_3 &= \begin{cases} 1 & \text{for } 4 \le x < 7\,, \\ 0 & \text{otherwise,} \end{cases} \\ w_4 &= \begin{cases} 1 & \text{for } 7 \le x < 9\,, \\ 0 & \text{otherwise,} \end{cases} \\ w_5 &= \begin{cases} 1 & \text{for } 9 \le x \le 12\,, \\ 0 & \text{otherwise,} \end{cases} \end{aligned} \tag{5.121}$$

then the membership functions have the form:

$$\begin{aligned} \mu_{A1} &= w_1 - 0.5(x - 4)w_2\,, \\ \mu_{A2} &= 0.5(x - 2)w_2 + w_3 - 0.5(x - 9)w_4\,, \\ \mu_{A3} &= 0.5(x - 7)w_4 + w_5\,. \end{aligned} \tag{5.122}$$

The membership functions satisfy the condition of unity partition (5.123).

$$\sum_{i=1}^{3} \mu_{Ai}(x) = 1 \tag{5.123}$$

The model output is determined by the formula (5.124).

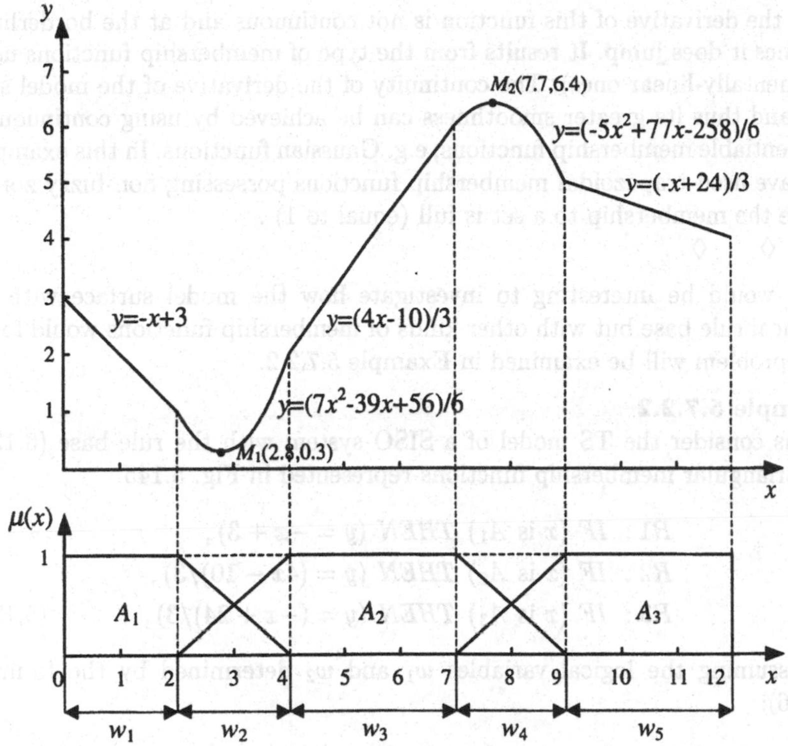

Fig. 5.144. The membership functions of the inputs of the exemplified Takagi-Sugeno model and the model surface

$$y = \sum_{i=1}^{3} \mu_{Ai}(x) f_i(x) = w_1(-x+3) + \frac{w_2(7x^2-39x+56)}{6} + \frac{w_3(4x-10)}{3} +$$
$$+ \frac{w_4(-5x^2+77x-258)}{6} + \frac{w_5(-x+24)}{3} \qquad (5.124)$$

As results from an analysis of the formula (5.124), the model surface is exactly as given by the rule conclusions (5.119) only in those input zones (w_1, w_3, w_5) where the grades of membership in the particular sets A_i satisfy the condition: $\mu_{Ai}(x) = 1$. In the zones where these grades are fractional (w_2, w_4) the model surface passes on from one linear form (determined by the appropriate conclusion) to another. The width of the transient zones depends on the width of the fractional zones $\mu_{Ai}(x) < 1$ of membership functions and the mathematical form of the transient function depends on the type of membership functions used.

In the example considered, transient functions have a quadratic form. The fact must be noted, that the function (5.124) defining the model is continuous and at the borderlines of particular zones it does not change jumpwise. However, the derivative of this function is not continuous and at the borderlines of zones it does jump. It results from the type of membership functions used (segmentally-linear ones). The continuity of the derivative of the model surface and thus its greater smoothness can be achieved by using continuously differentiable membership functions, e.g. Gaussian functions. In this example, we have used trapezoidal membership functions possessing non-fuzzy zones, where the membership to a set is full (equal to 1) .

◊ ◊ ◊

It would be interesting to investigate how the model surface with an identical rule base but with other kinds of membership functions would look. This problem will be examined in Example 5.7.2.2.

Example 5.7.2.2

Let us consider the TS model of a SISO system with the rule base (5.125) and triangular membership functions represented in Fig. 5.145.

$$\begin{aligned} &R1: \ IF\ (x \text{ is } A_1)\ THEN\ (y = -x + 3)\,, \\ &R2: \ IF\ (x \text{ is } A_2)\ THEN\ (y = (4x - 10)/3)\,, \\ &R3: \ IF\ (x \text{ is } A_3)\ THEN\ (y = (-x + 24)/3)\,, \end{aligned} \tag{5.125}$$

Assuming the logical variables w_1 and w_2 determined by the formula (5.126):

$$\begin{aligned} w_1 &= \begin{cases} 1 & \text{for } 0 \le x < 5\,, \\ 0 & \text{otherwise,} \end{cases} \\ w_2 &= \begin{cases} 1 & \text{for } 5 \le x \le 12\,, \\ 0 & \text{otherwise,} \end{cases} \end{aligned} \tag{5.126}$$

one achieves the formula (5.127) determining the model surface.

$$\begin{aligned} y = &\frac{-0.2w_1(x-5)(-x+3)}{5} + \frac{0.2w_1x(4x-10)}{3} + \\ &+ \frac{-w_2(x-12)(4x-10)}{21} + \frac{w_2(x-5)(-x+24)}{21} = \\ = &\frac{w_1(7x^2-34x+45)}{15} + \frac{w_2(-5x^2+87x-240)}{21} \end{aligned} \tag{5.127}$$

As results from the formula (5.127) and Fig. 5.145, in no sector w_i does the model surface have a linear character determined in the rule conclusions (5.125). Even at the points where the membership to the sets A_1, A_2, A_2 is complete (equal to 1) the tangents to the model surface do not coincide with the linear functions given in the rules, Fig. 5.145.

For example, for $x = 0$ the membership to the set A_1 is complete and the tangent to the model surface at this point is described by the formula:

$$y = \frac{14x - 34}{15}\,,$$

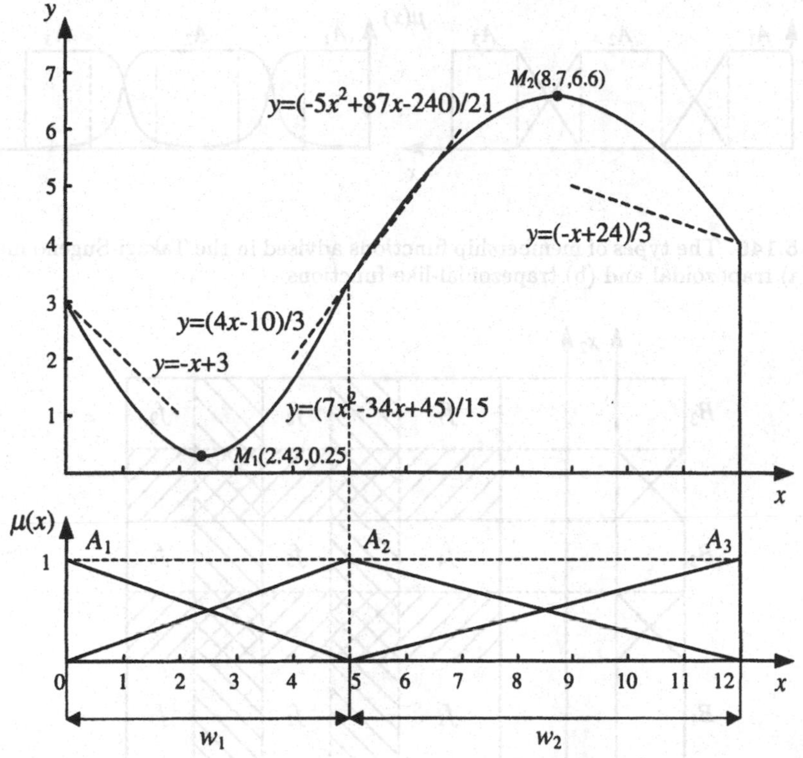

Fig. 5.145. An example of a Takagi-Sugeno model with triangular (non-trapezoidal) membership functions

whereas according to the conclusion of the rule $R1$ in the rule base (5.125) one would expect that it would be equal to:

$$y = -x + 3 .$$

A similar situation occurs at the points $x = 5$ and $x = 12$ where the membership in the sets A_2 and A_3 amounts to 1.

◊ ◊ ◊

As results from the presented example, the TS models should be applied mainly when membership functions have a trapezoidal character or a similar one, Fig. 5.146.

Notice that when we use trapezoidal-like functions with non-linear edges (e.g. Gaussian functions) we do not achieve (even in the zone where the membership to the set A_i is complete and amounts to 1) the model surface looking exactly as it is given by the appropriate rule Ri, but the surface which is more or less changed by the influence of the functions f_i of other rules. This results from the fact that the Gaussian membership functions have an

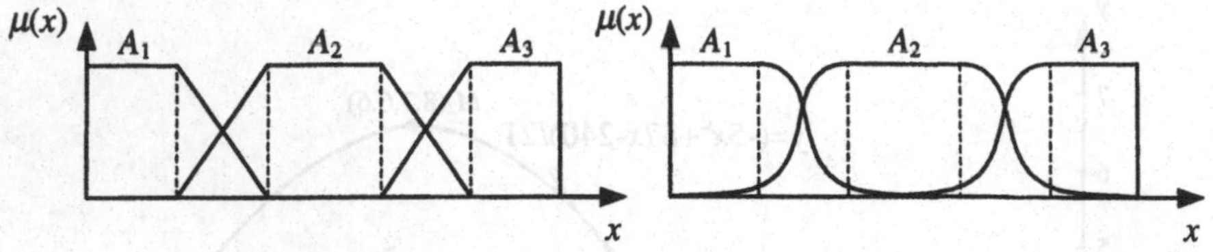

Fig. 5.146. The types of membership functions advised in the Takagi-Sugeno models: (a) trapezoidal and (b) trapezoidal-like functions

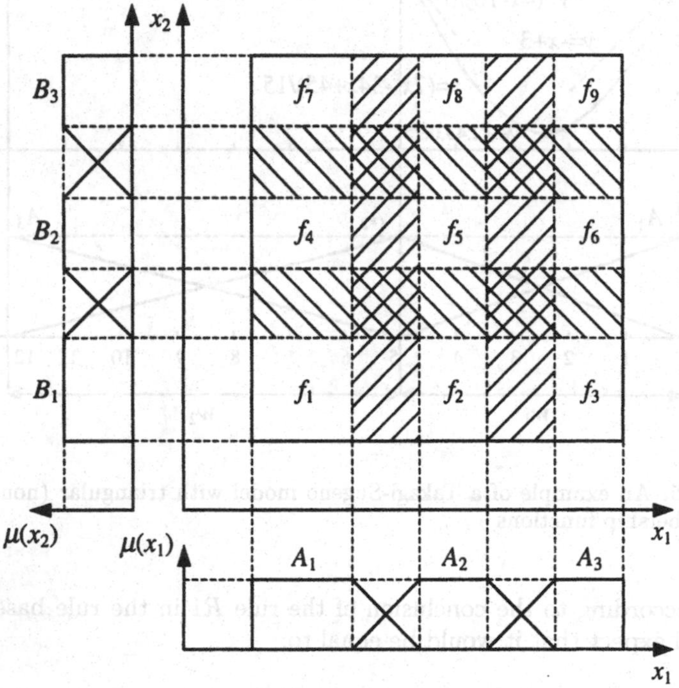

$$Rk: \; IF \; (x \text{ is } A_i \cap B_j) \; THEN \; (y = f_k(x)), \quad i,j = 1,\ldots,3, \quad k = 1,\ldots,9$$

Fig. 5.147. The recommended form of membership functions in a two-input model of the Takagi-Sugeno type enabling us to achieve rectangular sectors whose surfaces correspond exactly with the functions f_k given by the conclusions of the rules *Rk*. The lined areas are the transient zones between the particular conclusions f_k

infinitely large support, do not satisfy the condition of unity partition, and considerably extend the range of influence of the particular conclusions f_i.

In the case of systems with two (or more) inputs, trapezoidal membership functions allow for achieving (hyper-)rectangular sectors where the membership to the products of sets is equal to 1, Fig. 5.147.

It would be interesting to know what the relationship is between the Mamdani models and TS models and whether is it possible to map or convert one model into the other. For a SISO system this problem is discussed in (Babuška 1995a,b).

In order to map the TS model in a Mamdani model it is necessary to know the coordinates of the "important" points of the model surface. In the case of Takagi-Sugeno models the "important" points are the border points of the particular sectors of the inputs space, maxima and minima, and inflexion points of the model surface. For the "important" points of the surface we must calculate, on the basis of the TS model, the output values y and create Mamdani fuzzy rules characterizing the state of the model at these points as well as membership functions. The method is illustrated by Example 5.7.2.3.

Example 5.7.2.3

For the Takagi-Sugeno model from Example 5.7.2.2, the problem is to find the Mamdani model corresponding with it. In Fig. 5.148 the TS model with "important" points marked is represented.

For the "important" points of the TS model surface the Mamdani rules defining the state of the model are determined. The rule base of the Mamdani model has the form (5.128).

$$\begin{aligned}
&R1: \textit{IF } (x \text{ near to } 0) \textit{ THEN } (y \text{ near to } 3)\\
&R2: \textit{IF } (x \text{ near to } 2) \textit{ THEN } (y \text{ near to } 1)\\
&R3: \textit{IF } (x \text{ near to } 2.8) \textit{ THEN } (y \text{ near to } 0.3)\\
&R4: \textit{IF } (x \text{ near to } 4) \textit{ THEN } (y \text{ near to } 2)\\
&R5: \textit{IF } (x \text{ near to } 7) \textit{ THEN } (y \text{ near to } 6)\\
&R6: \textit{IF } (x \text{ near to } 7.7) \textit{ THEN } (y \text{ near to } 6.4)\\
&R7: \textit{IF } (x \text{ near to } 9) \textit{ THEN } (y \text{ near to } 5)\\
&R8: \textit{IF } (x \text{ near to } 12) \textit{ THEN } (y \text{ near to } 4)
\end{aligned} \tag{5.128}$$

The membership functions and surface of the Mamdani model are represented in Fig. 5.149.

As can be seen from the comparison of Fig. 5.148 with Fig. 5.149 the equivalent Mamdani model has more rules and fuzzy sets than the TS model. This results from the fact that in the TS models trapezoidal membership functions of long supports were used. These functions define large areas of the model surface functionally and not pointwise.

If we assume that the minima and maxima of transient zones are not essential for the model (in the example considered they are the points (2.8, 0.3) and (7.7, 6.4)), and we assume that only the border lines of the zones defined in the TS model conclusions (Fig. 5.148) are "important", then we achieve a simpler but a little less accurate Mamdani model represented in Fig. 5.150. The simplified model has the rule base (5.129).

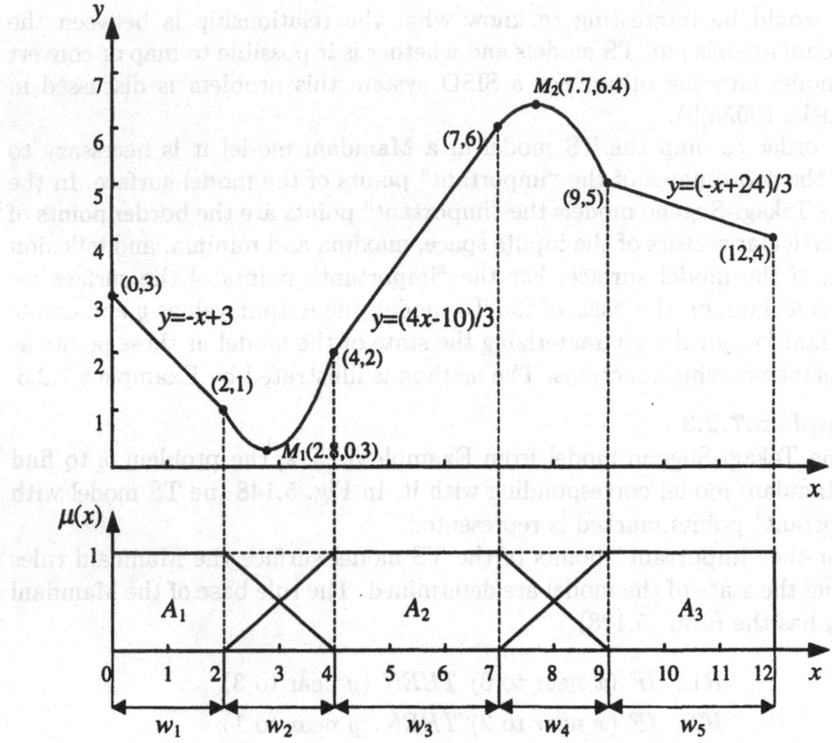

$$R1: \; IF \; (x \text{ is } A_1) \; THEN \; (y = -x + 3)$$
$$R2: \; IF \; (x \text{ is } A_2) \; THEN \; (y = (4x - 10)/3)$$
$$R3: \; IF \; (x \text{ is } A_3) \; THEN \; (y = (-x + 24)/3)$$

Fig. 5.148. The Takagi-Sugeno model with the "important" points of the $X \to Y$ mapping surface

$$\begin{aligned} R1 &: \; IF \; (x \text{ near to } 0) \; THEN \; (y \text{ near to } 3) \\ R2 &: \; IF \; (x \text{ near to } 2) \; THEN \; (y \text{ near to } 1) \\ R3 &: \; IF \; (x \text{ near to } 4) \; THEN \; (y \text{ near to } 2) \\ R4 &: \; IF \; (x \text{ near to } 7) \; THEN \; (y \text{ near to } 6) \\ R5 &: \; IF \; (x \text{ near to } 9) \; THEN \; (y \text{ near to } 5) \\ R6 &: \; IF \; (x \text{ near to } 12) \; THEN \; (y \text{ near to } 4) \end{aligned} \qquad (5.129)$$

The TS model from Fig. 5.144 has 3 rules and 3 fuzzy sets of the input. The simplified Mamdani model, which is equivalent to it, has 6 rules and 6 fuzzy sets of the input and so is more complex.

◊ ◊ ◊

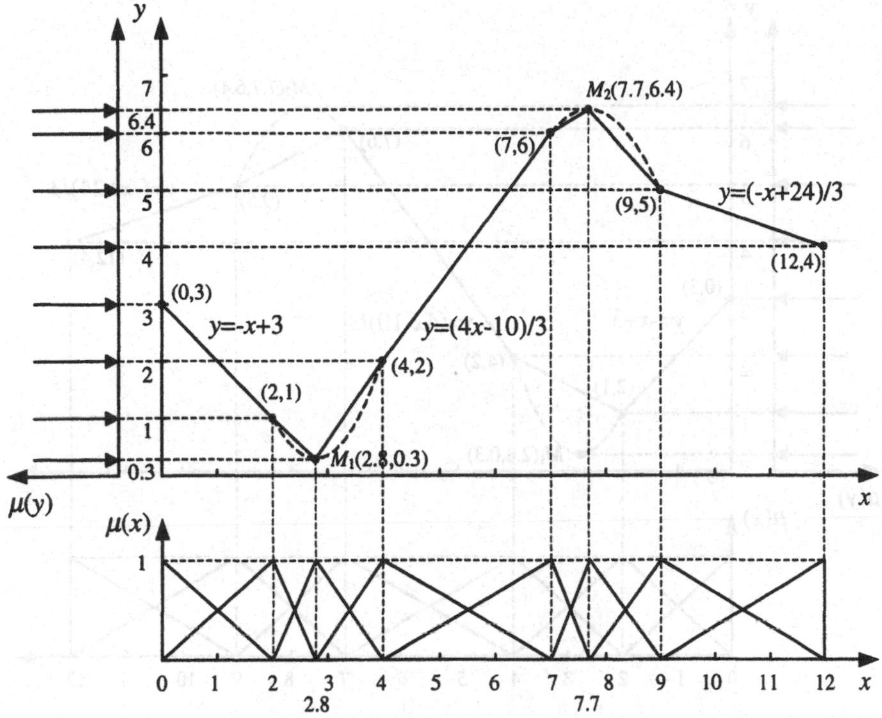

Fig. 5.149. The membership functions and surface of the Mamdani model approximately equivalent to the Takagi-Sugeno model from Fig. 5.148

The presented way of planning the Mamdani model equivalent to the given Takagi-Sugeno model on the basis of "important" surface points is relatively easy in the case of SISO systems. In multi-input systems the "important" points are the corner points of the (hyper-)rectangular sectors created by the products of fuzzy sets in the input space. The rules of the Mamdani models determine only the output state y of the model at these points, Fig. 5.147. The character of the function f_k over the given sector depends on the chosen type of operators accomplishing the *AND*, *OR* operation, the type of membership functions and the defuzzification method. For this reason, in practice, the exact conversion of the TS model onto the Mamdani model, particularly when the functions f_k in rule conclusions are more complicated, is very difficult or impossible.

What are the advantages and applications of the the Takagi-Sugeno model?

The TS models combine in themselves the system description based on linguistic rules with the traditional functional description of the system operation, which we are accustomed to, and which often, in the case of the actual

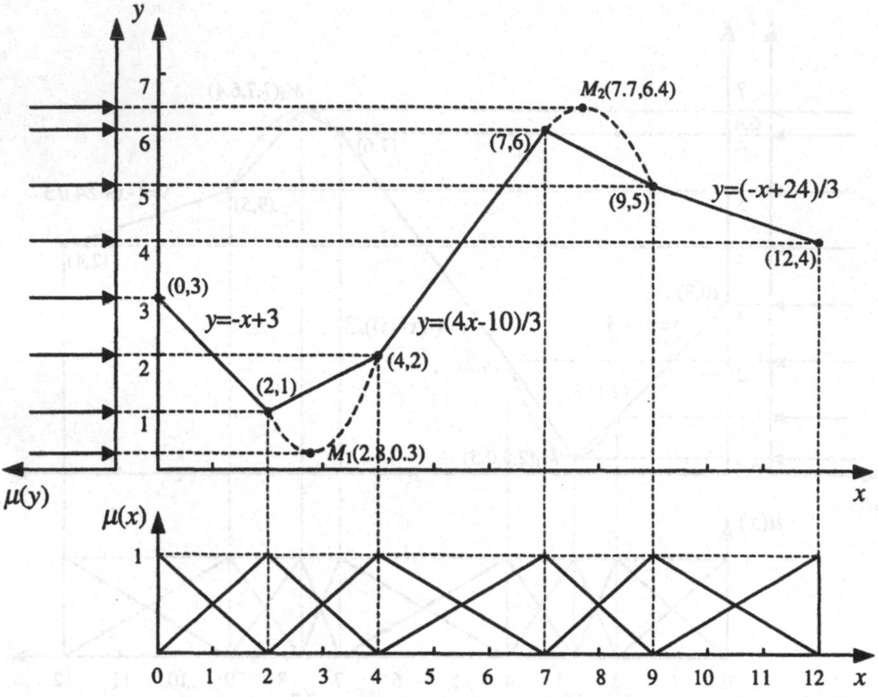

Fig. 5.150. The Mamdani model equivalent to the TS model from Fig. 5.144 with the rules determining the model state only at the border lines of the important zones in the TS model and disregarding the "important" points characterizing the transient zones $(2.8, 0.3)$ and $(7.7, 6.4)$

system, is well-known to us. The local description of the system operation, and in particular the linear description, is relatively easy to identify. It also has further advantages.

- Complicated, non-linear system surfaces can be approximated by a set of flat linear segments. Each such segment can be described by one rule in the TS model.
- The TS models are particularly suitable for describing controllers. At present, the design of linear controllers is a well-developed procedure. In the case of complex non-linear plants, optimal linear controllers can be designed for the most important working points (operating sectors) and then they can be combined to create one TS fuzzy controller (Cao 1997).
- The representation of a plant and controller in the TS form makes it easier to prove the stability of control systems owing to the locally linear structure of this form (Domański 1997).

The Mamdani model can be considered to be the simplest form of the TS model and the TS model to be the generalization of the Mamdani model.

This results from the comparison of the forms of the rules (5.130) and (5.131) used in both model types.

The Mamdani rule:

$$IF\ (x_1 \text{ is } A_1)\ AND\ \ldots\ AND(x_n \text{ is } A_n)\ THEN\ (y \text{ is about } c_0)\,. \quad (5.130)$$

The Takagi-Sugeno rule:

$$IF\ (x_1 \text{ is } A_1)\ AND\ \ldots\ AND(x_n \text{ is } A_n)\ THEN\ (y = a_0 + a_1x_1 + \ldots + a_nx_n)\,. \quad (5.131)$$

Because the conclusion in the TS model has a more complicated mathematical form it is not as easily comprehensible as the conclusion of the Mamdani model. The method of identification of local linear models is given in (Babuška 1995a).

5.7.3 Relational models

The relational models were introduced by Pedrycz in 1984 (Pedrycz 1984). They are also described in (Hung 1995; Pedrycz 1993). Their fundamental feature is that linguistic rules are not considered to be fully true but they are considered to be partly (more or less) true. Appropriate confidence coefficients are assigned to particular rules.

The rule base is represented by a fuzzy relation, and its identification and analysis the theory of relational equations is applied (Kacprzyk 1986; Pedrycz 1993). The idea of the confidence coefficients of rules will be explained in Example 5.7.3.1.

Example 5.7.3.1

The rule base (5.132) determines the relation between mental powers of a child and his school results.

*R*1 : A clever child learns well.

*R*2 : An ordinary child learns to an average degree.

*R*3 : A not-so-clever child learns poorly. (5.132)

Analyzing without insight we are ready to assume the rules (5.132) to be true. These rules can be presented in the form of the relational Table 5.22 using the confidence coefficients equal to 1 or 0 which join the input fuzzy sets with output sets.

But statistical investigations carried out at schools and experience of most parents show that there are clever children who get average or even poor school results as well as there being not-so-clever children who get average results. Statistically investigating groups of clever, average and not-clever pupils, we can calculate what percentage of children in these groups get good, average or poor results. On this basis we can determine the confidence coefficients in the relational table whose example is Table 5.23.

Table 5.22. The relational table representing the rule base (5.132) with the unique confidence coefficients of the rules

school results / level of mental powers	good	average	poor
clever	1	0	0
ordinary	0	1	0
not-so-clever	0	0	1

Table 5.23. The example table of the relations between the level of mental powers of a child and his school results with fuzzy confidence (truth) coefficients of rules

school results / level of mental powers	good	average	poor
clever	0.6	0.3	0.1
ordinary	0.1	0.7	0.2
not-so-clever	0	0.3	0.7

Table 5.23 corresponds with the rule base (5.133).

*R*1 : A clever child learns well (0.6)
to an average degree (0.3)
poorly (0.1)
*R*2 : An ordinary clever child learns well (0.1)
to an average degree (0.7)
poorly (0.2)
*R*3 : A not-so-clever child learns well (0)
to an average degree (0.3)
poorly (0.7)

(5.133)

The conclusions of the rules (5.133) are not unique but they reflect the reality more closely than rules (5.132).

◊ ◊ ◊

In relational fuzzy models before the reasoning process is carried out, the fuzzy sets of conclusions are usually defuzzified and replaced with singletons. After that the aggregation is accomplished by considering weighted confidence coefficients of the particular rules. The procedure of calculation will be illustrated by Example 5.7.3.2. It raises the question whether Mamdani fuzzy models are sufficient for practical purposes, and why (or when) should relational fuzzy models be used?

In problems of modeling we many a time encounter such cases where the membership functions of an output variable are defined a priori, on the basis of the knowledge or intuition of an expert (in the case of lack of knowledge of the system, the expert could assume the uniform distribution of membership functions). If the parameters of these functions, especially the modal values, are not optimal, then there is the possibility to increase the accuracy of the model. Example 5.7.3.2 presents a model where the uniform distribution of membership functions in the universe of discourse has been assumed.

Example 5.7.3.2
The universe of discourse of a SISO system is as follows:

$$X:\ 1 \le x \le 3, \quad Y:\ 5 \le y \le 15\,.$$

In the first phase of the work on the system, only the border points $P_1(1,5), P_2(3,15)$ of the universe of discourse are well known. We assume that the modal values of the membership functions are distributed symmetrically, as in Fig. 5.151. Let us assume the rule base (5.134).

$$\begin{aligned} &R1:\ IF\ (x \text{ is small})\ THEN\ (y \text{ is small})\\ &R2:\ IF\ (x \text{ is medium})\ THEN\ (y \text{ is medium})\\ &R3:\ IF\ (x \text{ is large})\ THEN\ (y \text{ is large}) \end{aligned} \qquad (5.134)$$

The rules (5.134) and assumed membership functions mean that the model surface is rectilinear and passes through the points P_1, P_2, P_3 in Fig. 5.151. This is the first coarse model of the system.

The rules (5.134) correspond with the Table 5.24 containing the confidence coefficients of the rules.

Further observation and better recognition of the system show that the output y has the value 12 and not 10 for the input $x = 2$, as the first version of the model gives. We can take this information into account by using the relational fuzzy model, Fig. 5.152, with the rule base (5.135) in which the conclusion of the rule $R2$ is not unique.

Table 5.24. The relational table of the rule base (5.134)

x \ y	S	M	L
S	1	0	0
M	0	1	0
L	0	0	1

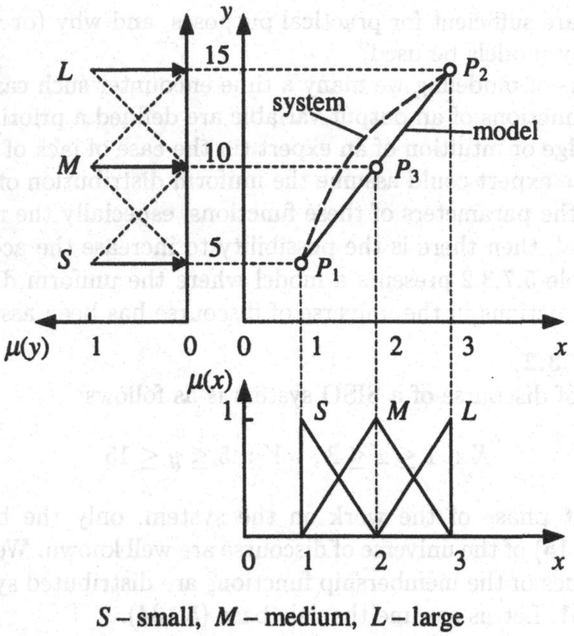

Fig. 5.151. The first version of a fuzzy model with uniformly distributed membership functions of its input and output

Table 5.25. The relational table of the rule base (5.135) containing confidence coefficients

x \ y	S	M	L
S	1	0	0
M	0	0.6	0.4
L	0	0	1

$$\begin{aligned} R1: &\ IF\ (x \text{ is small})\ THEN\ (y \text{ is small})\ (1) \\ R2: &\ IF\ (x \text{ is medium})\ THEN\ (y \text{ is medium})\ (0.6) \\ &\ \qquad\qquad\qquad\qquad\qquad (y \text{ is large})\ (0.4) \\ R3: &\ IF\ (x \text{ is large})\ THEN\ (y \text{ is large})\ (1) \end{aligned} \qquad (5.135)$$

The new rule $R2$ causes an apparent displacement of the singleton "medium" of the model output to the new position $y = 12$. This increases the accuracy of the model. The rule base (5.135) can be presented in the form of the relational Table 5.25.

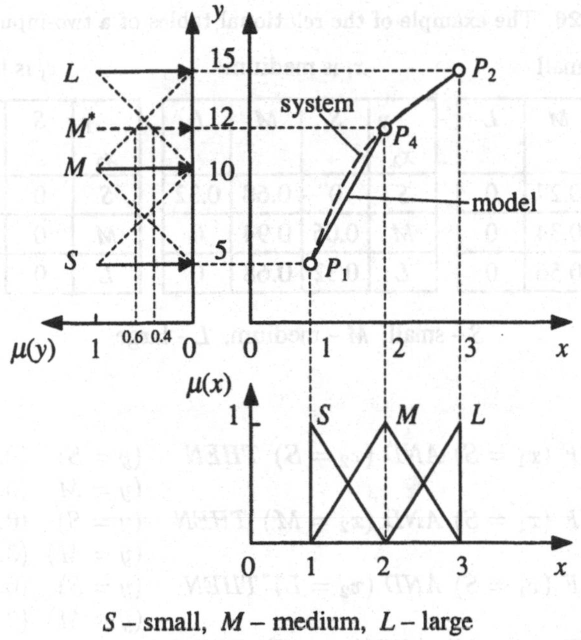

Fig. 5.152. The second version of the model, corresponding with the rule base (5.135)

The confidence coefficients result from the new position (M^*) of the singleton "medium" in Fig. 5.152.

◊ ◊ ◊

As the Examples 5.7.3.1 and 5.7.3.2 show, relational fuzzy models make it possible to increase the accuracy of a model by introducing the confidence coefficients of particular rules. The confidence coefficients enable us to express the conclusions of new rules as the convex combination of the adjacent fuzzy sets of the model output. The optimal values of the confidence coefficients can be determined on the basis of input/output measurements by using neuro-fuzzy networks, vector quantization methods (Pedrycz 1993) or LMS estimation (least mean squares) (Babuška 1995a).

The form of presentation of a relational fuzzy model becomes considerably complicated as the number of the inputs of the model increases. If there are two inputs, the relational table of the rule base is 3-dimensional and must be presented in the form of cuts. The example of the relational 2-input model is given in Tables 5.26.

The rule base (5.136) corresponds with the relational Tables 5.26.

Table 5.26. The example of the relational tables of a two-input model

x_1 is small

y / x_2	S	M	L
S	0.73	0.27	0
M	0.66	0.34	0
L	0.44	0.56	0

x_1 is medium

y / x_2	S	M	L
S	0	0.68	0.32
M	0.06	0.94	0
L	0.32	0.68	0

x_1 is large

y / x_2	S	M	L
S	0	0.7	0.3
M	0	0.33	0.67
L	0	0.4	0.6

S – small, M – medium, L – large

$$
\begin{array}{llll}
R1: & IF\ (x_1 = S)\ AND\ (x_2 = S)\ THEN & (y = S) & (0.73) \\
 & & (y = M) & (0.27) \\
R2: & IF\ (x_1 = S)\ AND\ (x_2 = M)\ THEN & (y = S) & (0.66) \\
 & & (y = M) & (0.34) \\
R3: & IF\ (x_1 = S)\ AND\ (x_2 = L)\ THEN & (y = S) & (0.44) \\
 & & (y = M) & (0.56) \\
R4: & IF\ (x_1 = M)\ AND\ (x_2 = S)\ THEN & (y = M) & (0.68) \\
 & & (y = L) & (0.32) \\
R5: & IF\ (x_1 = M)\ AND\ (x_2 = M)\ THEN & (y = S) & (0.06) \\
 & & (y = M) & (0.94) \\
R6: & IF\ (x_1 = M)\ AND\ (x_2 = L)\ THEN & (y = S) & (0.32) \\
 & & (y = M) & (0.68) \\
R7: & IF\ (x_1 = L)\ AND\ (x_2 = S)\ THEN & (y = M) & (0.7) \\
 & & (y = L) & (0.3) \\
R8: & IF\ (x_1 = L)\ AND\ (x_2 = M)\ THEN & (y = M) & (0.33) \\
 & & (y = L) & (0.67) \\
R9: & IF\ (x_1 = L)\ AND\ (x_2 = L)\ THEN & (y = M) & (0.4) \\
 & & (y = L) & (0.6)
\end{array}
\qquad (5.136)
$$

5.7.4 Global and local fuzzy models

The problem of the significance of global and local fuzzy models is more and more often discussed. Examples are the following publications (Babuška 1995c; Bossley 1995; Brown 1995a; Nelles 1997; Zeng 1996).

The first fuzzy models had a global character, i.e. they related to the whole universe of discourse. However, it was quickly noticed that the goal of high accuracy of a global model leads in the case of certain systems to immense complication, expressing itself in a large number of rules. It also appeared that global models are effective (tuning of parameters) when the

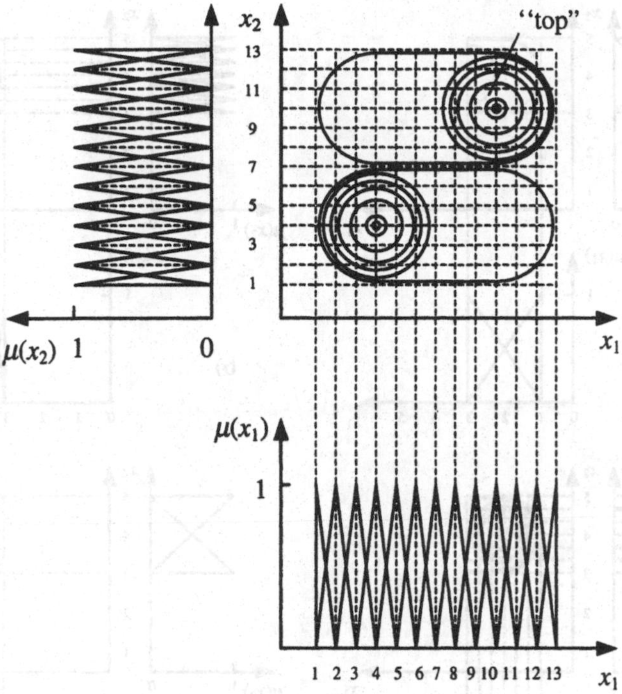

Fig. 5.153. The partition of the inputs' space $X_1 \times X_2$ in the global model of a system containing two "peaks" and two plateaus

number of inputs is relatively small $n \leq 4$ (Bossley 1995; Brown 1995a). Fig. 5.153 illustrates this problem.

To secure high accuracy of the mapping of "peaks" in Fig. 5.153, we require dense membership functions in the region of their occurrence. This way one obtains the partition grid of $13 \cdot 13 = 169$ nodes (support points). Because one fuzzy rule is assigned to each node, the model contains a large number (169) of rules. However, if a great density of nodes is needed in the regions of "peaks", then plateaus can be described sufficiently exactly using a small number of rules (nodes). Therefore, it would be reasonable to divide the global model into 4 local models represented in Fig. 5.154. In the local models the plateaus, Fig. 5.154a and d, are described with only 4 rules, whereas the "peaks" in Fig. 5.154b and c are described by means of a greater number (49) of rules. The total number of rules in the 4 local models amounts to 106 and is 63 less than in the global model.

In order to identify the model with a small number of rules it is sufficient to have a smaller amount of measurement data of the system inputs and output. This is a very important advantage of local modeling because there is often a lack of sufficient measurements made in the system to be modeled.

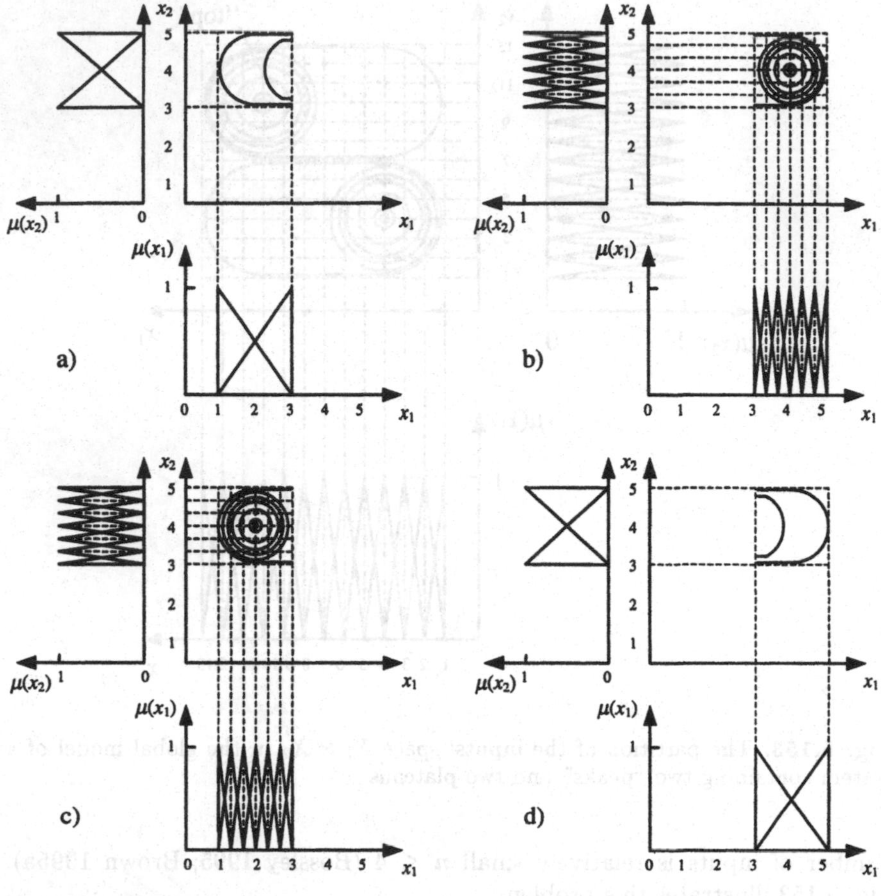

Fig. 5.154. 4 local models (a,b,c,d) replacing the global model from Fig. 5.153

The reason for the lack of information about the system can be high costs, labour consumption and time demand of measurements.

The required and desired feature of the set of local models replacing one global model is the continuity at the interfaces between the local models. Here, there should not be a drastic change in the output y. The method for securing this feature in the case of linear local models of the Takagi-Sugeno type is given in (Babuška 1995c). Another way applied both for Mamdani models and TS models is the use of a hierarchic structure which is represented in Fig. 5.155 for the case of 2 inputs.

In the hierarchic structure first one calculates the outputs of local fuzzy models LM_i and after that these outputs are aggregated using trapezoidal (or similar) membership functions having variable values at the zone boundaries. The method is illustrated by Example 5.7.4.1.

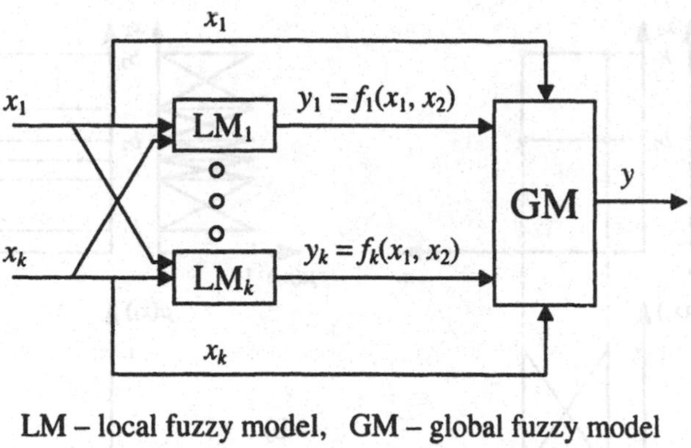

Fig. 5.155. Hierarchic structure of a fuzzy model enabling us to obtain a continuous global surface owing to fuzzy aggregation of local surfaces

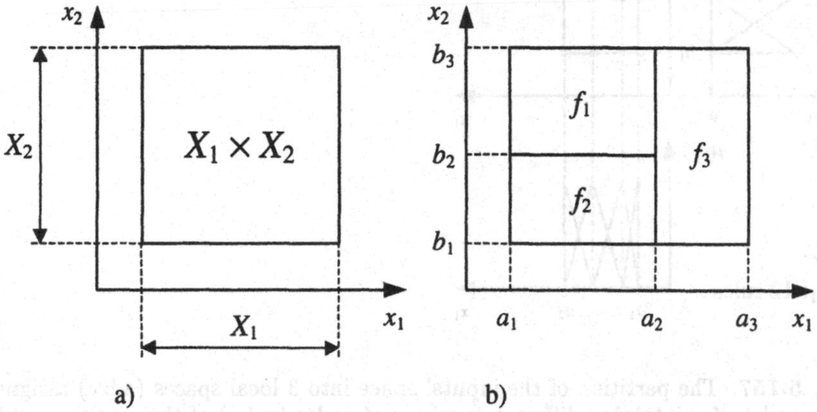

Fig. 5.156. The universe of discourse of the inputs of a global model and its partition into local sectors

Example 5.7.4.1
The universe of discourse of the model inputs has been divided into 3 sectors shown in Fig. 5.156.

For each of the sectors separate fuzzy models f_i have been elaborated. They are based on different local partitions of the input space, Fig. 5.157.
◊ ◊ ◊

As can be seen in Fig. 5.157, the density and distribution of the nodes of the partition grid and hence the number of rules and the parameters of membership functions are different in each sector, depending on the degree of complication of the system surface above this sector. Each of the local

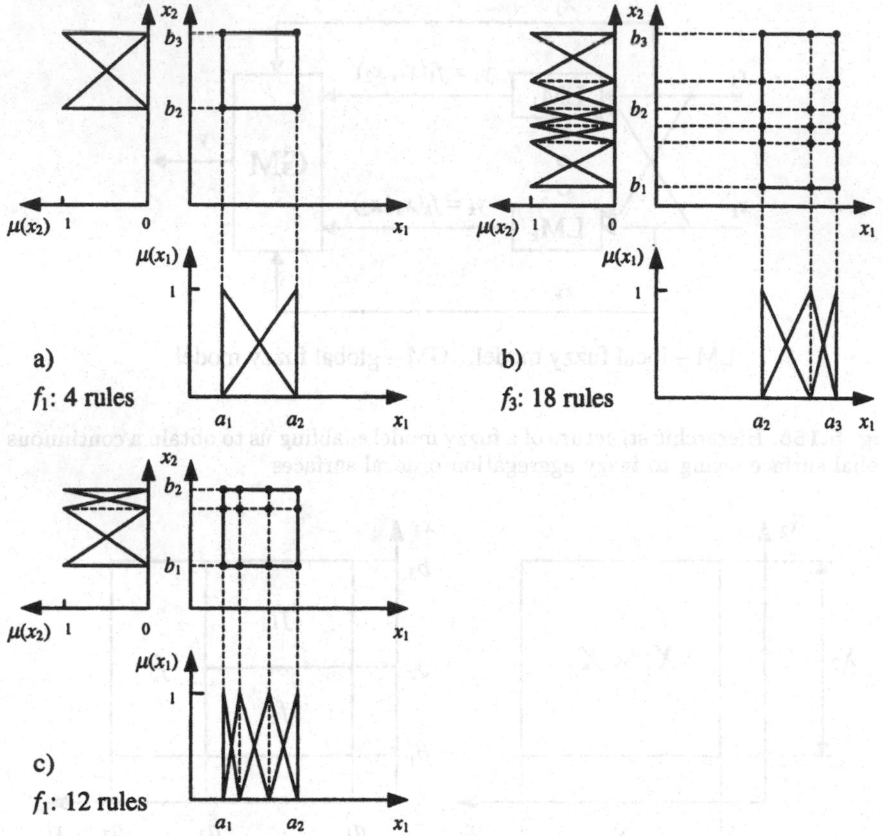

Fig. 5.157. The partition of the inputs' space into 3 local spaces (a,b,c) assigned to 3 models f_i containing different numbers of nodes (rules) of the partition grid

models f_i can be tuned separately, e.g. using neuro-fuzzy networks in order to determine optimal parameters.

The separate tuning of minor local models is much easier than the tuning of one big global model. This results from the fact that difficulties in tuning models increase exponentially with the number of rules (nodes of the partition grid). Also, the number of local minima of the surface of a model error, where the tuning process of the model can stop, increases considerably. After the separate tuning of local models their boundaries usually do not coincide and in passing from one model to another violent jumps of the output y can occur. Therefore the transient zones should be smoothed out. In order to accomplish this, one can apply a superordinated integrating global fuzzy model as represented in Fig. 5.158.

The width of the fuzzy zone around the boundaries of local zones (a_2 and b_2 in Fig. 5.158) can vary depending on the determined value of the "jump"

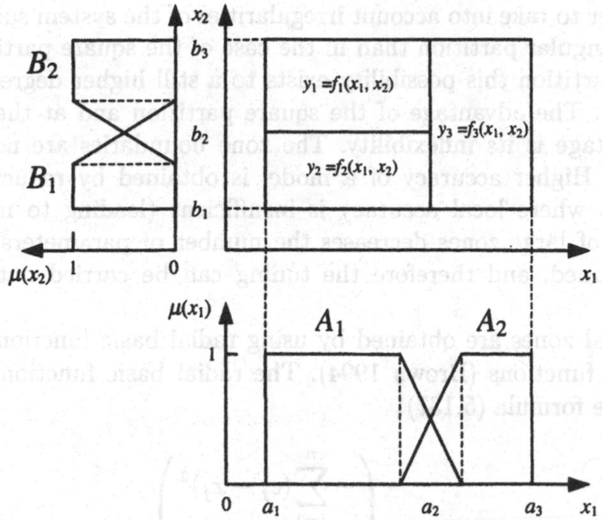

$R1$: *IF* $(x_1 = A_1)$ *AND* $(x_2 = B_1)$ *THEN* $(y = y_2)$
$R2$: *IF* $(x_1 = A_1)$ *AND* $(x_2 = B_2)$ *THEN* $(y = y_1)$
$R3$: *IF* $(x_1 = A_2)$ *THEN* $(y = y_3)$

Fig. 5.158. The superordinated global fuzzy model integrating local models f_i, which secures the smoothness of the transient zones

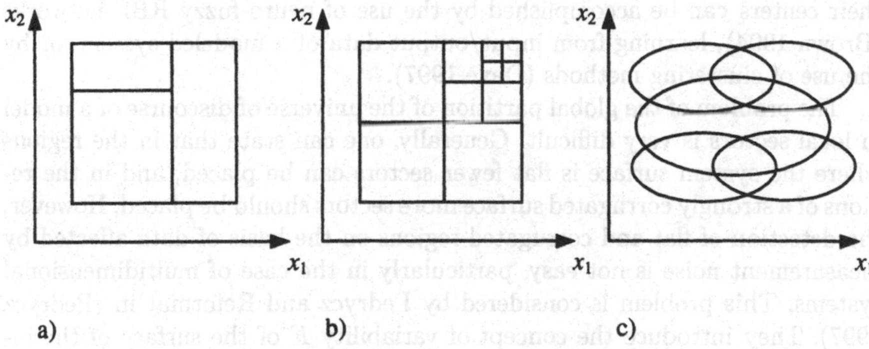

Fig. 5.159. The main methods of partition of the universe of discourse in local zones: a) rectangular zones, b) square zones, c) ellipsoidal zones

Δy at the interface of the models. If the "jump" value is large, then the width of the transition zone should be increased. Otherwise it should be decreased.

Very serious problems are the shape, size and distribution of the zones of particular local models. The partitions in rectangular, square and ellipsoidal zones are most often used, Fig. 5.159.

It is easier to take into account irregularities of the system surfaces in the case of rectangular partition than in the case of the square partition. In the ellipsoidal partition this possibility exists to a still higher degree (ellipsoids can overlap). The advantage of the square partition and at the same time its disadvantage is its inflexibility. The zone boundaries are not displaced (optimized). Higher accuracy of a model is obtained by reducing the size of the zones where local accuracy is insufficient (leading to more zones). Application of large zones decreases the number of parameters of a model which are tuned, and therefore the tuning can be carried out easier and quicker.

Ellipsoidal zones are obtained by using radial basic functions (RBF) as membership functions (Brown 1994). The radial basic functions are determined by the formula (5.137).

$$y = \exp\left(\frac{-\sum_{j=1}^{n}(c_j - x_j)^2}{2\cdot\sigma_j^2}\right), \tag{5.137}$$

where: c_j – the coordinate of the center of a function with respect to the axis of the variable x_j,
σ_j – mean square deviation with respect to the axis of the variable x_j.

The optimization of the sizes of ellipsoidal zones and the positioning of their centers can be accomplished by the use of neuro-fuzzy RBF networks (Brown 1994), learning from input/output data of a modeled system, or by the use of clustering methods (Davè 1997).

The problem of the global partition of the universe of discourse of a model in local sectors is very difficult. Generally, one can state that in the regions where the system surface is flat fewer sectors can be placed, and in the regions of a strongly corrugated surface more sectors should be placed. However, the detection of flat and corrugated regions on the basis of data affected by measurement noise is not easy, particularly in the case of multidimensional systems. This problem is considered by Pedrycz and Reformat in (Pedrycz 1997). They introduce the concept of variability F of the surface of the input/output mapping of a system to be modeled (5.138).

$$F = \int_{\mathbf{X}} \left|\frac{\partial f}{\partial \mathbf{X}}\right| \mathrm{d}\mathbf{X} = \int_{X_N}\int_{X_{N-1}}\dots\int_{X_1}\left(\sum_{i=1}^{N}\left|\frac{\partial f}{\partial x_i}\right| \mathrm{d}x_1\,\mathrm{d}x_2\,\dots\,\mathrm{d}x_N\right), \tag{5.138}$$

where: $\mathbf{X} = X_1 \times X_2 \times \dots \times X_N$ – the universe of discourse,
$f = f(x_1, x_2, \dots x_N)$ – the input/output mapping function of a system.

The method they propose consists in determining the regions of the universe of discourse where the variability F is less or greater and according to this the partition of the input space should be less or more dense, respectively. However, in the present version this method is only suitable for the fuzzy modeling of systems described by conventional mathematical functions. Unfortunately, it has no equivalent operating on the basis of noised measurement data.

It seems that the problem of optimal partition of the universe of discourse for such a case still requires further research. At present we can use preliminary knowledge of a system to be modeled as a basis and apply the trial-and error method or use our own intuition.

5.7.5 Fuzzy multimodels

The idea of multimodels was introduced by Pedrycz in (Pedrycz 1996). It is very important because the treating a multimodel system as a single-model system leads to erroneous interpretation of results. Furthermore, we deal with multimodel systems fairly often. Let us consider the system represented in Fig. 5.160.

This system is unique with respect to the input x: each value x corresponds with only one value of the output y. The system can be described with one fuzzy model shown in Fig. 5.160. Now compare this model with the model of a system of a chemical process (Pedrycz 1996) illustrating the interdependence between process temperature T and the Damkoeher number d characterizing certain features of the process, Fig. 5.161a.

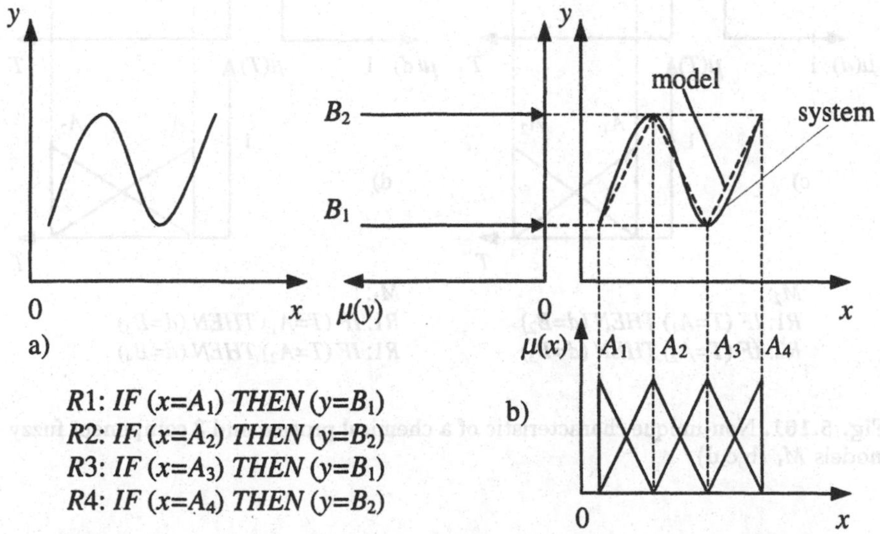

Fig. 5.160. The example of a unique system with respect to the input variable x (a) and its fuzzy model (b)

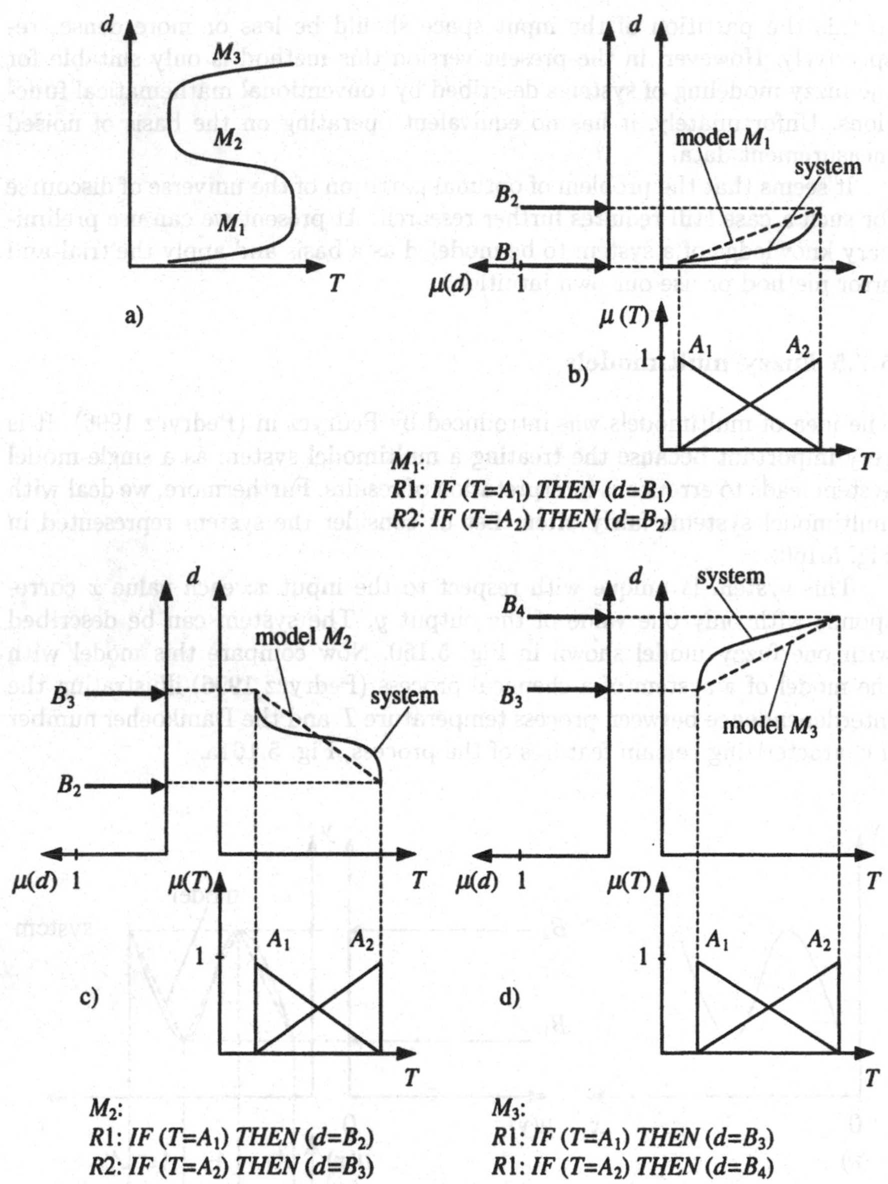

Fig. 5.161. Non-unique characteristic of a chemical process and 3 component fuzzy models M_i (b,c,d)

As results from Fig. 5.161a show, the process characteristic is not unique with respect to the input x. One value x corresponds with 3 (or 2) values of the output y. Which value y should be assigned to the current value of the input x of the model in this situation? Is this assignment unique in the real system?

The non-uniqueness of such characteristics as presented in Fig. 5.161a results from the fact that they are presented in the space of insufficient dimensionality. If the real system behaves non-uniquely, then there must exist in it an additional input deciding which part of the characteristic the system is at a given moment. In the case of a system with a non-unique 2-dimensional characteristic having the shape of the letter S, Fig. 5.161a, the output state d at the given moment t, $(d(t))$ can be determined uniquely only when we know additionally the values of two variables illustrating the direction of changes occurring in the process: $\dot{d}(t)$ as well as $\dot{T}(t)$.

The presentation of the process characteristic in the 4-dimensional system,

$$d(t) = f(T(t), \dot{d}(t), \dot{T}(t)),$$

which is graphically impossible, provides it with the feature of uniqueness.

The (as agreed upon) non-unique system is often encountered in engineering: the mechanical system of piston-cylinder-spring, having a hysteresis character caused by the presence of friction, is non-unique (Fig. 5.162).

A system of a hysteresis character can be presented uniquely in the 4-dimensional arrangement $x(t) = f(F(t), \dot{F}(t), \dot{x}(t))$, (Piegat 1995c). The problem of non-uniqueness is explained in Fig. 5.163.

If the model of a system does not include (or disregards) some inputs of it (because we don't known their values or we are not aware of their existence), then the state of the output can not be determined uniquely as the disregarded inputs affect the output too, changing it in a specified unknown manner. How should the problem of non-uniqueness be solved?

The first method is to detect the disregarded inputs and to present the model of a system in the extended space. The illustration of this method is Example 5.7.5.1 presenting a problem similar to the one in Example 5.7.3.1.

Example 5.7.5.1

Let us consider the relational fuzzy model (5.139) illustrating the relation between the mental powers of a child and his (her) results in school. The model is specified in two-dimensional space (mental powers, school results).

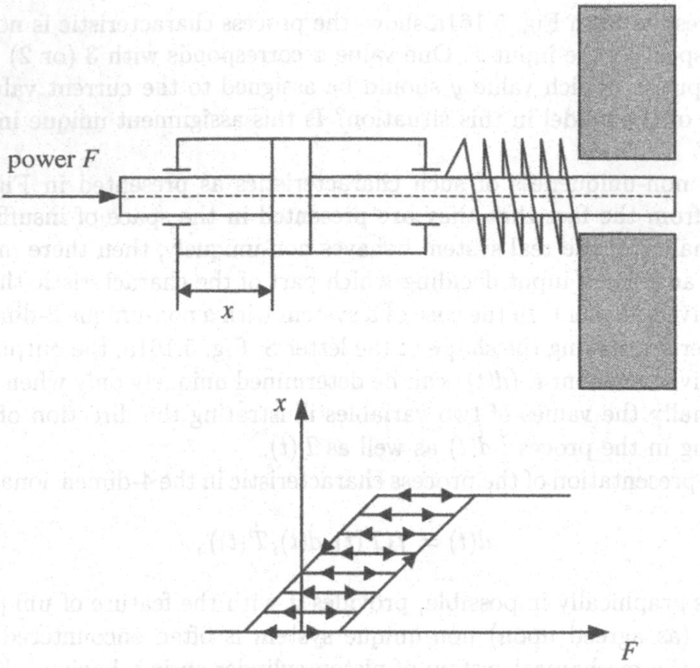

Fig. 5.162. A mechanical system and its 2-dimensional hysteresis characteristic

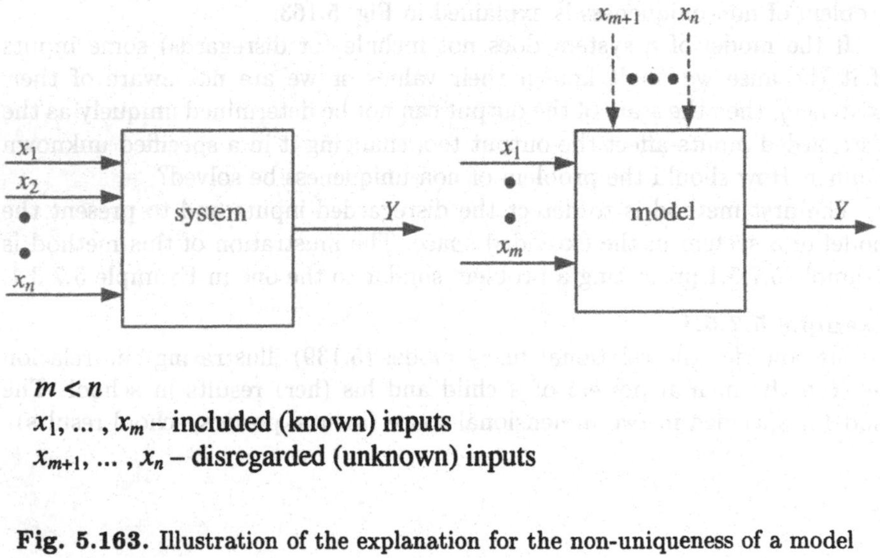

Fig. 5.163. Illustration of the explanation for the non-uniqueness of a model

*R*1 :	A clever child learns	well (0.6)
		to an average degree (0.3)
		poorly (0.1)
*R*2 :	An ordinary clever child learns	well (0.1)
		to an average degree (0.7)
		poorly (0.2)
*R*3 :	A not-so-clever child learns	well (0)
		to an average degree (0.3)
		poorly (0.7)

(5.139)

The model (5.139) has distinct features of a multimodel in which the state of its output is not unique and only its probable value can be given. The degree of probability in this case is expressed by confidence coefficients. However, if we take into account the second important input strongly affecting the school results of a child, namely studiousness (diligence), then we will be able to make this model (fairly) unique. The set of rules (5.140) is the rule base of the model in the enlarged space (mental powers, studiousness, school results).

*R*1 : A clever and hard-working child learns well.
*R*2 : A clever and average working child learns to an average degree.
*R*3 : A clever and non-diligent child learns poorly.
*R*4 : An average and hard-working child learns well.
*R*5 : An average and average working child learns to an average degree.
*R*6 : An average and non-diligent child learns poorly.
*R*7 : A not-so-clever and hard-working child learns to an average degree.
*R*8 : A not-so-clever and average working child learns poorly.
*R*9 : A not-so-clever and non-diligent child learns poorly. (5.140)

◊ ◊ ◊

One of the reasons for non-uniqueness can also be the insufficiently recognized input/output interdependence of a system resulting in our inability to determine the output state although the inputs are known (weather forecasting). If we don't know the cause-effect dependence of the system inputs and output or we don't know some inputs of the system, then we can apply the method of description of systems consisting in giving probabilities of various possible states of the output y at a given state of the input vector $\mathbf{X}$ of a model as is shown in Example 5.7.5.1, model (5.139).

Pedrycz defines in (Pedrycz 1996) the multimodel as "a set of models $M_1, M_2, \ldots, M_c$ equipped with a mechanism of appropriate switching from one model to another or, if required, a mechanism of aggregation of results provided by individual models". The switching mechanism operates on the

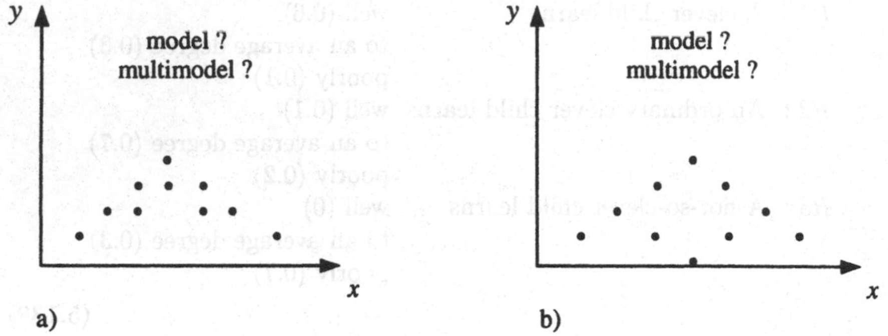

Fig. 5.164. Illustration of the problem of the degree of scatter of measurement data and of the evaluation of uniqueness of a model connected with this problem

basis of additional information about the system (inputs) while the mechanism of aggregation operates on the basis of confidence coefficients of individual models.

The next important problem connected with multimodels is distinguishing on the basis of measurement of input/output data if the given system is unique (i.e. it is a model) or non-unique (i.e. a multimodel). The problem is represented in Fig. 5.164. The measurements of input/output data of a system are usually noised (measurement noise) and/or affected by the inputs disregarded in the model. If the second case applies, the system should be represented in the form of a multimodel (if the influence of the disregarded inputs is "sufficiently" great). How it is possible to distinguish the first case from the second?

If the scatter of the measurements of the output is slight (within the error limits of a measurement device), then the system should be represented in the form of a (unique) model, Fig. 5.164a.

If the scatter is great, Fig. 5.164b, then it is appropriate to represent the system in the form of a multimodel. The qualifications "slight", "great" scatter are fuzzy and depend considerably on our intuition or preliminary knowledge of the modeled system. The difficulty in distinguishing the influence of measurement noise from the influence of disregarded inputs increases greatly as the dimensionality of the system (the number of inputs) becomes greater.

An interesting method of identification of multimodels based on directional clustering is presented by Pedrycz in (Pedrycz 1996). He also indicates there the directions of further studies and ways of making improvements of the method, which are necessary considering the complexity of the problem of the recognizability of multimodels.

5.7.6 Neuro-fuzzy models

To create a fuzzy model, all its elements: the rule base, number and kind of the membership functions of each of the model variables, parameters of the membership functions, type of logical operators and the like, must be determined.

The first fuzzy models were created on the basis of the expert's knowledge of the system to be modeled. Information about the system was acquired from the system expert and it was then converted by a fuzzy modeling expert into a model. This method is called the **acquisition of knowledge** (Preuss 1994a). It is effective if a system expert has complete knowledge about it, can formulate the possessed information in words and impart it. In practice the expert's knowledge is often incomplete, inaccurate and difficult to formulate. It can even include inconsistencies. This knowledge is subjective, i.e. individual persons can have different opinions of the operation of the same system. Taking this into account, it would be advantageous to base the modeling on objective information about a system. The measurement data of the system inputs and output is such information.

The acquisition of knowledge from this data is called the **extraction of knowledge**. Neural networks have such an ability. This observation caused the development of investigations of methods of converting fuzzy models into neural networks, which are called, considering their specificity, neuro-fuzzy networks. Various aspects of this interesting problem are described in (Altrock 1993; Brown 1994,1995a,b; Culliere 1995; Carpenter 1992; Eklund 1992; Feuring 1996; Gupta 1994; Higgins 1994; Hensel 1995; Hauptmann 1995; Halgamuge 1996; Horikawa 1992; Ishiguchi 1993; Kochlert 1994,1995; Lin 1991,1995; Nobre 1995; Osowski 1996; Piegat 1996,1997a; Pałęga 1996; Rutkowska 1997; Su 1995; Simpson 1992; Yao 1995).

A neuro-fuzzy network is a particular, equivalent form of a fuzzy model. For example, the fuzzy model with the segmentally-linear membership functions A_{ij} of the inputs x_1, x_2, singleton functions B_k of the output y and rule base (5.141),

$$\begin{aligned}
&R1: \ \mathit{IF}\ (x_1 = A_{11})\ \mathit{AND}\ (x_2 = A_{21})\ \mathit{THEN}\ (y = B_1)\,,\\
&R2: \ \mathit{IF}\ (x_1 = A_{11})\ \mathit{AND}\ (x_2 = A_{22})\ \mathit{THEN}\ (y = B_2)\,,\\
&R3: \ \mathit{IF}\ (x_1 = A_{12})\ \mathit{AND}\ (x_2 = A_{21})\ \mathit{THEN}\ (y = B_3)\,,\\
&R4: \ \mathit{IF}\ (x_1 = A_{12})\ \mathit{AND}\ (x_2 = A_{22})\ \mathit{THEN}\ (y = B_4)\,,
\end{aligned} \tag{5.141}$$

can be transformed into the neuro-fuzzy network represented in Fig. 5.165.

This network can be tuned with measurement samples of the inputs and output of a modeled system using the method of error back-propagation or other methods applied to neural networks.

The increasing popularity of neuro-fuzzy networks (NFN) results from their undeniable advantages.

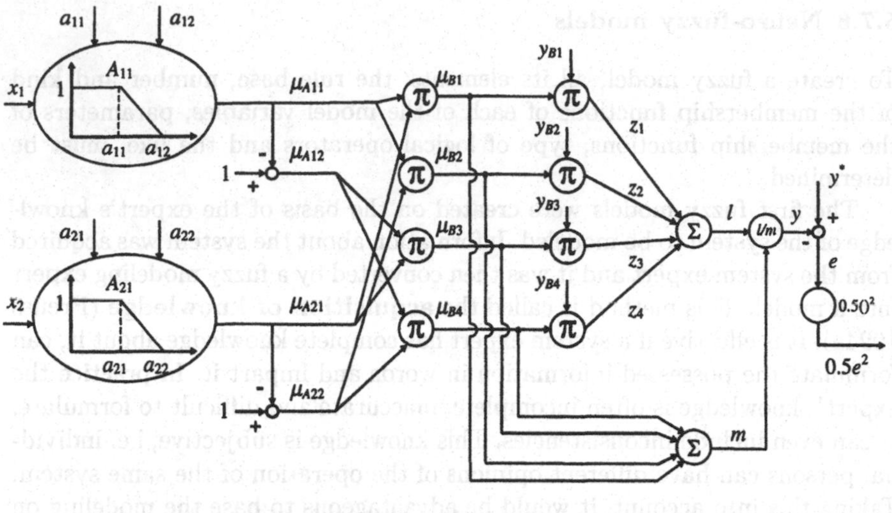

Fig. 5.165. Neuro-fuzzy network corresponding with the fuzzy model (5.141)

The advantages of neuro-fuzzy networks (NFNs)

1. NFNs make it possible to optimize (tune) the parameters of fuzzy models on the basis of measurement data of the inputs and output of real systems.
2. NFNs enable us to correct inaccurate fuzzy models formulated by experts.
3. NFNs enable us to complete fuzzy models formulated by experts in those regions of the inputs space where no expert's knowledge exists.
4. NFNs have structure and parameters intelligible to a person (sets: large, small, linguistic rules). They allow for generalizing the knowledge included in the noised measurement data of a system to be modeled and to present it in the form of linguistic rules (the extraction of knowledge) comprehensible to a person. In contrast the parameters of perceptron neural networks give nothing except numbers and the knowledge included in them is not comprehensible for a person. Therefore these networks are of a "black box" character.
5. If we have preliminary or partial knowledge of a modeled system, then it can be easily introduced in NFN. In the case of a perceptron network it is impossible or very difficult.
6. In the case of NFNs the structure of a network can be considerably more easily determined than in the case of a perceptron network where we usually must apply a trial-and-error method. Knowledge of the right structure of a network accelerates the learning process and decreases the influence of local minima of the error function.

The way of transforming a fuzzy model into a neuro-fuzzy network is rather complicated and depends on the kind of model. The transformation of Mamdani models into NFNs is described in Section 6.2.1.2 and of Takagi-Sugeno models in Section 6.2.1.3.

5.7.7 Alternative models

The Mamdani fuzzy models have the following features:

a) they accomplish the (hyper)-rectangular partition of the input space, Fig. 5.166,
b) the boundaries of rectangular sectors are usually linear,
c) the surfaces of local sector models are usually "weakly" non-linear (e.g. multi-linear).

The mentioned features can be both advantages and disadvantages of fuzzy models, depending on the chosen criterion of evaluation. The rectangular partition of the input space enables us to formulate a model in the form of rules comprehensible to a person. However, because membership functions are defined separately for each variable of a model, the trial of adding even one new function only causes a considerable increase in the number of rules. The next disadvantage of the rectangular partition of the input space is illustrated in Fig. 5.167a,b.

The surface $y = f(x_1, x_2)$ of the modeled system consists of two "peaks" and two plateaus, Fig. 5.167a. To exactly model the "peaks" a greater number of membership functions must be applied, Fig. 5.167b. The plateaus can be modeled accurately by means of far fewer rules and membership functions. However, the great density of membership functions imposed by the "peaks"

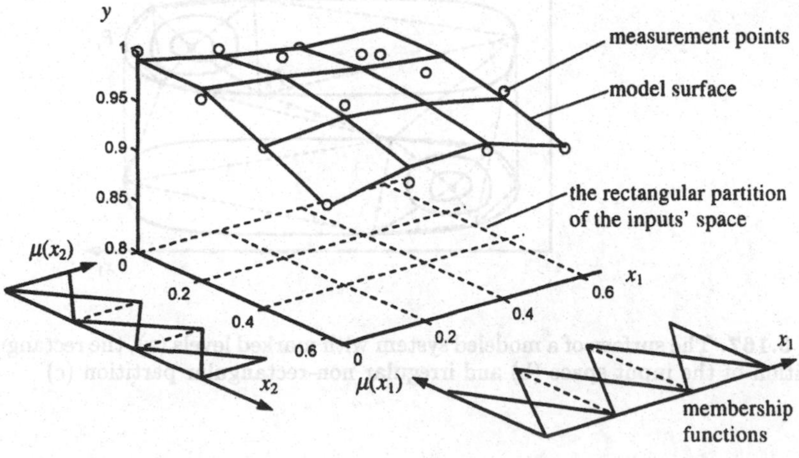

Fig. 5.166. Illustration of the features a), b), c) of a Mamdani fuzzy model

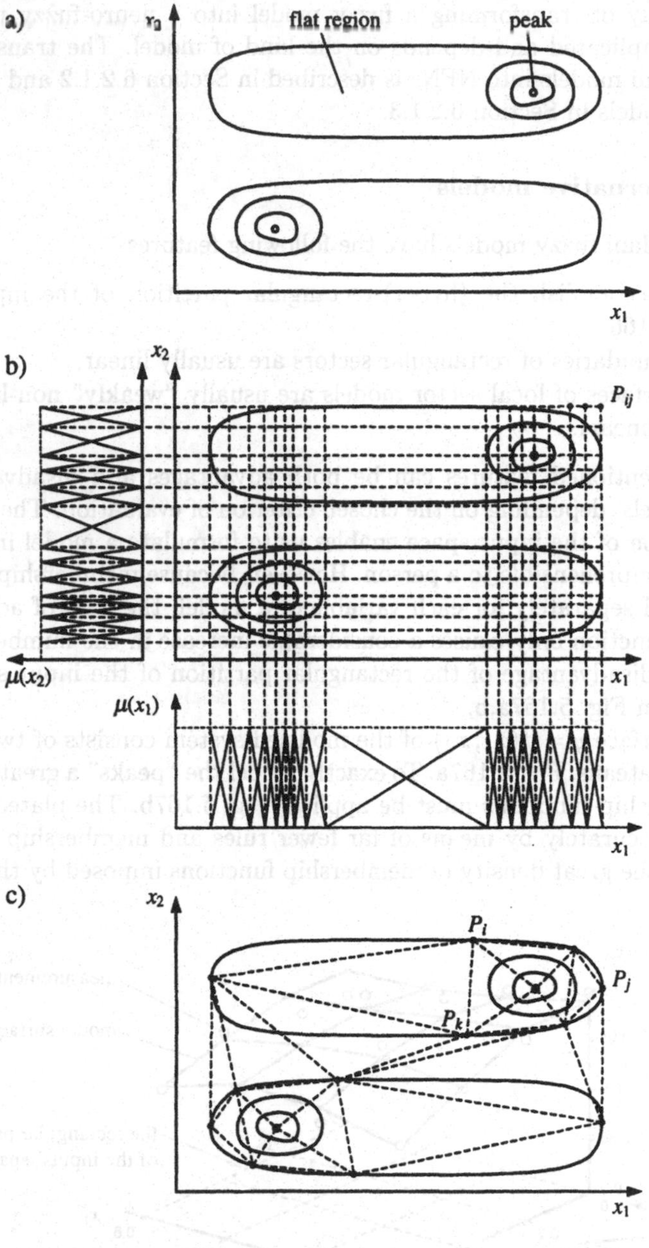

Fig. 5.167. The surface of a modeled system with marked levels (a), the rectangular partition of the input space (b) and irregular non-rectangular partition (c)

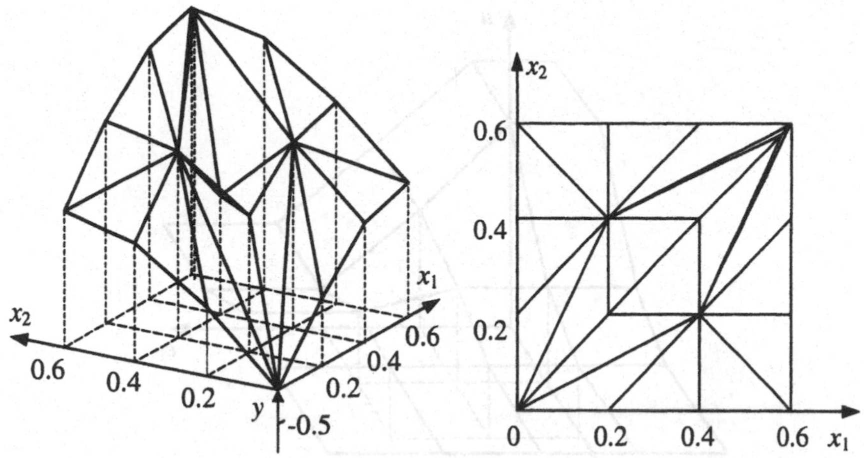

Fig. 5.168. The surface of a model (a) in the space $X_1 \times X_2 \times Y$ obtained by partition of the input space (b) $X_1 \times X_2$ in triangular sectors

is transferred onto the plateaus resulting in fact that the number of rules we must use to define them is unnecessarily great. Each node P_{ij} of the partition grid of the Mamdani model means one rule and three parameters, which must be tuned. Thus the great number of nodes requires that a respectively great amount of measurement information about the modeled system must be acquired. The increase in the number of parameters in a model causes its tuning process to become very difficult and brings the model to the limit of tuning possibility (curse of dimensionality).

The use of an irregular, e.g. triangular partition of the inputs space, Fig. 5.167c, makes it possible to considerably decrease the number of sectors and to adapt the density of their distribution to the grade of non-linearity of the region to be modeled. The number of triangular sectors must be greater about the "peaks". In plateau regions sufficient accuracy is obtained through the use of far fewer larger sectors. Fig. 5.168a shows, as an example, the model surface based on the triangular partition of the input space, Fig. 5.168b. For comparison, the surface of a fuzzy model based on the rectangular regular partition of the input space is represented in Fig. 5.169.

In the case of the rectangular partition, four input/output measurement data of a modeled system are needed to define the model surface over one rectangular sector. For the triangular partition of the input space three such data are needed. In the case of a system with three inputs, the fuzzy model is based on the cuboidal partition of the input space, Fig. 5.170a, and the alternative model based on the tetrahedral partition, Fig. 5.170b.

In a three input system eight input/output measurement data of the system are needed to define the model surface over one cuboidal sector of the input space, Fig. 5.170a. In the case of tetrahedral sectors, Fig. 5.170b, four

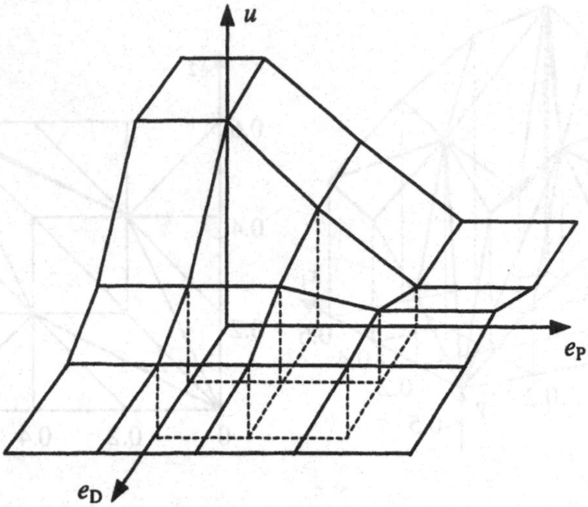

Fig. 5.169. The surface of a fuzzy model based on the regular rectangular partition of the input space

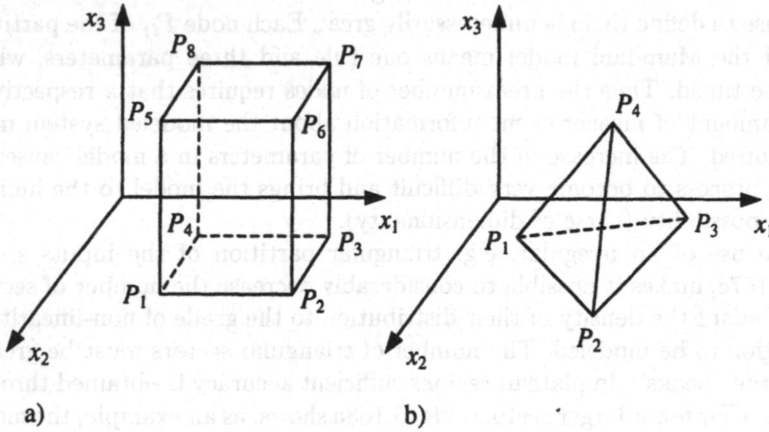

Fig. 5.170. A sector of the input space $X_1 \times X_2 \times Y$ at the regular hyper-rectangular partition (a) and at the irregular hyper-triangular partition of the input space (b)

measurement data are sufficient. With the increase in the number n of inputs of a modeled system, the difference of the number of required measurements increases drastically. For the hyper-rectangular partition of the input space the minimal number p of required measurement data amounts to:

$$p = 2^n \,, \tag{5.142}$$

whereas at the hyper-triangular partition only:

$$p = n + 1 \,, \tag{5.143}$$

measurements are needed. The relationship (5.142) is of a non-linear, exponential character, while the relationship (5.143) is linear.

In the case of a system having 10 inputs, $p = 1024$ measurement data are needed to define one hyper-rectangular sector, and only 11 measurement data to define one hyper-triangular sector. So, the difference is immense. The use of models with the hyper-triangular partition is particularly advantageous in the case of systems having a great number of inputs, systems determined by a lesser amount of measurement data and also when there is difficulty in making measurements, e.g. because of their costs, the lack of measurement equipment and so on. Lack of the required number of measurement data occurs particularly in economic systems. The great number of points defining the model surface over one hyper-rectangular sector – the formula (5.142) – makes it very difficult to tune a fuzzy model and is the cause of the "curse of dimensionality" (Bellmann 1961; Brown 1995a). In the case of the Mamdani fuzzy model one rule of the type (5.144) determining the output state y of the model corresponds with each node P_{ij} of the partition grid, Fig. 5.171.

$$\mathit{IF}\ (x_1 \text{ near } x_{1i})\ \mathit{AND}\ (x_2 \text{ near } x_{2j})\ \mathit{THEN}\ (y \text{ near } y_{ij}) \tag{5.144}$$

The position of the point P_{ij} defined by the rule is determined with the coordinates x_{1i}, x_{2j}, y_{ij}. Instead of the conventional fuzzy rule of the type (5.144), the rule of the type (5.145) can be applied, defining the surface $y = f_k(x_1, x_2)$ over the rectangular sector k of the input space in another manner.

$$\mathit{IF}\ (x_{1i} \le x_1 \le x_{1(i+1)})\ \mathit{AND}\ (x_{2j} \le x_2 \le x_{2(j+1)})\ \mathit{THEN}\ (y = f_k(x_1, x_2)) \tag{5.145}$$

In the case of the Mamdani models with the *PROD*-operator and segmentally-linear membership functions the surface f_k over the sector is of a multi-linear character (Zeng 1996) determined by the formula given in Fig. 5.171. The boundaries of each sector are straight segments. At the triangular partition of the input space, the model surface over an individual sector is described by the linear relationship given in Fig. 5.172. One rule of the model of type (5.146) corresponds with one triangular sector.

$$\mathit{IF}\ (x_2 \ge a_0^{ij} + a_1^{ij} x_1)\ \mathit{AND}\ (x_2 \ge a_0^{ik} + a_1^{ik} x_1)\ \mathit{AND}\ (x_2 \ge a_0^{jk} + a_1^{jk} x_1)$$
$$\mathit{THEN}\ (y = f_k(x_1, x_2)) \tag{5.146}$$

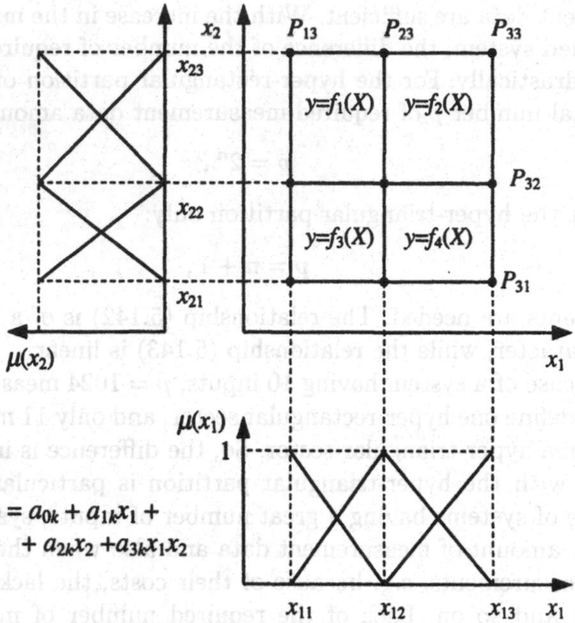

Fig. 5.171. The rectangular partition of the input space which is typical for the Mamdani fuzzy model

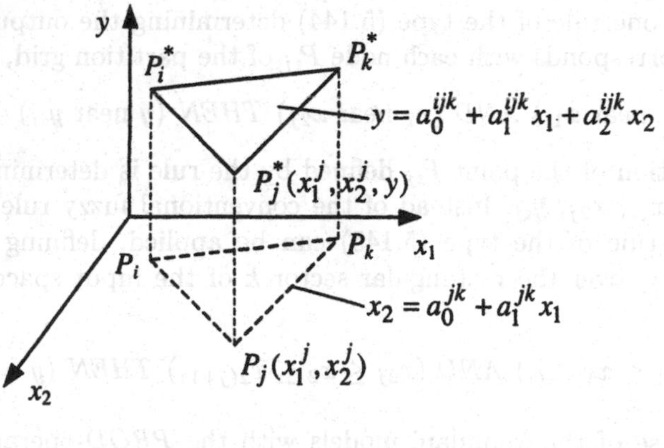

Fig. 5.172. A fragment of the model surface above the triangular sector of the input space

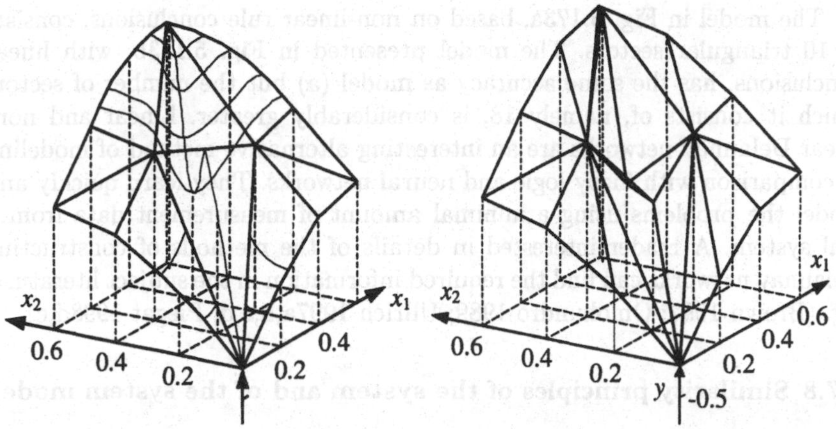

a) mean absolute error δ=0 b) mean absolute error δ=0

Fig. 5.173. The Delaunay model created with the use of non-linear conclusions (a) and linear conclusions (b)

The models with rules of type (5.146) which include linear conclusions, have been known under the name of "Delaunay networks" or "Delaunay triangularization" for the past few decades. Primarily they were used for geodesic descriptions of the areas of estates. Lately, new applications for the modeling of technical systems and control were found (Omohundro 1989; Ullrich 1997a,b,c). The Delaunay networks are self-organizing and self-tuning. There are quite a few construction methods. Some of them, as for example the method of placing new nodes of the networks at points of the maximal error of the model (Brown 1997; Piegat 1998d), do not show the disadvantage of stopping the learning process at the local minima of the error function, which is the typical disadvantage of neural networks. Therefore the learning process of the Delaunay networks is quick and ends in the proximity of the global minimum of the error function.

However, the investigations carried out by Ullrich (Ullrich 1997a) concerning Delaunay networks with linear conclusions, the formula (5.146), have shown that the number of triangular sectors (indispensable for obtaining the required accuracy of a model) they generate is often too great and with certain learning methods – when the inverting of matrices is required – they also show the feature of the "curse of dimensionality". The introduction of non-linear conclusions in the rules (5.146), i.e. the delinarization of the Delaunay networks (Piegat 1998d) enables us to obtain considerably higher accuracy of a model at an identical or a lesser number of triangular sectors. Fig. 5.173 represents the example of two models obtained with the use of the Delaunay networks.

The model in Fig. 5.173a, based on non-linear rule conclusions, consists of 10 triangular sectors. The model presented in Fig. 5.173b, with linear conclusions, has the same accuracy as model (a) but the number of sectors which it consists of, namely 18, is considerably greater. Linear and non-linear Delaunay networks are an interesting alternative method of modeling in comparison with fuzzy logic and neural networks. They learn quickly and model the problems using a minimal amount of measurement data from a real system. A reader interested in details of the methods of constructing Delaunay networks can find the required information in the subject literature, e.g. (Brown 1997; Omohundro 1989; Ullrich 1997a,b,c,d; Piegat 1998d,e).

5.7.8 Similarity principles of the system and of the system model

The model of an unique static system accomplishing the mapping $f_S : \mathbf{X} \rightarrow Y_S$, should satisfy best of all the two following similarity principles: the principle of similar mapping of the input vector $\mathbf{X}_S$ on the output Y_M and the principle of similar mapping of changes $\Delta\mathbf{X}_S$ of this vector on changes ΔY_M, Fig. 5.174 and 5.175.

1. The principle of similar mapping of the input vector $\mathbf{X}_S$ on the output Y_M This principle is determined by the formula (5.147).

$$Y_M(\mathbf{X}_S) \approx Y_S(\mathbf{X}_S) \tag{5.147}$$

   This means that at the same input vectors $\mathbf{X}_S$ of the system and its model the model output Y_M should be as close as possible to the system output Y_S (in terms of the assumed error evaluation) in the whole universe of discourse of the input vector.
2. The principle of similar mapping of changes $\Delta\mathbf{X}_S$ of the input vector on changes of the output ΔY at the same initial state of $(\mathbf{X}_S(0) = \mathbf{X}_M(0))$

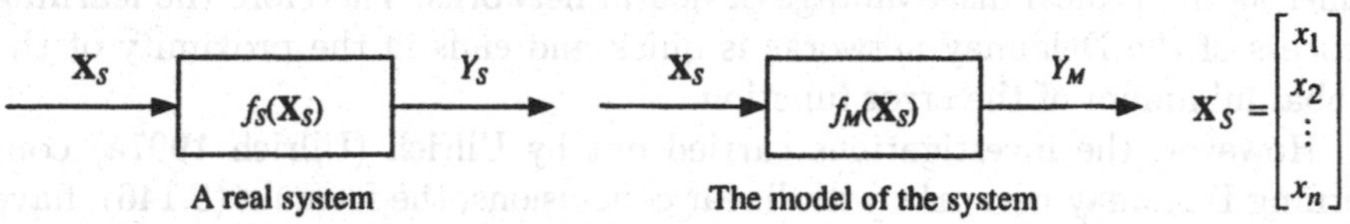

Fig. 5.174. The input/output mapping accomplished by the system and its model

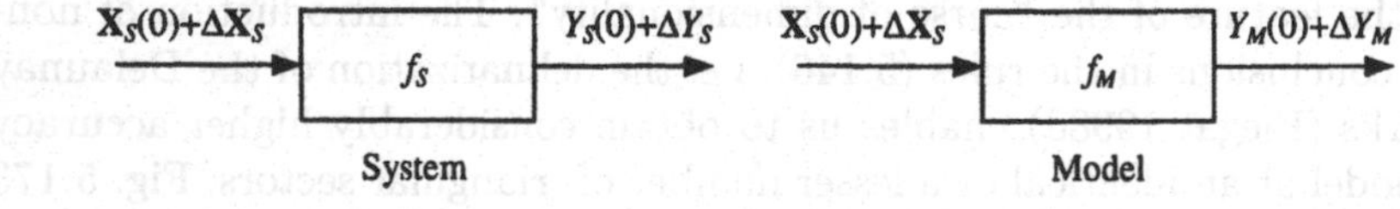

Fig. 5.175. The mapping of input changes on output changes by the system and its model

of the system and its model, Fig. 5.175. This principle is determined by the formula (5.148).

$$\Delta Y_M(\Delta \mathbf{X}_S, \mathbf{X}_S(0)) \approx \Delta Y_S(\Delta \mathbf{X}_S, \mathbf{X}_S(0)) \tag{5.148}$$

The second principle tells us that the system and its model are similar when the change $\Delta \mathbf{X}_S$ of the input vector (at the given initial state ($\mathbf{X}_S(0)$) of this vector) and the change ΔY_S of the system and ΔY_M of the model are similar to each other (in terms of the assumed error evaluation). The similarity principles should be satisfied in the whole universe of discourse of the system and its model.

Principle (2) of similar mapping of input changes results from principle (1) of the similar mapping of the inputs and it seems to be self-evident that if principle (1) is satisfied entirely then principle (2) should be satisfied automatically. Unfortunately, in practice, in fuzzy modeling, we usually obtain models which are only an approximation of the real system. Therefore principle (1) is not fulfilled perfectly. By regarding both principles when selecting the structure and elements of the model, we can obtain more accurate models.

For example, the application of the *MIN* operators for the aggregation of component premises, Section 5.1.2.1, or the realization of defuzzification using the Middle of Maxima method, Section 5.1.3, introduces insensitivity zones in the model where it does not react to changes of the inputs. This reduces the degree of satisfying principle (2) of mapping the input changes (also principle (1)).

5.7.9 Fuzzy classification

In the foregoing sections of Chapter 5 fuzzy models generating the mapping surface $y = f(X)$ of the causal relations existing in the modeled system, process or plant have been investigated.

The example here can be the dependence of the temperature inside a room T_{room} (°C) upon the temperature of the water flowing in a heater T_{water} (°C) and the outside temperature T_{outside} (°C), Fig. 5.176.

Fuzzy set theory is also very often used to solve classification problems. An example is the evaluation of customers trying to obtain credit at a bank. If a customer is more creditworthy, then the bank will more willingly grant him high credit. The bank can even reduce the rate of interest, i.e. the cost of credit. If a customer is not creditworthy, then the bank will not allow him credit. If a customer is partially creditworthy, then the bank will give him credit to a limited amount and perhaps increase the rate of interest.

Let us assume that we introduce the following classification of customers: creditworthy, partially creditworthy, and not creditworthy. These ideas are obviously fuzzy. In the past the bank granted credit to many customers. The customers who repaid their credit entirely and on time can be numbered

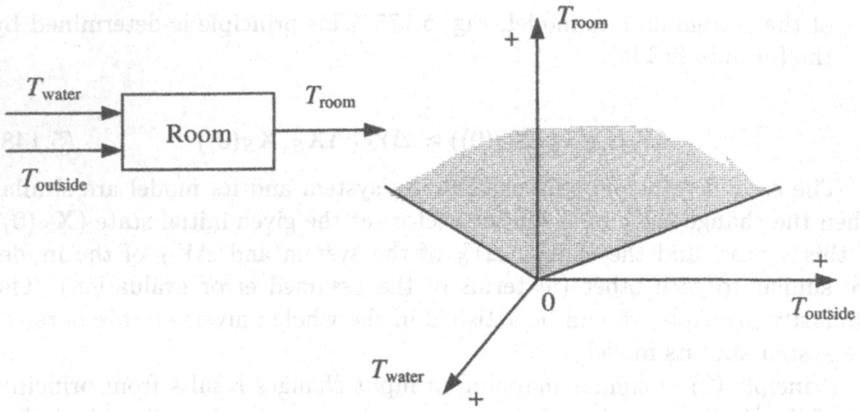

Fig. 5.176. The causal relation $T_{room} = f(T_{water}, T_{outside})$ of the heating of a room

among the creditworthy and those who did not repay the whole amount as not-creditworthy. The customers who did not repay it in the time allowed can be numbered among the partially creditworthy. The difference between a creditworthy customer and a partially creditworthy one is fuzzy, as the prolongation of the time of repayment or the amount of the credit repaid (losses to the bank) can be minor or greater.

Each of the customers trying to obtain credit had to give a series of data to the bank, for example:

x_1 – mean net earnings,
x_2 – the number of dependents,
x_3 – the amount of actual debts,
x_4 – the estimated value of their assets,
x_5 – the period of employment at the present place of employment,
x_6 – the period of employment at the previous place of employment,
x_7 – the sum of credit required,

and other data.

The data x_i can be called the features of a customer. They inform the bank whether the customer is capable of repaying the credit allowed to him (e.g. the earnings x_1 of the customer) or if it is possible to regain the allowed credit when it would not be repaid (e.g. the value of their assets x_4).

For each of the previous customers (i – the number of a customer) serviced by the bank up to the present the vector of features $(x_{1i}, x_{2i}, \ldots, x_{7i})$ can be prepared as well as for each of them the decisive classification μ_i (creditworthy – cr, partially creditworthy – p-cr, not-creditworthy – n-cr) made by a bank expert on the basis of the actual history of the repayment of credit. The example of the data base of customers serviced by the bank up to the present is given in Table 5.27.

Table 5.27. Database of bank customers

Customer no.	x_1 ($)	x_2	x_3 ($)	x_4 ($)	x_5	x_6	x_7 ($)	resulting credibility
1	5600	1	0	100000	26	0	10000	cr
⋮	⋮							⋮
500	1500	6	30000	2000	0.5	0.08	50000	n-cr
⋮	⋮							⋮
1000	3100	3	1000	50000	4	6	20000	p-cr

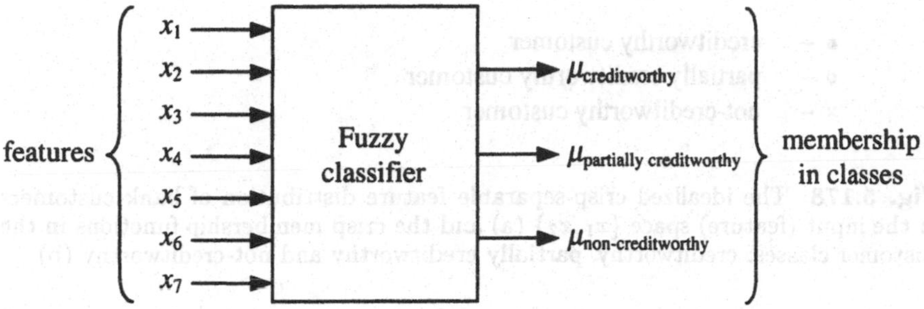

Fig. 5.177. The input/output scheme of the fuzzy classifier of bank customers

As the evaluation of creditworthiness is difficult considering the great amount of data (it is usually greater than 7), in order to accomplish this task an automatic classifier can be established, which can learn from the data of former bank customers, Table 5.27, and can then be used to estimate future customers. The input/output scheme of the classifier of the creditworthiness of customers is shown in Fig. 5.177.

On the basis of the data vector $\mathbf{X}_i$ of a potential customer the classifier calculates his membership in the sets of creditworthy customers (cr), partially creditworthy customers (p-cr), and not-creditworthy ones (n-cr). The greatest value of membership, which can assume fractional values in the interval $[0, 1]$, will decide which set the customer will be numbered amongst, i.e. how the customer will be classified.

Further examples of classification are:

- recognition of human beings on the basis of their faces: the features x_j can be here e.g. the distance between the eyes related to the head height, colour of eyes, the relative width of the mouth and so on;
- recognition of block or handwritten letters on the basis of the pixel notation or other specially selected features, e.g. the number of lines that intersect or the number of sharp bends in a letter;
- recognition of planes by their silhouettes;

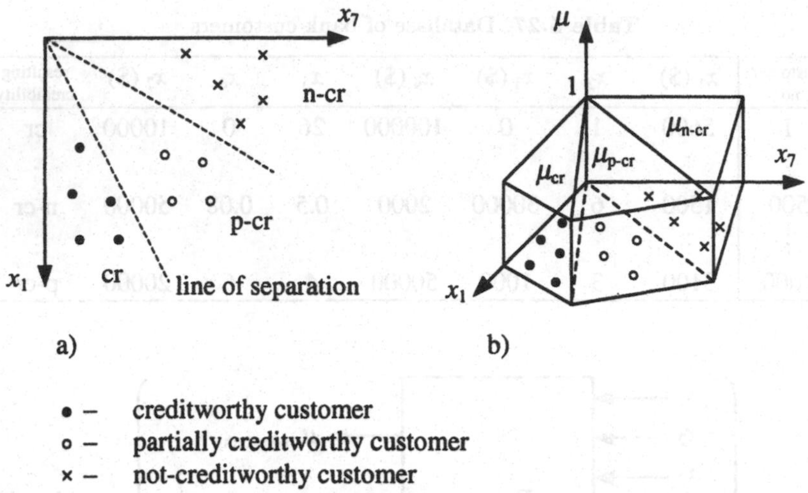

Fig. 5.178. The idealized crisp-separable feature distribution of bank customers in the input (feature) space $\{x_1, x_7\}$ (a) and the crisp membership functions in the customer classes: creditworthy, partially creditworthy and not-creditworthy (b)

- recognition of human beings by their voices;
- recognition of the type of tank by the sound of the running motor and taking an automatic decision to detonate an antitank mine.

The classification tasks are called pattern recognition because they consist in determining the similarity of the object (element) to the most typical element of the given class: its pattern.

An automatic classifier operates correctly if its membership functions in the particular classes are well tuned, i.e. correctly situated in the space of features $\{x_1, x_2, \ldots, x_n\}$. The problem will be presented using the example of a bank customer but – to simplify the task – the number of inputs (considered features of the customer) will be limited to two:

x_1 – the mean net salary of the customer,
x_7 – the amount of credit asked for.

Let us assume that the features of the bank customers up to the present are distributed in such a way as shown in Fig. 5.178a.

The example of the distribution of features represented in Fig. 5.178a is a crisp-separable problem. The particular classes of customers do not overlap and the lines of separation, which separate these classes uniquely, can be determined. Basing oneself on these lines one can easily create crisp functions of membership in the particular classes of customers, Fig. 5.178b. The crisp-separable sample distribution seldom occurs in practice. The particular classes usually overlap. For example, the distribution of the features of customers usually looks like that shown in Fig. 5.179a.

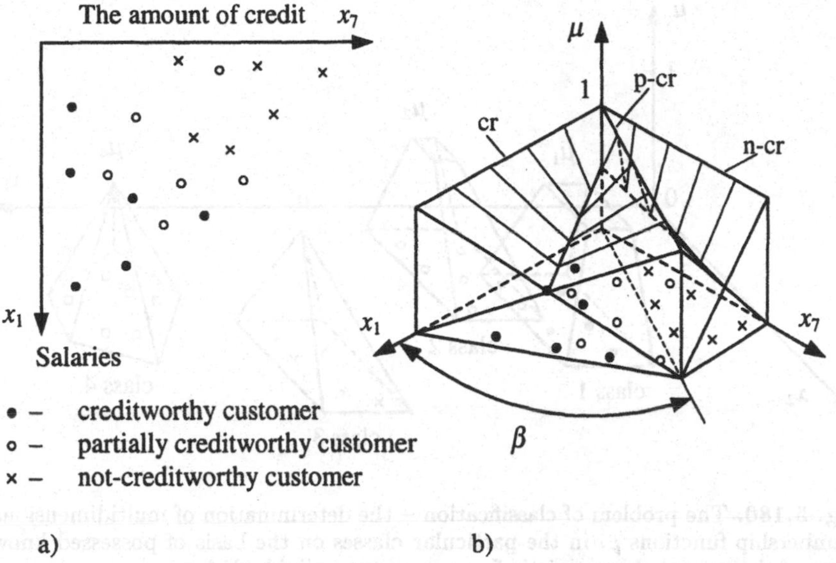

Fig. 5.179. The real, fuzzy-separable feature distribution of bank customers in the feature space $\{x_1, x_7\}$ (a) and the functions of membership in the customer classes: creditworthy (cr), partially creditworthy (p-cr), not-creditworthy (n-cr) (b)

The points corresponding to creditworthy customers are situated mainly in the zone of high earnings and low credit amounts, the points relating to not creditworthy customers lie in the zone of low earnings and high credit amounts, the points corresponding with partially creditworthy customers are situated in the intermediate zone. However, sometimes there are customers obtaining high earnings, who have not entirely repaid their credit, and also customers of low earnings who have repaid their credit completely and in due time, Fig. 5.179. For this reason the problem of estimation of creditworthiness is not crisp but fuzzy and each of the potential customers can be classified in two classes at least.

The tuning of the model in Fig. 5.179 on the basis of learning samples (features of the bank customers up to the present and the repayment by them of credit allowed by the bank) consists in selecting such a value of the angle β which enables us to evaluate customers as well as possible and assign them to the particular classes.

Generally, the problem of classification consists in determining the size, shape and position of membership functions in the particular classes in the feature space $X_1 \times X_2 \times, \ldots, \times X_n$. For the case of 2 features x_1, x_2 this problem can look like the one presented in Fig. 5.180.

The particular classes can overlap, as μ_1 and μ_2 in Fig. 5.180, or can be distinctly separate from each other, as the classes μ_3 and μ_4. In the latter case the classification is very easy.

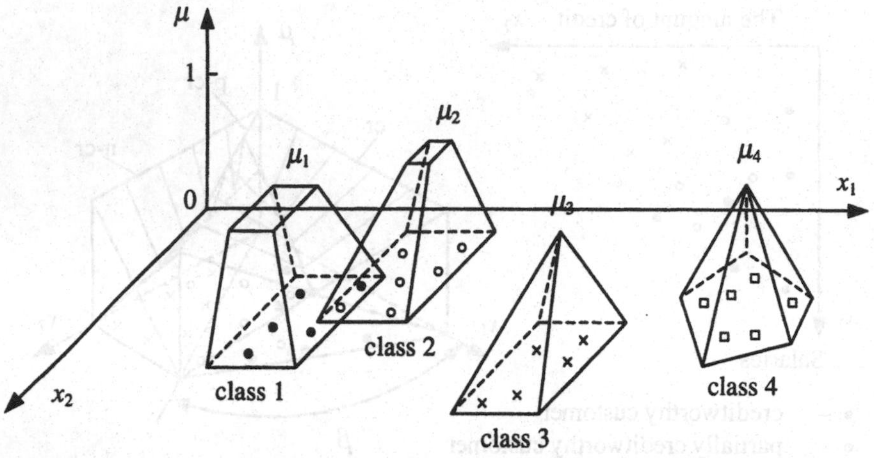

Fig. 5.180. The problem of classification – the determination of multidimensional membership functions μ_i in the particular classes on the basis of possessed knowledge of the typical characteristic features of the individual classes

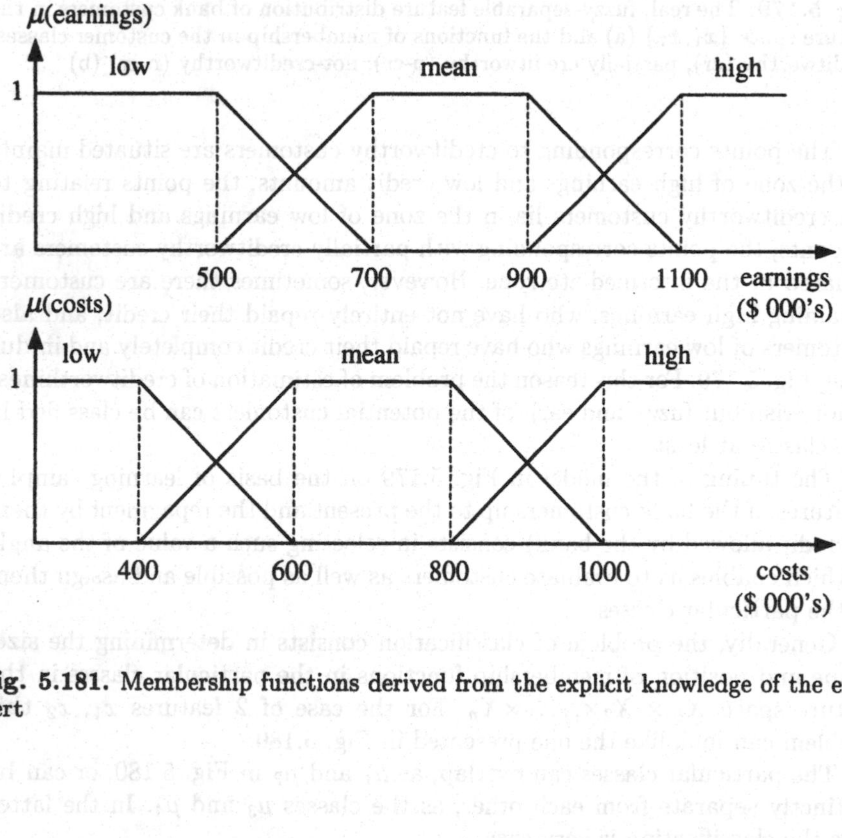

Fig. 5.181. Membership functions derived from the explicit knowledge of the expert

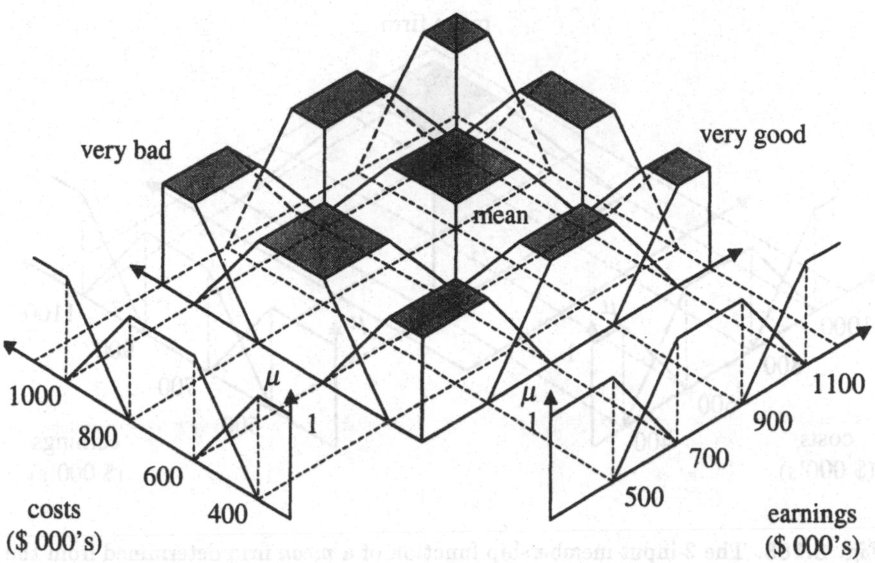

Fig. 5.182. 2-input membership functions based on the explicit expert-knowledge as *IF* (earnings high) *AND* (costs low) *THEN* (firm = very good)

The univariant membership functions are the type of functions most often used in fuzzy logic. Examples of such functions for variables: earnings and costs of a firm are presented in Fig. 5.181. By composing these univariant membership functions we can create bivariant membership functions which make possible the evaluation of a firm. The result of such composing, together with function names: very good (firm), mean, very bad, is presented in Fig. 5.182. Other functions denote intermediate evaluations as mentioned above.

The univariant membership functions, Fig. 5.181, are the only type of membership functions which can be given in words by the expert of the modeled system on the basis of his conscious knowledge of the system. The bivariant membership functions of the firm's evaluation, which can be created on their basis, are regular functions having rectangular supports. The sides of the supports are parallel to the coordinate axes, Fig. 5.183 and 5.184.

However, when taking decisions, human beings (experts) use not only their conscious but also subconscious knowledge, which they call "feeling", "intuition" or "the sixth sense". They can not express this knowledge with words in the form of rules. They are often not even aware of possessing such knowledge. The conscious knowledge also means emotions, likes and dislikes and various preferences, which affect decisions we make. We can suppose that real membership functions, which determine our decisions, can be not univariant but bivariant or even of a higher order. This means that they exist

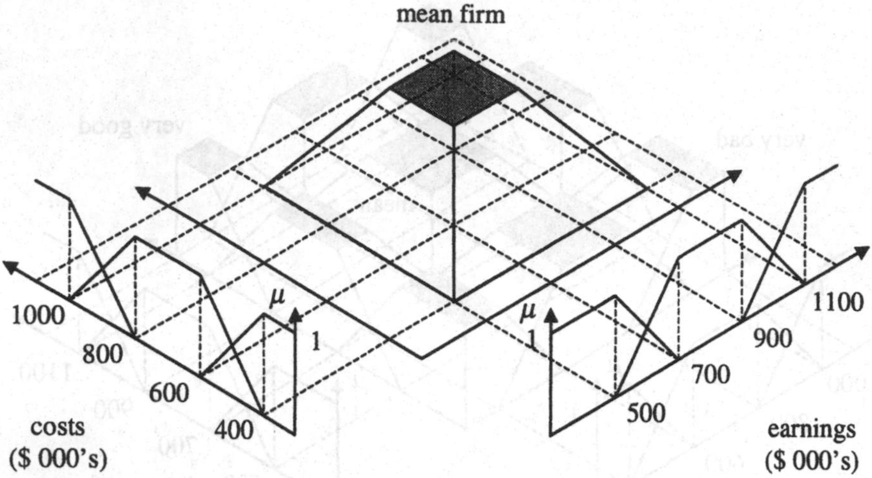

Fig. 5.183. The 2-input membership function of a *mean* firm determined from the expert knowledge in 3-dimensional space

or are defined in a space higher than the univariant space, are not parallel to the coordinate axes and their shape is not rectangular, Fig. 5.185 and 5.186.

If the expert has formed in his mind (conscious and subconscious knowledge) a bivariant rotated membership function A_{x1x2} as in Fig. 5.187, which he uses in taking decisions, then he can not express with words the information of the shape and rotation angle of this function. He can only give information on its univariant projections A_{x1} and A_{x2}, Fig. 5.187.

Creating a verbal rule (the composition of the functions A_{x1} and A_{x2}, we obtain a regular, rectangular bivariant membership function B_{x1x2}, which is different than the membership function A_{x1x2} used in reality by the expert, Fig. 5.188. A similar situation exists when the expert uses a concave membership function A_{x1x2} as presented in Fig. 5.189.

Giving verbal information of such a function in the form of the univariant functions A_{x1} and A_{x2}, he is not able to describe the concave shape of a bivariant function, either. The composition of two univariant membership functions A_{x1} and A_{x2} in a fuzzy model using the *AND* operation also gives in this case, a convex, regular, rectangular membership function, which is parallel to the coordinate axes x_1, x_2, Fig. 5.188.

As shown in the examples discussed above, fuzzy models composed on the basis of univariant membership functions are not able to form multidimensional membership functions of rule premises of irregular concave shape as well as functions rotated in relation to the coordinate axes. Such models are not able to map human knowledge (the expert's knowledge of the modeled system) exactly. So, if we want to increase the accuracy of fuzzy models,

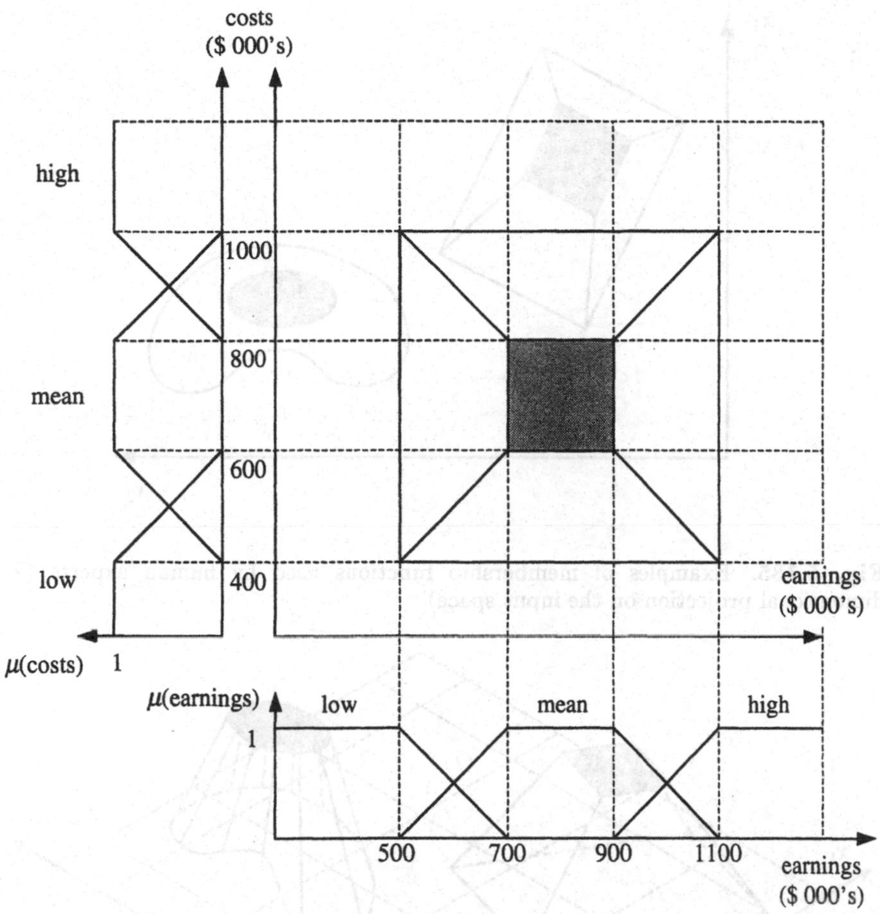

Fig. 5.184. The 2-dimensional projection of the membership function of a *mean* firm

we should use, in rule premises and fuzzification, multivariant and not only univariant membership functions.

Because the expert of a system is not able to describe with words the shape and angle position of such functions, the only way to determine them is to construct and tune them on the basis of premises corresponding in reality with the expert's decisions, i.e. measurement input/output samples of the expert. It enables us to investigate which membership functions are really used by the expert when he makes use of his complete (conscious and subconscious) knowledge. This method will be presented in Example 5.7.9.1.

The building and tuning of multidimensional irregular membership functions in the whole input (feature) space of the modeled system is possible but very difficult, particularly in situations when measurement samples of differ-

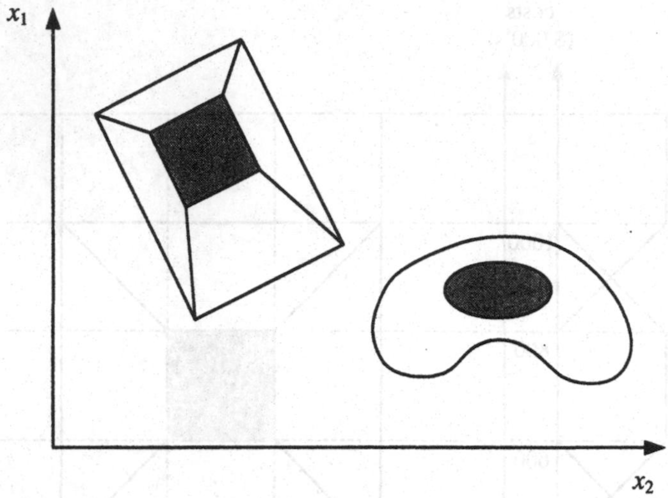

Fig. 5.185. Examples of membership functions used by human experts (2-dimensional projection on the input space)

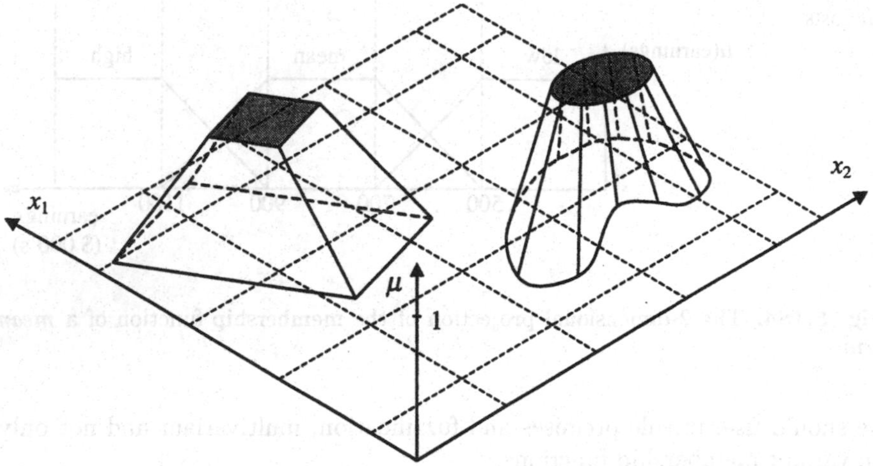

Fig. 5.186. Examples of membership functions used by human experts depicted in 3-dimensional space

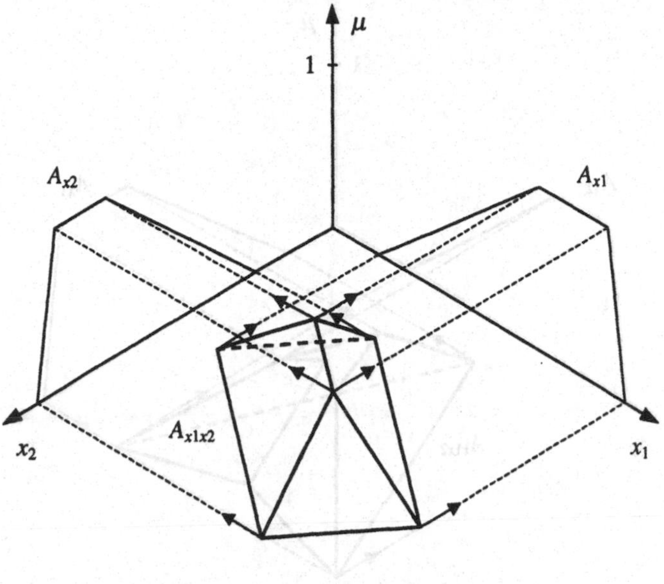

Fig. 5.187. The projection of a bivariant rotated fuzzy set A_{x1x2} from the space $X_1 \times X_2 \times M$ on the univariant spaces $X_i \times M$ resulting in fuzzy sets A_{x1} and A_{x2}

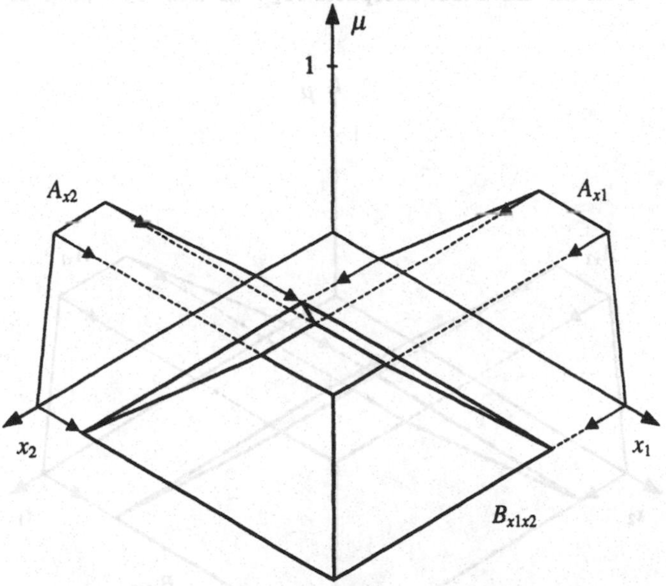

Fig. 5.188. Composition of a bivariant (convex) fuzzy set B_{x1x2} from two univariant fuzzy sets A_{x1} and A_{x2} achieved by projection of the rotated set A_{x1x2} in Fig. 5.187

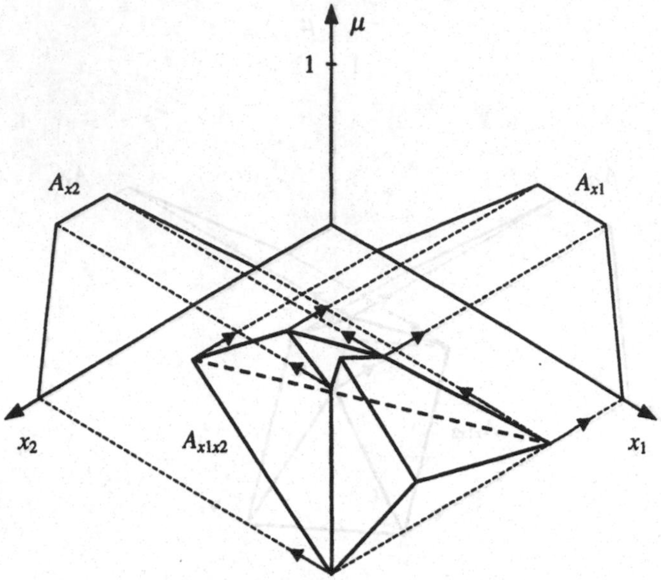

Fig. 5.189. The projection of a bivariant concave fuzzy set A_{x1x2} from the space $X_1 \times X_2 \times M$ on the univariant subspaces $X_1 \times M$ and $X_2 \times M$ (fuzzy sets A_{x1} and A_{x2})

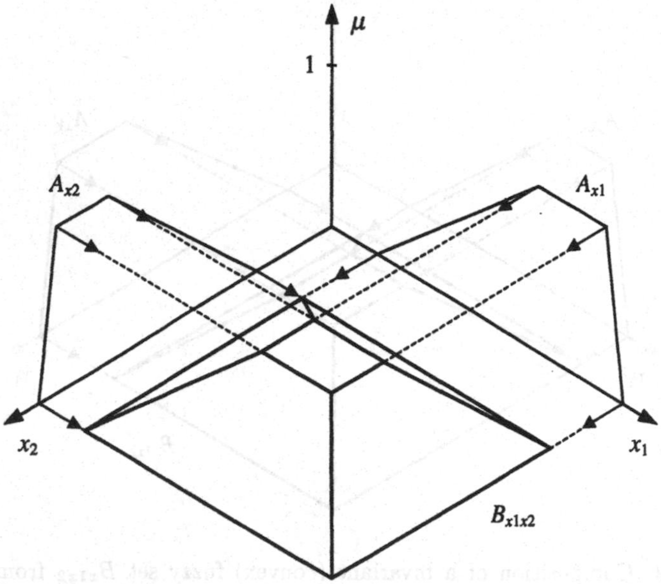

Fig. 5.190. Composition of a bivariant (convex) fuzzy set B_{x1x2} from two univariant fuzzy sets A_{x1} and A_{x2} achieved by projection of the concave set A_{x1x2} in Fig. 5.189

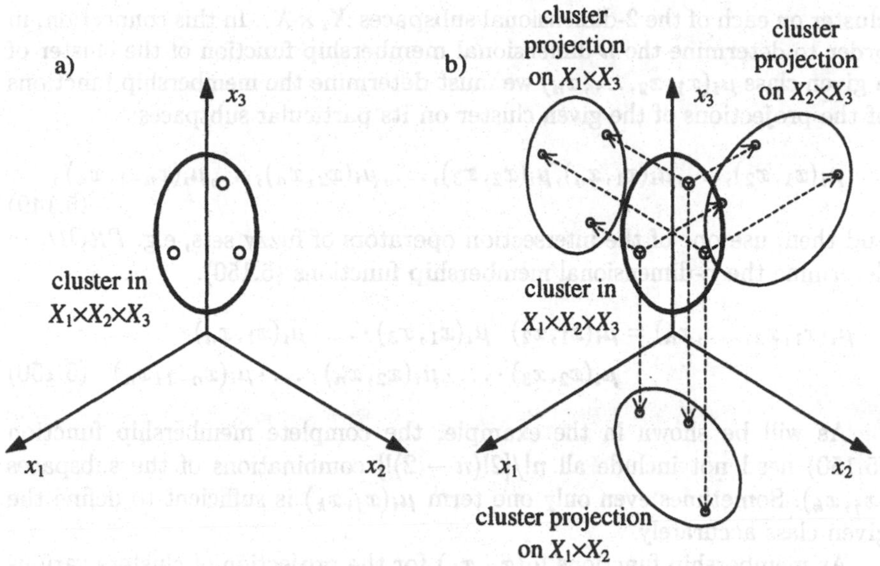

Fig. 5.191. The cluster of learning samples of the given class (a) and its projection on 2-dimensional subspaces (b)

ent classes are mixed. This task is an interesting field of future investigations in the domain of fuzzy logic. However, already, even now, one can progress in this domain by creating fuzzy models based on bivariant and not only on univariant membership functions. The advantage of bivariant membership functions is that one can see, imagine and graphically present them. The visualization of three- or more-variant membership functions is not possible.

In the following the method of fuzzy classification called the **method of two-dimensional projection of multi-dimensional clusters** will be presented. Although the author of the present book deduced this method himself in 1999, he thinks that it had been elaborated earlier by another researcher, because it is very simple and the idea suggests itself. Unfortunately, the author does not know the name of the original, former creator of this method and therefore he/she has not been acknowledged here.

Fuzzy classification by the two-dimensional projection of multi-dimensional clusters

The learning samples (e.g. the points determining features of the previous bank customers) create in the feature space $X_1 \times X_2 \times \ldots \times X_n$ the cluster of the given class. An example for the three-dimensional space is represented in Fig. 5.191a.

The basis of the method of 2-dimensional projection is the assumption that if any sample belongs to the cluster of the considered class in the n-dimensional space, then its projection belongs also to the projections of this cluster on each of the 2-dimensional subspaces $X_i \times X_j$. In this connection, in order to determine the n-dimensional membership function of the cluster of a given class $\mu_i(x_1, x_2, \ldots, x_n)$ we must determine the membership functions of the projections of the given cluster on its particular subspaces:

$$\mu_i(x_1, x_2), \ldots, \mu_i(x_1, x_n), \mu_i(x_2, x_3), \ldots, \mu_i(x_2, x_n), \ldots, \mu_i(x_{n-1}, x_n), \tag{5.149}$$

and then, use one of the intersection operators of fuzzy sets, e.g. *PROD*, to determine the n-dimensional membership functions (5.150).

$$\begin{aligned} \mu_i(x_1, x_2, \ldots, x_n) = {} & \mu_i(x_1, x_2) \cdot \mu_i(x_1, x_3) \cdot \ldots \cdot \mu_i(x_1, x_n) \cdot \\ & \mu_i(x_2, x_3) \cdot \ldots \cdot \mu_i(x_2, x_n) \cdot \ldots \cdot \mu_i(x_{n-1}, x_n) \end{aligned} \tag{5.150}$$

As will be shown in the example, the complete membership function (5.150) need not include all $n!/[2!(n-2)!]$ combinations of the subspaces (x_j, x_k). Sometimes even only one term $\mu_i(x_j, x_k)$ is sufficient to define the given class accurately.

As membership functions $\mu_i(x_j, x_k)$ for the projection of clusters various types of functions can be applied e.g. generalized, rotatable, non-symmetric Gaussian functions (5.151):

$$\begin{aligned} x_i^* &= (x_i - m_i) \cos \alpha_{ij} - (x_j - m_j) \sin \alpha_{ij} , \\ x_j^* &= -(x_i - m_i) \sin \alpha_{ij} + (x_j - m_j) \cos \alpha_{ij} , \\ \mu(x_i^*, x_j^*) &= \exp \left[- \left| \frac{x_i^*}{v_{ij} \cdot c_{i1} + (1 - v_{ij}) \cdot c_{i2}} \right|^{l_{ij1}} - \right. \\ & \qquad \left. - \left| \frac{x_j^*}{w_{ij} \cdot c_{j1} + (1 - w_{ij}) \cdot c_{j2}} \right|^{l_{ij2}} \right] , \end{aligned} \tag{5.151}$$

where:

- m_i – the center coordinate of the Gaussian function in relation to the x_i axis,
- m_j – the center coordinate of the Gaussian function in relation to the x_j axis,
- α_{ij} – the angle on the main axis of a cut (contour line) of the Gaussian function, see Fig. 5.193,
- l_{ij1}, l_{ij2} – exponents,
- v_{ij}, w_{ij} – the logical variables (0 or 1) activating different widths c_{i1}, c_{i2}, c_{j1}, c_{j2} of the non-symmetric Gaussian function, see Fig. 5.193,
- c_{i1}, c_{i2} – different widths of the non-symmetric Gaussian function.

In Fig. 5.192 the 3-dimensional Gaussian membership function is depicted. The 3-dimensional membership function can be shown on the 2-dimensional input-plane with the use of cuts (contour lines) made at different heights, Fig. 5.193.

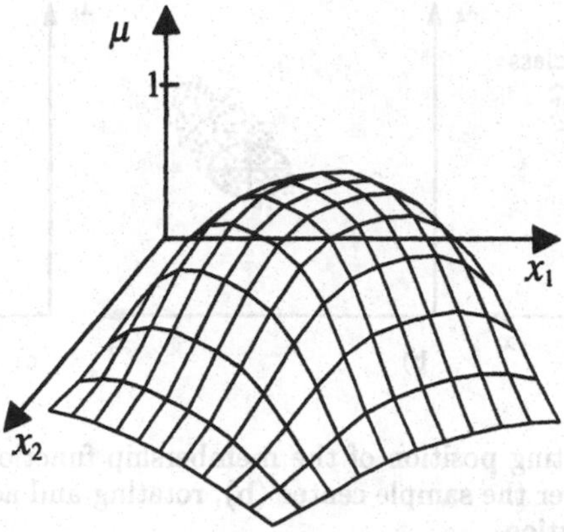

Fig. 5.192. 3-dimensional Gaussian membership function

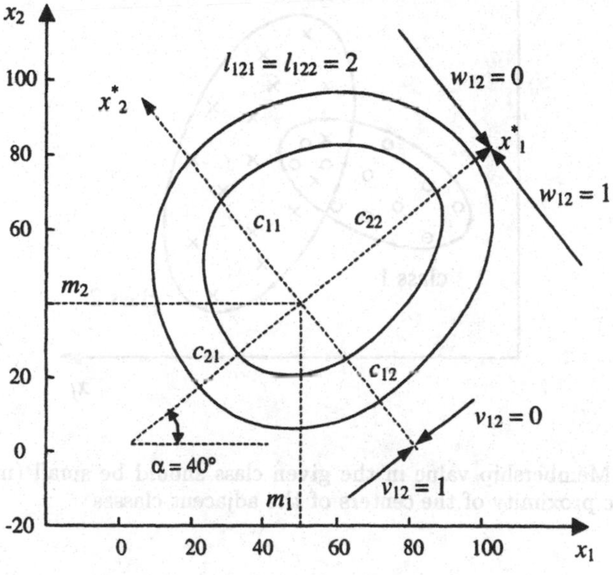

Fig. 5.193. Cuts (contour lines) of the generalized, non-symmetric Gaussian function projected on the subspace $X_i \times X_j$, for $i = 1$, $j = 2$

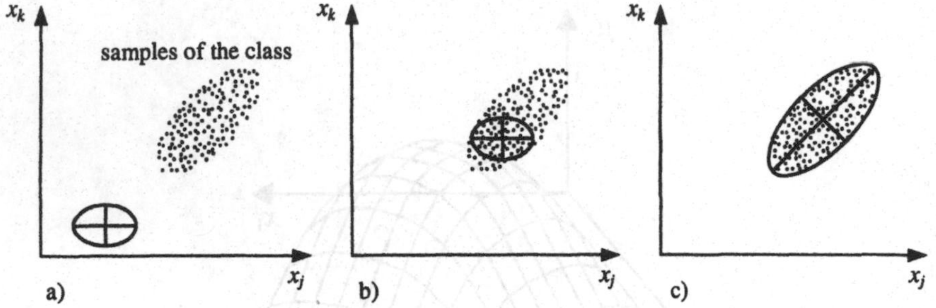

Fig. 5.194. The starting position of the membership function tuned (a), putting the function center over the sample center (b), rotating and adjusting the function to the sample distribution

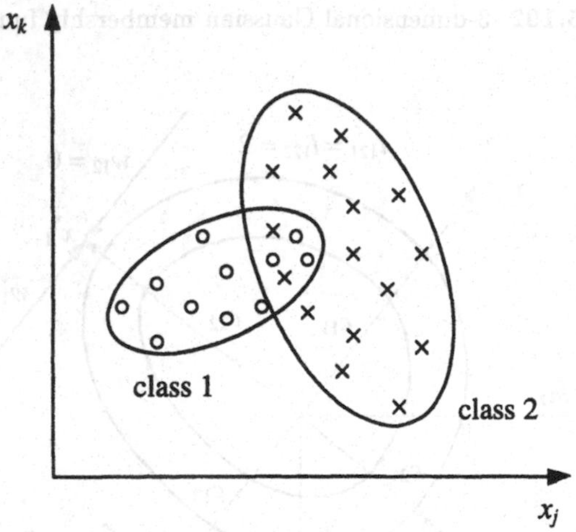

Fig. 5.195. Membership value in the given class should be small (much smaller than 1) in the proximity of the centers of the adjacent classes

The variation of the parameters m_i, m_j causes the Gaussian function to move along the axes, the variation of α_{ij} causes the rotation of the function, the parameters c_{i1}, c_{i2} change the function width asymetrically, and l_{ij1}, l_{ij2} change its shape. Owing to these parameters the Gaussian function can be visually or automatically moved in the center of the measurement samples of the given class and then rotated and shaped to encircle all or most samples of the class, Fig. 5.194.

When tuning the widths a_{jk}, b_{jk} of the membership function we must follow the principle that the function of membership in the given class should have a small value (e.g. 0.1) at the centers and cores of the membership functions of the adjacent classes, Fig. 5.195.

Sometimes it is not possible to keep this principle. This is because both the classes in the projection on the space $X_j \times X_k$ are very similar to each other or it is not possible to separate them. But the classes can differ in the projection on another pair of spaces, e.g. $X_{j+1} \times X_{k+2}$, and in such a case they can be separated.

If the separation of the classes in the n-dimensional space is not possible, then it is recommended to increase the order (dimension) of the space, i.e. the number of considered features, or change the coordinate system.

The classification with the method of 2-dimensional projection will be illustrated by the example (Piegat 2000a) below.

Example 5.7.9.1 Evaluation of 49 steel industry firms with the 2-dimensional projection method

Economic decisions have to take into account many specific data, which however, must finally be generalized to one complex measure giving the basis for the final decision. As complex, multicriterial evaluations are mostly not of a crisp but of a fuzzy character, Fuzzy Sets Theory suits here perfectly.

In the example, the application of the method of 2-dimensional projection for the complex evaluation of firms on the basis of 3 indices informing us about their financial standing:

$$x_1 = \frac{\text{gross profit} + \text{amortization}}{\text{total obligations}},$$
$$x_2 = \frac{\text{balance sum}}{\text{total obligations}},$$
$$x_3 = \frac{\text{turnover}}{\text{balance sum}},$$

will be presented. The values of these indices, normalized to the interval $[0, 1]$, for some of 49 companies of the Polish iron and steel industry, which have been investigated in the example, for the period from 1 June, 1994 till 31 December, 1997 are given in Table 5.28. In the last column the complex evaluation of the given company made by the expert with the use of 4 linguistic terms: *weak*, *mean*, *good* and *very good* is given. Fig. 5.196 shows the positions of the membership function *weak* in particular subspaces $X_1 \times X_2$, $X_1 \times X_3$, $X_2 \times X_3$ tuned first.

In the second step, the membership function *mean* is tuned, Fig. 5.197 It should have, if possible, small values at the center of the class *weak*.

As can be seen in Fig. 5.197 the similarity of the classes *weak* and *mean* is considerable. Certain samples belong to both the classes to a high degree, which can make distinguishing them difficult. In the next modeling steps the membership functions *good* and *very good* are tuned, Fig. 5.198.

In Figs. 5.198a and 5.198b a very interesting feature can be observed. The function *very good* is distinctly separated from the other classes. Therefore this class can be sufficiently well described by only one 2-input membership function, e.g. in the subspace $X_1 \times X_3$, (5.152).

Table 5.28. The normalized values of 3 significant economic indices x_i and the expert's evaluation y of example firms

No.	x_1	x_2	x_3	Class
1	0.0540	0.0845	0.5643	weak
⋮	⋮	⋮	⋮	⋮
19	0.2029	0.4047	0.2714	mean
⋮	⋮	⋮	⋮	⋮
36	0.1457	0.0287	0.9714	good
⋮	⋮	⋮	⋮	⋮
49	1.0000	0.7997	0.3119	very good

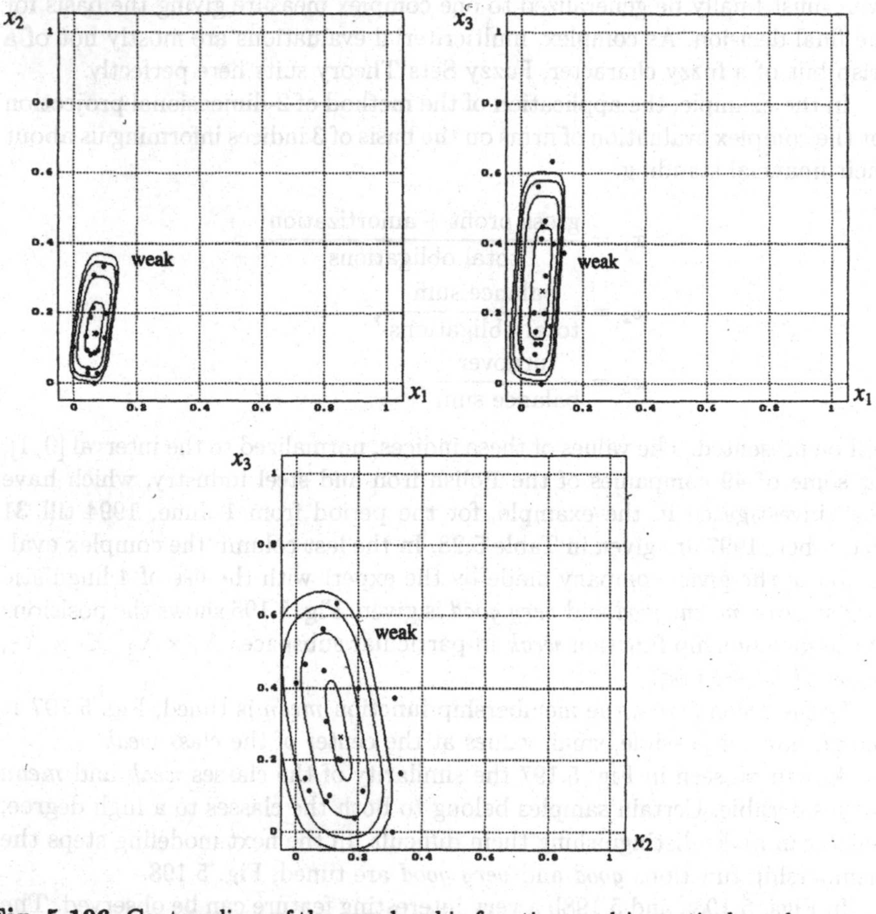

Fig. 5.196. Contour lines of the membership function *weak* in particular subspaces. The cross denotes the class center

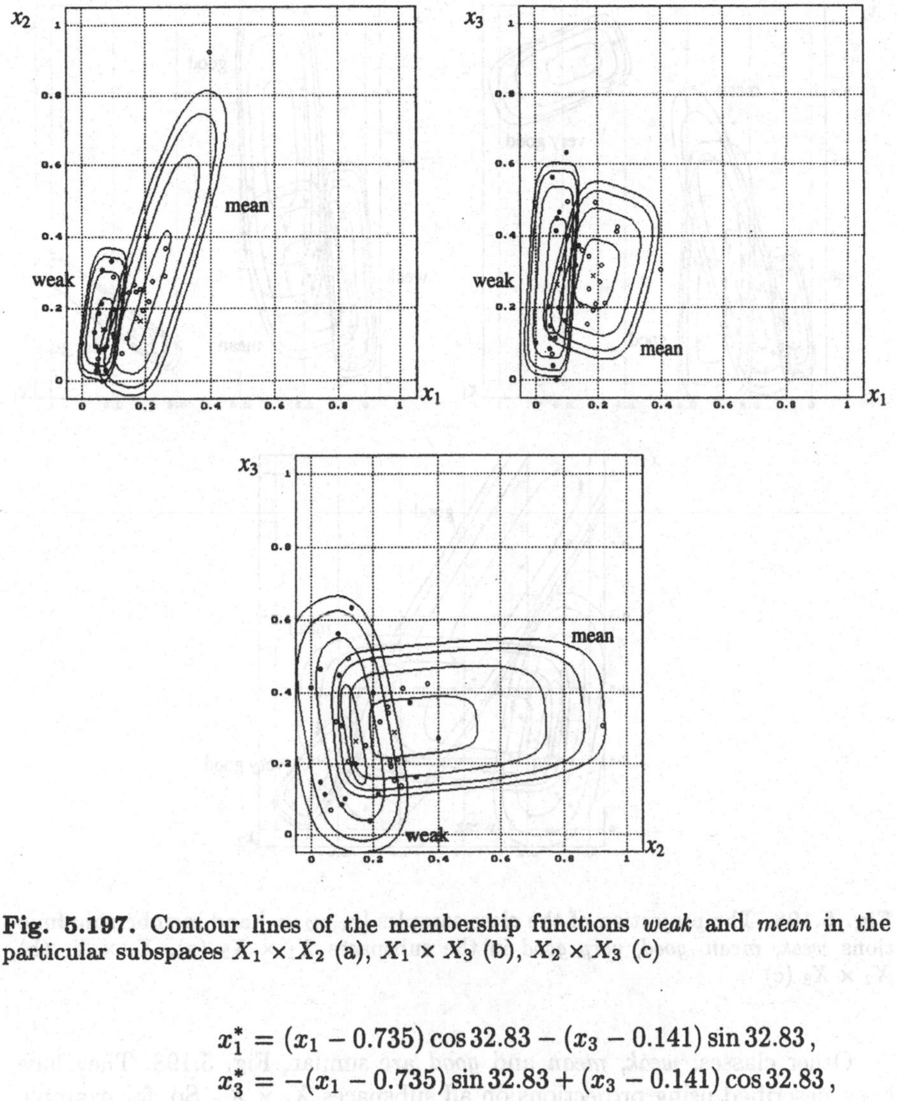

Fig. 5.197. Contour lines of the membership functions *weak* and *mean* in the particular subspaces $X_1 \times X_2$ (a), $X_1 \times X_3$ (b), $X_2 \times X_3$ (c)

$$x_1^* = (x_1 - 0.735)\cos 32.83 - (x_3 - 0.141)\sin 32.83\,,$$
$$x_3^* = -(x_1 - 0.735)\sin 32.83 + (x_3 - 0.141)\cos 32.83\,,$$
$$\mu_{\text{verygood}}(x_1^*, x_3^*) = \exp\left[-\left|\frac{x_1^*}{v_{13}\cdot 0.112 + (1 - v_{13})\cdot 0.249}\right|^4 - \right.$$
$$\left. - \left|\frac{x_3^*}{w_{13}\cdot 0.032 + (1 - w_{13})\cdot 0.031}\right|^4\right], \tag{5.152}$$

This example shows that we do not have to take the projections on all subspaces $X_i \times X_j$ to construct the class membership function in the full input space $X_1 \times X_2 \times \ldots \times X_n$ but only the projections on such subspaces which allow sufficiently distinct characterization of the class.

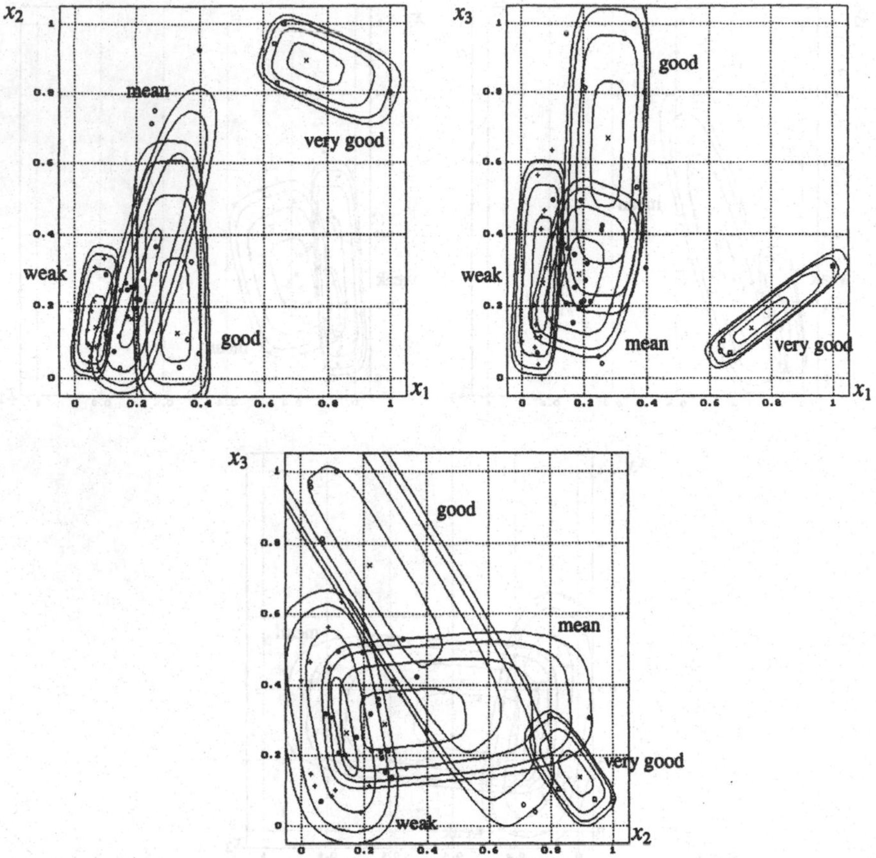

Fig. 5.198. The projection of the class samples $[x_1, x_2, x_3]$ and membership functions *weak*, *mean*, *good*, *very good* in the subspaces $X_1 \times X_2$ (a), $X_1 \times X_3$ (b), $X_2 \times X_3$ (c)

Other classes: *weak*, *mean* and *good* are similar, Fig. 5.198. They have been described using projections on all subspaces $X_i \times X_j$. So, for example, the membership function of the class *good* is given by (5.153).

$$\mu_{\text{good}}(x_1, x_2, x_3) = \mu_{\text{good}}(x_1, x_2) \cdot \mu_{\text{good}}(x_1, x_3) \cdot \mu_{\text{good}}(x_2, x_3) \tag{5.153}$$

$$x_1^* = (x_1 - 0.329)\cos 1.68 - (x_2 - 0.128)\sin 1.68$$

$$x_2^* = -(x_1 - 0.329)\sin 1.68 + (x_2 - 0.128)\cos 1.68$$

$$\mu_{\text{good}}(x_1^*, x_2^*) = \exp\left[-\left|\frac{x_1^*}{v_{12}\cdot 0.094 + (1 - v_{12})\cdot 0.067}\right|^4 - \right.$$

$$\left. -\left|\frac{x_2^*}{w_{12}\cdot 0.136 + (1 - w_{12})\cdot 0.363}\right|^4\right]$$

$$x_1^* = (x_1 - 0.279)\cos(-2.99) - (x_3 - 0.669)\sin(-2.99)$$
$$x_3^* = -(x_1 - 0.279)\sin(-2.99) + (x_3 - 0.669)\cos(-2.99)$$
$$\mu_{\text{good}}(x_1^*, x_3^*) = \exp\left[-\left|\frac{x_1^*}{v_{13}\cdot 0.091 + (1 - v_{13})\cdot 0.084}\right|^4 - \right.$$
$$\left. - \left|\frac{x_3^*}{w_{13}\cdot 0.334 + (1 - w_{13})\cdot 0.283}\right|^4\right]$$

$$x_2^* = (x_2 - 0.218)\cos 3.47 - (x_3 - 0.739)\sin 3.47$$
$$x_3^* = -(x_2 - 0.218)\sin 3.47 + (x_3 - 0.739)\cos 3.47$$
$$\mu_{\text{good}}(x_2^*, x_3^*) = \exp\left[-\left|\frac{x_2^*}{v_{23}\cdot 0.088 + (1 - v_{23})\cdot 0.155}\right|^4 - \right.$$
$$\left. - \left|\frac{x_3^*}{w_{23}\cdot 0.592 + (1 - w_{23})\cdot 0.542}\right|^4\right]$$

Fuzzy classification of 49 firms with the projection method classified 43 firms identically to the expert and 6 firms differently. It is a good result if we take into account that 3 classes: *weak*, *mean* and *good* are very similar and relatively difficult to distinguish as well as the fact that the expert could have made errors in his evaluation.

◊ ◊ ◊

A very important and interesting problem is the **determination of the zone affected by membership functions in particular classes in the regions not covered by measurement samples** (regions of lacking information). Let us assume that we investigate the problem of the evaluation of firms only on the basis of their earnings and costs of activity and use only 2 linguistic terms: *good* and *weak* (firm) for the evaluation. The measurement samples of the expert evaluation (Fig. 5.199a) and the result of the tuning of the membership function (Fig. 5.199b) are given.

Now, the membership functions *weak* and *good* firm tuned with the expert measurement samples can be used to evaluate new firms unaided by the expert. Assume that we have to evaluate 4 new firms whose measurement samples are distributed as in Fig. 5.200a.

Firm no. 2 will be evaluated as *good* (although its membership in the class *good* is very small, smaller than 0.01). Firm no. 4 will be evaluated by membership functions as *weak.* Both the evaluations agree with common sense because firm no. 2 is situated in the zone of high earnings and low costs and firm no. 4 has high costs and low earnings.

In the case of firms no.1 and no. 3 we can not trust the results of the classification carried out by means of the determined membership functions because these functions (Fig. 5.199b) are only tuned exactly in the region of

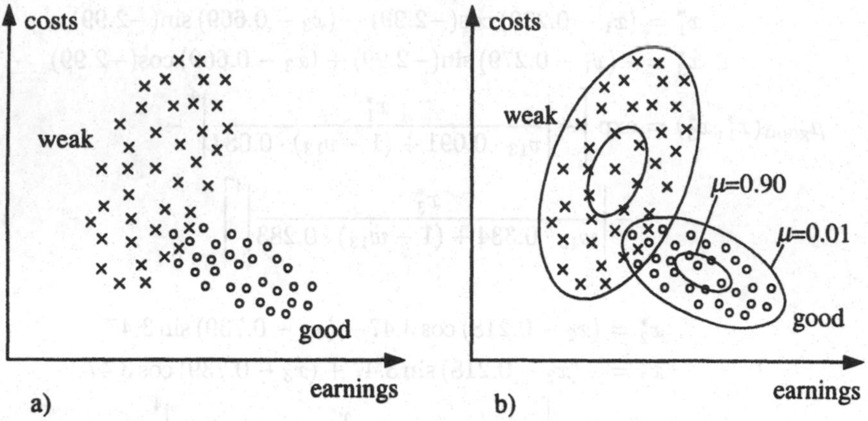

Fig. 5.199. The measurement samples of evaluations of firms made by an expert (a) and the membership functions *weak* and *good* firm tuned with them, given by contour lines (b)

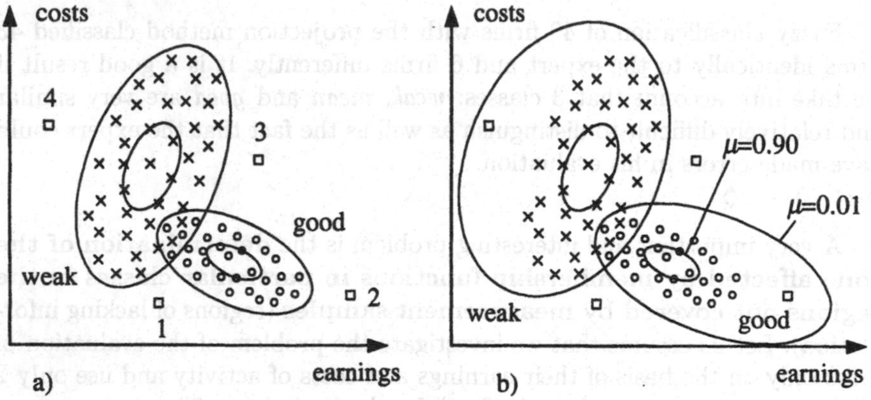

Fig. 5.200. The position of measurement samples of 4 new firms not participating in the tuning of the membership functions *weak* and *good* (a) and the membership functions *weak* and *good* with widened affected zones (b)

samples. Outside this region the evaluation can be questionable, particularly if it refers to the "controversial" region situated at a more or less equal distance from the cores (centers) of both the membership functions. In such doubtful cases (1 and 3) an expert should be charged with the evaluation of the new firms, and the result of this evaluation should be used to improve the possessed membership functions, i.e. to appropriately widen their influence zones in the region unknown at the present. But, if on the basis of our own knowledge we can draw a conclusion as to which side to widen the zone

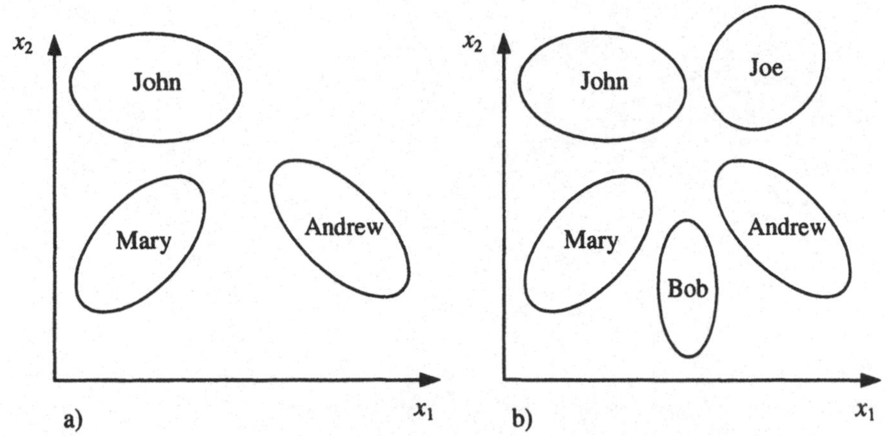

Fig. 5.201. Recognition of persons on the basis of their features as an example of a problem with an open, unlimited number of classes

of the given membership function, then we can make it without additional measurement samples.

In the investigated example the membership function *good* firm can certainly be widened in the region of high earnings and low costs and the membership function *weak* firm in the region of high costs and low earnings, as is shown in Fig. 5.200b. The widening of the influence zone is carried out by changing parameters of membership functions, mainly their widths c_{i1}, c_{i2} (5.151). The above example shows to what extent non-symmetric membership functions are necessary in classification problems.

The decision, which side of the membership function should be widened, is not always so obvious as in the case of the evaluation of firms. In complicated problems sound logic can disappoint. However, we must try to widen membership functions in unknown regions. It can considerably improve the effectiveness of fuzzy classification methods. The widening of membership functions can be made more safely in the case of **problems with a closed** and limited **number of classes**, for example when we use only two classes *weak* and *good* firm and when we are sure that new classes e.g. *mean* firm will not be introduced in the classification.

The widening of the influence zone should be carried out very carefully in **problems with an open number of classes**. An example here can be the recognition of human beings on the basis of pictures of their faces or finger-marks. Each class then denotes one human being. On a given day we can have membership functions recognizing 3 persons (Fig. 5.201a) but on the next day it can be necessary to introduce membership functions of 2 new persons (Fig. 5.201b). For the introduction of new membership functions free space of features is usually required.

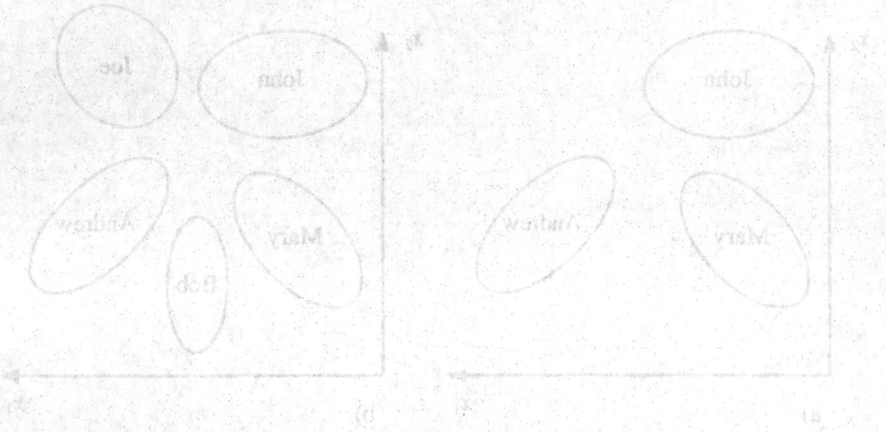

6. Methods of Fuzzy Modeling

Chapter 6 presents three methods of fuzzy modeling, i.e. building up the fuzzy models of real systems:

a) fuzzy modeling based on the system expert's knowledge,
b) creation of self-tuning fuzzy models based on input/output measurement data of the system,
c) creation of self-organizing and self-tuning fuzzy models based on input/output measurement data of the system.

The term "**self-tuning fuzzy model**" is to be understood, according to Driankov's concept (Driankov 1993,1996), as a model described with a base of fixed rules and a fixed number of fuzzy sets where only the membership function parameters are subject to tuning, Fig. 6.1, and presumably also the scaling coefficients of the model inputs and output.

During the tuning process the parameters of membership functions are modified (the modification of modal values is usually the most significant). The changes cause a displacement of the nodes P_i of the model surface defined with linguistic rules. The purpose of tuning is to position the nodes in such a way that the shape of the model surface approximates the shape of the system surface as closely as possible, in terms of minimizing the modeling criterion, where the latter can be a mean absolute error or a mean square error or a maximum error, etc. During the tuning process neither the number of fuzzy sets of inputs or output nor the number of rules (number of nodes P_i) of the model is altered. Moreover, the form of the rules or the number or type of the model inputs is not altered. The model structure remains fixed.

The term "tuning" may additionally denote the changes in the type of logical operators (*AND*, *OR*), type of membership function (segmentally linear, Gaussian type, etc.) or the changes in methods of inference or defuzzification. The alterations in the elements considered above cause the changes in type and size of surface curvature of the model sectors (interpolation type changes) among the nodes P_i defined with the rules, Fig. 6.1.

The term "**self-organizing fuzzy model**" is to be understood (Driankov 1993,1996) as a model employing its own automatic procedures to determine an optimum number and form of rules and fuzzy sets describing each of the model variables (inputs, output). The term "self-organization" also denotes the process of determining significant model inputs and a model structure

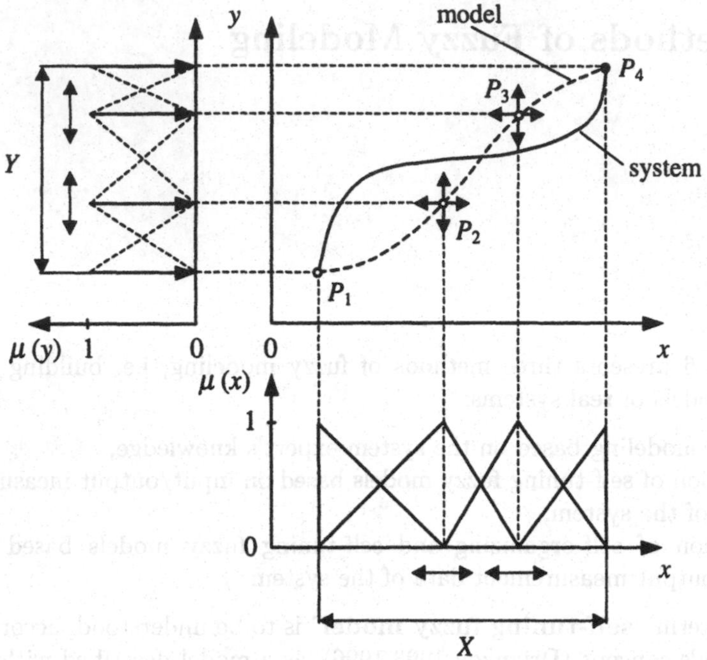

Fig. 6.1. Illustration of model tuning process

(e.g. the division of a global model into local models, defining the relationships in a hierarchic model, defining the component models in a multi-model structure). The effect of changes in the number of fuzzy sets and rules is presented in Fig. 6.2.

An increase in the number of fuzzy sets and rules enables us to achieve better accuracy of a model (provided it has been tuned well). However, the greater number of fuzzy sets and rules cause a rapid expansion of difficulties in the model's tuning, particularly in the case of a large number of inputs. If the model happens to be very complicated, its tuning may be practically unfeasible. Moreover, such a complicated model may appear to have insufficient capability to perform a generalization of the measurement results of the system inputs and output. Contrarily, it may result in unnecessary modeling of measurement noise and disturbances, Fig. 6.3, particularly in the case of a small number of measurements or their extensive scatter. Therefore, the model structure should not be unreasonably expanded and complicated and the number of rules and fuzzy sets should be reasonably limited. The optimization of the structure is a difficult but feasible task accomplished by self-organizing models.

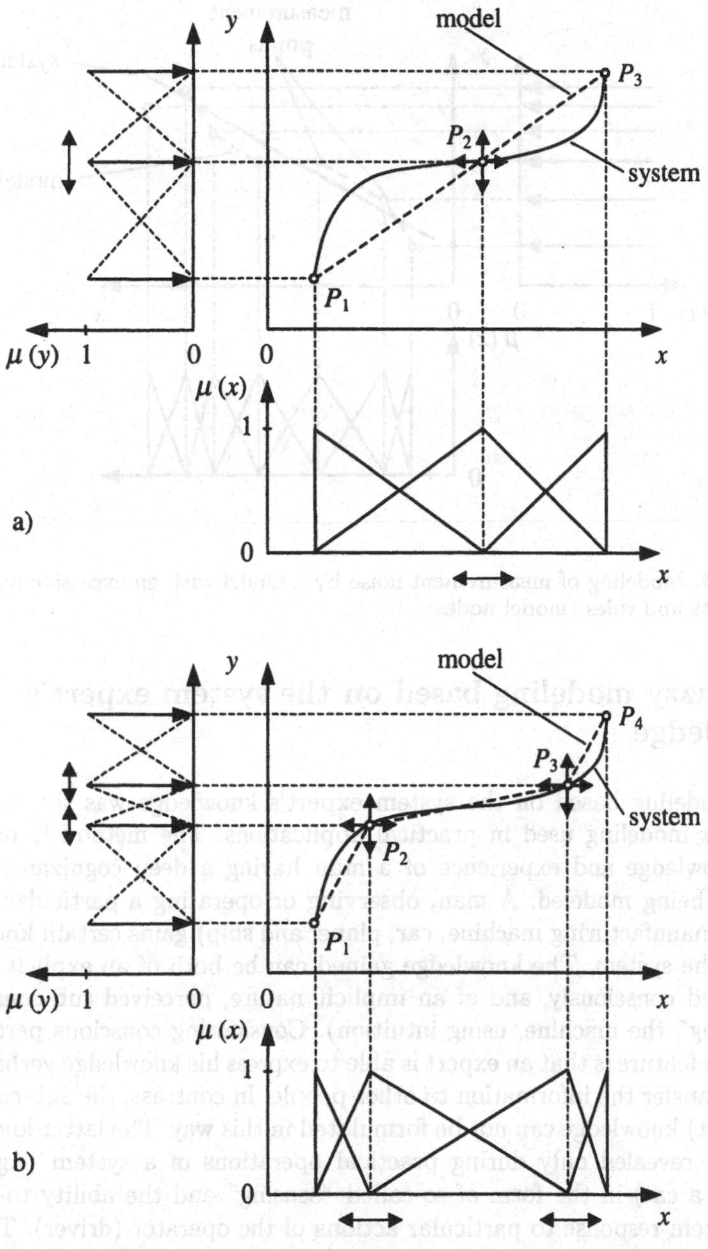

Fig. 6.2. **The influence of the number of fuzzy sets and rules (number of nodes P_i) on the potential accuracy of a model**

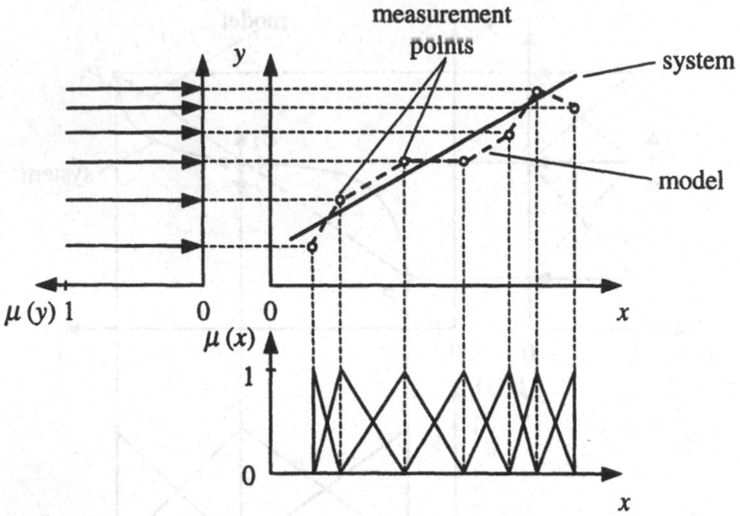

Fig. 6.3. Modeling of measurement noise by a model with an excessive number of fuzzy sets and rules (model nodes)

6.1 Fuzzy modeling based on the system expert's knowledge

The modeling based on the system expert's knowledge was the first type of fuzzy modeling used in practical applications. The method is based on the knowledge and experience of a man having a deep cognizance of the system being modeled. A man, observing or operating a particular system (e.g. a manufacturing machine, car, plane, and ship) gains certain knowledge about the system. The knowledge gained can be both of an explicit nature, perceived consciously, and of an implicit nature, perceived sub-consciously ("sensing" the machine, using intuition). Considering conscious perception, its main feature is that an expert is able to express his knowledge verbally and so to transfer the information to other people. In contrast, the sub-conscious (implicit) knowledge can not be formulated in this way. The latter knowledge may be revealed only during practical operations of a system (e.g. while driving a car) in the form of so-called "sensing" and the ability to predict the system response to particular actions of the operator (driver). The sum of conscious and sub-conscious knowledge about a real system accumulated in the expert's mind is recognized as a **mental model** (Babuška 1995b). An interview with an expert allows us to obtain only the explicit part of his knowledge about a system, expressed in the form of verbal rules describing the input/output relationships, e.g. (6.1):

IF (gas pedal is pushed heavily) *AND* (high gear) *THEN* (high speed), or in general:

$$IF\ (x_1 \text{ is } A_i)\ AND\ (x_2 \text{ is } B_j)\ THEN\ (y \text{ is } C_k)\,, \tag{6.1}$$

where: x_1, x_2 – system inputs,
y – system output,
A_i, B_j, C_k – fuzzy sets used by an expert to provide the linguistic evaluation of the inputs and output of a system.

An expert can also provide us with information on the linguistic values applied, e.g.:

- "By saying that the gas pedal is pushed heavily I mean that the pedal position is over 80% of pedal operation range",
- "By saying that the gear is high I mean it is in 4th or 5th gear".

Based on the above examples it can be said that the linguistic rules defining the system input/output relationships can be expressed more precisely by the expert, whereas the information concerning the linguistic values used is usually expressed less precisely, since the latter information is strongly related to the expert's "sensing" the system and the sub-conscious part of his knowledge.

A set of **verbally** expressed rules describing the system input/output relationships and the verbal information on linguistic values used by an expert is called a **verbal model** of a system. A verbal model is usually poorer than a mental model since the verbal model does not contain the implicit, subconscious knowledge about the system, the part the expert is not able to transfer to others. Moreover, the system expert is not able to transfer his knowledge about the inference processes taking place is his mind nor the type (shape) of membership functions used to describe the linguistic values nor the type of logical operators employed during the mental processing of information, etc.

All the above pieces of information, necessary to create a **fuzzy linguistic model** of a given system, have to be proposed by supposition or by intuition by the person building the model (modeler) who may be called an **expert in fuzzy modeling**. The information flow occurring during the process of creating a fuzzy linguistic model has been presented in Fig. 6.4.

An example of a fuzzy model is presented in the following.

Example 6.1.1

Let us assume that a system operator is observing a system with two inputs x_1, x_2 and a single output y; simultaneously the operator is recording the system behavior on the basis of its inputs and output measurements. The list of system state records is presented in Table 6.1.

It is the operator's task to determine such values of input signals x_1, x_2 for which the output y assumes characteristic values, i.e. minimum, maximum, mean values, etc. That type of knowledge is gained by an operator after a

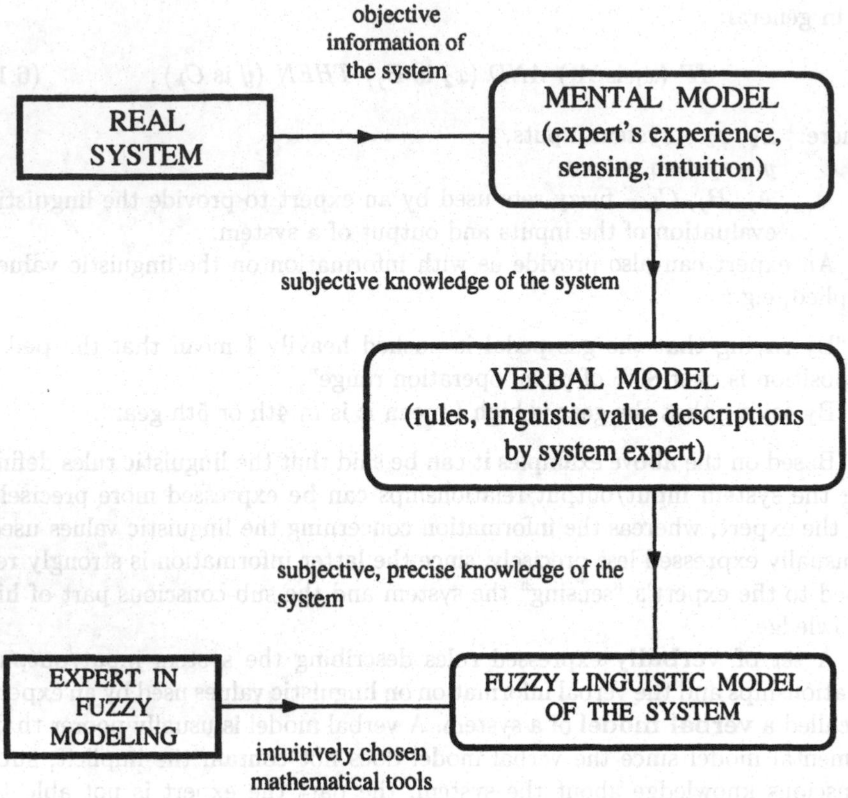

Fig. 6.4. Process of the creation of a fuzzy linguistic model of a real system

Table 6.1. Measurement data of real system states

x_1	55555	66666	77777	88888	99999
x_2	01234	01234	01234	01234	01234
y	96569	63236	52125	03236	96569

sufficiently long period of observing and operating the system – sufficiently long so that all the system specific states have occurred.

On the basis of data in Table 6.1 it can be seen that there are 4 states of system inputs (printed in bold) for which the output assumes a maximum value (9) and a single state of system inputs for which the system output assumes a minimum value (1). The information of the system presented in Table 6.1 constitutes the objective knowledge derived on the basis of measurements. In the case where precise measuring devices are not available and

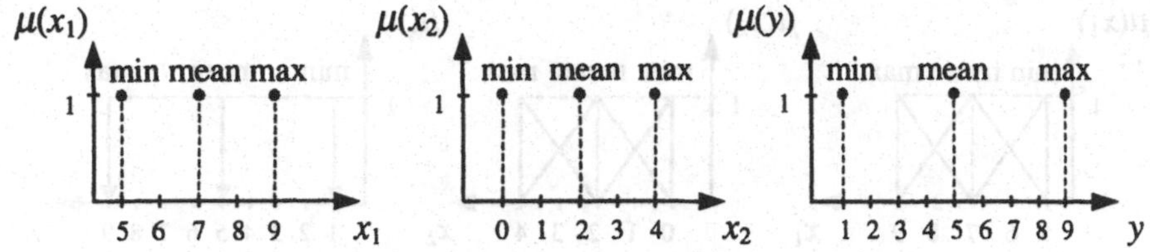

Fig. 6.5. Modal values of linguistic evaluations presented by a system expert

the inputs x_1, x_2 are adjusted manually, (e.g. using a lever) an operator is able to provide only such fuzzy information as, for example:

- "minimum position of lever no. 1",
- "maximum position of lever no. 2",
- "nearly central position of lever no. 1",
- "a bit lower than maximum position of lever no. 2", etc.

Considering the system described with data in Table 6.1, an expert can choose the following linguistic values:

$$x_1: \text{minimum, mean, maximum,}$$
$$x_2: \text{minimum, mean, maximum,}$$
$$y: \text{minimum, mean, maximum.}$$

Modal values of particular linguistic evaluations expressed by the system expert are presented in Fig. 6.5.

On the basis of objective knowledge of the system being modeled, presented in Table 6.1, an expert can define the rules (6.2) describing the system's operation.

$$
\begin{aligned}
&R1)\ IF\ (x_1 = \text{min})\ AND\ (x_2 = \text{min})\ THEN\ (y = \text{max})\\
&R2)\ IF\ (x_1 = \text{min})\ AND\ (x_2 = \text{mean})\ THEN\ (y = \text{mean})\\
&R3)\ IF\ (x_1 = \text{min})\ AND\ (x_2 = \text{max})\ THEN\ (y = \text{max})\\
&R4)\ IF\ (x_1 = \text{mean})\ AND\ (x_2 = \text{min})\ THEN\ (y = \text{mean})\\
&R5)\ IF\ (x_1 = \text{mean})\ AND\ (x_2 = \text{mean})\ THEN\ (y = \text{min})\\
&R6)\ IF\ (x_1 = \text{mean})\ AND\ (x_2 = \text{max})\ THEN\ (y = \text{mean})\\
&R7)\ IF\ (x_1 = \text{max})\ AND\ (x_2 = \text{min})\ THEN\ (y = \text{max})\\
&R8)\ IF\ (x_1 = \text{max})\ AND\ (x_2 = \text{mean})\ THEN\ (y = \text{mean})\\
&R9)\ IF\ (x_1 = \text{max})\ AND\ (x_2 = \text{max})\ THEN\ (y = \text{max}) \qquad (6.2)
\end{aligned}
$$

The information concerning the linguistic values presented in Fig. 6.5 and the set of rules (6.2) constitute a **verbal model** of the system. In the above case the mental model of the system is constituted by the expert's knowledge

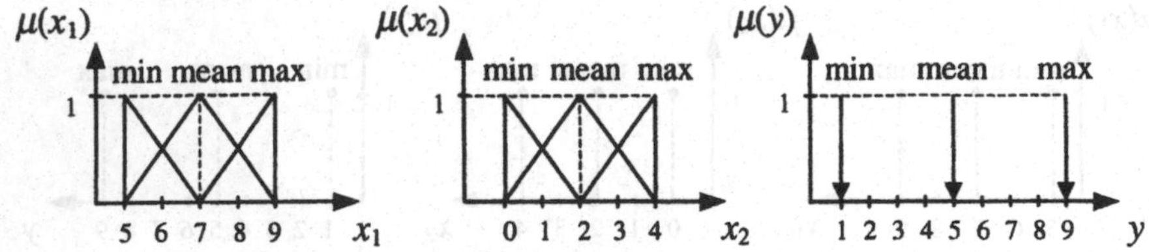

Fig. 6.6. The first variant of membership functions chosen for model inputs and output

about the states of the output, which the system expert is able to memorize and recall, as well as his "sensing" the system, i.e. his ability to generate certain system output states with methods he is not fully aware of.

An ordinary person is not able to remember more than 5 – 9 system states, i.e. much fewer than the quantity presented in Table 6.1. Moreover, a person cannot present the type of fuzzification (membership function shape and parameters), the method of premise aggregation in the rules (type of operator *AND* applied) or the type of defuzzification process in the person's mind. Nonetheless, the information mentioned above is necessary to build up a fuzzy linguistic model of the system. The task of fuzzy linguistic model creation has to be accomplished by an expert in fuzzy modeling, Fig. 6.4. The expert, working on the basis of an "intuition principle" (experience, intuition, knowledge), has to complete the verbal model with the missing elements of fuzzy logic apparatus. Depending on the aptness of the choice of element the model obtained is of a higher or lower accuracy.

Assuming that an expert in fuzzy modeling has chosen triangular membership functions to perform fuzzification for the inputs x_1, x_2 and singleton functions for the output y, (Fig. 6.6) and the operator *PROD* to perform *AND*-operations, then the fuzzy model will calculate the output y values presented in Table 6.2.

Table 6.2. Comparison between the output y of a real system and the output y^* of a fuzzy model with triangular membership functions for its inputs (according to Fig. 6.6), and with the operator *PROD*

x_1	5 5 5 5 5	6 6 6 6 6	7 7 7 7 7	8 8 8 8 8	9 9 9 9 9
x_2	0 1 2 3 4	0 1 2 3 4	0 1 2 3 4	0 1 2 3 4	0 1 2 3 4
y	9 6 5 6 9	6 3 2 3 6	5 2 1 2 5	6 3 2 3 6	9 6 5 6 9
y^*	9 7 5 7 9	7 5 3 5 7	5 3 1 3 5	7 5 3 5 7	9 7 5 7 9
$\|y-y^*\|$	0 1 0 1 0	1 2 1 2 1	0 1 0 1 0	1 2 1 2 1	0 1 0 1 0

mean absolute error of the model $|y-y^*|_{mean} = 0.8$

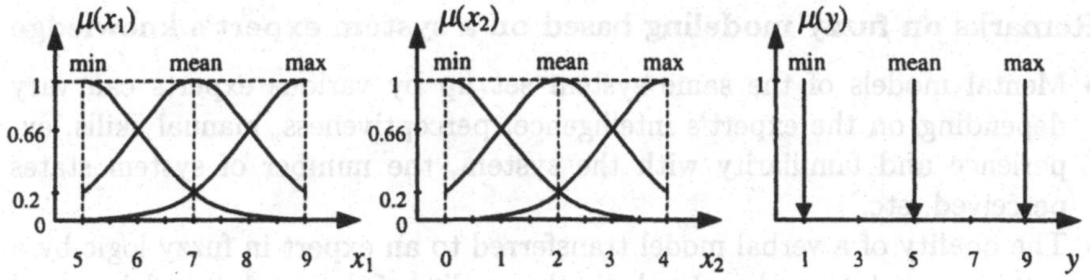

$$\mu_{\min}(x_1) = \exp[-((x_1 - 5)/1.5765)^2] \qquad \mu_{\min}(x_2) = \exp[-((x_2 - 0)/1.5765)^2]$$

$$\mu_{\mathrm{mean}}(x_1) = \exp[-((x_1 - 7)/1.5765)^2] \qquad \mu_{\mathrm{mean}}(x_2) = \exp[-((x_2 - 2)/1.5765)^2]$$

$$\mu_{\max}(x_1) = \exp[-((x_1 - 9)/1.5765)^2] \qquad \mu_{\max}(x_2) = \exp[-((x_2 - 4)/1.5765)^2]$$

Fig. 6.7. **Membership functions chosen for the intputs x_1, x_2 and output y in the second variant of the model**

Table 6.3. **Comparison between the output y of a real system and the output y^{**} of a fuzzy model with Gaussian membership functions according to Fig. 6.7, and with the operator *MIN***

x_1	5	5	5	5	5	6	6	6	6	6	7	7	7	7	7	8	8	8	8	8	9	9	9	9	9
x_2	0	1	2	3	4	0	1	2	3	4	0	1	2	3	4	0	1	2	3	4	0	1	2	3	4
y	9	6	5	6	9	6	3	2	3	6	5	2	1	2	5	6	3	2	3	6	9	6	5	6	9
y^{**}	7	6	5.4	6	7	6	5.4	4.9	5.4	6	5.4	5.5	4.7	5.5	5.4	6	5.4	4.9	5.4	6	7	6	5.4	6	7
$\lvert y-y^{**}\rvert$	2	0	0.4	0	2	0	2.4	2.9	2.4	0	0.4	3.5	3.7	3.5	0.4	0	2.4	2.9	2.4	0	2	0	0.4	0	2

mean absolute error of the model $\lvert y-y^{**}\rvert_{\mathrm{mean}} = 1.43$

On the other hand, if the Gaussian function has been chosen for the fuzzification process for the inputs x_1, x_2 and singleton functions for the output y, Fig. 6.7, and the operator *MIN* to perform *AND*-operation, then the fuzzy model will calculate the output y^{**} values presented in Table 6.3.

An analysis of the results in Tables 6.2 and 6.3 proves that each of the two models calculates the output in a different way and with a different accuracy and thus provides different mapping of the inputs onto the output.

Summarizing the above example we can conclude that the accuracy of a system set up on the basis of an expert's knowledge depends on the following two factors:

a) thorough recognition of a real system by a system expert and his capability to convert his knowledge into rules and linguistic values,
b) experience, knowledge and intuition of an expert in fuzzy modeling whose task is to complete the model delivered by a system expert, with the appropriate elements of fuzzy logic apparatus.

◊ ◊ ◊

Remarks on fuzzy modeling based on a system expert's knowledge

- Mental models of the same system set up by various experts can vary depending on the expert's intelligence, perceptiveness, manual skills, experience and familiarity with the system, the number of system states perceived, etc.
- The quality of a verbal model transferred to an expert in fuzzy logic by a system expert depends not only on the quality of the mental model created in the system expert's mind but also on his capability and skills to transfer his knowledge precisely and adequately.
- A verbal model transferred by a system expert can contain an incomplete set of rules, contradictory rules or incomplete information on linguistic values. Therefore, a verbal model should be subject to thorough verification and, if necessary, to adjustment or completion.
- A verbal model containing a set of rules without the information on linguistic value parameters can also be used for modeling, if combined with other methods that would enable us to determine the missing parameters, e.g. trial-and-error method, method of tuning the neuro-fuzzy network with system input/output data or others.
- One can manage to build up sufficiently precise verbal models for mechanical or electrical systems (Isermann 1995) but in the case of thermal or chemical systems the verbal models are usually less precise. Biological systems fall into the third class of precision. Economic and sociological systems are claimed to have the least precise models. The difficulties appear in formulating the terms and their fuzzy nature, in taking measurements and determining significant inputs of the systems as well as carrying out experiments and ensuring their long duration to provide sufficient data.
- The qualitative verbal models can be constructed only for systems of a low number of dimensions, mainly for systems with one or two inputs. Human perception abilities exclude the capacity to memorize the states of a greater number of inputs. In that case a man can give only a fragmentary knowledge of the system concerned.
- The possibility to recognize a system and to formulate its qualitative model depends on the speed of changes occurring in the system. In the case of rapidly changing processes, the task of system modeling may appear unfeasible even for a single input system.
- Actually, possessing only fragmentary qualitative knowledge about a system may appear very beneficial; still the system model structure (or a fragment of the structure) can be defined so that the work needed for system identification and use of other methods of fuzzy modeling can be significantly reduced.

The method of fuzzy modeling based on a system expert's knowledge enables us to build up the models of Mamdani type whereas the models of Takagi-Sugeno type can be constructed only with adaptation methods on the basis of system input/output measurement data.

6.2 Creation of fuzzy, self-tuning models based on input/output measurement data of the system

The first fuzzy models were built up on the basis of a system expert's knowledge. Practical experience gained during that time proved that there were several inconveniences in that approach – the most significant being the following:

a) difficulties in precise determination of fuzzy set parameters used by an expert (contribution of expert's subconscious in the creation of system mental model),
b) difficulties or inability to learn how to operate (control) the systems with a greater (more than 2) number of inputs, rapidly changing systems or systems performing complicated input/output mappings, and thus the expert's inability to formulate the fuzzy model of such a process.

The reasons listed above have made scientists develop the fuzzy self-tuning models. There are many significant publications which discuss the subject (Babuška 1995a,b,c,d,e; Baldwin 1995b; Bossley 1995; Brown 1994,1995a,b; Cao 1997a; Carpenter 1992; Cipriano 1995; Cho 1995; Davè 1997; Delgado 1995,1997; Driankov 1993,1996; Eklund 1992; Feuring 1996; Fiordaliso 1996; Fukumoto 1995; Furuhashi 1995; Gonzales 1995; Gorrini 1995; Gupta 1994; Hajek 1994,1995; Halgamuge 1996; Hanss 1996a; Hauptman 1995; Hensel 1995; Higgins 1994; Horikawa 1992; Ishibuchi 1993,1995; Jäckel 1997; Kahlert 1995; Kandel 1994; Katebi 1995; Kiriakidis 1995; Krone 1996a,c; Kwon 1994; Langari 1995; Lin 1991; Lin 1996; Locher 1996a,b; Magdalena 1995; Männle 1996; Murata 1995; Narazaki 1993,1995; Nelles 1996,1997; Nobre 1995; Nomura 1994; Osowski 1996; Otto 1995; Park 1995; Pedrycz 1997; Piegat 1996; Preus 1994a,b,1995; Preut 1995; Rovatti 1996; Rutkowska 1996,1997; Simpson 1992; Su 1995; Takagi 1985; Tan 1995; Wakabayashi 1995; Yao 1995; Zhou 1995).

As will be considered in this chapter, the tuning of a fuzzy model is mainly the process of determining the parameters of the membership functions for the inputs and an output where the parameters should allow us to minimize the model error with regard to the system modeled, in terms of employing the methods of error estimation (mean square error, mean absolute error or maximum error). Moreover, it is assumed that the model structure is known and is not subject to changes.

To tune a model, i.e. to optimize the model parameters, the following methods are most often used:

I. methods of neuro-fuzzy networks,
II. searching methods,
III. clustering-based methods,
IV. methods using non-fuzzy neural networks,
V. heuristic methods.

The first three methods are the most significant in practical applications. Method I is the method which converts a fuzzy model into a neuro-fuzzy network (Bossley 1995; Brown 1994,1995a,b; Gupta 1994; Hauptmann 1995; Nelles 1997; Osowski 1996; Piegat 1996; Yao 1995) and applies network training methods based on system input/output measurement data to tune the model parameters. This method, considering its practical and cognitive value, will be described in detail in Section 6.2.1.

Method II is the method of direct searching for optimum parameters of a fuzzy model. The searching process can be organized or not organized (trial-and-error method). The most often used method of organized searching is the genetic algorithm method (Murata 1995; Nobre 1995) presented in Section 6.2.2.

The methods based on the clustering process combine the tuning of model parameters and model structuring. These methods will be considered in Section 6.3. The methods making use of neural networks for fuzzy model tuning are infrequently used. The readers interested in these methods may find adequate examples in relevant publications (Carpenter 1992; Hauptmann 1995; Ishibuchi 1993; Narazaki 1995). The above remark applies to the heuristic methods as well (Eklund 1992; Gorrini 1995; Simpson 1992). Various interesting aspects of fuzzy modeling and relevant examples can be found in (Beigy 1995; Bartolan 1997; Culliere 1995; Hathaway 1996; Krone 1996b; Lofti 1996; Pütz 1996; Shmilovici 1996; Wang 1995a; Žižka 1996).

6.2.1 Application of neuro-fuzzy networks for fuzzy model parameter tuning

The neural networks can be trained on the basis of the input/output measurements of the system modeled. Nowadays, there are numerous methods of training the networks; the methods are extensively described in the professional literature (Haykin 1994; Masters 1993; Zell 1994). A fuzzy model can be presented in the form of a special neural network and one of the training methods can be applied to model parameter tuning. To understand the concept of a network constructed that way, referred to as a neuro-fuzzy network (Bossley 1995; Brown 1995a,b; Gupta 1994; Hauptmann 1995; Horikawa 1992; Nelles 1997; Osowski 1996; Preuss 1994a; Yao 1995), it seems essential to study the general concept of neural networks.

6.2.1.1 Structuring and training of neural networks

The basic structure of an artificial neural network (ANN) is presented in Fig. 6.8. The neurons of its input layer transfer the input signals to the neurons of a hidden layer. Networks with a single hidden layer are most often used since that type of network provides a sufficiently precise modeling of numerous real systems. The neurons in the hidden layers and output layer

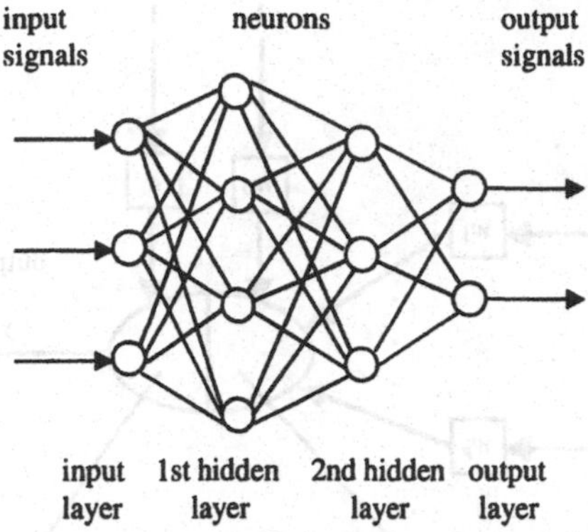

Fig. 6.8. Artificial neural network structure

provide processing of information transferred from the input layer. The multi-layer perceptron networks (MPN) and the neural networks comprising radial basis functions, so-called RBF networks (Preuss 1994a) are the most frequently used ANNs. The main element of the perceptron network is a neuron presented in Fig. 6.9.

The propagation function enables us to calculate a weighted sum s of all neural inputs and enables us to insert this sum to a non-linear element with an activation function $f(s)$, also called a threshold function, where the element mentioned generates an output signal y of the neuron. A coefficient value w_0 is called a threshold. At relatively large threshold values w_0 even small values of inputs x_i excite the neuron to generate an output signal. However, in the case of a small threshold value larger input values are required. In most cases a sigmoidal function, presented in Fig. 6.10, is used as the activation function $f(s)$.

The coefficient c affects the sigmoidal function inclination. When $c \to \infty$ the neuron is excited rapidly, otherwise when $c \to 0$ the excitation is slower. The MPNs usually consist of 3 layers, Fig. 6.11.

The process of training the neural network causes such a gradual change in its weight values w, v, c that a minimum or sub-minimum of the training criterion is reached. The criterion is usually a mean or total square error of the network output y with regard to the output y^* of the system modeled. The network output y is a function of network inputs and trained parameters (6.3).

$$y = F(\ldots, w_{ij}, \ldots, v_k, \ldots, c_l, \mathbf{x}), \qquad (6.3)$$

where: $\mathbf{x} = [x_1, \ldots, x_p]^{\mathrm{T}}$.

If a single training pair of input/output, obtained on the basis of measurements of the system modeled, is denoted as (6.4):

$$\{\mathbf{x}^{*i}, y^{*i}\}, \qquad (6.4)$$

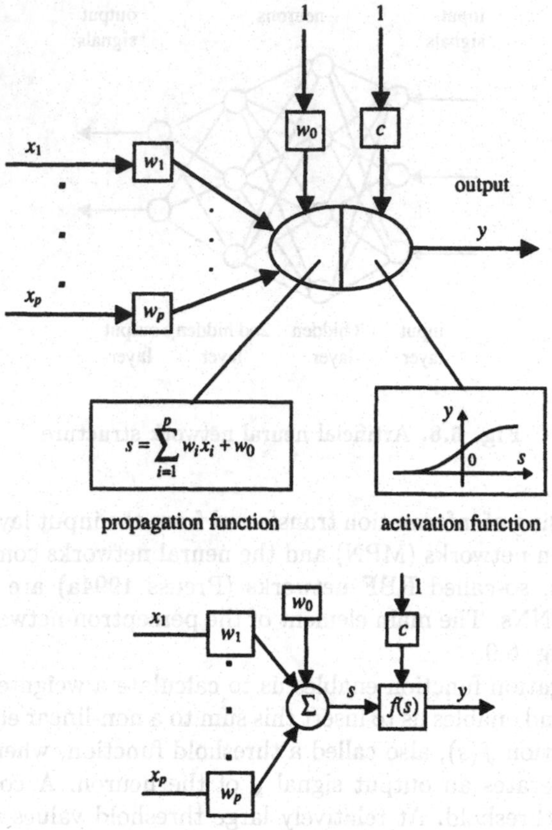

Fig. 6.9. Diagram of artificial neuron used in multi-layer perceptron networks

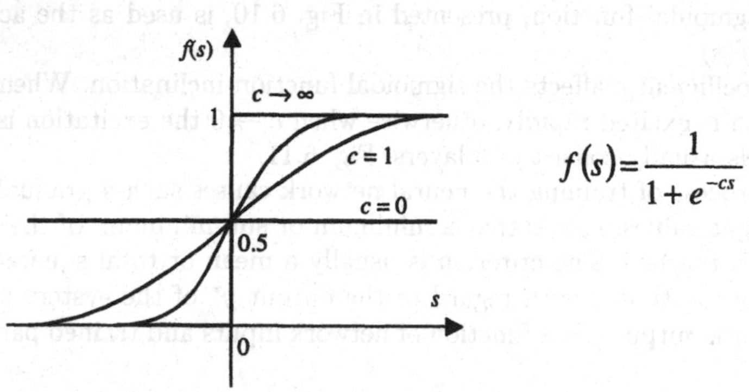

Fig. 6.10. Sigmoidal function

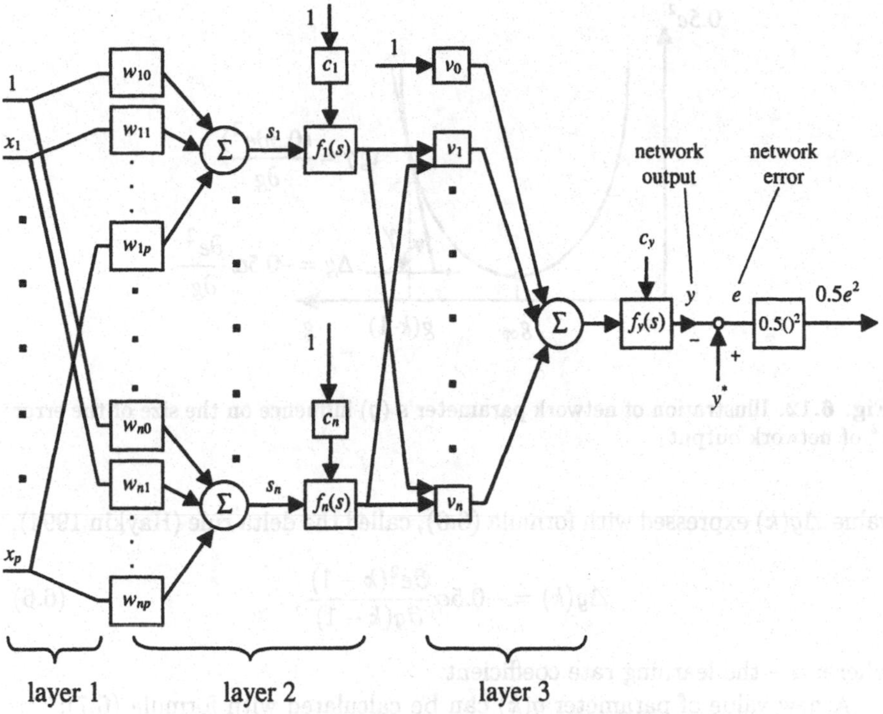

Fig. 6.11. Three-layer perceptron network with a single output

then the total square error of the network, for the entire training cycle containing m input/output training pairs, multiplied by 0.5 (for convenience of further calculations), is expressed with formula (6.5).

$$E = 0.5 \sum_{i=1}^{m} (y^{*i} - y^i)^2 = 0.5 \sum_{i=1}^{m} e_i^2 \tag{6.5}$$

The total error E will decrease if the errors e_i^2 of individual training pairs are reduced. The total network error, just as the output y, is a function of network parameters w_{ij}, v_k, c_l, which are the variables of the training process and are usually referred to as degrees of freedom of the network. The number of degrees is usually high to enable the most complicated input/output mappings to be performed. The network parameter tuning is usually performed following the principle of error back-propagation and using the gradient methods (Haykin 1994; Masters 1993; Zell 1994). If g is an arbitrary network parameter under tuning then its influence on the network error depends on the derivative $\partial e^2/\partial g$, Fig. 6.12.

Fig. 6.12 shows that to approximate the parameter g to its optimum value g_{opt}, the parameter has to be relocated towards the negative gradient by a

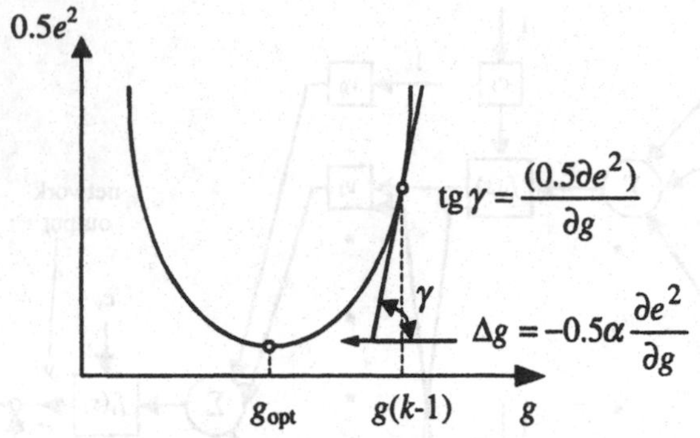

Fig. 6.12. Illustration of network parameter's (g) influence on the size of the error e^2 of network output

value $\Delta g(k)$ expressed with formula (6.6), called the delta rule (Haykin 1994).

$$\Delta g(k) = -0.5\alpha \frac{\partial e^2(k-1)}{\partial g(k-1)}, \tag{6.6}$$

where: α – the learning rate coefficient.

A new value of parameter $g(k)$ can be calculated with formula (6.7):

$$g(k) = g(k-1) + \Delta g(k), \tag{6.7}$$

where: k – the step number.

The correction value $\Delta g(k)$ depends on the value chosen of the coefficient α. The tuning process can be initiated starting from $\alpha = 0.1$ or similar values. If the value α is high then the approximation to the parameter optimum will proceed quickly. However, there is a danger that the optimum value may be "jumped over" and then oscillations around the optimum value would take place, Fig. 6.13. The above phenomenon can be avoided if the parameter correction is calculated with the formula (6.8) containing the momentum coefficient μ.

$$\Delta g(k) = -0.5\alpha \frac{\partial e^2(k-1)}{\partial g(k-1)} + \mu \Delta g(k-1) \tag{6.8}$$

The momentum coefficient μ is presumed to fall within the interval $[0, 1]$, usually to 0.9. The effect of the momentum coefficient is presented in Fig. 6.13.

Let us assume that the parameter g, tuned without the momentum, oscillates around the optimum value. The oscillation amplitude $\Delta^* g$, Fig. 6.13a, and its sign changes. If we initiate the correction calculations using the formula (6.8) with the momentum coefficient at a moment $(k-1)$, where the parameter value is $g(k-1)$, then the next correction will be immediately reduced by $0.9\Delta^* g(k-1)$ and the new parameter value $g(k)$ will not jump over

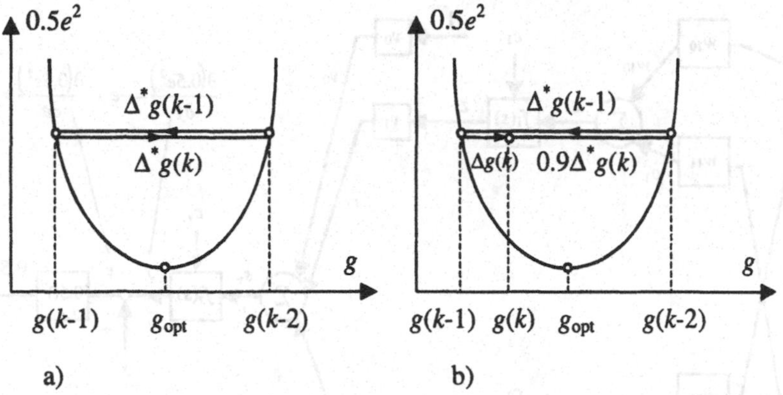

Fig. 6.13. Illustration of the momentum coefficient's μ influence on elimination of oscillations

the optimum. Further steps provide gradual approximation to the optimum value.

A key task during the correction process is the calculation of the derivatives $\partial e^2/\partial g$ of the parameter under tuning. The calculation is performed by multiplying all the partial derivatives of the network input/output elements found between the signal $0.5e^2$ and the corrected parameter, Fig. 6.14. In the case where there are many paths leading to the signal $0.5e^2$ the derivative products along each path have to be summed. The method of derivative calculation has been illustrated with an example based on Fig. 6.14.

To provide the correction of coefficient c_y in the lowest layer of the network the derivative $\partial(0.5e^2)/\partial c_y$ has to be calculated with formula (6.9).

$$\frac{\partial(0.5e^2)}{\partial c_y} = \frac{\partial(0.5e^2)}{\partial e} \cdot \frac{\partial e}{\partial y} \cdot \frac{\partial y}{\partial c_y} = e \cdot (-1) \cdot \frac{\partial f_y(s)}{\partial c_y} = -e \cdot \frac{\partial f_y(s)}{\partial c_y} \tag{6.9}$$

Formula (6.9) contains the network error e. The error "moves backwards" from the network output and appears in each derivative. That is why the method is called "error back-propagation".

To correct the coefficient v_0 in the lowest layer the derivative $\partial(0.5e^2)/\partial v_0$, formula (6.10), is applied.

$$\frac{\partial(0.5e^2)}{\partial v_0} = \frac{\partial(0.5e^2)}{\partial e} \cdot \frac{\partial e}{\partial y} \cdot \frac{\partial f_y(s)}{\partial s_y} \cdot \frac{\partial s_y}{\partial v_i} = -e \cdot \frac{\partial f_y(s)}{\partial s_y} \tag{6.10}$$

The correction of coefficient v_1 is performed on the basis of the derivative determined with formula (6.11).

$$\frac{\partial(0.5e^2)}{\partial v_1} = \frac{\partial(0.5e^2)}{\partial e} \cdot \frac{\partial e}{\partial y} \cdot \frac{\partial f_y(s)}{\partial s_y} \cdot \frac{\partial s_y}{\partial r_1} \cdot \frac{\partial r_1}{\partial v_1} = -z_1 \cdot e \cdot \frac{\partial f_y(s)}{\partial s_y} \tag{6.11}$$

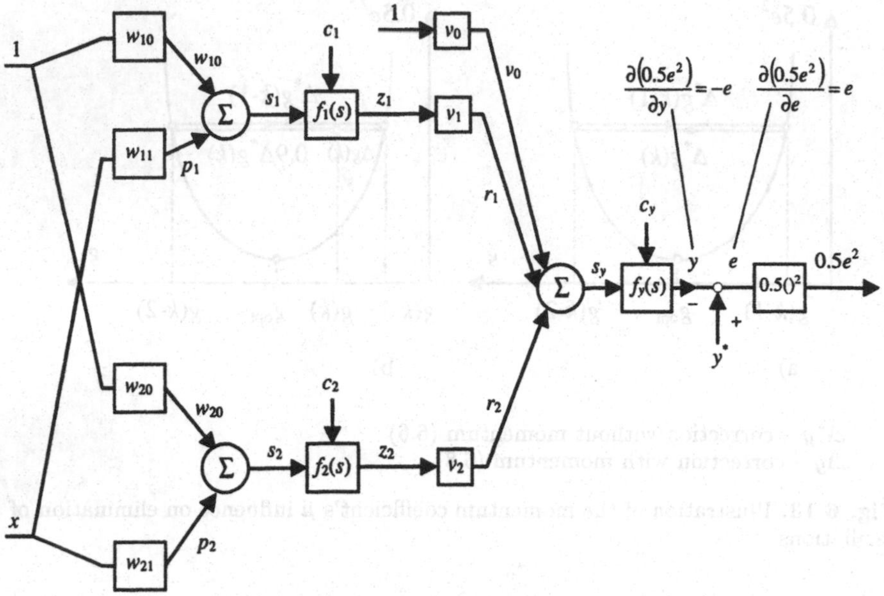

Fig. 6.14. Example illustrating the method of error back-propagation

The correction of coefficient v_1 depends not only on error e but also on the current value of input signal z_1 of element v_1, Fig. 6.14. To correct the coefficient w_{11} in the hidden layer the derivative calculated with formula (6.12) has to be used.

$$\frac{\partial(0.5e^2)}{\partial w_{11}} = \frac{\partial(0.5e^2)}{\partial e} \cdot \frac{\partial e}{\partial y} \cdot \frac{\partial y}{\partial s_y} \cdot \frac{\partial s_y}{\partial r_1} \cdot \frac{\partial r_1}{\partial z_1} \cdot \frac{\partial z_1}{\partial s_1} \cdot \frac{\partial s_1}{\partial p_1} \cdot \frac{\partial p_1}{\partial w_{11}} =$$
$$= -x \cdot e \cdot \frac{\partial f_1(s)}{\partial s_1} \cdot \frac{\partial f_y(s)}{\partial s_y} \cdot v_1 \qquad (6.12)$$

An advantage of sigmoidal activation functions, Fig. 6.10, is the simplicity of derivative calculation methods; see formulas (6.13) and (6.14).

$$y = f(s) = \frac{1}{1+e^{-cs}}$$
$$\frac{\partial f(s)}{\partial s} = \frac{c \cdot e^{-cs}}{(1+e^{-cs})^2} = \frac{c \cdot e^{-cs} + c - c}{(1+e^{-cs})^2} = c(y - y^2) \qquad (6.13)$$
$$\frac{\partial f(s)}{\partial c} = s(y - y^2) \qquad (6.14)$$

One of the difficulties found in neural network training is the problem of defining the network structure, i.e. the number of neurons in its intermediate layers and interdependencies among the neurons of particular layers. Since

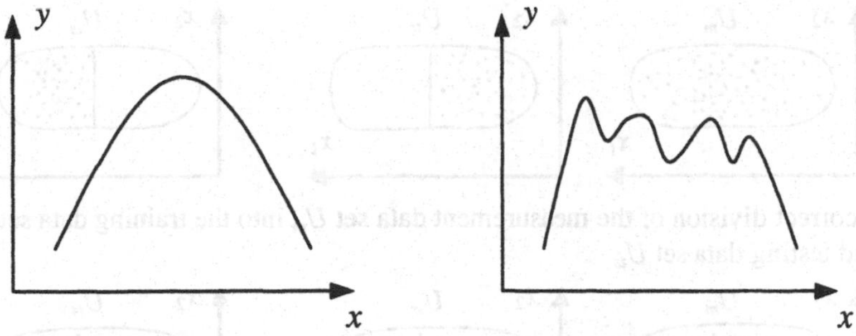

Fig. 6.15. Dependencies between the complexity of the modeled system $\mathbf{X} \rightarrow Y$ mapping surface and the number of neurons in the neural network

the most often used structures contain 3 layers, including one hidden layer, the problem of structure description is in fact the problem of determining the number of neurons n in that hidden layer. There are numerous valuable suggestions in many publications concerning the above problem, e.g. in (Haykin 1994; Osowski 1996; Zell 1994). However, these suggestions do not guarantee that we will choose an optimal or even good structure. The reasons will be considered in due course.

The tuning (training) process of a neural network aims at achieving a surface approximation of the input/output mapping of the modeled system using the measurement data representing the system and carrying information on system operation. The following general statement on the network structure remains true: "the more complicated the surface of system $\mathbf{X} \rightarrow Y$ mapping, the more neurons in the hidden layer are necessary for surface modeling", Fig. 6.15.

Unfortunately, while modeling the real systems there is often insufficient knowledge about the surface complication level, particularly in multi-input systems. We usually only have measurement value information describing the input and output states corresponding one to another. Additionally, the information is burdened with measurement errors and affected by inputs not considered during the modeling process. In the case of systems with more than 2 inputs the model cannot be presented in a graphic form to enable us to determine the surface complexity. That is the reason the trial method is usually applied. When initiating a modeling process we may assume the number n of neurons as a geometric average of inputs p and outputs l (Osowski 1996), formula (6.15). Later on, depending on the accuracy of the network set up, the quantity of neurons should be modified as needed.

$$n \cong \sqrt{pl} \tag{6.15}$$

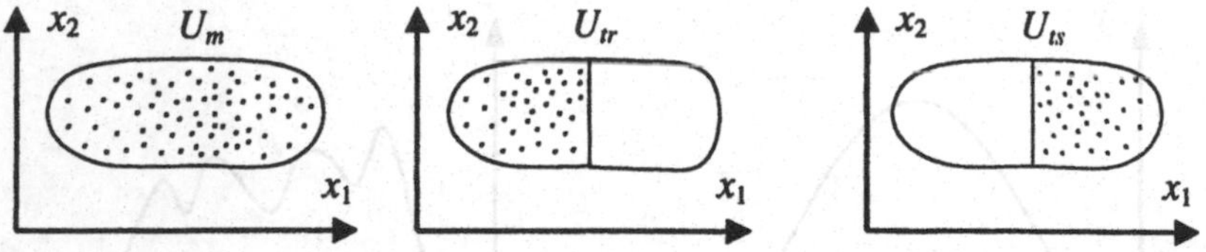

a) incorrect division of the measurement data set U_m into the training data set U_{tr} and testing data set U_{ts}

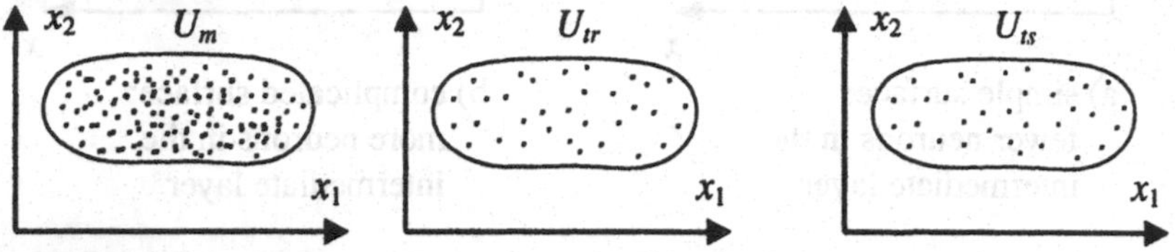

b) correct division of the measurement data set U_m into the training data set U_{tr} and testing data set U_{ts}

Fig. 6.16. Illustration of the problem of measurement data set division

Before the network training process is initiated, a set U_m, containing the measurement data $\mathbf{X}/Y$ of the system, has to be divided into two subsets: U_{tr} – a training data set and U_{ts} – a testing data set:

$$U_m = U_{tr} + U_{ts}, \quad U_{tr} \neq U_{ts}. \tag{6.16}$$

The testing data set U_{ts} is used to test the network trained by means of the training data set U_{tr}. The testing data set U_{ts} should contain the selected data from the measurement data set U_m, following the principle of uniform data allocation in the space of inputs $\mathbf{X}$, Fig. 6.16.

Considering the case in Fig. 6.16a, the network after tuning will provide accurate modeling of the system in the left-side part of the input space. The testing data located in the right-side part of the space will show a significantly large (very large) error. In this case the network has had no opportunity to "acquaint" itself with the system under modeling in the entire space of the system operation. In the case presented in Fig. 6.16b the network is trained using the data located within the entire space of system operation and is tested using the testing data set of similar features. In the latter case the testing data set will generate an objective, correct evaluation of the network's accuracy.

On having examined the initial structure, the number of neurons n in the structure's hidden layer should be modified until the test results are improved, i.e. until the mean network error for the testing data set U_{ts} or for the entire measurement data set U_m has decreased. In several cases the training is done without the testing data set, particularly where the data set is small. The training process and the structure modifications continue until

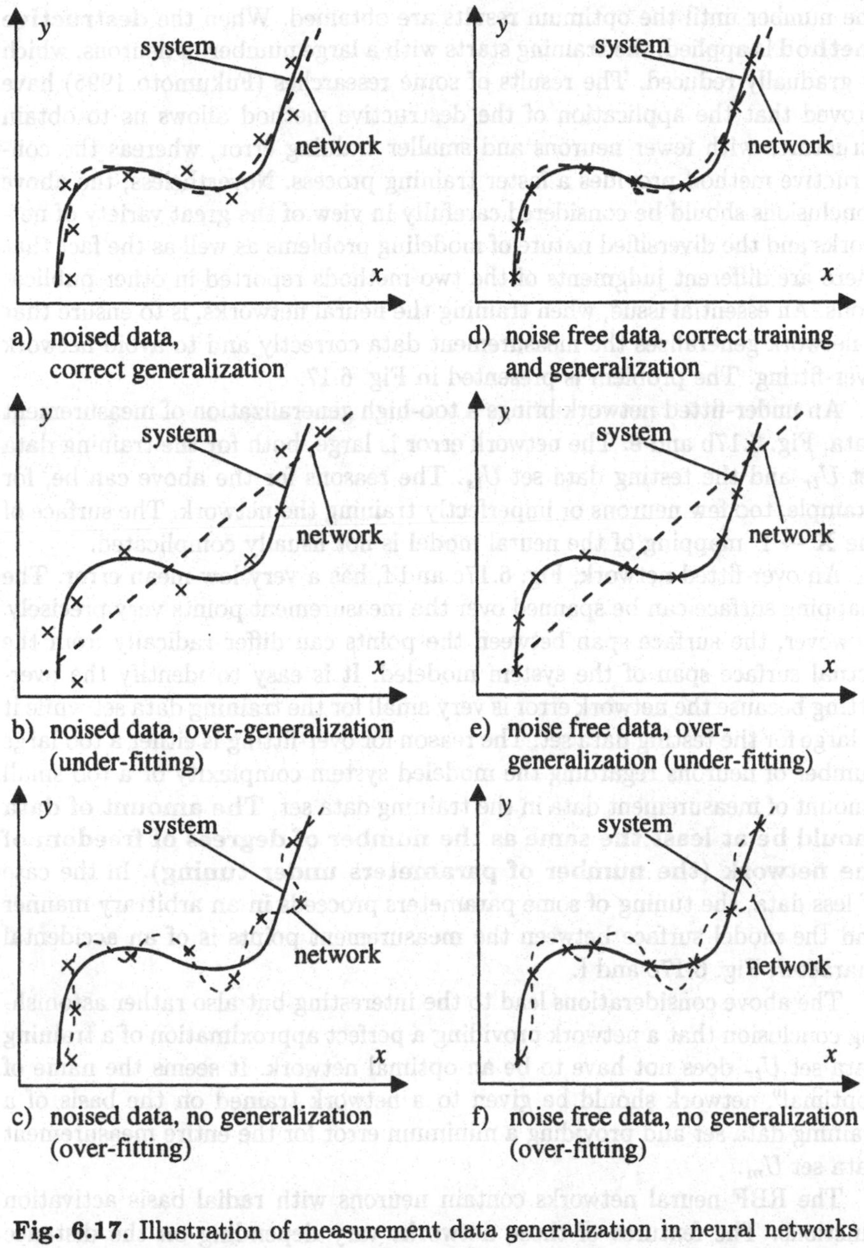

Fig. 6.17. Illustration of measurement data generalization in neural networks

the minimum error function is reached (assuming that the network has not been over-fitted, Fig. 6.17).

The structure modifications can be done using 2 methods: a constructive or destructive one. The **constructive method** means the initiation of the training process with a small number of neurons and the gradual increase of the number until the optimum results are obtained. When the **destructive method** is applied, the training starts with a large number of neurons, which is gradually reduced. The results of some researches (Fukumoto 1995) have proved that the application of the destructive method allows us to obtain structures with fewer neurons and smaller training error, whereas the constructive method provides a faster training process. Nevertheless, the above conclusions should be considered carefully in view of the great variety of networks and the diversified nature of modeling problems as well as the fact that there are different judgments of the two methods reported in other publications. An essential issue, when training the neural networks, is to ensure that a network generalizes the measurement data correctly and to avoid network over-fitting. The problem is presented in Fig. 6.17.

An under-fitted network brings a too-high generalization of measurement data, Fig. 6.17b and e. The network error is large, both for the training data set U_{tr} and the testing data set U_{ts}. The reasons for the above can be, for example, too few neurons or imperfectly training the network. The surface of the $\mathbf{X} \to Y$ mapping of the neural model is not usually complicated.

An over-fitted network, Fig. 6.17c and f, has a very low mean error. The mapping surface can be spanned over the measurement points very precisely. However, the surface span between the points can differ radically from the actual surface span of the system modeled. It is easy to identify the over-fitting because the network error is very small for the training data set while it is large for the testing data set. The reason for over-fitting is either a too large number of neurons regarding the modeled system complexity or a too small amount of measurement data in the training data set. **The amount of data should be at least the same as the number of degrees of freedom of the network (the number of parameters under tuning).** In the case of less data, the tuning of some parameters proceeds in an arbitrary manner and the model surface between the measurement points is of an accidental character, Fig. 6.17c and f.

The above considerations lead to the interesting but also rather astonishing conclusion that a network providing a perfect approximation of a training data set U_{tr} does not have to be an optimal network. It seems the name of "optimal" network should be given to a network trained on the basis of a training data set and providing a minimum error for the entire measurement data set U_m.

The RBF neural networks contain neurons with radial basis activation functions. The features of those networks vary depending on the distance $||\mathbf{x} - \mathbf{c}||$ between the point $\mathbf{x}$ belonging to the input space and the function center $\mathbf{c}$. The radial basis functions are defined with formula (6.17), (Brown 1994).

$$y = f(\mathbf{x}) = f(||\mathbf{x} - \mathbf{c}||)\,, \tag{6.17}$$

where: $\mathbf{x} = [x_1, \ldots, x_p]^{\mathrm{T}}$ – input vector,
$\mathbf{c} = [c_1, \ldots, c_p]^{\mathrm{T}}$ – vector of function center coordinates.

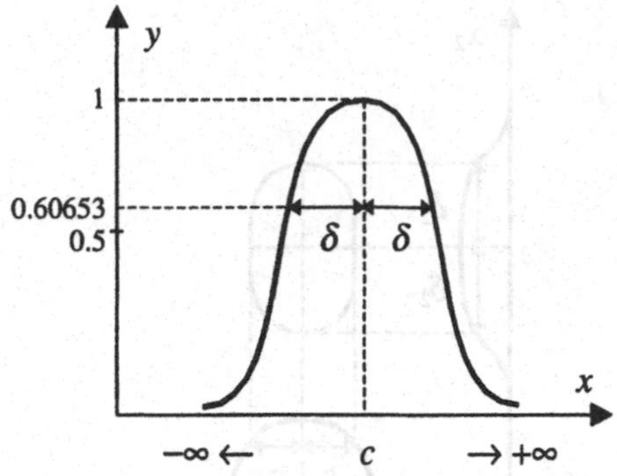

Fig. 6.18. Gaussian radial basis function (GRBF) in 2D space

The Gaussian radial basis function is the most often used radial basis function. The form of the above function in 2D space is presented in Fig. 6.18. A variance δ of the Gaussian radial basis function (abbreviated to GRB-function or GRBF) defines the deflection degree of the function arms, Fig. 6.18. The deflection value is equal to δ for the height:

$$y = \exp(-0.5) \cong 0.60653\,.$$

The GRB-function, commonly known as the Gaussian "bell" (Preuss 1994) has a certain useful feature. The function value is close to 1 only if the input value x is close to the center c.

The above feature makes the function particularly useful for application in the fuzzy rules defined by the formula (6.18),

$$\textit{IF } (x \text{ is close to } c) \textit{ THEN } (y \text{ is close to } y_0)\,, \tag{6.18}$$

where: $y = f(x) = y_0 \exp[-(x-c)^2/2\delta^2]$.

The multi-dimensional GRB-functions defined with formula (6.19) are used for modeling the multi-input systems.

$$y = f(\mathbf{x}) = \exp\left[-\sum_{i=1}^{p} \frac{(x_i - c_i)^2}{2\delta_i^2}\right], \tag{6.19}$$

where: $\mathbf{x} = [x_1, \ldots, x_p]^{\mathrm{T}}$ – input vector,
$\mathbf{c} = [c_1, \ldots, c_p]^{\mathrm{T}}$ – center coordinate vector of the function,
$\boldsymbol{\delta} = [\delta_1, \ldots, \delta_p]^{\mathrm{T}}$ – vector of variances (deflections) along particular axes of variables x_i.

The parameters of a two-input Gaussian "bell" are presented in Fig. 6.19. In the case of a multi-input GRB-function its output is close to 1 only if the input vector $\mathbf{x}$ is located close to the function center described by means of the vector $\mathbf{c}$, (6.19). A multi-input GRB-function provides good premise evaluation in fuzzy rules with complex premises described by (6.20).

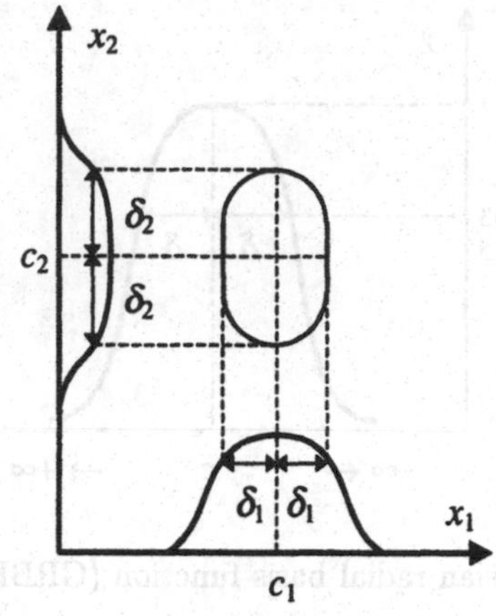

Fig. 6.19. Two-input GRB-function and its parameters

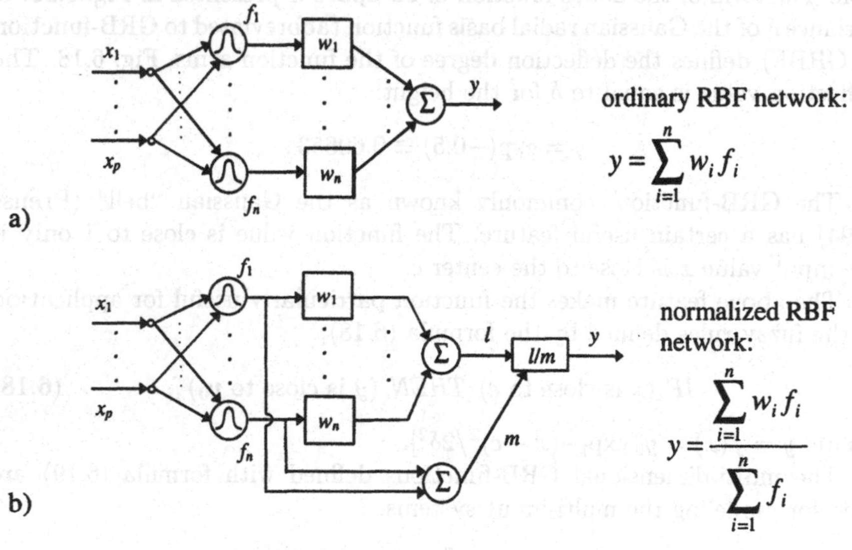

Fig. 6.20. Structures of RBF networks

$$IF\ (x_1 \text{ is close to } c_1)\ AND\ \ldots\ AND\ (x_p \text{ is close to } c_p)\ THEN\ (y \text{ is close to } y_0), \quad (6.20)$$

where: $y = f(\mathbf{x}) = \exp\left[-\sum_{i=1}^{p} \frac{(x_i - c_i)^2}{2\delta_i^2}\right]$.

Two types of RBF networks are used: ordinary (non-normalized) networks and normalized networks, Fig. 6.20.

A disadvantage of an ordinary RBF network, Fig. 6.20a, is the interpolation imperfection in cases where the arms of adjacent GRBF functions overlap

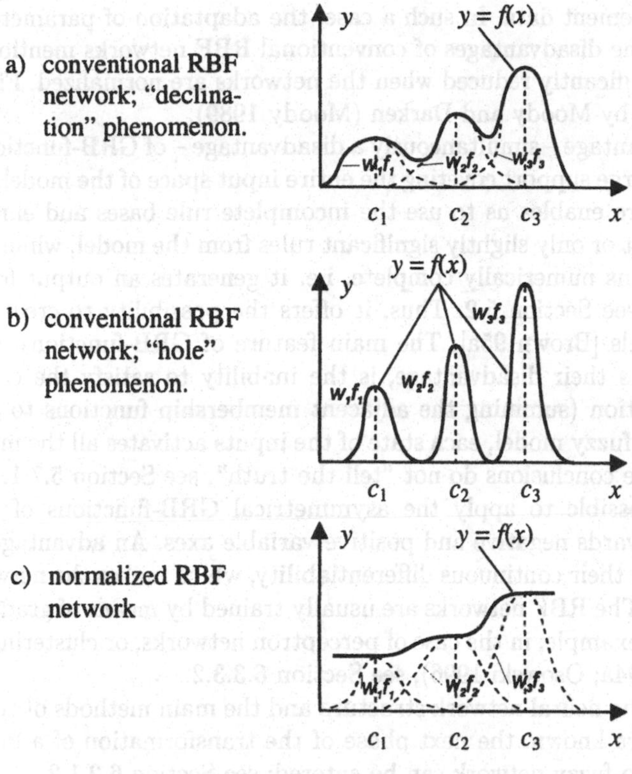

Fig. 6.21. Potential risks in system modeling using conventional RBF networks and their prevention through normalization of those networks

slightly. In that case the neural model surface can exhibit "declinations" between particular Gaussian "bells", Fig. 6.21a. The long distances between the centers of particular "bells" and the small span of their arms can stimulate the occurrence of "holes" in the model surface causing the model's local insensitivity to input changes, Fig. 6.21b.

The "declinations" and "holes" – attributes of the ordinary RBF networks – can be avoided if the arm span δ_i of adjacent functions is made dependent on the distance between their centers c_i. There are certain heuristic recommendations stating, for example, that δ should be equal to double the mean distance between the adjacent centers (Preuss 1994a). However, such "inflexibility" of the arm spans δ_i means excluding them as degrees of freedom in the tuning process. Then, during the training process only the centers c_i are tuned and span values δ_i are chosen depending on the center location. In the case where both the centers and spans are subject to tuning, then during the training of the conventional RBF network the "declinations" and "holes" will occur if at the start of the training process the span values δ_i chosen are too small and certain fragments of input space are not covered by the measurement data. In such a case, the adaptation of parameters will be difficult. The disadvantages of conventional RBF networks mentioned above can be significantly reduced when the networks are normalized, Fig. 6.20, as introduced by Moody and Darken (Moody 1989).

An advantage – simultaneously a disadvantage – of GRB-functions is their infinitely large support covering the entire input space of the model, Fig. 6.18. That feature enables us to use the incomplete rule bases and eliminate the insignificant or only slightly significant rules from the model, while the model itself remains numerically complete, i.e. it generates an output for each input state; see Section 5.2. Thus, it offers the possibility to create "sparse" fuzzy models [Brown 95a]. The main feature of GRB-functions, sometimes perceived as their disadvantage, is the inability to satisfy the condition of unity partition (summing the adjacent membership functions to 1). Therefore, in the fuzzy model, each state of the inputs activates all the model rules, but the rule conclusions do not "tell the truth", see Section 5.7.1.

It is possible to apply the asymmetrical GRB-functions of diversified spans δ towards negative and positive variable axes. An advantage of GRB-functions is their continuous differentiability, which makes the network training easier. The RBF networks are usually trained by means of gradient methods, as for example, in the case of perceptron networks, or clustering methods (Preuss 1994a; Osowski 1996), see Section 6.3.3.2.

When the neural network structure and the main methods of training the networks are known, the next phase of the transformation of a fuzzy model into a neuro-fuzzy network can be entered; see Section 6.2.1.2.

6.2.1.2 Transformation of a Mamdani fuzzy model into a neuro-fuzzy network

A fuzzy model of a MISO system is presented in Fig. 6.22. Further considerations in this chapter outline the methods of the transformation of various elements of fuzzy models into elements of neural networks and derivative calculation algorithms applied in gradient training methods.

Transformation of elements in Fuzzification block

Fig. 6.23 illustrates a transformation of a segmentally-linear membership function into a fragment of neural network. During training of the network, the membership function parameters a_i are tuned. In that case it is necessary to determine the derivatives of fuzzification block outputs with respect to parameters mentioned above. The membership functions and their derivatives are defined with formulas (6.21) – (6.30).

$$\mu_S(x) = \begin{cases} 1 & \text{if } x < a_1\,, \\ \frac{a_2 - x}{a_2 - a_1} & \text{if } a_1 \le x < a_2\,, \\ 0 & \text{if } x \ge a_2\,. \end{cases} \tag{6.21}$$

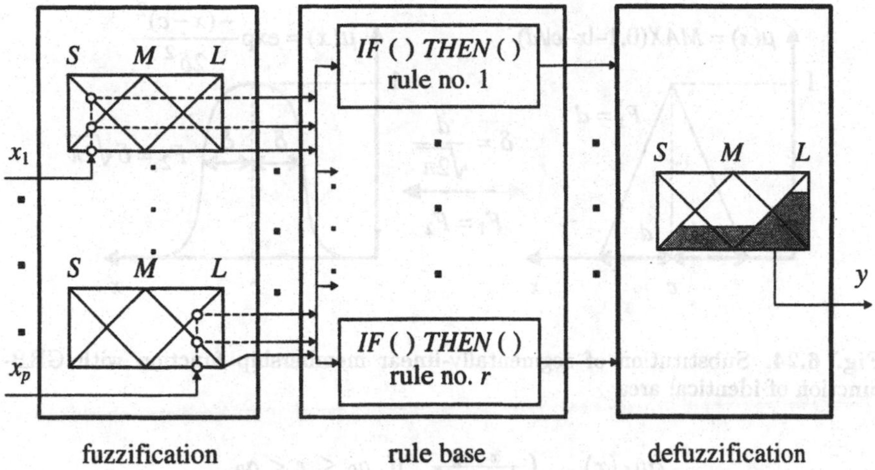

Fig. 6.22. General diagram of fuzzy model

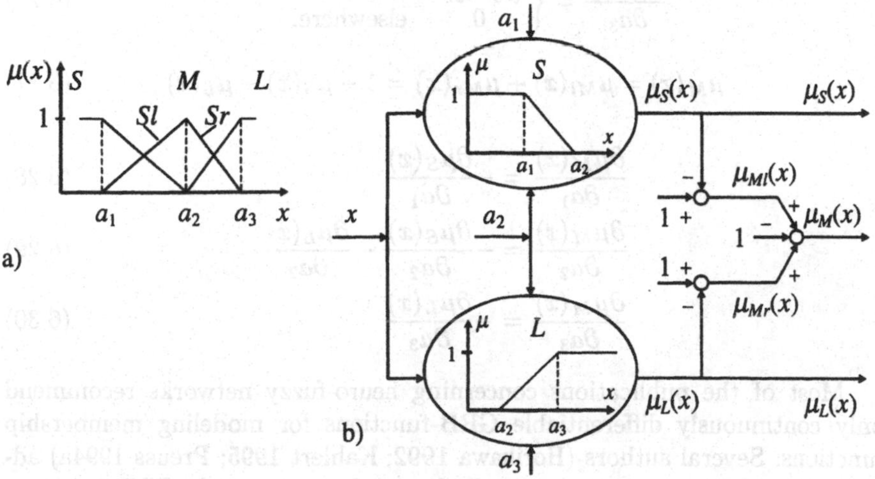

Fig. 6.23. Transformation of segmentally-linear membership functions (a) into a fragment of neural network (b)

$$\frac{\partial \mu_S(x)}{\partial a_1} = \begin{cases} \frac{a_2 - x}{(a_2 - a_1)^2} & \text{if } a_1 \le x < a_2\,, \\ 0 & \text{elsewhere.} \end{cases} \tag{6.22}$$

$$\frac{\partial \mu_S(x)}{\partial a_2} = \begin{cases} \frac{x - a_1}{(a_2 - a_1)^2} & \text{if } a_1 \le x < a_2\,, \\ 0 & \text{elsewhere.} \end{cases} \tag{6.23}$$

$$\mu_L(x) = \begin{cases} 0 & \text{if } x < a_2\,, \\ \frac{x - a_2}{a_3 - a_2} & \text{if } a_2 \le x < a_3\,, \\ 1 & \text{if } x \ge a_3\,. \end{cases} \tag{6.24}$$

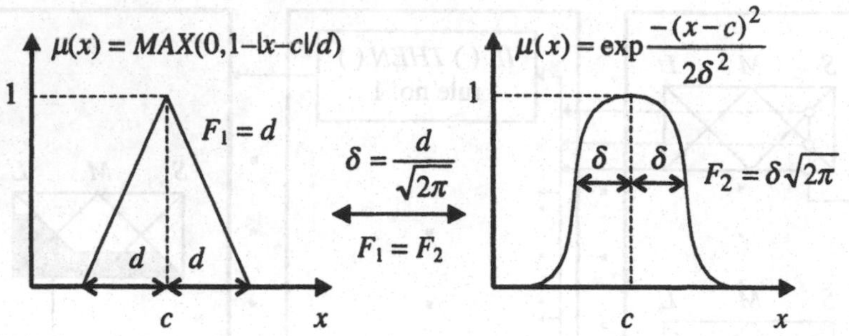

Fig. 6.24. Substitution of segmentally-linear membership function with GRB-function of identical area

$$\frac{\partial \mu_L(x)}{\partial a_2} = \begin{cases} \frac{x-a_3}{(a_3-a_2)^2} & \text{if } a_2 \le x < a_3\,, \\ 0 & \text{elsewhere.} \end{cases} \tag{6.25}$$

$$\frac{\partial \mu_L(x)}{\partial a_3} = \begin{cases} \frac{a_2-x}{(a_3-a_2)^2} & \text{if } a_2 \le x < a_3\,, \\ 0 & \text{elsewhere.} \end{cases} \tag{6.26}$$

$$\mu_M(x) = \mu_{Ml}(x) + \mu_{Mr}(x) = 1 - \mu_S(x) - \mu_L(x) \tag{6.27}$$

$$\frac{\partial \mu_M(x)}{\partial a_1} = -\frac{\partial \mu_S(x)}{\partial a_1} \tag{6.28}$$

$$\frac{\partial \mu_M(x)}{\partial a_2} = -\frac{\partial \mu_S(x)}{\partial a_2} - \frac{\partial \mu_L(x)}{\partial a_2} \tag{6.29}$$

$$\frac{\partial \mu_M(x)}{\partial a_3} = -\frac{\partial \mu_L(x)}{\partial a_3} \tag{6.30}$$

Most of the publications concerning neuro-fuzzy networks recommend only continuously differentiable GRB-functions for modeling membership functions. Several authors (Horikawa 1992; Kahlert 1995; Preuss 1994a) advise the substitution of segmentally-linear functions with GRB-functions (maintaining the equality of areas of functions being substituted, Fig. 6.24), and to build up the neuro-fuzzy networks exclusively on the basis of GRB-functions. The GRBF neuron is presented in Fig. 6.25. Calculation of derivatives $\partial\mu/\partial c$ and $\partial\mu/\partial\delta$ is very simple in this case.

The substitution of segmentally-linear membership functions with GRB-functions introduces so-called transformation error (Preuss 1995), caused by the fact that shapes and supports of both functions are different. Since the fuzzy models usually employ the asymmetrical triangular membership functions they should be substituted by asymmetrical GRB-functions of different left-side and right-side spreads. That makes the problem more complicated. Nevertheless, the researches carried out by the author of this book (Piegat 1996) as well as by Hensel, Holzmann and Pfeifer (Hensel 1995), have proved

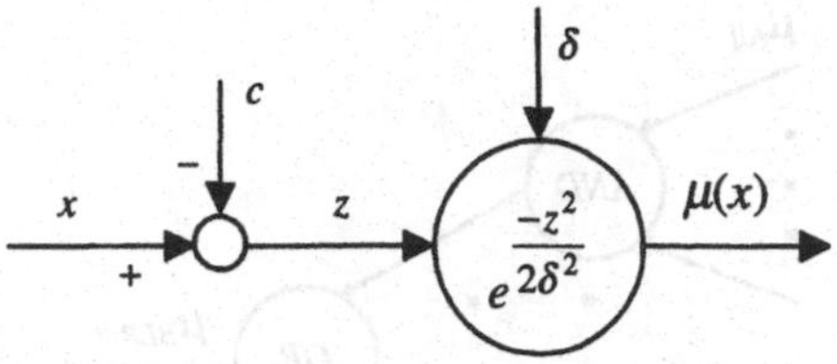

Fig. 6.25. GRBF neuron

that a neuro-fuzzy network with segmentally-linear membership functions learns as effectively as a network with continuously differentiable membership functions. The only condition restricting the application of a certain membership function in neuro-fuzzy networks is the possibility to calculate a derivative of that function and a finite value of the derivative.

Transformation of elements in a Rule Base block

The outputs of a Fuzzification block are the membership grades of particular inputs x_i to fuzzy sets A_{ij} of their linguistic universe of discourse. The outputs are simultaneously measures of satisfaction of component premises included in the part *IF* ( ) of fuzzy rules, (6.31).

$$
\begin{aligned}
&IF\ (x_1 = A_{11})\ AND\ (x_2 = A_{21})\ AND, \ldots,\ AND\ (x_p = A_{p1})\\
&OR\ (x_1 = A_{12})\ AND\ (x_2 = A_{22})\ AND, \ldots,\ AND\ (x_p = A_{p2})\\
&OR\ \ldots\\
&\vdots\\
&THEN\ (y = B_1)
\end{aligned}
\tag{6.31}
$$

The Rule Base block determines the grade of fulfillment for a complete complex premise (rule antecedent) on the basis of the fulfillment grades of component premises of segment *IF* ( ) in the rules, and on the basis of the types of logical operators (*AND*, *OR*) occurring in that segment. This grade, in turn, activates the membership function B_l appearing in the conclusion (consequent) of the rule. Usually a *MAX-MIN* inference method is used. A rule premise can be presented as a fragment of a neural network, Fig. 6.26. The operations *AND* or *OR* can also be accomplished by means of t-norms or s-norms or other operators. Each rule activates one of the sets of the output B_l. Since several rules *Rm* can activate the same output set B_l, their conclusions $\mu_{BlRm}(y)$ are combined, usually by means of a *MAX*-operator, to achieve the final form of activated membership function $\mu_{Bl}(y)$ of the fuzzy set B_l of the model output. The general neural diagram of the rule base usually has the form as shown in Fig. 6.27.

A network representing a rule base contains neurons performing logical operations. Usually, it is no problem to calculate input/output derivatives of those neurons except for the *MAX*- or *MIN*-operators existing independently

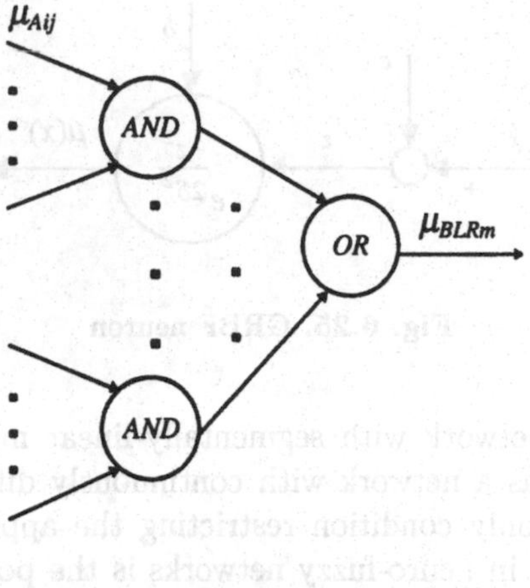

Fig. 6.26. Neural diagram of the premise of the rule *Rm* with neurons *AND* and *OR* activating the set B_l of the model output *y*

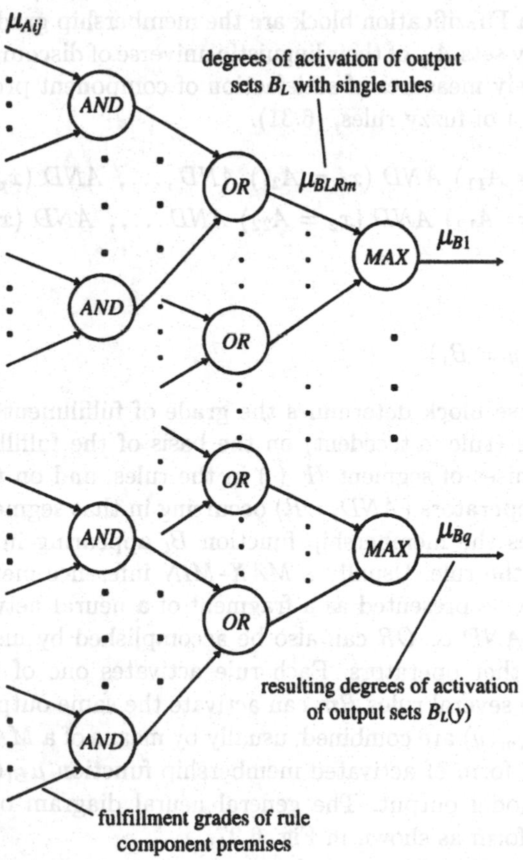

Fig. 6.27. General neural diagram of the Rule Base

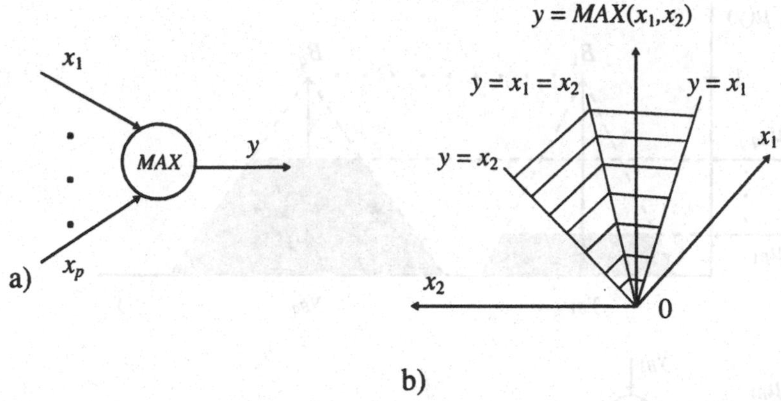

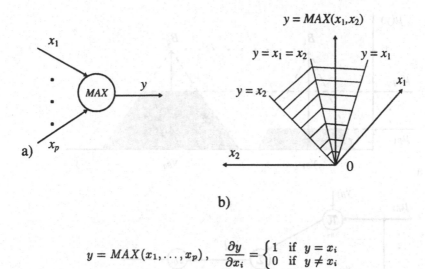

$$y = MAX(x_1, \ldots, x_p)\,, \quad \frac{\partial y}{\partial x_i} = \begin{cases} 1 & \text{if } y = x_i \\ 0 & \text{if } y \neq x_i \end{cases}$$

Fig. 6.28. *MAX*-neuron (a), the formula to calculate input/output derivatives and the surface of function $MAX(x_1, x_2)$, (b)

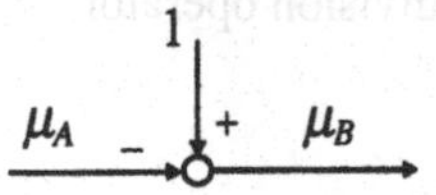

$$B = \overline{A}, \quad \mu_B = 1 - \mu_A, \quad \frac{\partial \mu_B}{\partial \mu_A} = -1$$

Fig. 6.29. Transformation of a negation operator to an element of a neuro-fuzzy network

in the networks or incorporated in other operators. Fig. 6.28 illustrates the multi-input *MAX*-operator.

The output of the *MAX*-neuron responds to the change Δx_i of input x_i with a change of Δy only when $x_i = MAX(x_1, \ldots, x_p)$. The changes in other inputs generate no response. Thus, the derivative value is equal to 1 for the above input but 0 for all other inputs, Fig. 6.28. Similarly, the derivatives of the *MIN*-neuron are calculated with formula (6.32).

$$y = MIN(x_1, \ldots, x_p)\,, \quad \frac{\partial y}{\partial x_i} = \begin{cases} 1 & \text{if } y = x_i \\ 0 & \text{if } y \neq x_i \end{cases} \tag{6.32}$$

If several outputs have the same values and simultaneously they are maximum (minimum) values then the derivative values are equal to 1 for each of the outputs. Similarly, modeling of a negation operator does not show problems; the transformation is presented in Fig. 6.29.

Transformation of Defuzzification block

The inputs to the Defuzzification block are the activation degrees $\mu_{Bl}(y)$ of individual fuzzy sets B_l of the model output. The neural network structure

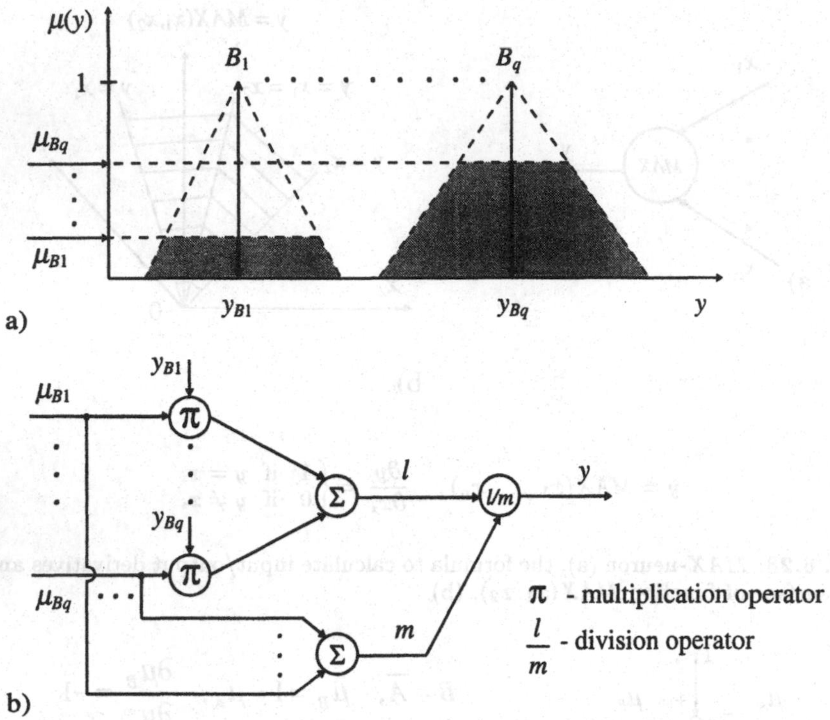

Fig. 6.30. Neural network representing a Defuzzification block

representing the block depends on the defuzzification method chosen and the level of method complexity. The easiest, very effective and frequently used defuzzification method is a singleton method where the singletons replace the output sets. Singletons are positioned either in membership function peaks or their centroids y_{bl}. In such cases the model output is calculated with formula (6.33).

$$y = \frac{\sum_{l=1}^{q} \mu_{Bl} \cdot y_{Bl}}{\sum_{l=1}^{q} \mu_{Bl}} \tag{6.33}$$

A neural network performing the singleton defuzzification is presented in Fig. 6.30.

The computation of network input/output derivatives appears to be quite simple. If the total of the membership grades y_{Bl}, occurring in the denominator of formula (6.33), always gives a sum of 1 then $m = 1$, and the lower branch of the network as well as the division operator l/m, Fig. 6.30, are

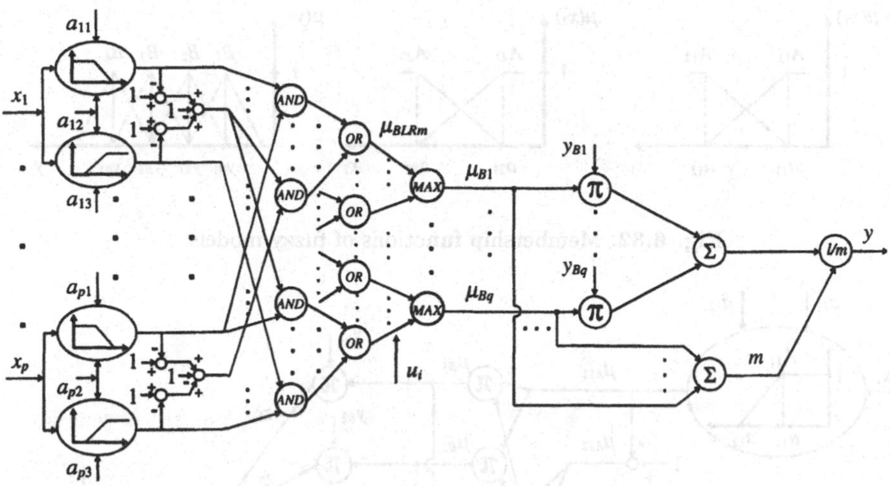

Fig. 6.31. Neuro-fuzzy network representing a fuzzy model

redundant. Finally, the neural network representing the entire fuzzy model has the structure presented in Fig. 6.31.

Let us assume we intend to test the significance of individual rules of the model. To accomplish the task we have to introduce the confidence coefficients u_i, within the range $[0, 1]$, to the network, i.e. to those network branches where the signal μ_{BlRm} (rule outputs) is transferred. If, on having trained the network with system input/output data, some of the coefficients appear to be close to 0, then the rules corresponding to those coefficients can be removed as being insignificant. The above procedure enables us to tune the relational models.

The neuro-fuzzy network diagram presented in Fig. 6.31 is of a general character. A network structure depends on a particular fuzzy model structure and particular elements. Example 6.2.1.2.1 below demonstrates a fuzzy model and its corresponding neuro-fuzzy network respectively.

Example 6.2.1.2.1

Let us consider a fuzzy model containing the rule base (6.34) and the membership functions presented in Fig. 6.32. The operation *AND* is to be performed by a *PROD*-operator. The *MAX-MIN* method is used to perform the inference while the defuzzification is done with the method of singletons located in the membership function peaks.

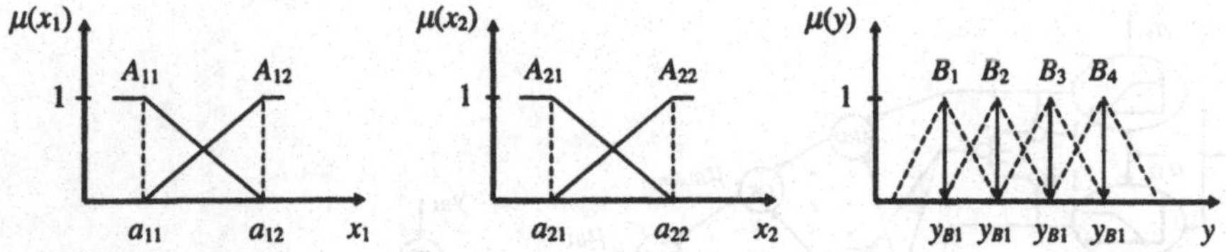

Fig. 6.32. Membership functions of fuzzy models

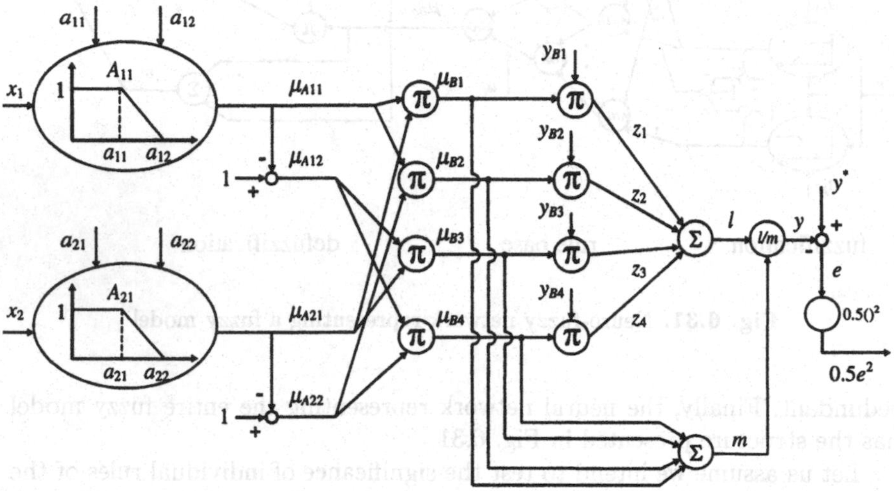

Fig. 6.33. Neuro-fuzzy network corresponding to fuzzy model (6.34) and Fig. 6.32

$$
\begin{aligned}
R1: &\ IF\ (x_1 = A_{11})\ AND\ (x_2 = B_{21})\ THEN\ (y = B_1)\\
R2: &\ IF\ (x_1 = A_{11})\ AND\ (x_2 = B_{22})\ THEN\ (y = B_2)\\
R3: &\ IF\ (x_1 = A_{12})\ AND\ (x_2 = B_{21})\ THEN\ (y = B_2)\\
R4: &\ IF\ (x_1 = A_{12})\ AND\ (x_2 = B_{22})\ THEN\ (y = B_3)
\end{aligned}
\qquad (6.34)
$$

The neuro-fuzzy network corresponding to the above model is presented in Fig. 6.33.

A derivative used to correct the parameters y_{B1} has the form expressed by (6.35).

$$
\begin{aligned}
\frac{\partial(0.5e^2)}{\partial y_{B1}} &= \frac{\partial(0.5e^2)}{\partial e}\cdot\frac{\partial e}{\partial y}\cdot\frac{\partial y}{\partial l}\cdot\frac{\partial l}{\partial z_1}\cdot\frac{\partial z_1}{\partial y_{B1}} =\\
&= e\cdot(-1)\cdot\left(\frac{1}{m}\right)\cdot(1)\cdot\mu_{B1} = \frac{-e\cdot\mu_{B1}}{m} = \frac{-e\cdot\mu_{B1}}{\mu_{B1}+\mu_{B2}+\mu_{B3}+\mu_{B4}}
\end{aligned}
\qquad (6.35)
$$

Formulas to calculate derivatives of parameters a_{ij} of neurons of the fuzzification layer are determined in a similar way. However, their mathematical form exhibits greater complexity.

◊ ◊ ◊

A neuro-fuzzy network can be trained by means of the following procedures.

1. Tuning of defuzzification layer parameters y_{Bl} exclusively, provided the parameters of membership functions of inputs a_{ij} remain constant (e.g. uniform distribution of the functions in the universe of discourse). The tuning process is usually short and the model precision is often high.
2. Simultaneous tuning of all parameters of the entire network. The tuning process can be time-consuming and difficult due to a large number of parameters and the phenomena of local minima of the error function. In the case of extremely large numbers of parameters the training of the network can appear practically unfeasible. This method allows us to obtain (potentially) the best tuning results. The initial parameters can be chosen using the genetic algorithm method (genetic-parametrical representation of a network).
3. Alternative tuning of parameters y_{Bl} of the defuzzification layer and the fuzzification layer. Since a single tuning cycle provides the tuning only for some of the model parameters, the effect of occurrences of the local error function minimum is slighter and a satisfactory set of network parameters can be acquired faster. The network training process can be supported with the genetic search method.

6.2.1.3 Transformation of a Takagi-Sugeno fuzzy model into a neuro-fuzzy network

The main difference between Mamdani models and Takagi-Sugeno models is the form of their rule conclusions. A typical rule *Ri* in a Takagi-Sugeno model can be expressed with formula (6.36).

$$\begin{aligned}
&IF\ (x_1 = A_{11})\ AND\ (x_2 = A_{21})\ AND, \ldots,\ AND\ (x_p = A_{p1}) \\
&OR\ (x_1 = A_{12})\ AND\ (x_2 = A_{22})\ AND, \ldots,\ AND\ (x_p = A_{p2}) \\
&OR\ \ldots \\
&\vdots \\
&THEN\ (y = b_{i0} + b_{i1}x_1 + \ldots + b_{ip}x_p),
\end{aligned} \tag{6.36}$$

where: i – rule number.

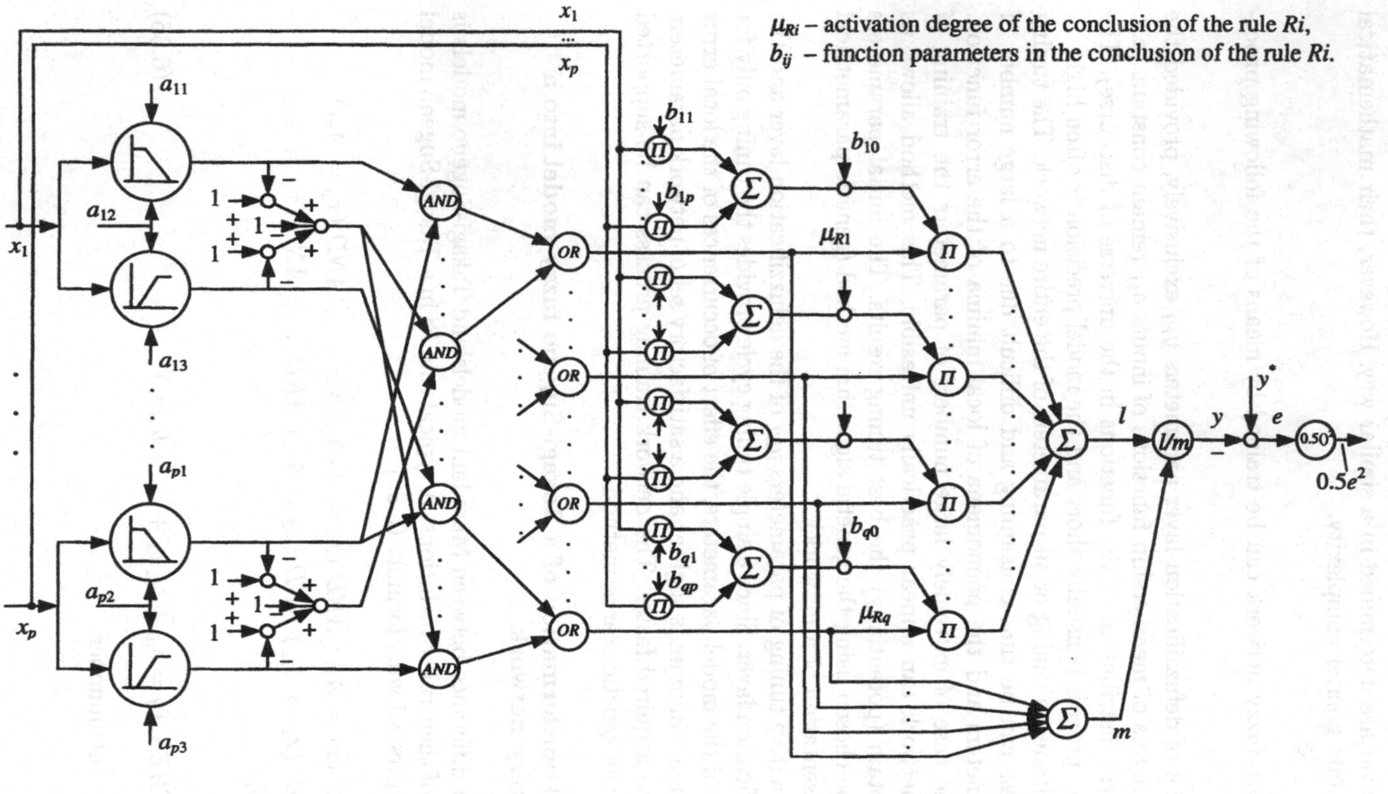

Fig. 6.34. General diagram of a neuro-fuzzy network representing a Takagi-Sugeno fuzzy model

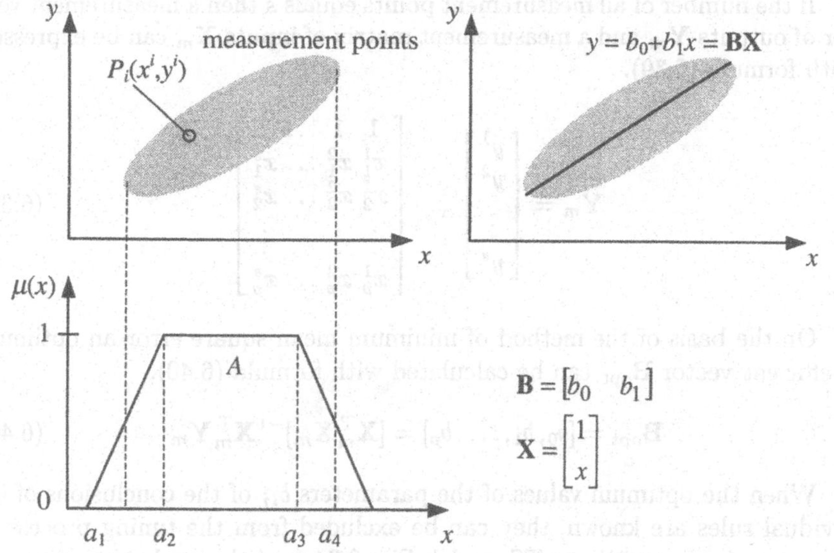

Fig. 6.35. Approximation of measurement results in a region defined with membership $\mu_A(x)$ using a local linear model

In most cases the linear functions are applied in the rule conclusions of Takagi-Sugeno models, formula (6.36). However, any arbitrary non-linear functions can be used as well. There are no fuzzy sets in rule conclusions but mathematical functions describing the model's local surface. The general diagram of a neuro-fuzzy network constituting a representation of a Takagi-Sugeno model is presented in Fig. 6.34.

A network representing the TS model can be trained using the same methods as in the case of a network representing a Mamdani model, see Section 6.2.1.2. There is, however, an opportunity to identify the parameters b_{ij} earlier in the formulas $y = f(x_1, \dots, x_p)$ of the rule conclusions described with formulas (6.36) (Cho 1995; Cipriano 1995; Park 1995; Takagi 1985).

In some cases the sector of input space over which the system surface can be approximated using a single local model $y = f(\mathbf{X})$, Fig. 6.35, is known.

If a local model is defined in the multi-dimensional space with formula (6.37):

$$y = b_0 + b_1x_1 + \ldots + b_px_p\,, \tag{6.37}$$

where: p – number of inputs, then the matrix of the model has the form (6.38):

$$y = \mathbf{BX}, \tag{6.38}$$

where: $\mathbf{B} = [b_0, b_1, \dots, b_p]$, $\mathbf{X}^{\mathrm{T}} = [1, x_1, x_2, \dots, x_p]$.

The coordinates of the measurement point P_i are equal to y^i, $\mathbf{X}^i$, where:

$$\mathbf{X}^{i\mathrm{T}} = [1, x_1^i, x_2^i, \dots, x_p^i]\,.$$

If the number of all measurement points equals s then a measurement vector of outputs $\mathbf{Y}_m$ and a measurement matrix of inputs $\mathbf{X}_m$ can be expressed with formula (6.39).

$$\mathbf{Y}_m = \begin{bmatrix} y^1 \\ y^2 \\ \vdots \\ y^s \end{bmatrix}, \quad \begin{bmatrix} 1 & 1 & \dots & 1 \\ x_1^1 & x_1^2 & \dots & x_1^s \\ x_2^1 & x_2^2 & \dots & x_2^s \\ \vdots & \vdots & & \vdots \\ x_p^1 & x_p^2 & \dots & x_p^s \end{bmatrix} \tag{6.39}$$

On the basis of the method of minimum mean square error an optimum coefficient vector $\mathbf{B}_{\text{opt}}$ can be calculated with formula (6.40).

$$\mathbf{B}_{\text{opt}} = [b_0, b_1, \dots, b_p] = [\mathbf{X}_m^{\mathrm{T}} \mathbf{X}_m]^{-1} \mathbf{X}_m^{\mathrm{T}} \mathbf{Y}_m \tag{6.40}$$

When the optimum values of the parameters b_{ij} of the conclusions of individual rules are known, they can be excluded from the tuning process of the network representing a TS model, Fig. 6.34, and then only tuning on the parameters a_{ij} of the membership function of the fuzzification layer need be accomplished. In this way, the network training process can be essentially simplified. However, the method of overall tuning applied for all model parameters is potentially more precise.

Various interesting methods of tuning and organizing the TS models can be found in (Babuška 1995c; Fiordaliso 1996; Männle 1996; Pedrycz 1997). Below an example of a TS fuzzy model and its corresponding neural network is presented.

Example 6.2.1.3.1

Let us consider a TS fuzzy model comprising the membership functions and rules presented in Fig. 6.36. Fig. 6.37 presents a neuro-fuzzy network corresponding to the model in Fig. 6.36.

◊ ◊ ◊

6.2.2 Tuning of fuzzy model parameters with the genetic algorithm method

Genetic algorithms allow us to tune the parameters of the most complicated fuzzy models; but it is a remarkably time-consuming process. Application examples can be found in (Bastian 1996; Jäkel 1997; Murata 1995). The genetic algorithm method considers each universe of discourse for each variable as a space divided into a finite number of intervals, Fig. 6.38. If a given interval incorporates a membership function peak then the interval is coded with digit 1, otherwise with digit 0. A coding string of a single variable membership function is called a chromosome (10110001011) and its elements (0 or 1) are

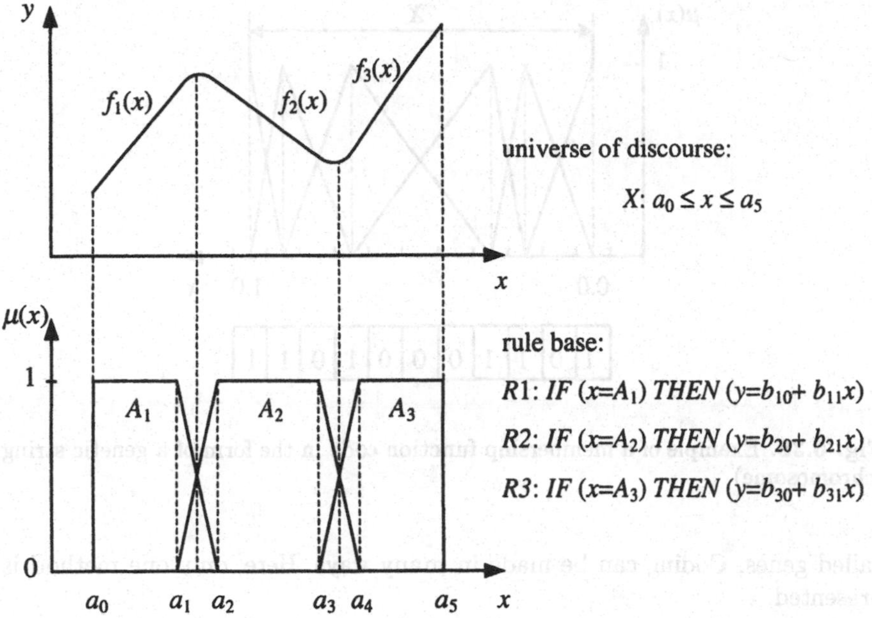

Fig. 6.36. Surface $y = f(x)$, membership functions and rules of a Takagi-Sugeno fuzzy model

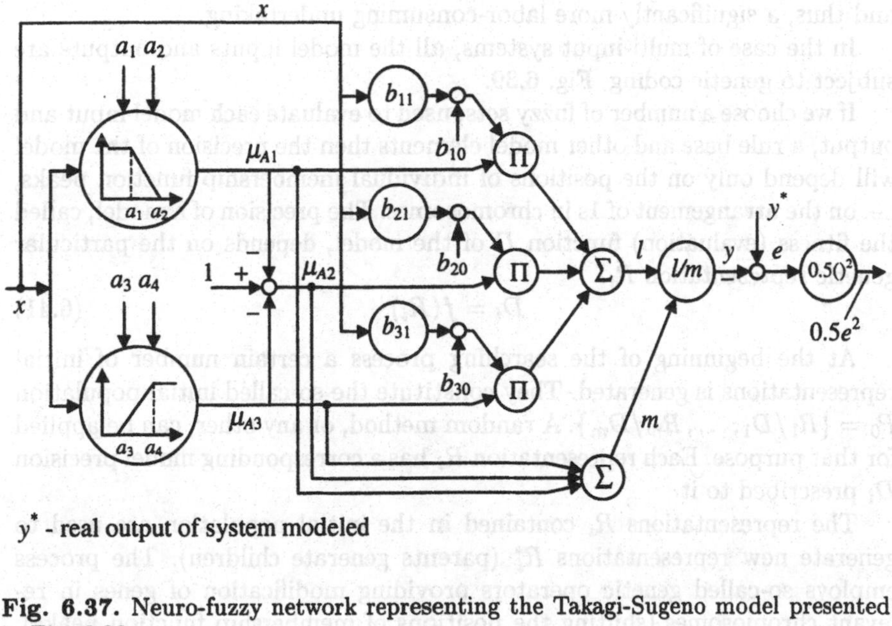

y^* - real output of system modeled

Fig. 6.37. Neuro-fuzzy network representing the Takagi-Sugeno model presented in Fig. 6.36

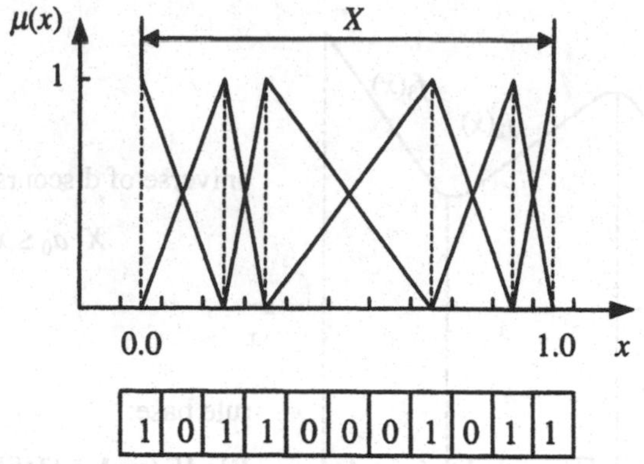

Fig. 6.38. Example of a membership function code in the form of a genetic string (chromosome)

called genes. Coding can be made in many ways. Here, only one method is presented.

An increment in the number of genes in a chromosome (number of intervals in the universe of discourse) increases the resolution of searching for an optimum solution. However, the higher the division resolution the more rapid is the growth in the number of potential solutions ("combinatorial explosion") and thus, a significantly more labor-consuming undertaking.

In the case of multi-input systems, all the model inputs and outputs are subject to genetic coding, Fig. 6.39.

If we choose a number of fuzzy sets, used to evaluate each model input and output, a rule base and other model elements then the precision of the model will depend only on the positions of individual membership function peaks, i.e. on the arrangement of 1s in chromosomes. The precision of a model, called the fitness (evaluation) function D of the model, depends on the particular genetic representation R_i.

$$D_i = f(R_i) \tag{6.41}$$

At the beginning of the searching process a certain number of initial representations is generated. They constitute the so-called initial population $P_0 = \{R_1/D_1, \ldots, R_m/D_m\}$. A random method, or any other, can be applied for that purpose. Each representation R_i has a corresponding model precision D_i prescribed to it.

The representations R_i contained in the initial population are used to generate new representations R_i^* (parents generate children). The process employs so-called genetic operators providing modification of genes in relevant chromosomes (shifting the positions of membership function peaks). Mutation and crossover constitute the basic genetic operators.

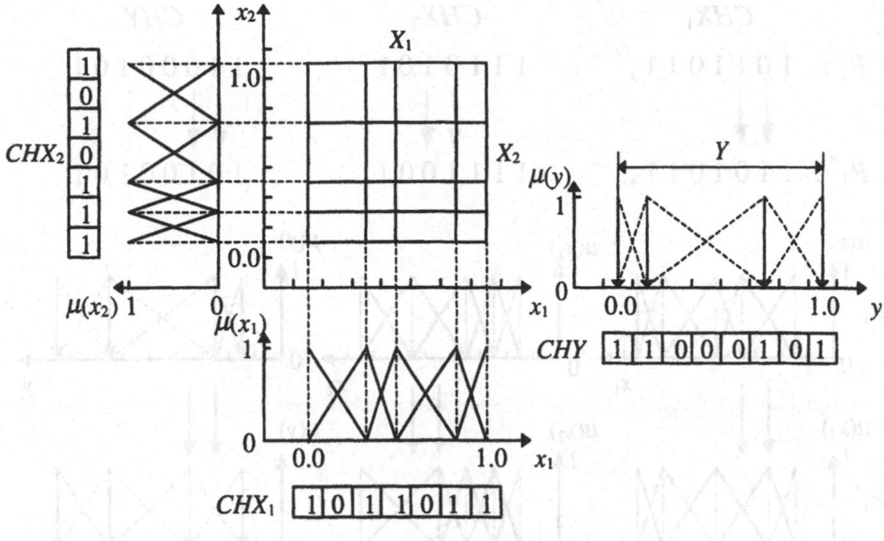

Fig. 6.39. Example of genetic representation R for a two-input fuzzy model described with a fixed number of sets for each variable and with a fixed rule base

Mutation – denotes the process of the creation of a progeny R_i^* of a given, individual representation R_i through modification of one or more of its genes ($1 \rightarrow 0$ or $0 \rightarrow 1$), Fig. 6.40. If we consider the tuning process exclusively limited to membership functions, the mutation should proceed in such a way that the number of 1-genes (fuzzy sets for a given variable) remains constant.

Crossover – denotes the process of exchange of one or more genes between two chromosomes (parents) to obtain new chromosomes (children), Fig. 6.41. When the tuning concerns the membership functions exclusively, the crossover should advance in such a way that the number of 1-genes in each type of chromosome (CHX_1, CHX_2 or CHY) remains unchanged.

Each new representation R_j^* is assigned its corresponding fitness grade D_j^* and incorporated in the initial population, giving rise to a new population P_1. The new population is reviewed and the worst representations i.e. those of lowest fitness grades (selection process) are rejected. The remaining representations are subject to genetic operations to generate new, better children. The process described above is continued until such a representation is found among the newly generated ones that its fitness satisfies the requirements imposed on model precision.

Summing up, the genetic algorithm-related operations are the following:

- coding of a problem to obtain its genetic representation R and determination of fitness function $D(R)$ of the representation,
- generation of initial population P_0,

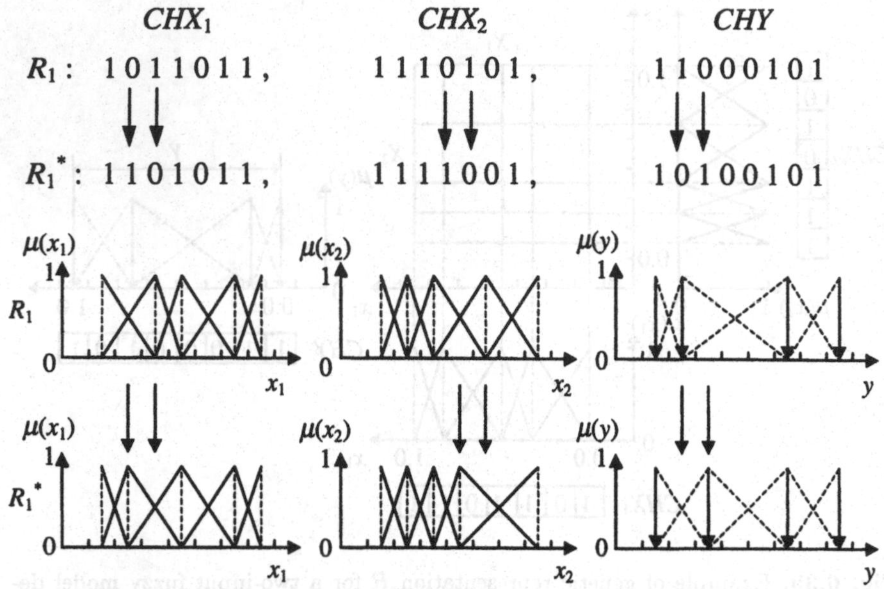

Fig. 6.40. Example of mutation operator application in generating a new representation of a fuzzy model

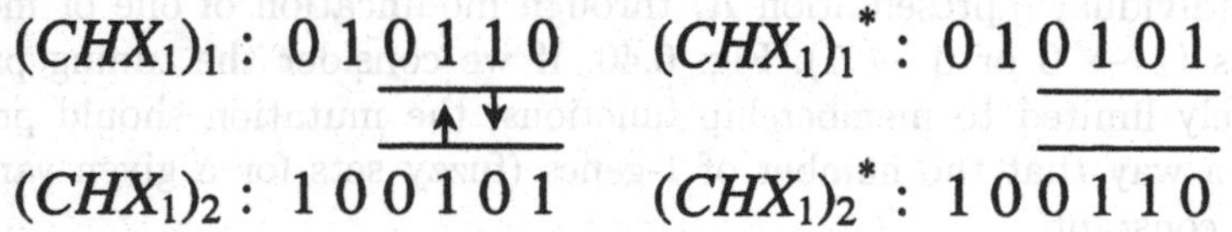

Fig. 6.41. Crossover operation: $(CHX_1)_1$ and $(CHX_1)_2$ – parents; $(CHX_1)_1^*$ and $(CHX_1)_2^*$ – children

- propagation and selection of the population acquired until such a representation R_0 is generated that it satisfies the fitness grade requirement (e.g. $D_0 \leq D_{min}$).

There are numerous methods of problem solving where genetic algorithms are employed. Nowadays genetic algorithms constitute an extensive and continuously developing science and any attempt to present a detailed description would require a separate monograph. As there are many books dealing with the subject, the author of this work would recommend them to the readers wishing to increase their knowledge (Davis 1991; Kinnebrock 1994; Kahlert 1995; Michalewicz 1996; Mitchell 1996).

The genetic algorithm method enables us to search for optimum solutions in cases of extremely difficult problems, including the problems with restrictions imposed on them. For that reason the method has become very popular and the number of its applications has been growing constantly. However, the method has a certain disadvantage, i.e. the necessity to divide a space of model variable solutions into a finite number of intervals to ensure the genetic representation contains a finite number of bits. Discretization of solution space usually disables us from acquiring as good solutions as the solutions potentially obtainable in the case of applying the methods operating in a continuous space, e.g. neuro-fuzzy networks. Yet, if the appropriate resolution of discretization is provided, the disadvantage mentioned becomes practically meaningless. Thus, the genetic algorithm method is worthy of recommendation. Moreover, an additional advantage of this method is its applicability in testing the influence of various model elements on the model precision; such elements as the types of *AND*- or *OR*-operators, inference methods, defuzzification methods, types of membership functions used, etc.

Let us assume we wish to include the type of *AND*-operator as a model variable where the solution space of *AND*-operator is a set containing seven t-norm operators:

$$\{AND\} = \{MIN, \text{drastic product, Łukasiewicz } AND\text{-operator,} \\ \text{Einstein product, Hamacher product, algebraic product,} \\ \text{Yager } AND\text{-operator}\}, \tag{6.42}$$

then, the model representation should be extended by a 7-bit chromosome:

$$CHAND = 0000010 \tag{6.43}$$

having its 1-gene at a position corresponding to the selected operator number in the set $\{AND\}$, (6.44).

$$\begin{array}{lcccc} & CHX_1 & CHX_2 & CHY & CHAND \\ R1: & 1011011, & 1110101, & 11000101, & 0000010 \end{array} \tag{6.44}$$

One should remember, however, that the extension of representation with new chromosomes enables us to find a better model but, on the other hand, requires far more computation.

6.3 Creation of self-organizing and self-tuning fuzzy models based on input/output measurement data of the system

A **self-organizing** model is to be understood as the model which defines its significant inputs itself, determines an optimum number of fuzzy sets for its inputs and output, and defines a form and number of rules. A **self-tuning** model is the model which is capable of defining optimum membership function parameters itself, to determine confidence coefficients of rules, etc.

Self-organizing fuzzy models are described in e.g. (Baldwin 1995b; Delgado 1995; Furuhashi 1995; Gonzales 1995; Gorrini 1995; Gupta 1994; Hajek 1994; Ishibuchi 1995a; Katebi 1995; Krone 1996a,c; Locher 1996a,b; Rovatti 1996; Wakabayashi 1995). The models which are simultaneously self-organizing and self-tuning are described in e.g. (Cho 1995; Cipriano 1995; Delgado 1997; Fiordaliso 1996; Fukumoto 1995; Halgamuge 1996; Hanss 1996a; Horikawa 1992; Kwon 1994; Langari 1995; Lin 1991,1996; Magdalena 1995; Meanle 1996; Murata 1995; Narazaki 1993,1995; Nobre 1995; Osowski 1996; Park 1995; Preuss 1995; Su 1995; Tan 1995; Yao 1995; Zho 1995).

Simultaneous application of both the self-organization and tuning processes usually provides higher precision of the fuzzy model. The main objective of using model self-organization methods is to obtain the minimum number of rules, fuzzy sets and parameters undergoing tuning, while maintaining satisfactory model precision. A simple model structure makes it easier to comprehend and tune the model; it also allows us to reduce the amount of measurement information necessary for tuning, (information which is taken from the system modeled but is not always easily accessible). Construction of a complicated model, and so the expansion of the number of its rules and fuzzy sets, can potentially increase model precision (in the case of a continuous function defined for a closed set in an input space (Lin 1995)). However, the difficulties in parameter tuning increase to such an extent that on having gone beyond a certain level of model complexity the tuning becomes practically unfeasible. Therefore, a tendency to simplify the model structures seems a reasonable approach.

The most beneficial method of model simplification is the removal of all meaningless inputs from the model. In such a case the number of rules decreases geometrically. The next section describes the methods of determining insignificant model inputs.

6.3.1 Determination of significant and insignificant inputs of the model

The determination of significant inputs of the model is not a trivial problem. While modeling the dynamic SISO systems, we usually consider what the order of the model should be. Depending on the model order, various signals may be considered as the model inputs, e.g.

$$y(kT-T),\ y(kT-2T),\ y(kT-3T),\ \text{etc.}$$

While modeling the static multi-input MISO systems we have to determine a set of inputs x_i "suspected" of having a significant effect on the output y. While modeling the dynamic multi-input systems, the number of significant "candidates" essentially increases:

$$x_1(kT),\dots,x_1(kT-n_1T),\dots,x_2(kT),\dots,x_2(kT-n_2T),\dots \\ \dots,y(kT),\dots,y(kT-mT)\,.$$

The problem of determining significant inputs is particularly difficult for economic systems. For example: what affects the number of unemployed in a country? Is the unemployment level affected by the unemployment in neighboring countries? What effect has the size of social support benefits? How great is the influence of the number of students graduating from universities every year? What effect has the "restrictiveness" of regulations forcing the unemployed to undertake jobs? How great is the influence of the size of a "grey" economic zone (how to measure it)? There are plenty of similar examples that could be presented. The determination of insignificant inputs in the system being modeled not only simplifies its essential model structure thus making its parameter tuning easier, but also reduces the cost incurred in acquiring measurement information. In certain cases the determination of insignificant inputs can determine whether we can model the process at all. The measurement data for certain inputs might have never been collected in the past and there is no possibility to acquire such data nor their estimates (e.g. measurement of average monthly temperature in past periods, before the invention of the thermometer, determination of amber picked up monthly by holiday makers on the Baltic coast, etc).

The problem of elimination of insignificant model inputs is often disregarded and all the "candidates" are considered as inputs. Another confirmation that the problem is underestimated is the fairly small number of publications concerning this subject. The assessment of input significance for non-linear models can be done with the following methods:

I. trial-and-error method,
II. mean fuzzy curve method (Lin 1995).

The trial-and-error method can be used as either a constructive or destructive process. For the purpose of the application of the constructive version a model with a minimum number of inputs is constructed and its precision is determined. The next step is to expand the model with a new "candidate" input and to determine the precision of the expanded model. Depending on the result obtained the new input is accepted or rejected, and so on. To apply the destructive version of the trial-and-error method the maximum number of inputs (all "candidates") is chosen in the initial model and its precision is evaluated. In the following steps the inputs are eliminated one by one and the precision of successive, simplified models is determined and compared. The trial-and-error method is fairly time-consuming and uncertain. The objectivity of the method depends on the extent (on the precision) to which the model has been tuned at a given number of inputs. The more inputs that have to be considered the more difficult the tuning process becomes. Eventually, the tuning may appear practically unfeasible. Moreover, if that method is to be used, an identical number of measurement samples for each analyzed input is necessary. In practice, however, we can have more data for some inputs but less for others. An improved, less time-consuming, version

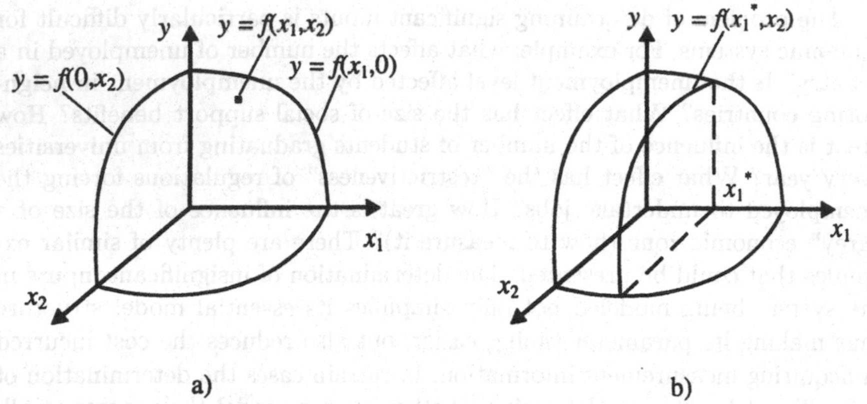

Fig. 6.42. Surface of system $y = f(x_1, x_2) = a^2 - x_1^2 - x_2^2$ (a), cross-section across surface $y = f(x_1^*, x_2)$ at the point $x_1 = x_1^*$ (b)

of the constructive method of determining significant inputs can be found in (Sugeno 1993).

Another method allowing us to evaluate the significance of model inputs, a method free from disadvantages described above, is the mean fuzzy curve method developed by Lin and Cunningham (Lin 1995). An advantage of the latter method is that it examines the influence of each input x_i on the output y separately. It provides visualization of the results on 2D curves y–x_1, y–x_2, ..., y–x_p. It also enables the comparative evaluation of significant inputs (e.g. input x_1 is more significant than input x_2) and the evaluation of the tendency of the influence of a given input x_i on the output y (if x_i increases then y increases). Moreover, the mean fuzzy curve method allows us to define the model structure and select "good" initial parameters to achieve fast tuning. The concept of the above method will be presented below.

Let us consider a continuous function: $y = f(x_1, x_2) = a^2 - x_1^2 - x_2^2$, Fig. 6.42a. If for $x_1 = x_1^*$ the function is sampled in regular intervals Δx_2 then a cross-section described with a set of discrete points $y_j(x_1^*, x_{2j})$ will be obtained, Fig. 6.43b.

The points projected on the surface x_1–y, Fig. 6.43b, are arranged unevenly, depending on the curvature of function $y = f(x_1, x_2)$. A mean value $y_{\text{mean}}(x_i^*)$ at the point $x_1 = x_1^*$ can be calculated for the entire cross-section of the function. The position of the mean value depends also on the curvature of the function surface in that cross-section. As well as calculating the mean values y_{mean} for cross-sections located on axes x_1 and x_2, we obtain the curves of mean values of the functions $y_{\text{mean1}} = f_{\text{mean}}(x_1)$ and $y_{\text{mean2}} = f_{\text{mean}}(x_2)$, Fig. 6.43c, d. That provides us with the information as to whether the increase in individual variables x_1, x_2 causes the function to increase or decrease (on average).

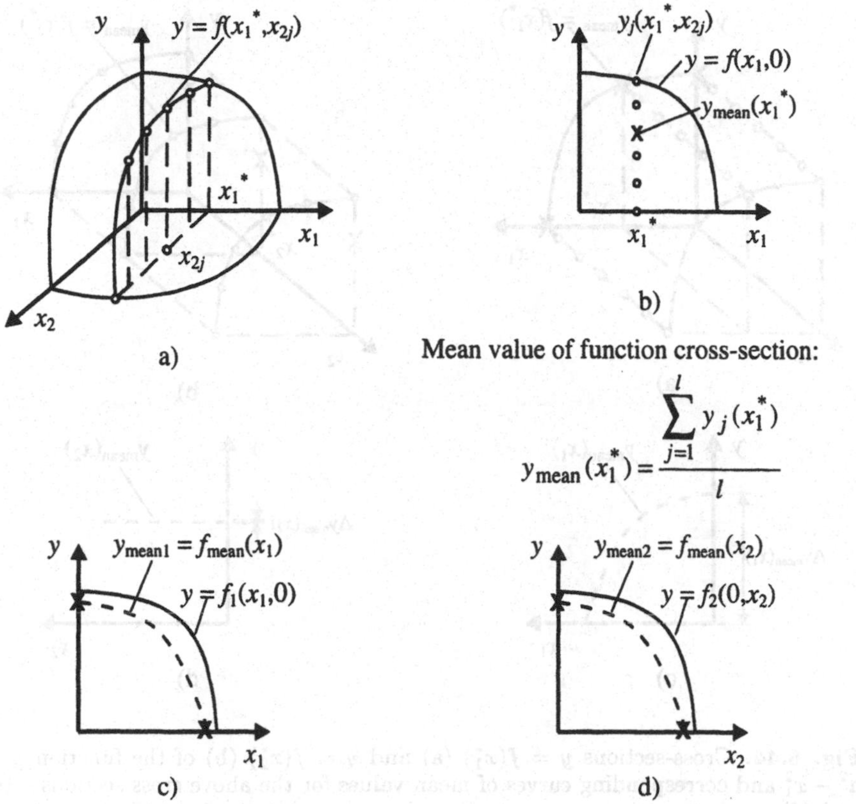

Fig. 6.43. Discrete cross-section $y_j(x_1^*, x_{2j})$ (a) and its projection on the surface x_1–y (b), curves of mean values of function cross-sections for variables x_1 and x_2 – (c) and (d)

Let us consider now that the system modeled performs the mapping $\mathbf{X} \to Y$, described with a function $y = a^2 - x_1^2$. Thus, the model output does not depend on the variable x_2. However, if the existence of such dependence has been presumed inappropriately, and the system surface has been sampled, then the result presented in Fig. 6.44 is obtained.

An analysis of Fig. 6.44 shows that if the function value y depends on a given variable x_i then, as well as changes of x_i, the function mean value also changes for the cross-sections $y_{\mathrm{mean}}(x_i)$, Fig. 6.44c. On the other hand, if the function value is independent from a given variable then the cross-section mean value of the function remains constant, Fig. 6.44d.

The above heuristic approach applied by Lin and Cunningham in their analysis of the significance of input x_i was verified in numerous experiments, the results of which can be found in (Lin 1995). **A relevantly large number of measurement samples and their possibly uniform space distri-**

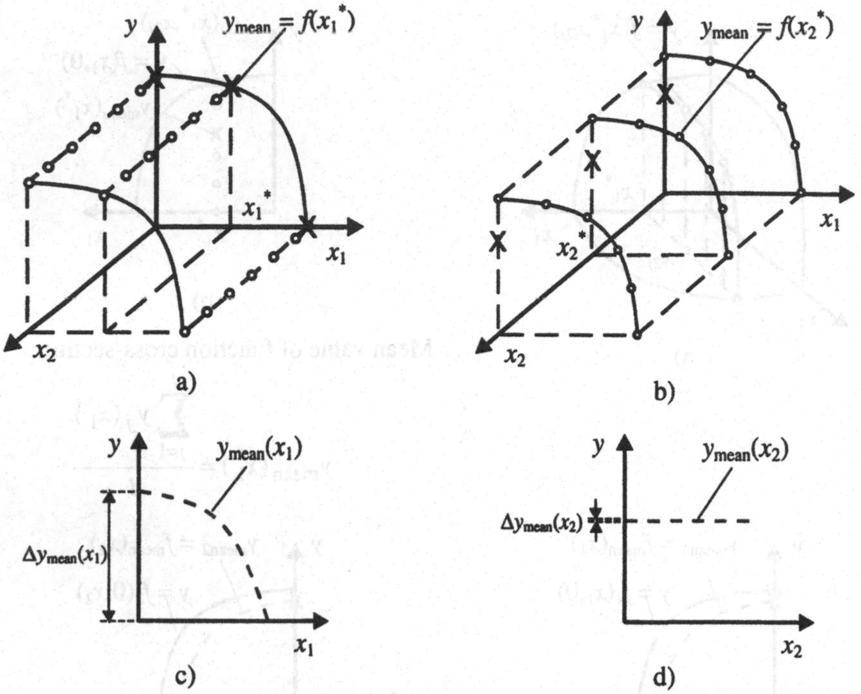

Fig. 6.44. Cross-sections $y = f(x_1^*)$ (a) and $y = f(x_2^*)$ (b) of the function $y = a^2 - x_1^2$ and corresponding curves of mean values for the above cross-sections – (c) and (d)

bution constitute an efficiency criterion of the method considered. The more remarkable the dependence of a function on a given variable, the greater, according to (Lin 1995), the range of variation of cross-section mean values $\Delta y_{\text{mean}}(x_i)$. The range extent makes a good basis for reasoning on the significance of variable x_i or comparison of various variables and determining their sequential order.

6.3.2 Determining of fuzzy curves

In the case of real systems, we usually do not know the function $y = f(x_1, \ldots, x_p)$ describing the mapping $\mathbf{X} \rightarrow Y$ performed by a given system. The measurement samples of inputs and output are often the only available information on the system. They are unevenly arranged within the space $\mathbf{X} \times Y$ and distorted with measurement noise, Fig. 6.45a.

On having projected the measurement samples on the planes $X_i \times Y$ of particular system inputs, Fig. 6.45b,c they might be perceived as a chaotic set of points. Can they constitute a basis to draw the curves of cross-section mean values $y_{\text{mean}}(x_i)$ of the mapping $\mathbf{X} \rightarrow Y$, for particular system inputs?

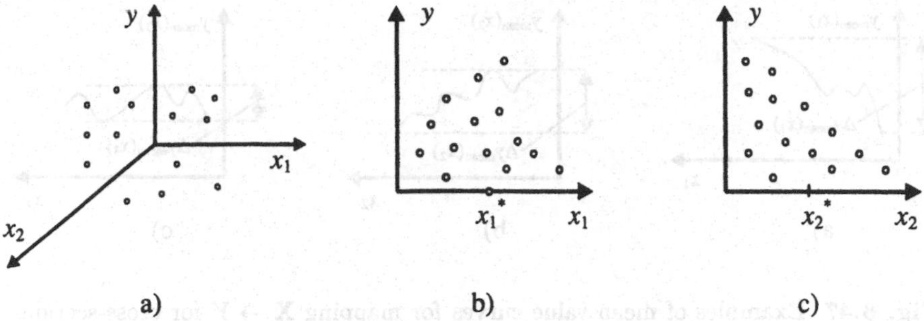

Fig. 6.45. Measurement samples of system inputs and output in space $\mathbf{X} \times Y$ (a) and sample projections on planes $X_1 \times Y$ (b) and $X_2 \times Y$ (c)

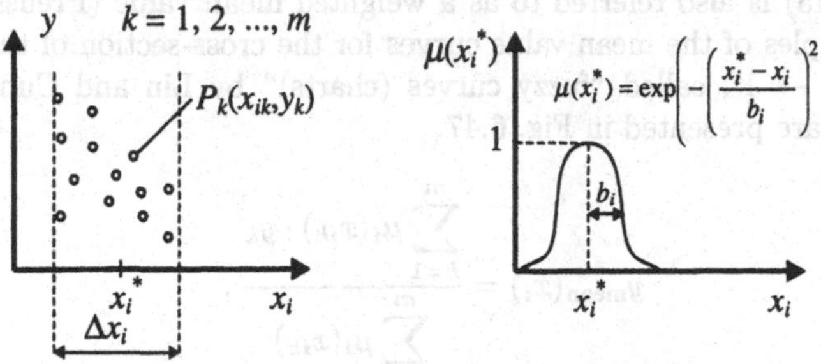

Fig. 6.46. Membership function of a fuzzy set the "closest" neighborhood of x_i^*

Let us point out again that the measurement points are usually arranged unevenly and there are many values x_i having no corresponding samples (or only a single sample, also distorted with measurement error). In such circumstances the only solution is to calculate a mean value $y_{\text{mean}}(x_i^*)$ for a given cross-section x_i^* on the basis of the measurement samples found in the closest neighborhood (it is a kind of filtering).

The term the "closest" neighborhood of x_i^* is a fuzzy term and can be defined with a membership function $\mu_i(x_i^*)$ for each input x_i, Fig. 6.46. The membership function $\mu_i(x_i^*)$ can be of any arbitrary type e.g. GRBF, triangular, trapezoid, etc. In the case of Gaussian functions, presented in Fig. 6.46, the width of the neighborhood zone for points x_i^* is determined by means of parameter b_i. According to Lin and Cunningham (Lin 1995) a recommended value b_i is equal to 20% of variability range Δx_i for a given input x_i.

It seems reasonable, however, to carry out an analysis of significant inputs for different values b_i, particularly in the case of uneven concentration of measurement points along the x_i axis. On applying the membership function of the "closest" neighborhood $\mu_i(x_i^*)$, the approximate mean value $y_{\text{mean}}(x_i^*)$ can be computed using formula (6.45) for an arbitrary input value x_i^* and its

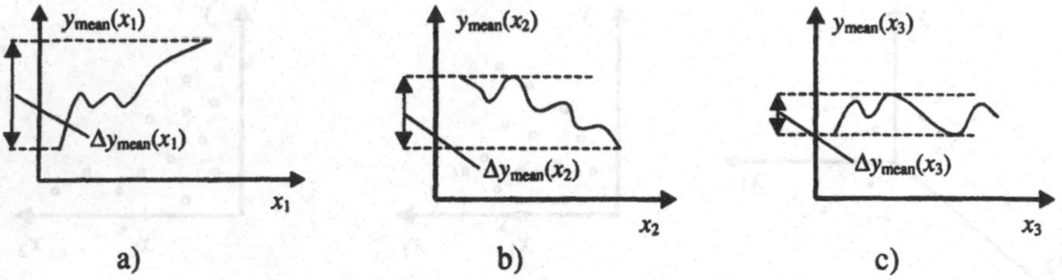

Fig. 6.47. Examples of mean value curves for mapping $\mathbf{X} \to Y$ for cross-sections along particular axes x_i

mean value curve can be outlined. In the literature, a mean value given by formula (6.45) is also referred to as a weighted mean value (Preuss 1994b). A few examples of the mean value curves for the cross-section of the system mapping $\mathbf{X} \to Y$, called "fuzzy curves (charts)" by Lin and Cunningham (Lin 1995), are presented in Fig. 6.47.

$$y_{\text{mean}}(x_i) = \frac{\sum_{k=1}^{m} \mu_i(x_{ik}) \cdot y_k}{\sum_{k=1}^{m} \mu_i(x_{ik})}, \tag{6.45}$$

where: i – input number,

k – measurement point number, Fig. 6.47.

If a mean value curve is nearly flat then a given input x_i has a slight influence on the output y. Since the analysis is based on the measurement samples distorted with errors or non-measurable inputs, then the flat mean value curves are hardly ever met in practice. Their "flatness" can be evaluated relatively, for example we can compare the variability ranges $\Delta y_{\text{mean}}(x_i)$ for individual inputs and determine the sequence of significant inputs. Then we can test only the models with the most significant inputs. It is easy to conclude that an input is insignificant if the differences $\Delta y_{\text{mean}}(x_i)$ for the inputs are clearly distinct, otherwise the decision is more difficult. If the latter is the case, the application of the trial-and-error method gives good results enabling us to reject successive inputs and analyze the model error.

Example 6.3.2.1

Let us consider a set of measurement samples of a system performing the mapping $y = x_1 + x_2^2$, Table 6.4.

The variables x_1, x_2, x_3, x_4 are considered the "candidate" inputs, where only the first two are significant inputs. The inputs x_3 and x_4 are insignificant – their values in particular samples have been generated with a random method. The task is to examine whether the method of mean fuzzy curves would provide correct determination of significant and insignificant

Table 6.4. Measurement samples of input/output of the system modeled

k	1	2	3	4	5	6	7	8	9
x_1	1	1	1	2	2	2	3	3	3
x_2	1	2	3	1	2	3	1	2	3
x_3	1	2	2	3	2	3	3	1	1
x_4	2	3	1	2	1	3	3	1	2
y	2	5	10	3	6	11	4	7	12

inputs. Fig. 6.48 illustrates the projections of measurement points on particular planes $X_i \times Y$.

Even now, merely a visual analysis of projections in Fig. 6.48 allows us to notice the significance of inputs x_1, x_2, i.e. their influence on output y of the system (an increase of x_1 and x_2 causes an increase of mean value y). The "close neighborhood" membership function for all the inputs is represented by the function (6.46).

$$\mu_i(x_i) = \exp\left[-\left(\frac{x_{ik} - x_i}{1}\right)^2\right] \tag{6.46}$$

The coefficient $b_i = 1$, describing the spread of the Gaussian function, is equivalent to 50% of the variability range of inputs x_i. The formula (6.47) has been used to calculate the mean fuzzy values for particular cross-sections,

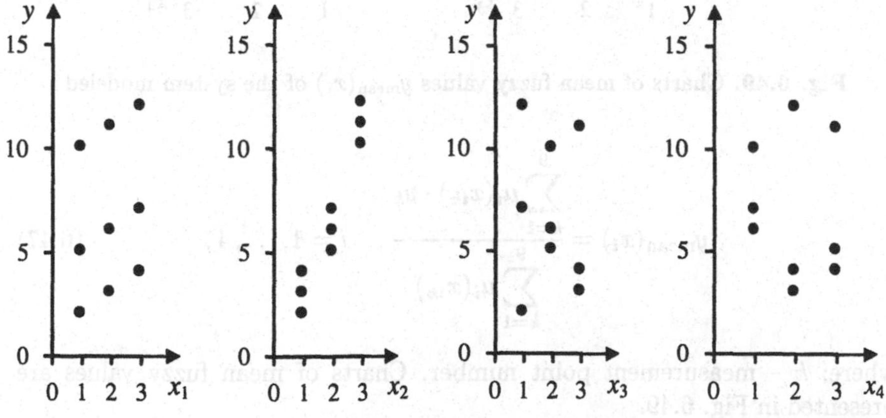

Fig. 6.48. Projections of measurement samples on planes $X_i \times Y$, $i = 1, \ldots, 4$

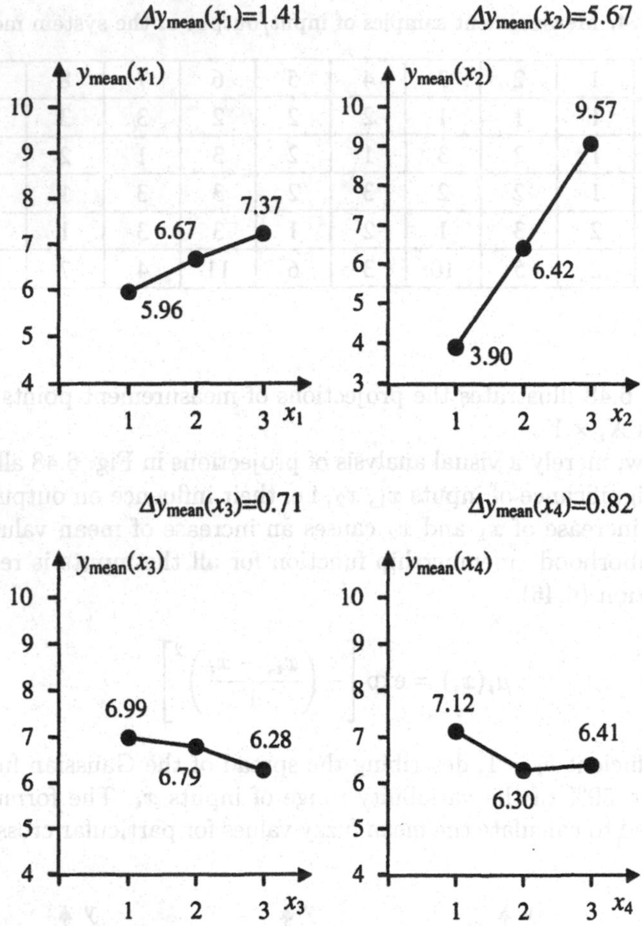

Fig. 6.49. Charts of mean fuzzy values $y_{\text{mean}}(x_i)$ of the system modeled

$$y_{\text{mean}}(x_i) = \frac{\sum_{k=1}^{9} \mu_i(x_{ik}) \cdot y_k}{\sum_{k=1}^{9} \mu_i(x_{ik})}, \quad i = 1, \ldots, 4, \tag{6.47}$$

where: k – measurement point number. Charts of mean fuzzy values are presented in Fig. 6.49.

Fig. 6.49 confirms that the input x_2 is the most significant one. The sequential order of significance (decreasing significance) for the inputs considered is the following:

$$x_2, x_1, x_4, x_3 .$$

The results of experiments have confirmed the actual state ($y = x_1 + x_2^2$). The inputs x_3, x_4 are the least significant.

◊ ◊ ◊

Research carried out by author of this book has shown that the method proposed by Lin and Cunnigham in (Lin 1995), which determines input significance on the basis of difference $\Delta y_{\text{mean}}(x_i)$ is an approximate method and is suitable for weakly non-linear systems with monotonic surfaces of input/output mapping. Significance evaluation of inputs on the basis of the difference $\Delta y_{\text{mean}}(x_i)$ is unreliable when the mapping $\mathbf{X} \rightarrow Y$ surface is strongly non-linear and wavy. Then, fuzzy curves are more wavy, Fig. 6.50a, than in the case of weakly non-linear or linear systems, Fig. 6.50b.

Fuzzy curve $y_{\text{mean}}(x_i)$ from Fig. 6.50a has the same difference $\Delta y_{\text{mean}}(x_i)$ as the curve from Fig. 6.50b. However, both fuzzy curves differ in their degree of waviness. Greater waviness of a fuzzy curve means grater dependence of the system output y on the given input x_i. In the example depicted in Fig. 6.50 the input x_i is more significant than the input x_j. A significance index, which takes into account waviness of a fuzzy curve, is index S_α. Its idea is explained in Fig. 6.51.

The wavier the fuzzy curve is, the more the system output is dependent on a particular input x_i. Next, greater waviness of a fuzzy curve means its greater length L_i. If a fuzzy curve were straightened as in Fig. 6.51, then at its greater length L_i we get a greater angle α_i. The angle, being the index of an average significance of input x_i, can be calculated from formula (6.48).

$$\alpha_i = \arccos \frac{x_{i\max} - x_{i\min}}{L_i}, \tag{6.48}$$

where:

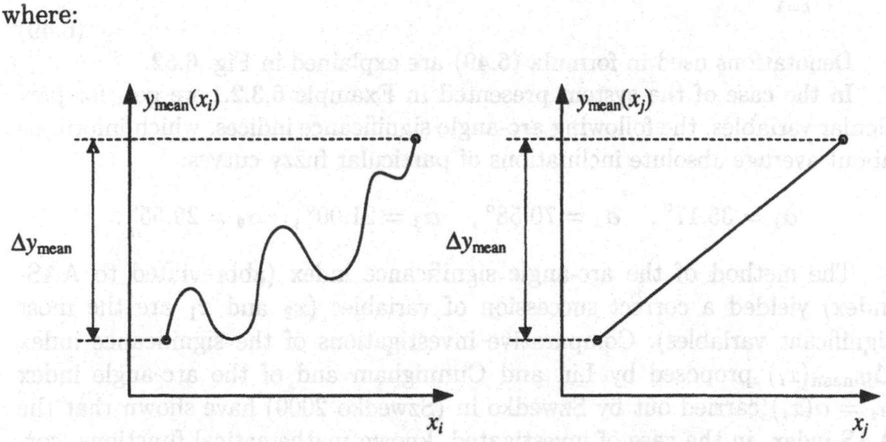

Fig. 6.50. An example of a wavy fuzzy curve (a) and of a non-wavy fuzzy curve (b)

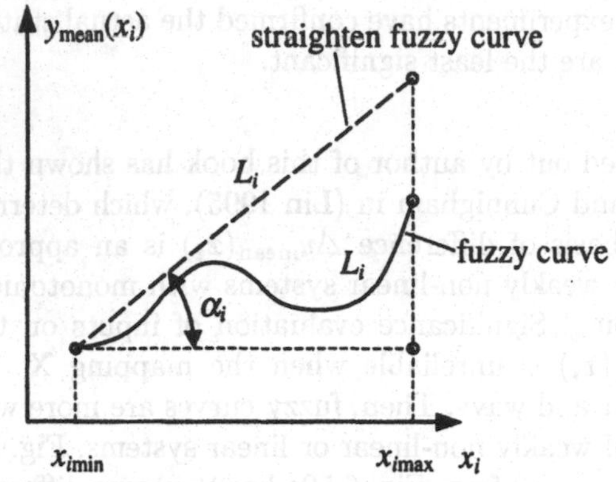

Fig. 6.51. Illustration of the idea of the arc-angle significance index S_α of input variable x_i (L_i – length of a fuzzy curve)

$$L_i = \int_{x_{imin}}^{x_{imax}} \sqrt{d^2 y_{mean} + d^2 x_i} = \int_{x_{imin}}^{x_{imax}} \sqrt{\left(\frac{dy_{mean}}{dx_i}\right)^2 + 1}\, dx_i \, .$$

In practical problems, the length L_i must be calculated numerically according to approximate formula (6.49), since we often don't know the exact mathematical expression describing the fuzzy curve, which is only given by a set of discrete points.

$$L_i = \sum_{k=1}^{n} \sqrt{[y_{mean}(x_{imin} + k\Delta x_i) - y_{mean}(x_{imin} + (k-1)\Delta x_i)]^2 + (\Delta x_i)^2} \tag{6.49}$$

Denotations used in formula (6.49) are explained in Fig. 6.52.

In the case of the system presented in Example 6.3.2.1 we get, for particular variables, the following arc-angle significance indices, which inform us about average absolute inclinations of particular fuzzy curves:

$$\alpha_1 = 35.17^\circ, \quad \alpha_2 = 70.58^\circ, \quad \alpha_3 = 21.00^\circ, \quad \alpha_4 = 29.55^\circ .$$

The method of the arc-angle significance index (abbreviated to AAS-index) yielded a correct succession of variables (x_2 and x_1 are the most significant variables). Comparative investigations of the significance index $\Delta y_{mean}(x_i)$ proposed by Lin and Cunnigham and of the arc-angle index $\alpha_i = \alpha(x_i)$ carried out by Szwedko in (Szwedko 2000) have shown that the AAS-index, in the case of investigated, known mathematical functions, correctly evaluated the significance of variables – on the average better than the difference index $\Delta y_{mean}(x_i)$.

The problem of the significance evaluation of system input variables requires further investigation. Significance indices discussed in this section are

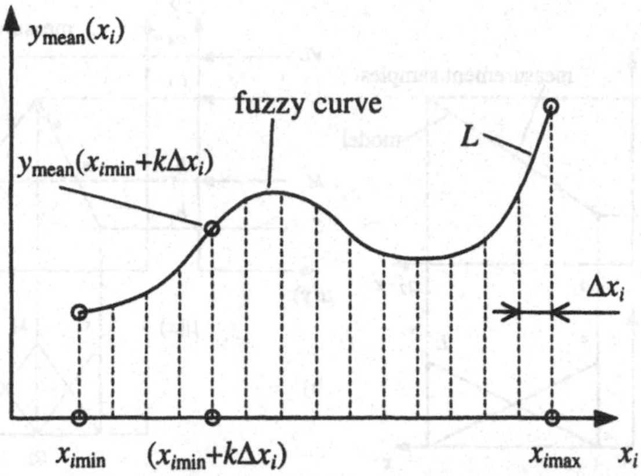

Fig. 6.52. Denotations used in the numerical calculation of length L_i of a fuzzy curve

only approximate evaluations. An interested reader can find new and more advanced methods based on fuzzy curves and fuzzy surfaces in (Lin 1998). The author of this book, with co-workers, is also developing methods based on the length of the fuzzy curve. Results of these investigations will be published in 2001.

6.3.3 Self-organization and self-tuning tuning of fuzzy model parameters

The fundamental elements of a fuzzy model structure are the rule base and the number of fuzzy sets assigned to particular inputs and the output of the model. It is assumed that the significant inputs of the model are known. In the literature we can often find methods determining a fuzzy model structure disregarding the distribution of measurement data representing a system being modeled. The model parameters are also often optimized disregarding structure optimization. The reference publications are numerous (Baldwin 1995b; Delgado 1995,1997; Furuhashi 1995; Gonzales 1995; Gorrini 1995; Gupta 1994; Hajek 1994; Ishibuchi 1995a,b; Katebi 1995; Krone 1996a; Locher 1996a,b; Rovatti 1996; Wakabayashi 1995).

The best results in modeling are achieved where the model structure optimization is synchronized with model parameter optimization. The above conclusion is explained with examples presented in Fig. 6.53. A lot of scientific publications consider the above problem, e.g. (Cho 1995; Cipriano 1995; Delgado 1997; Fiordaliso 1996; Fukumoto 1995; Halgamuge 1996; Hanss 1996a; Horikawa 1992; Kwon 1994; Langari 1995; Lin 1991,1996; Magdalena 1995; Mannle 1996; Murata 1995; Narazaki 1993,1995; Nobre 1995; Osowski 1996; Park 1995; Pedrycz 1997; Preuss 1995; Tan 1995; Yao 1995; Zhou 1995).

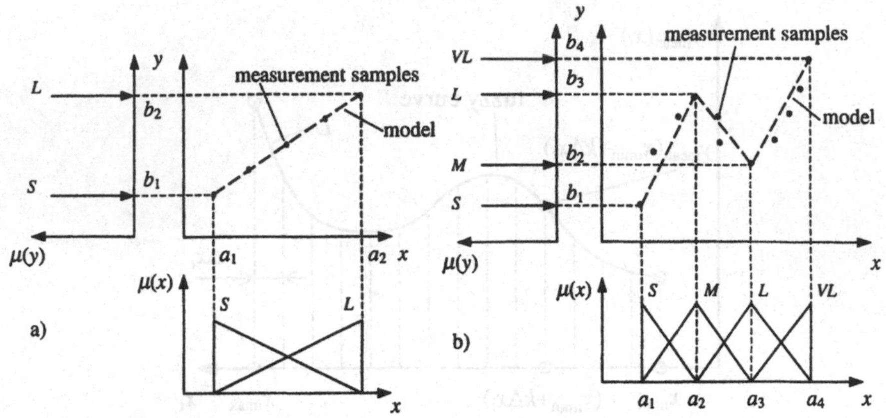

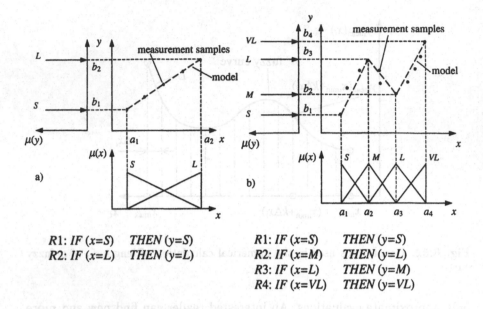

R1: IF (x=S) THEN (y=S)
R2: IF (x=L) THEN (y=L)

R1: IF (x=S) THEN (y=S)
R2: IF (x=M) THEN (y=L)
R3: IF (x=L) THEN (y=M)
R4: IF (x=VL) THEN (y=VL)

Fig. 6.53. Illustration of the relationship between model structure and membership function parameters and 2D distribution of measurement data of input and output of system modeled

If measurement data distribution indicates the linearity of the system modeled, Fig. 6.53a, then 2 rules are sufficient to construct a precise fuzzy model. If the measurement data distribution is non-linear, Fig. 6.53b, then more rules are required. It can also be seen that the position of internal membership functions of the sets M and L is not a random position but corresponds to the points (a_2, a_3, b_2, b_3) describing local extrema of the function $y = f(x)$, represented by the measurement samples. The number of rules and the parameters of membership functions depend on the input/output mapping performed by the system (plant) modeled. The number of rules and the parameters can not be determined in advance regardless of the type of system concerned. Moreover, they should not be determined separately one from another. The text below presents 3 basic methods (or groups of methods) of the self-organization and tuning of fuzzy models:

I. method of "important" points of system surface,
II. clustering methods,
III. searching methods.

Before the process of fuzzy model creation starts, one should be aware of the existence of a strong relationship between the quality of measurement samples and the quality and reliability of the resulting model. Fig. 6.54 presents a few cases of measurement sample arrangements in the space of inputs where the systems contain 2 inputs and 1 output each. The most

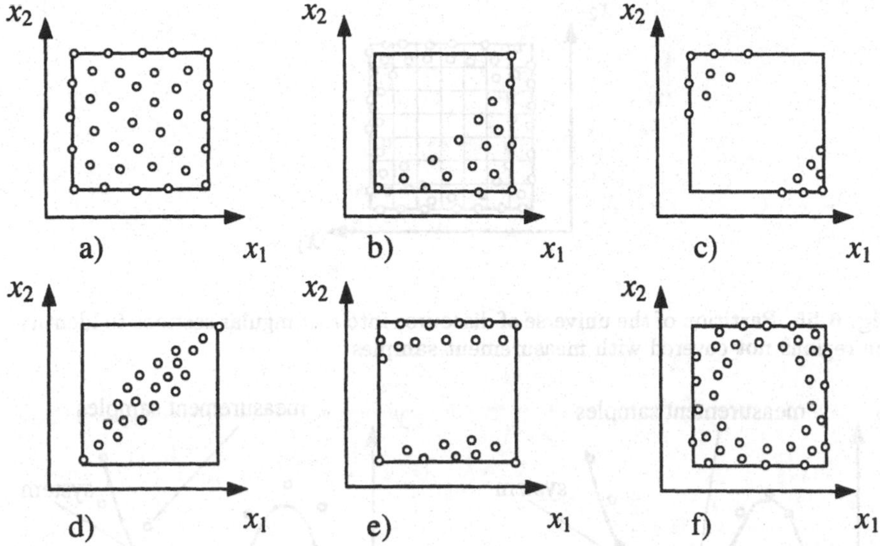

Fig. 6.54. Various distributions of measurement samples in the input universe of discourse $X_1 \times X_2$ of modeled system

beneficial distribution of samples is the uniform distribution presented in Fig. 6.54a. It provides a reliable surface approximation of the system mapping $\mathbf{X} \rightarrow Y$ for each region of the universe of discourse. The other cases either do not allow us to formulate any rules for the regions not covered with samples, or the model would be not reliable enough for the regions mentioned.

Testing of the model obtained, analogous to the modeling phase, should be carried out on data distributed in the entire space of inputs. The question is, however, what to do if data covering only a part of the universe of discourse are available. In such a case the regions of space not covered with measurement samples have to be identified and, if their size is essentially large, excluded from the modeling process. The "empty" regions do not require rules (defining model output states) to be specified. A future user of the model should be informed about the regions of input space where the model is reliable and effective for application. The regions not covered with measurement data can be detected, for example, by partitioning the universe of discourse into (hyper-) rectangular sections, counting the number of samples in each section, and then by searching for larger groups of empty sections adjacent one to another.

Another important factor affecting the model's precision is the noise or disturbance level (non-measurable inputs) distorting the data with measurement errors, Fig. 6.56. In the case of high level noise distorting the measurement samples of the system modeled, see Fig. 6.56b, a high precision of the model should not be expected. We often have to deal with such situations

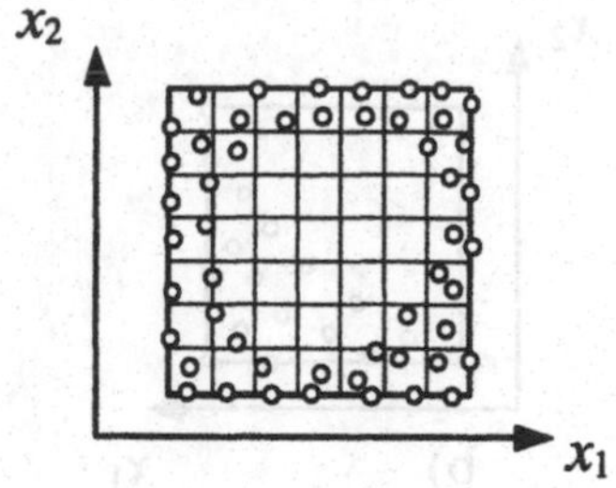

Fig. 6.55. Partition of the universe of discourse into rectangular sections to identify the regions not covered with measurement samples

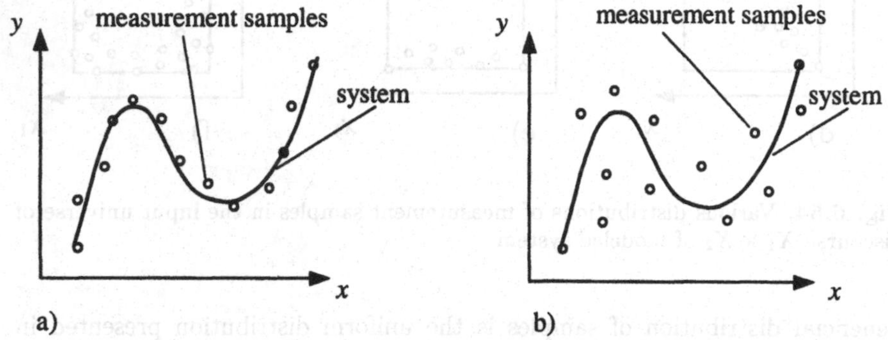

Fig. 6.56. Measurement data affected by low (a) and high (b) noise

in economic systems, where the figures coming from companies, concerning their production, profits, etc. can be significantly lower than they actually are.

6.3.3.1 Self-organization and tuning of fuzzy models with the geometric method of the maximum absolute error

Publications considering organization of fuzzy models often present an approach where the fuzzy model rules are recommended to be located in "important" points of the surface of system mapping $\mathbf{X} \rightarrow Y$, e.g. (Babuška 1995b; Lin 1995). The important points are usually the extrema of the system surface, Fig. 6.57.

In the case of multi-dimensional MISO systems, their extrema are the tops of "peaks" and bottoms of "troughs". If the system surface could be described with a continuous mathematical function $y = f(x_1, \ldots, x_p)$ then there would be no problem in finding the extreme surface points. However, while modeling the real systems, the available surface representation usually consists of point-like, discrete and noise-distorted measurement samples. In such situations, it is difficult to identify the surface "peaks" or "troughs".

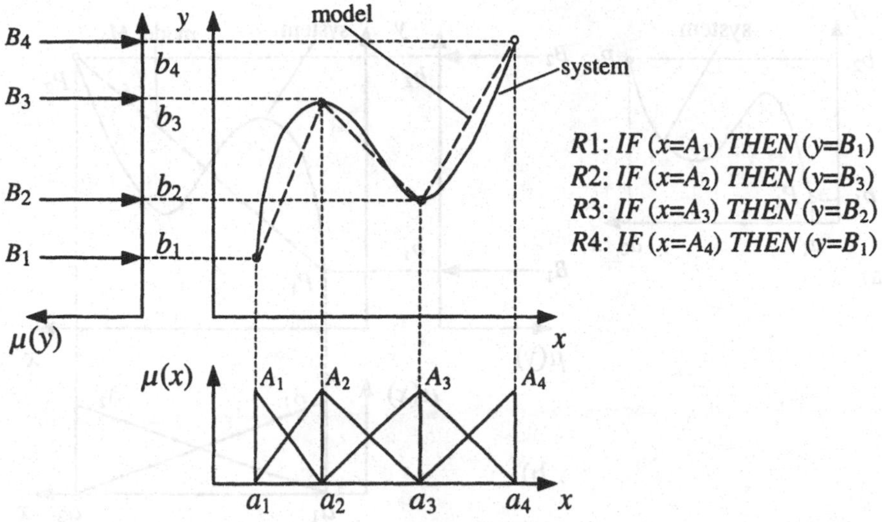

Fig. 6.57. Fuzzy model created by defining rules in system surface extrema

Higgins and Goodman in (Higgins 1994) proposed a certain method of structure organization and fuzzy model tuning. According to their method, the rules are located in "important" points of the system surface. However, the important points are here not the system surface extrema, but points of the maximum absolute error of the model in respect to the system.

Although a part of the rules located with the above method is found near to the local extrema, still the other part of the rules appears in other regions of the surface. The method described above represents a so called, "geometrical" approach to the problem of modeling (Babuška 1995), different from the frequency-related approach used in clustering methods. The method developed by Higgins and Goodman, enabling successive improvement of a single global model, leads to extensive growth in rule quantity and is affected with the "curse of dimensionality" phenomenon. Later in this section a method free from the above mentioned disadvantage is presented.

The method of maximum absolute error points, hereinafter referred to as the **MAEP method**, is an easily understandable and convincing method. However, **it can be applied mainly for modeling the systems represented by noise-free or slightly noise-affected measurement data.** In the case where the available data is highly distorted with noise, the data should be filtered, for example, with the method of "close neighborhood" membership functions presented in Section 6.3.2. First, a brief description of the MAEP method concept is presented below.

Let us consider a system surface represented by measurement samples. The continuous line in Fig. 6.58a represents the surface.

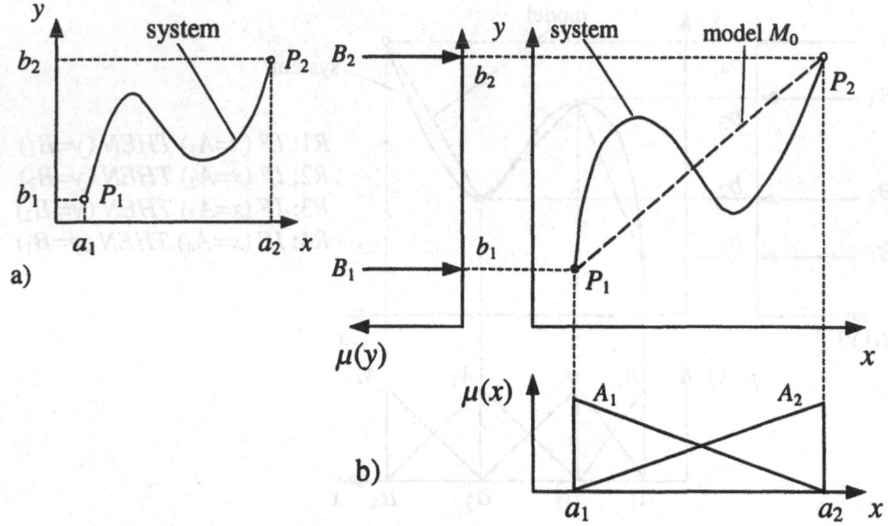

$R1_{M0}$: *IF* (x=A_1) *THEN* (y=B_1), $R2_{M0}$: *IF* (x=A_2) *THEN* (y=B_2).

Fig. 6.58. System surface (a) and base fuzzy model M_0 (b)

In the first step of the method the measurement points P_1 and P_2 are identified. The points represent a maximum and minimum value of input x (of the universe of discourse). There are two first rules $R1$ and $R2$ located at those points, which, at the same time, define the parameters a_i, b_i of the membership function (tuning of parameters).

The base model is the most approximate generalization of the system modeled. The rules of the base model describe the general tendencies encountered in the system. For example, if the rules presented in Fig. 6.58 are expressed with formula (6.50), it may be deduced that when the input x increases, the output y (on average) increases too.

$$\begin{aligned} R1 &: \textit{IF } (x \text{ is close to } 1) \textit{ THEN } (y \text{ is close to } 5) \\ R2 &: \textit{IF } (x \text{ is close to } 10) \textit{ THEN } (y \text{ is close to } 50) \end{aligned} \qquad (6.50)$$

Different types of membership functions can be applied in a base model. It sometimes allows us to improve the model's precision. However, it is difficult to predict what functions would be the most beneficial without computational experiments. The base model M_0 can be optimized. For that purpose the model can be converted into a neuro-fuzzy network and tuned on the basis of measurement samples of the system modeled. The reasoning for optimization is presented in Fig. 6.59.

The surface S of the system mapping $X \to Y$ can be symbolically expressed with formula (6.51).

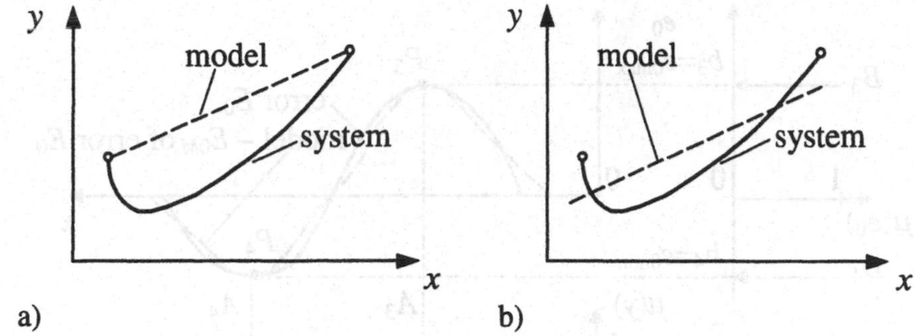

Fig. 6.59. Improvement of base model precision: (a) – input model, (b) – model after optimization

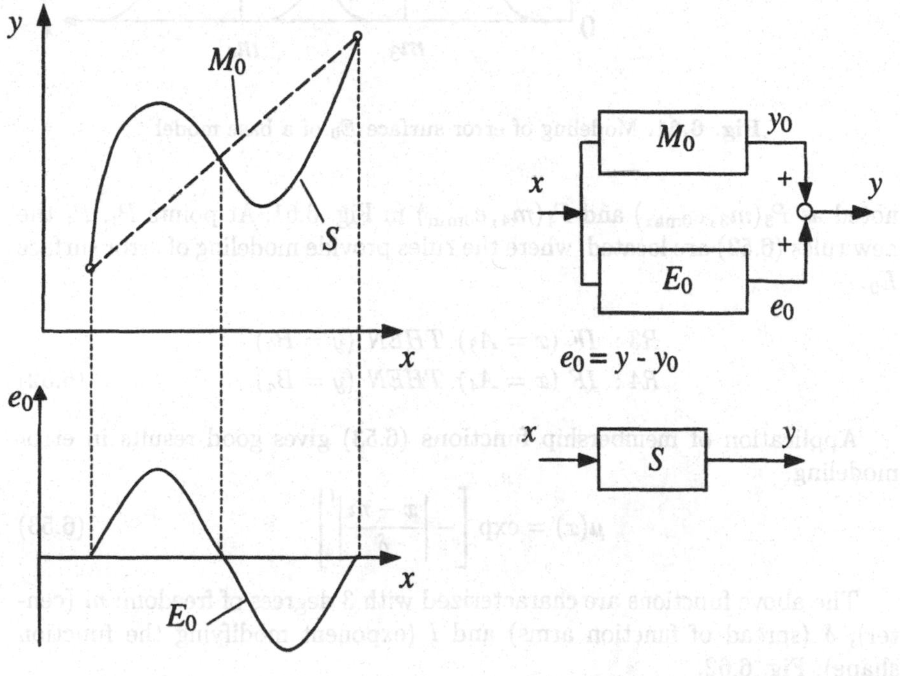

Fig. 6.60. Partition of system surface S into a base model surface M_0 and error surface E_0

$$S = M_0 + E_0 \, , \tag{6.51}$$

where: E_0 – error surface of the base model M_0 (in respect to the system surface S). The interpretation of formula (6.51) is explained in Fig. 6.60.

Now, the measurement points for which the error value e_0 reaches maximum ($e_{0\max}$) or minimum ($e_{0\max}$) should be identified. The points are de-

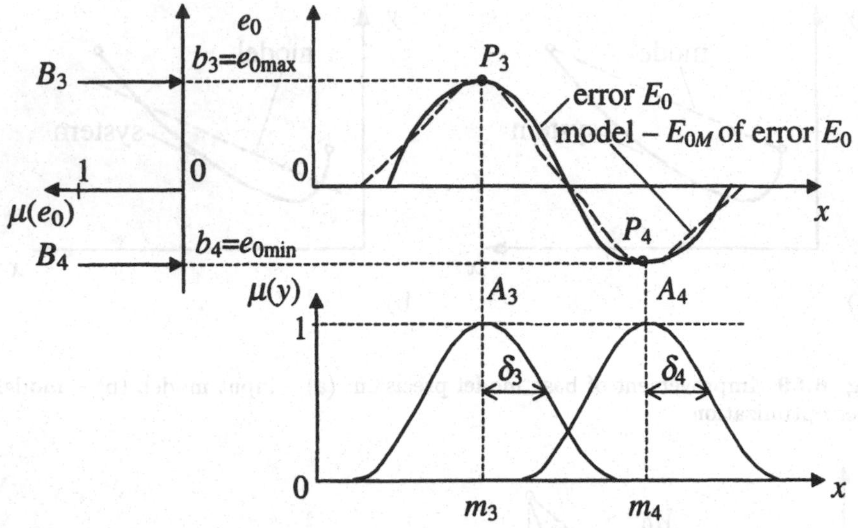

Fig. 6.61. Modeling of error surface E_0 of a base model

noted as $P_3(m_3, e_{0max})$ and $P_4(m_4, e_{0min})$ in Fig. 6.61. At points P_3, P_4 the new rules (6.52) are located, where the rules provide modeling of error surface E_0.

$$\begin{aligned} &R3: \; IF \; (x = A_3) \; THEN \; (y = B_3) \\ &R4: \; IF \; (x = A_4) \; THEN \; (y = B_4) \end{aligned} \tag{6.52}$$

Application of membership functions (6.53) gives good results in error modeling.

$$\mu(x) = \exp\left[-\left|\frac{x-m}{\delta}\right|^l\right] \tag{6.53}$$

The above functions are characterized with 3 degrees of freedom: m (center), δ (spread of function arms) and l (exponent modifying the function shape), Fig. 6.62.

Since the function (6.53) can vary in shape and size, it can be fitted quite accurately (on having been multiplied by the conclusion value e_{max} or e_{min}) to a local "peak" or "trough" of the error surface, Fig. 6.61. To achieve the above, the error model E_0 should be converted into a neuro-fuzzy network and the parameters m, δ, l of the input membership function and b_3, b_4 of the output membership function should be tuned on the basis of error samples e_0. The coordinates of maximum and minimum error points P_3 and P_4, Fig. 6.61, should be used as initial values for the parameters being tuned. The initial values of parameter δ, describing the spread of function arms, should be small at the beginning (a fraction of the distance between the centers m_3 and m_4).

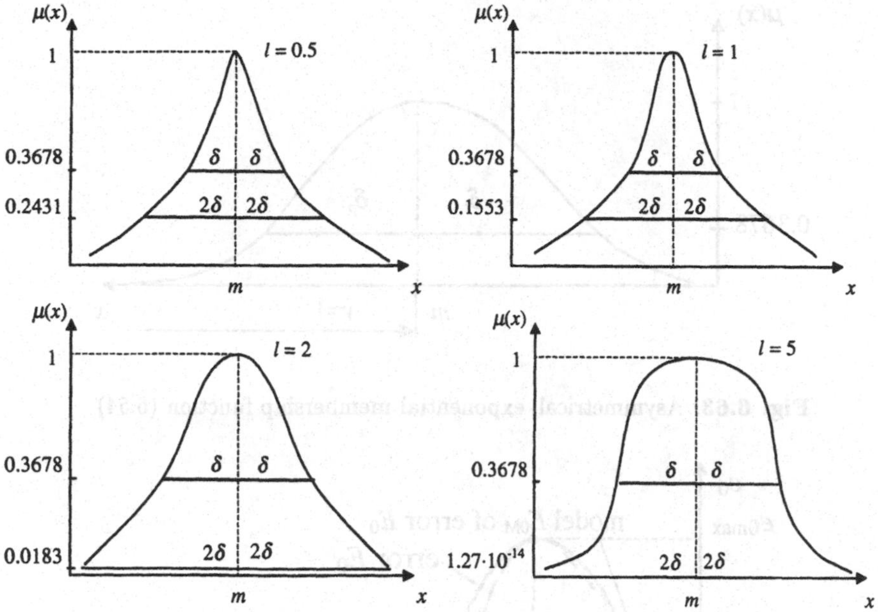

Fig. 6.62. Illustration of impact of parameters of function $\mu(x) = \exp\left[-\left|\frac{x-m}{\delta}\right|^{l}\right]$ on the shape and position of the function

During the tuning process the arms of the function are opened out and the membership function shape is fitted to the shape of the "peak" or "trough" of the error surface. Since the function curve sections can be asymmetrical it is advantageous to use the asymmetrical, exponential membership functions (6.54), Fig. 6.63.

$$\mu(x) = \exp\left[-\left|\frac{x-m}{v\delta_p + (1-v)\delta_l}\right|^{l}\right], \tag{6.54}$$

where:

$$v = \begin{cases} 1 & \text{for } x \geq m\,, \\ 0 & \text{otherwise.} \end{cases}$$

In the next step of the method the model error surface E_{0M} is subtracted from the error surface E_0 of the base model, giving in result the surface residuum E_1, Fig. 6.64.

On having determined the residuum E_1 of the modeling error, its mean absolute value or mean square value should be calculated. If the value obtained is appropriately small the modeling process can be terminated. Otherwise, the subsequent fuzzy model of error residuum is created and added to the previous models; this way the overall precision of the models is improved. In

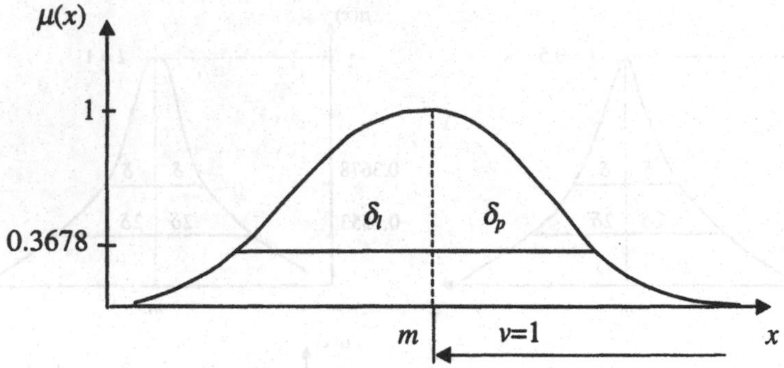

Fig. 6.63. Asymmetrical, exponential membership function (6.54)

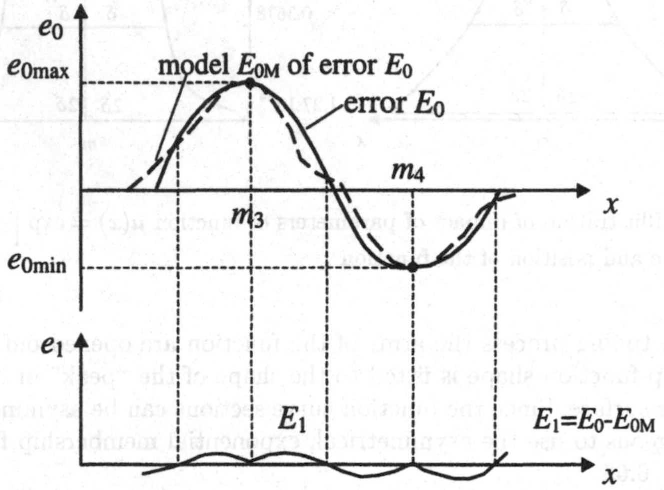

Fig. 6.64. Determination of residuum E_1 of model error surface

consequence, a set of parallel fuzzy models is obtained where each model has a simple structure, contains a small number of rules and fuzzy sets, and is easy to tune, Fig. 6.65.

The partition of a global model into a base model and error residuum models is, in fact, the partition of the global model into a single, simple base model covering the entire universe of discourse and a set of local models modeling exclusively the individual "peaks" and "troughs" of error residua. Since the local models only have a local range, they can be very simple. This method of modeling allows us to avoid an extensive increase in the number of rules that are specific to modeling the "peaks" and "troughs" in a global model. The problem is demonstrated in Fig. 6.66.

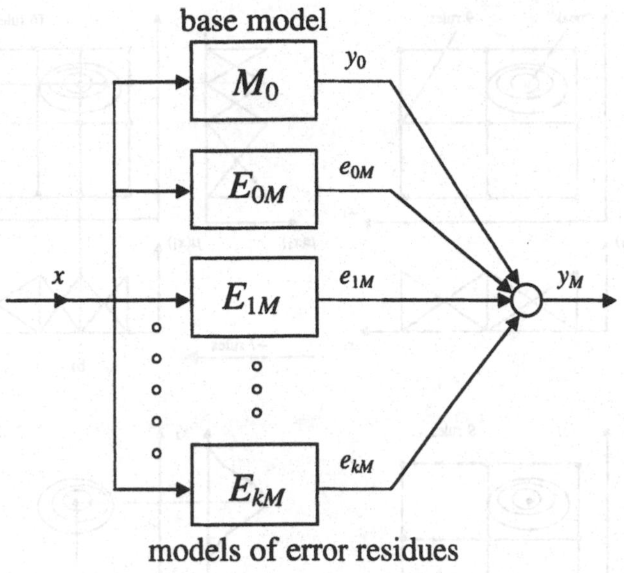

Fig. 6.65. Concept of partition of a global model into a base model and models of error residues

The steps I–VIII below present a simplified version of a modeling algorithm with the maximum error point method.

I. Definition of a base model M_0 of the system.
II. Tuning of the base model using the measurement samples of the system.
III. Checking the precision of the base model. If its precision is adequate – terminate the modeling, otherwise – skip to step VI.
IV. Determination of the base model error E_0.
V. Identification of the points of the maximum and minimum error of the base model E_0 ($e_{0\max}$ and $e_{0\min}$).
VI. Positioning 2 rules in error extrema – error model E_{0M}.
VII. Tuning the membership function parameters of error model E_{0M} on the basis of error samples of base model E_0.
VIII. Addition of the base model M_0 to the error model E_{0M}. The union of the two above models constitutes a new model M_1. Checking the model precision. If it is adequate – terminate the modeling, otherwise – calculate an error residuum E_1 and continue the modeling until satisfactory precision is reached.

Various improvements and versions of the above algorithm are possible. For example, in those steps where the simultaneous elimination of the highest error "peaks" and "troughs" is performed, the separate successive elimination can be applied, i.e. elimination of a protrusions, which show the highest absolute value. Another possibility is to identify such protrusions on the

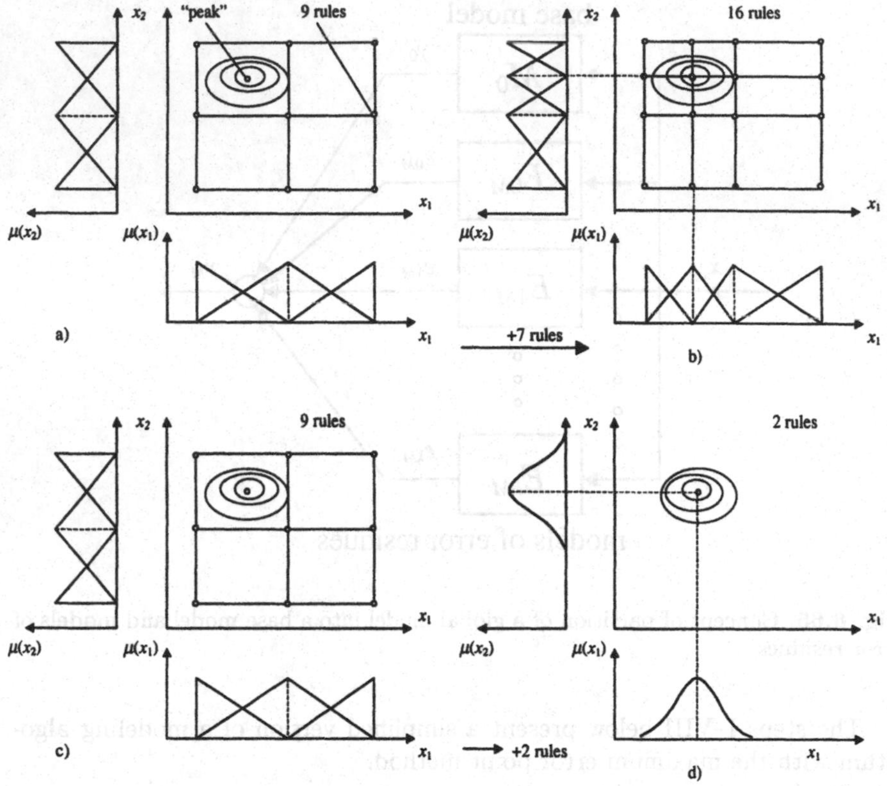

Fig. 6.66. Illustration of rule number increase in modeling a "peak" of a system surface in one global model (a) and (b) and in modeling with two parallel models (c) and (d)

error surface, which are not the highest but cover a relatively large region of input space or a region described with many samples. Elimination of such protrusions should improve the model precision considerably, Fig. 6.67.

Tuning of error models E_i can be performed in two ways. Each subsequent error model can be tuned jointly with the base model and all previously determined error models or it can be tuned individually on the basis of error residuum samples only.

Application of multi-dimensional membership functions (Lin 1996) is beneficial during modeling of multi-input systems. Examples of such functions are presented in Fig. 6.68. Multi-dimensional membership functions provide an advantageous possibility to abandon the regular grid-type partition of the input space and to decide on irregular partition instead, Fig. 6.69. Multi-dimensional membership functions enable us to reduce the number of model rules significantly.

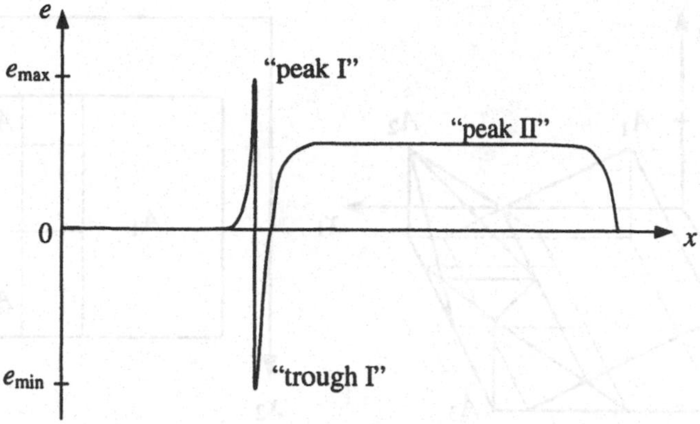

Fig. 6.67. Illustration of a case where elimination of a lower error "peak" (peak II) results in a more effective increase in model precision than elimination of a higher error "peak" (peak I)

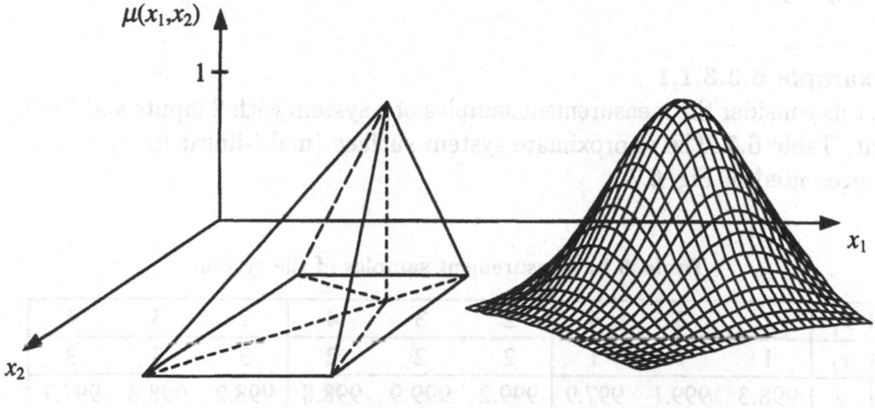

a) sector-linear membership function b) exponential membership function

Fig. 6.68. Examples of multi-dimensional membership functions

In the case of functions presented in Fig. 6.66, the complete rule base comprises as many rules as the product of all input membership functions. In the case of multi-dimensional membership functions, the minimum number of rules equals the number of membership functions only. Application of multi-dimensional membership functions for systems of less complicated surfaces of the mapping $\mathbf{X} \rightarrow Y$ enables us to construct a model comprising only two rules, regardless of the number of system inputs! In that way the problem of the "curse of dimensionality" can be overcome.

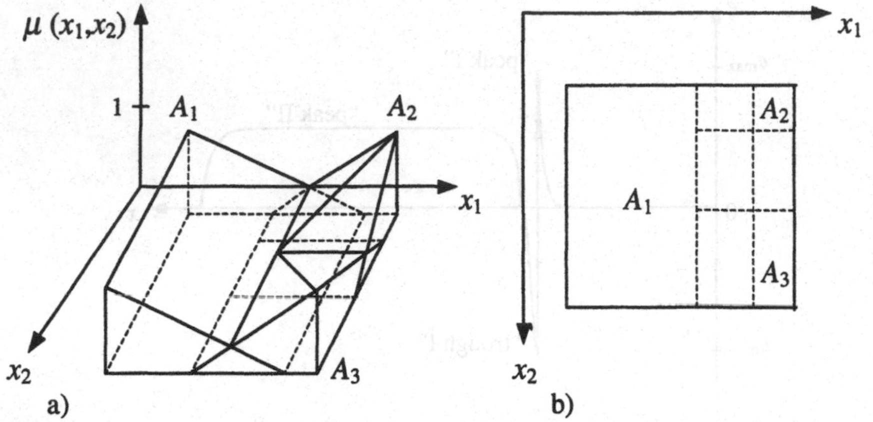

Fig. 6.69. Multi-dimensional membership functions arranged irregularly in the input space (a) and the partition of input space into receptive regions (influence zones) of particular functions

Example 6.3.3.1.1

Let us consider the measurement samples of a system with 2 inputs and 1 output, Table 6.5. The approximate system surface (multi-linear interpolation) is presented in Fig. 6.70.

Table 6.5. Measurement samples of the system

x_1	2	3	4	2	3	4	2	3	4
x_2	1	1	1	2	2	2	3	3	3
y	998.3	999.1	997.9	999.2	999.9	998.8	998.2	998.8	997.7

The system surface presented in Fig. 6.70 can be modeled with two rules (6.55) using only one multi-dimensional membership function A of the inputs and a singleton membership function B of the output, Fig. 6.71. The second, derivative rule $R1N$ fulfills only an auxiliary task and does not contain new information about the system modeled. It only allows for the fuzzy model to function, as it must have at least two rules. A model with only one rule $R1$ would always calculate a constant output value.

$$\begin{aligned} R1: &\ IF\ ((x_1, x_2) = A)\ THEN\ (y = B) \\ R1N: &\ IF\ ((x_1, x_2) = NOT\ A)\ THEN\ (y = 0) \end{aligned} \tag{6.55}$$

Special features of exponential functions allow us to represent the membership function in Fig. 6.71 with formula (6.56).

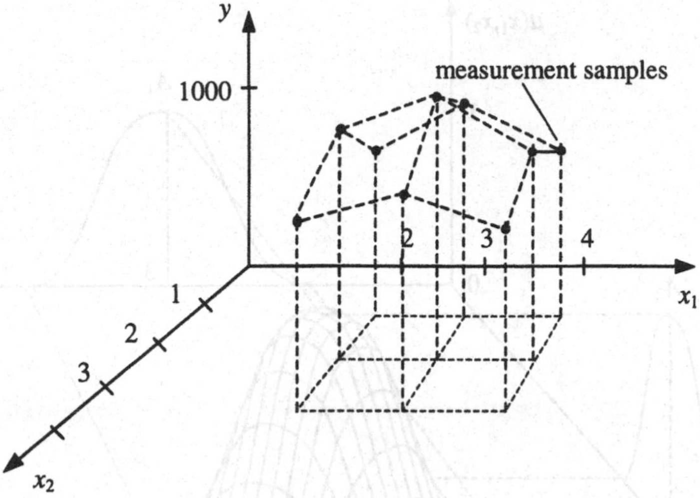

Fig. 6.70. Discrete measurement representation of system surface $\mathbf{X} \rightarrow Y$

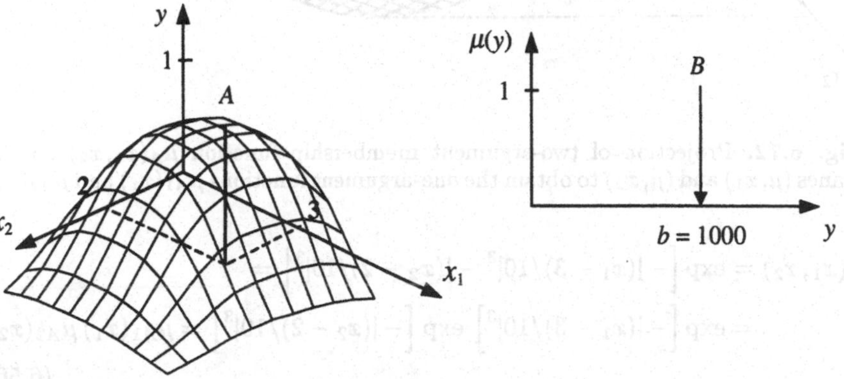

$$\mu_A(x_1, x_2) = \exp\left[-|(x_1-3)/10|^3 - |(x_2-2)/10|^3\right]$$

$$\mu_B(y) = \begin{cases} 1 & \text{for } y = 1000, \\ 0 & \text{otherwise.} \end{cases}$$

Fig. 6.71. Membership functions of input and output fuzzy set for the rule (6.55)

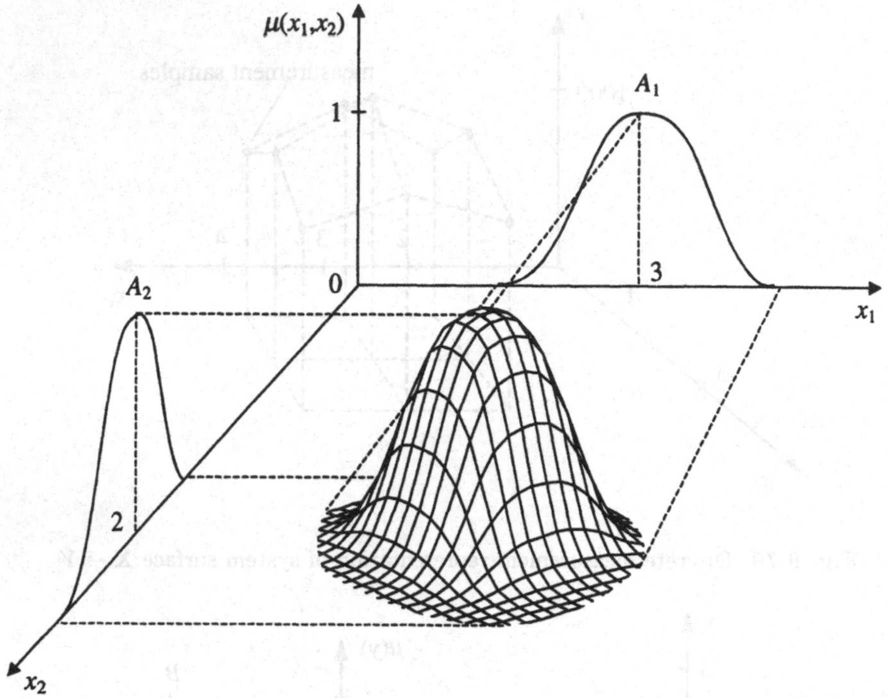

Fig. 6.72. Projection of two-argument membership function $\mu_A(x_1, x_2)$ on the planes (μ, x_1) and (μ, x_2) to obtain the one-argument functions $\mu_{A1}(x_1)$ and $\mu_{A2}(x_2)$

$$\mu(x_1, x_2) = \exp\left[-|(x_1 - 3)/10|^3 - |(x_2 - 2)/10|^3\right] =$$
$$= \exp\left[-|(x_1 - 3)/10|^3\right] \exp\left[-|(x_2 - 2)/10|^3\right] = \mu_{A1}(x_1)\,\mu_{A2}(x_2) \tag{6.56}$$

Moreover, the rule $R1$ (6.55) can be expressed in the form of a conjunction (6.57) containing the single-dimensional sets A_1 and A_2 in its premises, provided the *PROD*-operator is used to perform the *AND*-operation.

$$R1^*: \; IF\ (x_1 = A_1)\ AND\ (x_2 = A_2)\ THEN\ (y = B) \tag{6.57}$$

The other t-norms applied in the above case would not ensure the equivalence between the rule $R1^*$ (6.57) and the rule $R1$ (6.55). The membership functions $\mu_{A1}(x_1)$ and $\mu_{A2}(x_2)$, constituting the projection of multi-dimensional membership function $\mu_A(x_1, x_2)$ on the planes (μ, x_1) and (μ, x_2), are presented in Fig. 6.72.

The decomposition of multi-argument membership functions into one-argument functions makes fuzzy rules more understandable. It is easier to understand the premises in the form which uses one-argument membership

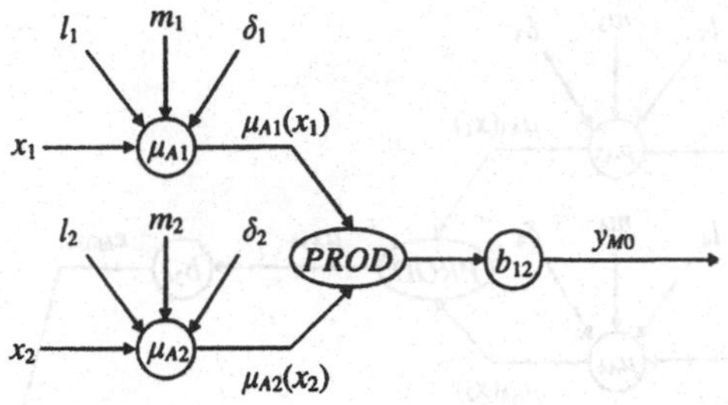

$$\mu_{Ai} = \exp\left[-|(x_i - m_i)/\delta_i|^{l_i}\right], \quad i = 1, 2, \ldots$$

Fig. 6.73. Neuro-fuzzy base model M_0 of the system modeled

functions:

$$IF\ (x_1 \text{ is close to } 3)\ AND\ (x_2 \text{ is close to } 2),$$

than the same premises in the form with two-argument functions:

$$IF\ (x_1, x_2) \text{ is close to } (3, 2).$$

Except for ensuring more understandable forms, the decomposition of multi-argument membership functions into one-argument functions brings no other benefits. However, a reverse operation, that is the creation of multi-argument membership functions from one-argument functions, and using them in a fuzzy model, allows us to reduce the number of rules in comparison with a conventional fuzzy model based on an input space rectangular grid resulting from the space partition.

The results presented in Table 6.6 provide a comparison between the system outputs y and model outputs y_{M0} and the base model error E_0.

Table 6.6. Results of system modeling using the rule (6.55) or (6.56)

x_1	2	3	4	2	3	4	2	3	4
x_2	1	1	1	2	2	2	3	3	3
y	998.3	999.1	997.9	999.2	999.9	998.8	998.2	998.8	997.7
y_{M0}	998.0	999.0	998.0	999.0	1000.0	999.0	998.0	999.0	998.0
E_0	0.3	0.1	- 0.1	0.2	- 0.1	-0.2	0.2	- 0.2	-0.3

The model M_0 (6.57) can be presented in the form of a neuro-fuzzy network, Fig. 6.73.

After tuning the network, the following results were obtained: $l_1 = 3$, $m_1 = 3$, $\delta_1 = 10$, $l_2 = 3$, $m_2 = 2$, $\delta_2 = 10$. The mean absolute error of the

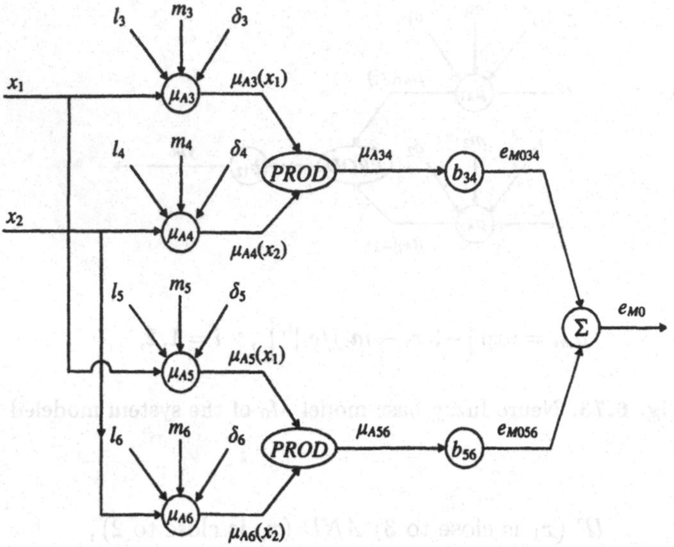

$$\mu_{Ai} = \exp\left[-|(x_i - m_i)/\delta_i|^{l_i}\right], \quad i = 3,4,5,6, \quad j = 1,2\,.$$

Fig. 6.74. Neuro-fuzzy network modeling the error surface E_{MO}

model is:

$$\sum_{i=1}^{9} |e_{0i}|/9 = 0.189\,.$$

If the model precision is found to be unsatisfactory, the process of modeling the base model error shown in Table 6.6 can be initiated. Maximum error value is equal to 0.3 and occurs at the point $(x_1, x_2) = (2, 1)$; minimum error value is equal to -0.3 and occurs at the point $(x_1, x_2) = (4, 3)$. New membership functions have modal values (centers) located at the above points, (6.58).

$$\begin{aligned} \mu_{A34}(x_1, x_2) &= \exp\left[-|(x_1 - m_3)/\delta_3|^{l_3} - |(x_2 - m_4)/\delta_4|^{l_4}\right], \\ \mu_{A56}(x_1, x_2) &= \exp\left[-|(x_1 - m_5)/\delta_5|^{l_5} - |(x_2 - m_6)/\delta_6|^{l_6}\right], \\ m_3 &= 3,\ m_4 = 1,\ m_5 = 4,\ m_6 = 3\,. \end{aligned} \tag{6.58}$$

To tune the parameters δ_i and l_i, a neuro-fuzzy network is created that provides modeling of error surface E_0, Fig. 6.74.

After having the network tuned, the following parameters of membership functions have been obtained: $m_3 = 2$, $l_3 = 3$, $\delta_1 = 0.3$, $m_4 = 1$, $l_4 = 0.5$, $\delta_4 = 10$, $m_5 = 4$, $l_5 = 2$, $\delta_5 = 1$, $m_6 = 3$, $l_6 = 1$, $\delta_6 = 2$. Table 6.7 presents the results of modeling: the output of error model e_{0M} and the error residuum $e_1 = e_0 - e_{0M}$.

Table 6.7. Results of modeling error E_0

x_1	2	3	4	2	3	4	2	3	4
x_2	1	1	1	2	2	2	03	3	3
e_0	0.3	0.1	- 0.1	0.2	- 0.1	- 0.2	0.2	- 0.2	- 0.3
e_{M0}	0.3	-0.041	-0.11	0.218	-0.067	-0.182	0.192	-0.11	-0.3
e_1	0	0.141	0.01	-0.018	-0.033	-0.018	0.008	-0.09	0

$$\text{mean square error: } \sum_{i=1}^{9} |e_{1i}|/9 = 0.035$$

As long as the model precision appears to be unsatisfactory, modeling of error residua can be continued. Let us assume, however, that the precision obtained (mean absolute error is 0.0035) is satisfactory. Thus, the model $M_1 = M_0 + E_{M0}$ is represented with formulas (6.59) and (6.60).

$$y_{M1} = y_{M0} + e_{M0}$$

Base model M_0:

$$\begin{aligned}
&R1 : IF\ ((x_1, x_2) = A_{12})\ THEN\ (y_{M0} = B_{12}) , \\
&R1N : IF\ ((x_1, x_2) = NOT\ A_{12})\ THEN\ (y_{M0} = 0) , \\
&\mu_{A12} = \exp\left[-|(x_1 - 3)/10|^3 - |(x_2 - 2)/10|^3\right] , \\
&\mu_{NOT\,A12} = 1 - \mu_{A12} , \\
&\mu_{B12} = \begin{cases} 1 & \text{for } y = 1000 , \\ 0 & \text{otherwise.} \end{cases}
\end{aligned} \tag{6.59}$$

Error model E_{M0}:

$$\begin{aligned}
&e_{M0} = e_{M034} + e_{M056} , \\
&R2 : IF\ ((x_1, x_2) = A_{34})\ THEN\ (e_{M034} = B_{34}) , \\
&R2N : IF\ ((x_1, x_2) = NOT\ A_{34})\ THEN\ (e_{M034} = 0) , \\
&\mu_{A34} = \exp\left[-|(x_1 - 2)/0.3|^3 - |(x_2 - 1)/10|^0 .5\right] , \\
&\mu_{NOT\,A34} = 1 - \mu_{A34} , \\
&\mu_{B34} = \begin{cases} 1 & \text{for } e_{M034} = 0.3 , \\ 0 & \text{otherwise,} \end{cases}
\end{aligned}$$

$$R3: IF\ ((x_1, x_2) = A_{56})\ THEN\ (e_{M056} = B_{56}),$$
$$R3N: IF\ ((x_1, x_2) = NOT\ A_{56})\ THEN\ (e_{M056} = 0),$$
$$\mu_{A56} = \exp\left[-|(x_1 - 4)/1|^2 - |(x_2 - 3)/2|^1\right],$$
$$\mu_{NOT\,A56} = 1 - \mu_{A56},$$
$$\mu_{B56} = \begin{cases} 1 & \text{for } e_{M056} = -0.3, \\ 0 & \text{otherwise.} \end{cases} \tag{6.60}$$

The neuro-fuzzy network corresponding to the model M_1 is presented in Fig. 6.75.

◊ ◊ ◊

A neuro-fuzzy network representing a model created with the maximum error method resembles an RBF neural network. They differ in the types of neurons used (the activation functions are not typical Gaussian functions) and the training method applied. The neuro-fuzzy network is trained in small fragments, one by one. The base model M_0 is trained first. Then, depending

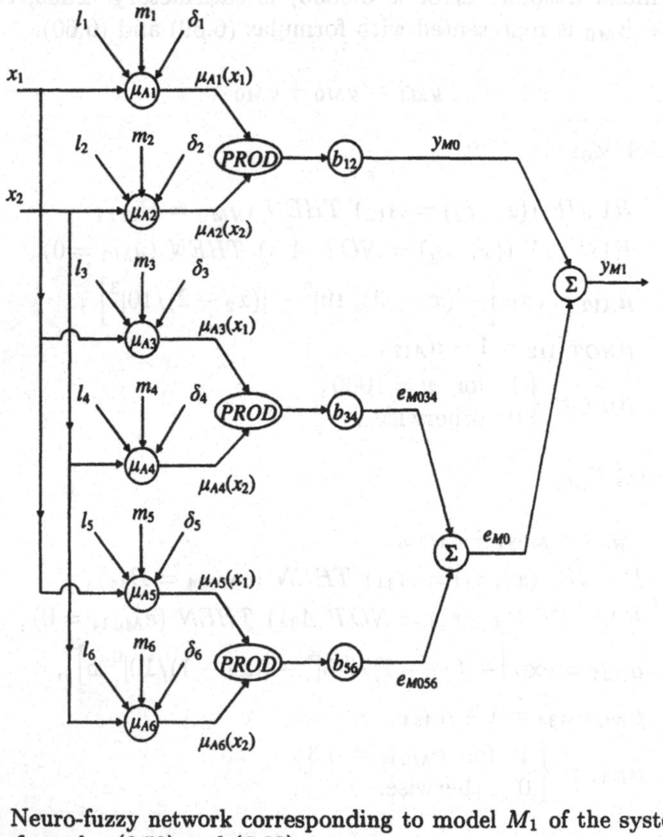

Fig. 6.75. Neuro-fuzzy network corresponding to model M_1 of the system represented with formulas (6.59) and (6.60)

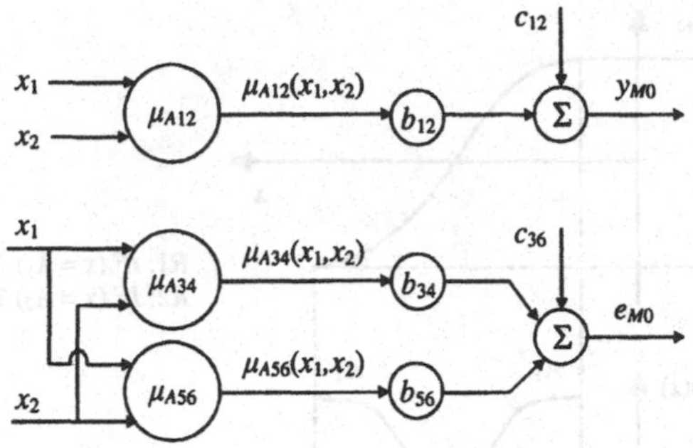

Fig. 6.76. Additional degree of freedom c performs shifts of particular models along the ordinate y-axis

on the results, the error model E_{M0} is trained, and so on. Since only a fragment of the network is trained at a time, the training is simpler and easier than in the case of entire network training. The latter situation may cause individual neurons to compete with one another and thus restrain the adaptation process.

The training process of both the base model M_0 and error models E_{Mi} can be made even more effective if their inputs are assigned an additional degree of freedom c, which allows the model surface to shift upwards or downwards along the ordinate y-axis, Fig. 6.76.

The application of the above degree of freedom may facilitate further model simplification and reduction of the number of rules in multi-dimensional fuzzy models. For example, if an error surface has a shape as in Fig. 6.77a, then it can be modeled with two rules and two membership functions. If the surface has been shifted upwards by a constant value c it is enough to use one membership function, which requires tuning, and its negation, which does not require any tuning (afterwards, value c has to be subtracted from the model output). The advantage of that approach strongly increases in high-dimensional models.

Another solution to increase the precision of a fuzzy model is to apply the elliptic membership functions with rotatable main axes, Fig. 6.78. A two-argument exponential elliptic membership function has the form (6.61) and has 4 degrees of freedom (m_1, m_2, a, b), whereas a rotatable function (6.62) has 5 degrees of freedom (m_1, m_2, a, b, α). Introduction of such an additional degree of freedom, however, causes particular consequences, i.e. it becomes more difficult to calculate the derivatives and it results in a greater number of parameters to be trained.

$$\mu(x_1, x_2) = \exp\left[-\left(\frac{x_1 - m_1}{a}\right)^2 - \left(\frac{x_1 - m_2}{b}\right)^2\right] \qquad (6.61)$$

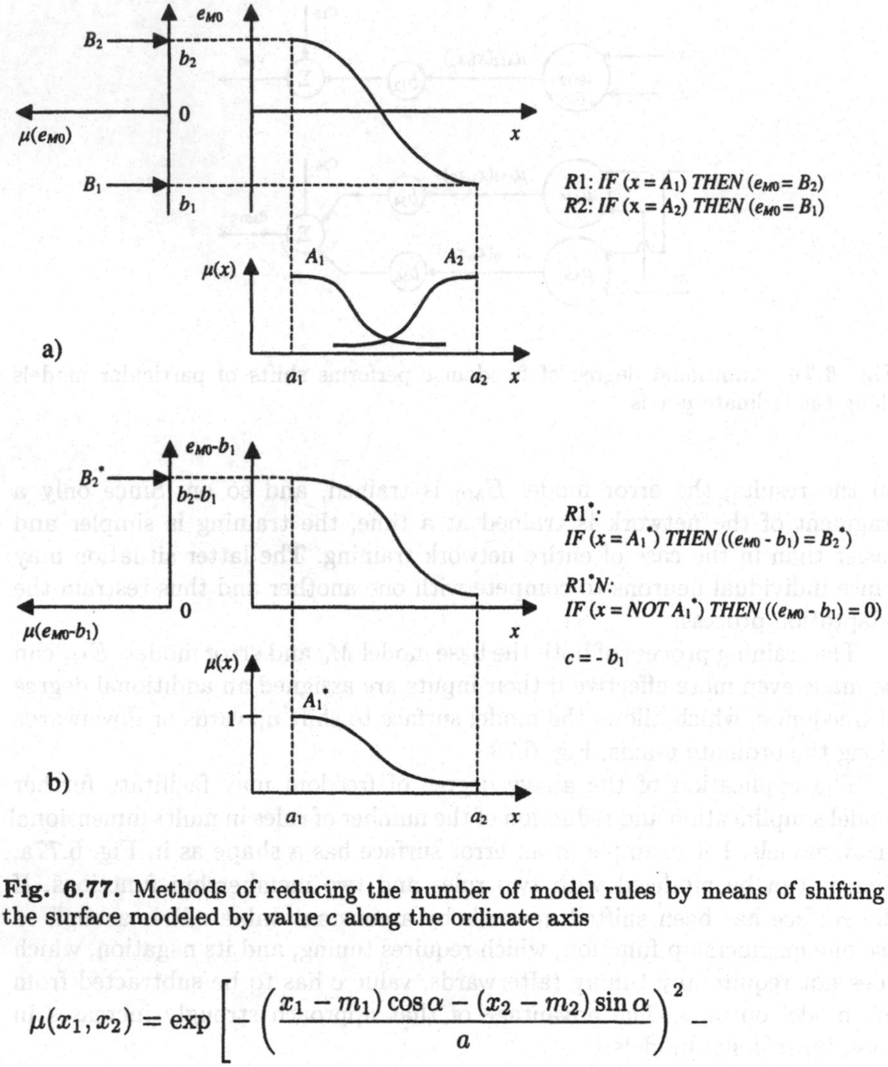

Fig. 6.77. Methods of reducing the number of model rules by means of shifting the surface modeled by value c along the ordinate axis

$$\mu(x_1, x_2) = \exp\left[-\left(\frac{(x_1 - m_1)\cos\alpha - (x_2 - m_2)\sin\alpha}{a}\right)^2 - \right.$$

$$\left. - \left(\frac{(x_1 - m_1)\sin\alpha + (x_2 - m_2)\cos\alpha}{b}\right)^2\right] \tag{6.62}$$

An exponent can be applied as an additional degree of freedom. The non-rotatable, elliptic, two-argument function (6.61) can be expressed with a general formula (6.63).

$$\mu(x_1, x_2) = \exp\left[-(a_1x_1^2 + a_2x_2^2 + a_3x_1 + a_4x_2 + a_5)\right] \tag{6.63}$$

In the general formula, which corresponds to the function rotated by an angle α (6.62), a product $b_3x_1x_2$ (6.64) appears.

$$\mu(x_1, x_2) = \exp\left[-(b_1x_1^2 + b_2x_2^2 + b_3x_1x_2 + b_4x_1 + b_5x_2 + b_6)\right] \tag{6.64}$$

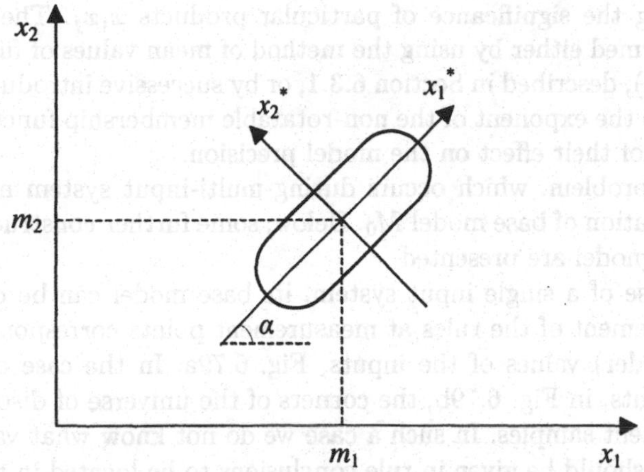

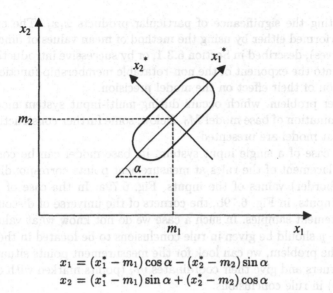

$$x_1 = (x_1^* - m_1)\cos\alpha - (x_2^* - m_2)\sin\alpha$$
$$x_2 = (x_1^* - m_1)\sin\alpha + (x_2^* - m_2)\cos\alpha$$

Fig. 6.78. Section across the elliptic exponential function of axes rotated by an angle α in respect to the system of coordinates (x_1, x_2)

In the case of three model inputs, a rotatable elliptic function will have 3 product elements (6.65).

$$\mu(x_1, x_2, x_3) = \exp\left[-(c_1 x_1^2 + c_2 x_2^2 + c_3 x_3^2 + c_4 x_1 x_2 + \right.$$
$$\left. + c_5 x_1 x_3 + c_6 x_2 x_3 + c_7 x_1 + c_8 x_2 + c_9 x_3 + c_{10})\right] \quad (6.65)$$

In the case of a larger number of model inputs n, the number of product elements m can be calculated with formula (6.66),

$$m = \frac{n!}{2!(n-2)!}, \quad (6.66)$$

while the number k of independent elements, occurring in the non-rotatable elliptic function (6.61) can be calculated with formula (6.67).

$$k = 2n \quad (6.67)$$

For 10 model inputs the non-rotatable function comprises $k = 20$ independent elements and parameters which need tuning. Thus, the rotatable function contains $(k + m) = 65$ parameters – that makes 45 more! Further on, for 100 model inputs we obtain $k = 200$ and $(k + m) = 5150$. The difference in the number of parameters to be tuned is as great as 4950! So, the application of a rotatable membership function makes us face the problem of the "curse of dimensionality". The extent of the phenomenon can be reduced by evaluating the significance of particular products $x_i x_j$. The evaluation can be performed either by using the method of mean values of function cuts (fuzzy curves), described in Section 6.3.1, or by successive introduction of the elements into the exponent of the non-rotatable membership function and an examination of their effect on the model precision.

Another problem, which occurs during multi-input system modeling, is the determination of base model M_0. Below, some further construction methods for that model are presented.

In the case of a single input system, its base model can be constructed through placement of the rules at measurement points corresponding to extremum (border) values of the inputs, Fig. 6.79a. In the case of a system with two inputs, in Fig. 6.79b, the corners of the universe of discourse show no measurement samples. In such a case we do not know what values of the coordinate y should be given in rule conclusions to be located in the corners. To solve the problem, we can look for the measurement points situated nearest the corners and give their coordinates y_{Pi} (points marked with crosses in Fig. 6.79b) in rule conclusions.

Let us consider the following points:

$$P_1(x_{1P1}, x_{2P2}, y_{P1}) \quad \ldots \quad P_4(x_{1P4}, x_{2P4}, y_{P4}),$$

then the base model rules are in the form (6.68).

$$\begin{aligned}
R1: &\ IF\ (x_1 = A_{11})\ AND\ (x_2 = A_{21})\ THEN\ (y \text{ is close to } y_{P1})\\
R2: &\ IF\ (x_1 = A_{12})\ AND\ (x_2 = A_{21})\ THEN\ (y \text{ is close to } y_{P2})\\
R3: &\ IF\ (x_1 = A_{12})\ AND\ (x_2 = A_{22})\ THEN\ (y \text{ is close to } y_{P3})\\
R4: &\ IF\ (x_1 = A_{11})\ AND\ (x_2 = A_{22})\ THEN\ (y \text{ is close to } y_{P4})
\end{aligned} \quad (6.68)$$

The parameters y_{Pi} in the rule conclusions (6.68) can be treated as the initial values in the process of tuning the neuro-fuzzy network representing the base model. The initial values y_{Pi} can also be determined randomly. However, that would make the network tuning process more difficult (in a general case). A disadvantage of locating the base model rules in the corners of the output universe of discourse is the phenomenon of the "curse of dimensionality".

In the above case the number of rules in the base model equals:

$$r = 2^n,$$

and increases rapidly when the number n of inputs increases. The extent of the problem can be reduced if **the universe of discourse of the model is expanded beyond the universe of discourse of the system being modeled**, Fig. 6.80. It is then presumed that the universe of discourse for the model is triangular. The coordinates of triangle corners P_i are the following:

$$P_1(a_{11}, a_{21}, y_{P1}), \quad P_2(a_{12}, a_{21}, y_{P2}), \quad P_1(a_{11}, a_{23}, y_{P3}).$$

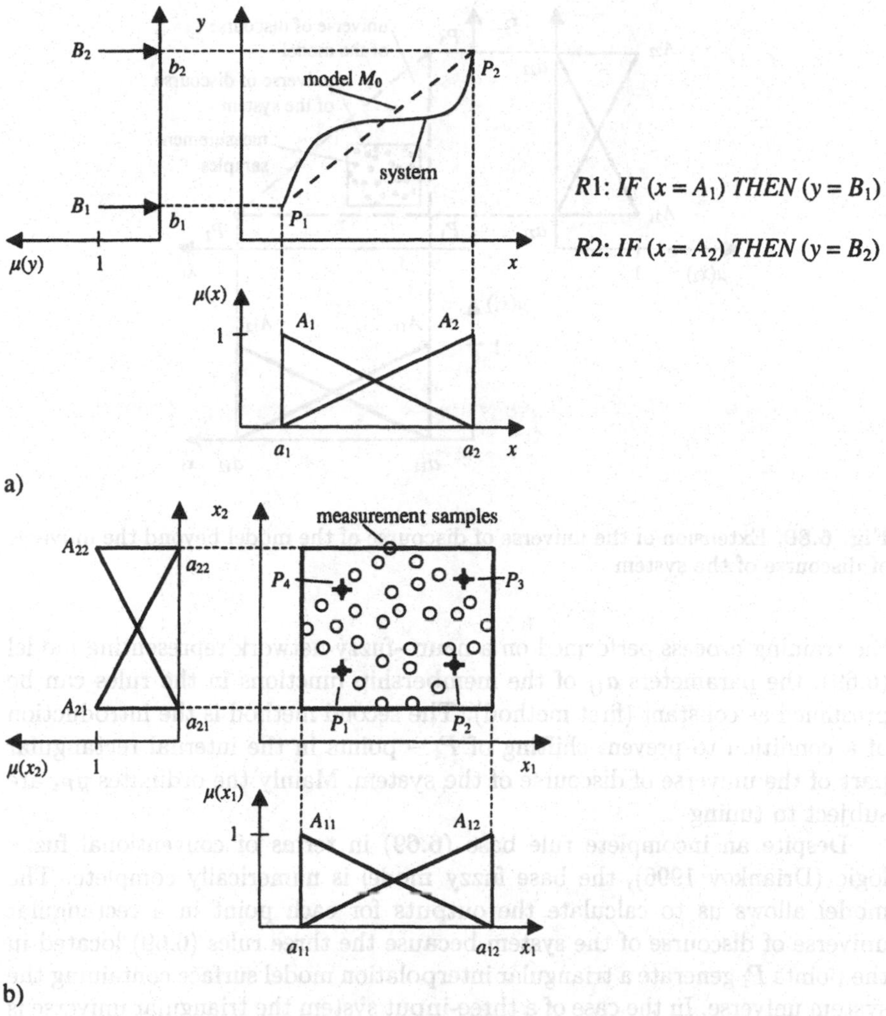

Fig. 6.79. Base model M_0 of a system with a single input (a) and illustration of the lack of samples in the corners of the universe of discourse of a system with two inputs (b)

The base model rules are located in the corners of the triangle. The rules are expressed with formula (6.69).

$$
\begin{aligned}
&R1: \ IF\ (x_1 = A_{11})\ AND\ (x_2 = A_{21})\ THEN\ (y \text{ is close to } y_{P1}) \\
&R2: \ IF\ (x_1 = A_{12})\ AND\ (x_2 = A_{21})\ THEN\ (y \text{ is close to } y_{P2}) \\
&R3: \ IF\ (x_1 = A_{11})\ AND\ (x_2 = A_{23})\ THEN\ (y \text{ is close to } y_{P3}) \qquad (6.69)
\end{aligned}
$$

At the starting point of base model training, the values y_{Pi} can be presumed on the basis of the closest measurement points or at random. During

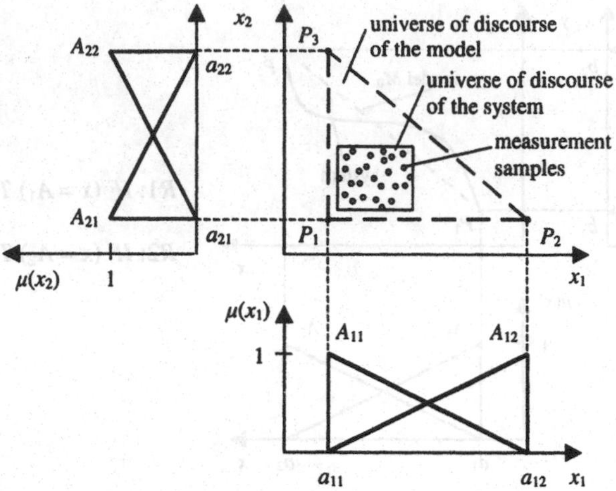

Fig. 6.80. Extension of the universe of discourse of the model beyond the universe of discourse of the system

the training process performed on a neuro-fuzzy network representing model (6.69), the parameters a_{ij} of the membership functions in the rules can be presumed as constant (first method). The second method is the introduction of a condition to prevent shifting of P_i – points in the internal rectangular part of the universe of discourse of the system. Mainly the ordinates y_{Pi} are subject to tuning.

Despite an incomplete rule base (6.69) in terms of conventional fuzzy logic (Driankov 1996), the base fuzzy model is numerically complete. The model allows us to calculate the outputs for each point in a rectangular universe of discourse of the system because the three rules (6.69) located in the points P_i generate a triangular interpolation model surface containing the system universe. In the case of a three-input system the triangular universe is substituted (replaced) with a tetrahedron comprising the cuboidal universe of discourse of the system, Fig. 6.81. In the above case, the minimum quantity of rules required to define the base model in a cubicoid universe of discourse of the system amounts to 8, while in the tetrahedral universe of discourse, expanded beyond the system universe, it amounts to only 4.

In the case of n model inputs, the number r of rules of a hypertetrahedral base model equals $r = n+1$. For 100 inputs a conventional base model would require:

$$r \cong 1.2676506 \cdot 10^{30} ,$$

rules and the hypertetrahedral model:

$$r = 101$$

rules. The difference needs no comment.

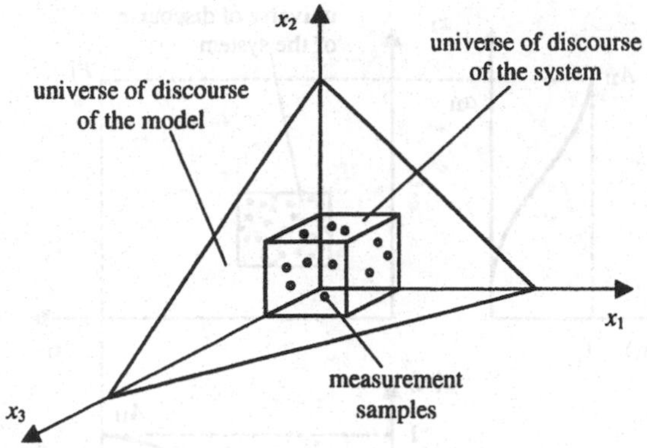

Fig. 6.81. Universe of discourse of the system with 3 inputs (8 corners) and universe of discourse of the model (4 corners)

The expansion beyond universe of discourse of the system allows us to limit the extent of the "curse of dimensionality". The number of rules of a base model can be reduced even more in certain circumstances, Fig. 6.82. The surface generated by a model with two rules can be sufficiently precise – provided the surface parameters are tuned correctly with system measurement samples. A model with two rules is considered numerically complete, provided the receptive fields of input membership functions are sufficiently large. The fields (supports) have to be so large as to cover the universe of discourse of the system as shown in Fig. 6.82.

If the condition is fulfilled, the rule $R1$ can sometimes also be located inside the rectangular system input space, Fig. 6.82a, depending on the system surface shape. The expansion beyond the universe of discourse allows us, depending on the case, to construct the base model with two rules for multi-input systems. Such a solution facilitates the construction of parsimonious fuzzy models (Brown 1995). The modeling method based on locating the rules in points of model error extrema gives the best effects when the measurement samples are noise-free or only slightly affected with noise. To increase the effectiveness of the method in conditions of stronger noise, filtration should be performed. The weighted mean method, discussed in (Preuss 1994b) and Section 6.3.2, can be applied for that purpose.

Let us assume the data consist of noise-affected measurement samples of the system modeled, where the sample distribution is as presented in Fig. 6.83a. The measurement sample (x_i, y_i) is a noise-affected representation of the model surface. Since the measurements of both the input x_i and the output y_i can be burdened with error, the determination of the most reliable value of the output y_i^* for the point $x = x_i$ only on the basis of the sample (x_i, y_i) would not be objective enough. The result would be more objective if

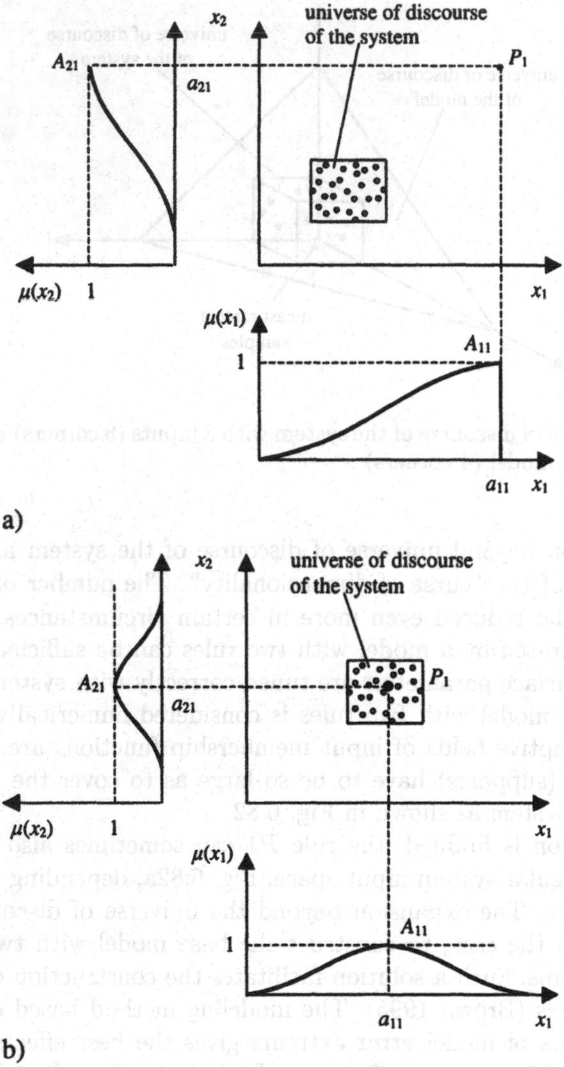

R1: *IF* $(x_1 = A_{11})$ *AND* $(x_2 = A_{21})$ *THEN* (y is close to y_{P1})

R1N: *IF* $(x_1 = NOT\, A_{11})$ *AND* $(x_2 = NOT\, A_{21})$ *THEN* $(y = 0)$

P_1: (a_{11}, a_{21}, y_{P1})

$\mu_A = 1 - \mu_{NOT\,A}$

Fig. 6.82. Base model with two rules

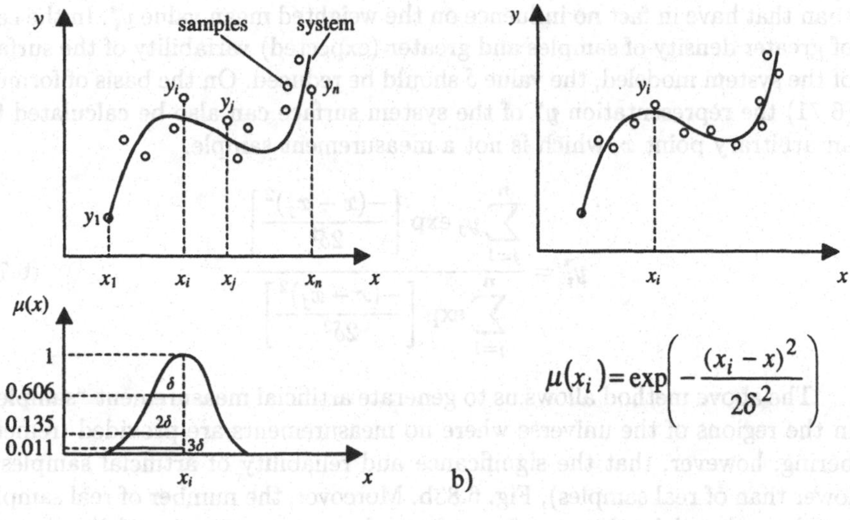

Fig. 6.83. Noise-affected measurement samples before filtration (a) and after filtration (b)

the representation of the output y_i^* is calculated with regard to other samples in the closest neighborhood of the sample (x_i, y_i).

The neighborhood can be defined using the membership function "close to x_i". Various types of membership function can be applied for that purpose. The Gaussian membership function, Fig. 6.83, seems beneficial for the purpose considered, as it has an infinitely large receptive field (support). That feature ensures that even in the case of very distant samples the adjacent neighboring samples are taken into account. The latter can not be obtained when a triangular membership function with an improperly chosen and too short support is applied.

A representation y_i^* of the system surface with a single input $x = x_i$ can be calculated with a weighted mean method using the formula (6.70).

$$y_i^* = \frac{\sum_{j=1}^{n} y_j \exp\left[\frac{-(x_i - x_j)^2}{2\delta^2}\right]}{\sum_{j=1}^{n} \exp\left[\frac{-(x_i - x_j)^2}{2\delta^2}\right]} \tag{6.70}$$

The measurement sample x_j located closer to the point x_i influences the value of representation y_i^* to a greater extent than the farther samples. The range of the nearest neighborhood can be adjusted through modification of the deviation δ of the membership function. At a distance 3δ from the point x_i the membership function value equals 0.011, i.e. the samples located farther than that have in fact no influence on the weighted mean value y_i^*. In the case of greater density of samples and greater (expected) variability of the surface of the system modeled, the value δ should be reduced. On the basis of formula (6.71) the representation y^* of the system surface can also be calculated for an arbitrary point x, which is not a measurement sample.

$$y_i^* = \frac{\sum_{j=1}^{n} y_j \exp\left[\frac{-(x-x_j)^2}{2\delta^2}\right]}{\sum_{j=1}^{n} \exp\left[\frac{-(x-x_j)^2}{2\delta^2}\right]} \tag{6.71}$$

The above method allows us to generate artificial measurement "samples" in the regions of the universe where no measurements are provided (remembering, however, that the significance and reliability of artificial samples is lower than of real samples), Fig. 6.83b. Moreover, the number of real samples can be reduced in those regions where they occur too plentifully. A large number of samples (x_i, y_i) can be replaced with a smaller number of representations (x_k^*, y_k^*). Thus, the weighted mean method enables us to reduce the volume of information. In the case of a system where the quantity of inputs is equal to p, the calculation formula for the weighted mean of system representation y_i^* in point x_i has the form (6.72):

$$y_i^*(x_{1i}, \ldots, x_{ki}, \ldots x_{pi}) = \frac{\sum_{j=1}^{n} y_j \exp\left[-\sum_{k=1}^{p} \frac{(x_{ki}-x_{kj})^2}{2\delta_k^2}\right]}{\sum_{j=1}^{n} \exp\left[-\sum_{k=1}^{p} \frac{(x_{ki}-x_{kj})^2}{2\delta_k^2}\right]}, \tag{6.72}$$

where: k – input number, j – measurement sample number.

Below, an example of the application of a geometrical approach to the fuzzy modeling of a 3-dimensional system $y = f(x_1, x_2)$ is presented. The approach uses a graphical visualization of the measurement sample distribution on a computer monitor screen.

Example 6.3.3.1.2

Measurement samples of the modeled system with two inputs x_1, x_2 and one output y, normalized to the interval $[0, 100]$ are given in Table 6.8.

Measurement samples presented in the form of a table are a mass of abstract numbers. Less abstract and more picturesque is their presentation in the form of points in the input space $X_1 \times X_2$ as depicted in Fig. 6.84.

The Gaussian function is represented by contour lines made at different levels, e.g. 1/3 and 2/3 of the full height. Before starting the tuning process, the starting parameters of the function have to be chosen. They determine the starting position, and its width and shape (exponent). In the example the starting parameters were chosen as shown in Fig. 6.84. In the course of tuning

Table 6.8. Measurement samples of the inputs x_1, x_2 and the corresponding output y – values of the modeled system

x_1	0	0	0	0	0	0	20	20	20	20	20	20
x_2	0	20	40	60	80	100	0	20	40	60	80	100
y	35.3	31.3	25.7	18.4	13.5	8.7	67.0	59.7	48.7	36.8	25.7	16.5

x_1	40	40	40	40	40	40	60	60	60	60	60	60
x_2	0	20	40	60	80	100	0	20	40	60	80	100
y	92.3	81.9	67.1	51.2	36.0	23.3	92.3	82.2	69.5	60.1	50.0	37.2

x_1	80	80	80	80	80	80	100	100	100	100	100	100
x_2	0	20	40	60	80	100	0	20	40	60	80	100
y	67.0	60.1	55.5	62.5	65.7	42.2	35.3	31.6	28.2	28.8	28.2	18.1

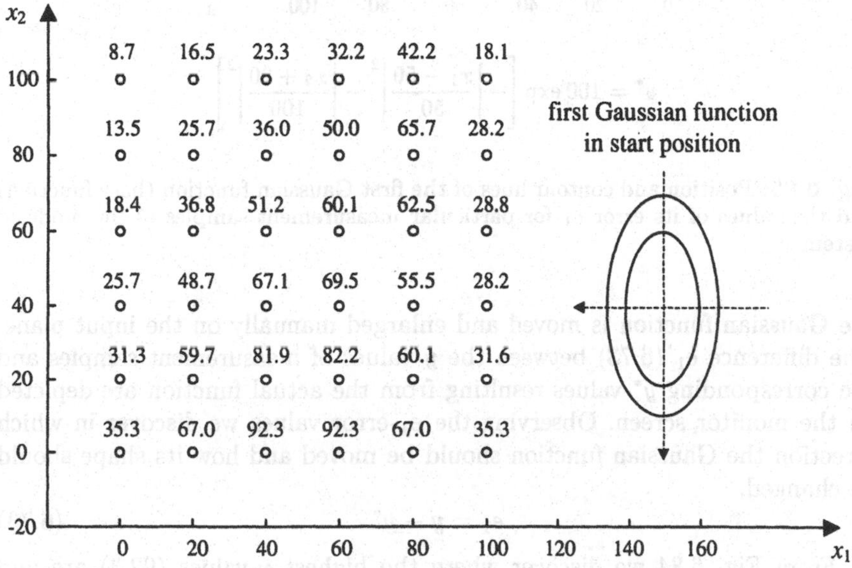

$$y^* = 2\exp\left[-\left|\frac{x_1 - 150}{20}\right|^2 - \left|\frac{x_2 - 40}{30}\right|^2\right]$$

Fig. 6.84. Measurement samples of the modeled system in the input space and an example of the starting position of the first bivariant Gaussian function

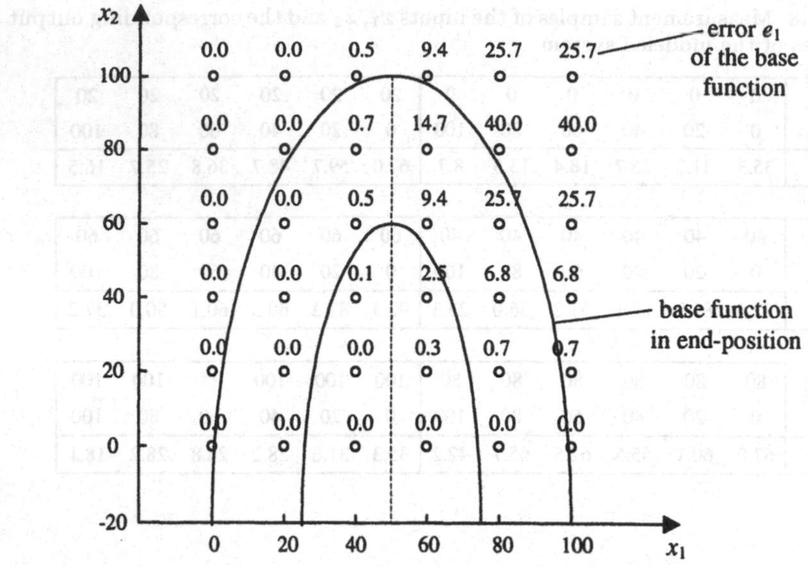

$$y^* = 100 \exp\left[-\left|\frac{x_1 - 50}{50}\right|^2 - \left|\frac{x_2 + 20}{100}\right|^2\right]$$

Fig. 6.85. Position and contour lines of the first Gaussian function (base function) and the values of its error e_1 for particular measurement samples of the modeled system

the Gaussian function is moved and enlarged manually on the input plane. The difference e_1 (6.73) between the y-values of measurement samples and the corresponding y^*-values resulting from the actual function are depicted on the monitor screen. Observing the e_1-error values we discover in which direction the Gaussian function should be moved and how its shape should be changed.

$$e_1 = y - y^* \tag{6.73}$$

From Fig. 6.84 we discover where the highest y-values (92.3) are and there we place the centrum of the Gaussian function. The function is moved, enlarged and/or diminished until citerion K_0 (being the mean absolute error of all measurement samples (6.74), where m denotes number of samples) has been minimized as much as possible.

$$K_0 = \frac{\sum_{i=1}^{m} |e_{1i}|}{m} \tag{6.74}$$

Fig. 6.85 depicts the optimal position of the Gaussian function found after a few trials, its shape (contour lines) and its error in relation to particular

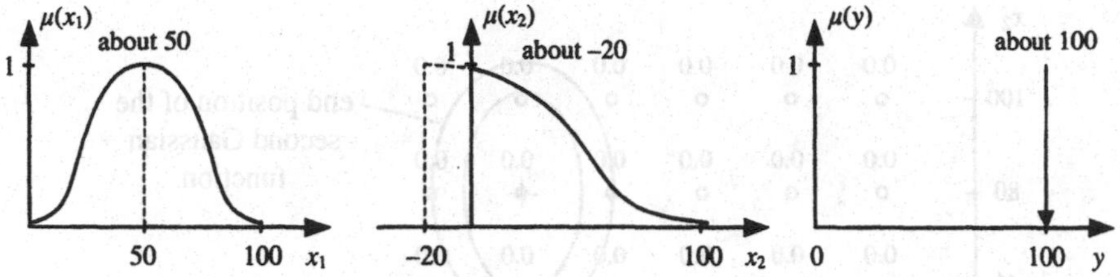

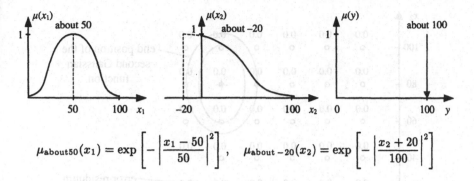

$$\mu_{\text{about50}}(x_1) = \exp\left[-\left|\frac{x_1 - 50}{50}\right|^2\right], \quad \mu_{\text{about}-20}(x_2) = \exp\left[-\left|\frac{x_2 + 20}{100}\right|^2\right]$$

Fig. 6.86. Univariant membership functions of the base model M_0

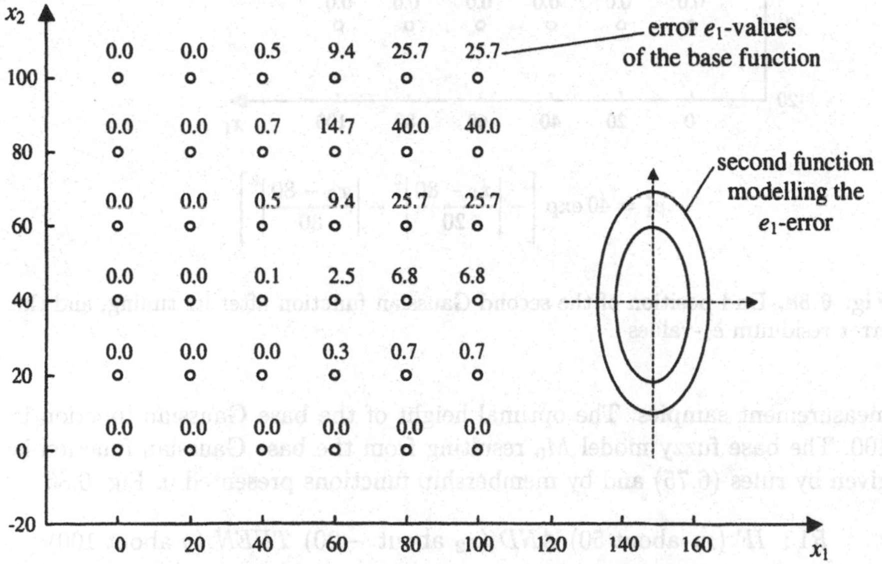

$$e_1^* = 10\exp\left[-\left|\frac{x_1 - 150}{20}\right|^2 - \left|\frac{x_2 - 10}{30}\right|^2\right]$$

Fig. 6.87. Values of the e_1-error of the base model M_0 and the starting position of the second Gaussian function

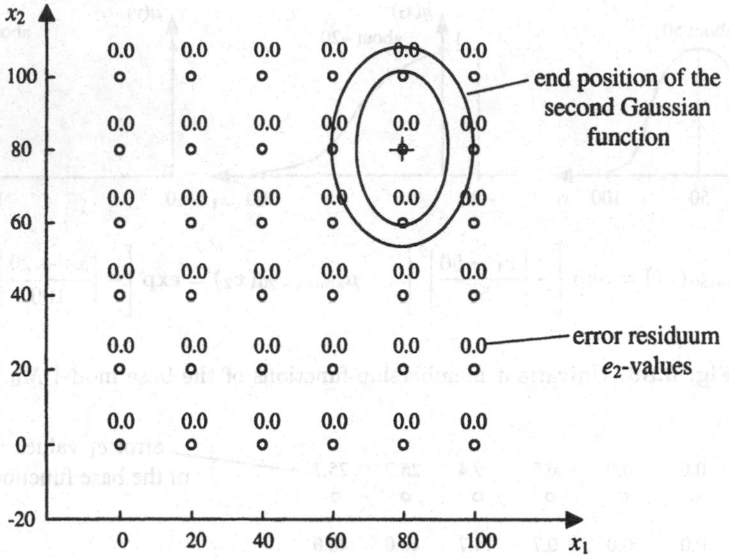

$$e_1^* = 40\exp\left[-\left|\frac{x_1-80}{20}\right|^2-\left|\frac{x_2-80}{30}\right|^2\right]$$

Fig. 6.88. End position of the second Gaussian function after its tuning, and the error residuum e_2-values

measurement samples. The optimal height of the base Gaussian function is 100. The base fuzzy model M_0 resulting from the base Gaussian function is given by rules (6.75) and by membership functions presented in Fig. 6.86.

$$\begin{aligned} R1:&\ IF\ (x_1 \text{ about } 50)\ AND\ (x_2 \text{ about } -20)\ THEN\ (y \text{ about } 100)\\ R1N:&\ IF\ (x_1\ NOT \text{ about } 50)\ OR\ (x_2\ NOT \text{ about } -20)\\ &\ THEN\ (y \text{ about } 0) \end{aligned} \tag{6.75}$$

The mean value of the absolute error of the base model M_0 equals 6.55. If this accuracy is insufficient then a second Gaussian function modeling the e_1-error can be added to the base model. Fig. 6.87 depicts error samples of the base model and the starting position of the second Gaussian function.

The second Gaussian function is displaced and enlarged or lessened by its parameter changes until its error e_2 has been minimized. Error of an error will be referred to as error residuum (6.76).

$$e_2 = e_1 - e_1^* \tag{6.76}$$

The optimal position and parameters of the second Gaussian function are depicted in Fig. 6.88.

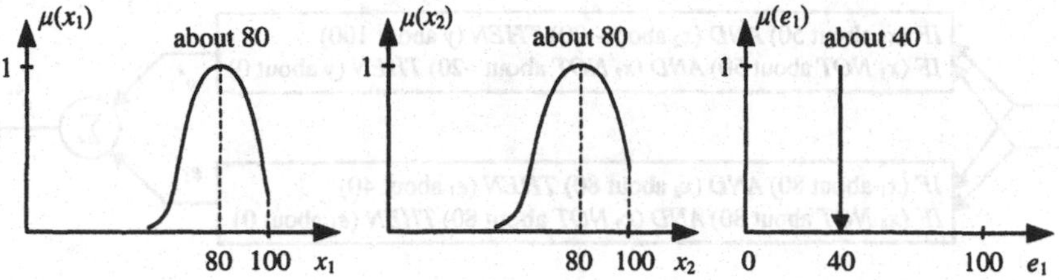

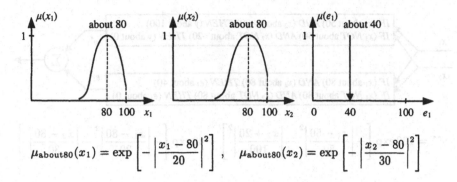

$$\mu_{\mathrm{about80}}(x_1) = \exp\left[-\left|\frac{x_1-80}{20}\right|^2\right], \quad \mu_{\mathrm{about80}}(x_2) = \exp\left[-\left|\frac{x_2-80}{30}\right|^2\right]$$

Fig. 6.89. Univariant membership functions of the error model E_{M0}

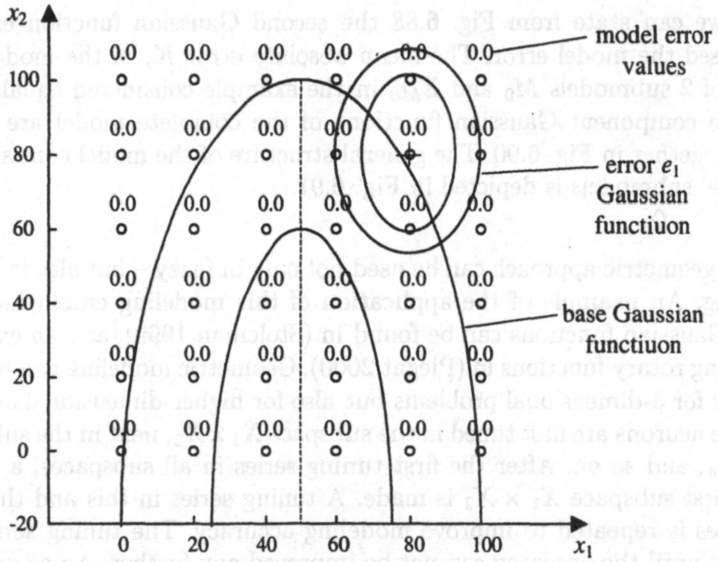

$$y^{**} = 100\exp\left[-\left|\frac{x_1-50}{50}\right|^2 - \left|\frac{x_2+20}{100}\right|^2\right] + 40\exp\left[-\left|\frac{x_1-80}{20}\right|^2 - \left|\frac{x_2-80}{30}\right|^2\right]$$

Fig. 6.90. Two component Gaussian functions of the complete model

A fuzzy error model E_{M0} of the base model M_0 is given by rules (6.77) and by membership functions, which are shown in Fig. 6.89.

$$\begin{aligned} R2:&\ \mathit{IF}\ (x_1 \text{ about } 80)\ \mathit{AND}\ (x_2 \text{ about } 80)\ \mathit{THEN}\ (e_1 \text{ about } 40)\\ R2N:&\ \mathit{IF}\ (x_1\ \mathit{NOT} \text{ about } 80)\ \mathit{OR}\ (x_2\ \mathit{NOT} \text{ about } 80)\\ &\ \mathit{THEN}\ (e_1 \text{ about } 0) \end{aligned} \tag{6.77}$$

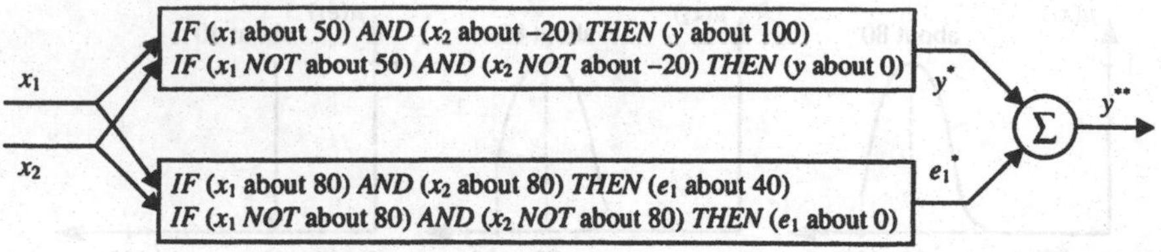
IF (x_1 about 50) AND (x_2 about −20) THEN (y about 100)
IF (x_1 NOT about 50) AND (x_2 NOT about −20) THEN (y about 0)
IF (x_1 about 80) AND (x_2 about 80) THEN (e_1 about 40)
IF (x_1 NOT about 80) AND (x_2 NOT about 80) THEN (e_1 about 0)
x_1
x_2
y^*
e_1^*
Σ
y^{**}

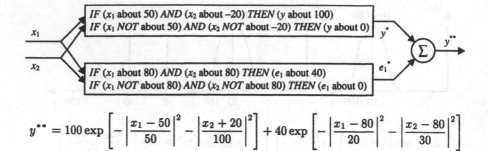

$$y^{**} = 100 \exp\left[-\left|\frac{x_1-50}{50}\right|^2 - \left|\frac{x_2+20}{100}\right|^2\right] + 40 \exp\left[-\left|\frac{x_1-80}{20}\right|^2 - \left|\frac{x_2-80}{30}\right|^2\right]$$

Fig. 6.91. Structure of the complete model of the system

As we can state from Fig. 6.88 the second Gaussian function entirely suppressed the model error. The mean absolute error K_1 of the model consisting of 2 submodels M_0 and E_{M0} in the example considered equals zero. The two component Gaussian functions of the complete model are represented together in Fig. 6.90. The general structure of the model consisting of 2 parallel submodels is depicted in Fig. 6.91.

◊ ◊ ◊

The geometric approach can be used not only in fuzzy – but also in neural modeling. An example of the application of this modeling employing nonrotary Gaussian functions can be found in (Stolcman 1999), and an example employing rotary functions in (Piegat 2000). Geometric modeling can be used not only for 3-dimensional problems but also for higher-dimensional ones. In this case neurons are first tuned in the subspace $X_1 \times X_2$, next in the subspace $X_3 \times X_4$, and so on. After the first tuning series in all subspaces, a return to the first subspace $X_1 \times X_2$ is made. A tuning series in this and the next subspaces is repeated to improve modeling accuracy. The tuning series are repeated until the accuracy can not be improved any further. An example of fuzzy geometrical modeling in 4-dimensional space is given in (Korzeń 2000) and of neural-geometric modeling in (Rzepka 2000).

6.3.3.2 Self-organization and self-tuning of fuzzy models with clustering methods

Beside the method of locating the model rules in "important" points of the system surface, described in Section 6.3.3.1, there are other methods, which can be called frequency methods. There are numerous phenomena for which the rules (regularity, generalization) can be discovered on the basis of the relative evaluation of the frequency of different variants of phenomenon behavior. The rules presented below describe certain examples:

R1: *IF* (the sunset is red) *THEN* (it will be nice weather the next day),
R2: *IF* (a student studies harder) *THEN* (the student will get a better job),
R3: *IF* (a person practices sports) *THEN* (he/she will live longer).

The above rules are of a statistical nature and the probability they are true is greater than 50% (that is why they have been recognized and formulated) for a reasonably large number of measurement samples (observations). The rules can appear false for if only a single sample or a few samples are taken. In a particular case, an individual practicing sports can live shorter than another doing no sports and smoking. A certain good student can get a worse job than his/her schoolmate who has poor results at school, etc. Still, we should remember that a rule is based on the statistical majority of cases and not on a single (or a few) individual cases. The genuineness of frequency rules can be evaluated on the basis of confidence coefficients (factor) reflecting the relative number of observations confirming a given rule. The question is whether the rules can be formulated with frequency methods in the case of modeling the surfaces of a system mapping $\mathbf{X} \to Y$. Let us make an analysis of the example in Fig. 6.92.

If the system surface is sampled in points located uniformly along the x-axis at intervals Δx, Fig. 6.92a, then on the basis of the sample projections on the y-axis the approximate location of the surface maximum point can be found. The highest density of sample projections on the y-axis occurs around the maximum. The membership function B_2 is placed at that point; the function will allow us to formulate a rule defining the maximum.

On the other hand, if the system surface is sampled along the y-axis at uniform interval Δy, Fig. 6.92b, then the density of those sample projections on the x-axis will be the lowest around the maximum. At that point the membership function A_2 is placed. The two membership functions B_2 and A_2 can be used to formulate the rule $R2$ defining the neighborhood of the maximum. The remaining two rules $R1$ and $R3$ define the boundaries of the universe of discourse X.

On the basis of Fig. 6.92 it can be stated that a single dimensional analysis (projections on each axis separately) of the frequency of sample occurrences, presuming the uniform sampling Δx, Δy, can provide us with information useful in formulating the rules for non-complicated $\mathbf{X} \to Y$ surfaces. Unfortunately, the measurement samples in real systems are rarely arranged uniformly in space. In some space regions the system can operate more frequently than in others. For instance, a vessel usually sails at a so-called economical speed, which, depending on the vessel type, can be, for example, 14, 15 or 16 knots. Examples of non-uniform sample configurations are presented in Fig. 6.93a.

The case illustrated in Fig. 6.93a shows the samples concentrated around the point P_2. On the right of that point the maximum of the system surface is found. However, as there are no measurement samples in that region it is

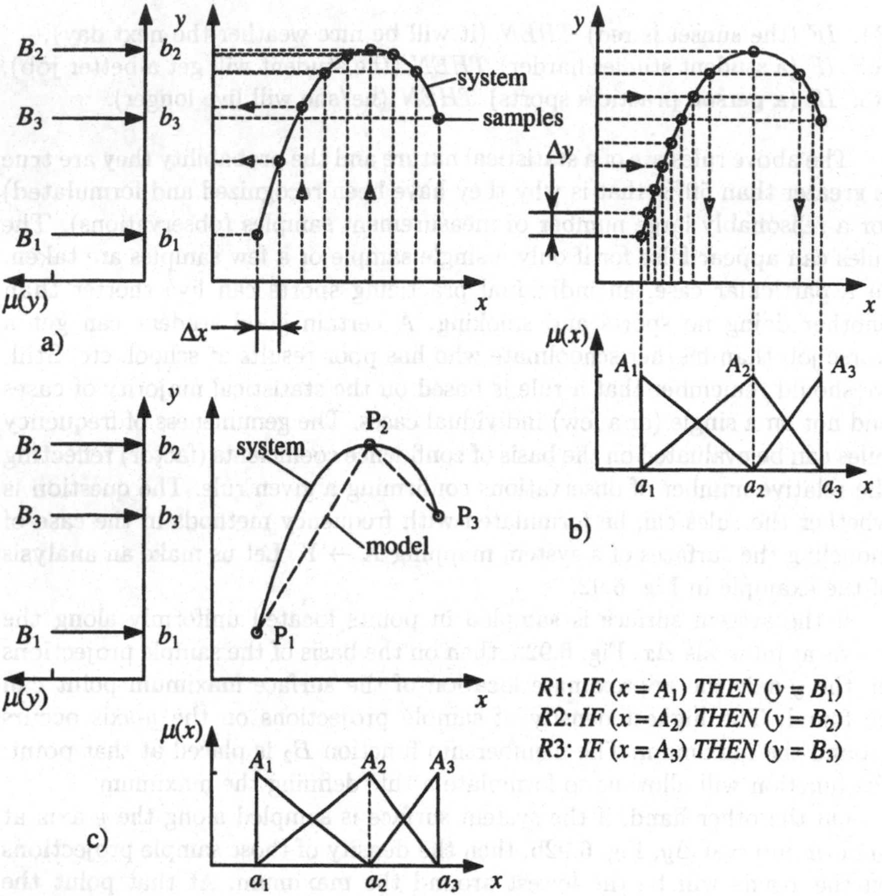

Fig. 6.92. Fuzzy model (c) defined on the basis of measurement sample frequency in a projection on the y-axis (a) and x-axis (b)

impossible to determine the coordinates of that "important" point. On the other hand, since the system never operates in the neighborhood of the maximum, the model surface without that point should be sufficient for practical purposes.

The case illustrated in Fig. 6.93b shows the greatest sample concentration around the point P_2 as well. In other regions the samples are arranged more or less uniformly. Thus, it is possible to determine the coordinates of the maximum and to place a rule there. On the other hand, it may appear beneficial to locate the rule at point P_2; such an operation enables us to reduce the mean error within the neighborhood of this point significantly as most of the measurement samples occur there. An imprecise location of the rule would contribute essentially to an increase of the model mean error.

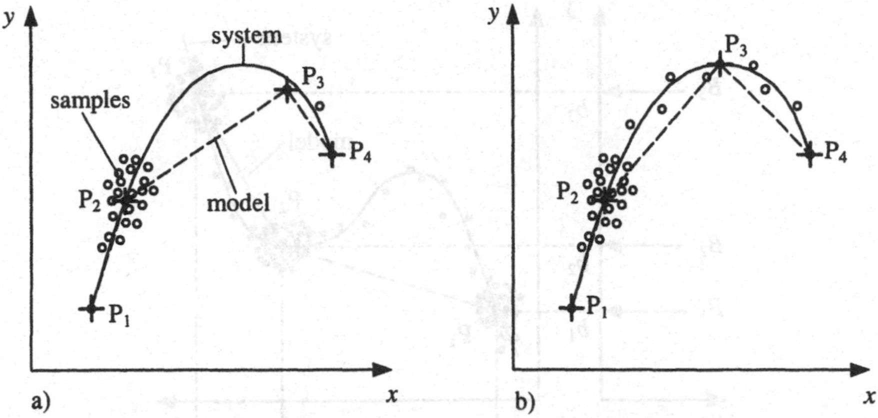

Fig. 6.93. Examples of non-uniform configurations of measurement samples in system input space

Frequency methods consider the points of maximum density of measurement samples, not the extrema of the model error, as the "important" points of the system. The rules are then located at the "important" points. The determination of a rule base, membership functions and their parameters depends on the capability to detect the dense measurement clusters and determine their centroids, Fig. 6.94.

It should be emphasized once more that even if the sample clusters were not located in the neighborhoods of system surface extrema the model constructed with the clustering method may have very high precision if the system operates mainly in areas close to clusters and the other states occur only sporadically. The samples obtained from system input and output measurements constitute a non-transparent mass of numbers; in the case of multi-input systems it is often difficult to provide a graphic representation of such a data mass and distinguish the clusters visually. Then, special mathematical methods detecting clusters automatically have to be applied. The easiest method of cluster detection is the **single dimension clustering method** (Preuss 1995). The method provides a projection of measurement samples on each axis separately, where the axes are $(x_1, \ldots, x_p, y)$, and allows us to identify one-dimensional clusters.

Fig. 6.95 presents the projection of 3D-samples $(X_1 \times X_2 \times Y)$ onto a 2D-plane $(X_1 \times X_2)$ and also onto axes X_1 and X_2. The membership function vertices a_{1i}, a_{2j} of sets A_{1i}, A_{2j} are located in the centroids of one-dimensional clusters. The next step is to complete the base of "candidate" rules:

$$Rr: \ IF\ (x_1 = A_{1i})\ AND\ (x_2 = A_{2j})\ THEN\ (y = B_k), \tag{6.78}$$

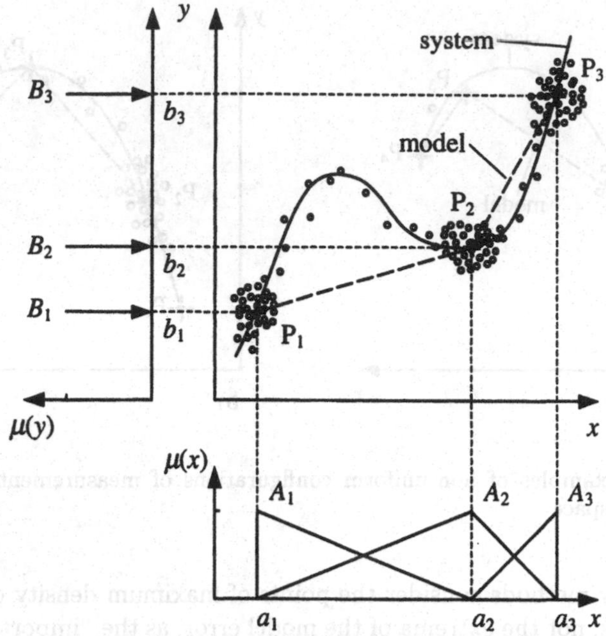

Fig. 6.94. Construction of a fuzzy model through the location of rules and peaks of membership functions in cluster centroids

where: $i = 1, \ldots m_1$ – number of fuzzy sets for input x_1,
$j = 1, \ldots m_2$ – number of fuzzy sets for input x_2,
$k = 1, \ldots m_3$ – number of fuzzy sets for input y.

The above base contains $m_1 \cdot m_2 \cdot m_3$ rules (all possible combinations of $A_{1i} \times A_{2j} \times B_k$). Part of the rules cannot be formulated because they occur in the input space regions where no measurement data are available. It may mean that the system never operates in those regions. The rules can be formulated only in regions covered by samples. If the sample P_l has coordinates (x_{1l}, x_{2l}, y_l) then the degree μ_r to which the sample confirms the veracity of rule Rr can be calculated with formula (6.79).

$$\mu_r(x_{1l}, x_{2l}, y_l) = MIN(\mu_{A1i}(x_{1p}), \mu_{A2j}(x_{2p}), \mu_{Bk}(y_p)) \tag{6.79}$$

The *MIN* operator can be replaced with the *PROD* operator or other t-norms. Particular measurement samples confirm the veracity of particular "candidate" rules to different degrees. If the condition (6.80) is met:

$$MAX \mu_r(x_{1l}, x_{2l}, y_l) > 0\,, \tag{6.80}$$

where: $l = 1, \ldots, m$ – measurement sample number, then rule Rr can potentially qualify to be included in the rule base. Fulfillment of the above condition means that the rule is located in such a sector of the universe of

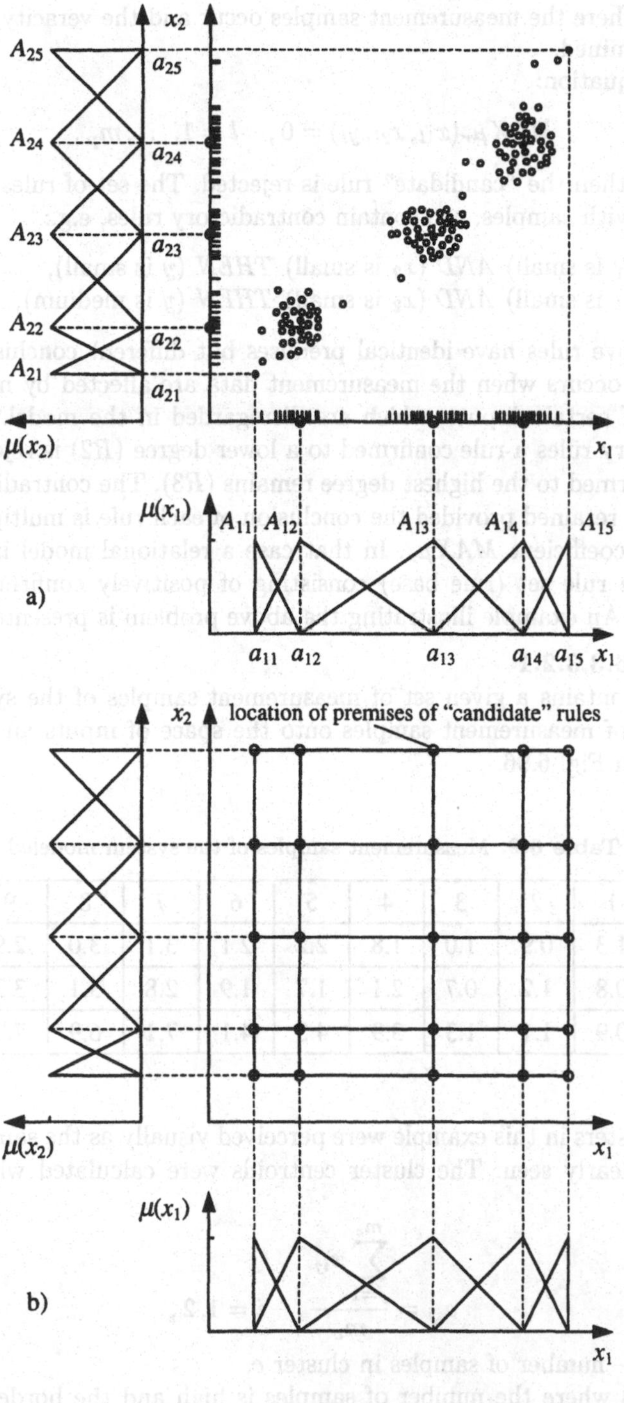

Fig. 6.95. Single dimension clustering (a) and formation of cluster membership functions (b)

discourse where the measurement samples occur and the veracity of the rule can be examined.

If the equation:

$$MAX\,\mu_r(x_{1l}, x_{2l}, y_l) = 0\,, \quad l = 1, \ldots, m_p\,,$$

is satisfied then the "candidate" rule is rejected. The set of rules, positively confirmed with samples, can contain contradictory rules, e.g.:

$$R1: IF\ (x_1 \text{ is small})\ AND\ (x_2 \text{ is small})\ THEN\ (y \text{ is small}), \quad \mu_1 = 0.9,$$
$$R2: IF\ (x_1 \text{ is small})\ AND\ (x_2 \text{ is small})\ THEN\ (y \text{ is medium}), \quad \mu_2 = 0.7.$$

The above rules have identical premises but different conclusions. Such a situation occurs when the measurement data are affected by noise or the influence of certain inputs, which are disregarded in the model. In case of contradictory rules a rule confirmed to a lower degree ($R2$) is rejected while a rule confirmed to the highest degree remains ($R3$). The contradictory rules can also be retained provided the conclusion of each rule is multiplied by its confidence coefficient $MAX\,\mu_r$. In that case a relational model is obtained. Finally, the rule set (rule base) consisting of positively confirmed rules is completed. An example illustrating the above problem is presented below.

Example 6.3.3.2.1

Table 6.9 contains a given set of measurement samples of the system. The projection of measurement samples onto the space of inputs and output is presented in Fig. 6.96.

Table 6.9. Measurement samples of the system modeled

l	1	2	3	4	5	6	7	8	9	10
x_1	1.3	0.9	1.0	1.8	2.2	2.1	3.1	3.0	2.9	2.7
x_2	0.8	1.2	0.7	2.1	1.7	1.9	2.8	3.1	3.3	2.9
y	0.9	1.1	1.3	3.9	4.3	4.1	7.1	6.9	7.2	4.5

The clusters in this example were perceived visually as the sample groups could be clearly seen. The cluster centroids were calculated with formula (6.81):

$$a_{ij} = \frac{\sum_{j=1}^{m_c} x_{ij}}{m_c}\,, \quad i = 1, 2\,, \tag{6.81}$$

where: m_c – number of samples in cluster c.

In cases where the number of samples is high and the borders between their groups are not sufficiently distinct, special methods of automatic searching for clusters are applied, e.g. method of c-means, method of k-means or

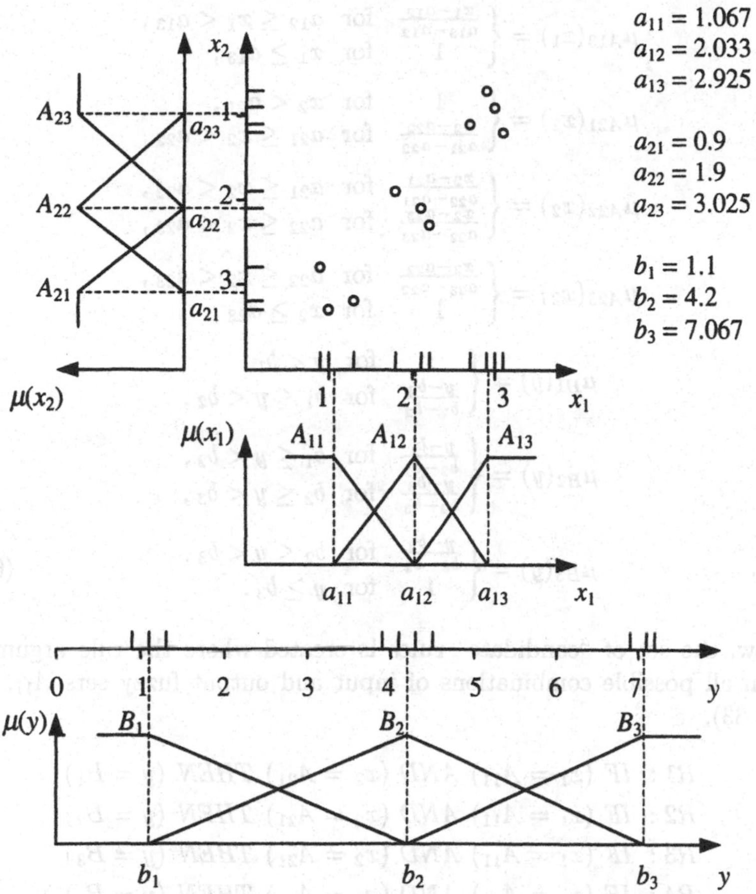

Fig. 6.96. Projections of measurement samples onto particular axes, the coordinates of clusters and the membership functions located in cluster centroids

others (Babuška 1995e; Cho 1995; Cipriano 1995; Delgado 1996,1997; Halgamuge 1996; Hanss 1996a; Kwon 1994; Langari 1995; Narazaki 1993; Osowski 1996; Rutkowska 1997; Su 1995; Yao 1995; Zhou 1995). The methods mentioned are described later in this chapter.

The modal values of membership functions are located in cluster centroids a_{ij}, Fig. 6.96. The membership functions are defined with formulas (6.82).

$$\mu_{A11}(x_1) = \begin{cases} 1 & \text{for } x_1 < a_{11}\,, \\ \frac{x_1 - a_{12}}{a_{11} - a_{12}} & \text{for } a_{11} \le x_1 < a_{12}\,, \end{cases}$$

$$\mu_{A12}(x_1) = \begin{cases} \frac{x_1 - a_{11}}{a_{12} - a_{11}} & \text{for } a_{11} \le x_1 < a_{12}\,, \\ \frac{x_1 - a_{13}}{a_{12} - a_{13}} & \text{for } a_{12} \le x_1 < a_{13}\,, \end{cases}$$

$$\mu_{A13}(x_1) = \begin{cases} \frac{x_1 - a_{12}}{a_{13} - a_{12}} & \text{for } a_{12} \le x_1 < a_{13}\,, \\ 1 & \text{for } x_1 \ge a_{13}\,, \end{cases}$$

$$\mu_{A21}(x_2) = \begin{cases} 1 & \text{for } x_2 < a_{21}\,, \\ \frac{x_2 - a_{22}}{a_{21} - a_{22}} & \text{for } a_{21} \le x_2 < a_{22}\,, \end{cases}$$

$$\mu_{A22}(x_2) = \begin{cases} \frac{x_2 - a_{21}}{a_{22} - a_{21}} & \text{for } a_{21} \le x_2 < a_{22}\,, \\ \frac{x_2 - a_{23}}{a_{22} - a_{23}} & \text{for } a_{22} \le x_2 < a_{23}\,, \end{cases}$$

$$\mu_{A23}(x_2) = \begin{cases} \frac{x_2 - a_{22}}{a_{23} - a_{22}} & \text{for } a_{22} \le x_2 < a_{23}\,, \\ 1 & \text{for } x_2 \ge a_{23}\,, \end{cases}$$

$$\mu_{B1}(y) = \begin{cases} 1 & \text{for } y < b_1\,, \\ \frac{y - b_2}{b_1 - b_2} & \text{for } b_1 \le y < b_2\,, \end{cases}$$

$$\mu_{B2}(y) = \begin{cases} \frac{y - b_1}{b_2 - b_1} & \text{for } b_1 \le y < b_2\,, \\ \frac{y - b_3}{b_2 - b_3} & \text{for } b_2 \le y < b_3\,, \end{cases}$$

$$\mu_{B3}(y) = \begin{cases} \frac{y - b_2}{b_3 - b_2} & \text{for } b_2 \le y < b_3\,, \\ 1 & \text{for } y \ge b_3\,. \end{cases} \tag{6.82}$$

Now, the set of "candidate" rules is created where the rule arguments contain all possible combinations of input and output fuzzy sets A_{1i}, A_{2j}, B_k, (6.83).

$$\begin{aligned}
&R1: \ IF\ (x_1 = A_{11})\ AND\ (x_2 = A_{21})\ THEN\ (y = B_1) \\
&R2: \ IF\ (x_1 = A_{11})\ AND\ (x_2 = A_{21})\ THEN\ (y = B_2) \\
&R3: \ IF\ (x_1 = A_{11})\ AND\ (x_2 = A_{21})\ THEN\ (y = B_3) \\
&R4: \ IF\ (x_1 = A_{11})\ AND\ (x_2 = A_{22})\ THEN\ (y = B_1) \\
&R5: \ IF\ (x_1 = A_{11})\ AND\ (x_2 = A_{22})\ THEN\ (y = B_2) \\
&R6: \ IF\ (x_1 = A_{11})\ AND\ (x_2 = A_{22})\ THEN\ (y = B_3) \\
&\vdots \\
&R27: \ IF\ (x_1 = A_{13})\ AND\ (x_2 = A_{23})\ THEN\ (y = B_3)
\end{aligned} \tag{6.83}$$

The rule set contains 27 "candidate" rules. For each measurement sample the confidence coefficient for each particular rule is calculated. Then the maximum μ_r is determined for all the rules. For example, the coefficient μ_r for $R1$ by the sample (1.3, 0.8, 0.9) is calculated using the formula (6.84).

$$\begin{aligned}
\mu_1(1.3, 0.8, 0.9) &= MIN(\mu_{A11}(1.3), \mu_{A21}(0.8), \mu_{B1}(0.9)) = \\
&= MIN((1.3 - a_{12})/(a_{11} - a_{12}), 1, 1) = \\
&= MIN(0.759, 1, 1) = 0.759
\end{aligned} \tag{6.84}$$

The degree to which rule $R3$ has been confirmed with the same sample is calculated with formula (6.85).

$$\mu_3(1.3, 0.8, 0.9) = MIN(\mu_{A11}(1.3), \mu_{A21}(0.8), \mu_{B3}(0.9)) = \\ = MIN(0.759, 1, 0) = 0 \qquad (6.85)$$

Upon calculation of the confidence coefficient of each "candidate" rule for each sample, the maximum coefficients for each rule respectively are determined. The calculation results for the example considered are presented in Table 6.10.

Table 6.10. Confidence coefficient of "candidate" rules (6.83)

Rule no. r	1	2	3	4	5	6	7	8	9
$MAX\,\mu_r$	0.935	0.064	0	0.3	0.241	0	0.097	0.178	0
r	10	11	12	13	14	15	16	17	18
$MAX\,\mu_r$	0	0	0	0.097	0.925	0.105	0.097	0.252	0.105
r	19	20	21	22	23	24	25	26	27
$MAX\,\mu_r$	0	0	0	0.032	0.748	0.2	0	0.105	0.972

On the basis of results in Table 6.10 it is seen that 15 rules have been evaluated positively while 12 rules, with confidence coefficients $MAX\,\mu_r = 0$, are classified for rejection from the rule set. Thus, the new rule set containing 15 "candidate" rules was determined. However, it comprises contradictory rules of identical premises but different conclusions. Those rules are grouped in Table 6.11. To simplify their form the rules:

$$IF\ (x_1 = A_{1i})\ AND\ (x_2 = A_{2j})\ THEN\ (y = B_k)\,,$$

are presented in abbreviated form as (A_{1i}, A_{2j}, B_k).

Table 6.11. Groups of contradictory rules

Rule no. r	$A_{11}A_{21}B_1,\ A_{11}A_{21}B_2$	$A_{11}A_{22}B_1,\ A_{11}A_{22}B_2$	$A_{11}A_{23}B_1,\ A_{11}A_{23}B_2$
$MAX\,\mu_r$	0.935 , 0.064	0.3 , 0.241	0.097 , 0.178

Rule no. r	$A_{12}A_{22}B_1,\ A_{12}A_{22}B_2,\ A_{12}A_{22}B_3$	$A_{12}A_{23}B_1,\ A_{12}A_{23}B_2,\ A_{12}A_{23}B_3$
$MAX\,\mu_r$	0.097 , 0.925 , 0.105	0.097 , 0.252 , 0.105
Rule no. r	$A_{13}A_{22}B_1,\ A_{13}A_{22}B_2,\ A_{13}A_{22}B_3$	$A_{13}A_{23}B_2,\ A_{13}A_{23}B_3$
$MAX\,\mu_r$	0.032 , 0.748 , 0.2	0.105 , 0.972

The problem of contradictory rules can be solved in two ways. The first method is to create a relational rule base (6.86) where the conclusions are multiplied by the confidence coefficients.

$$\begin{aligned}
R1:&\ IF\ (x_1 = A_{11})\ AND\ (x_2 = A_{21})\\
&\ THEN\ (y = 0.935B_1)\ OR\ (y = 0.064B_2)\\
R2:&\ IF\ (x_1 = A_{11})\ AND\ (x_2 = A_{22})\\
&\ THEN\ (y = 0.3B_1)\ OR\ (y = 0.241B_2)\\
R3:&\ IF\ (x_1 = A_{11})\ AND\ (x_2 = A_{23})\\
&\ THEN\ (y = 0.097B_1)\ OR\ (y = 0.178B_2)\\
R4:&\ IF\ (x_1 = A_{12})\ AND\ (x_2 = A_{22})\\
&\ THEN\ (y = 0.097B_1)\ OR\ (y = 0.925B_2)\ OR\ (y = 0.105B_3)\\
R5:&\ IF\ (x_1 = A_{12})\ AND\ (x_2 = A_{23})\\
&\ THEN\ (y = 0.097B_1)\ OR\ (y = 0.252B_2)\ OR\ (y = 0.105B_3)\\
R6:&\ IF\ (x_1 = A_{13})\ AND\ (x_2 = A_{22})\\
&\ THEN\ (y = 0.032B_1)\ OR\ (y = 0.748B_2)\ OR\ (y = 0.2B_3)\\
R7:&\ IF\ (x_1 = A_{13})\ AND\ (x_2 = A_{23})\\
&\ THEN\ (y = 0.105B_2)\ OR\ (y = 0.972B_3)
\end{aligned} \tag{6.86}$$

The conclusion in the form of, for example $(y = 0.935B_1)$, denotes that during the defuzification process the activation degree $\mu_{B1}(y)$ of the conclusion set B_1 should be multiplied by the confidence coefficient of the conclusion $\mu_r = 0.935$. The relational model (6.86) is more complicated, but it usually provides higher precision.

The second method of dealing with contradictory rules is the method of selecting the dominating conclusions, i.e. such conclusions that have the highest confidence coefficient, while the conclusions of lower confidence coefficients are rejected. This method provides model simplification and improvement of its clarity. The relational model (6.86) simplified with this method can be expressed with formula (6.87).

$$\begin{aligned}
R1:&\ IF\ (x_1 = A_{11})\ AND\ [(x_2 = A_{21})\ OR\ (x_2 = A_{22})]\ THEN\ (y = B_1)\\
R2:&\ IF\ (x_1 = A_{11})\ AND\ (x_2 = A_{23})\ THEN\ (y = B_2)\\
R3:&\ IF\ (x_1 = A_{12})\ AND\ [(x_2 = A_{22})\ OR\ (x_2 = A_{23})]\ THEN\ (y = B_2)\\
R4:&\ IF\ (x_1 = A_{13})\ AND\ (x_2 = A_{22})\ THEN\ (y = B_2)\\
R5:&\ IF\ (x_1 = A_{13})\ AND\ (x_2 = A_{23})\ THEN\ (y = B_3)
\end{aligned} \tag{6.87}$$

The simplified model (6.87) is usually less precise than the corresponding relational model. The smaller the difference between confidence coefficients of the retained and rejected rules, the greater the drop in model precision. For example, let us consider rule $R2$ from rule base (6.86):

$$R2:\ IF\ (x_1 = A_{11})\ AND\ (x_2 = A_{22})\ THEN\ (y = 0.3B_1)\ OR\ (y = 0.241B_2)\,, \tag{6.88}$$

the dominance of $(y = 0.3B_1)$ over $(y = 0.241B_2)$ is small. Acceptance of the conclusion $(y = 0.3B_1)$ and rejection of the conclusion $(y = 0.241B_2)$ may

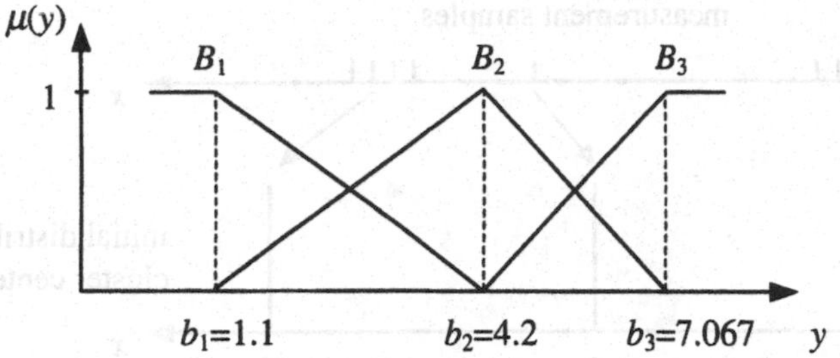

Fig. 6.97. Significant difference in modal values of the sets B_1 and B_2 of the conclusions

result in a significant drop in model precision, particularly if the modal value of the set B_2 is essentially higher than of the set B_1, Fig. 6.97.

In this example, the set B_2 has a greater effect on the defuzzification process than the set B_1. The product of confidence coefficient μ_r and modal value b_1 equals $0.3 \cdot 1.1 = 0.33$ and for the set B_2 it is equal to $0.241 \cdot 4.2 = 1.01$. **If the chosen criterion of conclusion dominance is not the confidence coefficient $\mu_r(B_k)$ but the product of this coefficient and the modal value b_k of the set B_k $(\mu_r(B_k) \cdot b_k)$ then usually a lower drop in model precision results** and a different form of rules is obtained. Thus, the simplified rule (6.88) can be represented by:

$$R2: \; IF \; (x_1 = A_{11}) \; AND \; (x_2 = A_{22}) \; THEN \; (y = 0.241 B_2),$$

but not by:

$$R2: \; IF \; (x_1 = A_{11}) \; AND \; (x_2 = A_{22}) \; THEN \; (y = 0.3 B_1).$$

◊ ◊ ◊

The methods of rule base creation presented above are not free from disadvantages, though. An important disadvantage results from the fact that the confidence coefficients of "candidate" rules are evaluated on the basis of measurement samples, which can be significantly affected by noise. Regarding the latter the evaluation itself can also be noise affected and therefore not accurate either. The resulting model can be therefore considered as an initial model that should be transformed into a neuro-fuzzy network and further tuned with the system measurement samples. Those operations often enable us to achieve a significant increase in the precision of the model. The method of elimination of "candidate" rules on the basis of their confidence coefficients can be used to build a rule base without clustering. The universe of discourse of each input x_i and output y of the system can be divided into an arbitrarily (intuitively) defined number of fuzzy sets. The sets can be distributed (arranged), for example, uniformly. In the next step a complete base of "candidate" rules should be created – the base should contain all combinations of

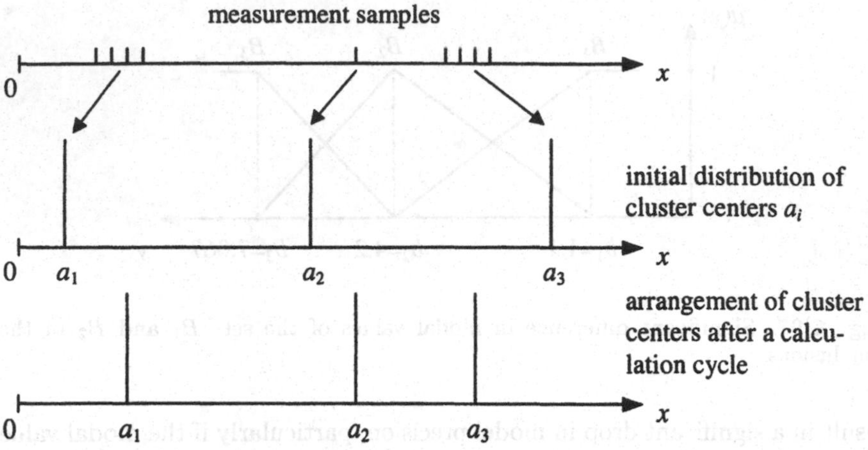

Fig. 6.98. Illustration of the k-means clustering method

input and output sets. Then each combination (candidate rule) should have its confidence coefficient evaluated on the basis of measurement samples.

Single dimension clustering can be carried out with various methods. One of the most popular methods is the **k-means method** (Preuss 1995; Osowski 1996). The algorithm of the above method **for a one-dimensional case** is presented below.

I. Define the number of clusters k and the initial coordinates $a_1, \ldots, a_k$ of their centers.
II. Assign each measurement sample to the nearest cluster.
III. Calculate new center coordinates $a_1, \ldots, a_k$ of the clusters on the basis of the samples assigned to them.
IV. Check whether any changes in cluster locations occurred, compared to the previous calculation cycle (in terms of a mean or maximum relative translocation). If yes: return to step II and repeat the calculation cycle. If no: end of calculation.

The method of k-means is illustrated in Fig. 6.98. The version of the method presented here is very general and provides information on the basic concept of the method that can be modified in many ways, for example two adjacent clusters can be merged into one cluster. Another concept might be the analysis of a number of samples assigned to each cluster and then the elimination of clusters with very few samples.

The process of center updating can be performed with the method of successive presentation of individual samples or all the samples concurrently (Osowski 1996). The assigning of samples to the clusters can be done on the basis of their Euclidean distance to the cluster centers or, in a fuzzy version, on the basis of the membership functions assigning the samples to a

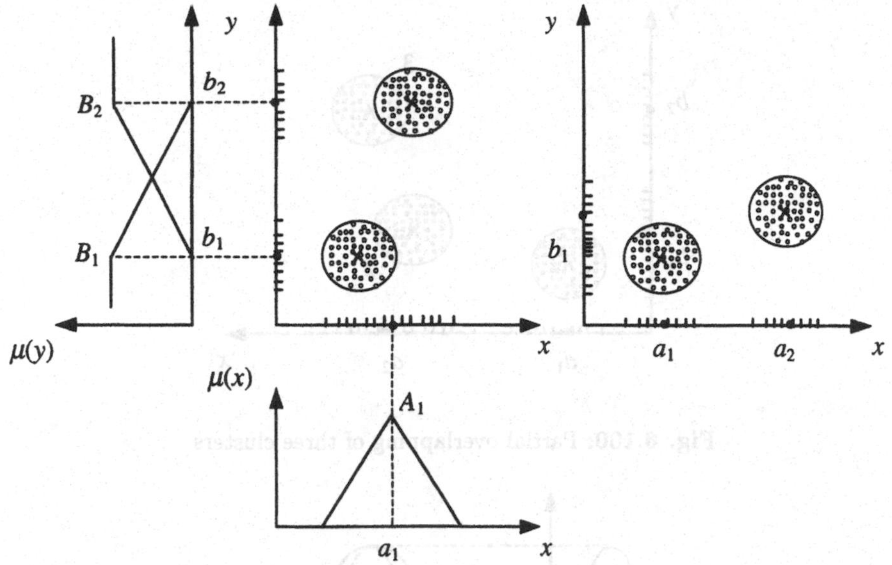

Fig. 6.99. Cluster overlapping in the projection on the x-axis (a) and the y-axis (b)

cluster. In that case a single measurement sample may be assigned to several clusters at a time, with a different membership grade in each cluster, and thus influence the cluster center translocation according to the membership grade. Another specific disadvantage of the one-dimensional clustering method is the potential occurrence of so called **pseudo-clusters**. This occurs as the result of individual projections of samples on each coordinate axis. The problem is illustrated in Fig. 6.99.

The two clusters, distinctly separated in 2D space $X \times Y$, overlap when projected on the x-axis. The one-dimensional analysis provides information only about the existence of a single cluster that allows us to construct only one fuzzy set A_1 in space X. The center a_1 does not match any of the centers of the two 2D clusters. A similar case of partially or entirely overlapping clusters may refer to the y-axis. The overlapping clusters prevent us from determining the number or coordinates of their centers correctly. Fig. 6.100 illustrates a case where the separate sample projections on the x-axis and the y-axis allow us to identify 2 clusters on each axis while actually 3 clusters exist.

Similar misleading conclusions may arise in multi-dimensional cases when the clusters are identified separately in the input space $\mathbf{X} = X_1 \times \ldots \times X_p$ and output space Y, Fig. 6.101. In that case only one cluster in the input space would be identified and its center would not match any center of the two actual clusters. The problem of cluster overlapping increases with the increase in the number of system inputs. The examples presented prove that the most

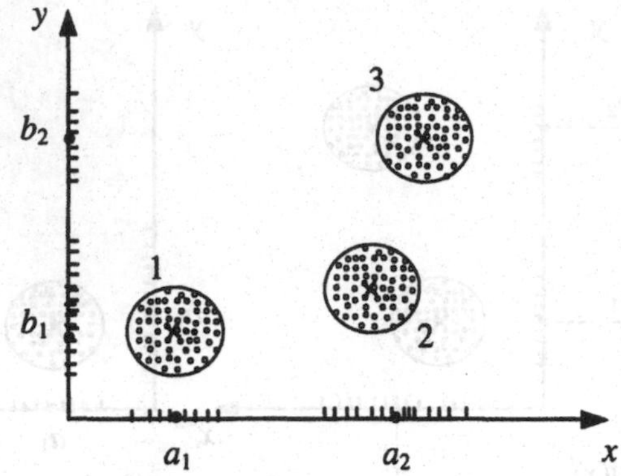

Fig. 6.100. Partial overlapping of three clusters

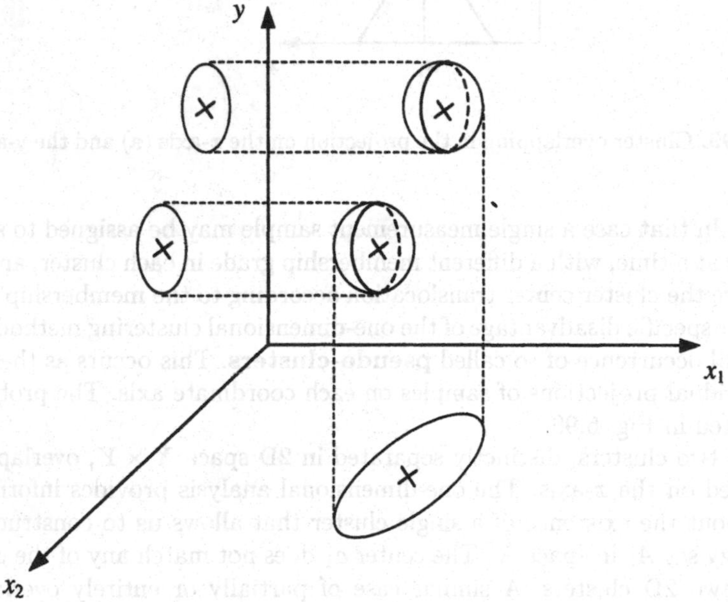

Fig. 6.101. Clusters overlapping in the projection on input space $X_1 \times X_2$

secured clustering is the full-dimensional clustering in system space $\mathbf{X} \times Y$, where $\mathbf{X} = X_1 \times \ldots \times X_p$ is the input space. Below we present an **algorithm of full-dimensional non-fuzzy clustering with the k-means method**, where the system output y is described with the symbol $x_{(p+1)}$ as a simplified form of y $(y = x_{(p+1)})$.

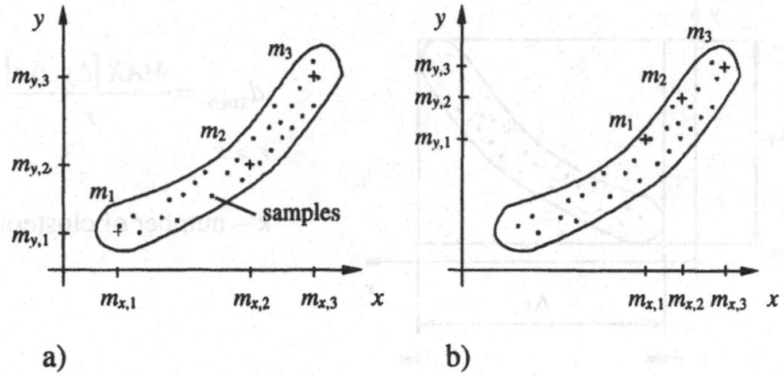

Fig. 6.102. Advantageous (a) and disadvantageous (b) location of initial cluster centers $\mathbf{m}_i$ generated at random

I. Initialization of clusters

Let us define the number of clusters k and the initial vectors of their center coordinates $\mathbf{m}_i(0), i = 1, \ldots, k$:

$$\mathbf{m}_1 = [m_{x1,1}, \ldots, m_{xl,1}, \ldots, m_{x(p+1),1}]^{\mathrm{T}},$$

$$\vdots$$

$$\mathbf{m}_i = [m_{x1,i}, \ldots, m_{xl,i}, \ldots, m_{x(p+1),i}]^{\mathrm{T}},$$

$$\vdots$$

$$\mathbf{m}_k = [m_{x1,k}, \ldots, m_{xl,k}, \ldots, m_{x(p+1),k}]^{\mathrm{T}},$$

where: p – number of system inputs.

The number of clusters k can be determined intuitively or generated at random. Another way is to perform clustering successively, beginning with a small quantity of clusters and expanding it gradually, testing the precision of successive models (Davè 1997). The cluster centers $\mathbf{m}_i$ can be determined with the method of random selection of k measurement samples $(x_1, \ldots, x_{(p+1)})$ from among all N samples. The random selection can be more or less advantageous for the further course of the modeling process, Fig. 6.102.

The recommended minimum number of clusters k should be equal to the dimensional size of the space $\mathbf{X}^p \times Y$ of the system $(p+1)$ to enable the model to form an interpolation surface of identical, full dimensional size. Lucky location of initial clusters reduces the computation efforts and facilitates finding the optimum (sub-optimum) solution. Unlucky location, Fig. 6.102b, may give an undesired result: the algorithm might easily get stuck in a local optimum. As a rule, the more uniform the location of initial clusters in the universe of discourse, the faster the algorithm operation.

To avoid the situation where the randomly selected cluster centers $\mathbf{m}_i$ may appear too close to each other, a condition may be imposed on the selection

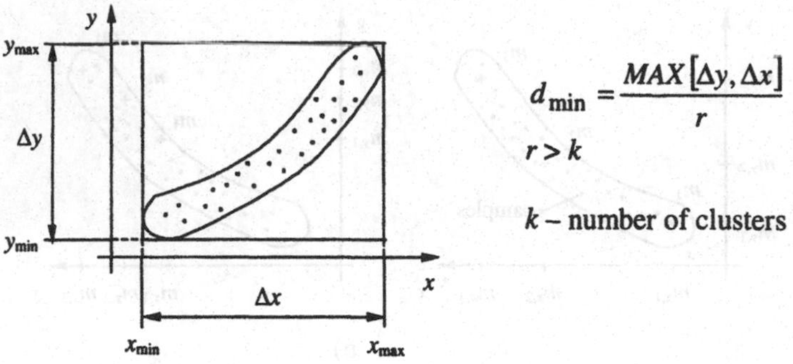

Fig. 6.103. One of the possible methods of determination of the minimum distance $d_{\min}$ for initial clusters

process to accept only those samples which are separated by at least a certain minimum distance $d_{\min}$ from each of the previously selected centers $\mathbf{m}_i$. The distance d_{ij} between the measurement sample $\mathbf{x}_j$ and cluster center $\mathbf{m}_i$ can be calculated with formula (6.89).

$$d_{ij} = \|\mathbf{x}_j - \mathbf{m}_i\| = \sqrt{\sum_{l=1}^{p+1} (x_{l,j} - m_{xl,i})^2} \tag{6.89}$$

The minimum distance $d_{\min}$ can be determined with the heuristic method if the following is known: the number of clusters k and the range of the universe of discourse defined with values $x_{l\min}$, $x_{l\max}$, $y_{l\min}$, $y_{l\max}$, for particular axes of the coordinate system, Fig. 6.103.

II. Assignment of measurement samples

Each measurement sample j has to be assigned to the closest cluster i. The distance d_{ij} from a cluster is calculated with formula (6.89).

III. Determination of new cluster centers

New cluster centers are calculated as the centroids of measurement samples assigned to them. If N_i denotes a number of points assigned to a cluster i then the coordinates x_l, y, $l = 1 + p$, of the centers are calculated with formula (6.90):

$$m_{xl,i}(t+1) = \frac{1}{N_i} \sum_{j=1}^{N_i} x_{l,j}(t), \quad l = 1, \ldots, (p+1), \tag{6.90}$$

where: t – number of clustering cycle.

The determination of new cluster centers on the basis of all points assigned to a cluster is known as a cumulative approach (Osowski 1996). There is also a direct approach where updating of cluster centers is done gradually, after presentation of individual samples. The direct approach method provides updating of only one center in each step, it is the center located at the closest position to the sample considered. The center is moved toward the sample by the value calculated with formula (6.91), depending on the learning rate coefficient $\eta \ll 1$.

$$m_{xl,i}(t+1) = m_{xl,i}(t) + \eta[x_l - m_{xl,i}(t)], \quad l = 1, \ldots, (p+1) \tag{6.91}$$

The direct approach, showing slighter changes in center translocations in successive steps, has a slightly better convergence than the cumulative approach and therefore is more frequently used in practice.

IV. Verification of cluster translocation Δm_i

The translocation of cluster centers in relation to their positions in the preceding clustering cycle $(t-1)$ is calculated with formula (6.92).

$$\Delta m_i(t) = \|\mathbf{m}_i(t) - \mathbf{m}_i(t-1)\| \tag{6.92}$$

If the minimum translocation Δm_i satisfies condition (6.93):

$$\min \Delta m_i(t) \le \varepsilon, \quad i = 1, \ldots, k, \tag{6.93}$$

where: ε – threshold value, then the updating of cluster centers $\mathbf{m}_i$ is completed and the procedure moves onto step V. Otherwise, repeat from step II.

V. Determination of cluster membership functions

The clusters were defined within the entire space $\mathbf{X} \times Y$ of the system. This fact allows us to more accurately distinguish the groups of samples and to avoid overlapping different clusters in projections on the space $\mathbf{X}$. In practical applications, however, it is usually indispensable to determine system output y for a given input vector $\mathbf{x}$. That is why the membership functions of fuzzy sets should be determined separately for inputs and output. Since the cluster centers m_{ix} are usually arranged unevenly in the space, it appears beneficial to apply exponential, Gaussian membership functions. The centers m_{ix} of those functions match the projections of cluster centers $\mathbf{m}_i$ on the input space $\mathbf{X}$ and their widths σ_i can be chosen to be 1/3 of the distance D_{ij} to the closest cluster, Fig. 6.104.

In the case of the Gaussian function the membership is equal to 0.011 at a distance 3σ from the center. This fact prevents one rule conclusion from disturbing others. It also makes the rules "tell the truth", i.e. the model output for vector $\mathbf{x}$, which perfectly matches the center of the membership function occurring in the rule premises, is (nearly) equal to the value of the output given in the rule conclusion.

Below we present an example of full-dimensional clustering with the k-means method according to the cumulative approach.

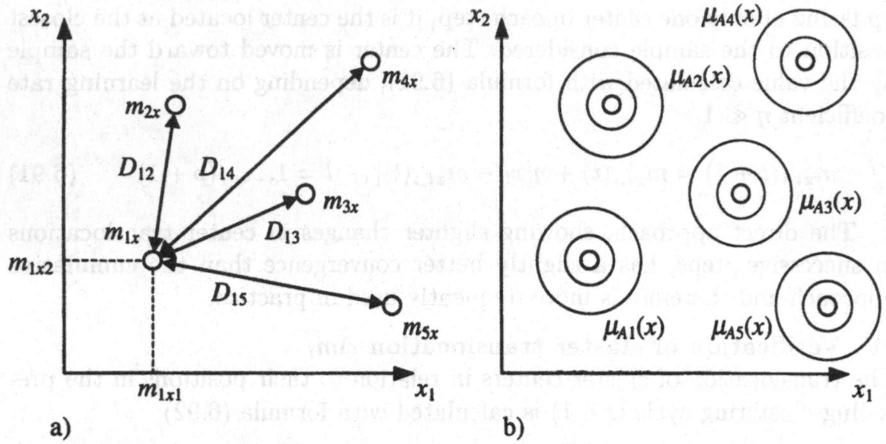

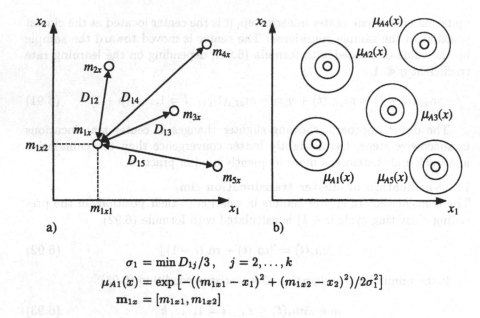

$$\sigma_1 = \min D_{1j}/3\,, \quad j = 2,\ldots,k$$
$$\mu_{A1}(x) = \exp\left[-((m_{1x1} - x_1)^2 + (m_{1x2} - x_2)^2)/2\sigma_1^2\right]$$
$$\mathbf{m}_{1x} = [m_{1x1}, m_{1x2}]$$

Fig. 6.104. Illustration of the method of determination of parameters $\mathbf{m}_i$ and σ_i of the Gaussian membership function (a) and the cross-sections of membership function in the input space (b)

Example 6.3.3.2.2

Let us consider a set of measurement data of the system modeled, Table 6.12.

The distribution of the measurement samples in $\mathbf{X} \times Y = X_1 \times X_2$ is presented in Fig. 6.105. Fig. 6.105a allows us to visually distinguish three clusters $K1$, $K2$, $K3$ in the space $X \times Y$. If clustering was performed only in the input space X then the projections of clusters $K2$ and $K3$ on the x-axis would overlap and the clusters would be indistinguishable.

Step I. Initialization of clusters

Let us consider 3 initial clusters with the following centers:

$$\mathbf{m}_1(0) = [0,0]^{\mathrm{T}}\,,$$
$$\mathbf{m}_2(0) = [3,0]^{\mathrm{T}}\,,$$
$$\mathbf{m}_3(0) = [3,3]^{\mathrm{T}}\,. \tag{6.94}$$

Table 6.12. System measurement samples

P_j	1	2	3	4	5	6	7	8	9
$x = x_1$	0.7	1.4	1.2	1.8	2.5	2.3	2.4	3.0	2.1
$y = x_2$	0.8	0.9	1.3	0.8	0.9	1.6	2.4	2.7	3.0

Step II. Assignment of measurement samples to the clusters
On the basis of formula (6.89) the Euclidean distances d_{ij} between particular samples P_j and the centers $\mathbf{m}_i$ are calculated (6.95).

$$d_{ij} = \|\mathbf{x}_j - \mathbf{m}_i\| = \sqrt{\sum_{l=1}^{2}(x_{j,1} - m_{xl,1})^2} \tag{6.95}$$

The distances calculated are presented in Table 6.13.

Table 6.13. The distances between the measurement samples P_j and initial cluster centers $\mathbf{m}_i(0)$ calculated with formula (6.95)

P_j	1	2	3	4	5	6	7	8	9
d_{1j}	1.06	1.66	1.77	1.97	2.66	2.80	3.39	4.04	3.66
d_{2j}	2.43	1.84	2.22	1.44	1.03	1.75	2,4	2.70	3.13
d_{3j}	3.18	2.64	2.48	2.51	2.16	1.56	0.85	0.30	0.90

On the basis of distances listed in Table 6.13 the closest cluster center for each sample P_j is determined (the darkened fields in Table 6.13). The assignment in the form of formula (6.96) and Fig. 6.105b demonstrates the results obtained.

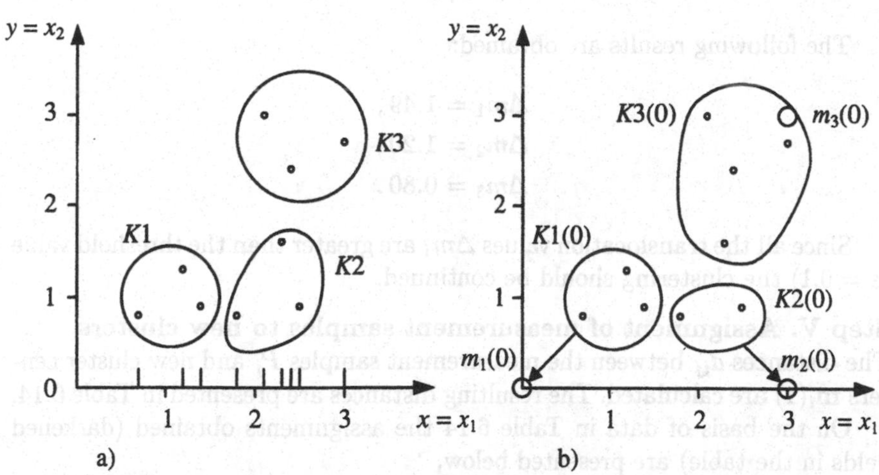

Fig. 6.105. Distribution of measurement samples in the space $X \times Y$ (a) and clusters initiating the algorithm (b)

$$K1(0) = (P_1, P_2, P_3)\,, \quad N_1 = 3\,,$$
$$K2(0) = (P_4, P_5)\,, \quad N_2 = 2\,,$$
$$K3(0) = (P_6, P_7, P_8, P_9)\,, \quad N_3 = 4\,. \tag{6.96}$$

Step III. Determination of new positions of cluster centers
New cluster centers $\mathbf{m}_i(1)$ are computed with the cumulative method using formula (6.97).

$$m_{xl,i} = \frac{1}{N_i}\sum_{j=1}^{N_i} x_{lj}(0) \tag{6.97}$$

The result obtained is presented in (6.98).

$$\mathbf{m}_1(1) = [1.1, 1.0]^T\,,$$
$$\mathbf{m}_2(1) = [2.15, 0.85]^T\,,$$
$$\mathbf{m}_3(1) = [2.45, 2.425]^T\,. \tag{6.98}$$

Step IV. Calculation of cluster translocation distance Δm_i
This step provides calculation of the distances by which the cluster centers $\mathbf{m}_i$ have moved. The results form the basis for making a decision whether it is necessary to continue the clustering process. Formula (6.99) is used to calculate cluster translocations.

$$\Delta m_i = \|\mathbf{m}_i(0) = \mathbf{m}_i(1)\| \tag{6.99}$$

The following results are obtained:

$$\Delta m_1 = 1.49\,,$$
$$\Delta m_2 = 1.20\,,$$
$$\Delta m_3 = 0.80\,.$$

Since all the translocation values Δm_i are greater than the threshold value ($\varepsilon = 0.1$) the clustering should be continued.

Step V. Assignment of measurement samples to new clusters
The distances d_{ij} between the measurement samples P_j and new cluster centers $\mathbf{m}_i(1)$ are calculated. The resulting distances are presented in Table 6.14.

On the basis of data in Table 6.14 the assignments obtained (darkened fields in the table) are presented below.

$$K1(1) = (P_1, P_2, P_3)\,, \quad N_1 = 3\,,$$
$$K2(1) = (P_4, P_5, P_6)\,, \quad N_2 = 3\,,$$
$$K3(1) = (P_7, P_8, P_9)\,, \quad N_3 = 3\,.$$

Table 6.14. The distances between samples P_j and cluster centers $\mathbf{m}_i(1)$

P_j	1	2	3	4	5	6	7	8	9
d_{1j}	0.45	0.32	0.32	0.73	1.40	1.26	1.91	2.46	2.24
d_{2j}	1.45	0.75	1.05	0.35	0.35	0.76	1.57	2.04	2.15
d_{3j}	2.39	1.85	1.68	1.75	1.53	0.84	0.06	0.61	0.67

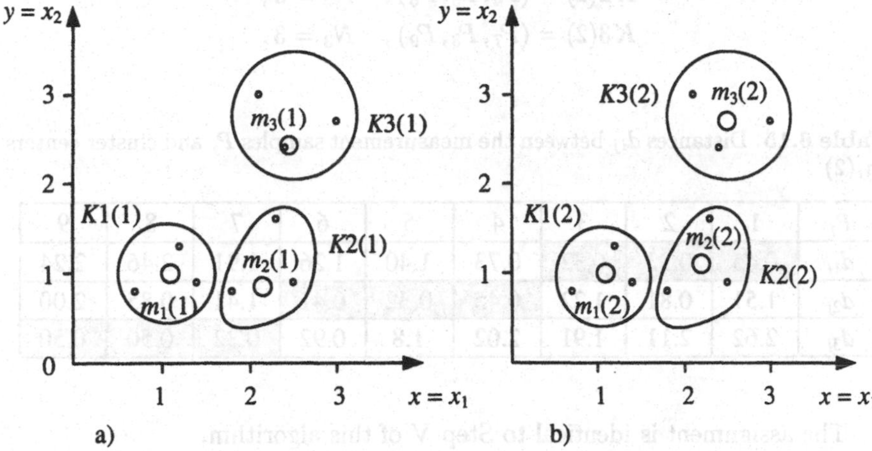

Fig. 6.106. Clusters obtained in Step II and IV of the algorithm

Step VI. Determination of new positions of cluster centers

The new positions of cluster centers, determined on the basis of the measurement samples assigned to them, have the following co-ordinates:

$$\mathbf{m}_1(2) = [1.1, 1.0]^T,$$
$$\mathbf{m}_2(2) = [2.2, 1.1]^T,$$
$$\mathbf{m}_3(2) = [2.5, 2.7]^T.$$

The clusters Ki and their centers $\mathbf{m}_i$ are presented in Fig. 6.106.

Step VII. Determination of cluster translocation size Δm_i

The cluster center translocations $\Delta m_i = ||\mathbf{m}_i(1) = \mathbf{m}_i(2)||$ are equal to:

$$\Delta m_1 = 0,$$
$$\Delta m_2 = 0.255,$$
$$\Delta m_3 = 0.279.$$

Since the translocation values Δm_2 and Δm_3 are greater than the threshold value ($\varepsilon = 0.1$) chosen, the process of clustering is continued.

Step VIII. Assignment of measurement samples to translocated clusters
Table 6.15 contains the distance values d_{ij} between the samples P_j and new cluster centers $\mathbf{m}_i(2)$. The assignment of samples to clusters is executed on the basis of the data in the table (the darkened fields in the table).

$$K1(2) = (P_1, P_2, P_3), \quad N_1 = 3,$$
$$K2(2) = (P_4, P_5, P_6), \quad N_2 = 3,$$
$$K3(2) = (P_7, P_8, P_9), \quad N_3 = 3.$$

Table 6.15. Distances d_{ij} between the measurement samples P_j and cluster centers $\mathbf{m}_i(2)$

P_j	1	2	3	4	5	6	7	8	9
d_{1j}	0.45	0.32	0.32	0.73	1.40	1.26	1.91	2.46	2.24
d_{2j}	1.51	0.81	1.22	0.45	0.32	0.41	1.41	0.85	2.00
d_{3j}	2.62	2.11	1.91	2.02	1.8	0.92	0.32	0.50	0.50

The assignment is identical to Step V of this algorithm.

Step IX. Determination of new positions of cluster centers
New positions of cluster centers $\mathbf{m}_i(3)$, determined on the basis of measurement samples assigned to them, have co-ordinates as follows:

$$\mathbf{m}_1(2) = [1.1, 1.0]^T,$$
$$\mathbf{m}_2(2) = [2.2, 1.1]^T,$$
$$\mathbf{m}_3(2) = [2.5, 2.7]^T.$$

Step X. Determination of cluster translocation size Δm_i
The cluster translocation values $\Delta m_i = \|\mathbf{m}_i(2) - \mathbf{m}_i(3)\|$, $i = 1, 2, 3$, are the following:

$$\Delta m_1 = 0,$$
$$\Delta m_2 = 0,$$
$$\Delta m_3 = 0.$$

Since all the values of cluster shiftings are lower than the threshold value ($\varepsilon = 0.1$), updating of cluster centers can be terminated. Now, the other parameters of the membership functions of clusters should be determined.

Step XI. Determination of the type and parameters of cluster membership function
The clusters have been defined in the entire space $X \times Y$ of the system. It allows for their full identification and prevents the clusters $K2$ and $K3$

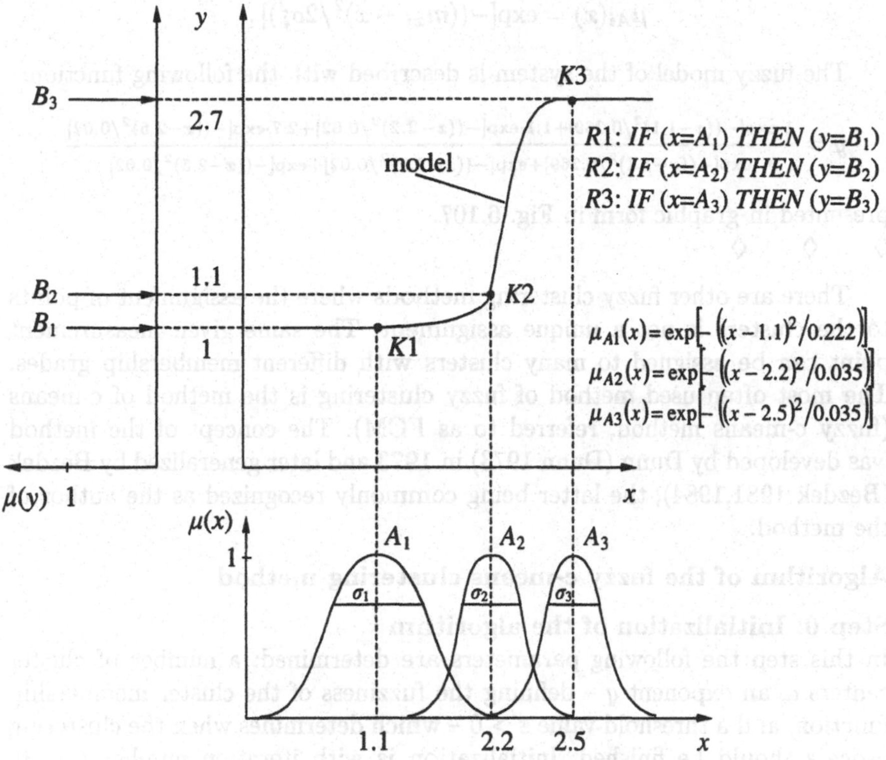

Fig. 6.107. Fuzzy model obtained in the clustering process

from overlapping in their projection on the input space X. However, the task which is usually to be solved during model construction is to calculate an output y corresponding to a given input vector $\mathbf{x}$. Thus, the membership functions of fuzzy sets should be defined separately for the input space and the output space. In this example the input space is the one-dimensional space and segmentally-linear functions might be applied. Regarding the irregular arrangement of cluster centers in the case of multi-dimensional input space, it is beneficial to use exponential membership functions having their centers identical with the cluster centers and having their widths σ dependent on the distances between the adjacent functions. In this example the functions presented in Fig. 6.107 have been chosen.

The width values σ_i chosen are equal to one third of the distances $|m_{xi} - m_{x(i+1)}|$ between the adjacent cluster centers in the input space:

$$\sigma_1 = |2.2 - 1.1|/3 = 0.367\,,$$
$$\sigma_2 = \sigma_3|2.5 - 2.2|/3 = 0.1\,.$$

The particular membership functions are Gaussian functions:

$$\mu_{Ai}(x) = \exp[-((m_{xi} - x)^2/2\sigma_i^2)] \, .$$

The fuzzy model of the system is described with the following function:

$$y = \frac{1 \cdot \exp[-((x-1.1)^2/0.269] + 1.1 \cdot \exp[-((x-2.2)^2/0.02] + 2.7 \cdot \exp[-((x-2.5)^2/0.02]}{\exp[-((x-1.1)^2/0.269] + \exp[-((x-2.2)^2/0.02] + \exp[-((x-2.5)^2/0.02]} \, ,$$

presented in graphic form in Fig. 6.107.

◊ ◊ ◊

There are other fuzzy clustering methods where the assignment of points to the clusters is not a unique assignment. The same given measurement point can be assigned to many clusters with different membership grades. The most often used method of fuzzy clustering is the method of c-means (fuzzy c-means method, referred to as FCM). The concept of the method was developed by Dunn (Dunn 1973) in 1973 and later generalized by Bezdek (Bezdek 1981,1984), the latter being commonly recognized as the author of the method.

Algorithm of the fuzzy c-means clustering method

Step 0. Initialization of the algorithm

In this step the following parameters are determined: a number of cluster centers c, an exponent q – defining the fuzziness of the cluster membership function, and a threshold value $\varepsilon > 0$ – which determines when the clustering process should be finished. Initialization is with iteration number $t = 0$. Moreover, an initial matrix $\Theta(0) = (\mu_{ij}(0))$ of membership grades describing membership of points j to clusters i is determined at random.

Step I. Updating of cluster centers $\mathbf{m}_i$, $i = 1, \ldots, c$

Updating of successive coordinates of centers c of clusters $m_{xl,i}$ is carried out with formula (6.100):

$$m_{xl,i} = \frac{\sum_{j=1}^{N} \mu_{ij}^q(t) \cdot x_{l,j}}{\sum_{j=1}^{N} \mu_{ij}^q(t)} \, , \quad \forall i = 1, \ldots, c \, , \tag{6.100}$$

where: $l = 1, \ldots, (p+1)$, $(x_{p+1} = y)$.

The cluster center co-ordinates are as follows:

$$\mathbf{m}_i = [m_{x1,i}, \ldots, m_{xp,i}, m_{x(p+1),i}] \, .$$

Step II. Updating of membership grade matrix Θ

Each updating cycle (t+1) has to have new values $\mu_{ij}(t+1)$ of the membership of measurement samples j to clusters i calculated with formula (6.101):

$$\mu_{ij}(t+1) = \frac{1}{\sum_{k=1}^{c}\left[\frac{d_{ij}(t)}{d_{kj}(t)}\right]^{\frac{2}{q-1}}}, \quad \forall i = 1,\ldots,c, \text{ and } \forall j = 1,\ldots,N, \tag{6.101}$$

where $d_{ij}(t)$ denotes the distance between a sample j and a cluster i, for which the membership is calculated. The values $d_{kj}(t)$, $k = 1,\ldots,c$, denote the distances between the sample and all the clusters. The distance $d_{ij}(t)$ is calculated with formula (6.102):

$$d_{ij}(t) = \|\mathbf{x} - \mathbf{m}_i(t)\|_A = \sqrt{(\mathbf{x}_j - \mathbf{m}_i(t))^{\mathrm{T}}\mathbf{A}(\mathbf{x}_j - \mathbf{m}_i(t))}, \tag{6.102}$$

where: $\mathbf{x}_j$ – vector of coordinates of sample j,
$\mathbf{A}$ – positive definite symmetrical matrix of dimension $(p+1)\times(p+1)$, where: p – number of inputs inducing the type of norm $\|\ldots\|$ used to calculate the distance d_{ij}.

On having presumed $\mathbf{A} = \mathbf{I}$ (identity matrix) the Euclidean norm (Euclidean distance) is obtained and formula (6.102) can be expressed in the form of (6.103).

$$d_{ij}(t) = \sqrt{\sum_{l=1}^{p+1}(x_{l,j} - m_{xl,i}(t))^2}, \quad l = 1,\ldots,(p+1) \tag{6.103}$$

The membership function $\mu_{ij}(t+1)$ is symmetrical function in all directions and a section across the cluster is the (hyper-) circle. On having presumed the matrix $\mathbf{A} = \mathbf{C}$, where $\mathbf{C}$ is a positive definite, symmetrical correlation matrix of vector $\mathbf{x}_j$ ($\mathbf{C} = [\mathrm{diag}(\sigma_j^2)]^{-1}$ which induces a Mahalanobis norm (Zimmermann 1994a)) and σ_j^2 denotes a variance of the sample j, then the membership functions $\mu_{ij}(t+1)$ of (hyper-) elliptic cross-sections are obtained. It usually allows us to obtain higher model precision. The value of exponent q in formulas (6.100) and (6.101) expresses the grade of fuzziness of the membership function for cluster i.

If $q = 1$ then clustering is not a fuzzy process any longer (**"hard" version** of algorithm) and each single sample j is assigned uniquely to only one cluster i. A disadvantage of "hard" clustering is its higher sensitivity to measurement noise which usually affects the samples. An increase in the exponent q ($q > 1$) results in an extension of the cluster fuzziness – larger spread of membership function arms (**"soft" version** of algorithm). The latter reduces the influence of measurement noise on the model and increases the model generalization grade (generalization of measurement information). When $q \to \infty$ the cluster centers come closer to a single point, being the centroid of all samples (maximum generalization of information) and the model computes identical outputs for each input vector; the outputs are the averages of all the samples.

The value q should be optimized with respect to the degree to which the measurement samples are affected by noise. For greater values q the samples located farther from a cluster i have a more significant effect on the location of cluster centers; for a smaller value q the influence of more distant samples decreases while the influence of those located closer increases. The process of clustering usually starts with $q = 2$.

Step III. Checking the condition of clustering termination
Comparison between the new membership matrix $\boldsymbol{\Theta}(t+1)$ and the previously iterated matrix $\boldsymbol{\Theta}(t)$ can be done with different methods. Three of the methods are presented with formulas (6.104) – (6.106):

$$IF \;\; \|\boldsymbol{\Theta}(t+1) - \boldsymbol{\Theta}(t)\| \leq \varepsilon_1 \; THEN \text{ end of clustering,} \tag{6.104}$$

$$IF \; \sum_{i=1}^{c}\sum_{j=1}^{N} |\mu_{ij}(t+1) - \mu_{ij}(t)| \leq \varepsilon_2 \; THEN \text{ end of clustering,} \tag{6.105}$$

$$IF \; \frac{1}{cN}\sum_{i=1}^{c}\sum_{j=1}^{N} |\mu_{ij}(t+1) - \mu_{ij}(t)| \leq \varepsilon_3 \; THEN \text{ end of clustering,} \tag{6.106}$$

where $|\ldots|$ denotes an absolute value of membership μ_{ij} increments for the previous iteration and $c \times N$ denotes the number of elements of the matrix $\boldsymbol{\Theta}$. The values ε are the arbitrary threshold values constituting conditions for clustering termination. In case the threshold values are smaller, a greater number of iterations is needed.

It seems that the threshold value ε_3 in condition (6.106) is the easiest one to be determined by intuition. The value is an absolute mean change of membership of the sample j to the cluster i. If the changes in membership μ_{ij} in the consecutive iterations are sufficiently small then the cluster centers $\mathbf{m}_i$ move insignificantly and clustering can be terminated. If the condition of clustering termination has not been fulfilled, the process is repeated from Step I. Otherwise proceed with Step IV.

Criteria of c-means algorithm convergence

a) For a single measurement sample the sum of memberships of all c clusters has to be equal to 1:

$$\sum_{i=1}^{c} \mu_{ij} = 1 \, , \forall j = 1, \ldots, N \, ,$$

b) the membership value has to be limited to an interval [0,1]:

$$\mu_{ij} \in [0,1] \, , \quad \forall i = 1, \ldots, c \, , \text{ and } \forall j = 1, \ldots, N \, .$$

Step IV. Determination of a fuzzy model
In the result of clustering, the cluster centers $\mathbf{m}_i$ for particular clusters c have been determined in the full system space $\mathbf{X} \times Y$. It allows us to avoid overlapping two (or a few) clusters in their projection on input space $\mathbf{X}$. However, since the model target is to compute an output y for a given vector input $\mathbf{x}$ then the centers of the clusters which have been found have to be projected on the input space, the clusters in this space have to be created, and an appropriate model performing the mapping $\mathbf{X} \rightarrow Y$ has to be constructed.

Clustering in the space $\mathbf{X} \times Y$ results in the creation of c clusters $\mathbf{m}_i$ of coordinates expressed with (6.107):

$$\mathbf{m}_i = \left[m_{x1,i}, \ldots, m_{xp,i}, m_{x(p+1),i}\right], \tag{6.107}$$

where: p – number of system inputs,
$x_{(p+1)} = y$ – system output.

The projection of cluster i on the input space has the center $\mathbf{m}_i^X$ of coordinates expressed with formula (6.108).

$$\mathbf{m}_i^X = \left[m_{x1,i}, \ldots, m_{xp,i}\right] \tag{6.108}$$

The membership $\mu_{ij} = \mu_i(x_j)$ of an arbitrary sample j (vector $\mathbf{x}_j = [x_{1j}, \ldots, x_{pj}]^{\mathrm{T}}$ in the input space) of the cluster i in the same space $\mathbf{X} \times Y$ is to be determined with formula (6.101) in the same way as for the full space; the only difference is that the distance d_{ij} between the vector $\mathbf{x}_j$ and the center of cluster $\mathbf{m}_i^X$ is determined only in the input space, with formula (6.109).

$$d_{ij}(t) = \|\mathbf{x} - \mathbf{m}_i^X(t)\|_A = \sqrt{(\mathbf{x}_j - \mathbf{m}_i^X(t))^{\mathrm{T}} \mathbf{A} (\mathbf{x}_j - \mathbf{m}_i^X(t))}, \tag{6.109}$$

The dimension of matrix $\mathbf{A}$ is equal to $p \times p$. The output y for a given vector $\mathbf{x}_i$ in the input space is calculated with formula (6.110).

$$y(\mathbf{x}_j) = \frac{\sum\limits_{i=1}^{c} m_{x(p+1),i} \cdot \mu_i(\mathbf{x}_j)}{\sum\limits_{i=1}^{c} \mu_i(\mathbf{x}_j)} \tag{6.110}$$

Instead of computing the membership $\mu_i(x_j)$ with formula (6.101) the Gaussian membership functions can be used. The centers of Gaussian functions should be located in the centers of clusters identified where the widths σ_i are defined as 1/3 of the distance to the closest cluster in the input space. The Gaussian membership functions provide a method with far less time-consuming computations.

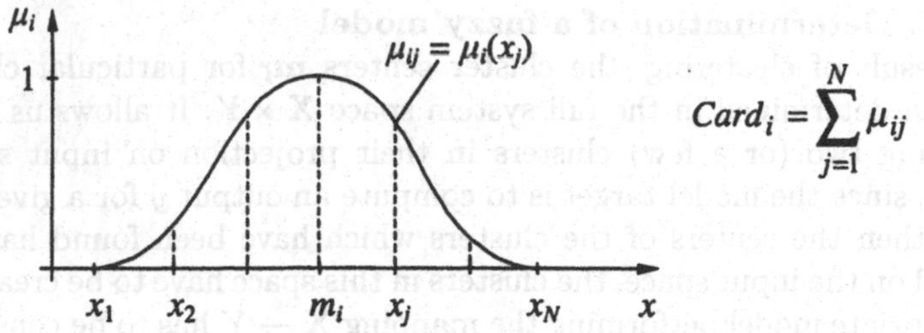

Fig. 6.108. Illustration of the notion of "cluster cardinal number"

Remarks on the c-means method

a) The c-means method allows us to find the cluster centers $\mathbf{m}_i$, minimizing the criterion of the weighted sum of square Euclidean distances of measurement samples j from the clusters i, expressed with formula (6.111).

$$\min Q = \sum_{i=1}^{c} \sum_{j=1}^{N} (\mu_{ij})^q \|\mathbf{x}_j - \mathbf{m}_i\|^2 \tag{6.111}$$

b) The solution obtained is usually only the **local optimum** (sub-optimum), not the global one and is dependent on initialization parameters. Various sets of parameters for the initiation of clustering should be used.
c) If a suitable number c of clusters has been selected and the data has a good clustering structure, then the c-means method usually generates similar clusters (solution stability).
d) Searching for an appropriate number c of clusters – this can be done with a trial-and-error method. The less significant clusters can be rejected after having evaluated their significance. The evaluation can be carried out on the basis of comparisons of certain measures, e.g. cardinal number Card_i or inclusion grade f of samples within the cluster i or validity Val_i of the cluster, (Davè 1997). The higher the total membership of all N samples to the cluster, the higher the cardinal number of cluster i, Fig. 6.108.

The inclusion grade of samples j within cluster i is calculated with formula (6.112).

$$f_i = \frac{\text{Card}_i}{N} \tag{6.112}$$

where: N – number of samples. The **inclusion grade** describes the mean membership of the cluster. The **validity** Val_i **of the cluster** i can be calculated with the formula presented in Fig. 6.109.

The problem of determination of cluster validity, described in (Davè 1997) is neither unique nor simple to formulate. The further the samples are located from the cluster center $\mathbf{m}_i$ and the higher the membership grade, the greater the value Val_i is.

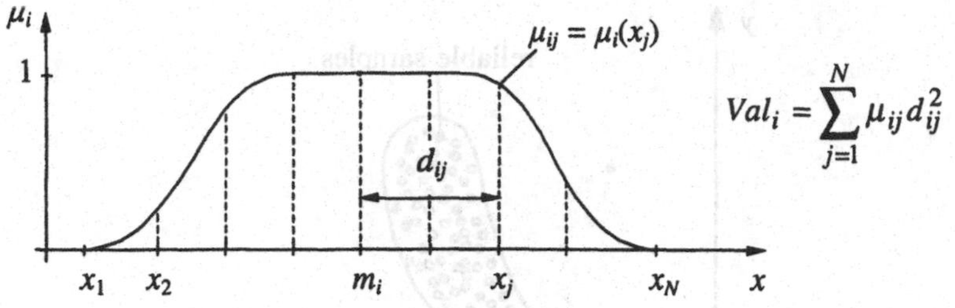

Fig. 6.109. Illustration of the notion of "cluster i validity"

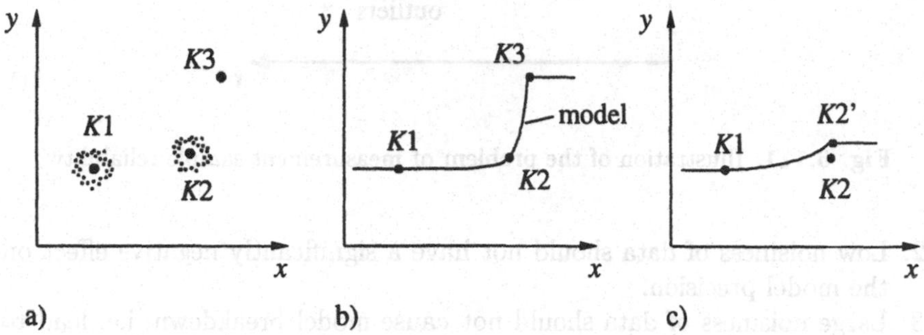

Fig. 6.110. Clusters representing great diversity of inclusion grades (a) and the effect of rejection of one of the clusters (c) on model precision (b)

On having applied one of the cluster validity measures, the clusters can be arranged in order of decreasing validity (significance) and an attempt to reject the least significant clusters can be undertaken. Clustering can be repeated again, however, with the fewer clusters. Rejection of clusters may be related to a certain danger, i.e. some essential information on the system modeled can get lost, Fig. 6.110. The cluster $K3$ has a very low grade of inclusion, represented by only one sample, in contrast to clusters $K1$ and $K2$. However, the cluster $K3$ carries information about the significant leap of the model surface in that region of input space, Fig. 6.110b. Removal of cluster $K3$ and assignment of its measurement sample to the cluster $K2$ will generate the model presented in Fig. 6.110c, essentially different from the more complicated model in Fig. 6.110b.

The methods of **clustering should be robust.** Clustering is recognized as robust (Davè 1997) if the criteria 1–3 below are fulfilled.

1. A system model obtained by the clustering process shows (reasonably) good precision. If the measurement samples are "pure" (not affected by noise) then the model obtained should give results falling within the limits of a certain assumed tolerance margin.

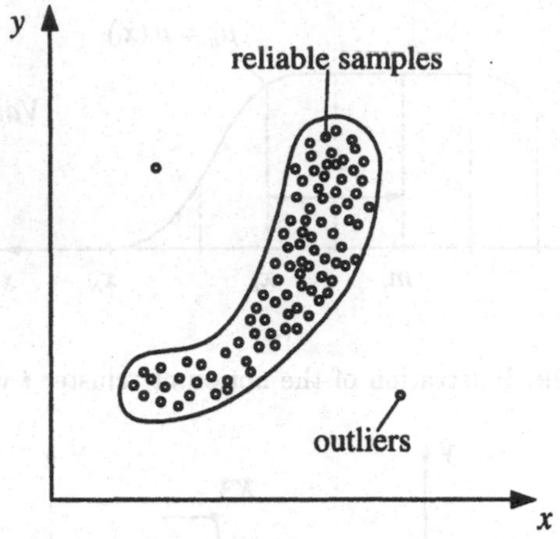

Fig. 6.111. Illustration of the problem of measurement sample reliability

2. Low noisiness of data should not have a significantly negative effect on the model precision.
3. Large noisiness of data should not cause model breakdown, i.e. lead to such behavior of the model which does not at all correspond to the behavior of the system modeled.

The third condition, in terms of practical applications, means that it is hardly possible to construct a good and robust model if the noise (error) of measurements exceeds a certain limit (theoretically 50%). On the basis of excessively noise-affected samples we can create several models of similar accuracy, which however, will be significantly different one from another and will generate absolutely different or even adverse results. If that is the case, which of the models is the appropriate (the best) one can hardly be evaluated.

The accomplishment of a resistant clustering method is a very difficult theoretical task, which has not been solved so far. Research on the robustness of particular clustering methods and the formulation of good measures of robustness has been constantly carried out. The results obtained so far indicate that one of the most significant conditions of achieving robust clustering is rejecting or ignoring such subsets of measurement samples which are located far from the center of a given cluster in so-called "regions of doubt" (Davè 1997). However, to formulate an optimum numerical measure defining the regions mentioned appears very difficult. It is possible to formulate only a certain general suggestion that the model robustness would increase if so-called "big errors" (improbable or slightly probable samples – **"outliers"**, Fig. 6.111), have been detected and removed from the measurement sample set.

The term "sample reliability" is a fuzzy term, difficult to define. The definition depends on the subjective approach presented by a person who constructs (formulates) the model and the problem itself.

The method of c-means is sensitive to measurement noise (Davè 1997), just as other methods based on the minimization of total square error. The noise can be the factor creating additional local minima where the process of clustering gets stuck. A method that should be recognized as a highly robust method, according to Davè and Krishnapuram, is Hough's method of generalized clustering (Davè 1997). It should be mentioned that the method of c-means shows several other advantages, which makes it a very often-used clustering method. The example presented below illustrates the computational procedure using the c-means method.

Example 6.3.3.2.3. Clustering with the fuzzy c-means method
Let us consider a set of measurement samples of the system modeled, Table 6.16.

Table 6.16. Measurement samples of the system modeled

P_j	1	2	3	4	5	6	7	8	9
$x = x_1$	0.7	1.4	1.2	1.8	2.5	2.3	2.4	3.0	2.1
$y = x_2$	0.8	0.9	1.3	0.8	0.9	1.6	2.4	2.7	3.0

Step 0. Initialization of the algorithm
The initial parameters chosen:

- number of clusters: $c = 3$,
- exponent: $q = 2$,
- threshold value: $\varepsilon = 0.1$,
- initial matrix of membership grades of samples j to clusters i (defined in an arbitrary way):

$$\Theta(0) = (\mu_{ij}(0)) = \overset{i \rightarrow}{\begin{bmatrix} 0.9 & 0.1 & 0 \\ 0.9 & 0.1 & 0 \\ 0.9 & 0.1 & 0 \\ 0.1 & 0.9 & 0 \\ 0.1 & 0.9 & 0 \\ 0.1 & 0.9 & 0 \\ 0 & 0.1 & 0.9 \\ 0 & 0.1 & 0.9 \\ 0 & 0.1 & 0.9 \end{bmatrix}} \downarrow j$$

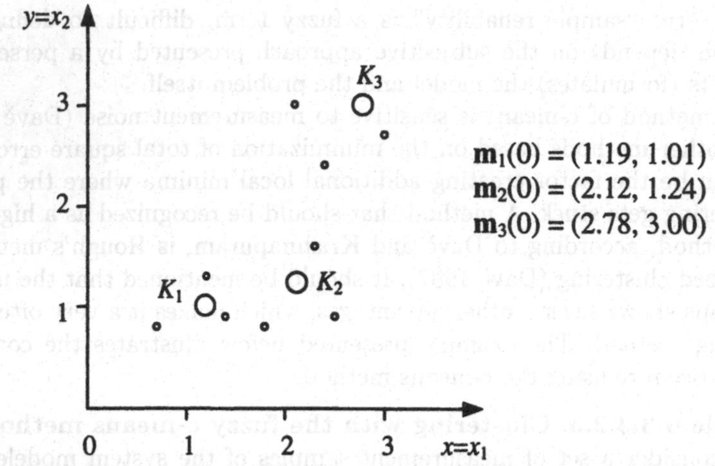

Fig. 6.112. Measurement samples and initial clusters

Step I. Updating of cluster centers $\mathbf{m}_i$, $i = 1, \ldots, c$
On the basis of matrix $\Theta(0)$ and formula (6.113) the cluster coordinates are determined.

$$m_{xl,i} = \frac{\sum_{j=1}^{9} \mu_{ij}^2 \cdot x_{l,j}}{\sum_{j=1}^{9} \mu_{ij}^2} \tag{6.113}$$

(6.3.3.2/36) The results obtained are presented in Fig. 6.112.

Step II. Updating of the membership grade matrix
On the basis of formula (6.114) the distances d_{ij}, between the samples j and particular clusters i, are calculated, Table 6.17.

$$d_{ij}(0) = \sqrt{\sum_{l=1}^{2} (x_{l,j} - m_{xl,i}(0))^2}, \quad l = 1, 2 \tag{6.114}$$

Formula (6.115) is used to calculate the elements of the matrix $\Theta(1)$, i.e. the membership grades μ_{ij} of particular samples of clusters.

$$\mu_{ij}(1) = \frac{1}{\sum_{k=1}^{3} \left[\frac{d_{ij}(0)}{d_{kj}(0)}\right]^2} \tag{6.115}$$

Table 6.17. Distances $d_{ij}(0)$ between the samples j and cluster centers $\mathbf{m}_i(0)$

$i\backslash j$	1	2	3	4	5	6	7	8	9
1	0.533	0.237	0.29	0.645	1.315	1.257	1.843	2.478	2.188
2	1.487	0.796	0.922	0.544	0.510	0.402	1.193	1.705	1.760
3	3.028	2.513	2.321	2.408	2.119	1.48	0.71	0.372	0.68

$$\boldsymbol{\Theta}(1) = (\mu_{ij}(1)) = \overset{i \rightarrow}{\begin{bmatrix} 0.863 & 0.110 & 0.027 \\ 0.845 & 0.081 & 0.074 \\ 0.897 & 0.081 & 0.014 \\ 0.404 & 0.568 & 0.028 \\ 0.124 & 0.828 & 0.048 \\ 0.087 & 0.850 & 0.063 \\ 0.099 & 0.236 & 0.665 \\ 0.021 & 0.044 & 0.935 \\ 0.078 & 0.120 & 0.802 \end{bmatrix}} \downarrow j$$

$$\sum_{i=1}^{3} \mu_{ij} = 1, \quad \forall j = 1, \ldots, 9$$

Step III. Checking the criterion of clustering termination
Condition (6.116), corresponding to condition (6.106), has been chosen as the clustering termination criterion.

$$IF \ \frac{1}{3 \cdot 9} \sum_{i=1}^{3} \sum_{j=1}^{9} |\mu_{ij}(1) - \mu_{ij}(0)| \leq 0.1 \ THEN \text{ end of clustering} \qquad (6.116)$$

$$|\boldsymbol{\Theta}(1) - \boldsymbol{\Theta}(0)| = \begin{bmatrix} 0.863 & 0.110 & 0.027 \\ 0.845 & 0.081 & 0.074 \\ 0.897 & 0.081 & 0.014 \\ 0.404 & 0.568 & 0.028 \\ 0.124 & 0.828 & 0.048 \\ 0.087 & 0.850 & 0.063 \\ 0.099 & 0.236 & 0.665 \\ 0.021 & 0.044 & 0.935 \\ 0.078 & 0.120 & 0.802 \end{bmatrix}$$

$$\frac{1}{3 \cdot 9} \sum_{i=1}^{3} \sum_{j=1}^{9} |\mu_{ij}(1) - \mu_{ij}(0)| = 0.073 < \varepsilon_3 = 0.1$$

Since the mean change in sample membership of clusters is lower than the threshold value $\varepsilon = 0.1$, the process of clustering can be terminated. The new

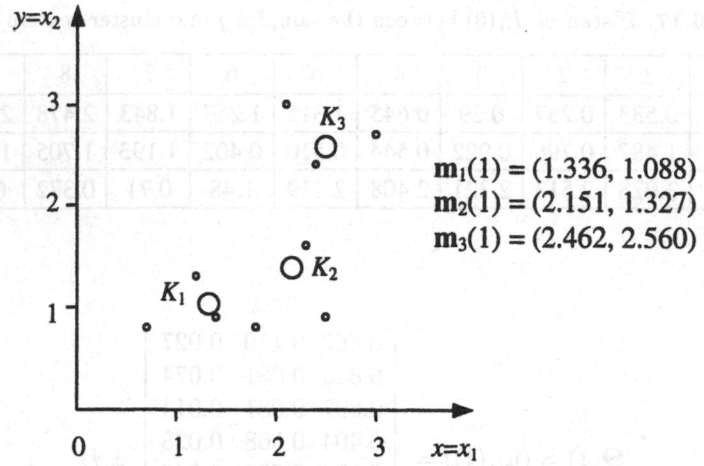

Fig. 6.113. Cluster centers obtained in iteration $t = 1$

centers of clusters, resulting from membership matrix $\Theta(1)$, are presented in Fig. 6.113.

The mean translocation of cluster centers Δm_i is equal to 0.267, compared with their position in the preceding iteration.

Step IV. Construction of fuzzy model X → Y
Particular clusters are projected on the input space X:

$$m_1^X = 1.336, \quad m_2^X = 2.151, \quad m_3^X = 2.462.$$

The membership functions μ_{ij}^X (defined in the input space) are located in the centers m_i^X – formula (6.117).

$$\mu_i^X(x) = \frac{1}{\sum_{k=1}^{3}\left[\frac{d_i^X(x)}{d_{kj}^X}\right]^2} = \mu_i^X(x_j), \quad i = 1,2,3, \quad j = 1,\ldots,9, \tag{6.117}$$

where d_i^X are expressed with formula (6.118).

$$d_i^X(x) = \|x - m_i^X\| = |x - m_i^X| \tag{6.118}$$

The output membership functions of the singleton type are located in the projections m_i^Y of cluster centers on the y-axis:

$$m_1^Y = 1.088, \quad m_2^Y = 1.327, \quad m_3^Y = 2.560.$$

The membership functions of the input and output of the model are presented in Fig. 6.114.

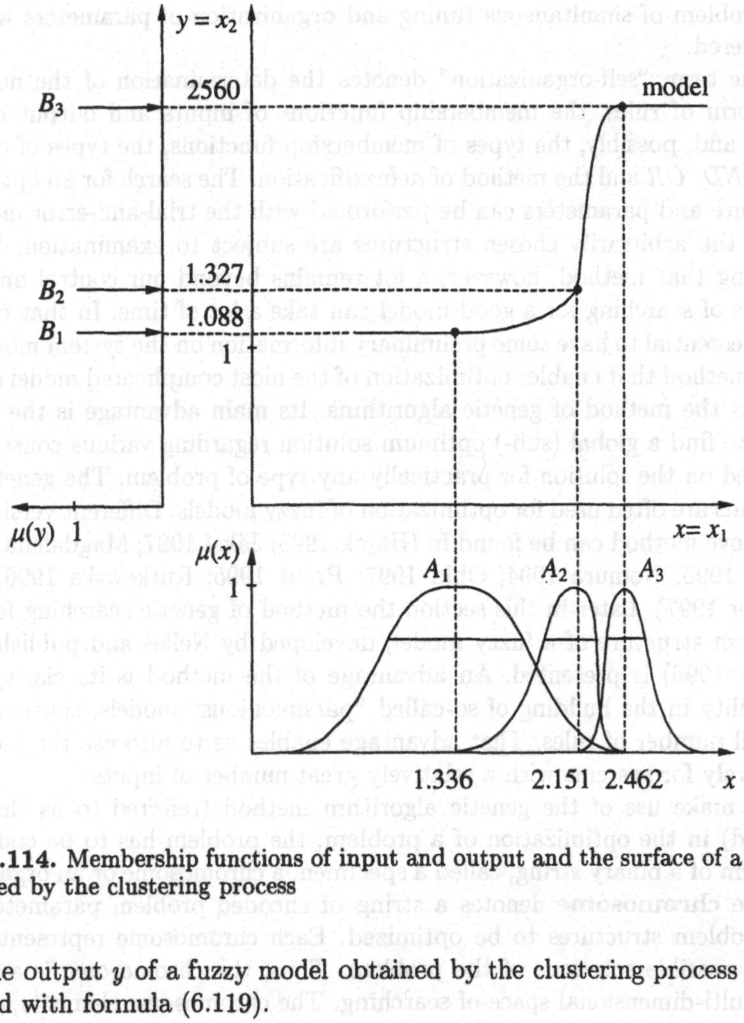

Fig. 6.114. Membership functions of input and output and the surface of a model obtained by the clustering process

The output y of a fuzzy model obtained by the clustering process is calculated with formula (6.119).

$$y(x) = \frac{\sum_{i=1}^{3} m_i^Y \cdot \mu_i^X(x)}{\sum_{i=1}^{3} \mu_i^X(x)} \tag{6.119}$$

◊ ◊ ◊

6.3.3.3 Self-organization and self-tuning of fuzzy models with the searching method

The problem of tuning fuzzy model parameters (parameters of input/output membership functions) has been presented in Section 6.2. In this section the problem of simultaneous tuning and organization of parameters will be considered.

The term "self-organization" denotes the determination of the number and form of rules, the membership functions of inputs and output of the model and, possibly, the types of membership functions, the types of operators *AND*, *OR* and the method of defuzzification. The search for an optimum structure and parameters can be performed with the trial-and-error method where the arbitrarily chosen structures are subject to examination. When applying that method, however, a lot remains beyond our control and the process of searching for a good model can take a lot of time. In that case it seems essential to have some preliminary information on the system modeled.

A method that enables optimization of the most complicated model structures is the method of genetic algorithms. Its main advantage is the capability to find a global (sub-) optimum solution regarding various constraints imposed on the solution for practically any type of problem. The genetic algorithms are often used for optimization of fuzzy models. Different versions of the above method can be found in (Hajek 1995; Jäkel 1997; Magdalena 1995; Nelles 1995; Nomura 1994; Ohki 1997; Preut 1995; Rutkowska 1996,1997; Wagner 1997). Later in this section the method of genetic searching for the optimum structure of a fuzzy model, developed by Nelles and published in (Nelles 1996) is presented. An advantage of the method is its clarity and suitability in the building of so-called "parsimonious" models, consisting of a small number of rules. That advantage enables us to also use the method effectively for systems with a relatively great number of inputs.

To make use of the genetic algorithm method (referred to as the GA method) in the optimization of a problem, the problem has to be coded in the form of a binary string, called a specimen, a chromosome or an organism.

The **chromosome** denotes a string of encoded problem parameters or the problem structures to be optimized. Each chromosome represents one of the possible solutions of the problem. Thus, the chromosome is a point in a multi-dimensional space of searching. The chromosome elements, called genes, represent individual elements of the problem under optimization. A set of chromosomes forms a **population**. The size of a population, where the best specimen is searched for, is defined by the user. Each chromosome incorporated in the population is evaluated on the basis of a criterion called the **fitness function**.

The procedure of the conventional GA method consists of the steps presented below.

1. **Initialization of algorithm.**
   Encoding of a problem in the form of genes and chromosomes.
   Determination of the fitness function.
   Determination of algorithm termination conditions (minimum acceptable value of fitness function).
   Selection of initial population of chromosomes.

2. **Evaluation of fitness grade** of particular chromosomes in a population.
3. **Checking termination conditions** of searching and, if fulfilled, determination of a chromosome (solution) satisfying the conditions (the end of searching). Otherwise, continuation of the process with item 4.
4. **Selection of chromosomes.** Selection of "better" and "worse" chromosomes to decide on the "candidates" to generate new chromosomes.
5. **Performance of genetic operations.** The "candidate" chromosomes selected are used to generate new chromosomes by means of special genetic operators.
6. **Creation of new population.** New population is subject to evaluation – return to Step 2.

The population of chromosomes (solutions) is improved and enhanced in each new generation, i.e. the iteration of the algorithm. The probability that a particular chromosome survives the selection, i.e. becomes an element of the subsequent population, is proportional to the chromosome fitness grade. Chromosome selection moves the population set forward to the regions of increasing fitness in the space of solutions. The most often used genetic operators are the operators of mutation and crossover.

The **crossover** operation cuts two binary strings of chromosomes at a randomly selected point and executes an exchange of cut-off fragments between the chromosomes.

The **mutation** operation changes the value of a bit in the chromosome string into the opposite value.

The probability of crossover or mutation to be performed on the chromosomes is defined by the user. As in natural conditions, the probability of crossover is usually a value close to 1 and of mutation – close to 0. It is not an easy problem to select an appropriate probability. In fact, not all the bits (genes) of a chromosome are equally significant in terms of optimization, for example, the chromosome 10000 corresponds to number 16 in the decimal system. The mutation performed on the first gene gives the chromosome 00000 corresponding to the number 0 in the decimal system. In that case the change resulting from the mutation is very significant. If the last gene of the chromosome 00001 is subject to mutation the resulting chromosome is also 00000. It corresponds to the change of 1 into 0 in the decimal system – such a change is not as significant as in the first case. Therefore, it seems advantageous to provide a separate value of the mutation probability for each individual gene. An example of crossover probability is 0.9 and the same for mutation is 0.2.

The main task in the process of searching for an optimum rule base is to design an appropriate encoding method. The first step should provide determination of a set of all possible rules for the model on the basis of previously assumed membership functions. Let us consider a system with 2 inputs and 1 output and assume there are 3 triangular membership functions assigned to each input and singleton membership functions for the output so

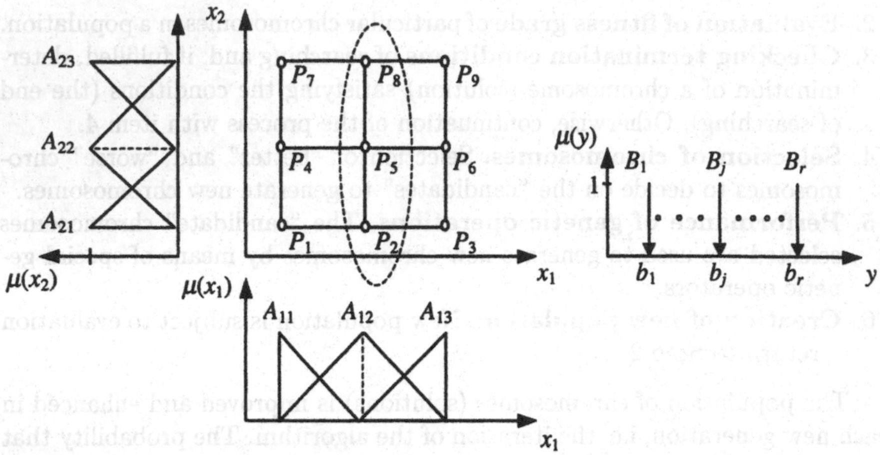

Fig. 6.115. Example of input space partition for the membership functions chosen

that a one-singleton conclusion corresponds to each rule. Fig. 6.115 represents the partition of input space **X** and the singletons B_j in the space Y.

The rule base can contain elementary rules and generalizing rules, the latter being the logical combinations of elementary rules.

An example of an **elementary rule** is presented in (6.120).

$$Rj: \ IF\ (x_1 = A_{13})\ AND\ (x_2 = A_{22})\ THEN\ (y = B_j) \tag{6.120}$$

The elementary rule assigns a single set B_j (singleton) of output space to the sets A_{13} and A_{22} of the input space. It means that a region around the point b_j, where the singleton B_j is located in the output space, corresponds to a region around the single point P_6, Fig. 6.115, in the input space.

An example of a **generalizing rule** is presented in (6.121).

$$Rj: \ IF\ (x_1 = A_{12})\ THEN\ (y = B_j) \tag{6.121}$$

The generalizing rule is generated as a logical combination of 3 elementary rules (6.122).

$$\begin{aligned} &R2: \ IF\ (x_1 = A_{12})\ AND\ (x_2 = A_{21})\ THEN\ (y = B_j) \\ &R5: \ IF\ (x_1 = A_{12})\ AND\ (x_2 = A_{22})\ THEN\ (y = B_j) \\ &R8: \ IF\ (x_1 = A_{12})\ AND\ (x_2 = A_{23})\ THEN\ (y = B_j) \end{aligned} \tag{6.122}$$

The above rules can be combined as follows:

$$\begin{aligned} &R2\ \ OR\ \ R5\ \ OR\ \ R8 = \\ &\quad IF\ (x_1 = A_{12})\ AND\ [(x_2 = A_{21})\ OR\ (x_2 = A_{22})\ OR\ (x_2 = A_{23})] \\ &\qquad\qquad THEN\ (y = B_j) = IF\ (x_1 = A_{12})\ THEN\ (y = B_j)\,. \end{aligned}$$

R1	R2	R3	R4	R5	R6	R7	R8	R9	R10	R11	R12	R13	R14	R15
1	0	1	0	0	0	0	0	0	1	1	1	0	0	0

Fig. 6.116. Chromosome corresponding to one of the possible rule bases

The single generalizing rule, e.g. (6.121), provides modeling of a large sector of the input space constituting the neighborhood of the points P_2, P_5, P_8, Fig. 6.115. Since the same particular conclusion of the above rule ($y = B_j$) refers to a larger region of the input space than in the case of the elementary rule, then the precision of the model with generalizing rules is generally lower than of the model with elementary rules. On the other hand, the possibility to cover a larger area with a single rule allows us to reduce the number of rules and obtain a "parsimonious" model. Thus, the method allows us to avoid the "curse of dimensionality" phenomenon.

An optimum fuzzy model can contain both elementary and generalizing rules. Therefore, before the encoding of the problem is initiated, the set Rj of all possible rules should be determined. The set Rj, considered in the example below, is described with formula (6.123).

$$\begin{aligned}
R1 &: IF\ (x_1 = A_{11})\ THEN\ (y = B_1)\\
R2 &: IF\ (x_1 = A_{12})\ THEN\ (y = B_2)\\
R3 &: IF\ (x_1 = A_{13})\ THEN\ (y = B_3)\\
R4 &: IF\ (x_1 = A_{21})\ THEN\ (y = B_4)\\
R5 &: IF\ (x_1 = A_{22})\ THEN\ (y = B_5)\\
R6 &: IF\ (x_1 = A_{23})\ THEN\ (y = B_6)\\
R7 &: IF\ (x_1 = A_{11})\ AND\ (x_2 = A_{21})\ THEN\ (y = B_7)\\
R8 &: IF\ (x_1 = A_{11})\ AND\ (x_2 = A_{22})\ THEN\ (y = B_8)\\
R9 &: IF\ (x_1 = A_{11})\ AND\ (x_2 = A_{23})\ THEN\ (y = B_9)\\
R10 &: IF\ (x_1 = A_{12})\ AND\ (x_2 = A_{21})\ THEN\ (y = B_{10})\\
R11 &: IF\ (x_1 = A_{12})\ AND\ (x_2 = A_{22})\ THEN\ (y = B_{11})\\
R12 &: IF\ (x_1 = A_{12})\ AND\ (x_2 = A_{23})\ THEN\ (y = B_{12})\\
R13 &: IF\ (x_1 = A_{13})\ AND\ (x_2 = A_{21})\ THEN\ (y = B_{13})\\
R14 &: IF\ (x_1 = A_{13})\ AND\ (x_2 = A_{22})\ THEN\ (y = B_{14})\\
R15 &: IF\ (x_1 = A_{13})\ AND\ (x_2 = A_{23})\ THEN\ (y = B_{15})
\end{aligned} \tag{6.123}$$

Each rule Rj of the set of possible rules has a single binary chromosome gene assigned to it, Fig. 6.116.

The positions occupied by 1-genes define a single, particular rule base. The base (6.124) corresponds to the chromosome presented in Fig. 6.116.

$$R1: \ IF\ (x_1 = A_{11})\ THEN\ (y = B_1)$$
$$R3: \ IF\ (x_1 = A_{13})\ THEN\ (y = B_3)$$
$$R10: \ IF\ (x_1 = A_{12})\ AND\ (x_2 = A_{21})\ THEN\ (y = B_{10})$$
$$R11: \ IF\ (x_1 = A_{12})\ AND\ (x_2 = A_{22})\ THEN\ (y = B_{11})$$
$$R12: \ IF\ (x_1 = A_{12})\ AND\ (x_2 = A_{23})\ THEN\ (y = B_{12}) \quad (6.124)$$

The base (6.124), despite having only 5 rules, covers the entire input universe of discourse. Otherwise, to cover the same universe with elementary rules would require $3 \cdot 3 = 9$ rules $R7 - R15$ of the base (6.123).

Another important task, constituting the condition to initialize the algorithm, is the determination of the fitness function. It would be beneficial to apply function F to evaluate the model structure, where the function should take the following two aspects into account: the number R of rules in the rule base and the model precision expressed, for example, by the mean absolute error E for measurement samples used in identification (6.125).

$$F = \frac{1}{E + \alpha \cdot R}, \quad (6.125)$$

where: $E = \frac{1}{N}\sum_{i=1}^{N} |y_i - y_i^m|$, N – number of measurement samples, y – system output, y^m – model output.

The **fitness function** F is an inverse of the **loss function** $S = E + \alpha R$, being the model "misfitness" measure. The coefficient α, determined arbitrarily by the user, changes the importance and significance of the number of rules to the form of the solution obtained. In the case of greater values α the solution will contain fewer rules and give lower precision of the model while lower values α will give adverse results. At the time of problem formulation a limit on the maximum number of rules may be defined. Then the models with a large number of rules would not be considered during the searching process.

If the fuzzy model is to be optimized regarding not only the number of rules but also other parameters, for example, the number of membership functions per each variable, the type of operator *AND* or *OR*, the implication method, the defuzzification method, etc. then the string of genes encoding the rules (6.121) should have additional encoding segments linked to them, concerning other model elements which are of interest, Fig. 6.117.

The chromosome containing the fragments encoding all parameters and elements of the fuzzy model structure, those the user is interested in, is called the **structure chromosome** or the code of **model structure**.

At the beginning of the searching process an initial population of structural chromosomes should be generated with an arbitrary method. Next, the population is subject to evolution, according to the genetic algorithm procedure. However, the evaluation of the fitness grade of each chromosome can be a problem. An individual chromosome provides the definition only of the model structure. It does not define the optimum parameters of the model.

a) The number of membership functions per each variable

2	3	4	5	. . .		
0	1	0	0			

b) Operator *AND*

MIN	*PROD*	*H.PROD.*	*MEAN*	. . .
0	1	0	0	

c) Operator *OR*

MAX	*L.SUM*	*E.SUM*	*H.SUM*	. . .
0	1	0	0	

d) Defuzzification method

*D*1	*D*2	*D*3	. . .		
0	0	1	0		

where: *D*1 – singleton method,
*D*2 – center of gravity defuzzification with inference *MAX-MIN*,
*D*3 – center of gravity defuzzification with inference *SUM-MIN*,
*D*4 – . . .

Fig. 6.117. Examples of genetic string fragments encoding various parameters of fuzzy model

Searching for the optimum parameters for each chromosome of the population appears to be a separate problem in itself. It may be identified as a local optimization within the structure defined by a given chromosome. The parameter optimization concerns the modal values, widths or other parameters of the membership functions of the inputs and output of the model. Optimization of that kind can be performed with either global methods, for example, genetic algorithms (see Section 6.2.2) or local methods, for example, neuro-fuzzy networks trained on the basis of measurement samples of system inputs and output or the method of the least square error sum.

Even though the GA methods allow us to find a good global solution in an arbitrarily large set of possible solutions and to avoid the problem of their combinatorial "explosion", empirical experiments (Nelles 1996) have proved that along with the increase in the solution set size the searching time also increases with that method. Therefore, it is recommended to limit the number of optimized elements of the model structure parameters to an indispensable minimum. Instead of the simultaneous optimization of membership function parameters of inputs and output, the membership function parameters of the inputs can be selected and the optimization can be performed only for membership function parameters of the output, e.g. singleton locations. In that case the optimization can be carried out quickly with the least square method.

Furthermore, the searching time increases when the optimization of input/output membership functions is accomplished with genetic algorithms. Despite the global character of the genetic method it considers only a limited set of discrete values of particular parameters: it operates within a discrete space and not within a continuous space as a neural network does. Thus, in practice, even though such a network performs local optimization, it can provide better solutions than the genetic algorithm method.

Concluding, it is recommended to optimize the model structure of a discrete nature with global methods such as genetic algorithms; while the optimization of parameters defined in the continuous universe of discourse ought to be performed with quicker methods such as the least square method or methods using neuro-fuzzy networks. The theory and details of genetic algorithms can be found in many books e.g. (Mitchell 1996; Goldberg 1995; Michalewicz 1996).

7. Fuzzy Control

7.1 Static fuzzy controllers

Static plants and some dynamic plants can be controlled by static controllers, transforming the control error e into the control signal u in accordance with the controller characteristic $u = F(e)$, Fig. 7.1.

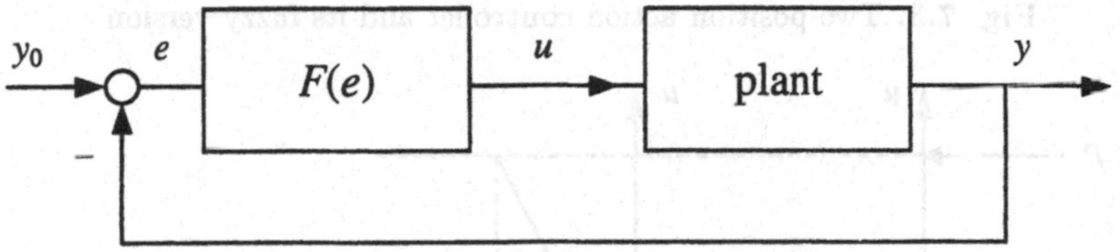

Fig. 7.1. Control system with a static controller

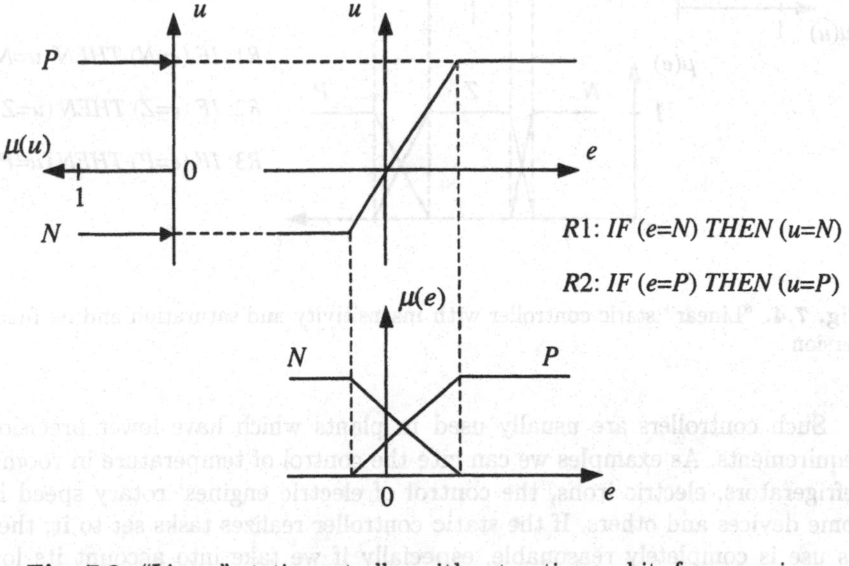

Fig. 7.2. "Linear" static controller with saturation and its fuzzy version

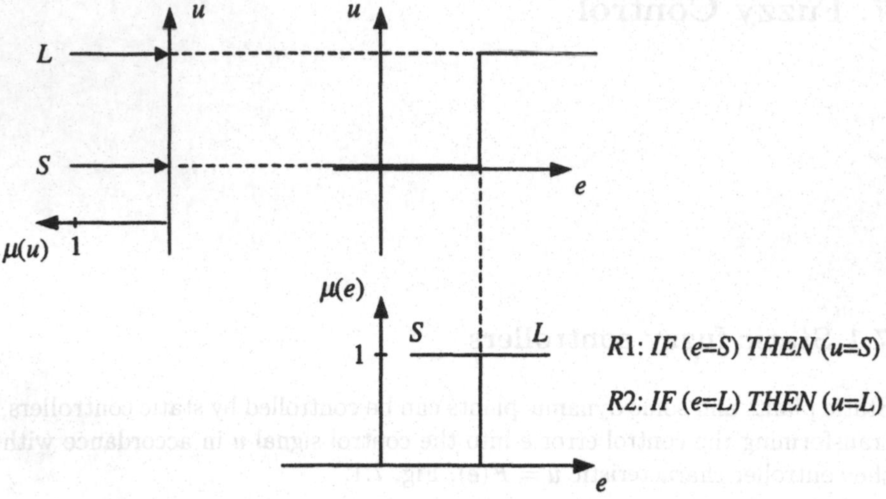

Fig. 7.3. Two-position action controller and its fuzzy version

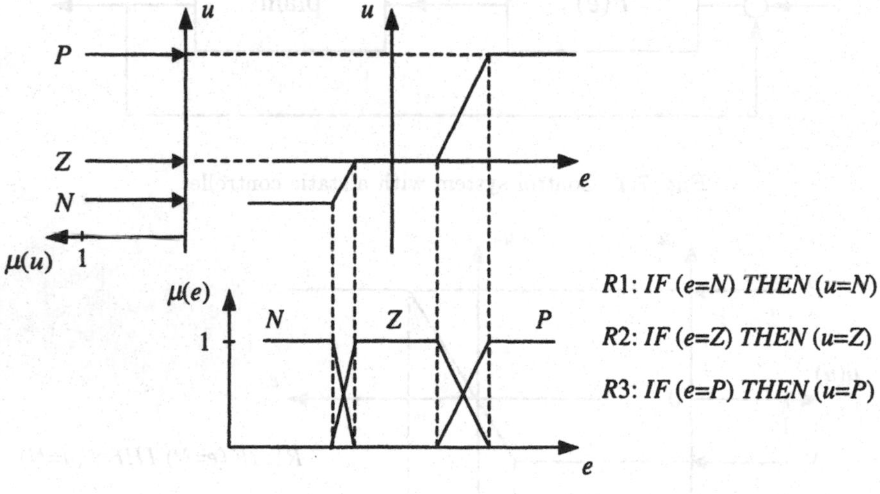

Fig. 7.4. "Linear" static controller with insensitivity and saturation and its fuzzy version

Such controllers are usually used in plants which have lower precision requirements. As examples we can cite the control of temperature in rooms, refrigerators, electric irons, the control of electric engines' rotary speed in some devices and others. If the static controller realizes tasks set to it, then its use is completely reasonable, especially if we take into account its low price. Figs. 7.2 – 7.7 show different conventional static controllers and their fuzzy versions.

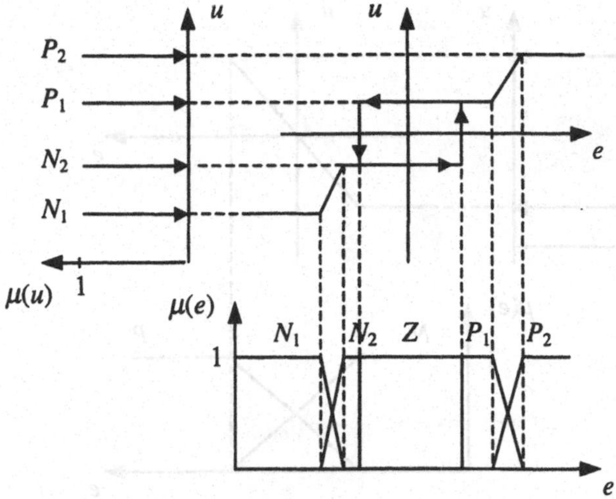

*R*1: *IF* (*e*(*t*)=N_1) *THEN* (*u*(*t*)=N_1)

*R*2: *IF* (*e*(*t*)=N_2) *THEN* (*u*(*t*)=N_2)

*R*3: *IF* (*e*(*t*)=*Z*) *AND* (*e*(*t*)-*e*(*t*-1)>0) *THEN* (*u*(*t*)=N_2)

*R*4: *IF* (*e*(*t*)=*Z*) *AND* (*e*(*t*)-*e*(*t*-1)=0) *AND* (*u*(*t*-1)=N_2) *THEN* (*u*(*t*)=N_2)

*R*5: *IF* (*e*(*t*)=*Z*) *AND* (*e*(*t*)-*e*(*t*-1)=0) *AND* (*u*(*t*-1)=P_1) *THEN* (*u*(*t*)=P_1)

*R*6: *IF* (*e*(*t*)=*Z*) *AND* (*e*(*t*)-*e*(*t*-1)<0) *THEN* (*u*(*t*)=P_1)

*R*7: *IF* (*e*(*t*)=P_1) *THEN* (*u*(*t*)=P_1)

*R*8: *IF* (*e*(*t*)=P_2) *THEN* (*u*(*t*)=P_2)

Fig. 7.5. Controller with hysteresis, saturation, and its fuzzy version

Figs. 7.2 – 7.7 show that all traditionally used static controllers can be represented in a fuzzy version. It concerns the controllers with hysteresis too, Fig. 7.5, but in this case the controller is not static but dynamic, because its current output $u(t)$ depends on the output state in the preceding moment $u(t-1)$. Fig. 7.6 shows that applying proper membership functions we can smooth down the controller characteristic. On the other hand, Fig. 7.7 shows that by increasing the number of input and output membership functions we can freely increase the degree of complexity of the characteristic.

Applying membership functions of different shapes, practically any non-linear characteristic of the controller can be obtained. Research made by specialists from the fuzzy logic domain, e.g. Wang (Wang 1994), have shown that fuzzy models are universal approximators of modeled systems, i.e. they allow for approximation of the given system with any precision. The difference

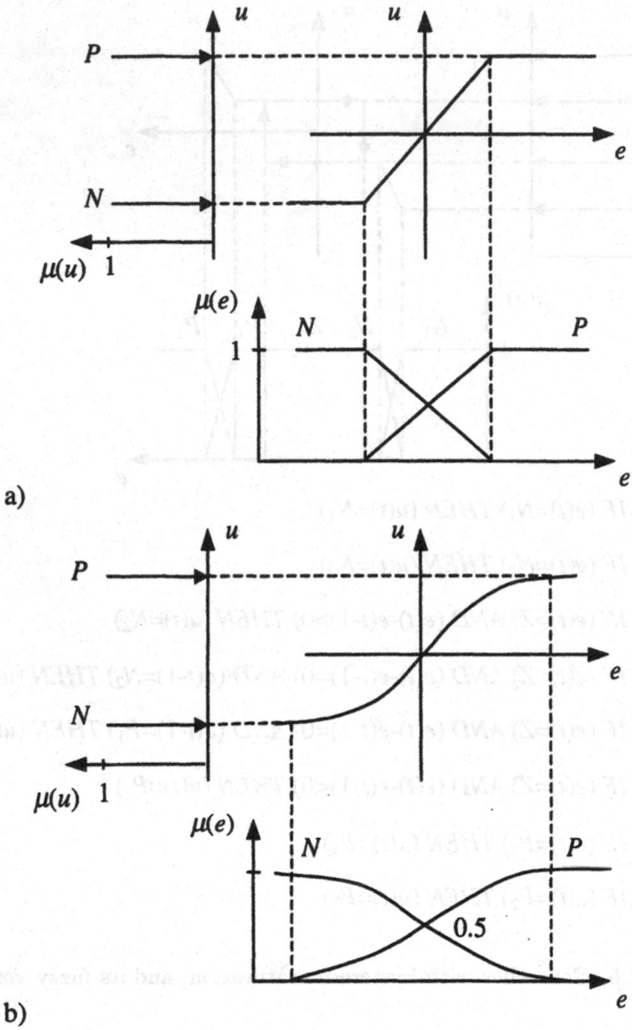

*R*1: *IF* (*e* = *N*) *THEN* (*u* = *N*) *R*2: *IF* (*e* = *P*) *THEN* (*u* = *P*)

Fig. 7.6. **Static controller with saturation of the input signal and its fuzzy version with segmentally-linear membership functions (a) and with smooth membership functions such as the Gaussian function or sin(x) (b)**

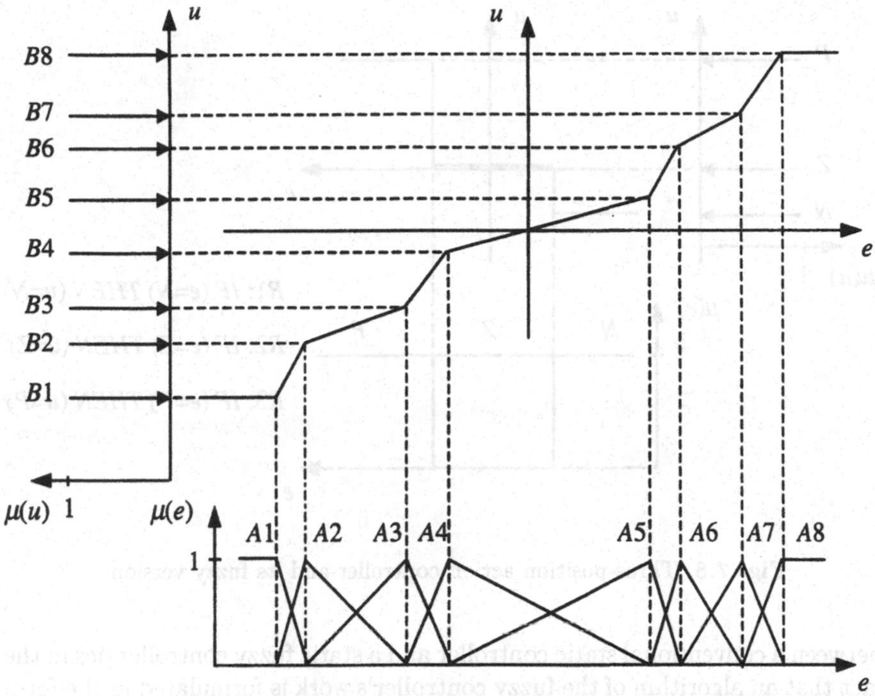

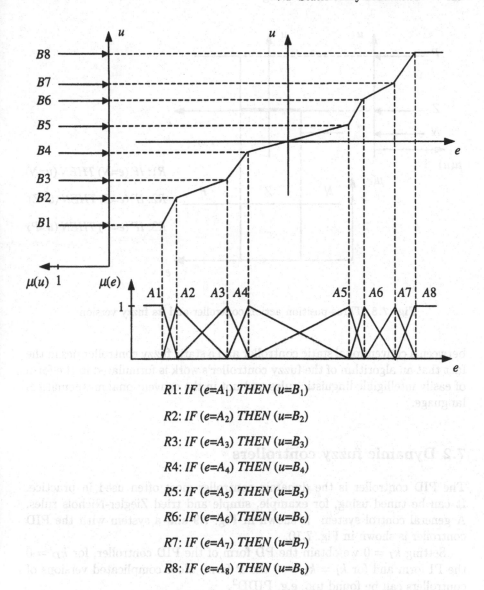

*R*1: *IF* (e=A_1) *THEN* (u=B_1)

*R*2: *IF* (e=A_2) *THEN* (u=B_2)

*R*3: *IF* (e=A_3) *THEN* (u=B_3)

*R*4: *IF* (e=A_4) *THEN* (u=B_4)

*R*5: *IF* (e=A_5) *THEN* (u=B_5)

*R*6: *IF* (e=A_6) *THEN* (u=B_6)

*R*7: *IF* (e=A_7) *THEN* (u=B_7)

*R*8: *IF* (e=A_8) *THEN* (u=B_8)

Fig. 7.7. Complicated static characteristic of the controller and its fuzzy version

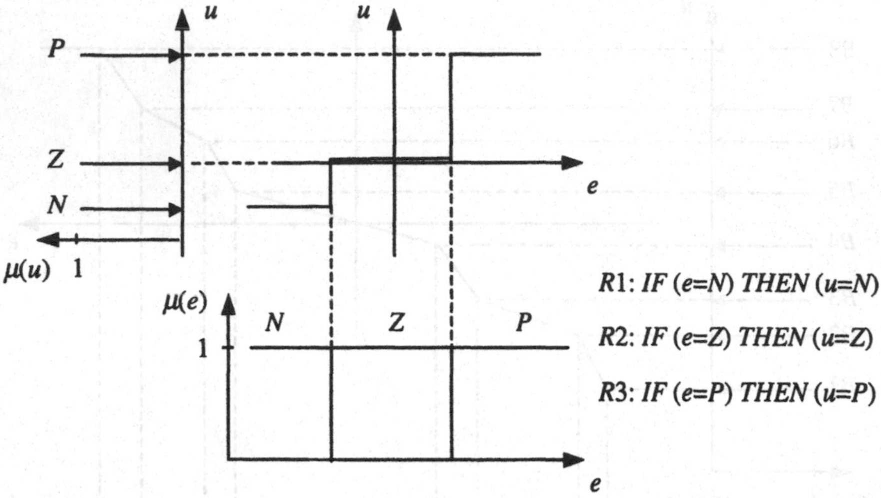

Fig. 7.8. Three-position action controller and its fuzzy version

between a conventional static controller and a static fuzzy controller lies in the fact that an algorithm of the fuzzy controller's work is formulated in the form of easily intelligible linguistic rules and not in the conventional mathematical language.

7.2 Dynamic fuzzy controllers

The PID controller is the dynamic controller most often used in practice. It can be tuned using, for example, simple and tried Ziegler-Nichols rules. A general control system is shown in Fig. 7.9 and a system with the PID controller is shown in Fig. 7.10.

Setting $k_I = 0$ we obtain the PD form of the PID controller, for $k_D = 0$ the PI form and for $k_I = k_D = 0$ the P form. More complicated versions of controllers can be found too, e.g. PIDD^2.

In digital versions differentiation is most often realized on the basis of equation (7.1):

$$e_D(k) = \frac{e(k) - e(k-1)}{T}, \tag{7.1}$$

where: T – sampling time, and integrating according to the equation (7.2):

$$e_I(k) = T \cdot \sum_{j=1}^{k} e(j), \tag{7.2}$$

or more exactly with the trapezoidal Tustin's method (7.3):

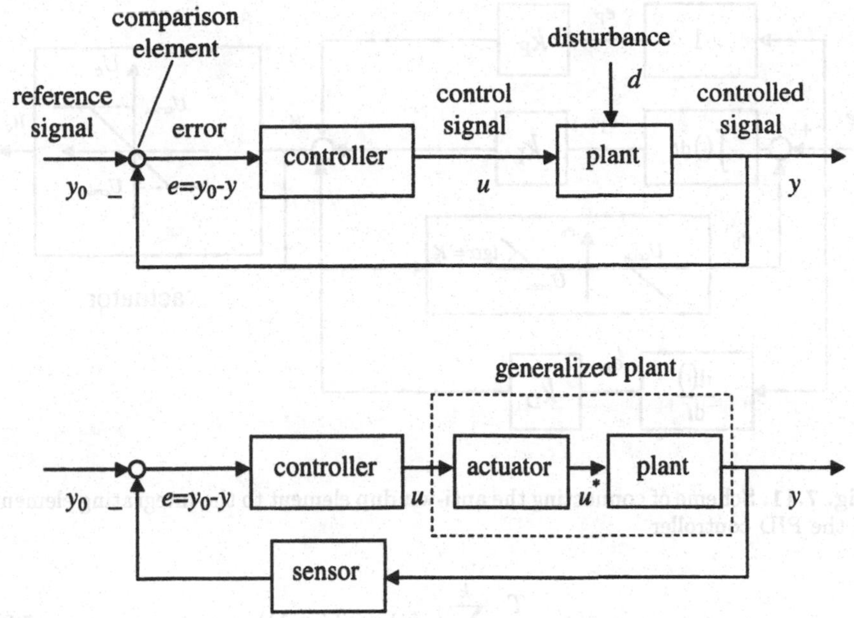

Fig. 7.9. General control system with the dynamic controller, notation

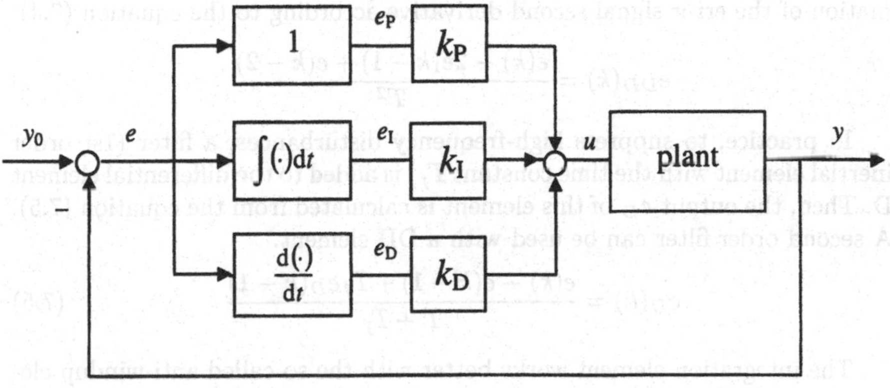

Fig. 7.10. Control system with conventional PID controller, k_P, k_I, k_D = const

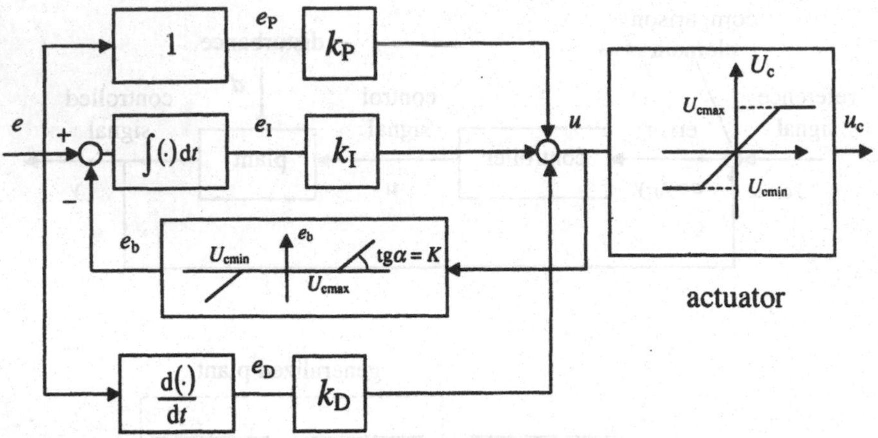

Fig. 7.11. Scheme of connecting the anti-windup element to the integrating element of the PID controller

$$e_I(k) = \frac{T}{2} \cdot \sum_{j=1}^{k} (e(j) + e(j-1)). \tag{7.3}$$

D^2 (DD) action can be obtained by calculation of the discrete approximation of the error signal second derivative according to the equation (7.4).

$$e_{DD}(k) = \frac{e(k) - 2e(k-1) + e(k-2)}{T^2} \tag{7.4}$$

In practice, to suppress high-frequency disturbances, a filter (1st order inertial element with the time constant T_f) is added to the differential element D. Then, the output e_D of this element is calculated from the equation (7.5). A second order filter can be used with a DD element.

$$e_D(k) = \frac{e(k) - e(k-1) + T_f e_D(k-1)}{T + T_f} \tag{7.5}$$

The integration element works better with the so-called anti-windup element (Franklin 1986) in accordance with the scheme shown in Fig. 7.11.

Actuators usually have limitations (saturations) U_{cmax} and U_{cmin} of generated signals. That is why they can not set such values of the control signal as the controller calculates in accordance with its algorithm. The anti-windup element prevents unnecessary loading of the integrating element and generating too large signals. In that way it decreases the overshoot and oscillations in the system and makes possible the setting of larger gains of the controller which increases the quality of the control.

Fuzzy PID controllers are most often realized in digital versions. Two versions are used here: the direct one shown in Fig. 7.12a and the incremental one shown in Fig. 7.12b.

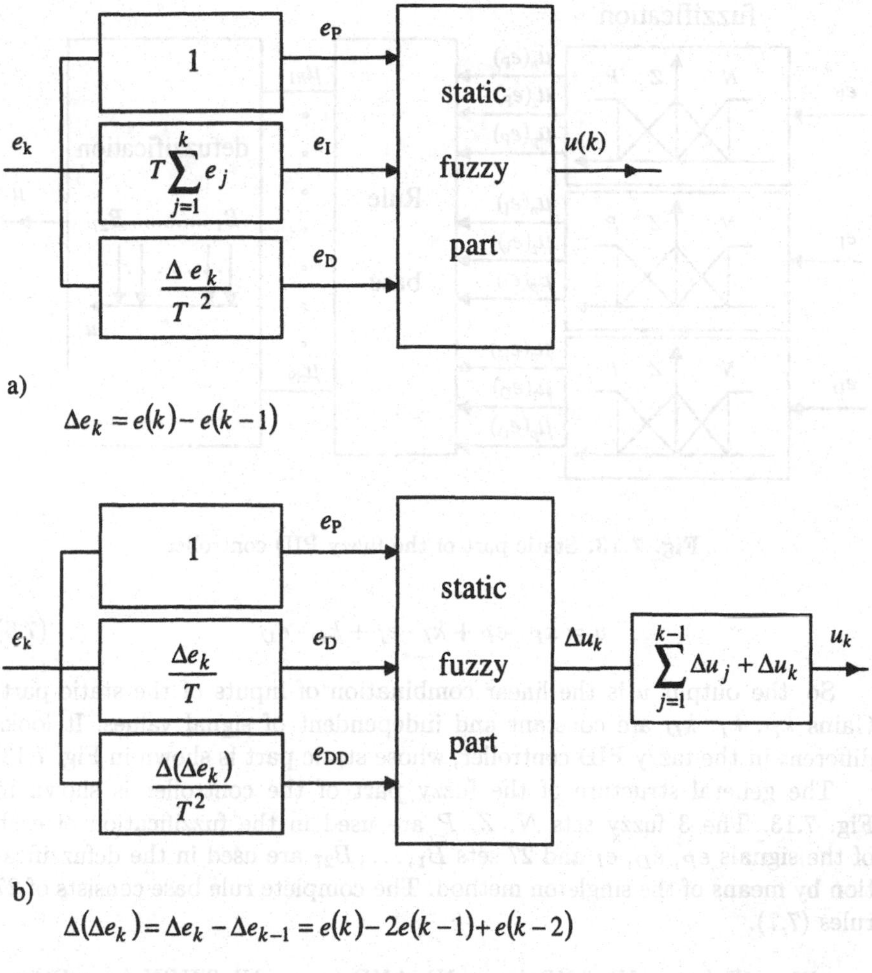

Fig. 7.12. Direct (a) and incremental (b) version of the fuzzy PID controller

In the direct version, the controller calculates the full value of the control signal u_k in each sampling step. In the incremental version, only the value Δu_k about which that signal has to be changed is calculated, after that the full value of the control signal is calculated by the connected adding element. It must also be noted that in the incremental version, in the dynamic part of the controller, only increments of inputs Δu_k and $\Delta(\Delta u_k)$ adequate to the first and second derivative of the error signal are calculated. The sum $e_I = T\sum_{j=1}^{k} e_j$ is not calculated here as in the direct version. The direct version is less sensitive to the noise of the input signal e. That is why it will be considered below. In the conventional PID controller, the controller output u is calculated as a sum of outputs of the dynamic part multiplied by corresponding gain coefficients, (7.6).

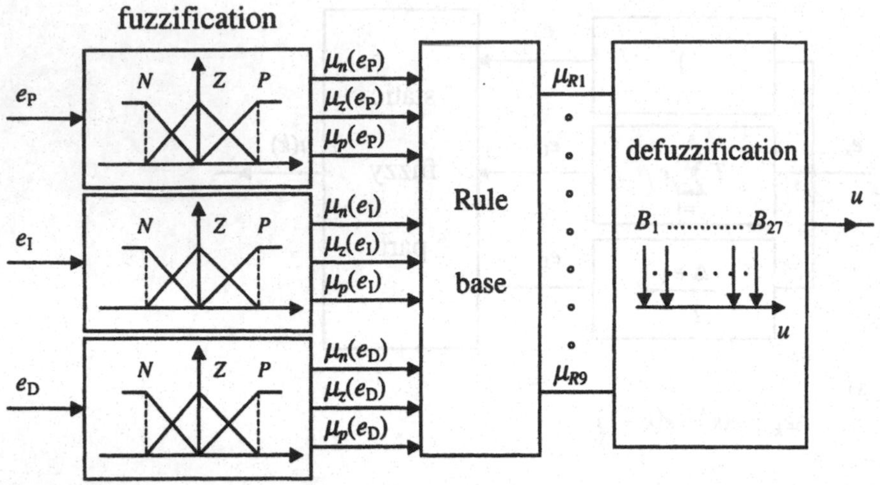

Fig. 7.13. Static part of the fuzzy PID controller

$$u = k_P \cdot e_P + k_I \cdot e_I + k_D \cdot e_D \tag{7.6}$$

So, the output u is the linear combination of inputs of the static part. Gains k_P, k_I, k_D are constant and independent of signal values. It looks different in the fuzzy PID controller, whose static part is shown in Fig. 7.13.

The general structure of the fuzzy part of the controller is shown in Fig. 7.13. The 3 fuzzy sets N, Z, P are used in the fuzzification of each of the signals e_P, e_D, e_I and 27 sets $B_1, \ldots, B_{27}$ are used in the defuzzification by means of the singleton method. The complete rule base consists of 27 rules (7.7).

$$\begin{aligned}
&R1: \ IF \ (e_P = N) \ AND \ (e_I = N) \ AND \ (e_D = N) \ THEN \ (u = B_1) \\
&R2: \ IF \ (e_P = N) \ AND \ (e_I = N) \ AND \ (e_D = Z) \ THEN \ (u = B_2) \\
&\vdots \\
&R27: \ IF \ (e_P = P) \ AND \ (e_I = P) \ AND \ (e_D = P) \ THEN \ (u = B_{27})
\end{aligned} \tag{7.7}$$

All the possible combinations of sets of inputs N, Z, P are expressed by premises, creating rules (7.7). Only one set of output B_i, $i = 1 \ldots 27$ corresponds to a given rule Ri.

The fuzzification realized on the basis of 2 fuzzy sets N and P leads to a less complicated version of the fuzzy PID controller – see Fig. 7.14.

The above situation means that the complete rule base is composed of 8 rules. So, there are numerous realizations of the fuzzy PID controller. The increasing number of fuzzy sets assigned to the input "generates" a larger

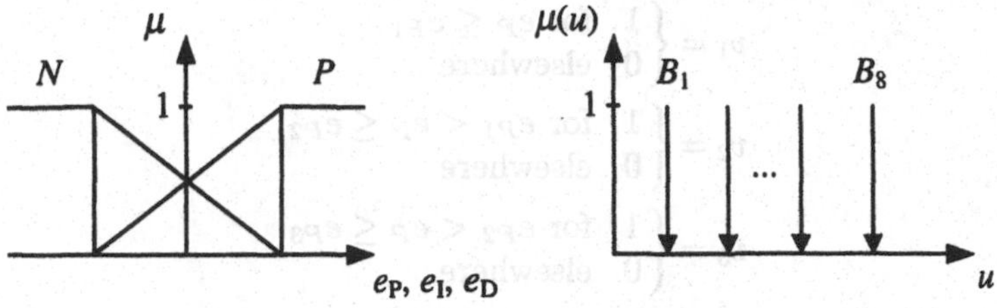

Fig. 7.14. The most simplified defuzzification for the fuzzy PID controller

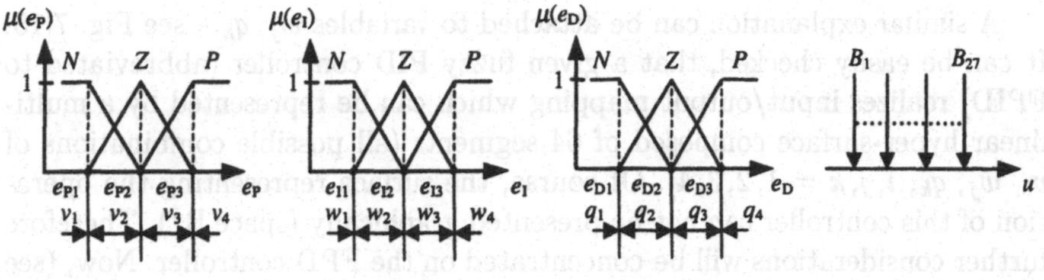

Fig. 7.15. The membership functions for inputs and output of a fuzzy PID controller

number of rules. Finally, it leads to a more complicated structure of the fuzzy controller.

The following questions arise:

1. How many fuzzy sets should be assigned to inputs and how many rules should be used in a fuzzy controller?
2. What are the differences between conventional PID controllers and fuzzy PID controllers?
3. Are there any reasons favoring applications of fuzzy PID controllers, if conventional PID controllers have been successfully used for several decades?

The equation (7.6) defines a linear operation realized by a conventional PID controller. The non-linear operation realized by a fuzzy PID controller, Fig. 7.15 (its fuzzification is based on 3 fuzzy sets, the logic operation *AND* is relized with *PROD* operator, the singleton method is applied to defuzzification) can be expressed in the form (7.8).

$$u = \sum_{i=1}^{4}\sum_{j=1}^{4}\sum_{k=1}^{4} v_i w_j q_k \big(a_{0ijk} + a_{2ijk}e_I + a_{3ijk}e_D + a_{4ijk}e_P e_I + a_{1ijk}e_P + a_{5ijk}e_P e_D + a_{6ijk}e_I e_D + a_{7ijk}e_P e_I e_D\big) \tag{7.8}$$

The logic variables v_i, w_j, q_k carry information about current membership of the input signals to a given sector of the input space. For example, the meaning v_i is given by relations (7.9).

$$v_1 = \begin{cases} 1 & \text{for } e_P \leq e_{P1} \\ 0 & \text{elsewhere} \end{cases}$$

$$v_2 = \begin{cases} 1 & \text{for } e_{P1} < e_P \leq e_{P2} \\ 0 & \text{elsewhere} \end{cases}$$

$$v_3 = \begin{cases} 1 & \text{for } e_{P2} < e_P \leq e_{P3} \\ 0 & \text{elsewhere} \end{cases}$$

$$v_4 = \begin{cases} 1 & \text{for } e_{P3} < e_P \\ 0 & \text{elsewhere} \end{cases} \quad (7.9)$$

A similar explanation can be attached to variables w_j, q_k – see Fig. 7.15. It can be easily checked, that a given fuzzy PID controller (abbreviated to FPID) realizes input/output mapping which can be represented by a multi-linear hyper-surface composed of 64 segments (all possible combinations of v_i, w_j, q_k, $i, j, k = 1, 2, 3, 4$). Of course, the surface representing the operation of this controller can not be presented graphically (space R^4). Therefore further considerations will be concentrated on the FPD controller. Now, (see Fig. 7.16) the hyper-surface representing input/output mapping is composed of 16 rectangular, multi-linear sectors.

The surface assigned to a conventional PD controller (abbreviated to CPD) is shown in Fig. 7.16a. It is an ordinary plane with 2 degrees of freedom k_P, k_D which can only change its inclination in relation to the co-ordinate system. The surface of the FPD controller is composed of 16 segments, which are represented by multi-linear surfaces separated by straight line segments (edges). There is a variety of possible locations of the segments in the space. Furthermore, the convexity of those segments can be considerably modified. The above adjustments can be done by choosing values for 15 degrees of freedom (the modal values of membership functions for inputs e_P, e_D and output u). The possibility of flexible shaping of the non-linear surface of that controller is a very important property ensuring high control performance. The possibilities of improving control performance indices are wider if the controller is more flexible. Of course, the greater the number of segments creating the controller surface, the more flexible the controller. The number of segments depends directly on the number of membership functions. Nevertheless, one should minimize the number of membership functions in order to avoid serious difficulties occurring when numerous controller parameters have to be tuned. Thus, the acceptable version of the controller should be as simple as possible for the required precision of plant control.

The superiority of a FPID controller can be confirmed by the following reasoning: the properly designed FPID controller (i.e. after suitable choice of *AND* operators, defuzzification, inference and membership functions) can "copy" the input/output mapping realized by any CPID controller (Voit 1994). The inverse task (CPID "copies" operation of any FPID) can not be realized. There are opinions, for instance (Isermann 1996), limiting the applications of fuzzy controllers to the control of non-linear plants, especially

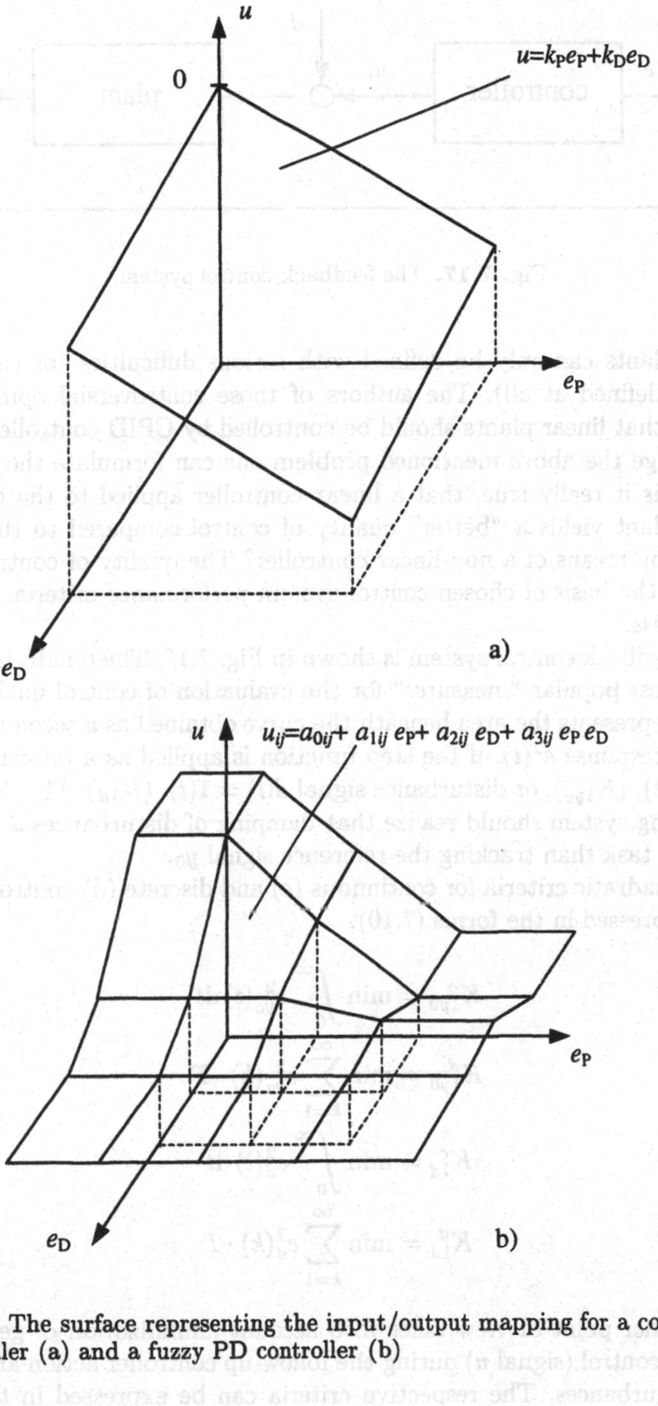

Fig. 7.16. The surface representing the input/output mapping for a conventional PD controller (a) and a fuzzy PD controller (b)

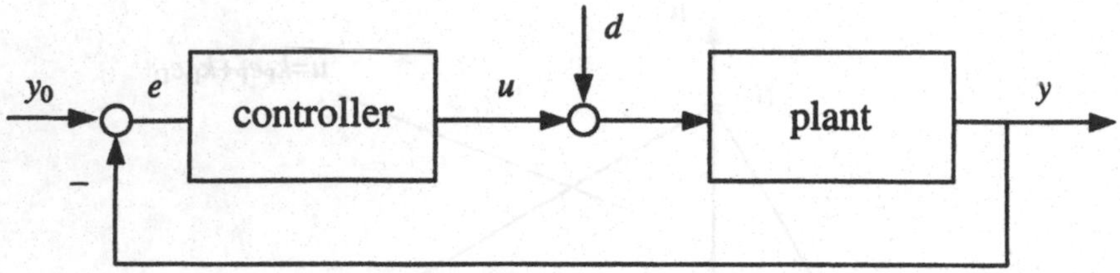

Fig. 7.17. The feedback control system

if those plants can only be defined with serious difficulties (or can not be precisely defined at all). The authors of those controversial opinions also maintain that linear plants should be controlled by CPID controllers.

To judge the above mentioned problem one can formulate the following question: is it really true, that a linear controller applied to the control of a linear plant yields a "better" quality of control compared to the quality obtained by means of a non-linear controller? The quality of control is evaluated on the basis of chosen control system performance criteria. What do they express?

The feedback control system is shown in Fig. 7.17. The quadratic criteria are the most popular "measures" for the evaluation of control quality. That measure represents the area beneath the curve obtained as a second power of unit step response $e^2(t)$, if the step function is applied as a reference signal $y_0(t) = 1(t)$, (K_{1y_0}), or disturbance signal $d(t) = 1(t)$, (K_{1d}). The designer of a stabilizing system should realize that damping of disturbances d is a more important task than tracking the reference signal y_0..

The quadratic criteria for continuous (c) and discrete (d) control systems can be expressed in the forms (7.10).

$$
\begin{aligned}
K^c_{1y0} &= \min \int_0^\infty e^2_{y_0}(t)\,\mathrm{d}t \\
K^d_{1y0} &= \min \sum_{k=1}^{\infty} e^2_{y_0}(k) \cdot T \\
K^c_{1d} &= \min \int_0^\infty e^2_d(t)\,\mathrm{d}t \\
K^d_{1d} &= \min \sum_{k=1}^{\infty} e^2_d(k) \cdot T
\end{aligned}
\tag{7.10}
$$

The other point of view takes into account minimization of generalized "costs" of control (signal u) during the follow-up controller action and damping of disturbances. The respective criteria can be expressed in the forms (7.11).

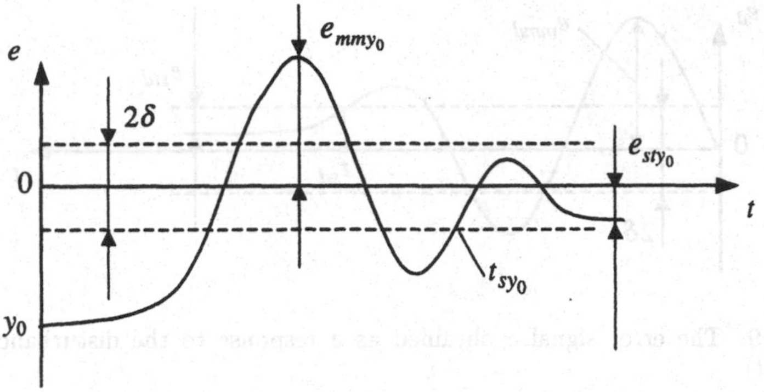

Fig. 7.18. The error signal e for the reference signal $y_0(t) = 1(t)$

$$K_{2y0}^{c} = \min \int_0^{\infty} u_{y_0}^2(t)\,\mathrm{d}t$$

$$K_{2y0}^{d} = \min \sum_{k=1}^{\infty} u_{y_0}^2(k) \cdot T$$

$$K_{2d}^{c} = \min \int_0^{\infty} u_d^2(t)\,\mathrm{d}t$$

$$K_{2d}^{d} = \min \sum_{k=1}^{\infty} u_d^2(k) \cdot T \tag{7.11}$$

There are control systems, where optimization should be aimed at minimization of the second power of the derivative (velocity) of the plant input signal u. By "optimization", in the above meaning, one ensures the "smooth" operation of the actuator (Kuhn 1994) and the minimization of the number of switches (+/−). Besides integral measures for the evaluation of control performance, one can find very useful criteria where the measure of system performance can be expressed as the length of characteristic segments defined for the error signal e, when the system is excited by reference signal $y_0(t) = 1(t)$ or disturbance signal $d(t) = 1(t)$ – see Fig. 7.18 and Fig. 7.19.

The tracking properties of a control system for $y_0(t) = 1(t)$ are usually evaluated on the basis of segments representing steady state error e_{sty_0}, overshoot ov_{y_0} and settling time t_{sy_0}. The respective conditions determining acceptable tracking properties state that the lengths of those segments should not be longer than admissible. The resistance of a control system to disturbances d can be measured by segments representing settling time t_{sd}, maximal amplitude of error e_{mmd} and steady state error e_{std}.

The amplitudes of error signal e for all $t > t_s$, where t_s is so-called settling time, satisfy condition $|e| < \delta$, where δ represents required accuracy – see Fig. 7.18 and 7.19. For example, δ can be assumed as 2% or 5% of amplitude of reference signal. The relative overshoot ov_{y_0} is defined by formula (7.12).

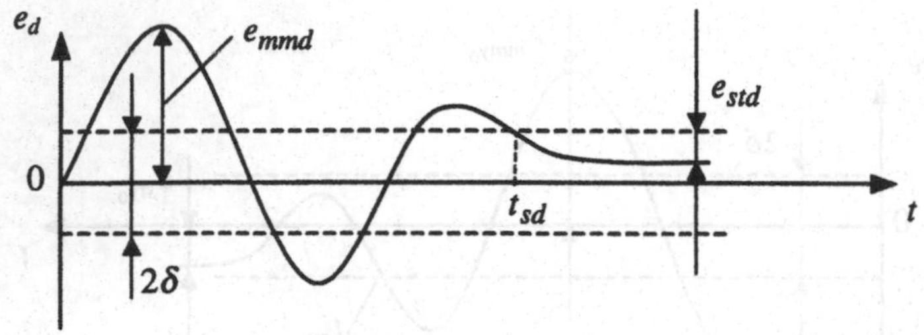

Fig. 7.19. The error signal e obtained as a response to the disturbance signal $d(t) = 1(t)$

$$ov_{y0} = \frac{e_{mmy0}}{y_0} \tag{7.12}$$

The steady state error e_{st} can be defined as the amplitude of the control system error signal after the end of the transient state caused by y_0 or d. Thus, the set of integral criteria K_1 and K_2 can be supplemented with "segmental" criteria (7.13).

$$\begin{aligned} K_{3y0} &= \min t_{sy0} \\ K_{3d} &= \min t_{sd} \\ K_{4y0} &= \min ov_{y0} \\ K_{4d} &= \min \frac{e_{mmd}}{d} \\ K_{5y0} &= \min e_{sty0} \\ K_{5d} &= \min e_{std} \end{aligned} \tag{7.13}$$

All the above criteria K_i represent non-linear functions of controller parameters, even in the case of linear plants and linear controllers. Thus, **if the quality of control is taken into account, then each control problem, irrespective of the type of controller (linear or non-linear), has to be classified as non-linear** (Hunt 1992).

The fuzzy controller, which in fact realizes the non-linear input/output mapping, "matches" much better to the optimization of non-linear criteria than linear controllers. Furthermore, the number of degrees of freedom (adjustable parameters) is usually greater for fuzzy controllers . For example, to the previously considered FPID controller, based on 3 fuzzy sets for each input, one can assign 15 degrees of freedom, whereas only 3 degrees of freedom (k_P, k_I, k_D) can be assigned to the linear CPID controller.

The meaningful role of the number of degrees of freedom at the disposal of the control system designer can also be perceived if one realizes that values of criteria are usually analyzed for various amplitudes of y_0 and d and, on the other hand, evaluation of control performance is often done on the basis of several criteria K_i simultaneously. It means that control system optimization has to be treated as a multi-criteria task (7.14),

$$K = \min(\alpha_1 K_1 + \alpha_2 K_2 + \ldots + \alpha_n K_n)\,, \tag{7.14}$$

where: α_i – weighting coefficients,
n – number of component criteria.

The widely used criterion taking into consideration error and cost of control (7.15)

$$K = \min \int_0^t \left[\alpha e^2(t) + (1-\alpha)u^2(t)\right] \, \mathrm{d}t\,, \tag{7.15}$$

can be treated as a good example of a multi-criteria problem.

The greater number of degrees of freedom assigned to a fuzzy controller means that component criteria for multi-criteria problems can be minimized much easier! The relatively small number of degrees of freedom assigned to a linear CPID controller makes the above task very difficult (interaction, contradictory requirements, etc.). Thus, there are important reasons favouring applications of fuzzy controllers for linear plants as well.

7.3 The determination of structures and parameters for fuzzy controllers (organization and tuning)

The following are approaches aimed at the synthesis of fuzzy controllers:

I. on the basis of expert knowledge,
II. by modeling the expert acting as a controller (on the basis of control signals generated by the expert),
III. on the basis of a model of the plant under control.

Fuzzy control seems to be one of the most important areas of application of fuzzy logic (Abdelnour 1992; Cao 1997b; Fisher 1996; Fischle 1997; Gupta 1991; Gorez 1996; Hanss 1996b; Hunt 1992; Iwasaki 1990; Isaka 1993; Isermann 1996; Jantzen 1997; Kouatli 1991; Koch 1994; Kang 1995; Kuhn 1996; Li 1995; Lewis 1996; Lopez 1997; Preuss 1992; Wang 1995b; Wu 1994; Ying 1993).

The models of plants under control are not unconditionally necessary for the synthesis of fuzzy controllers (Ying 1994). That fact has to be considered a huge advantage, because the design of conventional control systems using state space or frequency representations is based on approximate models of controlled plants. Of course, the above does not at all mean that knowledge about a controlled plant is useless in the case of fuzzy controllers. Formally speaking methods I and II do not use models of plants, because they are based on the knowledge of an experienced expert, who captured the "art" of plant control during many experiments and daily activities considered in the category called "man-made control of plant". It is obvious that some kind of intuitive model of a plant is hidden "inside" expert knowledge on the plant properties and its dynamic behavior. To successfully control car movement, the driver does not need to know the mathematical model of the car. He simply needs experience and practice, which gives him a good "feeling for" and "understanding" of a car behavior under different conditions (like different speed, road slope, etc.) and control activities (revolving the driving wheel, applying the brakes, clutch, gearbox lever, carburetor lever, etc.). It is a rule that even a good driver has to accustom himself to a new car, because only the "feel" of a car gives him fluency and perfection in its control. There are many fields where mathematical models of controlled plants can be replaced by the experience of an expert (piloting airplanes and vessels, "driving" cranes, supervision of industrial processes, etc.).

7.3.1 The design of fuzzy controllers on the basis of expert knowledge concerning plant under control

An expert's knowledge concerning the controlled plant can be used for designing a fuzzy controller. The designer of a fuzzy controller has to accumulate the necessary knowledge by interviewing the experienced plant operator. The expert knowledge can be expressed in the form of linguistic rules of the type:

$$IF\ (x = A)\ THEN\ (y = B)\,, \tag{7.16}$$

where y is a controlled signal and x is the signal set by the expert. A and B represent linguistic evaluations like, for example, small, large, slow, quick, etc. Besides linguistic rules, the expert should impart information on his linguistic evaluations. Such information is important for determining essential parameters of membership functions – especially their modal values.

Example 7.3.1.1
The designing of fuzzy controllers on the basis of expert knowledge can be illustrated with the example, dealing with the control of the travelling crab of a bridge crane (Watanabe 1991; Zimmermann 1994b).

During transportation of containers (see Fig. 7.20) it so happens that the container swings (gets out of plumb) and its angular displacements Θ can be quite large. The swinging has to be stopped before bringing the container to its destination. Otherwise, the neighboring containers would be damaged by the swinging container acting as a heavy hammer. The operator of the crane bridge controls the speed v of its travelling crab by means of a lever, which can be set in any position between its two terminal positions. The operator has to "feel" the responses of the crab when it is influenced by the weight of the container (the weights of particular containers are different) in order to undertake the control action using the lever. There are two possible simple ways of controlling the crab where expert knowledge is not necessary.

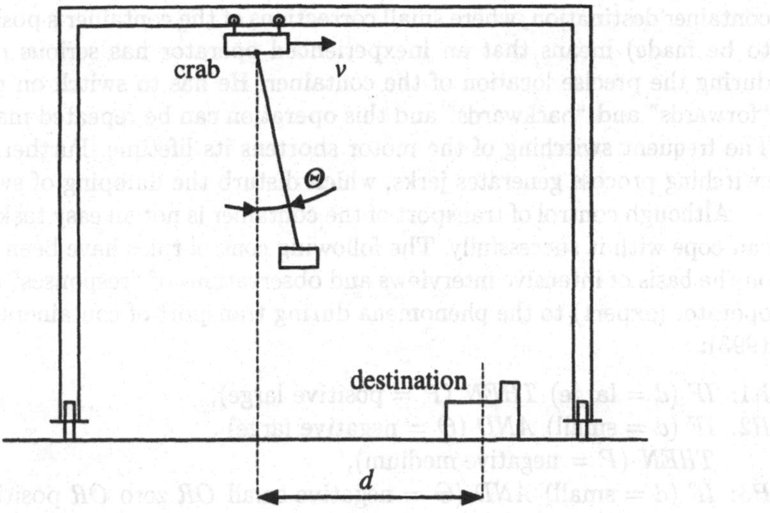

Fig. 7.20. The bridge crane transporting the container

The first control strategy can be to maintain a sufficiently low crab speed thereby avoiding swinging. The second simple way involves high-speed transportation of the container to the point above the destination. That operation would generate swings. Thus, after that speedy displacement the operator should wait until the container stops swinging. After that, the container could be slowly lowered to its destination. Both the above control strategies are time-consuming. An experienced operator of bridge cranes uses other control strategies. He can place the container in the destined area very quickly, efficiently damping swings of the container over the destined area. The conventional PID controller can not cope with the control of a crane bridge crab because of strong non-linearity of the process under control. If distance d between crab and destination is great, then swings of the container can be neglected and the operator is allowed to keep the high speed of the crab. When the distance to the destination diminishes, then the operator begins damping container swings. Finally, the angular displacements Θ of the container in the neighborhood of the destined area are almost completely damped. It means that the container can be quickly lowered to its destination. The experienced operator successfully modifies his control strategy in the presence of disturbances. The modification depends on the intensity and direction of the wind and on the weight of the container.

The presence of static friction makes the control of bridge crane movement more difficult. To begin crane movement the driving torque of an electric motor has to "overcome" the resistance caused by static friction. If crane movement begins, then resistance caused by static friction suddenly becomes small. This described phenomenon (especially in the neighborhood of the container destination, where small corrections of the container's position have to be made) means that an inexperienced operator has serious difficulties during the precise location of the container. He has to switch on the motor "forwards" and "backwards" and this operation can be repeated many times. The frequent switching of the motor shortens its lifetime. Furthermore, the switching process generates jerks, which disturb the damping of swings.

Although control of transport of the container is not an easy task, the man can cope with it successfully. The following control rules have been identified on the basis of intensive interviews and observations of "responses" of a crane operator (expert) to the phenomena during transport of containers (Alltrock 1993):

*R*1: *IF* (d = large) *THEN* (P = positive large),
*R*2: *IF* (d = small) *AND* (Θ = negative large)
THEN (P = negative medium),
*R*3: *IF* (d = small) *AND* (Θ = negative small *OR* zero *OR* positive small)
THEN (P = positive medium),
*R*4: *IF* (d = small) *AND* (Θ = positive large) *THEN* (P = positive large),
*R*5: *IF* (d = zero) *AND* (Θ = positive large *OR* small)
THEN (P = negative medium),
*R*6: *IF* (d = zero) *AND* (Θ = zero) *THEN* (P = zero),
*R*7: *IF* (d = zero) *AND* (Θ = negative small) *THEN* (P = positive medium),
*R*8: *IF* (d = zero) *AND* (Θ = negative large) *THEN* (P = positive large),

where: d – distance, Θ – angular displacement, P – electrical power supplying the motor (operator controls the power shifting the lever).

It has to be taken into account that another bridge crane operator can use different rules under the same circumstances. It simply depends on his experience, reflexes, temperament and maybe other individual features. The rules may also depend on the type of bridge crane, the technical performance of the electrical motor, maximal distance of container displacement, value of friction resistances, etc. Thus, the above control rules can not be treated as universal, but rather as an example.

To evaluate the particular "signals" or "quantities" (linguistic variables) the operator has used the following linguistic values:

d – large (L), small (S), zero (Z),
Θ – positive large (PL), positive small (PS), zero (Z), negative small (NS), negative large (NL),
P – negative large (NL), negative medium (NM), zero (Z), positive medium (PM), positive large (PL).

The membership functions of particular variables are shown in Fig. 7.21.

The power supply P is set by a lever. Thus, the lever position corresponding to the "forward" terminal point can be denoted by "1" and the position corresponding to the "backward" terminal point can be denoted by "−1", see Fig. 7.21. During interview the crane operator (expert) should supply the

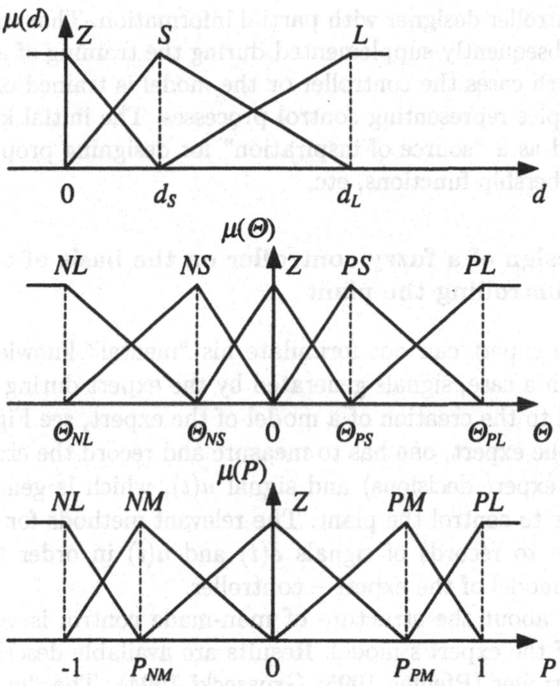

Fig. 7.21. The membership functions of the following linguistic variables: distance (d), angular displacement (Θ) and power supply (P)

controller designer with information about modal values of particular membership functions: d_S, d_L, Θ_{NL}, Θ_{NS}, Θ_{PS}, Θ_{PL}, P_{NM}, P_{PM}. It is a rather difficult problem, because the operator is "mostly driven by his intuition" and usually is not able to reveal the details of his "mental model for control". It means that the controller designer obtains rather rough information at his disposal. If information is too poor, then the parameters of a fuzzy controller have to be tuned using a neuro-fuzzy network, on the basis of input/output signals received and generated by the operator during real (actual) control of the plant.

◊ ◊ ◊

The synthesis of a fuzzy controller on the basis of expert knowledge is, in practice, narrowed to relatively slow, not very complicated plants equipped with not more than two control inputs. The man as "controller" is not able to control the multi-input plants (limitations caused by finite possibilities of perception of information and its processing) nor relatively "quick" plants (limitations caused by the rate of data processing). However, even the above disadvantageous circumstances do not eliminate the man's assistance in the process of collecting information on the control of the plant. The man can supply the controller designer with partial information. That initial information can be subsequently supplemented during the training of a controller or a model. In such cases the controller or the model is trained on the basis of measured samples representing control processes. The initial knowledge can also be treated as a "source of inspiration" for designing proper rule bases, choice of membership functions, etc.

7.3.2 The design of a fuzzy controller on the basis of a model of the expert controlling the plant

Sometimes the expert can not formulate his "mental" knowledge on plant control. In such a case, signals generated by the expert during plant control can be applied to the creation of a model of the expert, see Fig. 7.22.

To model the expert, one has to measure and record the error signal $e(t)$ (the basis for expert decisions) and signal $u(t)$, which is generated by the expert in order to control the plant. The relevant methods for model design can be applied to records of signals $e(t)$ and $u(t)$ in order to obtain the mathematical model of the expert – controller.

Knowledge about the structure of man-made control is very useful for the creation of the expert's model. Results are available describing features of man-as-controller (Pfeiffer 1995; Gruszecki 1994). The "man-controller" operates as a non-linear, adaptive controller. For small amplitudes of error signals, a man can be modeled as a PI controller. That is why a man eliminates the steady state errors. That feature means that a man can successfully stabilize the course of a ship or airplane. For greater values of error signal a man operates – approximately – as a PD controller. Especially for large error signals a man's control becomes "bang-bang" control (Pfeiffer 1995), i.e. control, where the control signal $u(t)$ is impetuously and frequently switched over from its maximal value (upper saturation) to its minimal value (lower saturation) and inversely.

The man adapts the method of control to the plant. For example, if a man controls the model of a helicopter (the transfer functions of that model have single or double poles equal to zero), then his operation is similar to the operation of a PD controller (integrating action of man-controller is not necessary, because it is realized by plant components represented in its transfer function by poles equal to zero, therefore steady state errors are eliminated

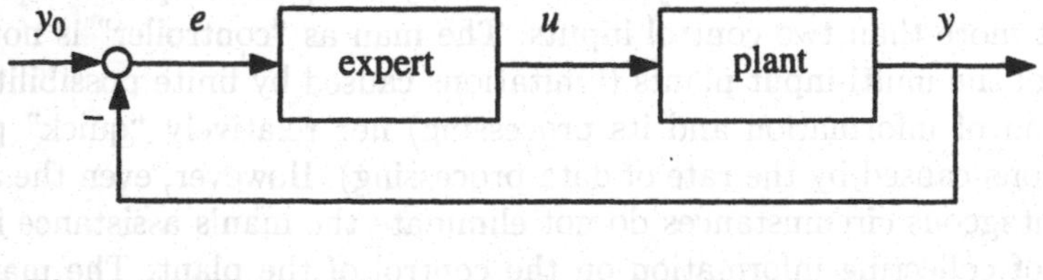

Fig. 7.22. The expert as controller in a feedback control system

without the integrating action of a man). The approximate transfer function of the man-controller can be written in the form (7.17).

$$G(s) = \frac{u(s)}{e(s)} = \frac{k_r e^{-sT_0}(1 + sT_d)}{(1 + sT_{nm})(1 + sT_r)} \tag{7.17}$$

The measurements of man delay time T_0 taken during the control of a model of a helicopter allow us to assume $T_0 = 0.1 - 0.2$ sec. However, the delay time T_0 increases if circumstances are more difficult and decisions have to be based on the processing of numerous data. The inertia of a nervous system (n) and a muscular system (m) is represented by time constant T_{nm}. The evaluation of T_{nm} yields $T_{nm} = 0.1 - 0.6$ sec. The gain k_r and time constant T_d representing derivative action as well as time constant T_r are variable and depend on the controlled plant.

The man controlling the cooling unit composed of piston blower and engine operates similarly to a PI controller. Taking into account the above facts one can conclude that man-as-controller can operate like a PID controller, which is "charged" by additional inertia and delay. The delay time T_0 can be neglected if a man controls slow plants, where sampling time $T > 0.25$ sec., and/or in the case when a man has extraordinary reflexes (short response time). The dynamics of a man controlling the plant or process can be approximated by transfer function (7.18).

$$G(s) = \frac{u(s)}{e(s)} = \frac{e^{-sT_0}}{(1 + sT_{nm})(1 + sT_r)} \left(k_P + k_D \cdot s + \frac{k_I}{s} \right) \tag{7.18}$$

It has to be kept in mind that gains k_P, k_D, k_I can vary (for example, when a man adjusts his control action to the parameters of a plant, type of reference signal, type of disturbances). Furthermore, under certain circumstances a man can operate a plant by "putting" $k_D = 0$ or $k_I = 0$. Those variable gains, depending on the reasons mentioned (amplitude of error signal, type of plant, etc.) , can be successfully modeled by fuzzy or neural controllers. The linear PID controller can not cope with the above task. The next example deals with the modeling of the man-controller by means of a neuro-fuzzy PID controller (Pałęga 1996).

Example 7.3.2.1

The control task lies in the course-change of a model of an underwater vehicle KRAB II (Technical University of Szczecin) shown in Fig. 7.23.

The torque M generated by two propellers changes the vehicle course Ψ. Because of propeller technical characteristics the values of the torque have to be in the range $[-18.3, 28.3]$, see Fig. 7.24. There are two different saturation levels for M: the upper one and the lower one. The lack of symmetry makes control more difficult. During the experiment, the operator observed the real course $\Psi(t)$ and compared it to the reference course Ψ_0. The error was compensated for by choice of joystick position. The joystick position was

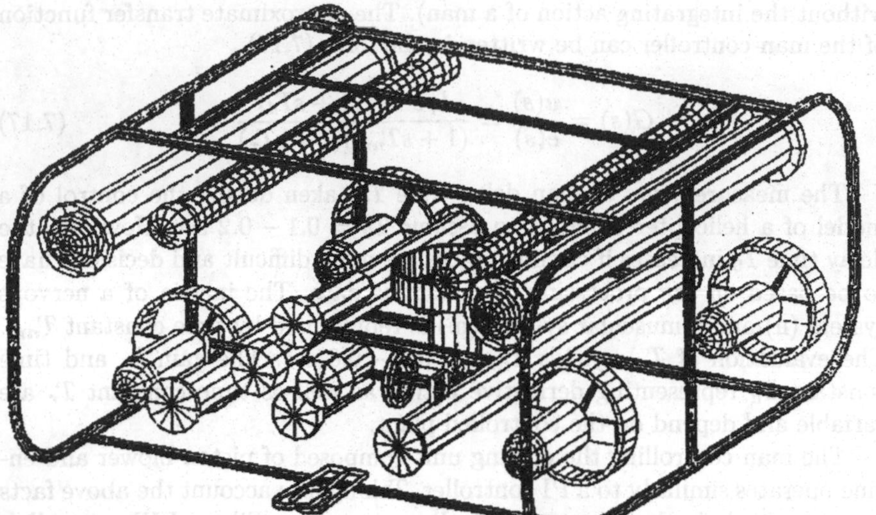

Fig. 7.23. The underwater vehicle as the plant under control

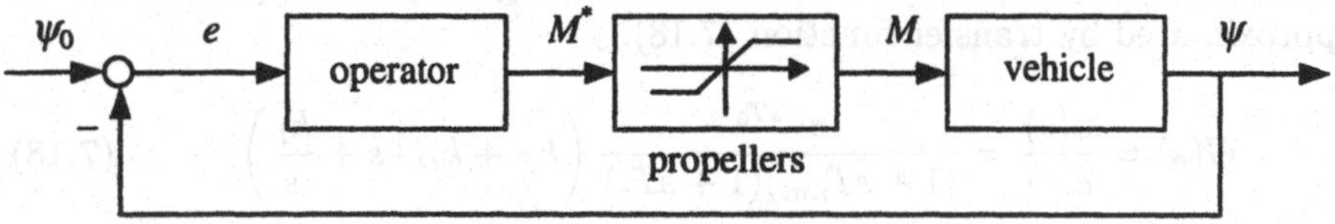

Fig. 7.24. The manual control of an underwater vehicle

transformed into the electrical signal M^* controlling the propellers. Finally, the thrusts of the two propellers generated the torque M turning the vehicle. The signals $M(t)$ and $\Psi(t)$ for several set values Ψ_0 were recorded during the experiment. Examples of records are shown in Fig. 7.25.

The recorded signals confirm the non-linearity of man as controller (impetuous changes of control signal, irregularities of course signal $\Psi(t)$). The signals $e(t)$ and $M(t)$ obtained during manual control were applied to the tuning of a neuro-fuzzy PID controller represented by 27 rules, see Fig. 7.26 (Piegat 1996). The operations *AND* of that controller were realized by means of operator *MEAN*. So, the sum of activations of all conclusions (defuzzification) was constant. That is why the defuzzification was simplified (the dividing element attached to the output of the network was redundant).

The tuned controller was introduced into the control system and its operation was examined. The comparison of the controlled signal (course $\Psi(t)$) obtained for the fuzzy controller to the same signal generated manually is

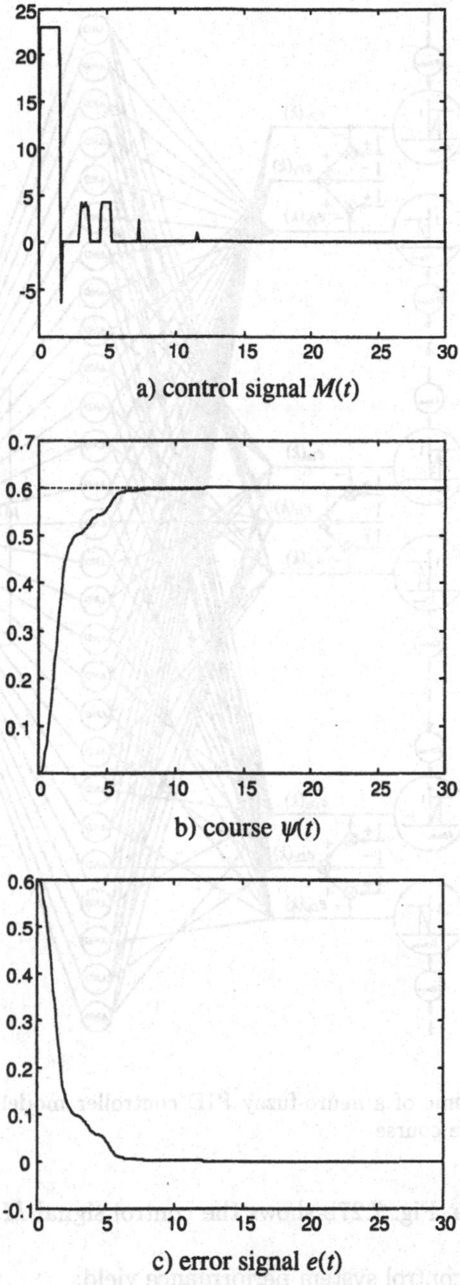

Fig. 7.25. Examples of signals representing torque $M(t)$ (a), vehicle course $\Psi(t)$ (b), and error signal $e(t)$ (c), obtained under manual control of vehicle

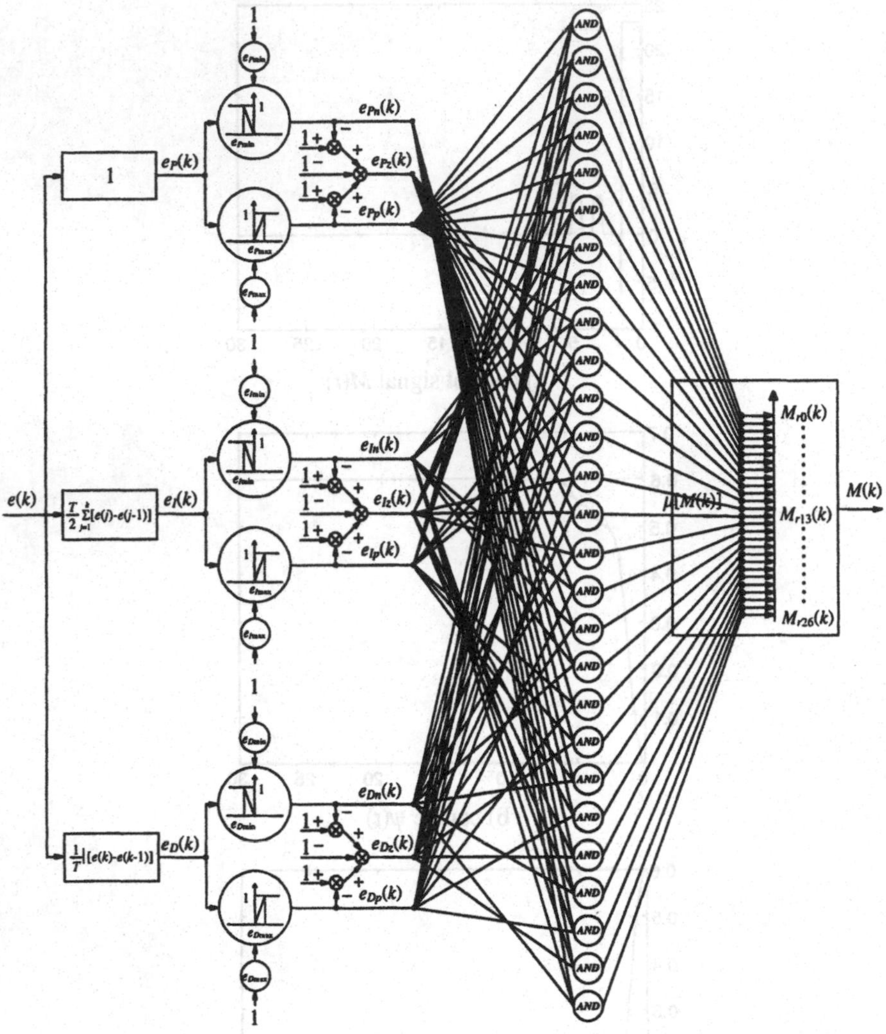

Fig. 7.26. The scheme of a neuro-fuzzy PID controller modeling manual control of underwater vehicle course

shown in Fig. 7.27a. Fig. 7.27b shows the control signal M generated by the fuzzy controller.

Evaluations of control system performance yield:

$$\int_0^{30\,(s)} |e|\,dt = 1.07\,, \qquad \int_0^{30\,(s)} |u|\,dt = 41.30\,,$$

for manual control and:

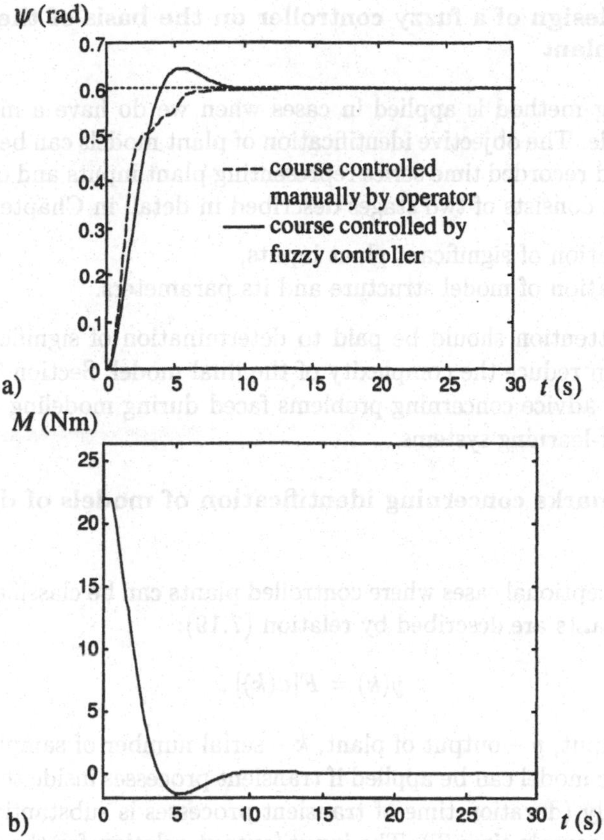

Fig. 7.27. The comparison of courses controlled manually and with the fuzzy controller (a) and control signal M generated by the fuzzy controller (b)

$$\int_0^{30\,(s)} |e|\, dt = 0.95\,, \qquad \int_0^{30\,(s)} |u|\, dt = 45.86\,,$$

for the fuzzy controller.

The values of performance criteria confirm that the fuzzy controller was slightly more accurate ($\int |e|\, dt$) but greater control cost ($\int |u|\, dt$) was the price paid for accuracy improvement. Comparing operations of both controllers (i.e. man and fuzzy controller) one can conclude that the operation of a fuzzy controller is softer and smoother – see Fig. 7.25a and 7.27b. The man often generated impetuous, sometimes redundant and random, movements of the joystick.

◊ ◊ ◊

7.3.3 The design of a fuzzy controller on the basis of the model of controlled plant

The following method is applied in cases when we do have a model of the plant available. The objective identification of plant models can be done using measured and recorded time series representing plant inputs and output. The identification consists of two stages described in detail in Chapter 6:

I. determination of significant plant inputs,
II. determination of model structure and its parameters.

Special attention should be paid to determination of significant inputs, because it can reduce the complexity of the final model. Section 7.3.3.1 contains general advice concerning problems faced during modeling of dynamic plants by self-learning systems.

7.3.3.1 Remarks concerning identification of models of dynamic plants

There are exceptional cases where controlled plants can be classified as static. The static plants are described by relation (7.19):

$$y(k) = F[u(k)]\,, \tag{7.19}$$

where: u – input, y – output of plant, k – serial number of sample.

The static model can be applied if transient processes inside the plant disappear quickly (duration time of transient processes is substantially shorter compared to sample time T). The input/output relation for the majority of plants can be represented by non-linear mapping F in the form (7.20).

$$y(k+1) = F[y(k), \ldots, y(k-n+1), u(k), \ldots, u(k-m+1)] \tag{7.20}$$

The state of plant output y for time instant assigned to sample number $(k+1)$ depends on former input states represented by "m" previous samples of u, and on former output states represented by "n" previous samples of y. There are different structures for the identification of the parameters of a neural model F^*. Some advice on how to use them is presented below.

If identified plant F is almost free of noise and disturbances, then the recommended scheme for identification of its model F^* is shown in Fig. 7.28 (Hunt 1992). The mathematical operator D denotes delay, for example:

$$v(k-1) = D(-1)(v(k))\,.$$

The plant model is determined on the basis of signals from a real plant. The plant input/output mapping can be represented by equivalent formulae:

$$y(k+n-m) = F[u(k), \ldots, u(k-m), y(k+n-m-1), \ldots, y(k-m)]\,,$$

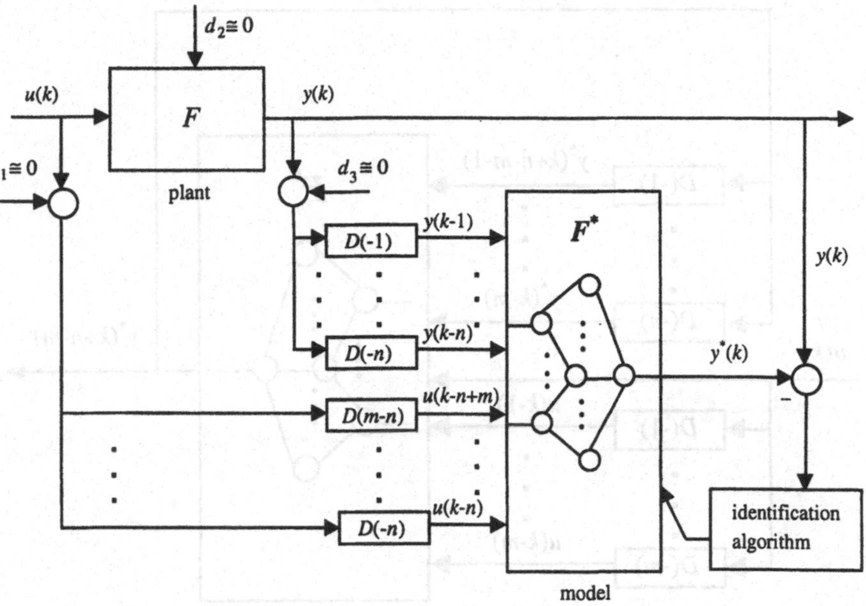

Fig. 7.28. The series-parallel model structure for identification of plant F (the structure is recommended for small measurement noises (d_1, d_3) and small disturbances (d_2) influencing the plant)

or:

$$y(k) = F[u(k-n+m), \ldots, u(k-n), y(k-1), \ldots, y(k-n)]\,. \qquad (7.21)$$

The input/output mapping realized during model identification (aimed at the determination of model structure and tuning of its parameters) can be represented by equivalent formulae (7.22):

$$y^*(k+n-m) = F^*[u(k), \ldots, u(k-m), y(k+n-m-1), \ldots, y(k-m)]\,,$$

or:

$$y^*(k) = F^*[u(k-n+m), \ldots, u(k-n), y(k-1), \ldots, y(k-n)]\,. \qquad (7.22)$$

The process of tuning model parameters for the series-parallel structure of a model (the model is "supplied" by plant input and output) is more stable compared to that for other structures (Hunt 1992; Narenda 1990a,b).

When identification is finished, the output of model $y^*(k+n-m)$ is connected to its input, instead of plant output $y(k+n-m)$. Thus, for practical purposes (control systems, etc.) the structure of model F^* is modified to the form shown in Fig. 7.29.

The formula (7.23) represents mapping realized by model F^* after its identification.

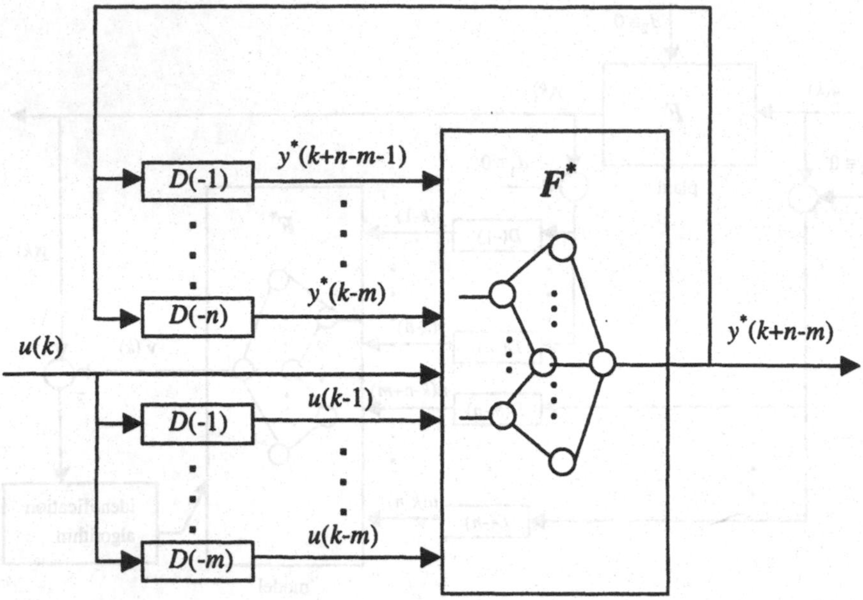

Fig. 7.29. The modified structure of a series-parallel model after identification

$$y^*(k+n-m) = F^*[u(k), \ldots, u(k-m), y^*(k+n-m-1), \ldots, y^*(k-m)] \quad (7.23)$$

If measurement noises (d_1, d_3) and disturbances (d_2) influencing the plant are large, then identification based on the parallel structure of the model (Fig. 7.30) is advised. Of course, the above attributes "large", "small" attached to the noise and disturbance levels have to be treated as fuzzy. So, they can not be precisely defined (attributes "large" and "small" have to be considered in relation to the sensitivity of plant under consideration). Usually experience and intuition are necessary for the correct choice of model structure (otherwise one has to carry out many experiments to "discover" a good structure).

Part of the inputs of the parallel model, Fig. 7.29 and 7.30, is free of noise and disturbances, because they are "taken" from model output $y^*(k+n-m)$. Thus, the resultant accuracy of identification can be higher. The model maps its inputs into output according to formula (7.23).

To obtain a good model F^*, which operates like the plant F, one has to synthesize it using all possible signals u influencing the plant under real conditions. Special attention should be paid to amplitude and spectral representation of signals u applied during model tuning. Nevertheless, even good practical models operate similarly to their primary "standards" F, if input signals u are represented by low-frequency spectra. The accuracy of a model decreases with the increase of higher frequency components in the spectral representations of signals u (Häck 1997).

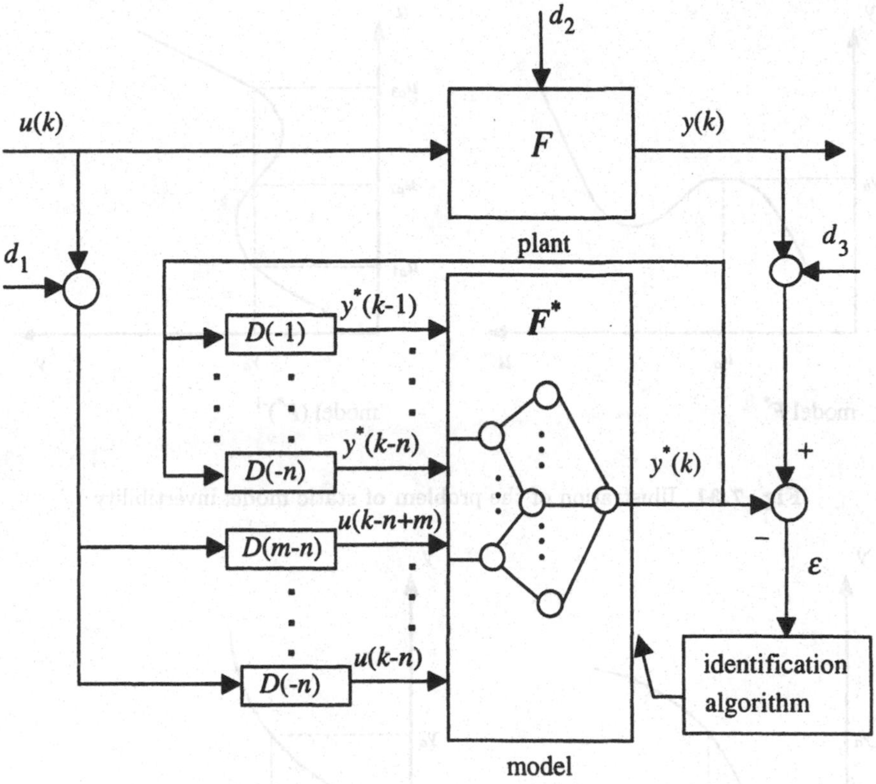

Fig. 7.30. The parallel model structure for identification of plant F (the structure is recommended for large measurement noise signals (d_1, d_3) and large disturbances (d_2) influencing the plant)

7.3.3.2 Some remarks concerning the identification of inverted models of dynamical plants

The inverted models F^*_{inv} of plants are often used in control algorithms (Häck 1997; Hunt 1992; Morari 1989). Such models, called "inverses", can be determined only for invertible models. The problem of model invertibility is illustrated in Fig. 7.31.

The static model F^* of plant is unique in relation to its input u. Each input value u_a is mapped onto one output value y_a. The inverse F^*_{inv} of that model does not represent unique mapping. There are such ranges of F^*_{inv} inputs (y), that one input value y_a is mapped onto two or three values of output (u_{ai}).

The model inverse can be obtained (Babuška 1995e) if the model represents a monotonically increasing or decreasing function realizing unique mapping in relation to output y – see Fig. 7.32.

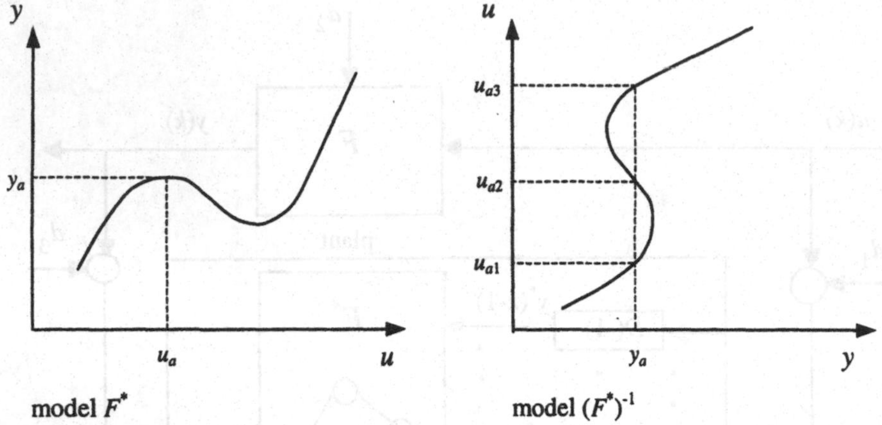

Fig. 7.31. Illustration of the problem of static model invertibility

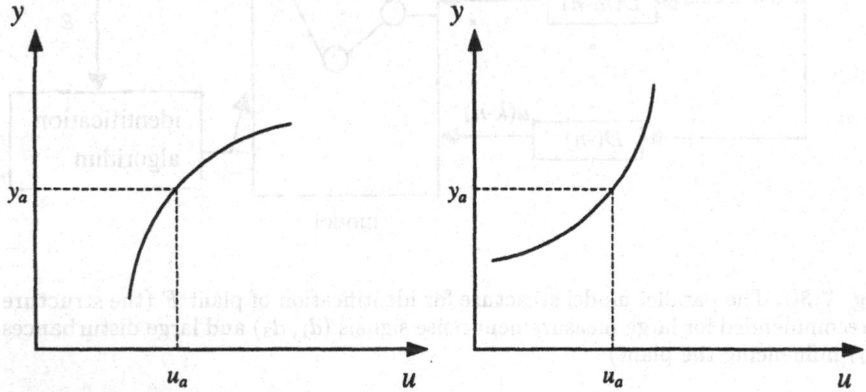

Fig. 7.32. An example of an invertible static model (representation of unique mapping in relation to output y)

The above explanations incline to the conclusion that the model F^* of plant F has to be determined before determination of its inverse F^*_{inv}. Model F^* has to be carefully examined in order to confirm or disprove its invertibility (mapping F^* has to be a monotonic one!). If such examination is neglected, then direct calculation of inverse F^*_{inv} can lead us to an averaged, untrue inverted model (Fig. 7.31).

Now, let us consider the problem of invertibility for dynamic models. What does inversion of a dynamic model mean?

The problem of invertibility of continuous models given in the form of transfer functions will be explained at the outset. The perfect inverse $G^*_{\text{inv}}(s)$ of model $G^*(s)$ should satisfy condition $G^*(s) \cdot G^*_{\text{inv}}(s) = 1$. Furthermore,

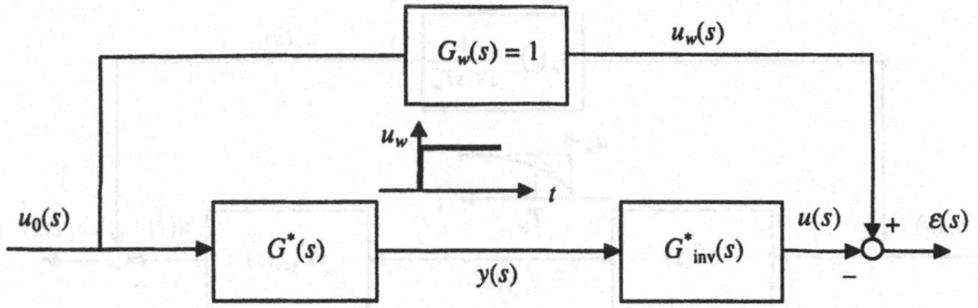

Fig. 7.33. The structure illustrating the problem of perfect invertibility of a linear model $G^*(s)$

the inverse obtained should be a minimum-phase transfer function. Thus, the invertibility problem can be modeled by means of the structure shown in Fig. 7.33.

For perfect inversion, signal $u(s)$ should be exactly equal to input signal $u_0(s)$. The structure shown in Fig. 7.33 compares the operation of models $G^*(s)$ and $G^*_{\text{inv}}(s)$ connected in series to the operation of reference model $G_w(s)$, where $G_w(s) = 1$. Hence, the transfer function of the perfect model inverse is given by (7.24).

$$G^*_{\text{inv}}(s) = \frac{G_w(s)}{G^*(s)} \tag{7.24}$$

Let us assume that $G^*(s)$ represents the inertia system of first order (7.25).

$$G^*(s) = \frac{1}{1+s} \tag{7.25}$$

Using (7.24) one obtains expression (7.26) representing a perfect inverse of model (7.25).

$$G^*_{\text{inv}}(s) = 1 + s \tag{7.26}$$

The result obtained contains a component realizing ideal differentiation. It is known that such an operation is physically unrealizable. Assuming reference model $G_w(s) = 1$ we require that all changes of input signal $u_0(s)$ have to be transmitted perfectly to the output of cascade $[G^*_{\text{inv}}(s)G^*(s)]$. Since perfect differentiation can not be realized, therefore perfect transmission of u_0 can also not be executed. It means that the perfect inverse of (7.25) does not exist. Let us try to "soften" the above requirements for the reference model. Let us assume that $G_w(s)$ represents the transfer function of the inertia element, with gain equal to 1 and time constant T_w, see Fig. 7.34.

Now, equation (7.24) yields (7.27).

$$G^*_{\text{inv}}(s) = \frac{1+s}{1+sT_w} \tag{7.27}$$

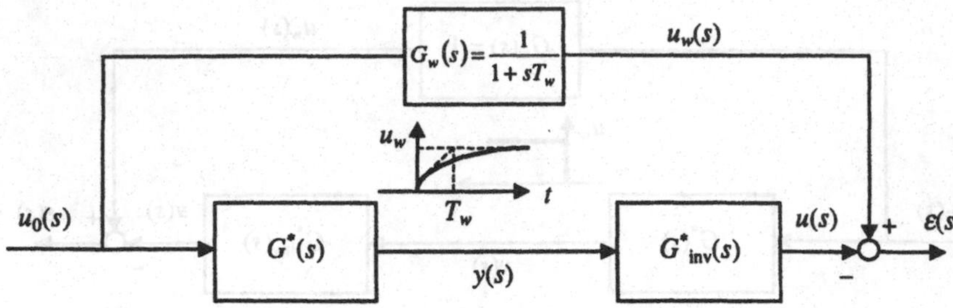

Fig. 7.34. So-called direct identification structure applied to the determination of approximate inverse $G^*_{\text{inv}}(s)$ of the model $G^*(s)$

The approximate inverse obtained (7.27) is realizable. The smaller the time constant T_w, the better the approximation of the perfect inverse $G^*_{\text{inv}}(s) = 1 + s$. For low frequency excitations u_0 the approximate inverse operates identically to the perfect inverse. Of course, the higher the frequency range assigned to spectral representation of excitation, the bigger the difference between operations represented by the perfect inverse and its realizable approximation. The current example poses the question: do such dynamic models exist, where application of reference model $G_w(s) = 1$ leads to realizable inverses?

Let us assume that the model of plant $G^*(s)$ is given by (7.28).

$$G^*(s) = \frac{b_0 + b_1 s + \ldots + b_n s^n}{a_0 + a_1 s + \ldots + a_n s^n} \tag{7.28}$$

The orders of denominator and numerator in (7.28) are identical and equal to n. The plants described by such transfer functions are called "proper" (Morari 1989).

Definition 7.3.3.2.1
The plant described by transfer function $G(s)$ is **proper**, if $G(s)$ satisfies the condition (7.29).

$$\lim_{s\to+\infty} |G(s)| = \text{finite value} \neq 0 \tag{7.29}$$

If $G(s)$ satisfies the condition (7.30), then the plant is called **strictly proper.**

$$\lim_{s\to+\infty} |G(s)| = 0 \tag{7.30}$$

The plants which are not proper are called improper. The transfer function $G(s)$ describes the improper plant, if order m of its numerator is higher than order n of its denominator. For proper plants $n = m$ and for strictly proper plants $n > m$. The proper plants as well as the strictly proper plants are physically realizable. The improper plants can not be realized physically.

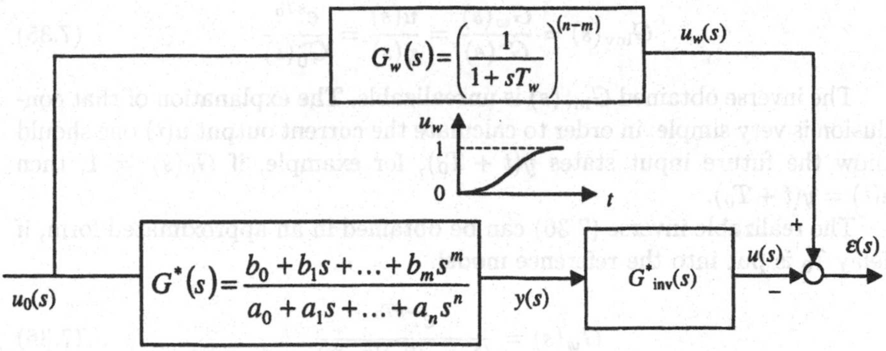

Fig. 7.35. The structure for calculation of realizable inverses of linear plant models on the basis of an inertia reference model

They could be realizable if the operation of perfect differentiation were possible. Of course, there is no such possibility because of physical constraints.

Within the meaning of the above definition, model (7.28) is a proper one. Its inverse, calculated from (7.24) for reference model $G_w(s) = 1$, is given by (7.31).

$$G^*_{inv}(s) = \frac{G_w(s)}{G^*(s)} = \frac{a_0 + a_1 s + \ldots + a_n s^n}{b_0 + b_1 s + \ldots + b_n s^n} \tag{7.31}$$

For proper models ($n = m$) one can apply reference model $G_w(s) = 1$. For strictly proper models $G_w(s) = 1$ can not be applied. Thus, if $n > m$, then inertia models of the appropriate order should be assumed as reference models. How to chose the correct order of inertia n_w for the reference model? There is a simple answer:

$$n_w = n - m, \quad n > m. \tag{7.32}$$

The structure enabling calculation of the inverse for an inertia reference model is shown in Fig. 7.35.

The structure shown in Fig. 7.35 yields the formula (7.33) representing the inverse of strictly proper models of plants.

$$G^*_{inv}(s) = \frac{a_0 + a_1 s + \ldots + a_n s^n}{(b_0 + b_1 s + \ldots + b_m s^m)(1 + sT_w)^{(n-m)}}, \quad n \geq m. \tag{7.33}$$

The next question is connected with the presence of delay T_0 in a model of a plant. Does the inverse for model (7.34) exist, if transfer function $G^*_0(s)$ is proper or strictly proper?

$$G^*(s) = e^{-sT_0} \cdot G^*_0(s) \tag{7.34}$$

The assumption of a perfect inverse ($G_w(s) = 1$) yields the following formula (7.35).

$$G_{\text{inv}}(s) = \frac{G_w(s)}{G^*(s)} = \frac{u(s)}{y(s)} = \frac{e^{sT_0}}{G_0^*(s)} \tag{7.35}$$

The inverse obtained $G_{\text{inv}}(s)$ is unrealizable. The explanation of that conclusion is very simple: in order to calculate the current output $u(t)$ one should know the future input states $y(t + T_0)$, for example, if $G_0(s) = 1$, then $u(t) = y(t + T_0)$.

The realizable inverse (7.36) can be obtained in an approximated form, if delay T_0 is put into the reference model.

$$G_w(s) = \frac{e^{-sT_0}}{(1 + sT_w)^{(n-m)}} \tag{7.36}$$

The above reference model yields an inverse in the form (7.33), which is identical to that obtained for the model of a plant without delay. For steady states, the inverse calculated by means of a delaying reference model (7.36) is exactly the same as the theoretical inverse (7.35). That property is conserved if very slow signals are taken into account. The error of approximate inverse increases rapidly if the frequency of input signal y becomes higher and higher.

The above discussion concerning inversion of models with delays clearly explains that approximate inverses of such models can be applied for steady or quasi-steady states. That exceptional and narrowed area of applications means that inverses for plant models with delays are used rather rarely (Häck 1997; Hunt 1992; Morari 1989). It has to be added that models with delays are readily used as simplified models of high-order inertia plants. Under such circumstances, problems concerning invertibility can be simplified by using high-order inertia models of plants (like Strejc's model (Żuchowski 1998)) instead of models with delays.

The idea of proper and improper models can be implemented for discrete linear models of plants.

Definition 7.3.3.2.2
The plant described by discrete transfer function $G(z)$ is a **proper** one, if condition (7.37) is satisfied (Morari 1989).

$$\lim_{z\to\infty} G(z) = \text{finite value} \neq 0 \tag{7.37}$$

If condition (7.38) is satisfied, then the plant is classified as **strictly proper**.

$$\lim_{z\to\infty} G(z) = 0 \tag{7.38}$$

The plant which is not proper is classified as improper. The transfer function $G(z)$ describes the improper plant, if order m of its numerator is higher than order n of its denominator. For proper plants $n = m$ and for strictly proper plants $n > m$. The proper plants as well as strictly proper plants are physically realizable. The improper plant can not be realized physically, because prediction of its input would be necessary for calculation of plant output, for example:

$$G(z) = \frac{y(z)}{u(z)} = \frac{z+1}{1} \quad \Leftrightarrow \quad y(k) = u(k+1) + u(k)\,,$$

where: y – plant output, u – plant input.

The reference model $G_w(z) = 1$ can be applied for the discrete proper model $G(z)$. For a strictly proper model ($n > m$) one can find the approximate inverse choosing reference model $G_w(z)$ in the form of single discrete pole of $(n-m)$ order (7.39).

$$G_w(z) = \left(\frac{1-c}{z-c}\right)^{(n-m)} \qquad 0 \leq c < 1 \tag{7.39}$$

The responses $u_w(k)$ of a single discrete pole (7.40):

$$G_w(z) = \frac{1-c}{z-c} \tag{7.40}$$

to step input signal $u_0(k) = 1(k)$, $k = 1, 2, \ldots, \infty$ are shown in Fig. 7.36.

Studying results presented in Fig. 7.36 one can conclude that inertia "charging" the reference model increases for increasing values of parameter c. Increasing inertia decreases the exactness of the inverse obtained. The most accurate inverse (acceptable for quick excitations) is obtained for $c = 0$. The reference model of the form (7.41):

$$G_w(z) = \left(\frac{1}{z}\right)^{(n-m)} \tag{7.41}$$

generates the most difficult demands in relation to the inverse. The structure for calculation of "quick" inverse $G^*_{\text{inv}}(z)$ for linear model $G^*(z)$ of plant is shown in Fig. 7.37.

The inverse resulting from Fig. 7.37 is expressed by formula (7.42):

$$G^*_{\text{inv}}(z) = \frac{G_w(z)}{G^*(z)} = \frac{u(z)}{y(z)} = \frac{a_0 + a_1 z + \cdots + a_n z^n}{(b_0 + b_1 z + \cdots + b_m z^m) z^{(n-m)}}\,, \quad n \geq m\,. \tag{7.42}$$

It has to be repeated once more that a perfect version of reference model $G_w(z) = 1$ can not be applied for strictly proper plants ($n > m$), i.e. plants containing inertia and/or delay. The greater the difference $(n-m)$, the higher the inaccuracy of the inverse. For large $(n-m)$ one obtains an inaccurate inverse, except in static states or quasi-static states (presence of very "slow" signals). The loss of accuracy accompanies the increasing frequency of input signal y.

The realizable inverse $G^*_{\text{inv}}(z)$, expression (7.42), always takes the form of a proper transfer function. It means that orders of plant denominator,

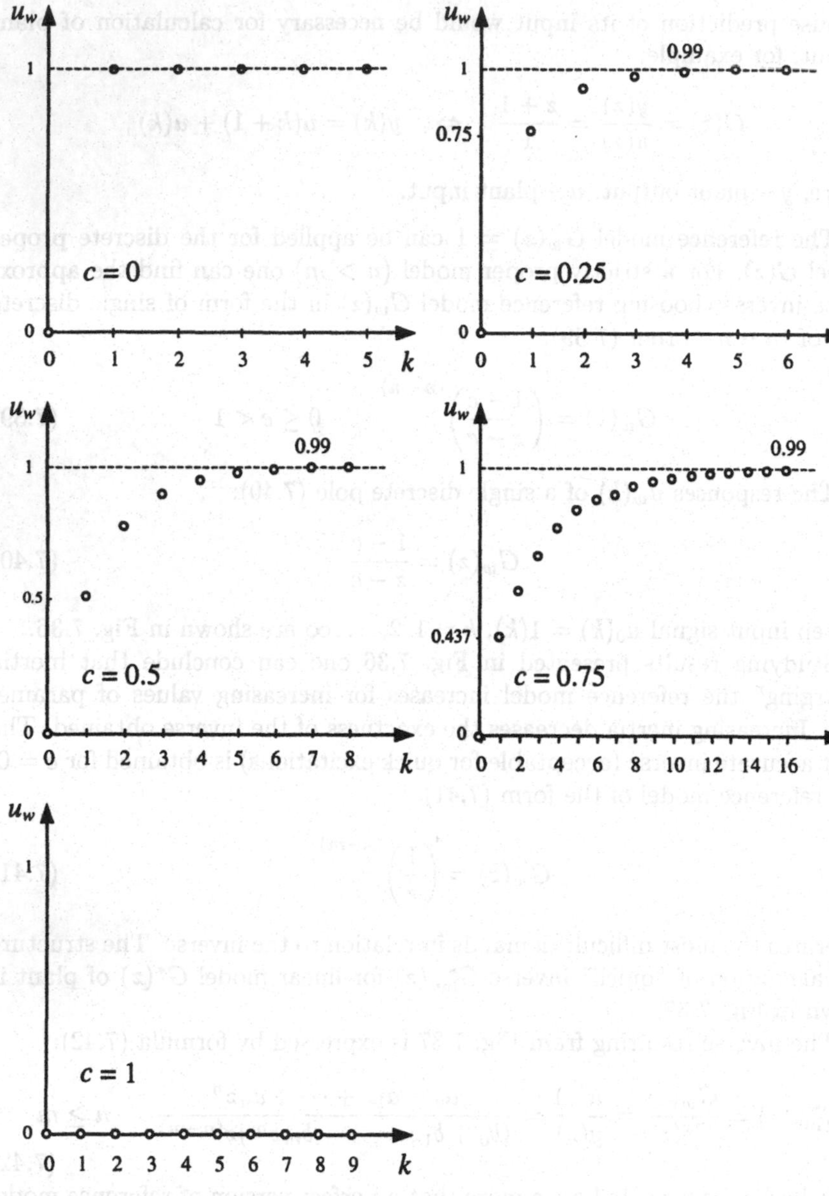

$$u_w(k) = (1-c)u_0(k-1) + cu_w(k-1), \quad u_0(k) = 1(k), \quad k = 0,1,\ldots,\infty .$$

Fig. 7.36. The step responses of reference model (7.40) for various values of parameter c

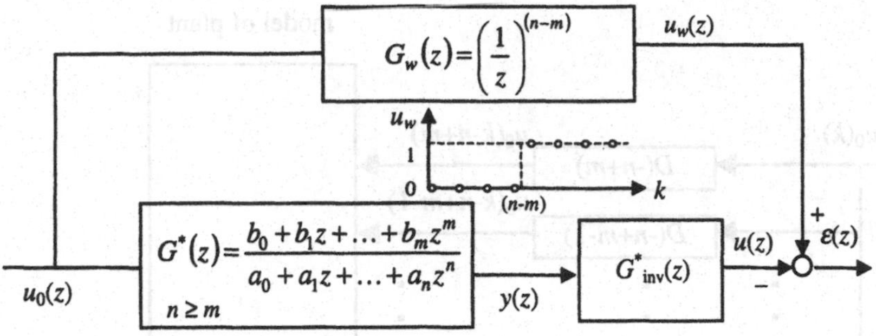

Fig. 7.37. The structure for calculation of "quick" inverse $G^*_{\mathrm{inv}}(z)$ for a linear model $G^*(z)$ of plant

inverse numerator and inverse denominator are exactly the same, equal to n. Taking into account problems concerning the perfect reference model and the approximate inverse, one can conclude that realizable inverse $G^*_{\mathrm{inv}}(z)$ is usually different compared to the reciprocal of the plant transfer function. The above is usually manifested by a different denominator of inverse in comparison to the numerator of the plant transfer function. The formula (7.43) transforms the model transfer function $G^*(z)$ based on operator z to the form based on operator z^{-1}. The resulting difference equation assigned to the transformed transfer function of the model completes the relation expressing the above transformation.

$$G^*(z) = \frac{y(z)}{u_0(z)} = \frac{b_0 + b_1 z + \cdots + b_m z^m}{a_0 + a_1 z + \cdots + a_n z^n} = \frac{b_0 z^{-n} + \cdots + b_{m-1} z^{-n+m+1} + b_m z^{-n+m}}{a_0 z^{-n} + \cdots + a_{n-1} z^{-1} + a_n}$$

$$\begin{aligned} a_n y(k) + \cdots + a_1 y(k-n+1) + a_0 y(k-n) = \\ = b_m u_0(k-n+m) + \ldots + b_1 u_0(k-n+1) + b_0 u_0(k-n) \end{aligned} \quad (7.43)$$

The vectors $\mathbf{y}$ and $\mathbf{u_0}$ given by (7.44) can be defined for the difference equation (7.43).

$$\begin{aligned} \mathbf{y}^{\mathrm{T}} &= [y(k), y(k-1), \ldots, y(k-n)] \\ \mathbf{u}_0^{\mathrm{T}} &= [u_0(k-n+m), u_0(k-n+m-1), \ldots, u_0(k-n)] \end{aligned} \quad (7.44)$$

The inputs and output of the model are shown in Fig. 7.38. The function F^* realizes linear input/output mapping defined by the equation of plant model (7.43). In general, the mapping F^* can be non-linear.

$$y(k) = F^*[u_0(k-n+m), \ldots, u_0(k-n), y(k-1), \ldots, y(k-n)] \quad (7.45)$$

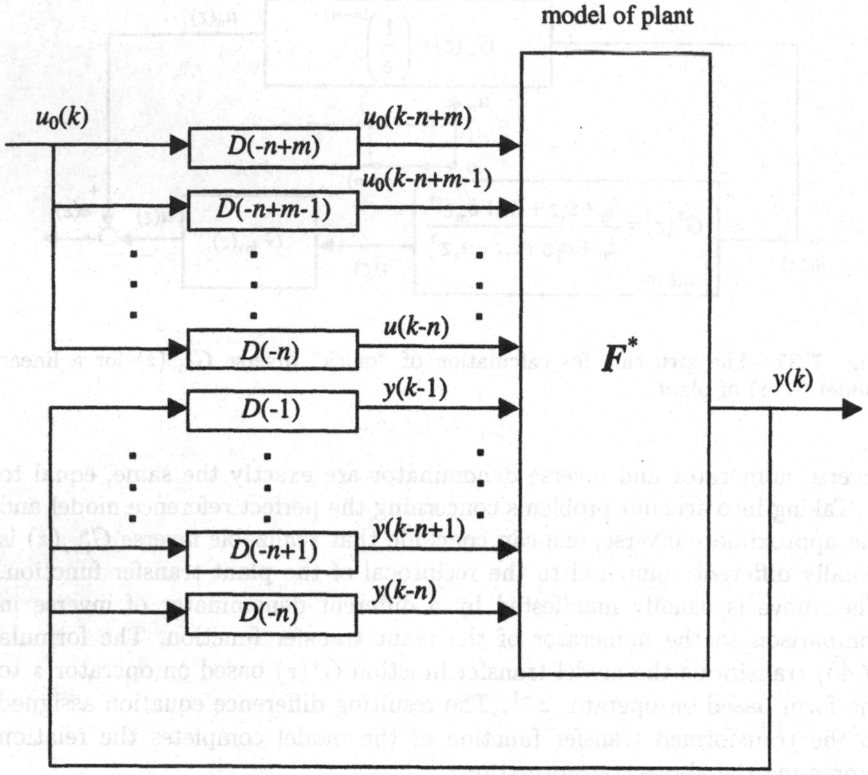

Fig. 7.38. The inputs and output of a plant model (operator $D(-n)$ denotes the time delay equal to nT)

If delays of inputs $u_0(k-n+m)$ and $u_0(k-n)$ are known, then values n and m can be determined. For realizable inverse $G^*_{\text{inv}}(z)$ of a linear model, one obtains the following relations (7.46).

$$G^*_{\text{inv}}(z) = \frac{u(z)}{y(z)} = \frac{a_0 + a_1 z + \cdots + a_n z^n}{b_0 z^{n-m} + b_1 z^{n-m+1} + \cdots + b_m z^m} =$$
$$= \frac{a_0 z^{-n} + \cdots + a_{n-1} z^{-1} + a_n}{b_0 z^{-m} + \cdots + b_{m-1} z^{-1} + b_m}$$

$$b_0 u(k-m) + \cdots + b_{m-1} u(k-1) + b_m u(k) =$$
$$= a_0 y(k-n) + \cdots + a_{n-1} y(k-1) + a_n y(k) \quad (7.46)$$

The vectors **y** and **u** for the inverse of a model are given by (7.47). For comparison, vectors **y** and $\mathbf{u}_0$ are given by relations (7.48).

The **inverse** F^*_{inv} of a plant model:

$$\mathbf{u}^{\mathrm{T}} = [u(k), u(k-1), \ldots, u(k-m)] \,,$$
$$\mathbf{y}^{\mathrm{T}} = [y(k), y(k-1), \ldots, y(k-n)] \,. \tag{7.47}$$

The **model** F^* of a plant:

$$\mathbf{u}_0^{\mathrm{T}} = [u_0(k-n+m), u_0(k-n+m-1), \ldots, u_0(k-n)] \,,$$
$$\mathbf{y}^{\mathrm{T}} = [y(k), y(k-1), \ldots, y(k-n)] \,. \tag{7.48}$$

Comparison of vectors $\mathbf{u}$ for the model and $\mathbf{u}_0$ for the inverse of the model yields the conclusion that the time shifts of respective elements of both vectors are identical $(-n+m=0)$ if $m=n$, (i.e. model is proper). If the model is improper $(-n+m>0)$, then the time shifts of respective elements of inverse vector $\mathbf{u}$ are smaller, which means that the inverse processes the earlier samples of signal u (for example, if $(-n+m)=1$, then $u(k)$ is the first element of signal vector $\mathbf{u}$ processed by the inverse, and $u_0(k-1)$ is the first element processed by the plant). The scheme of a model inverse is shown in Fig. 7.39.

The number of model inputs (see Fig. 7.38) and the number of inputs of its inverse (see Fig. 7.39) are identical, equal to $(n+m+1)$. It means that the dimensions of both input spaces are identical too.

If a linear model is strictly proper $(n>m)$, then the vector of model $\mathbf{u}_0$ is shifted by the time $(n-m)T$ in relation to the vector $\mathbf{u}$ of the inverse. In order to determine the inverse of such a model, one should transform it into the form of a proper model, shifting vector $\mathbf{u}_0$ by the time $(n-m)T$. After that transformation the inverse can be determined. The Example 7.3.3.2.1 elucidates the calculation method of the inverse.

Example 7.3.3.2.1
Let us assume that a proper model given in the form of transfer function $G^*(z)$ or respective difference equation (7.49) represents linear input/output mapping F^* realized by a plant.

$$G^*(z) = \frac{y(z)}{u_0(z)} = \frac{z+0.75}{z^2+z-0.25} \,,$$
$$y(k+1) + y(k) - 0.25y(k-1) = u_0(k) + 0.75u_0(k-1) \,,$$
$$m=1 \,, \quad n=2 \,, \quad n-m=1 \,. \tag{7.49}$$

Now, let us shift the vector $\mathbf{u}_0^{\mathrm{T}} = [u_0(k), u_0(k-1)]$ by the time $1T$. The above operation yields vector $\mathbf{u}^{\mathrm{T}} = [u(k+1), u(k)]$ of the inverse. Next, the vector of inverse obtained is inserted into the equation of the model (7.49). Hence the difference equation of inverse as well as the transfer function corresponding to this equation is obtained – see relations (7.50).

$$y(k+1) + y(k) - 0.25y(k-1) = u(k+1) + 0.75u(k)$$
$$(G^*(z))_{\text{inv}} = \frac{u(z)}{y(z)} = \frac{z^2+z-0.25}{z^2+0.75z} \tag{7.50}$$

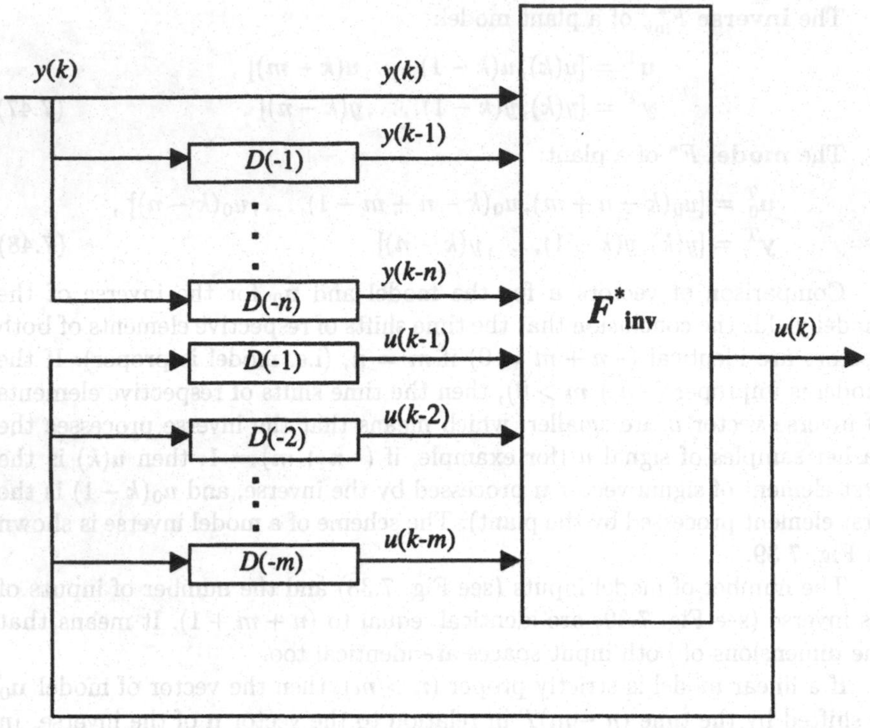

Fig. 7.39. The scheme of a model inverse (operator $D(-n)$ denotes the time delay equal to nT)

The output of model $y(k+1)$ is determined by equation (7.51) resulting from (7.49) and the output of the inverse is determined by equation (7.52) resulting from (7.50).

$$y(k+1) = u_0(k) + 0.75u_0(k-1) - y(k) + 0.25y(k-1) \qquad (7.51)$$
$$u(k+1) = y(k+1) + y(k) - 0.25y(k-1) - 0.75u(k) \qquad (7.52)$$

The cascade composed of model F^* and its inverse F^*_{inv}, both coupled in series, is shown in Fig. 7.40.

Putting (7.51) into equation (7.52) one obtains a relation of the type $u(k+1) = f[u_0(k)]$, which is, for the circumstances considered above, specifically of the form (7.53).

$$u(k+1) = u_0(k) + 0.75u_0(k-1) - 0.75u(k) \qquad (7.53)$$

The signal $u(k)$ is shown in Fig. 7.40. There is accordance between $u(k)$ and reference signal u_w generated by reference model $F_w : u_w(z)/u_0(z) = 1/z$ for excitation $u_0(k) = 1(k)$ and initial conditions equal to zero. The order of the reference model results from the difference $(n - m)$ determined by the

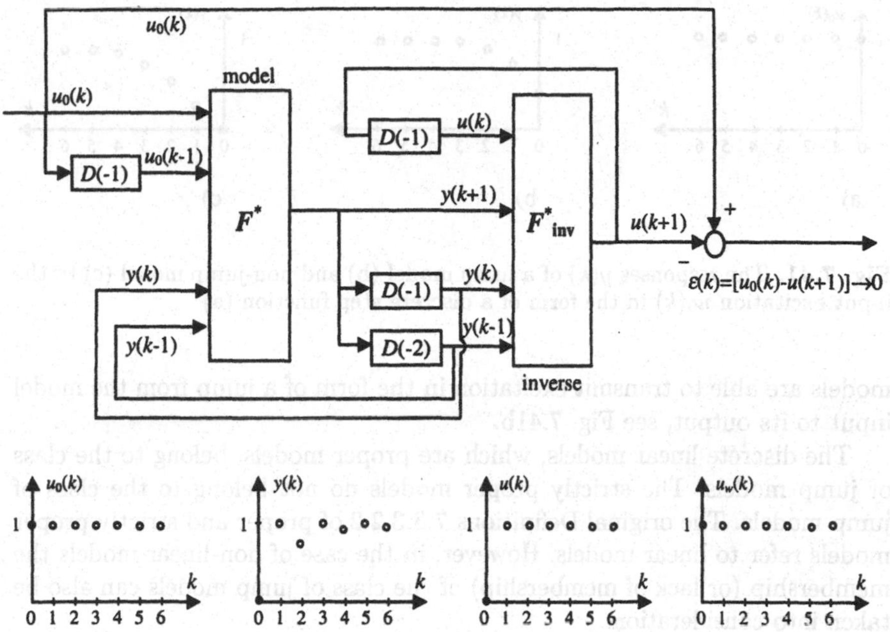

Fig. 7.40. The scheme of model F^*, its inverse F^*_{inv} and the reference model as well as important signals attached to the structure under consideration $u_w(k) = u_0(k)$

orders of denominator n and nominator m of model transfer function $G^*(z)$, see (7.49).

◊ ◊ ◊

The above example confirms that the inverse of a strictly proper, linear model can be obtained directly from plant model F^* by its formal reciprocation and the time shift of vector $\mathbf{u_0}$ ($\mathbf{u_0} \to \mathbf{u}$).

What about the inversion of fuzzy models of non-linear plants? Is it possible to apply the similar procedure?

The non-linear plant can be represented by a discrete model of the form (7.54).

$$y(k) = F^*[u_0(k-n+m), \ldots, u_0(k-n), y(k-1), \ldots, y(k-n)] \tag{7.54}$$

For $n - m = 0$ the above model obtains the form (7.55).

$$y(k) = F^*[u_0(k), \ldots, u_0(k-n), y(k-1), \ldots, y(k-n)] \tag{7.55}$$

The output variable of that model $y(k)$ depends directly on the variable $u_0(k)$. It means that input u_0 for instant kT directly influences (without any delay) the current output of model $y(k)$. The models "equipped" with that property are called "jump" models (sprungfähig) (Häck 1997). The jump

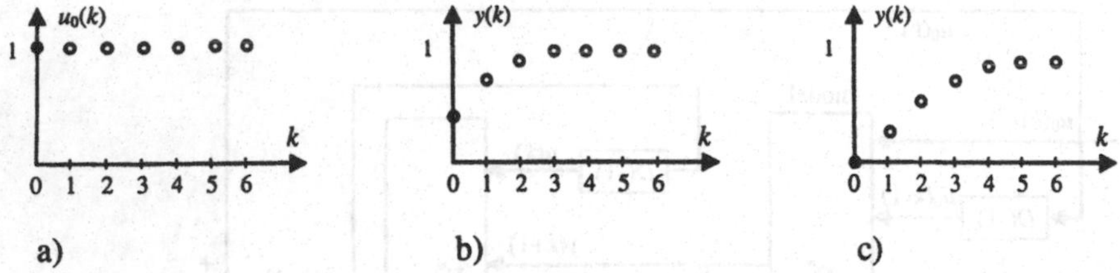

Fig. 7.41. The responses $y(k)$ of a jump model (b) and non-jump model (c) to the input excitation $u_0(k)$ in the form of a discrete step function (a)

models are able to transmit excitation in the form of a jump from the model input to its output, see Fig. 7.41b.

The discrete linear models, which are proper models, belong to the class of jump models. The strictly proper models do not belong to the class of jump models. The original Definitions 7.3.3.2.2 of proper and strictly proper models refer to linear models. However, in the case of non-linear models the membership (or lack of membership) of the class of jump models can also be taken into consideration.

If a fuzzy model of a plant is a standard Mamdani model with singleton type conclusions, and properties of that model are as follows:

- the premises of triangular membership functions of fuzzy sets satisfy the partition of unity condition ($\sum \mu = 1$),
- the rules are of the conjunction form and the operator *PROD* is applied to the realization of operation *AND*,
- the singleton method is used in deffuzification,

$$y = \frac{\sum_{j=1}^{K} \beta_j c_j}{\sum_{j=1}^{K} \beta_j}, \tag{7.56}$$

  where: β_j – denotes the grade of satisfying the premise of the rule distinguished by index j, and K – represents the consecutive number of rules in the rule base,
- the fuzzy model represents unique mapping in relation to output variable $y(k)$,
- the model belongs to the class of jump models and contains the rules of the form (7.57),

$$\begin{aligned} &IF\ (u_0(k) = A_1)\ AND\ \ldots AND\ (u_0(k-n) = A_{n+1}) \\ &AND\ (y(k-1) = B_1)\ \ldots\ AND\ (y(k-n) = B_n)\ THEN\ y(k) = c\,, \end{aligned} \tag{7.57}$$

where: $A_1, \ldots, A_{n+1}, B_1, \ldots, B_n$ are fuzzy sets and c is a singleton,

then the inverse of the whole model can be obtained by consecutive inversion of each separate rule, which means that component premise $(u_0(k) = A_1)$ of a given rule is replaced by its conclusion $(y(k) = c)$ and vice versa, (Babuška 1995e).

Thus, the inverse is represented by a fuzzy model containing rules of the form (7.58).

$$IF\ (u_0(k-1) = A_2)\ AND\ \ldots AND\ (u_0(k-n) = A_{n+1})\ AND\ (y(k) = C)$$
$$AND\ (y(k-1) = B_1)\ \ldots\ AND\ (y(k-n) = B_n)\ THEN\ u_0(k) = a_1\,, \tag{7.58}$$

where: a_1 – singleton with modal value equal to modal value of set A_1,
C – triangular fuzzy set with modal value equal to modal value of singleton c.

The following Example 7.3.3.2.2 illustrates the procedure of inverting a model belonging to the class of jump models.

Example 7.3.3.2.2

The fuzzy model of plant $y(k+1) = F^*[u_0(k), y(k)]$, where the rule-base is defined by relations (7.59) and membership functions for inputs and output are shown in Fig. 7.42, has been chosen in order to illustrate the procedure of inverting.

$$
\begin{aligned}
R1:\ & IF\ (u_0(k) \text{ is about } 0)\ AND\ (y(k) \text{ is about } 0)\\
& THEN\ (y(k+1) \text{ is about } 0)\\
R2:\ & IF\ (u_0(k) \text{ is about } 1)\ AND\ (y(k) \text{ is about } 0)\\
& THEN\ (y(k+1) \text{ is about } 1/3)\\
R3:\ & IF\ (u_0(k) \text{ is about } 2)\ AND\ (y(k) \text{ is about } 0)\\
& THEN\ (y(k+1) \text{ is about } 4/3)\\
R4:\ & IF\ (u_0(k) \text{ is about } 0)\ AND\ (y(k) \text{ is about } 1)\\
& THEN\ (y(k+1) \text{ is about } 1/6)\\
R5:\ & IF\ (u_0(k) \text{ is about } 1)\ AND\ (y(k) \text{ is about } 1)\\
& THEN\ (y(k+1) \text{ is about } 0.5)\\
R6:\ & IF\ (u_0(k) \text{ is about } 2)\ AND\ (y(k) \text{ is about } 1)\\
& THEN\ (y(k+1) \text{ is about } 1.5)\\
R7:\ & IF\ (u_0(k) \text{ is about } 0)\ AND\ (y(k) \text{ is about } 2)\\
& THEN\ (y(k+1) \text{ is about } 2/3)\\
R8:\ & IF\ (u_0(k) \text{ is about } 1)\ AND\ (y(k) \text{ is about } 2)\\
& THEN\ (y(k+1) \text{ is about } 1)\\
R9:\ & IF\ (u_0(k) \text{ is about } 2)\ AND\ (y(k) \text{ is about } 2)\\
& THEN\ (y(k+1) \text{ is about } 2)
\end{aligned}
\tag{7.59}
$$

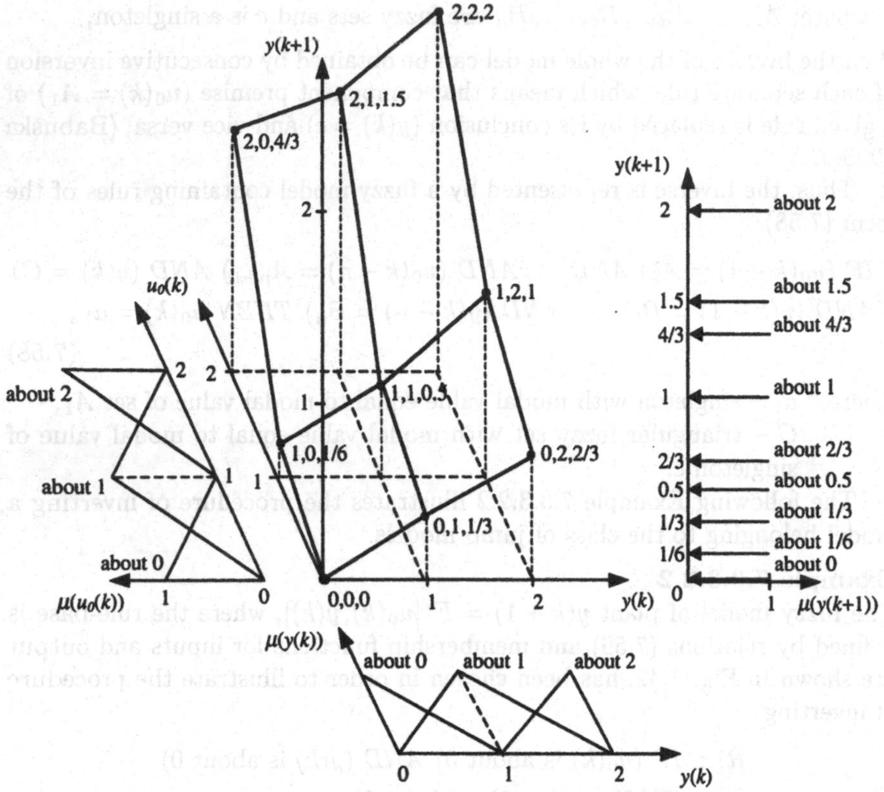

Fig. 7.42. The membership functions for inputs and output and the surface of a fuzzy model with rule base (7.59). The numbers assigned to surface nodes represent coordinates of those nodes

The model representation shown in Fig. 7.42 proves that fuzzy model F^* realizes a unique operation in relation to its output $y(k+1)$. It means that the model can be inverted. Although the edges of particular sectors are segments of straight lines, the surfaces enclosed by those straight segments are not flat at all. The general form of a multi-linear equation defining the surfaces of particular sectors is given by (7.60).

$$y(k+1) = a_0 + a_1 u_0(k) + a_2 y(k) + a_3 u_0(k) y(k) \tag{7.60}$$

At the second stage of model inversion the vector $\mathbf{u}_0(k) = u_0(k)$ is shifted and the value of that shift equals $(n - m)$. Thus, for the example considered, the shift is equal to 1. The above shift of vector $\mathbf{u}_0(k)$ transforms the non-jump model (7.59) into a jump model (7.61).

$$\begin{aligned}
R1: \;& IF\ (u_0(k+1) \text{ is about } 0)\ AND\ (y(k) \text{ is about } 0)\\
& THEN\ (y(k+1) \text{ is about } 0)\\
&\vdots\\
R9: \;& IF\ (u_0(k+1) \text{ is about } 2)\ AND\ (y(k) \text{ is about } 2)\\
& THEN\ (y(k+1) \text{ is about } 2)
\end{aligned} \tag{7.61}$$

At the third stage, the conclusions comprising the variable $y(k+1)$ are replaced by component promises comprising $u_0(k+1)$ and vice versa. Those exchanges yield the rule base (7.62).

$$\begin{aligned}
R1: \;& IF\ (y(k+1) \text{ is about } 0)\ AND\ (y(k) \text{ is about } 0)\\
& THEN\ (u_0(k+1) \text{ is about } 0)\\
R2: \;& IF\ (y(k+1) \text{ is about } 1/3)\ AND\ (y(k) \text{ is about } 0)\\
& THEN\ (u_0(k+1) \text{ is about } 1)\\
R3: \;& IF\ (y(k+1) \text{ is about } 4/3)\ AND\ (y(k) \text{ is about } 0)\\
& THEN\ (u_0(k+1) \text{ is about } 2)\\
R4: \;& IF\ (y(k+1) \text{ is about } 1/6)\ AND\ (y(k) \text{ is about } 1)\\
& THEN\ (u_0(k+1) \text{ is about } 0)\\
R5: \;& IF\ (y(k+1) \text{ is about } 0.5)\ AND\ (y(k) \text{ is about } 1)\\
& THEN\ (u_0(k+1) \text{ is about } 1)\\
R6: \;& IF\ (y(k+1) \text{ is about } 1.5)\ AND\ (y(k) \text{ is about } 1)\\
& THEN\ (u_0(k+1) \text{ is about } 2)\\
R7: \;& IF\ (y(k+1) \text{ is about } 2/3)\ AND\ (y(k) \text{ is about } 2)\\
& THEN\ (u_0(k+1) \text{ is about } 0)\\
R8: \;& IF\ (y(k+1) \text{ is about } 1)\ AND\ (y(k) \text{ is about } 2)\\
& THEN\ (u_0(k+1) \text{ is about } 1)\\
R9: \;& IF\ (y(k+1) \text{ is about } 2)\ AND\ (y(k) \text{ is about } 2)\\
& THEN\ (u_0(k+1) \text{ is about } 2)
\end{aligned} \tag{7.62}$$

The input space of the direct inverse of a model with rule base (7.62) as well as membership functions for inputs are shown in Fig. 7.43.

Fig. 7.43 shows that rule base (7.62) of the direct inverse of a model is incomplete. The information on values of outputs is available only for 9 nodes of the input space – it is the effect of the number of rules in the primary model (7.59). The incomplete rule-base does not allow us to calculate the outputs assigned to those sectors, where some nodes are not "equipped" with complete rules. To overcome that problem one has to spread out the rule base (7.62) to the complete rule base (7.63) containing $9 \times 3 = 27$ rules. The above means that additional values of output $u(k+1)$ have to be determined for all those nodes which remain without assigned rules.

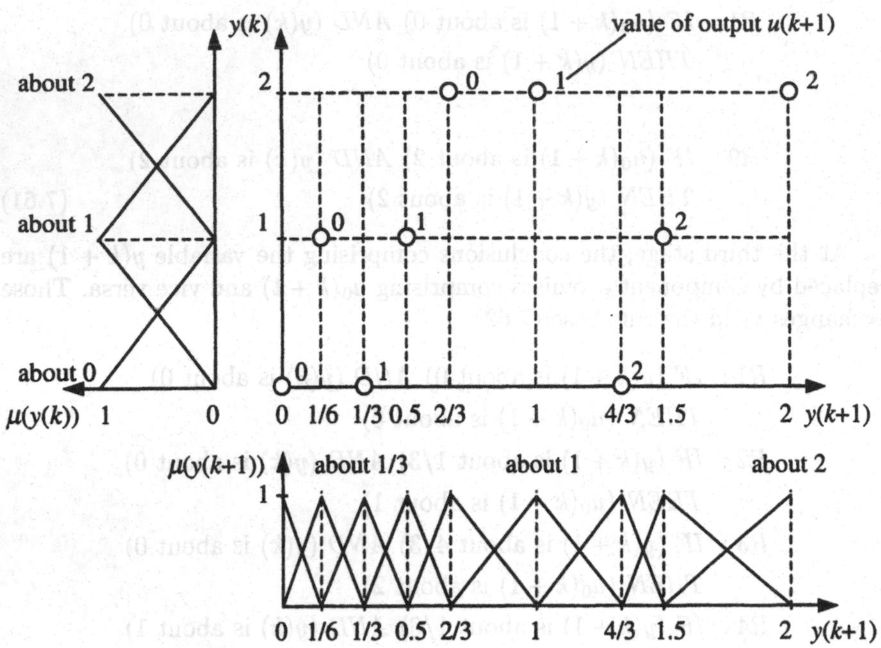

Fig. 7.43. The membership functions for inputs, the partition of input space and nodes determined by rules (7.62) resulting from the procedure of direct inversion of a model (7.59)

$$
\begin{aligned}
R1:\ & \mathit{IF}\ (y(k+1) \text{ is about } 0)\ \mathit{AND}\ (y(k) \text{ is about } 0) \\
& \mathit{THEN}\ (u_0(k+1) \text{ is about } a_1) \\
R2:\ & \mathit{IF}\ (y(k+1) \text{ is about } 0.5)\ \mathit{AND}\ (y(k) \text{ is about } 0) \\
& \mathit{THEN}\ (u_0(k+1) \text{ is about } a_1) \\
R3:\ & \mathit{IF}\ (y(k+1) \text{ is about } 1)\ \mathit{AND}\ (y(k) \text{ is about } 0) \\
& \mathit{THEN}\ (u_0(k+1) \text{ is about } a_3) \\
& \vdots \\
R27:\ & \mathit{IF}\ (y(k+1) \text{ is about } 2)\ \mathit{AND}\ (y(k) \text{ is about } 2) \\
& \mathit{THEN}\ (u_0(k+1) \text{ is about } a_{27})
\end{aligned}
\tag{7.63}
$$

The premises of the rule base (7.63) contain 27 (all possible) combinations defined by fuzzy sets assigned to inputs. The parameters of conclusions a_i represent the values (positions) of singletons. The values of 9 singletons in (7.63) are known due to direct inversion of the primary model, for example, $a_1 = 0$, $a_{27} = 2$. The values of the remaining singletons can be determined using the procedures described below.

I. The first procedure transforms the complete rule base (7.63) into a neuro-fuzzy network with membership functions for inputs identical to those in Fig. 7.43 and membership functions of the form of a singleton for output. The samples taken from plant inputs and output can be used for tuning a neuro-fuzzy network obtained according to the principles of the method of error back-propagation or, alternatively, for calculation of a_i by the application of the least square method.
II. The second procedure is quite similar to the first one. However, there is a difference. The neuro-fuzzy network is tuned on the basis of samples of inputs and output generated by means of the primary fuzzy model (7.59).

Procedure I is usually more accurate. Some advantageous and disadvantageous properties of both procedures will be discussed later on. The first procedure has been applied to find the values of singletons a_i. The values obtained have been collected in Table 7.1.

Table 7.1. The values of singletons a_i obtained as a result of tuning a complete inverse of a model (7.63)

i	1	2	3	4	5	6	7	8	9	10	11	12	13	14
a_i	0	0.5	1	1.167	1.333	1.667	2				0	0.833	1	1.167
i	15	16	17	18	19	20	21	22	23	24	25	26	27	
a_i	1.5	1.833	2						0	1	1.333	1.5	2	

There are empty fields in Table 7.1. That phenomenon can be easily explained. The measured samples taken from plant inputs and output have not been regularly distributed in the input space, which is determined here by inequalities:

$$0 \le u_0(k) \le 2\,, \quad 0 \le y(k) \le 2\,, \quad 0 \le y(k+1) \le 2\,.$$

The range of samples for training is also shown in Fig. 7.42. Because of the lack of training samples in some sectors of the input space for the direct inversion of the model (Fig. 7.43), the singletons assigned to nodes of such sectors have not been tuned. Thus, only singletons assigned to sectors "covered" with training samples (lined area in Fig. 7.44) can be tuned.

The real rule base of the model inverse containing 19 rules is shown in Table 7.2. The 19 values of conclusion singletons have been calculated on the basis of measurement samples taken from the plant and put to particular rules (7.63).

The membership functions as well as the surface of the model inverse with the rule base defined by Table 7.2 is shown in Fig. 7.45.

As is shown in Fig. 7.45, only one point of the surface of the model inverse is attached to each single output value $u(k+1)$ (uniqueness). The structure

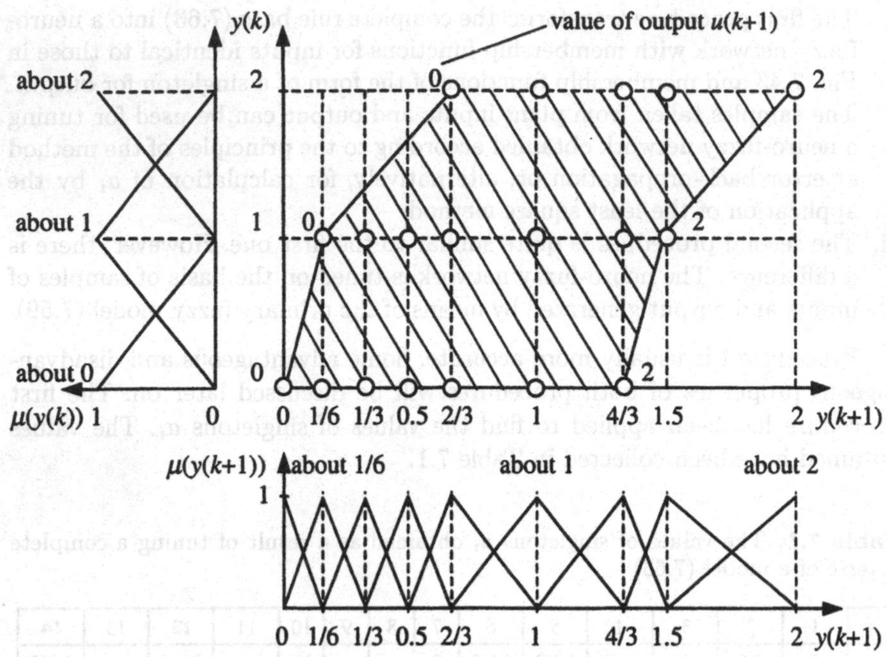

Fig. 7.44. The sub-space (lined area) of input space of the fuzzy model inverse where output of the inverse can be calculated on the basis of measurement samples taken from the plant

Table 7.2. The rule base of the model inverse (7.62)

$y(k+1)$ \ $y(k)$	about 0	about 1/6	about 1/3	about 0.5	about 2/3	about 1	about 4/3	about 1.5	about 2
about 0	0	0.5	1	1.167	1.333	1.667	2		
about 1		0	0.833	1	1.667	1.5	1.833	2	
about 2					0	1	1.333	1.5	2

$u(k+1)$

composed of fuzzy model F^* (7.59) and its inverse F^*_{inv} (Table 7.2) and responses generated by that structure to input excitation $u_0(k) = 0.5 \cdot 1(k)$ are shown in Fig. 7.46.

◊ ◊ ◊

Recapitulating the above: in order to determine the inverse F^*_{inv} of fuzzy model F^* one should follow the consecutive steps of the algorithm presented below:

I. determine the vectors $\mathbf{u}_0$ and $\mathbf{y}$ and delay $(n-m)T$,

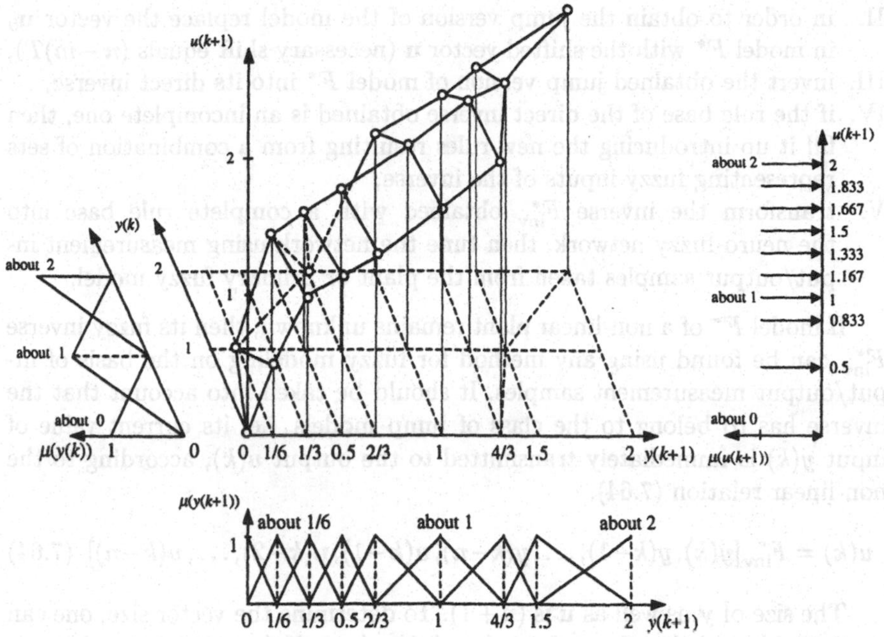

Fig. 7.45. The surface and membership functions of the model inverse (7.62) with the rule base defined by Table 7.2

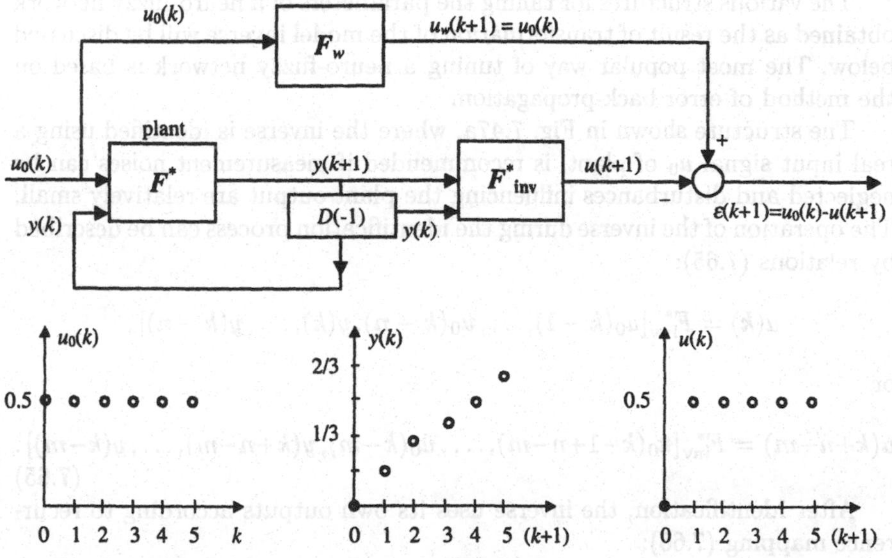

Fig. 7.46. The structure composed of model F^*, its inverse F^*_{inv}, reference model F_w and signals generated by that structure for a test input signal of step form $u_0(k) = 0.5 \cdot 1(k)$

II. in order to obtain the jump version of the model replace the vector $\mathbf{u}_0$ in model F^* with the shifted vector $\mathbf{u}$ (necessary shift equals $(n-m)T$),
III. invert the obtained jump version of model F^* into its direct inverse,
IV. if the rule base of the direct inverse obtained is an incomplete one, then fill it up introducing the new rules resulting from a combination of sets representing fuzzy inputs of the inverse,
V. transform the inverse F^*_{inv} obtained with a complete rule base into the neuro-fuzzy network, then tune the network using measurement input/output samples taken from the plant or primary fuzzy model.

If model F^* of a non-linear plant remains unknown, then its fuzzy inverse F^*_{inv} can be found using any method for fuzzy modeling on the basis of input/output measurement samples. It should be taken into account that the inverse has to belong to the class of jump models, i.e. its current value of input $y(k)$ is immediately transmitted to the output $u(k)$, according to the non-linear relation (7.64).

$$u(k) = F^*_{\text{inv}}[y(k), y(k-1), \ldots, y(k-n), u(k-1), u(k-2), \ldots, u(k-n)] \tag{7.64}$$

The size of $\mathbf{y}$ as well as $\mathbf{u}$ is $(n+1)$. To determine the vector size, one can use initial knowledge about the order of the plant. If that initial knowledge is not sufficient, then tests enabling the evaluation of the significance of inputs of the model inverse have to be carried out and, on that basis, the insignificant inputs can be rejected.

The various structures for tuning the parameters of a neuro-fuzzy network obtained as the result of transformation of the model inverse will be discussed below. The most popular way of tuning a neuro-fuzzy network is based on the method of error back-propagation.

The structure shown in Fig. 7.47a, where the inverse is identified using a real input signal u_0 of plant, is recommended if measurement noises can be neglected and disturbances influencing the plant output are relatively small. The operation of the inverse during the identification process can be described by relations (7.65):

$$u(k) = F^*_{\text{inv}}[u_0(k-1), \ldots, u_0(k-n), y(k), \ldots, y(k-n)]\,,$$

or

$$u(k+n-m) = F^*_{\text{inv}}[u_0(k-1+n-m), \ldots, u_0(k-m), y(k+n-m), \ldots, y(k-m)]\,. \tag{7.65}$$

After identification, the inverse uses its own outputs according to recurrence mapping (7.66):

$$u(k) = F^*_{\text{inv}}[u(k-1), \ldots, u(k-n), y(k), \ldots, y(k-n)]\,,$$

or

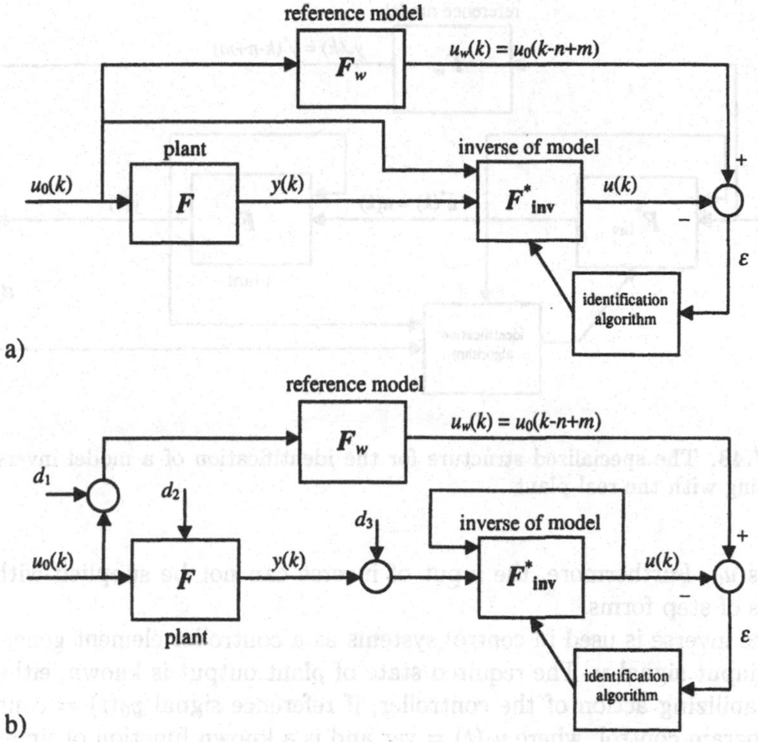

Fig. 7.47. The direct structure of identification of model inverse F^*_{inv} advised for a low level of measurement noise and negligible effects of disturbances influencing the plant – scheme (a), and scheme (b) advised for those cases where noises d_1 and d_3 as well as disturbances d_2 can not be neglected

$$u(k+n-m) = F^*_{inv}[u(k-1+n-m), \ldots, u(k-m), y(k+n-m), \ldots, y(k-m)]. \tag{7.66}$$

If signal samples are substantially charged by measurement noise and/or plant output is "contaminated" by disturbances, then the relevant identification structure is that shown in Fig. 7.47b, because – according to the mentioned scheme – model inverse F^*_{inv} uses its own, noise-free output samples $u(k)$. Of course, the mapping realized by the inverse is now given by equations (7.66). If the exact model F^* of the plant is known, then it can be used for identification, because mapping F realized by the plant (see Fig. 7.47) can be replaced by its model F^*.

There are advantages and disadvantages of using the direct structure for identification. The parameters of inverse can be tuned using well known algorithms of error ε back-propagation. Undoubtedly, such a possibility is very advantageous. However, there are difficulties in determining the best training

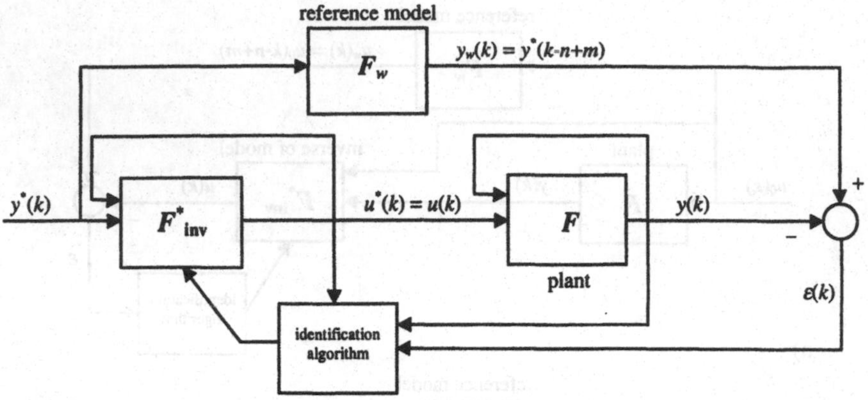

Fig. 7.48. The specialized structure for the identification of a model inverse co-operating with the real plant

signals u_0. Furthermore, the input of inverse can not be supplied with the signals of step forms.

The inverse is used in control systems as a controller element generating plant input signal u. The required state of plant output is known, either for the stabilizing action of the controller, if reference signal $y_0(t)$ = const. or for program control, where $y_0(t)$ = var and is a known function of time. The real value of plant output "oscillates" in the neighborhood of reference signal $y_0(t)$. The approximate frequency and range of those oscillations are usually known. There are difficulties in determining which of such input signals exciting the plant $u(t)$ would be transformed by the plant to the exact required form of plant output $y(t) = y_0(t)$. The solution of that problem, especially for complicated plants under control, is not an easy task. Because we do not know the most advantageous training signals $u_0(t)$, we must use a variety of them, in order to "cover" the wide ranges of such parameters like frequency, average value, amplitude. Besides the problem of numerous training signals the obtained inverse has a further weak point. It can usually be classified as "non-oriented towards the goal of control" (Hunt 1992). It negatively affects the control performance for the real range of plant operation.

The inverse "oriented towards the goal of control" can be obtained by the use of a so-called "specialized" structure for identification, shown in Fig. 7.48.

The identification of inverse realized on the basis of a specialized structure can be carried out by means of training signals $y^*(k)$ which are "close" to those generated by the plant in the neighborhood of its set point $y_0(t)$. Then, the inverse F^*_{inv} generates signals $u(k)$ controlling the plant, which keep the plant output close to the set point. Thus, the specialized structure allows us to recognize the amplitude and frequency ranges of the control signals $u(k)$ "oriented to the control goal".

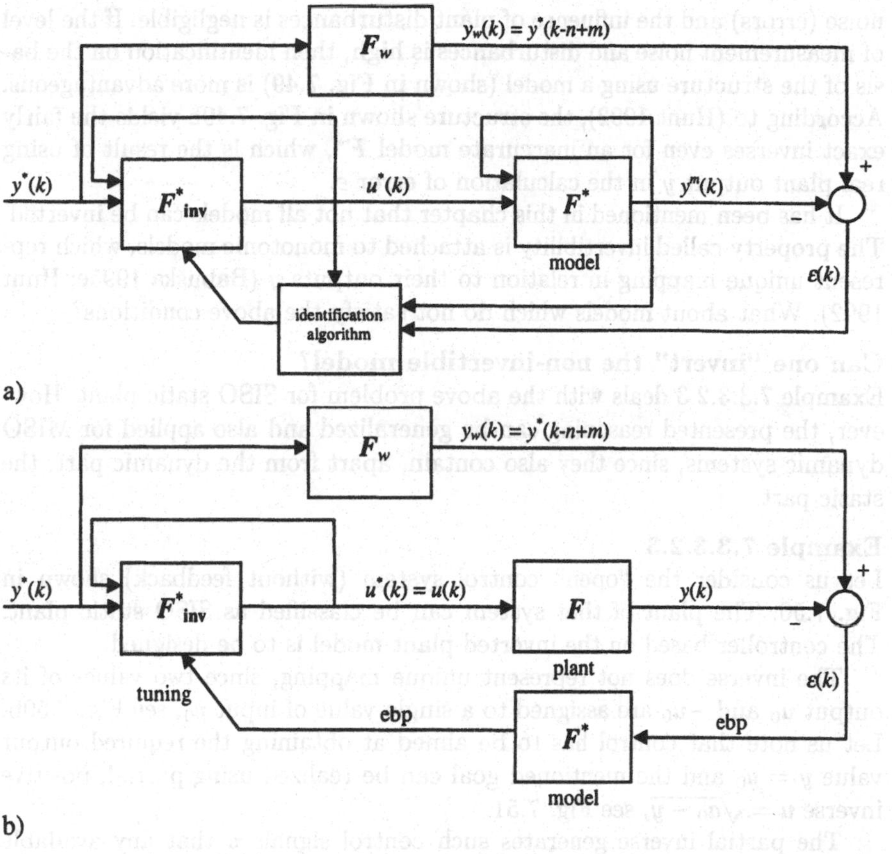

Fig. 7.49. The specialized structures for identification of inverse enabling the application of error back-propagation, a) the structure with plant model, b) the structure with plant and its model

The method of error back-propagation can not be applied for the specialized structure shown in Fig. 7.48, since plant output $y(k)$ has to be treated now as a component of the error signal in that structure. However, the plant F is unknown. It means that error can not be back propagated through the plant and identification has to be done on the basis of other methods described in Chapter 7. However, if a sufficiently accurate model F^* of plant is known, then the method of error back-propagation (as well as other methods) can be applied to the identification of the model inverse. The respective structures are shown in Fig. 7.49. The error ε of those structures can be propagated through the known model F^*, which is not tuned (the inverse of the model is the only subject of tuning).

The identification scheme using a plant, represented by Fig. 7.48, is recommended if measurement samples of signals are not charged by measurement noise (errors) and the influence of plant disturbances is negligible. If the level of measurement noise and disturbances is high, then identification on the basis of the structure using a model (shown in Fig. 7.49) is more advantageous. According to (Hunt 1992), the structure shown in Fig. 7.49b yields the fairly exact inverses even for an inaccurate model F^*, which is the result of using real plant output y in the calculation of error ε.

It has been mentioned in this chapter that not all models can be inverted. The property called invertibility is attached to monotonic models, which represent unique mapping in relation to their outputs y (Babuška 1995e; Hunt 1992). What about models which do not satisfy the above conditions?

Can one "invert" the non-invertible model?
Example 7.3.3.2.3 deals with the above problem for SISO static plant. However, the presented reasoning can be generalized and also applied for MISO dynamic systems, since they also contain, apart from the dynamic part, the static part.

Example 7.3.3.2.3
Let us consider the "open" control system (without feedback) shown in Fig. 7.50. The plant of that system can be classified as SISO static plant. The controller based on the inverted plant model is to be designed.

The inverse does not represent unique mapping, since two values of its output u_0 and $-u_0$ are assigned to a single value of input y_0, see Fig. 7.50b. Let us note that control has to be aimed at obtaining the required output value $y = y_0$ and the mentioned goal can be realized using partial, positive inverse $u = \sqrt{a_0 - y}$, see Fig. 7.51.

The partial inverse generates such control signals u that any available level of output value y can be obtained. The controller operating with the positive part of the inverse (Fig. 7.51b) generates only positive control signals u (if the negative part of the inverse was applied, then the controller would generate only negative signals u).

In the case of more complicated characteristics of SISO plants, one can invert only that part of the characteristic, which is necessary for the operation of the plant in practice (the chosen part of the characteristic should "cover" the necessary range Δy of output value y) – see Fig. 7.52. A similar way of determining the inverse can be applied for MISO plants, if they realize multi-valued mappings in relation to output y – see Fig. 7.53.

Fig. 7.53 shows that only part of a surface (that assigned to positive coordinates $x(k-1)$, $x(k)$, $y(k)$) has been chosen as the base for determining the partial inverse. There is the probability of facing "dangerous" situations, if partial inverse is implemented to the control algorithm. Fig. 7.54 explains the above problem.

The disturbances d change the plant output y. The small error $e = y_0 - y$ can be compensated for by introduction of feedback to the system shown in Fig. 7.54. However, if disturbances are intensive, then there is a risk that the state of the system will be transferred from the sub-space I to the sub-space

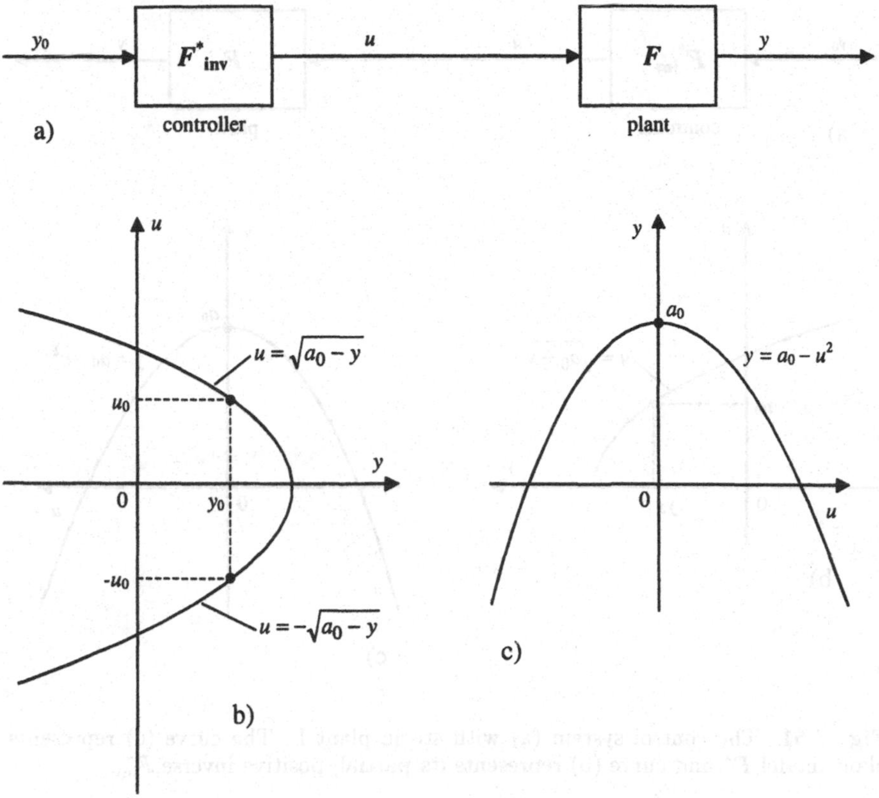

Fig. 7.50. The control system (a) containing static plant F. The curve (c) represents model F^* of the plant and the curve (b) represents inverse F^*_{inv} of model F^*

II (Fig. 7.54), where the negative control signal u is required. Of course, a controller based on a positive, partial inverse of plant $u = \sqrt{a_0 - y}$ (Fig. 7.54b will not cope under those circumstances. Its operation will be faulty. The above can be easily explained. If plant state is located in sub-space I, then decreasing values of plant output y are assigned to increasing values of input u (Fig. 7.54c). If plant state is located in sub-space II, then the above relation between increments of input and output signals is the opposite.

Thus, the partial inverse can be applied to control of plants operating without "significant" disturbances, i.e. disturbances which are able to change the sub-space of plant operation. If disturbances can be measured, then the necessary correction of controller output signal u can be calculated and executed using measurement results, even in the case of "significant" disturbances. If one is able to determine which input sub-space refers to the current state of plant operation ($u_r < 0$ or $u_r > 0$), then the influences of disturbances

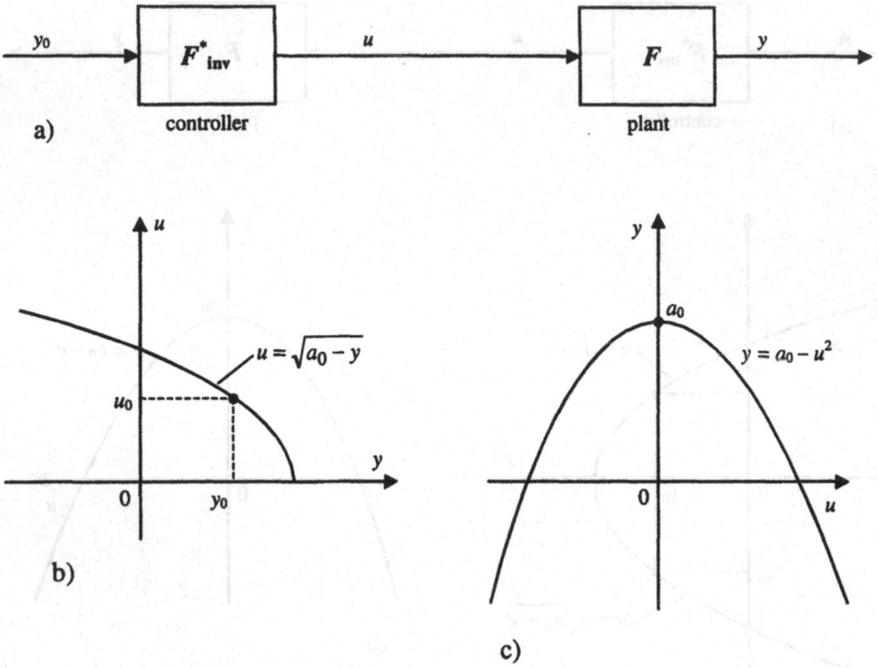

Fig. 7.51. The control system (a) with static plant F. The curve (c) represents plant model F^* and curve (b) represents its partial, positive inverse F^*_{inv}

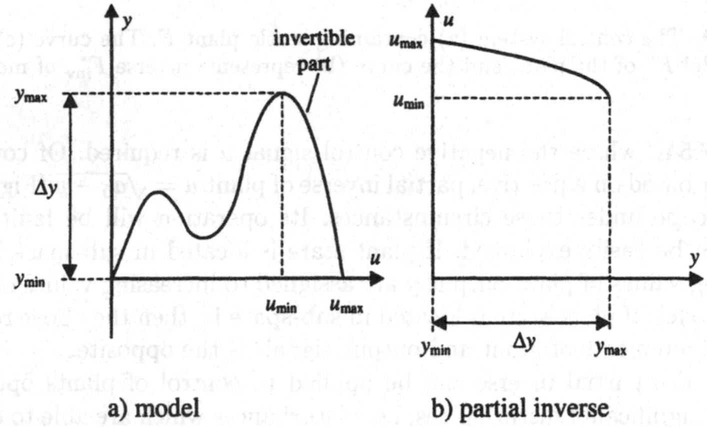

Fig. 7.52. The model of plant (a) and its partial inverse for control purposes (b)

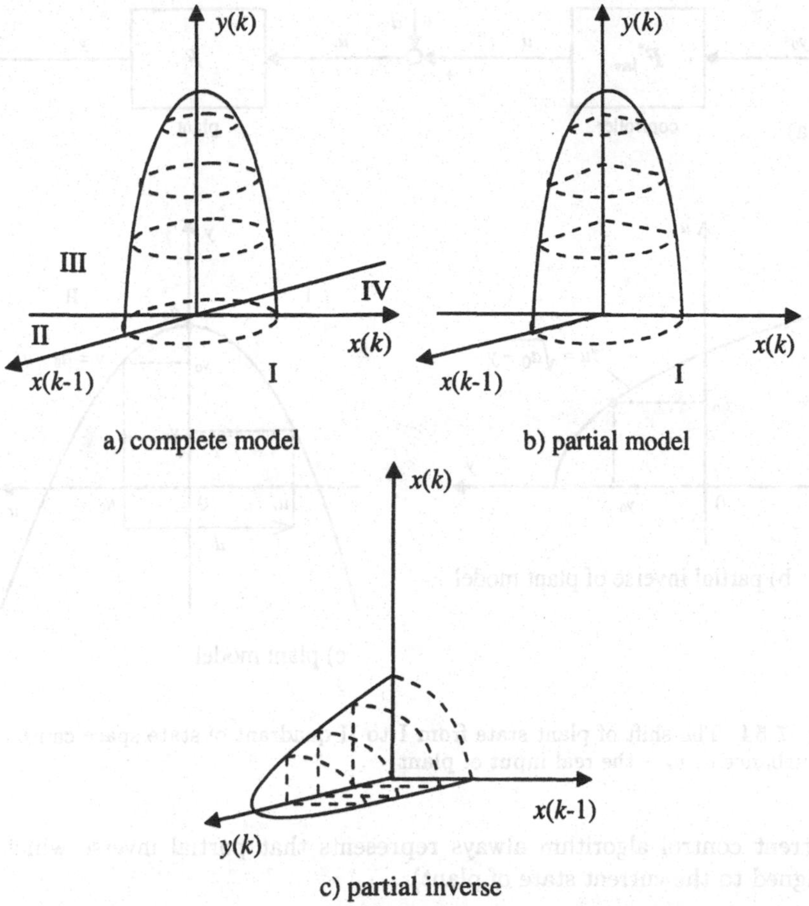

Fig. 7.53. The partial inverse of a dynamic MISO plant realizing multi-valued mapping in relation to output $y(k)$

can be compensated for by control action based on a modified partial inverse, given in the form $u = \mathrm{sgn}(u_r)\sqrt{a_0 - y - d}$.

◊ ◊ ◊

Now, let us come back to the original question "can one invert the non-invertible model?". The above considerations incline us to the following answer: the non-invertible model can not be totally inverted, but its invertible parts can be successfully inverted, especially if the inverse of its invertible part is sufficient for control purposes. If the real plant state can be observed continuously, then another solution can be proposed. The plant control can be realized by a polyzonal controller based on a certain number of partial inverses. The controller adjusts its control action to the current plant state

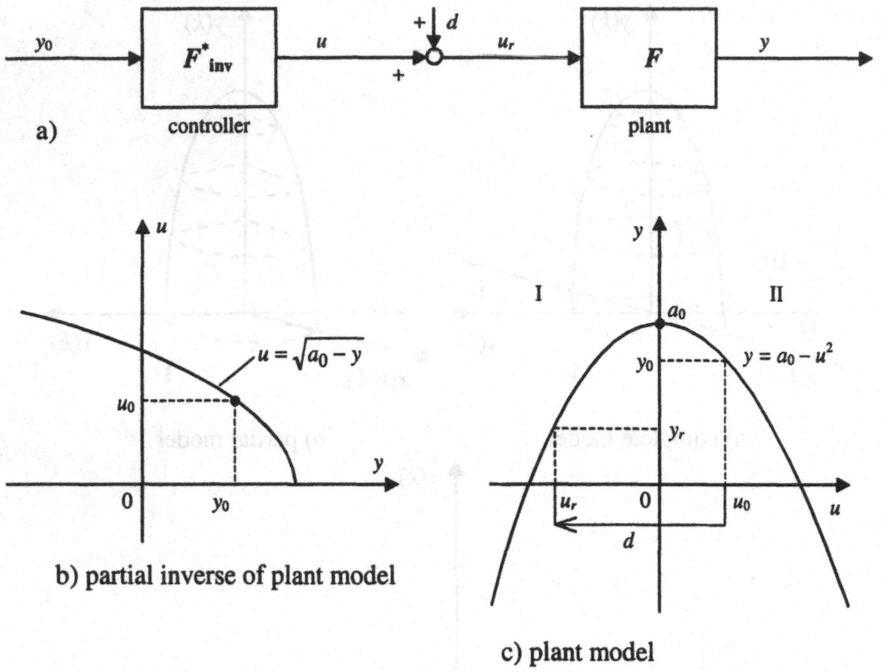

Fig. 7.54. The shift of plant state from I to II quadrant of state space caused by disturbance d, u_r – the real input of plant

(current control algorithm always represents that partial inverse which is assigned to the current state of plant).

7.3.3.3 Tuning a fuzzy controller with an a priori chosen structure

Results from long experience show that typical control algorithms (PID, PI, PD, P, PDD^2) can be successfully applied to the control of many plants.

The relation between control system performance and controller parameters is usually non-linear. Thus, introduction of variable controller parameters to "classic" algorithms (PID, PI, PD, P, PDD^2) can fit them to non-linear requirements generated by non-linear control performance criteria. The fuzzy controllers can satisfy the above non-linear requirements very well because of their non-linear character. The fuzzy PID-controllers are widely applied in practice. There are many papers and technical reports dealing with problems of fuzzy PID-controllers (Kahlert 1994,1995; Yager 1995; Brown 1994; Koch 1993,1996; Driankov 1996; Lichtenberg 1994; Altrock 1993; Kuhn 1994,1996; Pfeiffer 1995; Wang 1995; Isermann 1995,1996; Fisher 1996; Pułaczewski 1997; Piegat 1995,1996,1997,1997a,1997b; Pałęga 1996; Domański 1997).

If the type of controller is known (let us say PID), then a suitable number of fuzzy sets can be assumed for each controller input and output (the possible

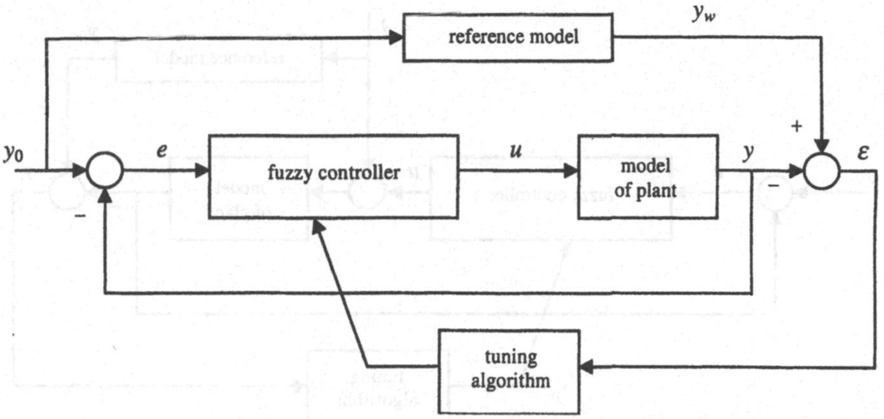

Fig. 7.55. The tuning of a fuzzy controller in a structure with a reference model representing the desired tracking of reference signal $y_0(t)$

"curvature" of the control surface depends directly on the number of fuzzy sets) and, after that, the parameters of the controller have to be tuned. If the proper type of controller is not known in advance, then the choice of controller type has to be done on an experimental basis.

The tuning of a controller can be done according to the scheme shown in Fig. 7.55. The reference model represents our preferences and requirements in relation to the control performance. The method of error back-propagation can be utilized to the tuning of the fuzzy controller.

Transfer function of order $(n - m)$ of an inertial element given by (7.67) is often chosen as a reference model for continuous systems:

$$G_w(s) = \left(\frac{1}{1 + sT_b}\right)^{n-m} \tag{7.67}$$

where: $(n - m)$ is the difference of orders of denominator and numerator of the resulting transfer function obtained as the product of transfer functions representing the plant (after its linearization) and the controller (so-called "open loop" transfer function). T_b is the time constant of the control system with a closed feedback loop.

For a discrete system the reference model of the form (7.68) is often used:

$$G_w(z) = \left(\frac{1 - c}{z - c}\right)^{n-m} \tag{7.68}$$

where: $0 \leq c < 1$.

The reference model $G_w(s) = 1$ or $G_w(z) = 1$ can be used only for static or jump plant models and controllers (see Section 7.3.3.2). Thus, these static reference models are used exceptionally, because most plants must be treated as dynamic ones.

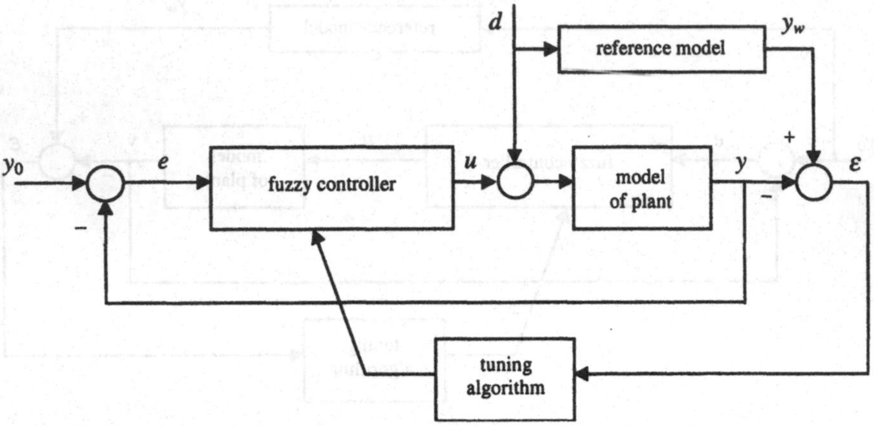

Fig. 7.56. The tuning of a fuzzy controller in a structure with the reference model representing the desired damping of disturbances

If the damping of disturbances has to be treated as a basic task of the controller, then tuning can be done in the structure shown in Fig. 7.56.

Now, for the above-mentioned controller task, the reference model can be chosen in the form of a differentiating element with inertia of order $(n - m)$:

$$G_w(s) = \frac{s}{(1 + sT_b)^{n-m}} \tag{7.69}$$

or its discrete equivalent. The reference model (7.69) constraints the influence of disturbances d till its complete suppression. That process is faster for a smaller time constant T_b.

Example 7.3.3.3.1

The mathematical model of the angular movement of the underwater vehicle Krab II is presented in (Piegat 1997a). The model expresses relations between vehicle course angle (heading angle) ψ and control signal M_r (the torque generated by the vehicle's lateral propellers). The model is represented by a recurrent neural network shown in Fig. 7.57, where z^{-1} denotes an operator expressing "unit" delay equal to sample time T, and a, b, $M_{1\min} = M_{\min}$, $M_{1\max} = M_{\max}$ are constant and known parameters of the model, the integral represents discrete integration realized according to formula (7.70):

$$\int_0^{Tk} \mathrm{d}t : \quad c(k) = \frac{T}{2}\sum_{i=1}^{k}[M_1(k) + M_1(k-1)] \,. \tag{7.70}$$

The neuro-fuzzy controller is shown in Fig. 7.58. The model of the angular movement of the underwater vehicle is non-linear. The linearization of that model for longitudinal velocity $U = 0$ yields:

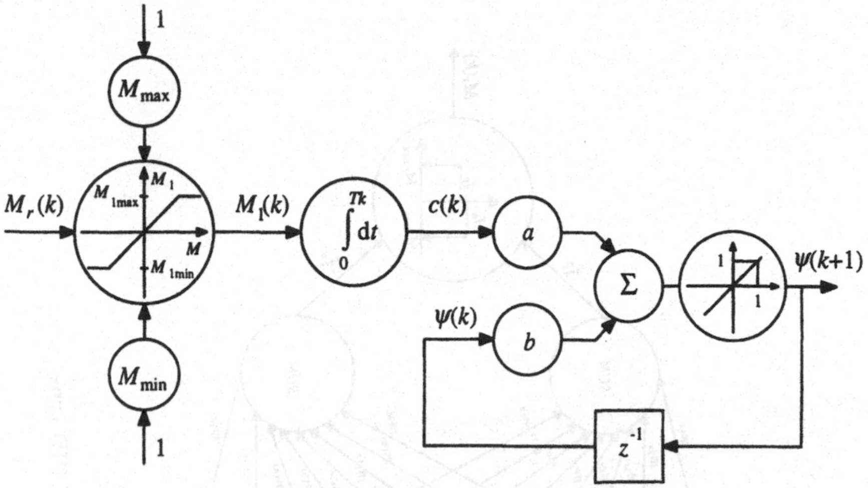

Fig. 7.57. The neural model of angular movement of the underwater vehicle

$$G^*(s) = \frac{\psi(s)}{M_1(s)} = \frac{k_0}{s \cdot (1 + sT_0)} . \tag{7.71}$$

The discretization of transfer function (7.71) (made for the "zero order hold" method of approximation of plant input signal) yields the neural model shown in Fig. 7.57. The rule base of the fuzzy controller is presented below by relations (7.72).

$$\begin{aligned}
R1:\ & IF\ (e_P = p)\ AND\ (e_I = p)\ AND\ (e_D = p) \\
& THEN\ [M_r = M_{\max}(k_{1p}\ OR\ -k_{1n})] \\
R2:\ & IF\ (e_P = p)\ AND\ (e_I = p)\ AND\ (e_D = n) \\
& THEN\ [M_r = M_{\max}(k_{2p}\ OR\ -k_{2n})] \\
R3:\ & IF\ (e_P = p)\ AND\ (e_I = n)\ AND\ (e_D = p) \\
& THEN\ [M_r = M_{\max}(k_{3p}\ OR\ -k_{3n})] \\
R4:\ & IF\ (e_P = p)\ AND\ (e_I = n)\ AND\ (e_D = n) \\
& THEN\ [M_r = M_{\max}(k_{4p}\ OR\ -k_{4n})] \\
R5:\ & IF\ (e_P = n)\ AND\ (e_I = p)\ AND\ (e_D = p) \\
& THEN\ [M_r = M_{\max}(k_{5p}\ OR\ -k_{5n})] \\
R6:\ & IF\ (e_P = n)\ AND\ (e_I = p)\ AND\ (e_D = n) \\
& THEN\ [M_r = M_{\max}(k_{6p}\ OR\ -k_{6n})] \\
R7:\ & IF\ (e_P = n)\ AND\ (e_I = n)\ AND\ (e_D = p) \\
& THEN\ [M_r = M_{\max}(k_{7p}\ OR\ -k_{7n})] \\
R8:\ & IF\ (e_P = n)\ AND\ (e_I = n)\ AND\ (e_D = n) \\
& THEN\ [M_r = M_{\max}(k_{8p}\ OR\ -k_{8n})]
\end{aligned} \tag{7.72}$$

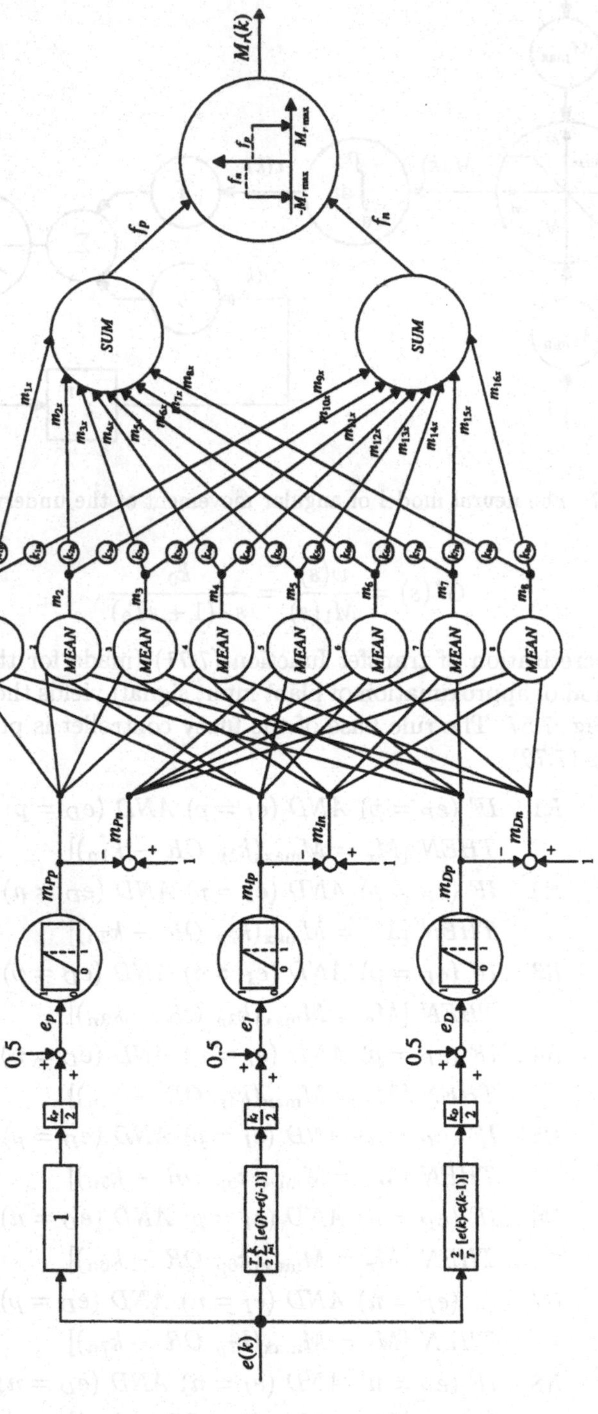

Fig. 7.58. The structure of a neuro-fuzzy controller of PID type

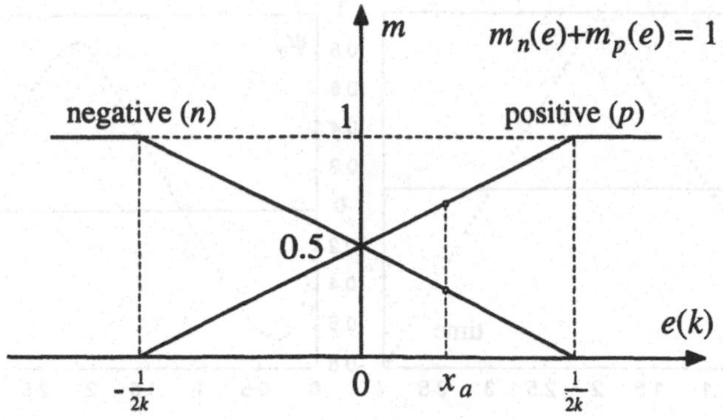

Fig. 7.59. The fuzzification of inputs in a neuro-fuzzy controller of PID type

The negative type (n) and positive type (p) membership functions for inputs are shown in Fig. 7.59. The defuzzification is realized according to expression (7.73):

$$M_r(k) = \left(\frac{f_p - f_n}{f_p + f_n}\right) \cdot M_{\max}\,, \tag{7.73}$$

where: f_p – the resultant activation grade of conclusion $M_r = M_{\max}$,
f_n – the resultant activation grade of conclusion $M_r = -M_{\max}$.

There are 11 parameters representing the neuro-fuzzy controller (K_P, K_I, K_D, $k_{1p} \ldots k_{8p}$). All those parameters can be tuned. However, application of additional conditions (7.74), ensuring the symmetry of the controller operation for positive and negative values of error signal e, means that only 6 parameters (K_P, K_I, K_D, k_{2p}, k_{3p}, k_{4p}) have to be tuned. Of course, because of the decrease of the parameters from 11 to 6 the duration time of the tuning process can be shortened.

$$\begin{aligned} k_{ip} &= 1 - k_{(9-i)p}\,, \\ k_{in} &= 1 - k_{ip}\,, \\ k_{1p} &= 1\,, \\ k_{8p} &= 0\,. \end{aligned} \tag{7.74}$$

The model of a discrete fuzzy controller is a jump model, and the discrete model of the vehicle (Fig. 7.57) is not a jump model ($n - m = 1$). Taking into account the former discussion concerning jump models, the following form of reference model G_w has been chosen, which represents the desired processing of set signal $\psi_0(t)$:

$$G_w(s) = \frac{\psi_w(s)}{\psi_0(s)} = \frac{1}{1 + 0.2s}\,. \tag{7.75}$$

The triangular desired signal $\psi_0(t)$ and corresponding reference signal $\psi_w(t)$ have been used in the tuning of the controller – see Fig. 7.60.

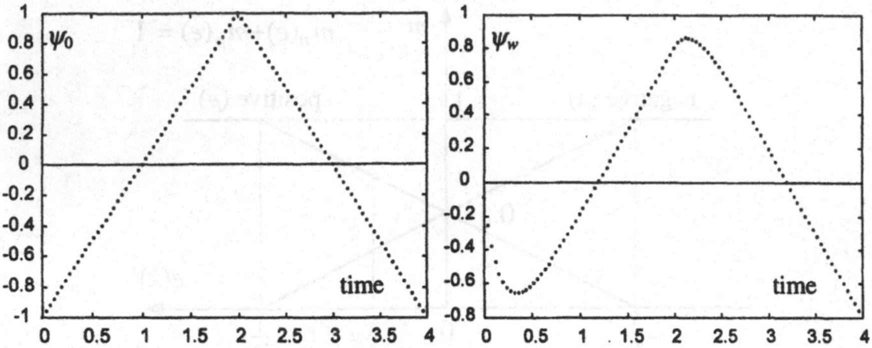

Fig. 7.60. The control system's desired signal $\psi_0(t)$ and corresponding output signal of reference model $\psi_w(t)$ utilized in the tuning of a PID controller

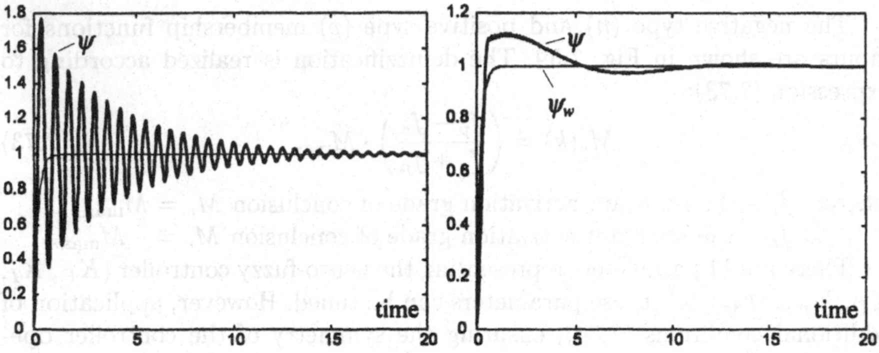

a) before tuning b) after tuning

Fig. 7.61. The response ψ of a control system with a neuro-fuzzy controller to excitation $\psi_0(t) = 1(t)$ and the response of the reference model to the same excitation

The method of error back-propagation and a structure with a reference model shown in Fig. 7.55 have been used for tuning the controller. The response of the control system ψ to the step excitation $\psi_0(t) = 1(t)$ (rad) shown in Fig. 7.61a has been obtained before tuning the controller, for a random choice of initial controller parameters. The response of the control system ψ shown in Fig. 7.61b has been obtained after tuning the controller. The reference signal ψ_w, necessary for calculation of error ε is also shown in Fig. 7.61b.

The responses of the control system ψ and of reference model ψ_w to the desired signals ψ_0 of a sine wave form and a triangular form are shown in Fig. 7.62.

The results of experiments presented in Figs. 7.61 and 7.62 prove that the controller, which is "in tune", copes successfully not only with triangular desired signals (those signals have been previously used for tuning), but

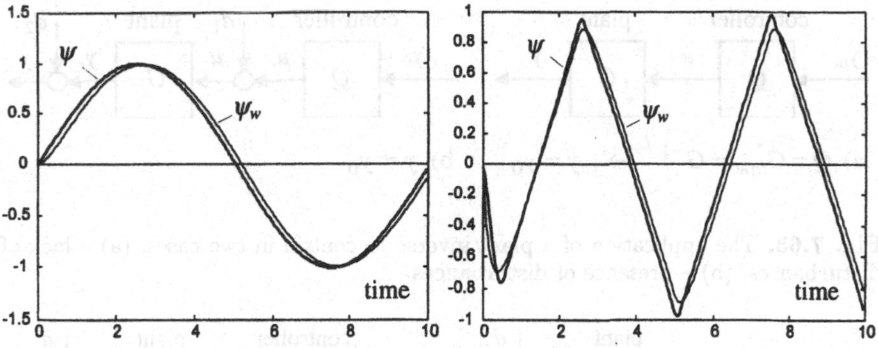

Fig. 7.62. The comparison of responses ψ_w – obtained from the reference model and ψ – from the control system with a tuned neuro-fuzzy controller for sine wave (a) and triangular (b) forms of the desired signal ψ_0

also with other forms of desired signals ψ_0 (sine wave signals, unit step signals, etc.). Experiments aimed at tuning the controller for reference model $G_w(s) = 1$ failed. The requirement of the perfect tracking of the desired signal ψ_0 was too difficult to achieve. The convergence of tuning was not attained at all. A quick destabilization of the tuning processes was observed.
◊ ◊ ◊

7.3.3.4 Fuzzy control based on the Internal Model Control Structure (IMC structure)

The control structure containing the internal model of the controlled plant (called the IMC structure) was probably "invented" in 1957, although it is difficult to find only one person "responsible" for that invention (Garcia 1982). The idea of the IMC structure has been spread mainly by Morari, Garcia and Zifiriou (Garcia 1982; Morari 1989).

The IMC structure is very useful, especially for the control of non-linear plants. Of course, it can be used for linear plants as well. The fuzzy logic and neural networks are very powerful tools for modeling non-linear plants. Thus, it was quite natural to apply it to the design of fuzzy and neural controllers (Hunt 1992; Edgar 1997; Häck 1997; Neumerkel 1992; Narenda 1990a,b; Ullrich 1997b; Piegat 1998b).

The idea of control on the basis of the IMC structure seems to be very interesting. Let us consider the open-loop control system shown in Fig. 7.63, where G and Q are linear operations (either discrete or continuous).

If $Q = G^{-1}$ and the operations G, Q are stable and disturbances do not influence the plant, then $y = y_0$. The real plant operation is affected by disturbances. Therefore the introduction of a feedback loop, see Fig. 7.64a, is necessary. The influence of both disturbances d_1, d_2 (Fig. 7.63) are represented now by the resultant, single disturbance $d = d_1 G + d_2$ (Fig. 7.64).

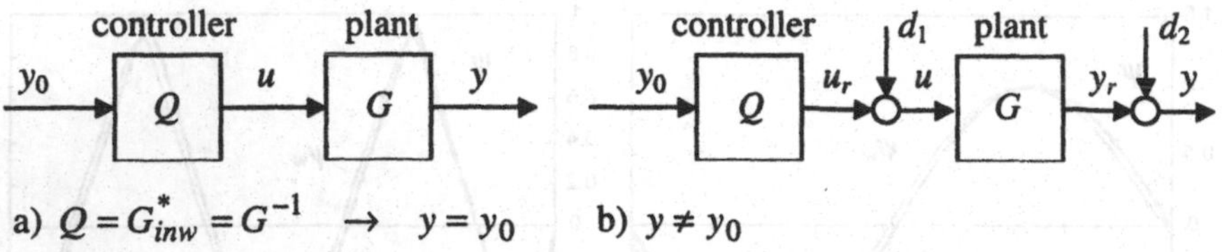

Fig. 7.63. The application of a plant inverse to control in two cases: (a) – lack of disturbances, (b) – presence of disturbances

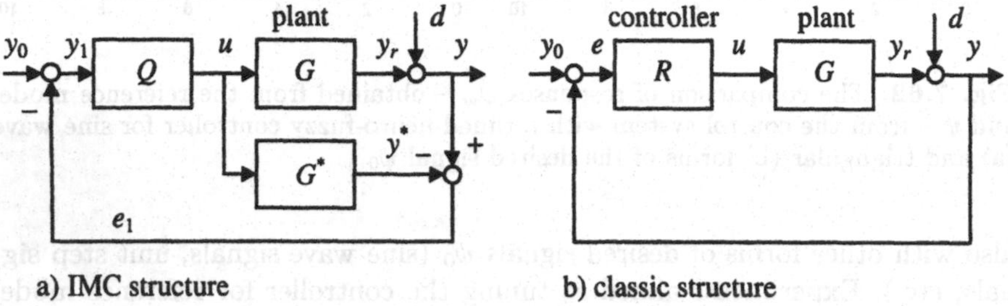

Fig. 7.64. The structure of a control system containing internal model G^* representing plant G (a) and the classic structure of a control system (b)

Remarks concerning IMC structures for linear plants

1. If plant model G^* is accurate ($G = G^*$) and disturbance $d = 0$, then feedback signal $e_1 = 0$. It means that the control system operates like an open-loop system shown in Fig. 7.63a. If $d \neq 0$, then $e_1 = d$.
2. If G^* represents an inaccurate model of a plant ($G \neq G^*$) and plant output is affected by a disturbance ($d \neq 0$), then the feedback signal is given by formula (7.76).

$$e_1 = (G - G^*)u + d \tag{7.76}$$

   So, signal e_1 depends on the error of the model ($G - G^*$) and disturbance d.
3. If plant model G^* is accurate ($G = G^*$), then the control system of an IMC type is stable, if G and Q are stable (Morari 1989). It means that one can separately examine the stability of component elements of the system instead of the complete system. That property simplifies the examination of system stability, which is of great importance, especially for non-linear control systems.
4. The IMC control structure can be transformed to an equivalent classic structure (Fig. 7.64). The controller R of the classic structure corresponding to the IMC structure should be calculated from (7.77).

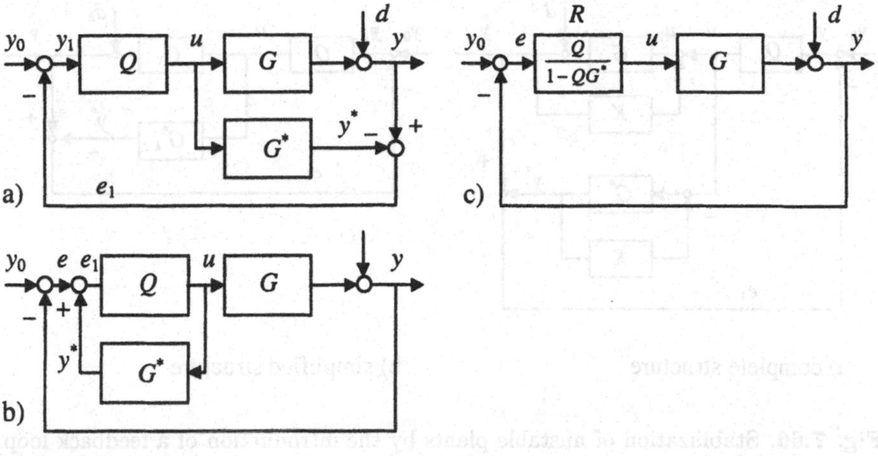

Fig. 7.65. The consecutive stages of transforming an IMC structure (a) into a classic structure (c)

$$R = \frac{Q}{1-QG^*} \tag{7.77}$$

The above transformation means that the plant model G^* becomes the internal element of controller R. It suggests the name "structure with internal model". If G and G^* are stable and the model is accurate ($G = G^*$), then the classic controller R obtained from (7.77) ensures stability of the classic control system.

5. The error e of an IMC structure (Fig. 7.65) depends on the difference $(y_0 - d)$ – see formula (7.78).

$$e = \frac{(1-QG)(y_0-d)}{1+Q(G-G^*)} \tag{7.78}$$

If plant model G^* is accurate ($G = G^*$) and Q represents the perfect inverse of the plant $Q = (G^*)^{-1}$, then $e = 0$ for any reference signal y_0 and disturbance d. Thus, the IMC structure becomes "perfect" (Morari 1989) and the controller R obtains the form (7.79).

$$R = \frac{u}{e} = \frac{Q}{1-QG^*} = \frac{G^{*-1}}{1-G^{*-1}G^*} = \infty \tag{7.79}$$

The expression (7.79) yields the controller gain equal to infinity, which can be treated as an explanation of "perfect" controller performance. The implementation of a "perfect" controller to a real control system is not possible. Even if error signal e were extremely small, then the controller would immediately generate the control signal u with an infinite amplitude (otherwise condition $e = 0$ would not be satisfied for all t). Because of the finite power of all real actuators, like servomotors, valves, rudders,

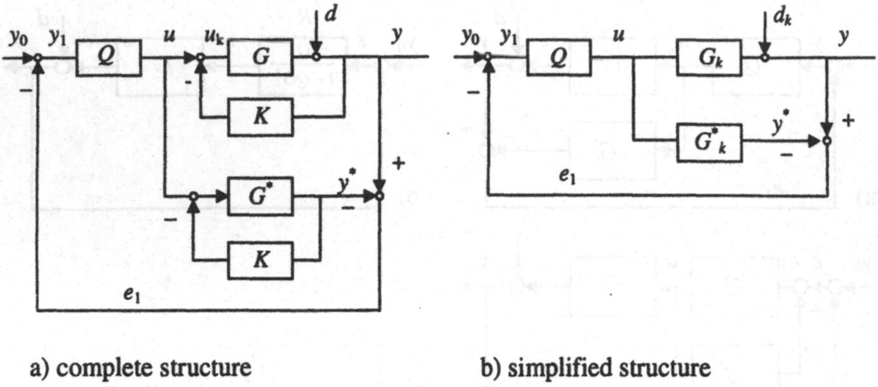

Fig. 7.66. Stabilization of unstable plants by the introduction of a feedback loop K to enable implementation of the IMC structure

etc., control signals with infinite amplitudes do not exist in reality. On the other hand, the perfect models of plants G^* and corresponding perfect inverses G^*_{inv} can not be usually obtained. So, we have in reality:

$$G^* \neq G, \quad G^*_{\text{inv}} \neq G^{*-1}.$$

However, it is obvious that a more accurate model and its inverse yield better control than a less accurate model and its inverse.

6. The IMC structure can be applied for minimum phase plants. If zeros of G^* are positive, then the poles of G^{*-1} are unstable. It means that the controller Q as well as the whole IMC system would be unstable for non-minimum phase plants.
7. The IMC structure is usually used for stable plants. However, it can also be used for unstable plants, but the design of such a controller is more complicated. The method of designing linear controllers Q is presented in (Morari 1989). For non-linear fuzzy or neural controllers and unstable plants one can apply the IMC structure introducing the additional feedback loop represented by K – see Fig. 7.66. The feedback loop K stabilizes the plant G and creates the "new", modified plant G_k, which is stable. The gains of correction element K should not be too large. Theoretically, the choice of large gains K means that amplitudes of plant input signal u_k (Fig. 7.66) are very large too. Thus, actuators of real control systems would not be able to generate the required input signals and the new plant G_k would be unstable. This loss of system stability would mean that the main task assigned for the correction K would not be realized.
8. The accuracy of model G^* decreases for increasing frequencies of signals (the models obtained are accurate for static states or slowly altering signals). The above feature violates fundamental assumptions (of the idea)

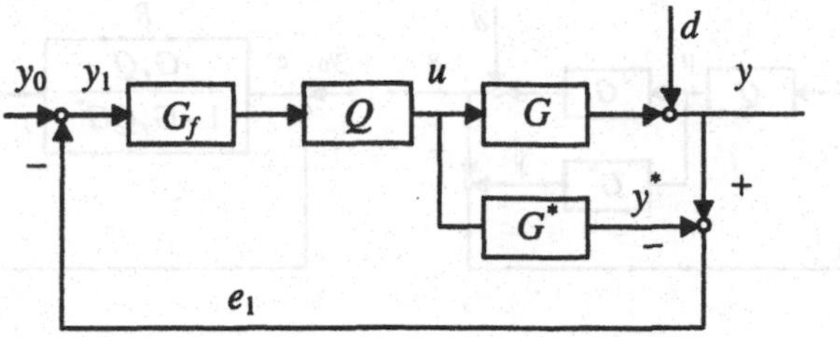

Fig. 7.67. The IMC structure with low-pass filter G_f improving its robustness

of IMC structure and leads to the loss of system stability. The additional low-pass filters G_f with static gains equal to 1 (see Fig. 7.67) allow us to overcome the above-mentioned problem. Those filters damp the high frequency components of signals. Thus, signals passing the system are slowed down. The filter damps the amplitudes of signals u generated by controller Q and decreases the rate of signal rise. It allows us to conserve the linearity of operation $Q(s)G(s)$ and the linearity of the whole system, if signal u does not attain its saturation levels. The filter G_f can also be helpful for the solution of other tasks, such as:

- elimination of system steady state error,
- asymptotic tracking of altering signals $y_0(t)$,
- simultaneous optimization and stabilization of system.

The inertial filter (7.80) or its discrete equivalent can be used as the simplest technical realization of G_f for a constant reference signal ($y_0 =$ const.) (Morari 1989).

$$G_f(s) = \frac{1}{(1+sT_f)^p} \tag{7.80}$$

The controller Q should be a proper one. It can be ensured by the choice of inertia of order p. It is a known property of a low-pass filter, that a greater T_f causes better damping of high frequency components of signals. If models G^* and inverses G^*_{inv} are relatively accurate and amplitudes of saturation for signal u are large, then a small T_f can be chosen. The filter given by (7.81) is advised for reference signals of the form $y_0(t) = at$.

$$G_f(s) = \frac{1+pT_f s}{(1+sT_f)^p} \tag{7.81}$$

If a filter is used in order to fulfill additional tasks influencing control system performance, then more degrees of freedom of the filter can be useful. Under such circumstances the following form of filter is advised:

$$G_f(s) = \frac{1+c_1 s+\ldots+c_r s^r}{(1+sT_f)^p}, \quad r \le p. \tag{7.82}$$

Further principles applied to the choice of filters are presented in (Morari 1989). The design of IMC structures can be divided into two stages. Dur-

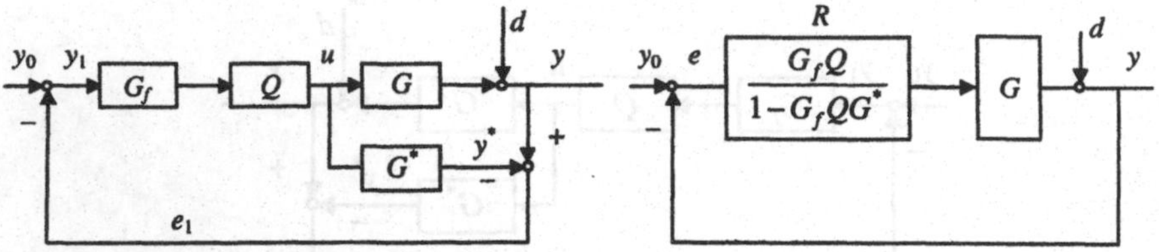

Fig. 7.68. The IMC structure containing low-pass filter G_f (a) and its transformation into the corresponding classic structure (b)

ing the first stage one designs an IMC structure neglecting the problem of filter G_f. Attention is concentrated on a high quality of control. During the second stage the previously obtained IMC structure is "equipped" with filter G_f. The filter which is "in tune", should guarantee the stability of a system for all possible conditions of the system operation. The control performance usually deteriorates (compared to performance achieved during stage 1) because of the filter inertia. Of course, an experienced designer can begin the design procedure at stage 2.

9. The error e for an IMC structure containing a filter (see Fig. 7.68) is given by formula (7.83):

$$e = \frac{(1 - G_fQG)(y_0 - d)}{1 + G_fQ(G - G^*)}\,. \tag{7.83}$$

If controller Q represents such an imperfect inverse of G^*_{inv} of a model, that $Q = G^*_{\text{inv}}$ only for static states of the control system and the whole system is stable, then the error signal asymptotically tends to zero for unit step form of excitations y_0 and d, if condition (7.84) is satisfied.

$$\lim_{t\to\infty} e(t) = 0 \text{ if } = \begin{cases} \lim_{s\to 0} G_f(s)Q(s)G(s) = 1 \\ \text{or } \lim_{z\to 1} G_f(z)Q(z)G(z) = 1 \end{cases} \tag{7.84}$$

The resultant static gain of the series connection of the filter of G_f, controller Q and plant G has to be equal to 1. If the static gain of the filter equals 1, then the static gain of inverse $G^*_{\text{inv}} = Q$ has to equal the reciprocal of static gain of plant G.

The above remarks 1 – 9 concern the design of IMC structures for linear plants. What about IMC structures for non-linear plants? The transfer functions $G(s)$, $G^*(s)$, $G^*_{\text{inv}}(s)$, $Q(s)$ assigned to linear IMC structures have to be replaced by non-linear operators.

$G \to F$ – plant
$G^* \to F^*$ – model of plant
$G^*_{\text{inv}} \to F^*_{\text{inv}}$ – inverse of plant model
$Q \to Q_n$ – IMC controller

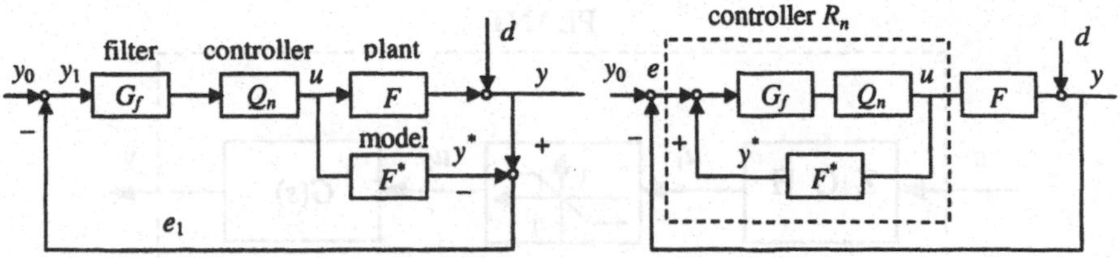

Fig. 7.69. The IMC structure represented by non-linear operators (a) and its classic equivalent containing non-linear controller R_n (b)

$R \rightarrow R_n$ – controller in classic structure
$G_f \rightarrow G_f$ – filter (the linear filters are used in non-linear IMC structures)

The non-linear IMC structure is shown in Fig. 7.69a. According to (Economou 1986; Häck 1997; Morari 1989) many properties of linear IMC structures hold in the case of non-linear IMC structures, if $d = 0$.

Remarks concerning non-linear IMC structures

1. If the model of plant F^* is accurate and controller Q as well as plant F are stable, then the IMC structure is stable.
2. If the exact plant inverse F^*_{inv} exists and it is used as a controller ($Q = F^*_{\text{inv}}$), then the IMC structure guarantees perfect control $y = y_0$.
3. If the static state of controller Q_n is equal to the static state of plant inverse F_{inv} and the IMC structure is stable as a whole, then static accuracy of control (elimination of steady state error) is obtained for constant excitations $y_0 =$ const. and $d =$ const.
4. The perfect control, as in a linear IMC structure, can be obtained only for static states. Worsening control performance accompanies the increasing frequency of signals passing through the IMC structure.
5. The plant model F^*, its inverse F^*_{inv} and controller Q_n need not be represented by conventional mathematical expressions. They can also be fuzzy models or neural networks ("black boxes") with optionally defined structures and parameters, e.g. on the basis of input/output measurements.
6. If controller Q_n represents the exact inverse F^*_{inv} of non-linear plant F, then the series connection of plant and controller can be treated as a linear element (Neumerkel 1992). The non-linear components of plant are compensated for by the controller and the whole IMC structure obtains the properties of the linear structure, which means that performance of control processes can be improved. If the inverse can be treated as accurate only for static states, then static non-linearities of the plant are compensated for.
7. The usefulness of non-linear IMC structures with fuzzy and neural controllers has been confirmed by many practical applications (Morari 1989; Hunt 1992; Edgar 1997; Häck 1997; Neumerkel 1992; Narenda 1990a,b;

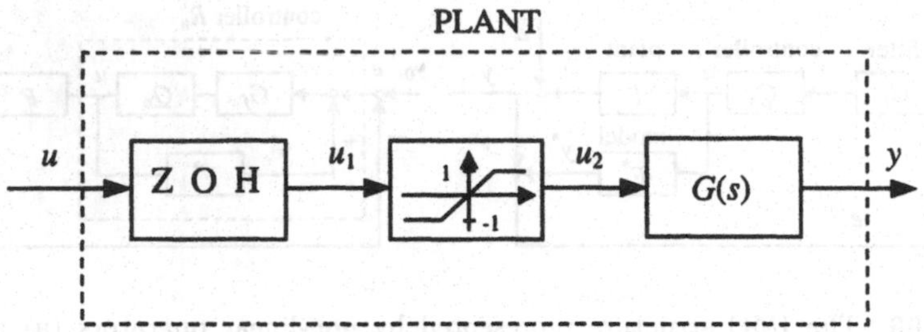

Fig. 7.70. The elements of the plant under control

Ullrich 1997b; Piegat 1998b). It means that theoretical considerations on the possibility of the transfer of some advantageous properties of linear IMC structures to non-linear IMC structures can be treated as proved and a useful source of information necessary for practical purposes.

Example 7.3.3.4.1
Let F be a non-linear operator describing the system shown in Fig. 7.70. The system F is composed of a ZOH element, actuator featuring saturation $-1 \le u_2 \le 1$ and plant under control $G(s)$. For simplicity, the whole system F will be called "plant".

The discrete model of a series connection of a ZOH element and $G(s)$, obtained for sample time $T = 0.1$ (s), is given by expression (7.85).

$$G(z) = \frac{0.9z^{-1} - 0.8z^{-2}}{0.6 - 0.5z^{-1}} \,. \tag{7.85}$$

The input/output mapping realized by the plant is described by relations (7.86) – (7.88).

$$y(k+1) = 0.83333y(k) + 1.5u(k) - 1.33333u(k-1) \text{ for } -1 \le u \le 1\,, \tag{7.86}$$

$$y(k+1) = 0.83333y(k) + 0.166667 \text{ for } u > 1\,, \tag{7.87}$$

$$y(k+1) = 0.83333y(k) - 2.83333 \text{ for } u < -1\,. \tag{7.88}$$

The response of the plant to step input signal $u(k) = 1(k)$, for all initial conditions equal to zero, is shown in Fig. 7.71.

The plant is symmetric. It means that plant responses to any signal u satisfy condition $y(u) = -y(-u)$. That is why the fuzzy model of plant F^* should be symmetric too. The results of model tuning, i.e. membership functions for model inputs $u(k)$, $u(k-1)$, $y(k)$ as well as the membership function for model output $y(k+1)$, are shown in Fig. 7.72.

The extrapolation truth (see Section 5.6) has been applied for variable $y(k)$ in order to avoid the saturation of model output $y(k+1)$, for $y(k) > 1$

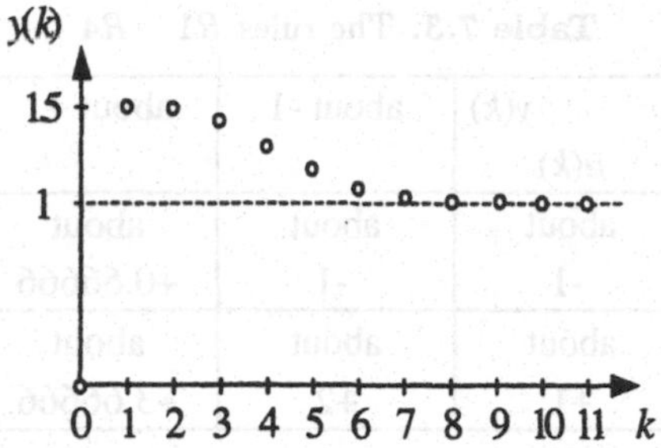

Fig. 7.71. The responses of the plant to step input signals of the form $u(k) = \alpha 1(k)$, obtained for $\alpha = 1$ and $\alpha > 1$

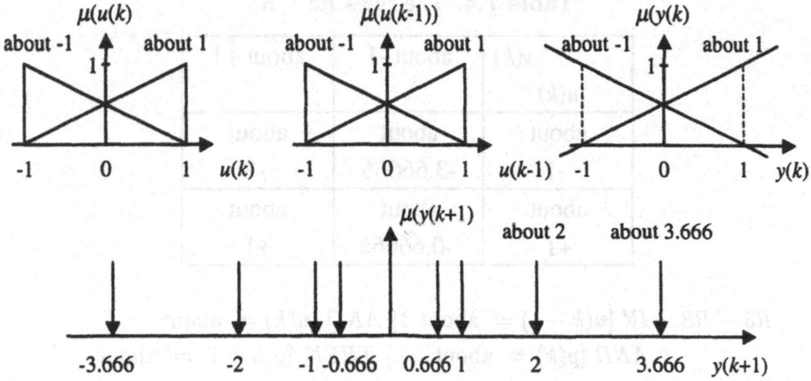

Fig. 7.72. The membership functions for inputs and output of plant model F^*

and $y(k) < -1$. The fuzzy model of the plant is represented by 8 rules Ri defined in Tables 7.3 and 7.4.

The operations *AND* have been realized by application of operator *PROD*. The block diagram illustrating the process of tuning the model inverse F^*_{inv} is shown in Fig. 7.73. The satisfying of requirement $y(k) = y_2(k)$ has not been possible. Thus, that unrealizable requirement has been weakened to the form $y(k) = y_2(k-1)$ yielding relations (7.89).

$$\begin{aligned} y(k+1) &= y_2(k) \\ y(k) &= y_2(k-1) \end{aligned} \tag{7.89}$$

In order to obtain the fuzzy inverse of a model one has to change the rules ($R1 - R8$) according to (7.89). Next, those modified rules are transformed into such forms that variable $u(k)$ representing the output of the inverse has to be inserted into the conclusion of each rule. For example, let us implement the above procedure for rule $R2$.

Table 7.3. The rules $R1$ – $R4$

$u(k)$ \ $y(k)$	about -1	about +1
about -1	about -1	about +0.66666
about +1	about +2	about +3.66666

$$R1 - R4: \; IF\ [u(k-1) = \text{ about } -1]\ AND\ [u(k) = \text{ about } \ldots] \\ AND\ [y(k) = \text{ about } \ldots]\ THEN\ [y(k+1) = \text{ about } \ldots]$$

Table 7.4. The rules $R5$ – $R5$

$u(k)$ \ $y(k)$	about -1	about +1
about -1	about -3.66666	about -2
about +1	about -0.66666	about +1

$$R5 - R8: \; IF\ [u(k-1) = \text{ about } 1]\ AND\ [u(k) = \text{ about } \ldots] \\ AND\ [y(k) = \text{ about } \ldots]\ THEN\ [y(k+1) = \text{ about } \ldots]$$

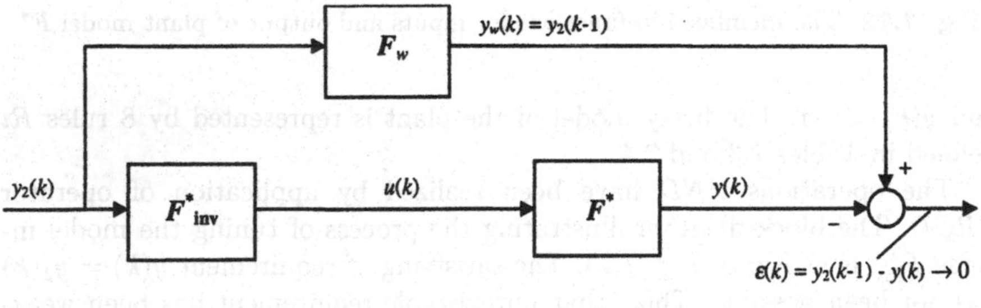

Fig. 7.73. The scheme of the system for tuning the model inverse F^*_{inv}

The primary rule:

$$R2: \; IF\ [u(k-1) = \text{ about } -1]\ AND\ [u(k) = \text{ about } -1] \\ AND\ [y(k) = \text{ about } +1]\ THEN\ [y(k+1) = \text{ about } 0.66666]$$

The primary rule after exchange of variables:

$$R2^*: \ IF\ [u(k-1) = \text{ about } -1]\ AND\ [u(k) = \text{ about } -1]$$
$$AND\ [y_2(k-1) = \text{about } +1]\ THEN\ [y_2(k) = \text{ about } 0.66666]$$

The rule of model inverse corresponding to rule $R2$:

$$R^*_{\text{inv}}: \ IF\ [u(k-1) = \text{ about } -1]\ AND\ [y_2(k-1) = \text{ about } +1]$$
$$AND\ [y_2(k) = \text{ about } 0.66666]\ THEN\ [u(k) = \text{ about} -1]$$

As a result of directly inverting 8 rules for model F^* we obtain 8 rules for its inverse. The obtained rules are listed in Tables 7.5 and 7.6.

Table 7.5. The rules $R1$ – $R8$ of the inverse of the plant model

$y_2(k)$ / $y_2(k$-1)	about -3.6666	about -2	about -1	about -0.6666	about +0.6666	about +1	about +2	about +3.6666
about -1			about -1				about +1	
about +1					about -1			about +1

$u(k)$

$$R1 - R8: \ IF\ [u(k-1) = \text{ about } -1]\ AND\ [y_2(k-1) = \text{ about} \dots]$$
$$AND\ [y_2(k) = \text{ about } \dots]\ THEN\ [u(k) = \text{ about } \dots]$$

Table 7.6. The rules $R9$ – $R16$ of the inverse of the plant model

$y_2(k)$ / $y_2(k$-1)	about -3.6666	about -2	about -1	about -0.6666	about +0.6666	about +1	about +2	about +3.6666
about -1	about -1			about +1				
about +1		about -1			about +1			

$u(k)$

$$R9 - R16: \ IF\ [u(k-1) = \text{ about} +1]\ AND\ [y_2(k-1) = \text{ about} \dots]$$
$$AND\ [y_2(k) = \text{about } \dots]\ THEN\ [u(k) = \text{ about } \dots]$$

Let us note that 32 rules are necessary to complete the inverse of the model. The additional rules came into existence because of the transfer of output variable $y(k+1)$ together with its 8 membership functions to the set of input variables. Because of incompleteness, the rule base defined by Tables 7.5 and 7.6 has to be filled up. There are two possible ways to make the rule base complete. The inverse can be tuned by application of the error back-propagation method to the scheme shown in Fig. 7.73, which contains the plant model. The tuning of the model inverse on the basis of the plant input/output data and the substitution $y(k+1) = y_2(k)$, $y(k) = y_2(k-1)$ can be treated as a second way aimed at filling the rule base up (see Section 7.3.3.2). Assuming saturation of the output of the inverse $-1 \leq u \leq +1$ one obtains the result of tuning the inverse in the form of rule bases defined by Tables 7.7 and 7.8.

Table 7.7. The rules $R1 - R16$ of inverse F^*_{inv} of plant model F^*

$y_2(k)$ $y_2(k\text{-}1)$	about -3.6666	about -2	about -1	about -0.6666	about +0.6666	about +1	about +2	about +3.6666
about -1	about -1	about -1	about -1	about -0.7777	about +0.1111	about +0.3333	about +1	about +1
about +1	about -1	about -1	about -1	about -1	about -1	about -0.7777	about -0.1111	about +1

$$
\begin{aligned}
R1 - R16: \ & IF\ [u(k-1) = \text{ about } -1]\ AND\ [y_2(k-1) = \text{ about } \ldots]\\
& AND\ [y_2(k) = \text{ about } \ldots]\ THEN\ [u(k) = \text{ about } \ldots]
\end{aligned}
$$

Table 7.8. The rules $R17 - R32$ of inverse F^*_{inv} of plant model F^*

$y_2(k)$ $y_2(k\text{-}1)$	about -3.6666	about -2	about -1	about -0.6666	about +0.6666	about +1	about +2	about +3.6666
about -1	about -1	about +0.1111	about +0.7777	about +1	about +1	about +1	about +1	about +1
about +1	about -1	about -1	about -0.3333	about -0.1111	about +0.7777	about +1	about +1	about +1

$$
\begin{aligned}
R17 - R32: \ & IF\ [u(k-1) = \text{ about } +1]\ AND\ [y_2(k-1) = \text{ about } \ldots]\\
& AND\ [y_2(k) = \text{ about } \ldots]\ THEN\ [u(k) = \text{ about } \ldots]
\end{aligned}
$$

The membership functions for variables of the inverse are shown in Fig. 7.74. To finish the synthesis of the IMC structure (Fig. 7.75) one should chose the filter G_f. Following advice given in (Morari 1989) the inertial filter (7.90) is chosen:

$$
G_f(s) = \frac{1}{1+sT_f}, \quad G_f(z) = \frac{1-e^{-T/T_f}}{z-e^{-T/T_f}}. \tag{7.90}
$$

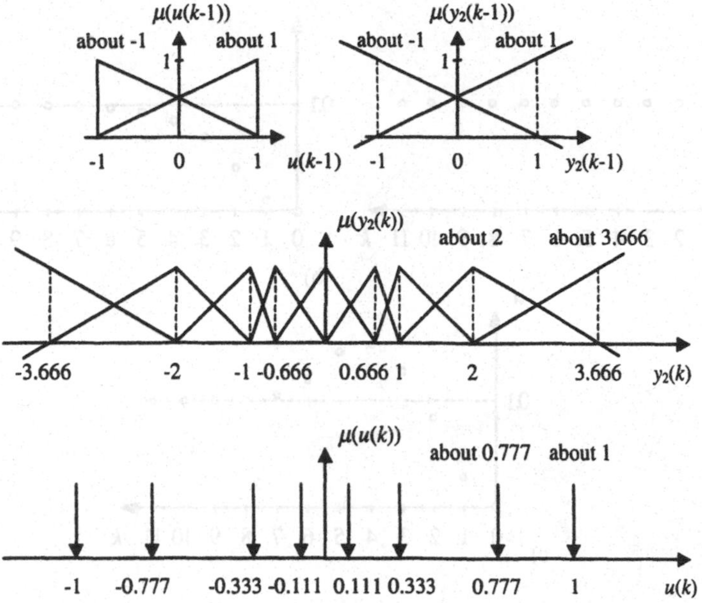

Fig. 7.74. The membership functions for inputs $u(k-1)$, $y_2(k-1)$, $y_2(k)$ and output $u(k)$ of fuzzy model inverse F^*_{inv} (defined for rule base presented in Tables 7.7 and 7.8)

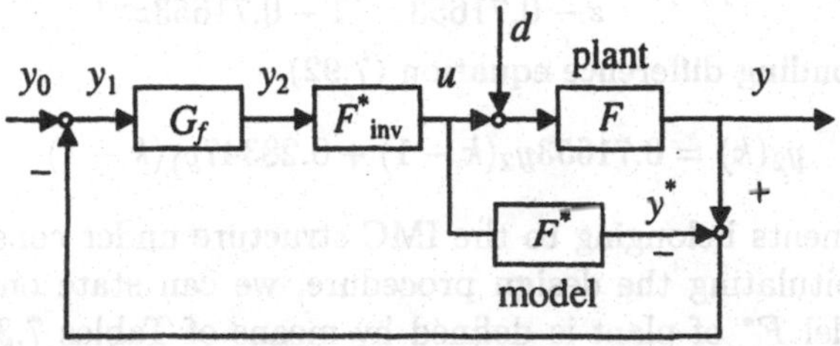

Fig. 7.75. The elements and signals of the designed IMC structure

Of course, the value of filter time constant T_f has to be chosen. The resultant time constant of a closed loop of the control system is approximately equal to $(T_f + T)$. The first choice is such, that T_f will be as small as possible. It accelerates the process of control. However, the above method of process speeding up is connected with the necessity of generating high amplitudes of control signal u, whereas the actuator imposes a restriction $-1 \leq u \leq 1$.

Taking the above into account, $T_f = 0.3$ (s) has been assumed. Hence, the discrete equivalent of the filter transfer function $G_f(s)$ is given by equation (7.91).

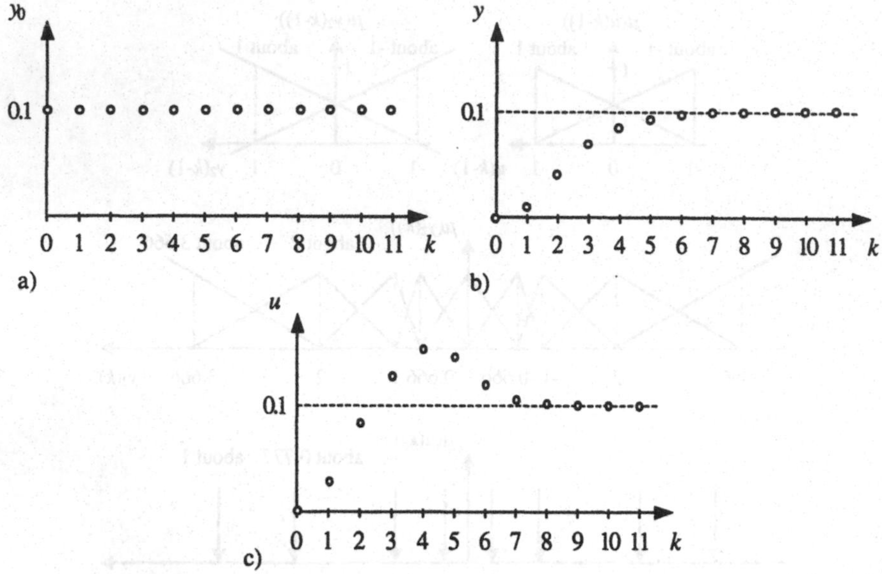

Fig. 7.76. Response y (b) of the IMC structure to step input signal $y_0(k) = 0.1(k)$, (a), and control signal $u(k)$ – the plant input (c)

$$G_f(z) = \frac{0.28347}{z - 0.71653} = \frac{0.28347z^{-1}}{1 - 0.71653z^{-1}} \tag{7.91}$$

or by corresponding difference equation (7.92).

$$y_2(k) = 0.71653y_2(k-1) + 0.28347y_1(k-1) \tag{7.92}$$

So, all elements belonging to the IMC structure under consideration are known. Recapitulating the design procedure, we can state once more that the fuzzy model F^* of plant is defined by means of Tables 7.3 and 7.4 and Fig. 7.72, the fuzzy inverse F^*_{inv} is defined by Tables 7.7 and 7.8 and Fig. 7.74, the filter is defined by equation (7.92).

Response y of the IMC structure to the "small" input signal $y_0(k) = 0.1(k)$ and control signal $u(k)$ generated by inverse F^*_{inv}, for initial conditions equal to zero, are shown in Fig. 7.76. It can be seen in Fig. 7.76 that $u(k)$ has not been saturated – this is because of a small input signal. The signals $y(k)$ and $u(k)$ for large reference signal $y_0(k) = 1(k)$ are shown in Fig. 7.77 – now, signal $u(k)$ attains the level of saturation $(-1 \le u \le 1)$.

If the amplitude of reference signal increases more and more or filter time constant T_f becomes shorter, then the control signal $u(k)$ "tends" to form a unit step. In a "terminal" case the control process can be classified as an "on – off" one.

The process of damping disturbance $d(k) = 0.1(k)$ influencing the plant input (see Fig. 7.75) is shown in Fig. 7.78. It shows that the IMC system

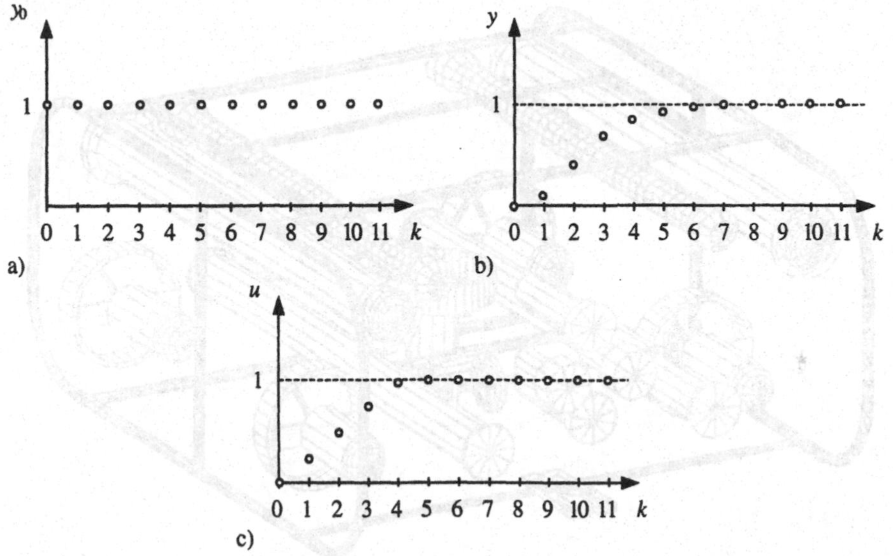

Fig. 7.77. Response y of the IMC structure (b) to step reference signal $y_0(k) = 1(k)$, (a), and control signal $u(k)$ – the plant input (c)

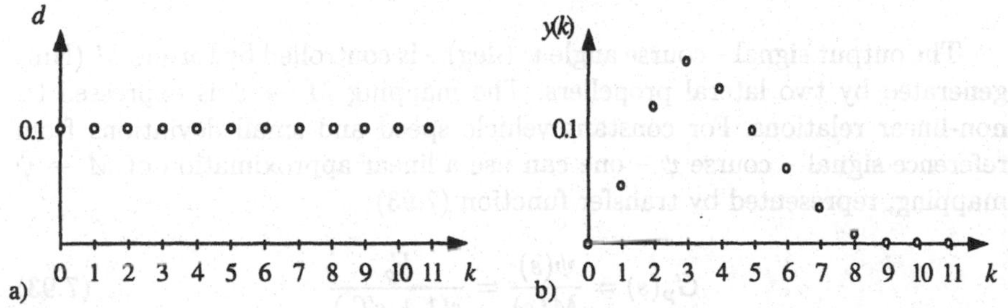

Fig. 7.78. The damping (b) of disturbance $d(k) = 0.1(k)$, (a), influencing the plant input

eliminates (in the sense of a complete static compensation, i.e. for $k \to \infty$) the influences caused by disturbances. If amplitudes of disturbances are greater, but still ($|d| < 1$), then the time necessary for complete compensation is considerably longer, which is caused by saturation of u ($-1 \leq u \leq 1$).

◊ ◊ ◊

Example 7.3.3.4.2

Let us consider the course control problem for the underwater vehicle "KRAB II" (see Fig. 7.79).

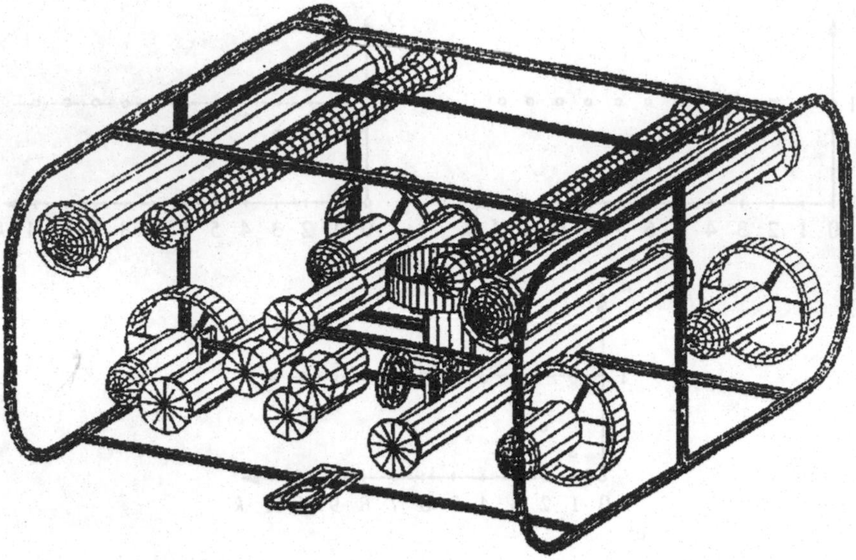

Fig. 7.79. The unmanned, remote controlled underwater vehicle "KRAB II" (Technical University of Szczecin)

The output signal – course angle ψ (deg) – is controlled by torque M (Nm) generated by two lateral propellers. The mapping $M \to \psi$ is expressed by non-linear relations. For constant vehicle speed and small deviations from reference signal – course ψ – one can use a linear approximation of $M \to \psi$ mapping, represented by transfer function (7.93).

$$G_p(s) = \frac{\psi(s)}{M(s)} = \frac{k_p}{s(1+sT_p)} \tag{7.93}$$

The experimental identification of parameters for average speed (Pluciński 1996) has yielded:

$$k_p = 0.04\,(\text{deg s/Nm})\,, \quad T_p = 1\,(\text{s})\,.$$

For other speeds, parameters cover the ranges:

$$0.015 \le k_p \le 0.063\,, \quad 0.5 \le T_p \le 11\,(s)\,.$$

One of the poles of the transfer function (7.93) equals zero. The non-linear static characteristic of lateral propellers (actuator) is shown in Fig. 7.80.

The dead zone of the propeller characteristic is $(-1.12 \le U \le 1.12)$. The propeller saturations are asymmetrical $(-18.20 \le M \le 28.31)$. The static correction of the propeller characteristic has been applied in order to eliminate the dead zone – see Fig. 7.81.

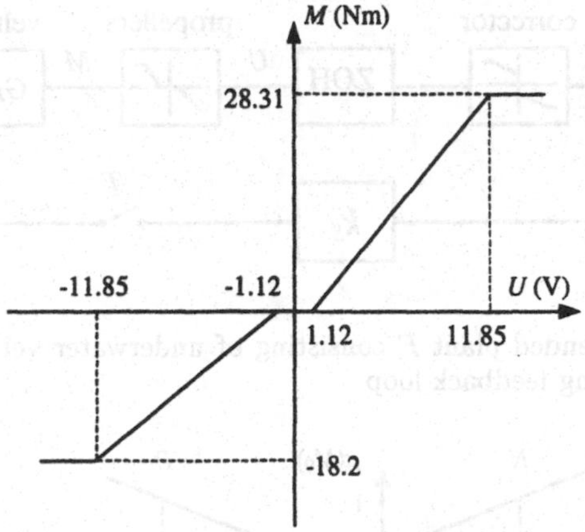

Fig. 7.80. The static characteristic of vehicle lateral propellers

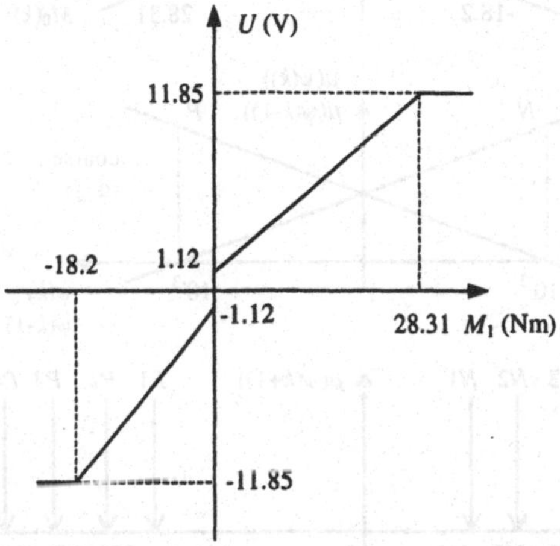

Fig. 7.81. The static characteristic of the propeller correction

The plant controlled by an IMC structure has to be stable. To fulfill that requirement, the plant has been "equipped" with stabilizing feedback represented by gain k_s.

Results from experimental examinations (Piegat 1998b) show that if the gain $k_s = 8.6$, it guarantees stability of the vehicle for all possible parameters k_p, T_p from the ranges defined above. The plant F, Fig. 7.82, is a non-linear one. The fuzzy model of F in the form $\psi(k+1) = F^*[\psi(k), \psi(k-1), M_0(k)]$ has been determined on the basis of measurement records of the output signal ψ and input signal M_0. The membership functions for model variables are shown in Fig. 7.83.

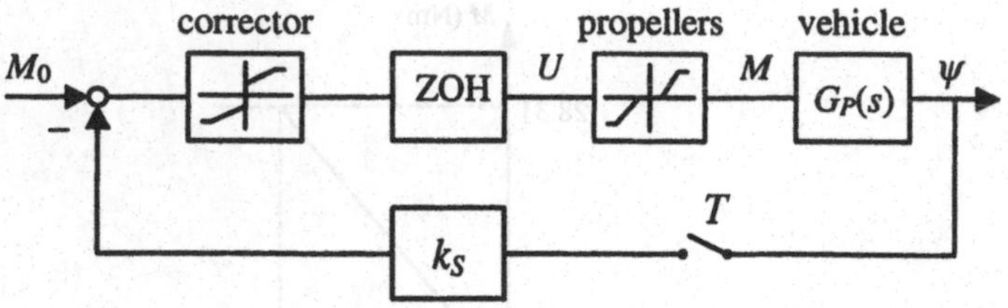

Fig. 7.82. The extended plant F consisting of underwater vehicle, propeller correction and stabilizing feedback loop

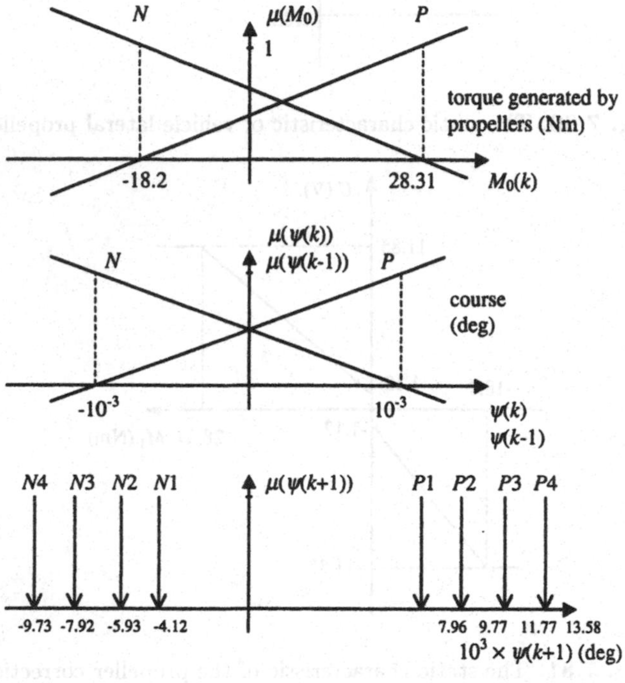

Fig. 7.83. The membership functions for variables representing fuzzy model F^* of the vehicle

The extrapolation truth described in (Piegat 1997c,1998a) and in Section 5.6 has been applied to membership functions. That is why the model of vehicle F^* can be turned through any course angle ψ (the membership functions are not restricted). Tables 7.9 and 7.10 define 8 rules representing the rule base for vehicle model F^*. The operation *AND* is realized by means of the *PROD*-operator.

An analysis of the rule base shows that plant model F^* is a monotonic one. So it can be inverted. The present section explains exactly the method of inverting F^*. The number of rules representing the inverse is greater than the

Table 7.9. Rules $R1 - R4$ representing fuzzy model F^* for $M_0(k) = N$

$\psi(k-1)$ \ $\psi(k)$	N	P
N	$N3$	$N1$
P	$N4$	$N2$

IF $[M_0(k) = N]$ *AND* $[\psi(k) = \ldots]$ *AND* $[\psi(k-1) = \ldots]$ *THEN* $[\psi(k+1) = \ldots]$

Table 7.10. Rules $R5 - R8$ representing fuzzy model F^* for $M_0(k) = P$

$\psi(k-1)$ \ $\psi(k)$	N	P
N	$P2$	$P4$
P	$P1$	$P3$

IF $[M_0(k) = P]$ *AND* $[\psi(k) = \ldots]$ *AND* $[\psi(k-1) = \ldots]$ *THEN* $[\psi(k+1) = \ldots]$

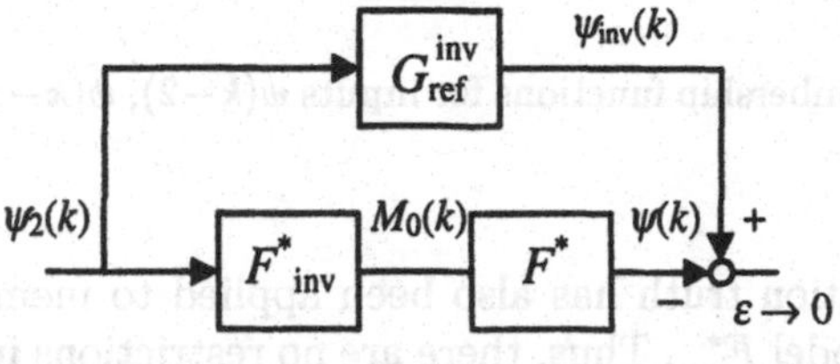

Fig. 7.84. The structure with reference model $G_{\mathrm{ref}}^{\mathrm{inv}}$ for tuning inverse F_{inv}^* of model F^*

number of rules representing the model. Thus, the tuning of those new membership functions has been realized using the structure shown in Fig. 7.84.

The reference model is defined by expressions (7.94).

$$G_{\mathrm{ref}}^{\mathrm{inv}} = \frac{\psi_{\mathrm{ref}}(z)}{\psi_2(z)} = z^{-1}$$
$$\psi_{\mathrm{ref}}(k) = \psi_2(k-1) \tag{7.94}$$

Hence, for $\varepsilon \to 0$, it is:

$$\psi(k+1) = \psi_2(k)\,,$$
$$\psi(k) = \psi_2(k-1)\,. \tag{7.95}$$

The above relations have been used to exchange variables during model inversion. The membership functions for model inverse are shown in Fig. 7.85.

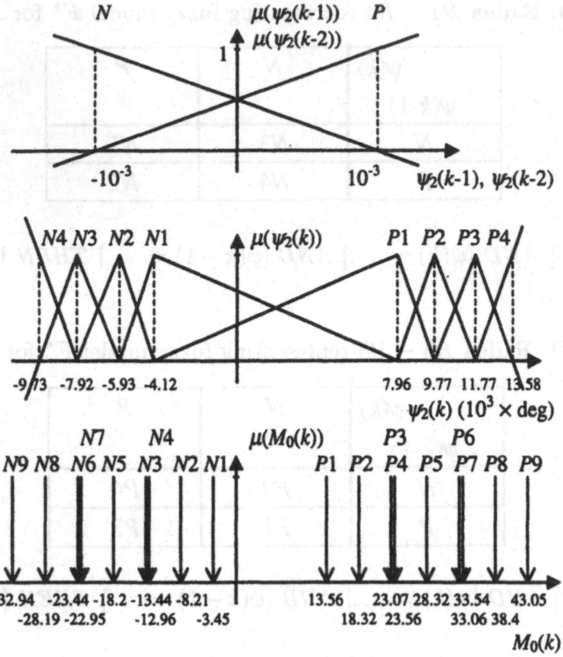

Fig. 7.85. The membership functions for inputs $\psi(k-2)$, $\psi(k-1)$, $\psi(k)$ and output $M(k)$ of inverse F^*_{inv}

The extrapolation truth has also been applied to membership functions of the inverted model F^*_{inv}. Thus, there are no restrictions imposed on output signal $M_0(k)$ of the inverse. It means that the underwater vehicle can be turned through any course angle ψ. The rule base composed of 32 rules is presented in Tables 7.11 and 7.12.

Taking into account Tables 7.11 and 7.12 and membership functions shown in Fig. 7.85 one can easily check that the model inverse, like the model itself, is monotonic. The complete IMC structure with fuzzy model F^* and its inverse F^*_{inv} is shown in Fig. 7.86.

The inertial filter G_f given by transfer function (7.96) has been applied:

$$G_f(z) = \frac{\psi_2(z)}{\psi_1(z)} = \frac{1-c}{z-c} \tag{7.96}$$

where: $c = \exp(-T/T_f)$, T_f – filter time constant, T – sample time 0.1 (s). In approximate terms, the approximate value of the resultant time constant of the whole system is equal to the filter time constant T_f. Thus, the value of T_f determines the system response time. The shortening of the response time by choosing smaller T_f means that amplitudes of control signal M_0 become greater. Finally, large control signals M_0 cause a saturation effect and the

Table 7.11. The rules $R1 - R16$ of inverse F^*_{inv} for $\psi_2(k-2) = N$

$\psi(k)$ / $\psi(k\text{-}1)$	N4	N3	N2	N1	P1	P2	P3	P4
N	N6	N5	N3	N2	P4	P5	P7	P8
P	N9	N8	N6	N5	P1	P2	P4	P5

$IF\ [\psi_2(k-2) = N]\ AND\ [\psi_2(k-1) = \ldots]\ AND\ [\psi_2(k) = \ldots]\ THEN\ [M_0(k) = \ldots]$

Table 7.12. The rules $R17 - R32$ of inverse F^*_{inv} for $\psi_2(k-2) = P$

$\psi(k)$ / $\psi(k\text{-}1)$	N4	N3	N2	N1	P1	P2	P3	P4
N	N5	N4	N2	N1	P5	P6	P8	P9
P	N8	N7	N5	N4	P2	P3	P5	P6

$IF\ [\psi_2(k-2) = P]\ AND\ [\psi_2(k-1) = \ldots]\ AND\ [\psi_2(k) = \ldots]\ THEN\ [M_0(k) = \ldots]$

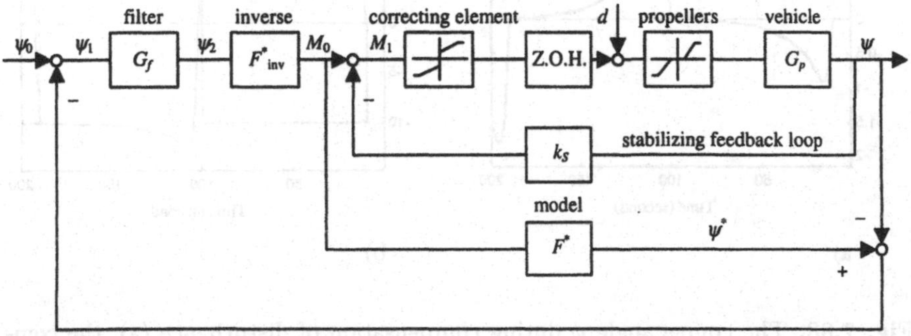

Fig. 7.86. The IMC structure for the control of underwater vehicle Krab II

control process becomes similar to an "on-off" process. Simultaneously, the difference between responses of real plant F and its fuzzy model F^* increases, which can lead to loss of system stability. For relatively large T_f the signals inside the system are rather slow and amplitudes of M_0 are too small for the development of the saturation effect. So, the operation of the system can be treated as linear.

The optimal value of filter time constant ($T_f = 17$ (s)) has been found experimentally. The response of the IMC structure $\psi(t)$ to the reference signal $\psi_0(t)$ given in the form of a rectangle wave with alternating amplitudes

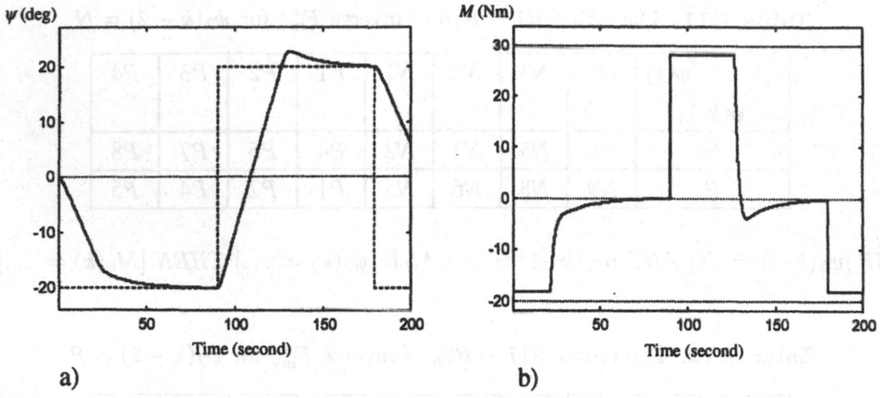

Fig. 7.87. The course angle ψ (continuous line) and reference course angle ψ_0 (dashed line) (a), the torque M generated by propellers (b)

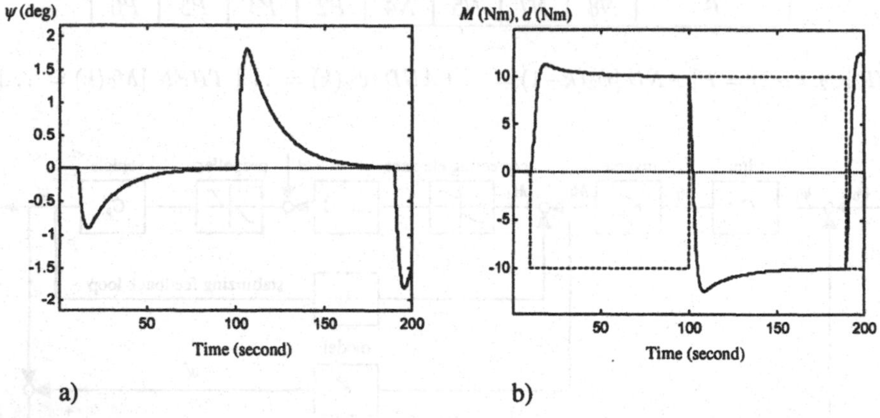

Fig. 7.88. The course angle ψ during compensation of disturbance (a), the compensating torque M (continuous line) as well as disturbance d in the form of a rectangular wave (dashed line) (b)

(+20°, −20°) as well as torque M (control signal) generated by propellers are shown in Fig. 7.87.

The signals shown in Fig. 7.87 confirm a good control performance. For the greater part of the signal period, torque M "holds" its extreme values (maximum or minimum). That is why the state $\psi_0(t) = \psi(t)$ is attained quickly and overshoots are very small (if the characteristic of the propellers had been symmetric, the overshoots would have been eliminated completely). The propellers operate without any redundant switching activity.

The process of damping disturbance d is shown in Fig. 7.88. The perturbations of course angle $\psi(t)$ caused by disturbance d in the form of a rectangular wave is shown in Fig. 7.88a, whereas Fig. 7.88b shows signal d (dashed line) and torque M (continuous line). The curves in Fig. 7.88 confirm that compensation of disturbance d is done fluently and quickly, without introducing oscillatory components to signal ψ. The propellers operation is also quick and without unnecessary switching.

The present example supports the statement that an IMC structure can also be successfully applied to the control of unstable plants. However, in the beginning, the plant must be adjusted to co-operate with an IMC structure by the introduction of an external, stabilizing feedback loop.

◊ ◊ ◊

7.3.3.5 Fuzzy control structure with an inverse of a plant model (InvMC structure)

There are theoretical possibilities of "perfect" control by using control system structures considered in Section 7.3.3.4 (Morari 1989). If model G^* of plant G and its inverse G^*_{inv} are perfectly accurate, i.e. $G = G^*$ and $G^*_{\text{inv}}G = 1$ ($G^*_{\text{inv}} = G^{-1}$), then the resultant transfer function of the control system equals 1, which means that the reference signal is always equal to the plant output signal and disturbances are damped immediately, see Fig. 7.89.

Now, let us consider the structure shown in Fig. 7.90, which contains plant model inverse G^{*-1}, but does not contain the plant model. If the considered system contains a perfect inverse (ideal case), then it exactly follows the reference signal ($y_0 = y$). So, results of its operation are the same as the results of the operation of a perfect IMC structure, shown in Fig. 7.89. Nevertheless, that identical result of operation is achieved by using a smaller number of elements in the control loop. The only requirement is inverse G^{*-1}, whereas the IMC structure "requires" the plant model G^* and inverse.

Both the structures, i.e. perfect IMC as well as a structure with perfect plant inverse, generate infinitely large control signals u. In the case of perfect control, the reference model (see Fig. 7.91) ought to be represented by transfer function $G_w = 1$.

There are many technical reasons which make implementation of that perfect control ($G_w = 1$) impossible. Thus, one has to assume a more realistic reference model. What kind of reference models G_w can be "imitated" by a realizable control loop? Let us consider the structure shown in Fig. 7.92.

Let the plant transfer function $G(s)$ be expressed by (7.97).

$$G(s) = \frac{y(s)}{u(s)} = \frac{b_0 + b_1 s + \ldots + b_{m-1}s^{m-1} + b_m s^m}{a_0 + a_1 s + \ldots + a_{n-1}s^{n-1} + a_n s^n}, \quad n \geq m \qquad (7.97)$$

Hence, the perfect plant inverse $G^{-1}(s)$ obtains the form (7.98). Let us note that all remarks from Section 7.3.3.4 referring to model inversion are still valid.

$$G^{-1}(s) = \frac{a_0 + a_1 s + \ldots + a_{n-1}s^{n-1} + a_n s^n}{b_0 + b_1 s + \ldots + b_{m-1}s^{m-1} + b_m s^m} \qquad (7.98)$$

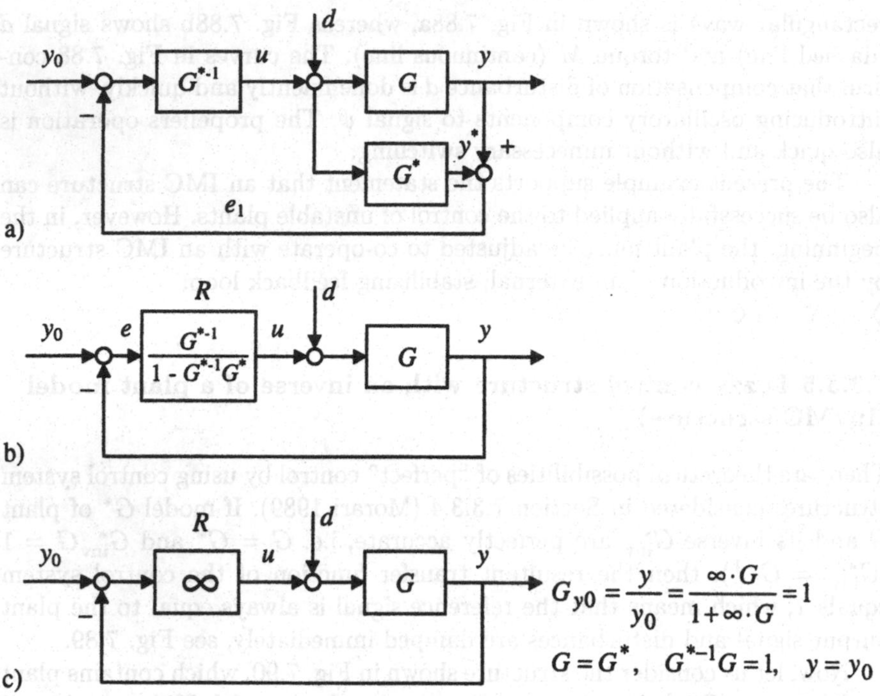

Fig. 7.89. The IMC structure with a perfect plant model G^* and its perfect inverse G^{*-1}

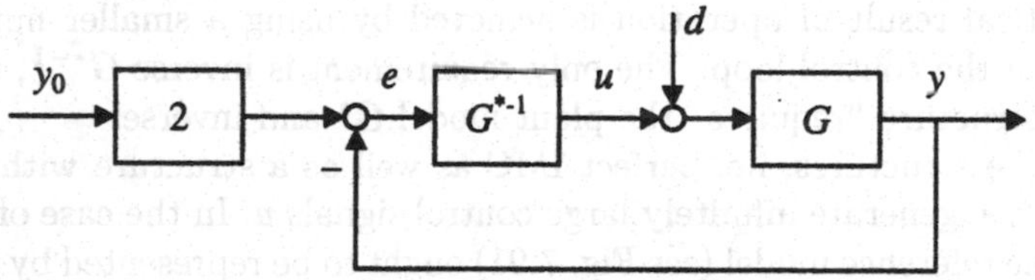

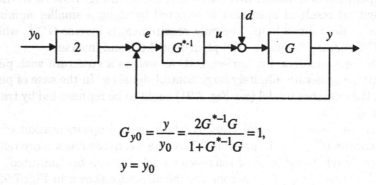

$$G_{y0} = \frac{y}{y_0} = \frac{2G^{*-1}G}{1 + G^{*-1}G} = 1,$$

$$y = y_0$$

Fig. 7.90. The control system with a controller represented by a perfect plant inverse G^{*-1}

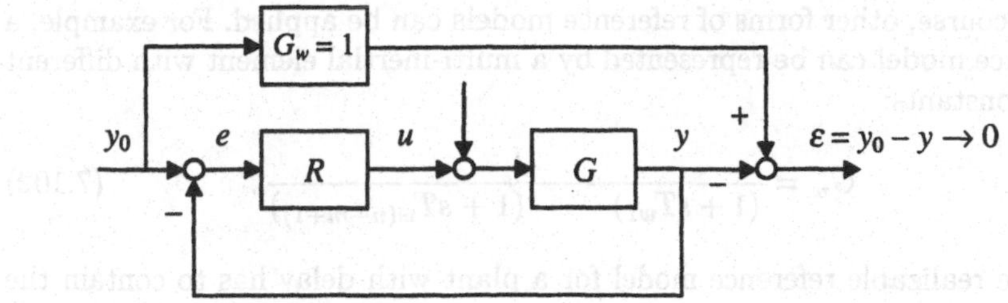

Fig. 7.91. The control structure with reference model G_w for perfect control ($y = y_0$)

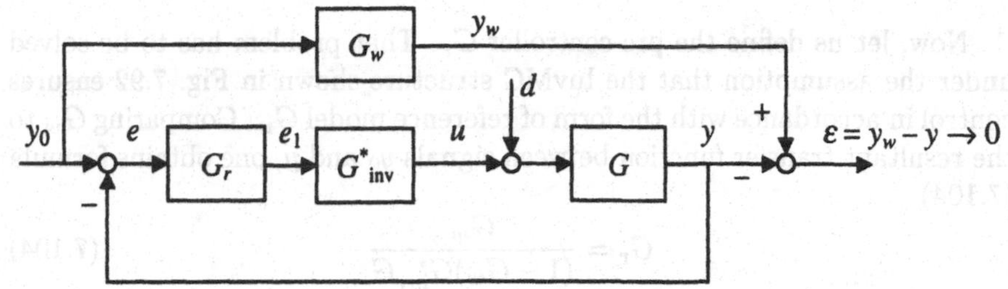

Fig. 7.92. The control structure containing an inverse of plant model G^*_{inv}, pre-controller G_r and reference model G_w (InvMC structure)

If perfect inverse $G^{*-1}(s)$ is not a proper one ($n > m$), then an additional, inertial element of order ($n-m$) has to be connected in series with that inverse in order to obtain the physically realizable inverse G^*_{inv} in the form (7.99).

$$G^*_{inv}(s) = \frac{a_0 + a_1 s + \ldots + a_{n-1}s^{n-1} + a_n s^n}{(b_0 + b_1 s + \ldots + b_{m-1}s^{m-1} + b_m s^m)(1 + sT_b)^{n-m}} \tag{7.99}$$

The resultant transfer function representing a series connection of realizable inverse G^*_{inv} and plant G (see Fig. 7.92) can be approximated (note that inverse G^*_{inv} is usually inaccurate) by means of inertial element of ($n - m$) order – formula (7.100).

$$G^*_{inv}(s) \cdot G(s) \cong \frac{1}{(1 + sT_b)^{n-m}} \tag{7.100}$$

For decreasing T_b, product $G^*_{inv} \cdot G$ tends to 1. It means that the control process is accelerated. It has to be added that larger amplitudes of the control signal u are generated, if T_b is shorter. So, the technical performance of actuators determines the choice of the minimum value of T_b.

Now, let us assume that inverse G^*_{inv} in the system shown in Fig. 7.92 is determined using (7.99) and reference model G_w is given by (7.101).

$$G_w = \frac{1}{(1 + sT_w)^{n-m+1}} \tag{7.101}$$

Of course, other forms of reference models can be applied. For example, a reference model can be represented by a multi-inertial element with different time constants:

$$G_w = \frac{1}{(1+sT_{w1})\cdot\dots\cdot(1+sT_{w(n-m+1)})}. \tag{7.102}$$

The realizable reference model for a plant with delay has to contain the delaying part, as in formula (7.103).

$$G_w = \frac{e^{-sT_0}}{(1+sT_w)^{n-m+1}} \tag{7.103}$$

Now, let us define the pre-controller G_r. That problem has to be solved under the assumption that the InvMC structure shown in Fig. 7.92 ensures control in accordance with the form of reference model G_w. Comparing G_w to the resultant transfer function between signals y_0 and y, one obtains formula (7.104).

$$G_r = \frac{G_w}{(1-G_w)G^*_{\text{inv}}G} \tag{7.104}$$

If the plant transfer function is given in the form (7.97) and the inverse is given by formula (7.99), then the difference $(1 - G_w)$ obtains the form (7.105) and quotient $G_w/(1-G_w)$ contains an integrating component, which is clearly visible in expression (7.106).

$$1-G_w(s) = 1-\frac{1}{(1+sT_w)^{n-m+1}} = \frac{s(c_1+c_2s+\dots c_{n-m+1}s^{n-m})}{(1+sT_w)^{n-m+1}}, \tag{7.105}$$

where: $(1+sT_w)^{n-m+1} = 1 + c_1s + c_2s^2 + \dots c_{n-m+1}s^{n-m+1}$.

$$\frac{G_w(s)}{1-G_w(s)} = \frac{1}{s(c_1+c_2s+\dots c_{n-m+1}s^{n-m})} \tag{7.106}$$

Putting (7.105) and (7.106) into (7.104) one obtains a pre-controller transfer function in the form (7.107).

$$G_r(s) = \frac{(1+sT_b)^{n-m}}{s(c_1+c_2s+\dots c_{n-m+1}s^{n-m})} \tag{7.107}$$

The coefficients depend on time constant T_w (or time constants T_{wi}) of the reference model (7.105). Thus, if they are chosen "wisely", then components of the numerator and denominator in (7.107) can be reduced. So, under friendly circumstances, one can obtain an extremely simplified form of the pre-controller (7.108).

$$G_r(s) = \frac{k_r}{s} \tag{7.108}$$

Recapitulating the above: the controller R is composed of two parts, i.e. the pre-controller G_r and model inverse G^*_{inv}. If (7.108) holds, then tuning the controller is simplified, because only a single parameter k_r has to be tuned.

The next important problem refers to the suppression of disturbances d. As results from Fig. 7.92 show the component of y created by disturbance d can be calculated using transfer function G_d in the form (7.109).

$$G_d(s) = \frac{y}{d} = \frac{G}{(1 + G_r G^*_{\text{inv}} G} \tag{7.109}$$

If G is given by (7.97), G^*_{inv} by (7.99), G_r by (7.104), then G_d is given in the form (7.110).

$$G_d(s) = G(s)[1 - G_w(s)] = \\ = \frac{(b_0 + b_1 s + \ldots + b_m s^m)s(c_1 + c_2 s + \ldots c_{n-m+1} s^{n-m})}{(a_0 + a_1 s + \ldots + a_n s^n)(1 + sT_w)^{n-m+1}} \tag{7.110}$$

The static gain k_d of that transfer function is equal to zero, because:

$$k_d = \lim_{s \to \infty} G_d(s) = 0 \tag{7.111}$$

Thus, the time-invariable components of disturbances d are perfectly damped, it is static damping (neglecting system transient state). Let us note that the above conclusion holds for proper plants. If plant G is an improper plant and contains, for example, the integrating part, as in expression (7.112), then a suitable controller G_r has to be designed in order to perfectly suppress (in the static meaning) the time-invariable components of disturbances.

$$G(s) = \frac{b_0 + b_1 s + \ldots + b_m s^m}{s(a_0 + a_1 s + \ldots + a_n s^n)} \tag{7.112}$$

The model inverse for the above plant transfer function is given by (7.113).

$$G^*_{\text{inv}}(s) = \frac{s(a_0 + a_1 s + \ldots + a_n s^n)}{(b_0 + b_1 s + \ldots + b_m s^m)(1 + sT_b)^{n-m+1}} \tag{7.113}$$

Let us assume that controller transfer function G_r is given by (7.114).

$$G_r(s) = \frac{k_{r0} + k_{r1} s}{s^2} \tag{7.114}$$

Putting (7.112), (7.113) and (7.114) into ((7.110)), one obtains G_d in the form (7.115).

$$G_d(s) = \frac{G(s)}{(1 + G_r(s) G^*_{\text{inv}}(s) G(s)} = \\ = \frac{(b_0 + b_1 s + \ldots + b_m s^m) s (1 + sT_b)^{n-m+1}}{(a_0 + a_1 s + \ldots + a_n s^n)[s^2(1 + sT_b)^{n-m+1} + k_{r1} s + k_{r0}]} \tag{7.115}$$

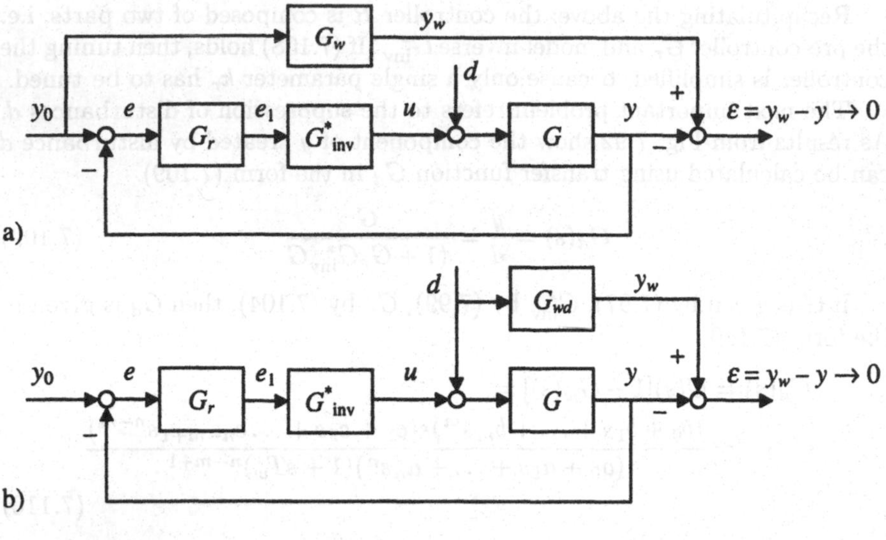

Fig. 7.93. The structure with reference model G_w representing desired tracking action (a) and a variant of this structure with reference model G_{wd} representing desired disturbance suppression (b)

Using final value theorem, it can easily be checked that:

$$k_d = \lim_{s \to \infty} G_d(s) = 0 \tag{7.116}$$

It means that the controller (7.114) damps perfectly the disturbances of a static "nature". Calculating the input/output transfer function G_{y0}, for G_r given by (7.114), one obtains formula (7.117).

$$G_{y0}(s) = \frac{y(s)}{y_0(s)} = \frac{G_r(s)G^*_{\text{inv}}(s)G(s)}{1 + G_r(s)G^*_{\text{inv}}(s)G(s)} = \frac{k_{r0} + k_{r1}s}{k_{r0} + k_{r1}s + s^2(1 + sT_b)^{n-m+1}} \tag{7.117}$$

Thus, the system stability as well as its control performance can be influenced by controller parameters k_{r0} and k_{r1} and the time constant of inverse T_b. The above adjustable parameters can be set as the result of a tuning process or by calculations made on the basis of structures shown in Fig. 7.93. It is obvious that reference transfer functions G_w and G_{wd}, see Fig. 7.93, have to be realizable.

Let us note that for a plant with the integrating component (7.112) the numerator of controller transfer function $R = G_r G^*_{\text{inv}}$ – formula (7.118) – has not got a root equal to zero. So, the unstable pole of plant ($s_1 = 0$), formula (7.112), is not reduced. Reduction of that type is forbidden (Morari 1989).

$$R(s) = G_r(s)G^*_{\text{inv}}(s) = \frac{(k_{r0} + k_{r1}s)(a_0 + a_1 s + \ldots + a_n s^n)}{s(b_0 + b_1 s + \ldots + b_m s^m)(1 + sT_b)^{n-m+1}} \tag{7.118}$$

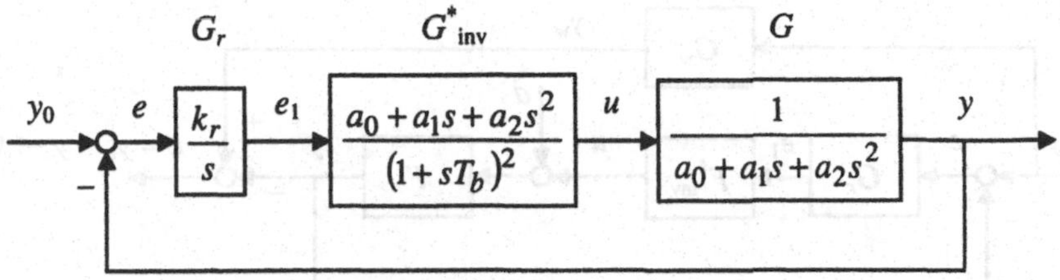

Fig. 7.94. The control system with model inverse for the second order plant

If the poles of a plant are located in the right half-plane of complex variable "s" (plant is unstable), then the plant can be stabilized by an additional stabilizing feedback loop. Some advice on how to stabilize the plant has been discussed in Section 7.3.3.4. After stabilization of the plant, the controller with an inverse can be designed. Analysis of InvMC structures leads to interesting observations. Example 7.3.3.5.1 discusses one such observation.

Example 7.3.3.5.1
Let the plant be either an oscillatory or inertial system of second order (7.119).

$$G(s) = \frac{1}{a_0 + a_1 s + a_2 s^2} \tag{7.119}$$

The control system with model inverse designed on the basis of rules discussed in the current chapter obtains the form shown in Fig. 7.94.

The controller R for the considered plant is represented by transfer function (7.120).

$$R(s) = G_r(s)G^*_{\text{inv}}(s) = k_r \left[\frac{a_1}{(1+sT_b)^2} + \frac{a_0}{s(1+sT_b)^2} + \frac{a_2 s}{(1+sT_b)^2} \right] \tag{7.120}$$

It is a typical, widely used and commonly known PID controller (incidently, the PID controller has been called "the draught horse for control engineering" (Franklin 1986)). If T_b tends to zero, then controller transfer function (7.120) tends to a standard form for the PID controller. Therefore the PID can be treated as an optimal controller for the second order plants. The above gives us a convincing explanation of the popularity of applications of PID controllers in industry. Similar reasoning for first order plants yields the conclusion that the PI controller ($a_2 = 0$) is an optimal one for those plants.

◊ ◊ ◊

The non-jump version of a pre-controller described by transfer function (7.121) is advised for discrete InvMC structures.

$$G_r(z) = \frac{e_1(z)}{e(z)} = \frac{k_r}{z-1} \tag{7.121}$$

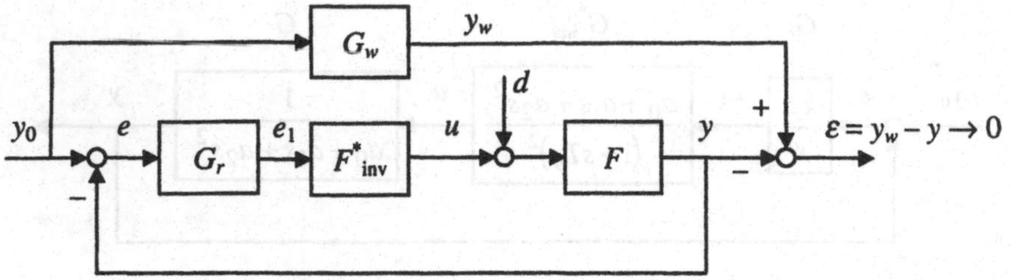

Fig. 7.95. The InvMC structure for non-linear plants

The difference equation (7.122) corresponds to the above transfer function form.

$$e_1(k) = k_r e(k-1) + e_1(k-1) \tag{7.122}$$

The impetuous, even step variations of reference signal y_0 do not generate step variations of controller output signal $e_1(k)$, because transfer function (7.121) describes a discrete version of integration. Due to such a property of the pre-controller, the amplitudes of control signal u, generated by the next element of controller R, i.e. inverse, are not too large for the actuators (let us note that the inverse usually contains differentiating components). Thus, the saturation state of the actuator can be avoided. If saturation states are eliminated, then loss of system stability seems to be less probable.

The above deep analysis of properties of InvMC structure inclines us to the conclusion that the considered structure is surely advantageous, if linear plants are taken into consideration. Using the InvMC concept the quasi-perfect controllers can be designed easily. Now, let us discuss the problem of control for non-linear plants in the context of the InvMC structure. The transfer function models of operations have to be substituted by non-linear operators defining input/output mappings.

$G \to F$ – plant
$G^* \to F^*$ – model of plant
$G^*_{inv} \to F^*_{inv}$ – inverse of plant model
$R \to R_n$ – controller composed of pre-controller and inverse

The InvMC structure for a non-linear plant is shown in Fig. 7.95. As in previous, linear cases, so also now, for non-linear plants, one can not find the perfect inverse imitating the reference model $G_w = 1$ – see Fig. 7.96.

Due to careful choice of inverse structure (fuzzy or neural) and accurate tuning of its parameters, one can ensure more or less exact compensation of non-linear, either static or dynamic, phenomena directly characterizing the model F^* and indirectly the plant F. The efficient compensation means that the relation between plant output y and input of inverse e_1 (see Fig. 7.95) can be approximately treated as a linear one. This will be assumed in the multi-inertial form (7.123) or (7.124).

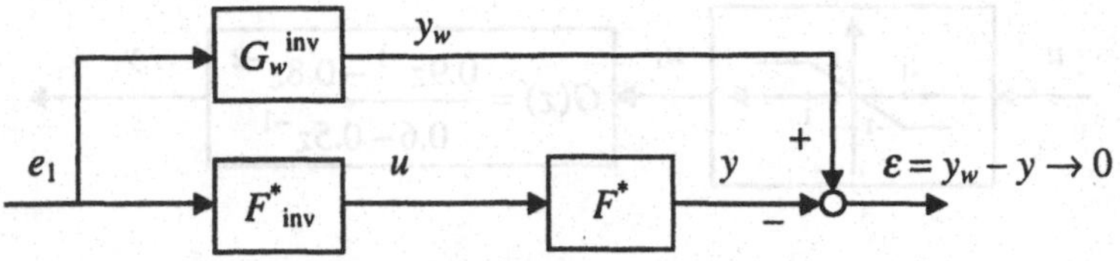

Fig. 7.96. The tuning of non-linear plant inverse F^*_{inv} in a system with reference model G_w^{inv}

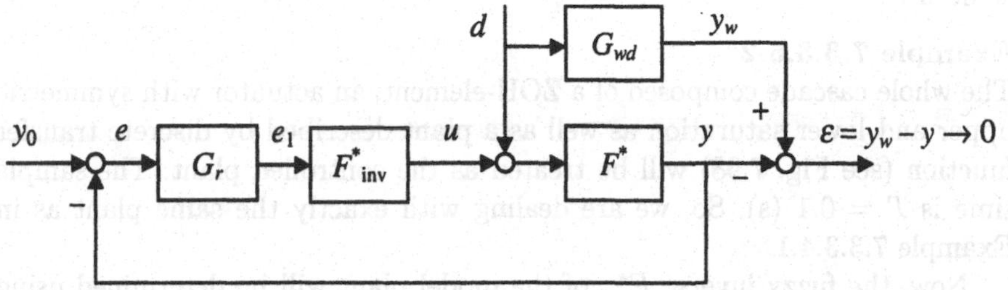

Fig. 7.97. The application of a structure with the reference model of disturbance rejection G_{wd} to the tuning of controller G_r

$$\frac{y(s)}{e_1(s)} \cong \frac{1}{(1+sT_b)^{n-m}} \tag{7.123}$$

$$\frac{y(s)}{e_1(s)} \cong \frac{1}{(1+sT_{b1}) \cdot \ldots \cdot (1+sT_{b(n-m)})} \tag{7.124}$$

Because static gains of both the above forms are equal to 1, after the transient state, $y = e_1$. The assumption (7.123) or (7.124) means that controller G_r as well as reference model G_w can be linear too. Thus, the designing rules obtained for linear plants can be applied for designing controller G_r, if the plant is non-linear. Of course, other methods for the synthesis of linear controllers can also be utilized for determining controller G_r. We must keep in mind that resultant properties of the cascade composed of F^*_{inv} and F can be treated as linear, if inverse F^*_{inv} does not generate too-large signals u, i.e. signals responsible for saturation of an actuator.

The compensation of disturbance d can not be treated as a linear operation, because d directly influences plant input, omitting the inverse – see Fig. 7.95. Exactly the same statement applies to the IMC structure. However, if one knows the exact model F^* of plant, then the plant model can be inserted into the InvMC structure (Fig. 7.95) instead of plant F. After that replacement one can try to suppress disturbance d effectively, by tuning controller G_r on the basis of the scheme shown in Fig. 7.97 (inverse F^*_{inv} can be tuned in a system shown in Fig. 7.96).

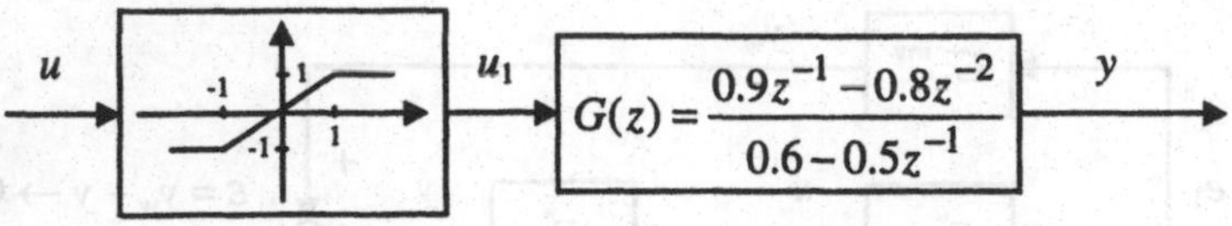

Fig. 7.98. The plant and actuator

Let us consider the example of fuzzy control on the basis of the InvMC structure.

Example 7.3.3.5.2
The whole cascade composed of a ZOH-element, an actuator with symmetric upper and lower saturation as well as a plant described by discrete transfer function (see Fig. 7.98) will be treated as the controlled plant. The sample time is $T = 0.1$ (s). So, we are dealing with exactly the same plant as in Example 7.3.3.4.1.

Now, the fuzzy inverse F^*_{inv} of the model plant will be determined using rule bases defined by Tables 7.13 and 7.14. The membership functions are shown in Fig. 7.99. The operation *AND* is realized using the *PROD* operator.

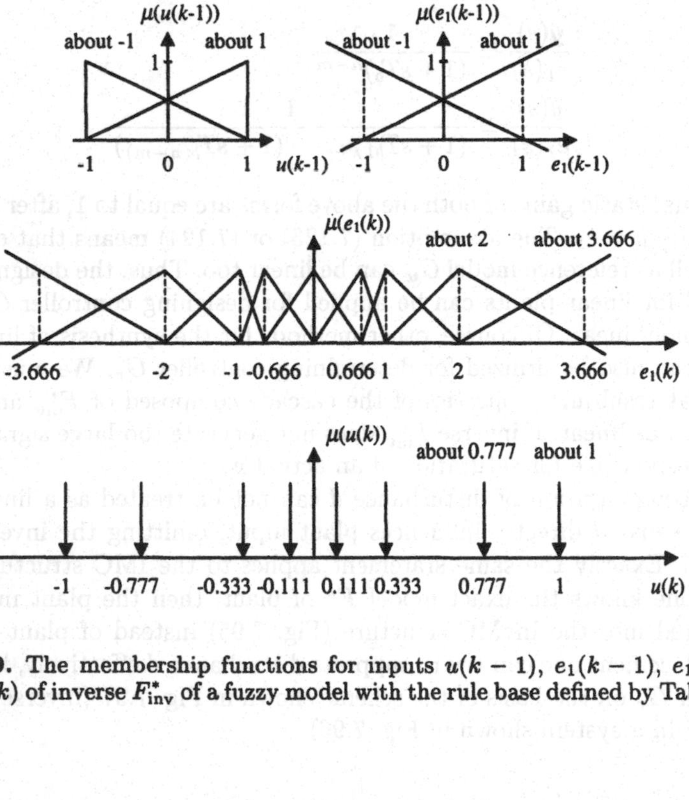

Fig. 7.99. The membership functions for inputs $u(k-1)$, $e_1(k-1)$, $e_1(k)$ and output $u(k)$ of inverse F^*_{inv} of a fuzzy model with the rule base defined by Tables 7.13 and 7.14

Table 7.13. The rules $R1$ – $R16$ of model inverse F^*_{inv}

$e_1(k)$ / $e_1(k\text{-}1)$	about -3.6666	about -2	about -1	about -0.6666	about +0.6666	about +1	about +2	about +3.6666
about -1	about -1	about -1	about -1	about -0.7777	about +0.1111	about +0.3333	about +1	about +1
about +1	about -1	about -1	about -1	about -1	about -1	about -0.7777	about -0.1111	about +1

$u(k)$

$$R1 - R16: \; IF\ [u(k-1) = \text{about } -1]\ AND\ [e_1(k-1) = \text{about } \ldots]$$
$$AND\ [e_1(k) = \text{about } \ldots]\ THEN\ [u(k) = \text{about } \ldots]$$

Table 7.14. The rules $R17$ – $R32$ of model inverse F^*_{inv}

$e_1(k)$ / $e_1(k\text{-}1)$	about -3.6666	about -2	about -1	about -0.6666	about 0.6666	about +1	about +2	about +3.6666
about -1	about -1	about +0.1111	about +0.7777	about +1	about +1	about +1	about +1	about +1
about +1	about -1	about -1	about -0.3333	about -0.1111	about +0.7777	about +1	about +1	about +1

$u(k)$

$$R17 - R32: \; IF\ [u(k-1) = \text{about } +1]\ AND\ [e_1(k-1) = \text{about } \ldots]$$
$$AND\ [e_1(k) = \text{about } \ldots]\ THEN\ [u(k) = \text{about } \ldots]$$

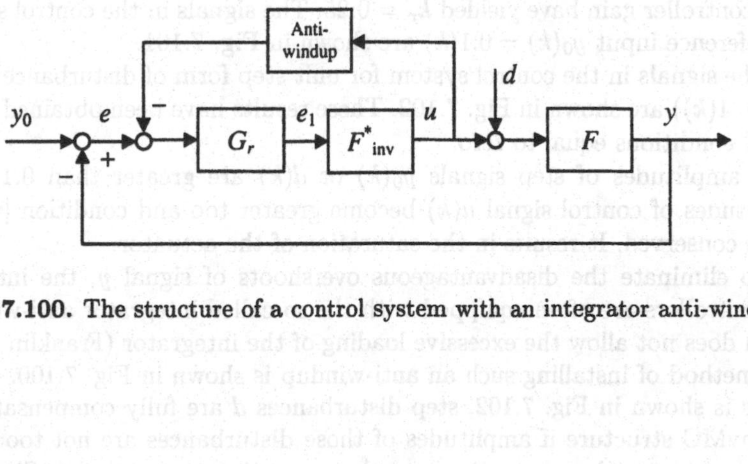

Fig. 7.100. The structure of a control system with an integrator anti-windup

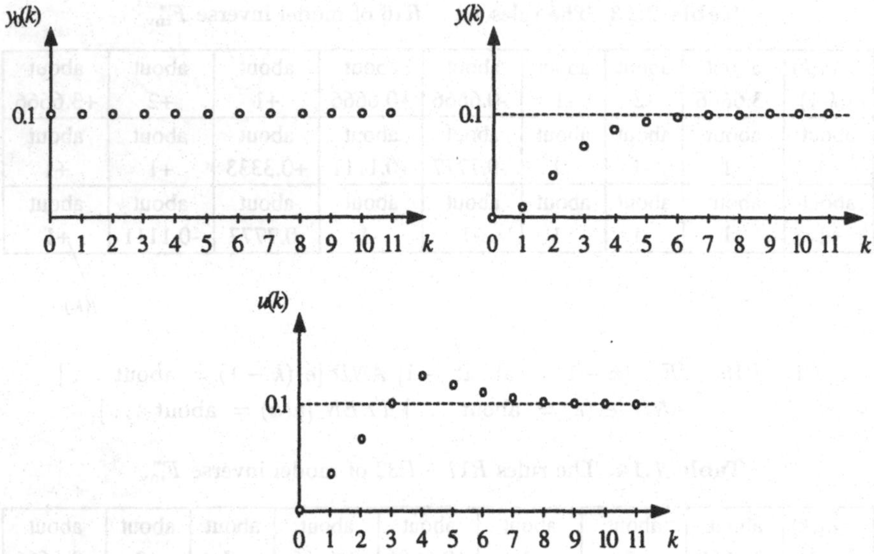

Fig. 7.101. The signals in the control system for reference signal $y_0(k) = 0.1(k)$

The choice of pre-controller G_r (see Fig. 7.100) is also included in the synthesis of control structure InvMC. The pre-controller has been chosen in the form (7.125). So, the recommendation (7.101) has been satisfied.

$$G_r(z) = \frac{e_1(z)}{e(z)} = \frac{k_r}{z-1} \tag{7.125}$$

Experimental examinations of system performance aimed at choosing the best controller gain have yielded $k_r = 0.25$. The signals in the control system for reference input $y_0(k) = 0.1(k)$ are shown in Fig. 7.101.

The signals in the control system for unit step form of disturbance signal ($d = 0.1(k)$) are shown in Fig. 7.102. These results have been obtained under initial conditions equal to zero.

If amplitudes of step signals $y_0(k)$ or $d(k)$ are greater than 0.1, then amplitudes of control signal $u(k)$ become greater too and condition $|u| \leq 1$ is not conserved. It results in the saturation of the actuator.

To eliminate the disadvantageous overshoots of signal y, the integrating controllers are often equipped with the so-called integrator anti-windup, which does not allow the excessive loading of the integrator (Franklin 1986). The method of installing such an anti-windup is shown in Fig. 7.100.

As is shown in Fig. 7.102, step disturbances d are fully compensated by the InvMC structure if amplitudes of those disturbances are not too great, i.e. the actuator does not enter into the saturation state $|u| \leq 1$. Similarly,

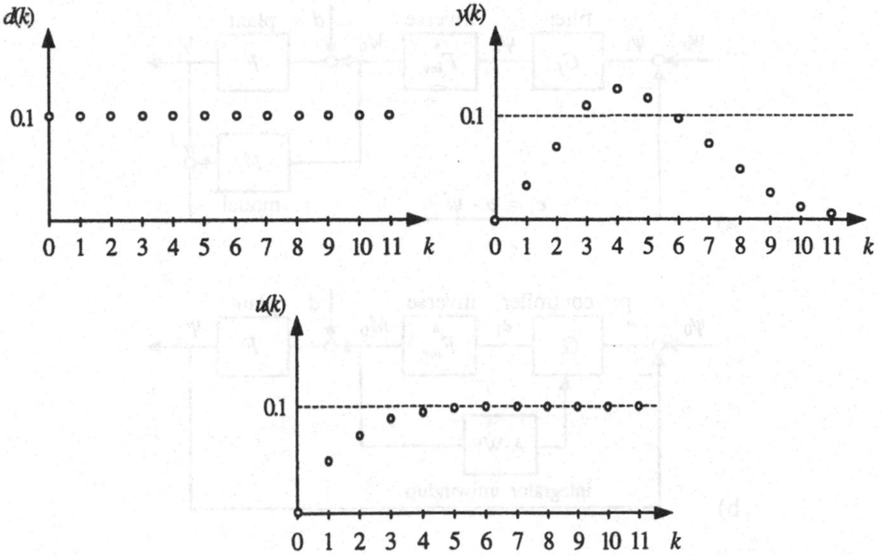

Fig. 7.102. The signals in the InvMC structure for step disturbance signal $d(k) = 0.1(k)$

for the same reason, the system follows reference signal $|y_0| \le 1$. Too-great gains of pre-controller k_r destabilize the system.

◊ ◊ ◊

A comparison of the results obtained during control of the same plant using the InvMC structure and the IMC structure is presented in Example 7.3.3.5.3.

Example 7.3.3.5.3

The InvMC structure has been applied to the control of an underwater vehicle's course. The IMC structure as well as the InvMC structure is shown in Fig. 7.103.

The plant F, its fuzzy model F^* and inverse F^*_{inv} are presented with details in Example 7.3.3.4.2. The controller of the InvMC structure is composed of a pre-controller G_r and an inverted model of the plant – see Fig. 7.103b. The formula (7.126) determines the pre-controller transfer function.

$$G_c(z) = \frac{e_1(z)}{e_2(z)} = \frac{k_c}{z-1} \tag{7.126}$$

Many simulation experiments prove that the best control performance is obtained if $k_c = 3$. The comparison of both structures during tracking of the rectangular reference signal is shown in Fig. 7.104.

The settling time is shorter for the InvMC structure. The responses of the InvMC structure to the rectangular wave are of identical shape – it refers

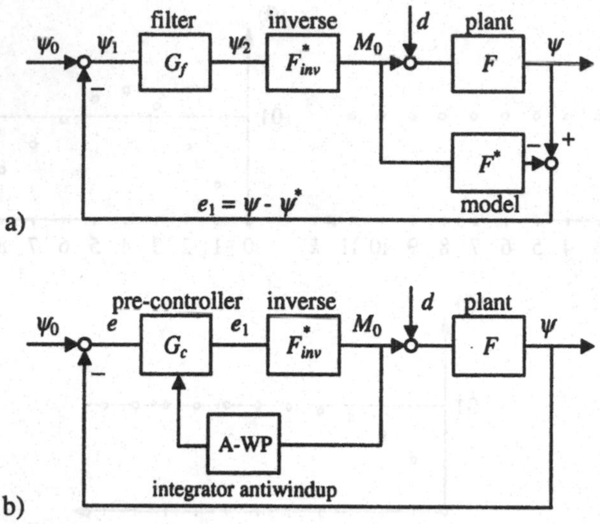

Fig. 7.103. Both structures under comparison: IMC (a) and InvMC (b)

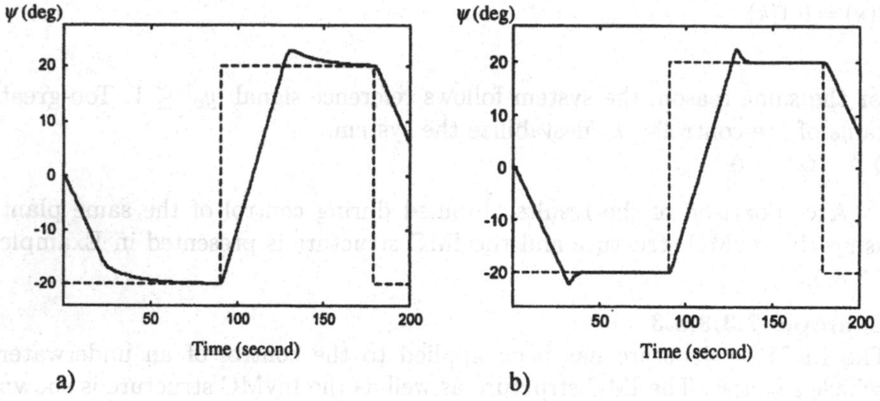

Fig. 7.104. Comparison of the responses $\psi(t)$ of the IMC structure (a) and the InvMC structure (b) to reference signal $\psi_0(t)$ in the form of a rectangular wave (dashed line)

to the positive and negative amplitudes of the reference signal. For the IMC structure we can observe that response $\psi(t)$ during tracking of the positive amplitude of the rectangular wave differs to the response obtained for the negative amplitude of the reference signal. In the first case we can observe the overshoot whereas in the second case the overshoot disappears.

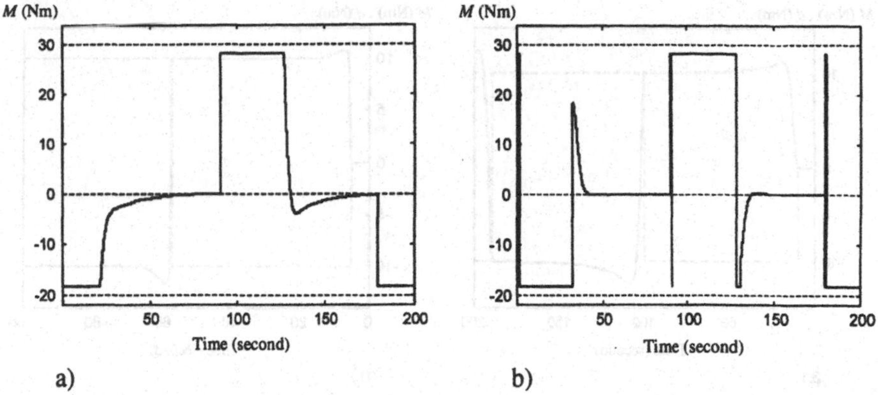

Fig. 7.105. Comparison of control signals M generated by IMC (a) and InvMC (b) structures for reference signal $\psi_0(t)$ as in Fig. 7.104

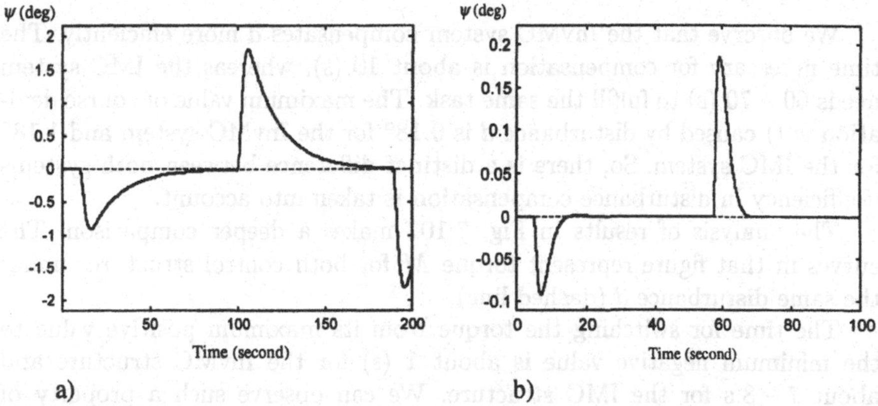

Fig. 7.106. Compensation of step disturbance d in IMC (a) and InvMC (b) structures

The torques M generated by propellers for both control structures are shown in Fig. 7.105. These curves have been obtained for reference signal $\psi_0(t)$ as in Fig. 7.104.

It can be observed that the curve representing the IMC structure torque is more gentle and smooth compared to the curve representing the InvMC structure. A large part of the reference $\psi_0(t)$ signal consists of the maximum or minimum available values of torque M which results in transient states caused by steps of reference signals being relatively short.

The responses $\psi(t)$ to the "step" disturbance d (-10 (Nm), $+10$ (Nm)) acting on the plant input are shown in Fig. 7.106.

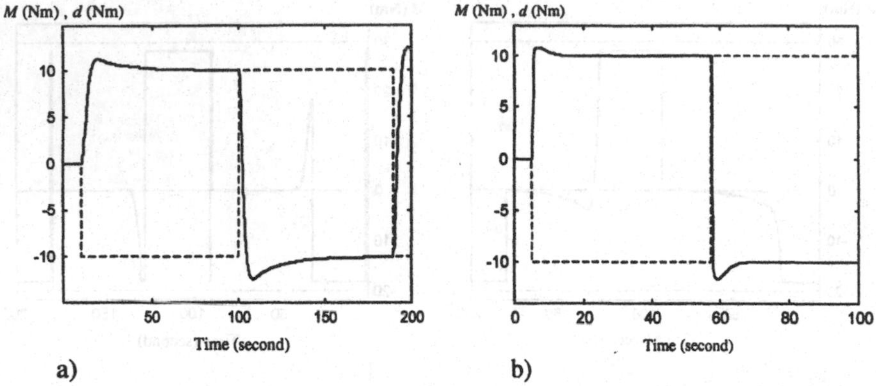

Fig. 7.107. The torque M generated in the IMC structure (a) and in the InvMC structure (b) during compensation of disturbance d (dashed line)

We observe that the InvMC system compensates d more efficiently. The time necessary for compensation is about 10 (s), whereas the IMC system needs 60 – 70 (s) to fulfill the same task. The maximum value of course deviation $\psi(t)$ caused by disturbance d is 0.18° for the InvMC system and 1.18° for the IMC system. So, there is a distinct difference between both systems if efficiency in disturbance compensation is taken into account.

The analysis of results in Fig. 7.107 makes a deeper comparison. The curves in that figure represent torque M for both control structures under the same disturbance d (dashed line).

The time for switching the torque from its maximum positive value to the minimum negative value is about 1 (s) for the InvMC structure and about 7 – 8 s for the IMC structure. We can observe such a property of both structures, that the greater part of the period of reference signal $\psi_0(t)$ consists of maximum or minimum available values of torque M which means that transient states caused by step disturbance signals are relatively short and course $\psi(t)$ deviations are eliminated quickly.

Thus, judging the above results, one can draw the conclusion that the InvMC structure is better for control of underwater vehicle course $\psi(t)$ – it is faster, because it does not contain the inertia of the plant model. Of course, it has not been proved that the above conclusion can be generalized. So, for other control tasks the IMC structure can be the better one. Results from experiments show that the InvMC structure considered for course control is stable for the following parameter ranges of the plant (vehicle) transfer function:

$$0.015 \leq k_p \leq 0.065\,, \quad 0.5 \leq T_p \leq 11\,.$$

Good results of application of the inverse model of a non-linear plant (distillation column) are reported in (Ławryńczuk 2000)

◊ ◊ ◊

7.3.3.6 Adaptive fuzzy control

The methods for identification of fuzzy models of plants as well as ways of determining model inverse on the basis of measurement samples representing plant input and output have been discussed in Section 6.3.3. If plant parameters vary, then on-line identification methods can be applied. On that basis the parameters of the fuzzy controller or its separated parts (like model F^* or the model inverse F^*_{inv} treated as elements of controller R – Sections 7.3.3.4, 7.3.3.5) can be incessantly tuned. Such adjustment of control action to current parameters of a plant is called "adaptive control" (Åström 1989).

If changes of plant parameters are very slow or they occur rather sporadically, then identification of the plant as well as tuning of the controller can be done periodically. The adaptation of elements of the controller for a fuzzy IMC structure (Section 7.3.3.4) can be realized on the basis of the scheme in Fig. 7.108. The choice of parameters of filter G_f ought to guarantee system stability for all possible parameters of plant F. From a practical point of view it means that time constants of the inertial filter should be enlarged.

The adaptive control scheme for an InvMC structure (Section 7.3.3.5) is shown in Fig. 7.109. The permanent actualization of parameters of fuzzy model F^* and its fuzzy inverse F^*_{inv} improves the quality of linearization of operators $F^*_{\text{inv}}F$ and $F^*_{\text{inv}}F^*$ included in structures shown in Figs. 7.108 and 7.109, if plant parameters drift. Due to the adaptation described above, one can neglect the actualization of parameters of filter G_f and pre-controller G_r. Accuracy of identification of parameters of model F^* and its inverse F^*_{inv} is not high especially for quick changes of plant parameters. That is why the time constants of filter G_f in IMC structures should be enlarged and the gain of controller G_r in InvMC structures ought to be decreased in comparison to the advised values of the mentioned parameters for plant F with fixed parameters. Following that advice makes the control system more robust.

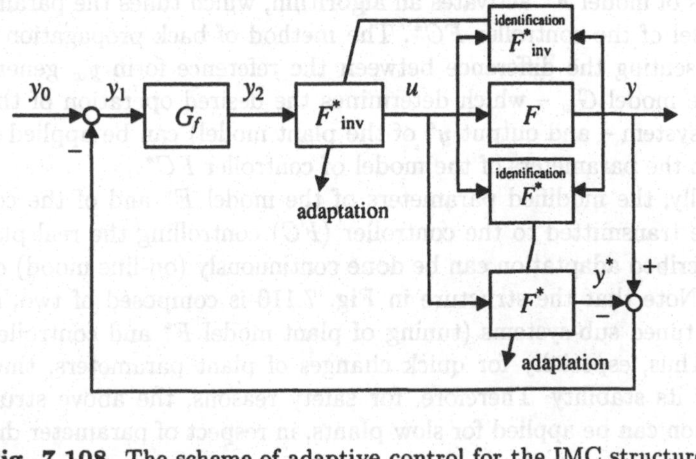

Fig. 7.108. The scheme of adaptive control for the IMC structure

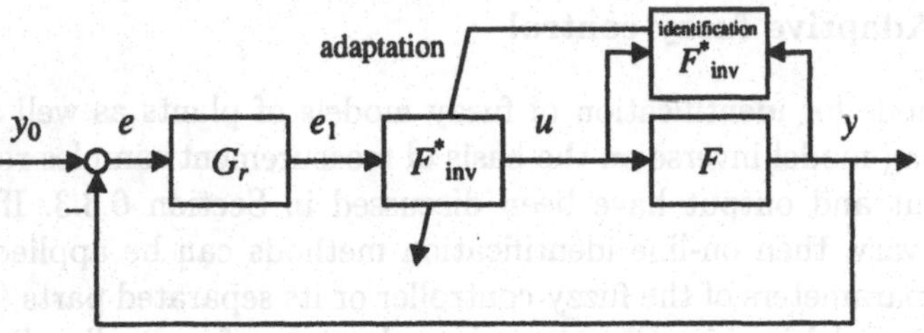

Fig. 7.109. The scheme of adaptive control for an InvMC structure

The adaptive IMC and InvMC structures shown in Fig. 7.108 and 7.109 can be designed in more complicated forms, if parameters of filter G_f and controller G_r are also tuned. On the other hand, there are serious difficulties in system stabilization, if the number of adapted parameters is too great (Åström 1989, Anderson 1986). Increasing the number of loops for adaptation of parameters results in many difficulties in analytical calculations aimed at the determination of the system stability region. Thus, the number of adapted parameters ought to be minimized.

Besides IMC and InvMC adaptive structures one can use a general adaptive structure, where the fuzzy controller (FC) obtains any form (for example, a fuzzy PID form). An example of the structure for the adaptation of a controller in such a case is shown in Fig. 7.110. The lower part of that structure represents a classic loop of the control system, composed of plant F and controller FC. The upper part of that structure represents the model of a real control system containing model F^* of the plant and model FC^* of the controller. The plant model F^* is tuned on the basis of current measurement samples of the input signal u and output signal y. The up-to-date parameters are transmitted to model F^* located in the model of the control system (adaptation of parameters). The modification (adaptation) of the parameters of model F^* activates an algorithm, which tunes the parameters of the model of the controller FC^*. The method of back propagation of error ε (representing the difference between the reference form y_w generated by reference model G_w – which determines the desired operation of the whole control system – and output y^* of the plant model) can be applied in order to adapt the parameters of the model of controller FC^*.

Finally, the modified parameters of the model F^* and of the controller FC^* are transmitted to the controller (FC) controlling the real plant (F). The described adaptation can be done continuously (on-line mood) or periodically. Note that the structure in Fig. 7.110 is composed of two, simultaneously tuned sub-systems (tuning of plant model F^* and controller model FC^*). Thus, especially for quick changes of plant parameters, the system can lose its stability. Therefore, for safety reasons, the above structure of adaptation can be applied for slow plants, in respect of parameter drift or in

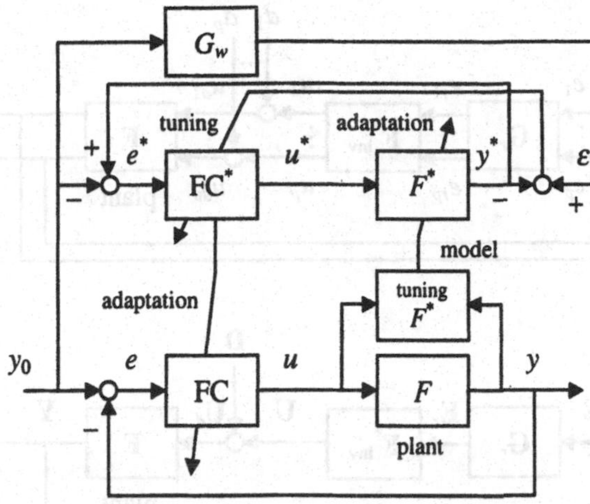

Fig. 7.110. The adaptive system with fuzzy controller FC

the case of sporadic changes of plant parameters. Cargo vessels are a good example of such plants, which distinguishably change the dynamic parameters sporadically, during loading in harbor. After departure from the harbor, the ship parameters drift only very slowly, because of consumption of fuel, drinkable water, etc.

Apart from three structures discussed in the current section, one can apply other adaptive structures containing fuzzy controllers. The examples of fuzzy, adaptive control can be found in (Brown 1994; Fischle 1997; Koch 1996; Serac 1996; Sousa 1995; Wang 1994a).

7.3.3.7 Multivariable fuzzy control (MIMO)

The block diagram of a multivariable InvMC control structure is shown in Fig. 7.111. The vectors and operators in Fig. 7.111 are defined as follows:

$$\begin{aligned}
\mathbf{Y}_0 &= [y_{01}, \ldots, y_{0p}]^{\mathrm{T}}, \\
\mathbf{E}_0 &= [e_1, \ldots, e_p]^{\mathrm{T}}, \\
\mathbf{E}_r &= [e_{r1}, \ldots, e_{rp}]^{\mathrm{T}}, \\
\mathbf{U} &= [u_1, \ldots, u_p]^{\mathrm{T}}, \\
\mathbf{Y} &= [y_1, \ldots, y_p]^{\mathrm{T}}, \\
\mathbf{D} &= [d_1, \ldots, d_p]^{\mathrm{T}}, \\
\mathbf{U}_d &= [u_{d1}, \ldots, u_{dp}]^{\mathrm{T}}, \\
\mathbf{G}_r &= \begin{bmatrix} G_{r11} & \ldots & G_{r1p} \\ \vdots & & \vdots \\ G_{rp1} & \ldots & G_{rpp} \end{bmatrix}
\end{aligned} \tag{7.127}$$

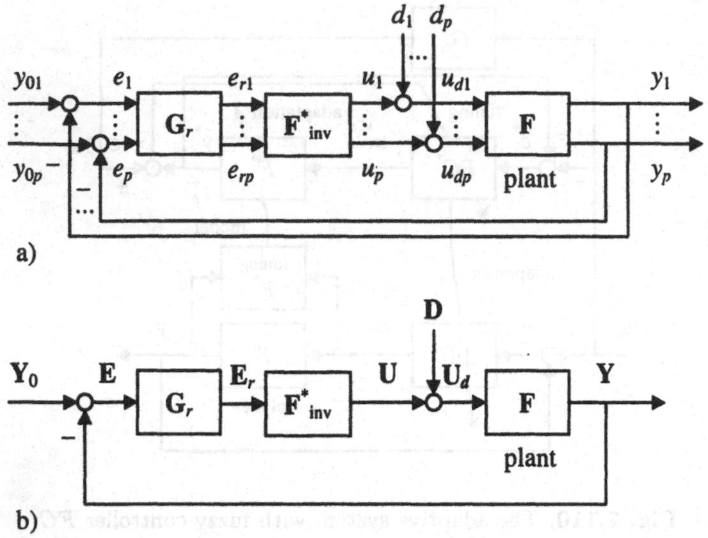

Fig. 7.111. Block diagram of multivariable (MIMO) control system with model inverse (a) and its simplified version (b)

The multivariable, non-linear operator representing model $\mathbf{F}^*$ of plant $\mathbf{F}$ describes the input/output mapping of the form (7.128).

$$\mathbf{Y}^*(k) = \mathbf{F}^*[\mathbf{U}_d(k-n+m), \dots, \mathbf{U}_d(k-n), \mathbf{Y}(k-1), \dots, \mathbf{Y}(k-n)] \quad (7.128)$$

Fig. 7.111 and associated relations (7.127) have been prepared for the same dimensions (equal to p) of the all vectors . However, there are many real plants where the number of inputs is different from the number of outputs. Designing a control system under the mentioned circumstances can be considerably more complicated.

The methods discussed in Section 7.3.3.1 can be successfully applied in order to determine the fuzzy model of plant $\mathbf{F}^*$. Assuming $\mathbf{U}_d = \mathbf{U}$, one can tune the plant model using the scheme shown in Fig. 7.112.

Similarly, the multivariable versions of methods discussed in Section 7.3.3.2 can be applied to determine the model inverse $\mathbf{F}^*_{\text{inv}}$. An example is shown in Fig. 7.113.

The input/output mapping realized by the inverse is determined by expression (7.129):

$$\mathbf{U}(k) = \mathbf{F}^*_{\text{inv}}[\mathbf{E}_r(k), \dots, \mathbf{E}_r(k-n), \mathbf{U}(k-1), \dots, \mathbf{U}(k-m)]\,. \quad (7.129)$$

The reference model for the inverse can be chosen in the form of a matrix of transfer functions (7.130).

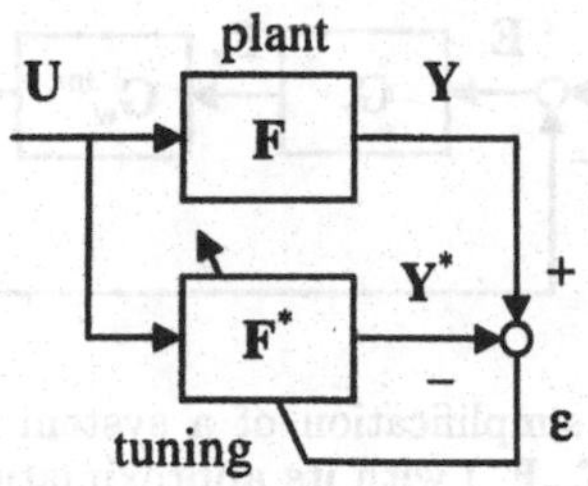

Fig. 7.112. The system for tuning plant model $\mathbf{F}^*$

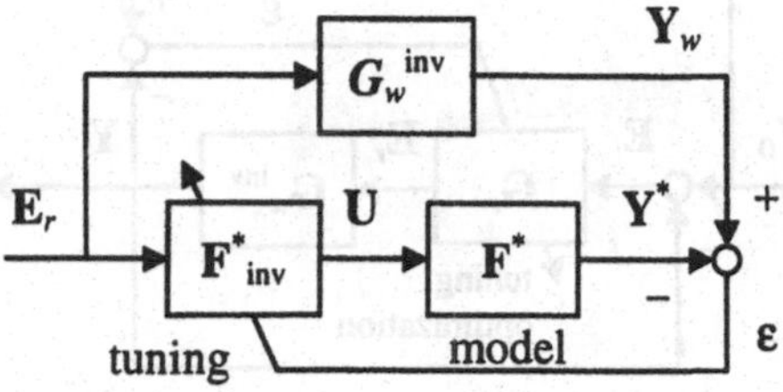

Fig. 7.113. Tuning inverse $\mathbf{F}^*_{\text{inv}}$ on the basis of a structure with reference model G_w^{inv}

$$\mathbf{G}_w^{\text{inv}} = \begin{bmatrix} G_{w11}^{\text{inv}} & 0 & \dots & 0 \\ 0 & G_{w22}^{\text{inv}} & \dots & 0 \\ \vdots & & & \vdots \\ 0 & \dots & 0 & G_{wpp}^{\text{inv}} \end{bmatrix} \tag{7.130}$$

The component transfer functions G_{wii}^{inv} can be represented by inertial elements of $(n-m)$ orders. The particular time constants of component transfer functions can be identical or different, depending on the designer's choice. The static gains of G_{wii}^{inv} should be equal to 1.

Thus, for the continuous version one can choose:

$$G_{wii}^{\text{inv}}(s) = \frac{1}{(1+sT_i)^{n-m}},$$

whereas for the discrete version:

$$G_{wii}^{\text{inv}}(z) = \left(\frac{1-c_i}{z-c_i}\right)^{n-m}, \tag{7.131}$$

where: $c_i = e^{-T/T_i}$, T – sample time, T_i – time constant, $i = 1, \dots, p$.

The identification of the multivariable inverse and its tuning is considerably more complicated compared to the same task for SISO systems (see also the note in Section 7.3.3.2). If a model $\mathbf{F}^*$ of plant $\mathbf{F}$ and its inverse $\mathbf{F}^*_{\text{inv}}$

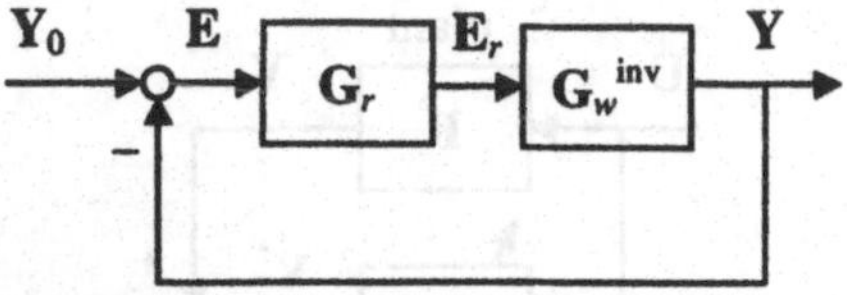

Fig. 7.114. The result of simplification of a system in Fig. 7.111b obtained by substitution of mapping $(\mathbf{F}^*_{\text{inv}}\mathbf{F}^*)$ with its approximation $\mathbf{G}_w^{\text{inv}}$

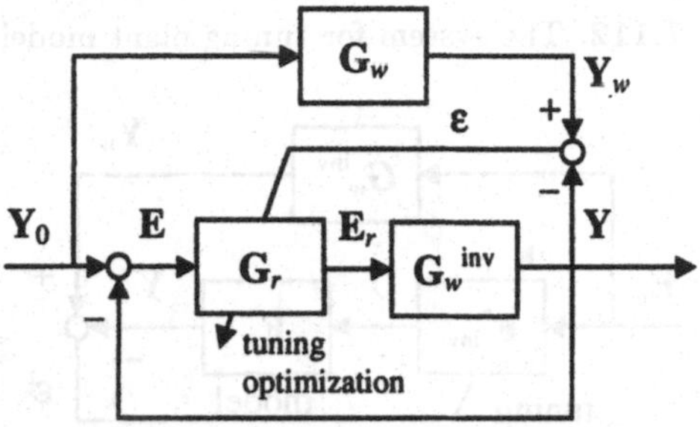

Fig. 7.115. The design of multivariable controller $\mathbf{G}_r$ on the basis of a structure with reference model $\mathbf{G}_w$

are sufficiently accurate, then series connection of $\mathbf{F}^*_{\text{inv}}$ and $\mathbf{F}^*$ (Fig. 7.113) can be approximated by linear reference model $\mathbf{G}_w^{\text{inv}}$. Assuming sufficiently accurate approximation in the meaning given above, one can simplify the structure in Fig. 7.111b to the form shown in Fig. 7.114.

The control system in Fig. 7.114 is a linear one. To determine controller $\mathbf{G}_r$ one can use well-known methods of designing controllers for linear multi-dimensional control systems. It can be, for example, a method aimed at the synthesis of a robust controller (Weinmann 1991), the method of control on the basis of a structure with reference model $\mathbf{G}_w$ (Fig. 7.115) and so on.

To avoid misunderstanding, one should realize that reference model $\mathbf{G}_w$ in Fig. 7.115 determines the desired dynamic properties of the whole control system, whereas reference model $\mathbf{G}_w^{\text{inv}}$ in Fig. 7.113 represents the chosen basis (a series connection $F^*_{\text{inv}}F^*$) for determination of plant model inverse $\mathbf{F}^*_{\text{inv}}$. The reference model for control system $\mathbf{G}_w$ is usually chosen as a diagonal matrix containing inertial elements $\mathbf{G}_{wii}$ (7.132). The static gains of those elements are equal to 1 (as for $\mathbf{G}_w^{\text{inv}}$) but inertia orders assigned to elements of $\mathbf{G}_w$ are higher compared to those assigned to $\mathbf{G}_w^{\text{inv}}$.

$$\mathbf{G}_w = \begin{bmatrix} G_{w11} & 0 & \dots & 0 \\ 0 & G_{w22} & \dots & 0 \\ \vdots & & & \vdots \\ 0 & \dots & 0 & G_{wpp} \end{bmatrix}$$

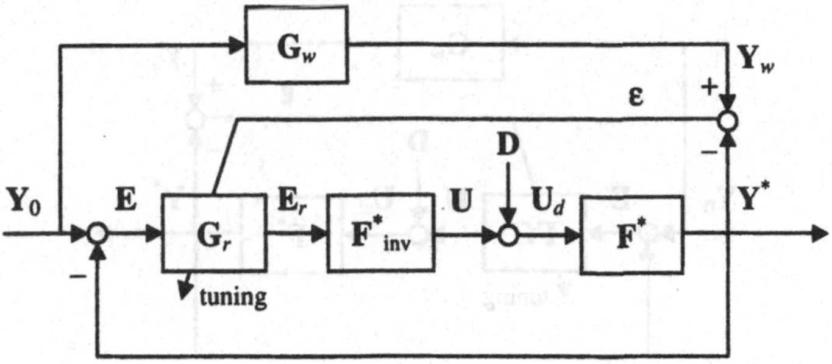

Fig. 7.116. The complete, non-linear structure with reference model $\mathbf{G}_w$ enabling the final tuning of multivariable controller $\mathbf{G}_r$

Thus, depending on the continuous or discrete version one can assume respectively:

$$G_{wjj}(s) = \frac{1}{(1+sT_{wj})^{n-m+1}},$$
$$G_{wjj}(z) = \left(\frac{1-c_j}{z-c_j}\right)^{n-m+1}. \tag{7.132}$$

where: $c_j = e^{-T/T_{wj}}$, T – sample time, T_{wj} - time constant, $j = 1, \ldots, p$.

The controller of a system shown in Fig. 7.115 can be determined from equation (7.133):

$$\mathbf{G}_r\mathbf{G}_w^{\text{inv}}(\mathbf{I} - \mathbf{G}_r\mathbf{G}_w^{\text{inv}})^{-1} = \mathbf{G}_w, \tag{7.133}$$

where: $\mathbf{I}$ – identity matrix of order p.

The general form of formulae defining component transfer functions G_{rij} of multivariable controller $\mathbf{G}_r$ (7.127) can be obtained from equation (7.133). Furthermore, the obtained transfer functions G_{rij} can be treated as the basis for an initial choice of approximate values of controller parameters. Let us note that a simplified control structure (Fig. 7.114 and 7.115) neglects disturbances $\mathbf{D}$ influencing the input of non-linear plant $\mathbf{F}$ and, as a linear structure, does not represent the constraints in the operation of actuators. Thus, after the preliminary choice of controller parameters, they have to be tuned once again in the non-simplified structure shown in Fig. 7.116.

The fuzzy model $\mathbf{F}^*$ and its inverse $\mathbf{F}^*_{\text{inv}}$ can be represented by a neuro-fuzzy network, and controller $\mathbf{G}_r$ by a neural network of a dynamic type. The method of back propagation of error ε can be used for controller tuning.

If one wants to apply the other structure of a fuzzy controller (**FC**), then the form of that structure has to be assumed in advance. The parameters assigned to the assumed controller **FC** can be tuned using the structure with reference model shown in Fig. 7.117.

Let us consider the multivariable IMC structure shown in Fig. 7.118. If one is able to determine a sufficiently accurate model $\mathbf{F}^*$ of plant and its

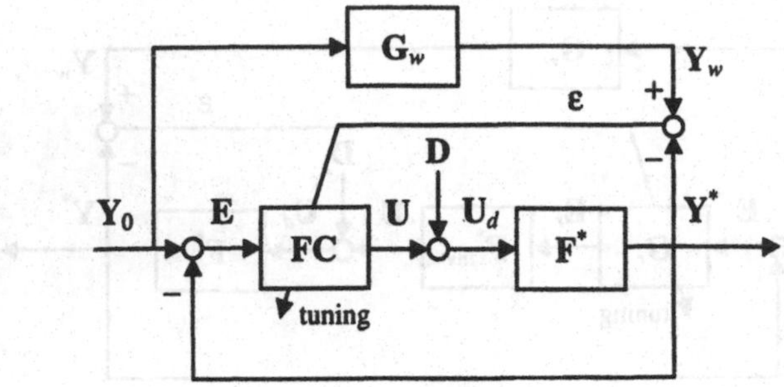

Fig. 7.117. Structure with reference model $\mathbf{G}_w$ for tuning fuzzy controller **FC** representing any controller structure assumed in advance

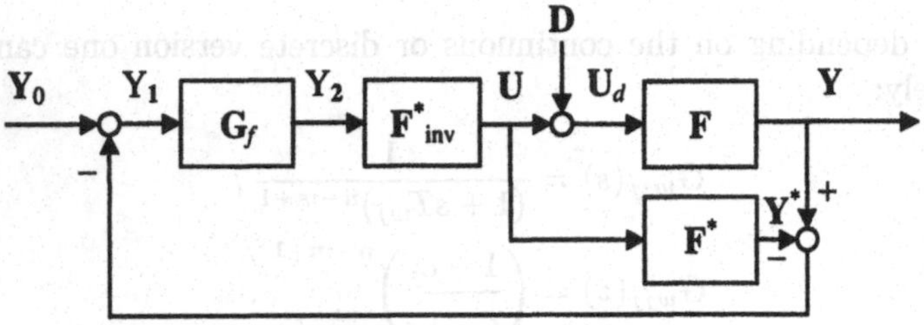

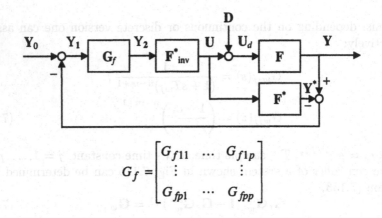

$$\mathbf{G}_f = \begin{bmatrix} G_{f11} & \cdots & G_{f1p} \\ \vdots & & \vdots \\ G_{fp1} & \cdots & G_{fpp} \end{bmatrix}$$

Fig. 7.118. Multivariable IMC control system containing low-pass filter $\mathbf{G}_f$

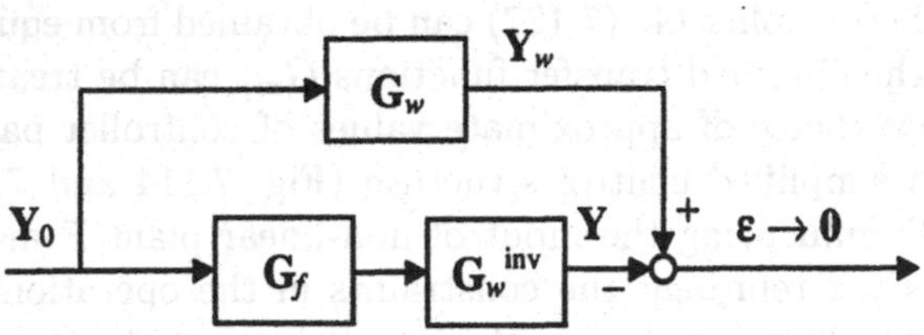

Fig. 7.119. System for determining the form of filter $\mathbf{G}_f$ and preliminary values of its parameters

inverse $\mathbf{F}^*_{\text{inv}}$ and series connection $(\mathbf{F}^*_{\text{inv}}\mathbf{F}^*)$ can be treated as equivalent to the reference model for inverse $\mathbf{G}^{\text{inv}}_w$ (Fig. 7.113), then the choice of component transfer functions G_{fij} of multivariable filter $\mathbf{G}_f$ and the initial tuning of filter parameters (for $\mathbf{D} = \mathbf{0}$) can be carried out on the basis of a structure with a reference model (for control) shown in Fig. 7.119.

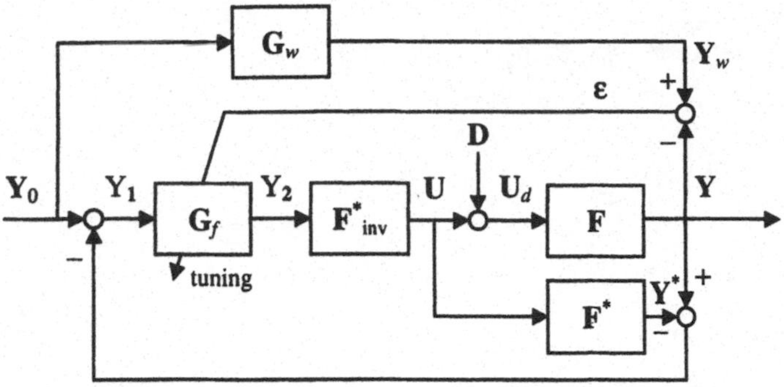

Fig. 7.120. Structure with multivariable reference model $\mathbf{G}_w$ advised for tuning parameters of filter $\mathbf{G}_f$

The following equations (7.134) and (7.135) result directly from Fig. 7.119.

$$\mathbf{G}_f\mathbf{G}_w^{\text{inv}} = \mathbf{G}_w \tag{7.134}$$

$$\mathbf{G}_f = \mathbf{G}_w(\mathbf{G}_w^{\text{inv}})^{-1} \tag{7.135}$$

Based on those equations one can chose the correct form of component transfer functions G_{fij} of filter $\mathbf{G}_f$ as well as determine the preliminary values of the filter parameters. The final tuning of filter parameters can be done on the basis of the complete IMC structure shown in Fig. 7.120, where non-linear constraints and disturbances are represented. The back propagation of error ε as well as other known methods can be used for tuning filter parameters.

The error ε ought to be propagated through the model $\mathbf{G}^*$ of plant, which can be represented by a neural network. $\mathbf{F}^*_{\text{inv}}$ and $\mathbf{G}_f$ can be represented similarly, i.e. also by neural networks.

Numerous examples of multivariable, fuzzy control systems are discussed in (Cormac 1997; Dias 1997; Lacrose 1997; Lopez 1997; Serac 1996; Soria 1996). They deal mostly with the control of plants for $p = 2$ (two inputs and two outputs). If the number of plant inputs and outputs increases, then one has to struggle with more and more difficult problems during fuzzy modeling of such plants while designing fuzzy controllers for them. A similar statement is also valid, if tuning is taken into account. That is why designing fuzzy controllers for a number of plant inputs greater than 4 is not recommended (Brown 1994,1995a). However, it does not mean that the problem for $p > 4$ can not be solved at all. There are special structures of neural controllers, which are useful in modeling multidimensional mappings, if dimensions are really large. Structures of that type are discussed with details in (Lin 1995). A brief presentation of those structures can be found in Section 6.3.

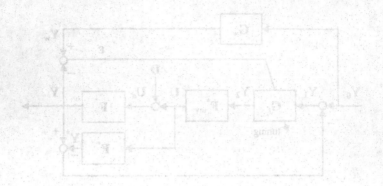

8. The Stability of Fuzzy Control Systems

According to obligatory industrial regulations issued by Authorities of many countries, the stability of a control system governed by a controller of a proposed type has to be proved. That requirement is treated as a necessary condition for use of the control system. There are many applications where "delivery" of proof of control system stability has to be perceived as a task of crucial importance. The control systems affect the safety of people (stabilizing airplane flight, etc.), govern costly plants and complicated industrial processes which are sensitive to loss of stability. The same regulations should be satisfied regardless of whether fuzzy or non-fuzzy controllers are considered.

The operations realized by fuzzy controllers can usually be classified as complicated non-linear mappings of input signals into output signals. There are many excellent methods for the investigation of the stability of linear control systems but, unfortunately, there are still difficulties with proving the stability of complicated, non-linear systems. Not long ago, in 1992 pessimistic opinions were published (Preuss 1992) stating that methods for proving the stability of fuzzy control systems had not been invented yet.

In 1996 the "latest" fuzzy controllers were presented during The German Technology Fairs aimed at applications of the results of contemporary research work. The presence of many prominent experts created the opportunity for serious discussion of important aspects of fuzzy control. Döbrich, reporting that discussion, wrote: "It is still an open question, if fuzzy controllers can be applied under critical conditions where the stability of a control system has to be guaranteed" (Döbrich 1996).

Difficulties with proving the stability of fuzzy systems narrow the applications of fuzzy controllers in practice. Despite those unfortunate results intense work is continuing in order to overcome difficulties. Thus, more or less advanced methods for proving the stability of fuzzy systems have been found since the end of the previous decade. Some of them have been adopted and developed from classic control theory. The most famous methods designed for proving the stability of fuzzy systems are listed below:

- Popov's stability criterion (Kahlert 1995; Böhm 1993; Opitz 1993; Bühler 1994; Cook 1986; Khalil 1992; Föllinger 1993),

- the circle criterion (Kahlert 1995; Driankov 1993; Opitz 1993; Cook 1986; Khalil 1992),
- direct Lyapunov's method (Kahlert 1995; Hung 1995; Böhm 1993; Kiendl 1993; Weinman 1991; Cook 1986; Khalil 1992; Föllinger 1993; Marin 1995; Sheel 1995; Tanaka 1990,1992),
- analysis of system stability in phase (state) space (Kahlert 1995; Driankov 1993; Cook 1986; Maeda 1991),
- analysis of stability using computer model of system
- the describing function method (Kahlert 1995; Kiendl 1992,1993; Cook 1986; Aracil 2000),
- method of stability indices and system robustness (bifurcation method) (Driankov 1993; Kahlert 1995),
- methods based on theory of input/output stability (small gain theorem) (Driankov 1993; Malki 1994; Noisser 1994; Cook 1986; Khalil 1992; Suykens 1995; Aracil 2000),
- conicity criterion (Driankov 1993; Aracil 1991,2000),
- methods based on Popov's hyperstability theory (Opitz 1986,1993; Böhm 1993; Schmitt 1996, Bindel 1995; Piegat 1997b,1997d; Han 1970; Li 1991; Popov 1963,1973; Anderson 1968),
- heuristic methods (Ying 1994; Wang 1996; Sommer 1993,1994; Rumpf 1994; Singh 1992; Aracil 2000).

The list of methods for the investigation of fuzzy systems stability is being supplemented by new proposals. Most of them can be classified as heuristic methods. It often happens that heuristic methods are based on very interesting intricate ideas. For example, Wang (Wang 1996) proposes the stabilization of a fuzzy system with a known fuzzy model of a plant in the following way: each inference rule assigned to the plant is compensated by a single, properly designed controller rule. A similar approach is proposed in (Aracil 2000) for Takagi-Sugeno fuzzy systems.

The large number of methods for the investigation of fuzzy system stability creates a kind of "information noise". This phenomenon is rather typical for these days. There is no reason to deal thoroughly with all existing methods. Thus, the usefulness of particular methods should be evaluated in order to select some of them. Finally, the "best" methods ought to be carefully studied. The following criteria assumed by the author of this book have been taken into account during evaluation of methods:

1. possibility of obtaining exact proof of system stability,
2. the "level" of difficulties while proving the system stability,
3. simplicity – making the method easily understandable,
4. generality of method (possibility of applications for investigation of various systems),
5. possibility of computer-aided support while struggling with proving the stability.

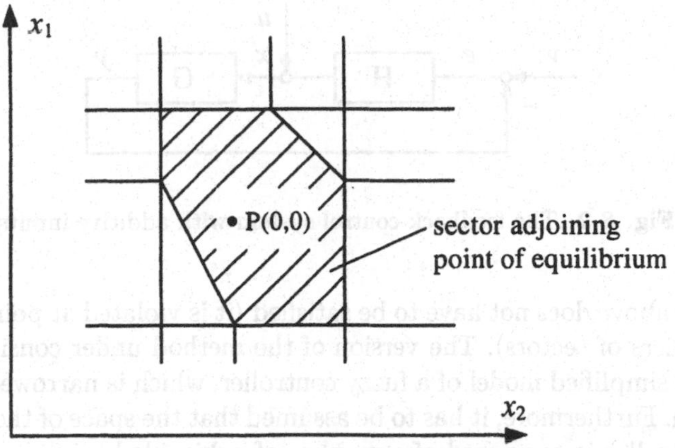

Fig. 8.1. The exemplary "split" of fuzzy system operation space into sectors (P – point of equilibrium)

The analysis of a system in phase space, the describing function method, analysis of stability using computer models of systems – these methods do not yield exact proofs of system stability but rather enable some kind of inspection of the system for precisely defined disturbances, excitations, initial conditions, etc. The analysis of stability using computer models of systems is very important, because it can be applied to investigations of very complicated, multi-dimensional systems, despite the number of signals exciting them. However, it is an experimental method, which means that its results can not be generalized. They simply confirm system stability for particular circumstances considered during simulations.

The widely known **Lyapunov's method** for the investigation of the stability of non-linear systems yields exact, mathematical proof of stability. There are some disadvantages of Lyapunov's method.

- To prove the stability one has to find the so-called Lyapunov function. This is a difficult, inventive task. The design of a computer program realizing a task of that type is also difficult.
- The Lyapunov's method allows us to prove the stability of a system for those sectors of the operation of the system, which are adjacent to the point representing the state of system equilibrium (Böhm 1993). What about remaining sectors?

According to (Böhm 1993), the Lyapunov's method can not be effectively applied outside the sector adjoining the equilibrium point. The Lyapunov function has to be continuously differentiable. That is why a quadratic form of the function is chosen – to represent generalized energy. The "softened" version of Lyapunov's method is proposed in (Kiendl 1993). The condition

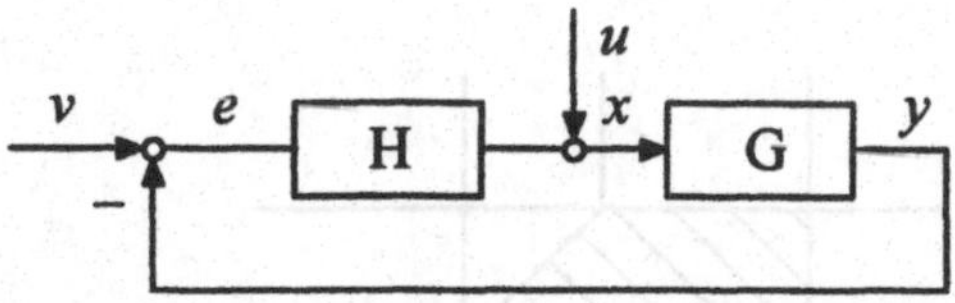

Fig. 8.2. The feedback control system with additive inputs

mentioned above does not have to be satisfied (it is violated at points belonging to borders of sectors). The version of the method under consideration is based on a simplified model of a fuzzy controller, which is narrowed to multilinear form. Furthermore, it has to be assumed that the space of the operation of the controller is composed of a number of cubicoids having parallel axes. The method seems to be interesting but the simplifications mentioned narrow its application field. The attempts aimed at improving the applicability of Lyapunov's method to investigation of fuzzy system stability are still continuing. In the near future, disadvantages of the method may be eliminated. Moreover, the existing versions of Lyapunov's method are said to be arduous and work consuming (Böhm 1995).

The methods based on the theory of **input/output stability** do not guarantee, that system steady state error is eliminated (Driankov 1993). The main results of those methods are based on the following theorem: the feedback control system is stable if the product of gains of feedback loop elements, i.e. plant $g(\mathrm{G})$ and controller $g(\mathrm{H})$, satisfies inequality:

$$g(\mathrm{G}) \cdot g(\mathrm{H}) < 1 \tag{8.1}$$

The analytic ways for determining the gains of feedback loop elements are difficult if a general case is taken into account. The procedure for determining plant gain $g(\mathrm{G})$ and controller gain $g(\mathrm{H})$ is relatively easy only in cases of linear, time-invariant plants governed by static controllers without hysteresis (see Fig. 8.2).

Then, the gains can be determined using formula (8.2):

$$\begin{aligned} g(\mathrm{H}) &= \sup \left\{ \tfrac{|\mathrm{H}(e)|}{|e|} \right\}, \quad |e| \neq 0\,, \\ g(\mathrm{G}) &= \sup_{\omega} \bar{\delta}\{\mathrm{G}(j\omega)\}\,, \end{aligned} \tag{8.2}$$

where $\bar{\delta}\{\mathrm{G}(j\omega)\}$ denotes the maximum singular value of matrix $\mathbf{A}$ assigned to the equation representing the state of the plant. Dealing with maximal gain $g(\mathrm{G})$ and $g(\mathrm{H})$ the theorem yields "conservative" results, ensuring a margin of system stability which is unnecessarily large. The gains of plant and controller do not have to be maximal simultaneously. Therefore, it sometimes happens that application of a considered method of investigation of system stability does not yield proof of stability, whereas the investigated system is undoubtedly stable.

The **bifurcation method** determines the "distance" of the system to the state where the system loses its stability. The method analyzes conditions responsible for loss of stability and ways of stabilizing the system in the neighborhood of unstable equilibrium points. The stabilization is realized by proper forming of controller characteristics. A deep theoretical analysis of system properties is necessary for successful application of this method. It means that automation of the method is extremely difficult. There is a version of this method (Driankov 1993) for non-linear systems but this method seems to be useful only for systems described by simple mathematical models. If a system can be represented in 2-dimensional space (phase plane) then the method is strongly recommended, since theoretical analysis of system properties and understanding of its operation appear relatively simple.

The **conicity criterion** has been developed using the small gain theorem. The disadvantages mentioned accompanying the idea of input/output stability are also applicable to the method using the conicity criterion.

The review of methods for the investigation of system stability incline to the opinion that Popov's criterion and the circle criterion seem to be worthy of recommendation for SISO systems, and methods developed from hyperstability theory are advised for MIMO systems. They yield exact, mathematical proofs of stability and can be supported by relatively uncomplicated computer algorithms. Special attention should be paid to the **theory of hyperstability**. It enables us to obtain proof of stability for systems with numerous equilibrium points. The stability can be proved not only in the neighborhood of those equilibrium points (as in the case of Lyapunov's method) but also globally, including sectors which do not adjoin equilibrium points. These features of the method result in its application in practice. Therefore, the hyperstability theory and circle criterion (as similar to but more simple than Popov's criterion) are presented in detail in Sections 8.2 and 8.3 respectively.

Section 8.1, dealing with the stability of fuzzy systems in cases where plant models remain unknown, precedes Sections 8.2 and 8.3. The fuzzy controller can be applied either if mathematical models of plants are known or unknown.

If the mathematical model of a plant is unknown, then the system stability can not be proved analytically. The necessary methods have not been "invented" yet (Ying 1994; Opitz 1993). However, they can appear in the near future. Buckley in (Hung 1995) proves conditions of stability, stating that one can design many fuzzy controllers (FC) ensuring stable operation of systems with plants of P-type, if this P-plant can be stably controlled by a controller C (for example, by man). Ying in (Ying 1994) presents the systematic algorithm for the design of fuzzy controllers ensuring stable control of plants with unknown models. During the first step of that algorithm one has to determine the parameters of the linear PI controller which enables the

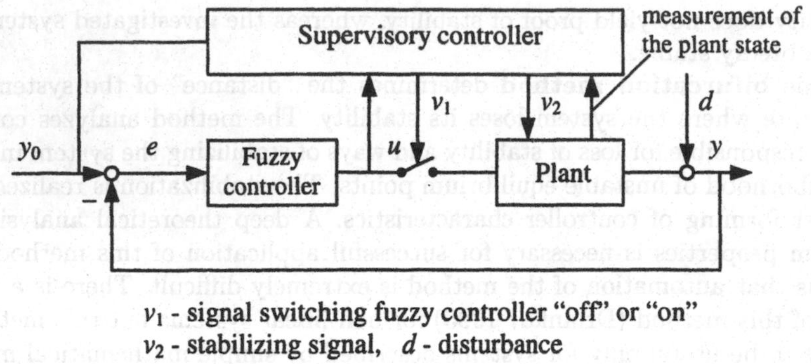

Fig. 8.3. The idea of the introduction of supervisory controllers to ensure the stability of fuzzy control systems

stable control of the plant. It means that even the Ziegler-Nichols method of controller tuning can be utilized to fulfill the above task.

The proposal of Wang published in (Wang 1994b) is also important and interesting. The paper presents the exact proof of stability of a control system if the model of a plant is unknown. The application of two controllers (the fuzzy controller R_1 and a supervisory one R_2) ensures the stability of a system under control. The problem of stabilizing fuzzy systems by means of supervisory controllers is presented in Section 8.1.

8.1 The stability of fuzzy control systems with unknown models of plants

The fuzzy controllers for plants with unknown mathematical models are designed on the basis of knowledge of a person, who achieved the sufficient knowledge on plant control during the learning process (an expert). There is still a risk that the knowledge of the expert is not complete. It sometimes happens that an operator has never dealt with rarely appearing critical states of a plant. That is why the supervisory controller shown in Fig. 8.3 (Opitz 1993) can be used to ensure system stability.

The supervisory controller observes the state and the input of the plant. If the state of the plant exceeds the admissible range (becoming the member of dangerous states) then the supervisory controller activates the stabilizing action v_2 in order to bring the plant to an area of safe operation. In case of poor results of the stabilizing action, the signal v_1 switches the input signal u off.

The method for realization of stabilization which ensures the stability of control systems containing certain classes of non-linear plants controlled by a

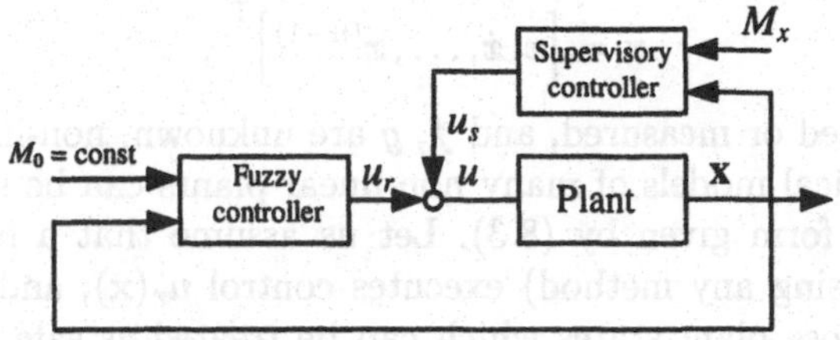

Fig. 8.4. Block diagram of a fuzzy control system with a supervisory controller ensuring stability of the system

fuzzy controller and a supervisory one is described in (Wang 1994b). The stability defined by Wang means that the system state does not exceed the range which has been previously determined by the designer. The introduction of a supervisory controller seems to be very advantageous, because many strong restrictions usually resulting from stability criteria can be omitted during the design process. It means that one can design a fuzzy controller with a bigger proportional gain and higher control performance which ensures the "proved stability" of the whole system. The stability criteria like the input-output method and Lyapunov's method enable us to prove the stability of control systems mostly for smaller gains; when the rate of the controller action increases, proof of stability becomes more and more difficult.

The operation of a supervisory controller that is considered in this section can be described as below:

- if the fuzzy controller itself controls the plant stably (plant state is kept inside the constraint set) then the supervisory controller stays in standby, idle mode,
- if the state of a system consisting of a plant and a fuzzy controller exceeds the constraints of safe operation then the supervisory controller begins a stabilizing operation to force the state back within the constraint set.

The considerations of the present section will concentrate on designing a supervisory controller according to the method proposed in (Wang 1994b). The block diagram of a system with a supervisory controller is shown in Fig. 8.4

Wang's supervisory controller ensures stability of fuzzy control for a certain class of non-linear plants described by equation (8.3):

$$x^{(n)} = f(\mathbf{x}) + g(\mathbf{x})u\,, \tag{8.3}$$

where $x \in R$ is the plant output and $u \in R$ is the plant input. It is assumed that state vector:

$$\mathbf{x} = \left[x, \dot{x}, \ldots, x^{(n-1)}\right]^{\mathrm{T}},$$

can be calculated or measured, and f, g are unknown, non-linear functions. The mathematical models of many non-linear plants can be successfully expressed in the form given by (8.3). Let us assume that a fuzzy controller (designed by using any method) executes control $u_r(\mathbf{x})$, and the boundary M_x limiting those plant states which can be treated as safe and stable has been defined.

The moment of activation of the supervisory controller depends on the value M_x given by (8.4):

$$M_x: \quad |\mathbf{x}(t)| \le M_x, \quad \forall t > 0, \tag{8.4}$$

where: $|\mathbf{x}(t)|$ – magnitude of state vector.

The input signal of the plant is given by expression (8.5):

$$u = u_r(\mathbf{x}) + I \cdot u_s(\mathbf{x}), \tag{8.5}$$

where the function of initiation of fuzzy controller action I is defined by (8.6):

$$I = \begin{cases} 1 & \text{if } |\mathbf{x}| \ge M_x, \\ 0 & \text{if } |\mathbf{x}| < M_x, \end{cases} \tag{8.6}$$

while u_s denotes the signal generated by the supervisory controller according to formula (8.7):

$$u_s = -\operatorname{sgn}\left(\mathbf{x}^{\mathrm{T}} \mathbf{P} \mathbf{b}_c\right) \left[\frac{1}{g_l}(f^u + |\mathbf{k}^{\mathrm{T}} \mathbf{x}|) + |u_r|\right], \tag{8.7}$$

where:

$$\mathbf{b}_c = \begin{bmatrix} 0 \\ 0 \\ \ldots \\ g \end{bmatrix}, \quad g > 0, \tag{8.8}$$

P represents a symmetric, positive definite matrix satisfying the Lyapunov equation (8.9):

$$\Lambda_c^{\mathrm{T}} \mathbf{P} + \mathbf{P} \Lambda_c = -\mathbf{Q}, \tag{8.9}$$

where $\mathbf{Q} > 0$ is specified by the designer of the system. The matrix Λ_c has the form (8.10):

$$\Lambda_c = \begin{bmatrix} 0 & 1 & 0 & 0 & \ldots & 0 & 0 \\ 0 & 0 & 1 & 0 & \ldots & 0 & 0 \\ \ldots & \ldots & \ldots & \ldots & \ldots & \ldots & \ldots \\ 0 & 0 & 0 & 0 & 0 & 0 & 1 \\ -k_n & -k_{n-1} & -k_{n-2} & -k_{n-3} & \ldots & -k_2 & -k_1 \end{bmatrix}, \tag{8.10}$$

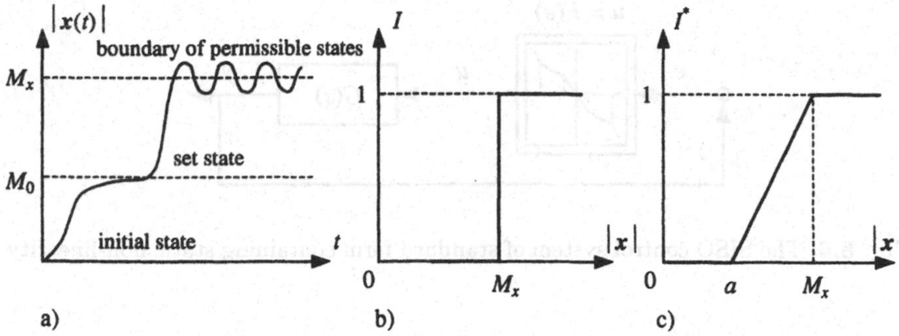

Fig. 8.5. The oscillations in a system (a) which can be generated if one uses the initiating function I (b) and the way of avoiding oscillations by the application of an alternative form of initiating function I^* (c)

where vector $\mathbf{k} = (k_n, \ldots, k_1)^{\mathrm{T}} \in R^n$ contains such coefficients k_i, that all roots s_i of polynomial $s^n + k_1 s^{n-1} + \cdots + k_n$ belong to the left-side half-plane of the complex plane "s".

The proof of stability of a system governed by a fuzzy controller and a supervisory controller according to rules given by (8.5),(8.6),(8.7) is presented in (Wang 1994b). To prove the stability of such a system, Wang assumes that the upper limit f^u of function f and the lower limit g_l of function g are known. The relations (8.11) define the requirements for functions f^u and g_l:

$$\begin{aligned} f^u(\mathbf{x}): &\quad |f(\mathbf{x}| \le f^u(\mathbf{x}) \\ g_l(\mathbf{x}): &\quad 0 < g_l(\mathbf{x}) \le g(\mathbf{x}) \end{aligned} \tag{8.11}$$

The function initiating the activity of supervisory controller I (8.6) has the form of a step function. The initiating function switches the signal u_s "on" and "off" if the state vector of the plant reaches the limit of admissible states $|\mathbf{x}| = M_x$. It means that the plant is treated roughly (Fig. 8.5b). That is why the periodical oscillations can appear in the neighborhood of state M_x – see Fig. 8.5a.

The use of initiating function I^* (Fig. 8.5c) protects the systems against oscillations. The function I^* is determined by formula (8.12):

$$I^* = \begin{cases} 0 & \text{if } |\mathbf{x}| < a\,, \\ \dfrac{|\mathbf{x}| - a}{M_x - a} & \text{if } a \le |\mathbf{x}| < M_x\,, \\ 1 & \text{if } |\mathbf{x}| \ge M_x\,, \end{cases} \tag{8.12}$$

where parameter a is specified by the designer. The small value of a (in relation to M_x) means that the supervisory controller influences the plant

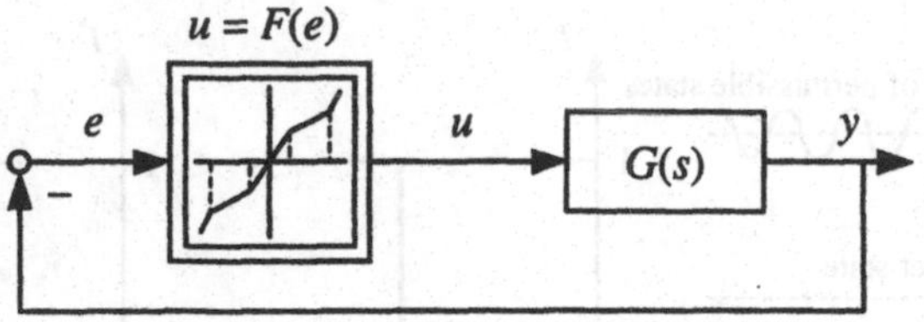

Fig. 8.6. The SISO control system of standard form containing static non-linearity

frequently but gently. If the value a is close to M_x, then the supervisory controller undertakes the control activity rather rarely but roughly. Thus, the probability of generating oscillations is higher compared to the previous choice (small a). The switching function I^* also ensures stability of the fuzzy system (in the meaning given above). Fig. 8.5 shows that the supervisory controller does not "keep" the plant state in the neighborhood of set state M_0 (usually, this task is realized successfully by the fuzzy controller) but protects the plant from operating within dangerous states. The presented idea of distribution of control tasks seems to be a good solution. It is obvious that the use of a supervisory controller would also be advantageous in cases of classic control systems governed by PID-controllers. Proof of stability for classic controllers is much easier to achieve. However, to prove the system stability under PID control, simplified, linear models of non-linear plants are usually assumed. It means that destabilizing phenomena like non-linearities, dead bands (friction), hysteresis, etc. are neglected. Under the above circumstances there is no proven answer if the PID-controller ensures the stability of a real system for all possible conditions of operation. The possibility of equipping fuzzy control systems with supervisory controllers can be considered as an important reason leading to wider application of fuzzy controllers in industry.

8.2 The circle stability criterion

The circle stability criterion (sometimes called the criterion of Kudrewicz and Cypkin) (Markowski 1985) is mostly applied to those types of SISO (single input – single output) systems which can be transformed to standard system represented by a block diagram shown in Fig. 8.6 (Driankov 1993; Opitz 1993; Kahlert 1995; Cook 1986; Khalil 1992).

The linear part $G(s)$ of the standard system shown in Fig. 8.6 satisfies the following conditions L1 and L2:

L1: $G(s)$ is rational (the denominator order is higher than nominator order),
L2: $G(s)$ is asymptotically stable (all poles are located in the left half-plane of "s", the imaginary axis is excluded as a place for pole location).

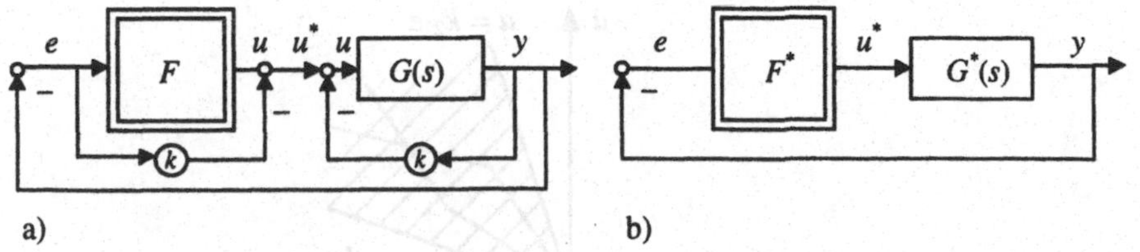

Fig. 8.7. The introduction of fictitious degree of freedom k to the primary system (a) and the secondary system after transformation to the standard form (b)

The non-linear part F satisfies conditions N1 – N3,

N1: $F(e)$ represents a static and unique relation (only one output value $F(e)$ can be assigned to a given input value e, i.e. $F(e)$ can not represent hysteresis and other relations of "memory type"),
N2: the characteristic $F(e)$ consists of straight line segments,
N3: $F(0) = 0$ (the characteristic crosses the intersection of the e-axis and $F(e)$-axis).

If primary transfer function $G(s)$ does not satisfy the L2 condition then fictitious degrees of freedom can be added to it in order to transform the original system into an equivalent, secondary system that does fulfill L2. The function of original inputs and outputs of the linear and non-linear parts of the primary system has to be conserved (not changed) during introduction of fictitious degrees of freedom – see Fig. 8.7.

The following relations (8.13) describe operations realized by elements of the system shown in Fig. 8.7b:

$$G^*(s) = \frac{G(s)}{1 + k \cdot G(s)}, \qquad F^*: \quad u^* = F(e) - k \cdot e. \tag{8.13}$$

By proper choice of coefficient k one can usually obtain the asymptotically stable form of transfer function $G^*(s)$. A later example will show this possibility .

The circle criterion of stability enables us to draw conclusions on system stability by consideration of the membership of a non-linear characteristic to the family of characteristics constrained by straight lines $u = k_2 \cdot e$ and $u = k_1 \cdot e$ – see Fig. 8.8.

The standard control system with a linear part and non-linear part fulfilling conditions Li and Ni is globally and asymptotically stable if the circle defined by its center located on the real axis in the point (8.14):

$$c = -0.5\left(\frac{1}{k_1} + \frac{1}{k_2}\right) \tag{8.14}$$

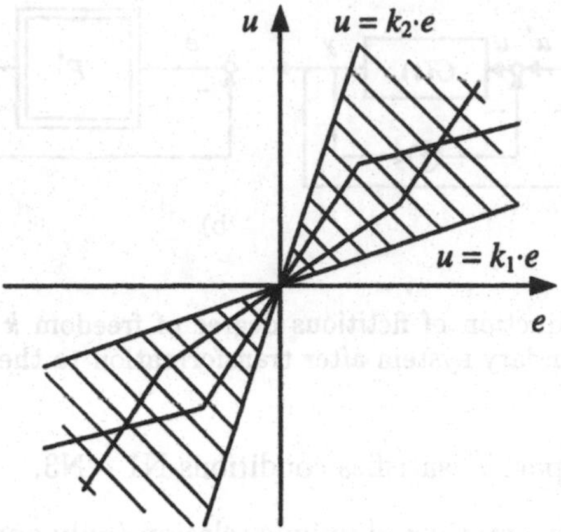

Fig. 8.8. The "upper" and "lower" constraints applied to the family of static characteristics

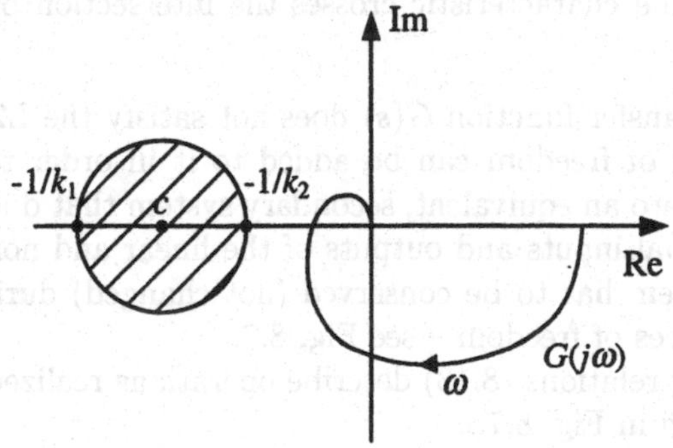

Fig. 8.9. Example of location of the circle representing constraints imposed on non-linear part in relation to the Nyquist plot of linear part $G(j\omega)$ for an asymptotically stable control system

with its radius (8.15):

$$r = 0.5\left(\frac{1}{k_1} - \frac{1}{k_2}\right) \tag{8.15}$$

is placed entirely (contactless) in the area adjacent to the left side of the frequency response (Nyquist plot) of the system linear part $G(j\omega)$, see Fig. 8.9.

The values "zero" and "infinity" can also be put as k_1 and k_2 (Markowski 1995). For example, if $k_1 = 0$, then the circle representing constraints is transformed into a half-plane shown in Fig. 8.10. For simplicity of form of the circle criterion it is assumed that $k_1 \geq 0$, $k_2 \geq 0$ (Driankov 1993).

Example 8.2.1

The fuzzy control system consists of a linear element:

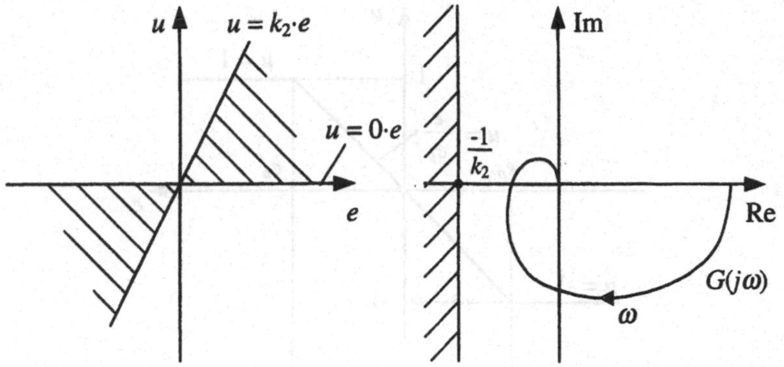

Fig. 8.10. The circle criterion for critical value of the lower constraint $k_1 = 0$

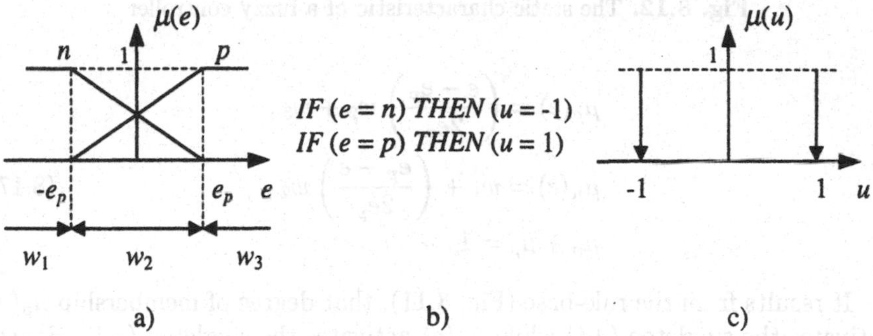

Fig. 8.11. The fuzzification(a), rule-base (b) and defuzzification (c) realized by a fuzzy-controller

$$G(s) = \frac{1}{s(1+s)},$$

and fuzzy controller $F(e)$. The fuzzification element characteristic, rule-base and defuzzification element characteristic are shown in Fig. 8.11. Let us consider the conditions for system stability.

Choosing logical variables w_i in the form (8.16):

$$w_1 = \begin{cases} 0 : e \le -e_p, \\ 1 : \text{otherwise}, \end{cases}$$

$$w_2 = \begin{cases} 0 : -e_p < e \le e_p, \\ 1 : \text{otherwise}, \end{cases} \tag{8.16}$$

$$w_3 = \begin{cases} 0 : e > e_p, \\ 1 : \text{otherwise}, \end{cases}$$

one can obtain the following membership functions (8.17) for fuzzification:

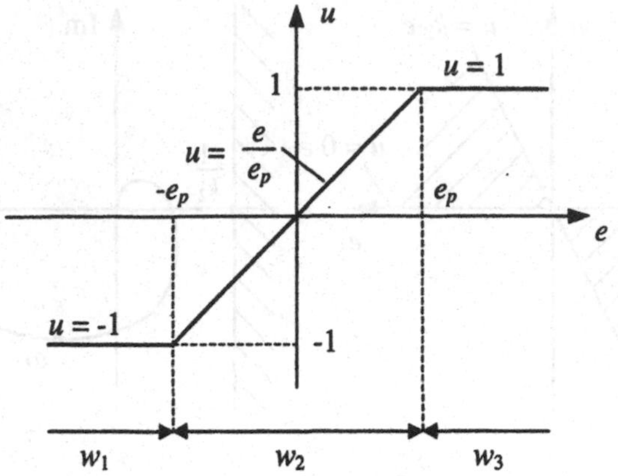

Fig. 8.12. The static characteristic of a fuzzy controller

$$\begin{aligned}
\mu_p(e) &= \left(\frac{e - e_p}{2e_p}\right) w_2 + w_3 \,, \\
\mu_n(e) &= w_1 + \left(\frac{e_p - e}{2e_p}\right) w_2 \,, \\
\mu_n + \mu_p &= 1 \,.
\end{aligned} \tag{8.17}$$

It results from the rule-base (Fig. 8.11), that degree of membership $\mu_p(e)$ activates the singleton (+1) while $\mu_n(e)$ activates the singleton (−1). Hence, the output u of the non-linear part is (8.18):

$$u = \frac{\mu_p \cdot 1 + \mu_n \cdot (-1)}{\mu_n + \mu_p} = -w_1 + \frac{e}{e_p} w_2 + w_3 \,. \tag{8.18}$$

The characteristic of the fuzzy non-linear part is shown in Fig. 8.12.

The linear part $G(s)$ has poles $s_1 = 0$ and $s_2 = -1$. It means that the linear part has to be stabilized because of the pole belonging to the imaginary axis ($s_1 = 0$). If the method of stabilization shown in Fig. 8.7 is applied, then, using expression (8.13), one obtains the secondary transfer function G^* in the form (8.19):

$$G^*(s) = \frac{G(s)}{1 + kG(S)} = \frac{1}{s^2 + s + k} \,. \tag{8.19}$$

Now, the poles of $G^*(s)$ are:

$$\begin{aligned}
s_1 &= -0.5 \cdot \left(1 + \sqrt{1 - 4k}\right) \,, \\
s_2 &= -0.5 \cdot \left(1 - \sqrt{1 - 4k}\right) \,.
\end{aligned}$$

If $k > 0$ then $G^*(s)$ is asymptotically stable. Thus, the first condition of choosing the value of k is:

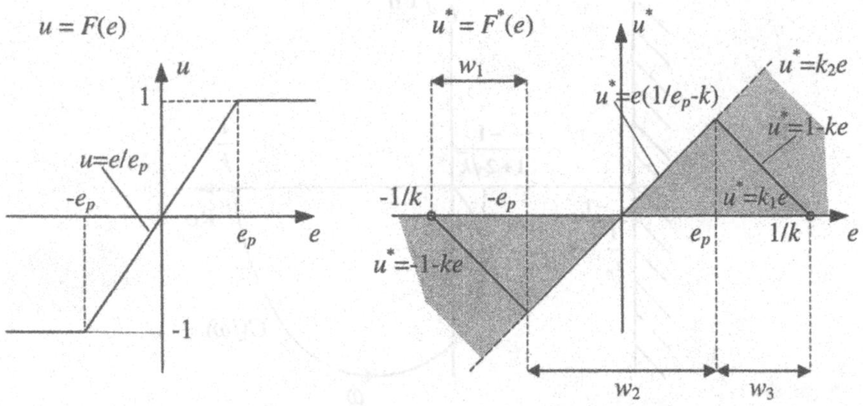

Fig. 8.13. The primary static characteristic of fuzzy-controller $F(e)$ and secondary one $F^*(e)$ obtained by introduction of fictitious degree of freedom k

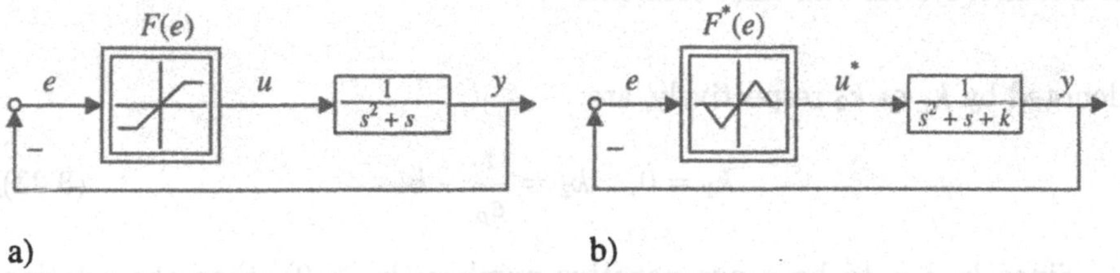

Fig. 8.14. The primary control system (a) and its equivalent system (b) created in order to apply the circle stability criterion

$$k > 0 \tag{8.20}$$

The secondary non-linear part F^* can be determined using (8.13). The result of calculations is expressed by (8.21):

$$u^* = -w_1 + \frac{e}{e_p} w_2 + w_3 - ke\,. \tag{8.21}$$

The primary and secondary characteristics of the non-linear part are shown in Fig. 8.13. The primary control system and its equivalent with a fictitious degree of freedom are shown in Fig. 8.14.

It has been mentioned that gains k_1 and k_2 defined by the circle criterion are usually considered as positive numbers or equal to 0. It means that the stability of the system can be proved for input range:

$$-\frac{1}{k} \leq e \leq \frac{1}{k}\,. \tag{8.22}$$

The smaller the value k fulfilling the stability condition the bigger the range (8.22). It results from Fig. 8.13 that lower and upper limits of gain,

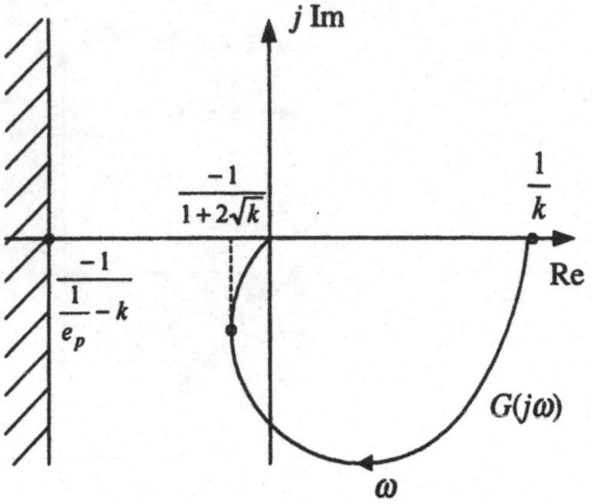

Fig. 8.15. Result of application of the circle stability criterion in the form of a graphic representation of condition (8.26) determining requirements for the stability of a control system with fuzzy controller

denoted by k_1 or k_2 respectively, are:

$$k_1 = 0, \quad k_2 = \frac{1}{e_p} - k. \tag{8.23}$$

Since k_2 has to be a non-negative number ($k_2 \geq 0$), then the relation between value k and the parameter determining the gain of fuzzy controller e_p can be expressed by inequality (8.24):

$$e_p \leq \frac{1}{k}. \tag{8.24}$$

The minimum of the real part of frequency response $G^*(j\omega)$ is:

$$\text{Re}_{\text{min}}[G^*(j\omega)] = -\frac{1}{1+2\sqrt{k}}. \tag{8.25}$$

Taking into account conditions (8.23) and (8.25) one obtains condition (8.26) determining stability of the considered system, Fig. 8.15.

$$\frac{-1}{\frac{1}{e_p} - k} < \frac{-1}{1+2\sqrt{k}} \tag{8.26}$$

The conjunction of conditions (8.24) and (8.26) can be written in the form (8.27):

$$\frac{1}{1+k+2\sqrt{k}} < e_p < \frac{1}{k} \tag{8.27}$$

If we assume a small, positive value of k, for example $k = 10^{-100}$, then approximate transformation of (8.27) yields (8.28):

$$1 < e_p < \infty . \tag{8.28}$$

The above condition determines those parameters e_p characterizing the fuzzification membership function $\mu(e_p)$, which guarantee the stable operation of a control system. The proof of stability refers to the range (see Fig. 8.13):

$$-\frac{1}{k} \leq e \leq \frac{1}{k} .$$

The above range tends to be infinitely large for very small, positive k. For bigger values of k, the gain of controller $(1/e_p)$ can also be bigger but stable operation of the system is narrowed to a smaller amplitude range of signal e.
◊ ◊ ◊

There are versions of the circle criterion for MIMO (Multi Input Multi-Output) systems (Driankov 1993) but experts maintain (Opitz 1993) that applicable results are obtained for relatively easy cases if multidimensional characteristic of the non-linear part can be brought (by means of a set of linear operations) to a simple resultant form (8.29):

$$u = f(\mathbf{e}) = f(\mathbf{k}^{\mathrm{T}} \cdot \mathbf{e}) , \tag{8.29}$$

where: **k** – vector of gain coefficients. If condition (8.29) is not fulfilled, then the circle criterion for MIMO systems "generates" conservative and narrowed results. It suggests that the usefulness of the MIMO version of the circle criterion for practical purposes is rather poor.

8.3 The application of hyperstability theory to analysis of fuzzy-system stability

The rudiments of the theory of hyperstability were created by Romanian mathematician V.M. Popov and published in (Popov 1963,1973). Popov's general formulation of the theory of hyperstability was not adjusted to the analysis of stability of all systems, but only systems fulfilling many initial requirements. However, obstacles have been gradually removed. Now, the hyperstability theory represents the "mature" form and can be efficiently applied to stability analysis of a wide class of non-linear systems, including fuzzy-control systems. Recent papers, like: (Schmitt 1996; Opitz 1986,1993; Böhm 1995; Bindel 1995; Föllinger 1993; Piegat 1997b,1997d), seem to be very interesting.

The recent results aimed at the application of hyperstability theory are judged by the author as attractive and useful. Methods derived from hyperstability theory can be treated as competitive to other methods of stability analysis. This section contains a detailed discussion of the above statement. There are many advantages connected with an approach based on hyperstability theory. The method enables systematic organization of stability

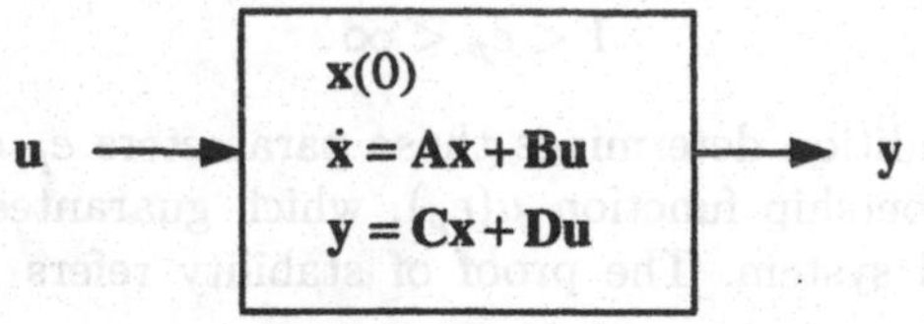

Fig. 8.16. The multidimensional, time-invariant linear system

verification and gives a better understanding of problems due to the possibility of its visualization. Furthermore, the chosen steps of analytic research can be substituted by numerical research realized with the aid of a computer (the method can be automated). What does hyperstability mean? The following definition has been given in (Anderson 1968):

"Hyperstability is such a property of a system which ensures that the system state vector will stay constrained if input quantities are constrained to a determined subset of all possible input quantities".

The mathematical formulation of hyperstability conditions for linear systems (Fig. 8.16) is the easiest task, compared to other cases. However, mathematical formulations of conditions of hyperstability are accessible for certain classes of non-linear systems (Han 1970), bilinear systems (Ionescu 1978) and system with distributed parameters (Jumaire 1983).

If the dimension of input vector $\mathbf{u}$ equals the dimension of output vector $\mathbf{y}$, and the system is completely controllable (which depends on matrices $\mathbf{A}$ and $\mathbf{B}$) and observable (which depends on matrices $\mathbf{C}$ and $\mathbf{A}$), then the system is hyperstable, if for all $u(t)$ fulfilling the integral inequality (8.30):

$$I = \int_0^t \mathbf{u}^{\mathrm{T}}(\tau)\mathbf{y}(\tau)\mathrm{d}\tau \le \beta_0^2, \quad \forall t > 0,\ \beta_0 > 0\,, \tag{8.30}$$

one can prove inequality (8.31) in the form:

$$\|\mathbf{x}(t)\| \le \beta_0 + \beta_1\|\mathbf{x}(0)\|\,, \quad \forall t > 0\,, \tag{8.31}$$

where $\mathbf{x}(0)$ denotes the initial condition of state vector $\mathbf{x}(t)$, β_0 and β_1 are any positive constants, $\|\ldots\|$ represents the Euclidean norm.

The system fulfilling (8.30) and (8.31) can be considered as hyperstable in the ordinary sense. If, for all input vectors $\mathbf{u}$ satisfying inequality (8.30), the inequalities (8.31) and (8.32) are also satisfied:

$$\lim_{t\to\infty} \mathbf{x}(t) = 0\,, \tag{8.32}$$

then the system is asymptotically hyperstable.

The idea of system hyperstability can be roughly explained considering a SISO (single input, single output) system. Let us suppose that the system is

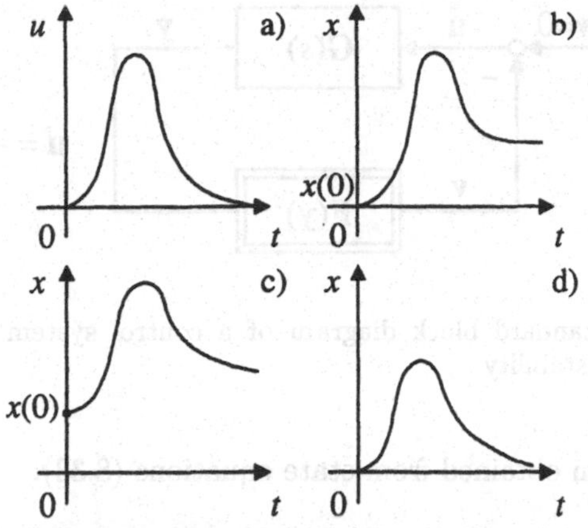

Fig. 8.17. Illustration of the hyperstability idea: input signal $u(t)$ (a), state variable of hyperstable system for $x(0) = 0$ (b) and $x(0) \neq 0$ (c), state variable for asymptotically hyperstable system (d)

activated with input signal u of finite duration time (Fig. 8.17a). The energy delivered to the system by the signal u is limited. Let us assume that u will cause the limited amplitude growth of state variables $\mathbf{x}$ (estimated with the norm $\|\mathbf{x}(t)\|$), see Fig. 8.17b. Of course, the amplitude growth of state variables accompanies the increase of system energy. If the increase of system energy depends exclusively on energy delivered with an input signal (which can be estimated by β_0) and eventually on system potential energy determined by initial conditions (estimated with the norm $\|\mathbf{x}(0)\|$), see Fig. 8.17c, then the system can be treated as hyperstable.

Thus, the hyperstable system does not contain internal sources of energy. It means that amplitude increase of state variables $\mathbf{x}$ generated by input signal $\mathbf{u}$ depends on energy delivered with an input signal. That is why state variables can not tend to infinity, if amplitude level of input is constrained. If system state $\mathbf{x}(t)$ tends to zero, fulfilling condition (8.32), Fig. 8.17d, then system is called asymptotically hyperstable .

Note that output signal $\mathbf{y} = \mathbf{Cx} + \mathbf{Du}$ is linearly dependent on the input signal and system state (Fig. 8.16). The constrained $\mathbf{x}$ and $\mathbf{u}$ yield constrained output $\mathbf{y}$.

8.3.1 The frequency domain representation of hyperstability conditions for control systems with a time invariant non-linear part

The conditions have been formulated for a standard form of control system shown in Fig. 8.18. The standard control system consists of a linear, time invariant subsystem described by a matrix of transfer functions $\mathbf{G}(s)$ or by

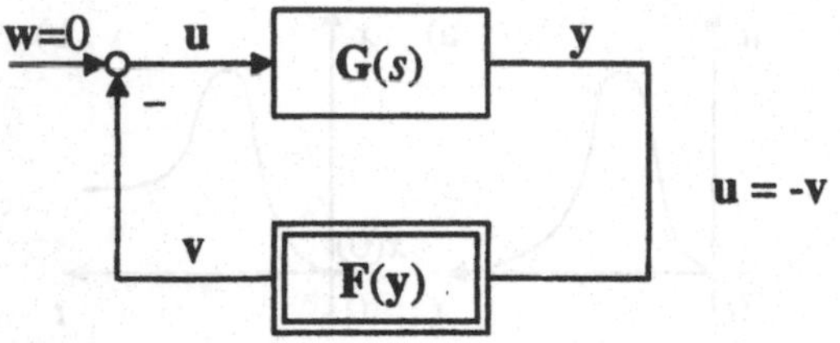

Fig. 8.18. The standard block diagram of a control system considered during discussion of hyperstability

an equivalent form obtained from state equations (8.33):

$$\mathbf{G}(s) = \mathbf{C}(s\mathbf{I} - \mathbf{A})^{-1}\mathbf{B} + \mathbf{D}, \tag{8.33}$$

(where: $\mathbf{I}$ – identity matrix) and static, non-linear subsystem $\mathbf{F}(\mathbf{y})$. The theory of hyperstability can also be extended to a form enabling investigation of system stability in cases of a time variant non-linear part.

The matrix $\mathbf{G}(s)$ represents plant and other linear elements co-operating with the plant. Those elements can be aggregated (using rules for equivalent transformation of block diagrams). Thus, one can treat $\mathbf{G}(s)$ as representing the resultant matrix of transfer functions of the linear subsystem. Similarly, $\mathbf{F}(\mathbf{y})$ can also be considered as the resultant set of static, non-linear operations. The theory of hyperstability can be applied to all systems which can be transformed to the standard form shown in Fig. 8.18.

In order to apply the hyperstability method the linear part of control system $\mathbf{G}(s)$ as well as its non-linear part have to fulfill some initial requirements.

Preliminary conditions for linear subsystem $\mathbf{G}(s)$

PL1) The matrix of transfer functions $\mathbf{G}(s)$ has to be a square one. It means that the number of inputs of the linear part (dimension of vector $\mathbf{u}$) has to be the same as the number of linear part outputs (dimension of vector $\mathbf{y}$). If the "primary" system does not fulfill the above requirement, then additional signals, always keeping values equal to zero, can be used in order to fit dimensions of matrix $\mathbf{G}(s)$ to the condition under consideration. An example of such procedure is shown in Fig. 8.19.

PL2) The linear part $\mathbf{G}(s)$ has to be completely controllable and observable. It happens that the controllable condition can be weakened (Popov 1973; Landau 1979). If all eigenvalues of matrix $\mathbf{A}$ which are not connected to its inputs are represented by complex numbers with negative real parts, and the partial matrix of transfer functions, describing that part of the linear subsystem which is completely controllable and observable, is strictly positive, then the system is hyperstable, if the rest of the remaining requirements is satisfied.

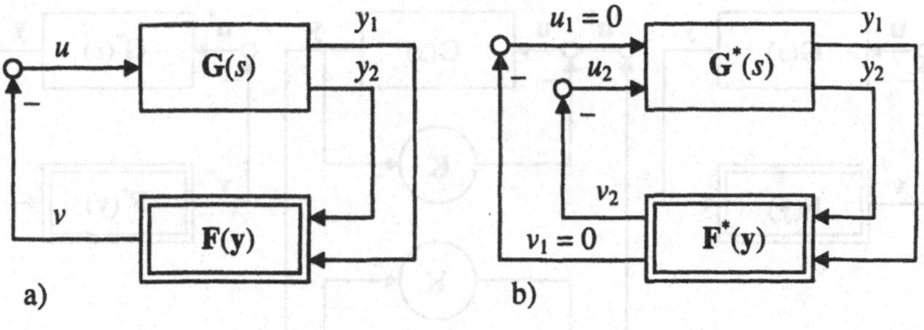

Fig. 8.19. The "primary" system (a) failing the requirement PL1 and equivalent "secondary" system (b) fulfilling PL1

PL3) All transfer functions $G_{ij}(s)$ belonging to matrix $\mathbf{G}(s)$ have to be stable (poles of all $G_{ij}(s)$ have to be represented by complex numbers with negative real parts). If the condition mentioned above is not fulfilled, then such fictitious degrees of freedom k_{ij} can be introduced to the primary system that inputs of primary linear parts are "conserved". The degrees of freedom are introduced in the form of such feedback loops around the linear part $\mathbf{G}(s)$ and parallel connections encircling the non-linear subsystem $\mathbf{F}(\mathbf{y})$, that signals leaving the introduced connections mutually reduce themselves in the system – see Fig. 8.20. The relation between the primary standard form of a system and its secondary standard form is given by (8.34).

$$\mathbf{G}^*(s) = [\mathbf{I} + \mathbf{G}(s)\mathbf{K}]^{-1}\mathbf{G}(s)\,,$$
$$\mathbf{F}^*(\mathbf{y}): \quad \mathbf{v}^* = \mathbf{v} - \mathbf{K}\mathbf{y} = \mathbf{F}(\mathbf{y}) - \mathbf{K}\mathbf{y}\,. \qquad (8.34)$$

The number of non-zero elements in matrix $\mathbf{K}$ should be sufficient to stabilize all unstable poles. However, sometimes the necessary number of non-zero elements is smaller than the number of unstable poles. Then the stability proof is more difficult. The range of available values of coefficients k_{ij} in matrix $\mathbf{K}$ seems to be quite large. But one has to keep in mind that values of those coefficients influence the transformation of primary operator $\mathbf{F}(\mathbf{y})$ into the secondary operator $\mathbf{F}^*(\mathbf{y})$, which should also fulfill certain requirements imposed by the theory of hyperstability. The fulfilling of these requirements can be impossible if coefficients of matrix $\mathbf{K}$ are chosen unluckily. Expert advice is: to chose the values of those coefficients after formulation of all conditions for system hyperstability, referring to both subsystems: the linear one $\mathbf{G}(s)$ and the non-linear $\mathbf{F}(\mathbf{y})$. The examples discussed in this section will explain the method of creation of matrix $\mathbf{K}$.

PL4) The order of the nominator of each component transfer function $G_{ij}(s)$ belonging to the matrix of transfer functions $\mathbf{G}(s)$ has to be not higher

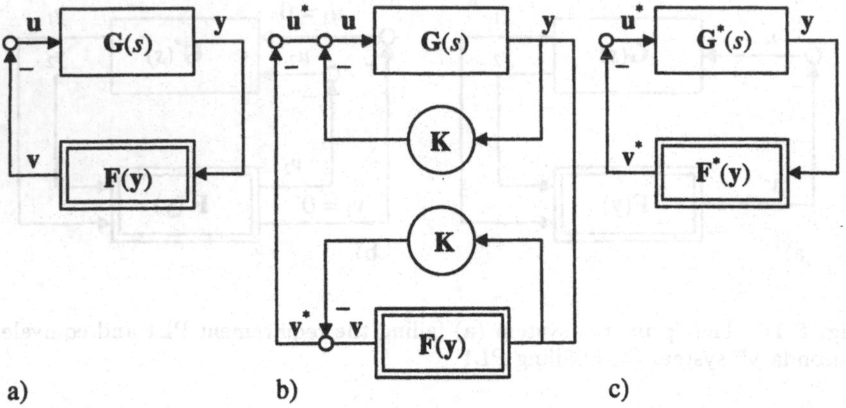

Fig. 8.20. The stabilization of unstable poles of a primary linear subsystem $\mathbf{G}(s)$ (a) by introduction of fictitious degrees of freedom $\mathbf{K}$ (b) and new, secondary, standard form of system (c) equivalent to the primary form (a)

than the order of its denominator. The real plants always satisfy this requirement.

The preliminary conditions for non-linear subsystem $\mathbf{F}(\mathbf{y})$

PN1) The operation $\mathbf{v} = \mathbf{F}(\mathbf{y})$ has to realize unique mapping of $\mathbf{y}$ into $\mathbf{v}$. The non-linear elements with a memory of a hysteresis type (Fig. 8.21) do not satisfy the considered condition.

PN2) The mapping $\mathbf{v} = \mathbf{F}(\mathbf{y})$ has to fulfill the zeroing condition (8.35), Fig. 8.22.

$$\mathbf{F}(\mathbf{0}) = \mathbf{0} \tag{8.35}$$

If requirements PL1 – PL4 for the linear part and PN1 – PN2 for the non-linear part are fulfilled, then the main conditions referring to linear part (ML) and non-linear part (MN) can be checked.

The main conditions for hyperstability (sufficient conditions)

The standard control system shown in Fig. 8.20a, and its equivalent systems, obtained by such introduction of fictitious degrees of freedom to the primary system that inputs and outputs of its linear and non-linear parts are left unchanged after this transformation, are asymptotically hyperstable, if:

ML) time-invariant linear part $\mathbf{G}(s)$ is strictly, positive real,

MN) the non-linear part $\mathbf{F}(\mathbf{y})$ satisfies Popov's integral inequality (8.36).

$$I = \int_0^t \mathbf{v}^{\mathrm{T}}(\tau)\mathbf{y}(\tau)\mathrm{d}\tau \geq -\beta_0^2, \quad \forall t > 0,\ \beta_0 > 0\,. \tag{8.36}$$

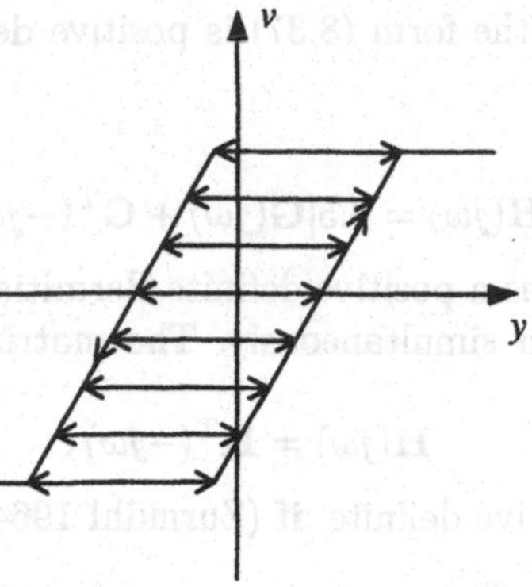

Fig. 8.21. The example of non-linearity called hysteresis – the elements with hysteresis do not realize unique mapping

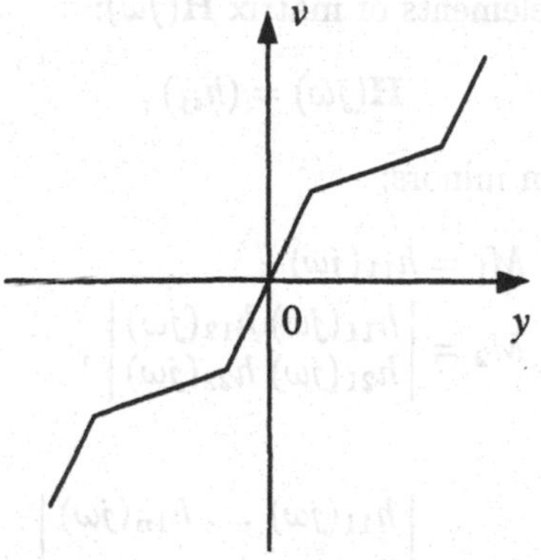

Fig. 8.22. The example of mapping $\mathbf{F}(\mathbf{y})$ which satisfies the zeroing requirement PN2: $\mathbf{v}(0) = \mathbf{F}(0) = 0$

If the non-linear part fulfills condition (8.36), then the linear part "automatically" fulfills hyperstability condition (8.30) (because $\mathbf{u} = -\mathbf{v}$). Thus, condition (8.36) determines the stability of the linear part of a standard system and finally the stability of the standard system as a whole. If application of condition ML to the linear part allows us to classify the linear part as positive definite (instead of strictly positive definite), then the standard system is referred to as normally stable or stable.

Explanations concerning the condition ML determining hyperstability of the linear part

The matrix of transfer functions $\mathbf{G}(s)$ is strictly positive definite, if:

- poles of all component transfer functions $G_{ij}(s)$ are located to the left of an imaginary axis of complex plane "s",

- the matrix given in the form (8.37) is positive definite and Hermitian for all $\omega \geq 0$.

$$\mathbf{H}(j\omega) = 0.5[\mathbf{G}(j\omega) + \mathbf{G}^{\mathrm{T}}(-j\omega)] \tag{8.37}$$

The matrix $\mathbf{H}(j\omega)$ is a positive definite Hermitian matrix, if it is positive definite and Hermitian simultaneously. The matrix is called Hermitian, if (Schmitt 1996):

$$\mathbf{H}(j\omega) = \mathbf{H}^{\mathrm{T}}(-j\omega)\,. \tag{8.38}$$

The matrix is positive definite, if (Zurmühl 1964):

$$Q = \mathbf{Z}^{\mathrm{T}}\mathbf{H}\mathbf{Z} > 0 \ \text{ for each vector } \mathbf{Z} \in \mathbf{R}^n\,. \tag{8.39}$$

The other way of checking if matrix is positive definite or not, is by examination of its main minors M_i by means of Sylvester's theorem (Opitz 1986). Let h_{ij} denote elements of matrix $\mathbf{H}(j\omega)$:

$$\mathbf{H}(j\omega) = (h_{ij})\,, \tag{8.40}$$

and M_i denote its main minors:

$$\begin{aligned} M_1 &= h_{11}(j\omega)\,, \\ M_2 &= \begin{vmatrix} h_{11}(j\omega) & h_{12}(j\omega) \\ h_{21}(j\omega) & h_{22}(j\omega) \end{vmatrix}\,, \\ &\vdots \\ M_n &= \begin{vmatrix} h_{11}(j\omega) & \dots & h_{1n}(j\omega) \\ \vdots & & \vdots \\ h_{n1}(j\omega) & \dots & h_{nn}(j\omega) \end{vmatrix}\,. \end{aligned} \tag{8.41}$$

The matrix $\mathbf{H}(j\omega)$ is positive definite if the following condition is fulfilled:

$$M_i(j\omega) > 0\,, \quad \forall\omega\,, \quad i = 1, \dots, n\,. \tag{8.42}$$

The condition (8.39) as well as equivalent conditions (8.41) and (8.42) are usually expressed in the form of a system of inequalities containing elements of stabilizing matrix $\mathbf{K}$. It mostly occurs that $\mathbf{H}(j\omega)$ is not positive definite. This fact should not be treated as pessimistic. There is still a possibility to use the theory of hyperstability. By introduction of fictitious degrees of freedom one can replace the primary system with such a secondary system that conditions for a positive definite matrix will be fulfilled. The system with linear part $\mathbf{G}(s)$ of SISO type can be treated as the easiest example, which explains the required properties of additional degrees of freedom. For the SISO system, the matrix $\mathbf{H}(j\omega)$ is positive definite if all poles of the linear part are stable and the matrix $\mathbf{H}(j\omega)$ obtained for $\mathbf{G}(j\omega) = G(j\omega)$ satisfies inequality:

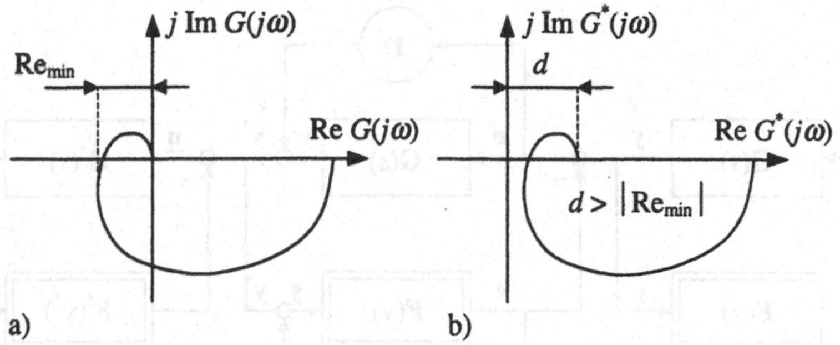

Fig. 8.23. The shift of the Nyquist plot (b) of the primary linear part $G(j\omega)$ as a result of adding a suitable constant d to a primary transfer function (a) (SISO case)

$$H(j\omega) = 0.5[G(j\omega) + G^{\mathrm{T}}(-j\omega)] = \mathrm{Re}(j\omega) > 0\,. \tag{8.43}$$

The condition (8.43) is fulfilled if representation of $G(j\omega)$ in complex plane (Nyquist plot) belongs entirely to the right half-plane – Fig. 8.23b. The uncorrected real systems usually do not fulfill the above requirement. To satisfy the requirement under consideration the Nyquist plot of the primary linear part $G(j\omega)$ has to be shifted right. Of course, the measure of shift has to satisfy inequality $d > |\mathrm{Re}\, G(j\omega)|$, for all $\mathrm{Re}\, G(j\omega) < 0$ – see Fig. 8.23.

The transformation of primary transfer function $G(j\omega)$ to secondary form $G^*(j\omega)$ has to be connected with the modification of primary non-linear part F. This modification has to be made so that input and output signals of linear and non-linear parts remain unchanged. The degrees of freedom for a MIMO system can be represented by matrix $\mathbf{D}$. The way of introducing matrix $\mathbf{D}$ in order to shift right all component transfer functions $G_{ij}(s)$ of matrix $\mathbf{G}(s)$ is shown in Fig. 8.24.

The relations between a primary system and a secondary system shown in Fig. 8.24 can be expressed in the form (8.45):

$$\mathbf{G}^*(s) = \mathbf{G}(s) + \mathbf{D}\,, \tag{8.44}$$

$$\mathbf{F}^*(\mathbf{y}^*):\quad \mathbf{v} = \mathbf{F}^*(\mathbf{y}^*) = \mathbf{F}(\mathbf{y} + \mathbf{D}\mathbf{v})\,. \tag{8.45}$$

The choice of coefficients d_{ij} of matrix $\mathbf{D}$ seems to be an easy task. The choice of very big coefficients, for example $d = 10^{1000}$, can be treated as the simplest idea for the creation of D. For big coefficients d one does not need to calculate the values $\mathrm{Re}_{\min}[G_{ij}(j\omega)]$, because shifts represented by such coefficients are undoubtedly sufficient to locate all component Nyquist plots $G_{ij}(j\omega)$ in the right half-plane.

The problem becomes more complicated if one realizes that matrix $\mathbf{D}$ also modifies the non-linear part $\mathbf{F}(\mathbf{y})$ to the secondary form $\mathbf{F}^*(\mathbf{y}^*)$ and that the secondary form obtained has to fulfill certain requirements. It is advised that

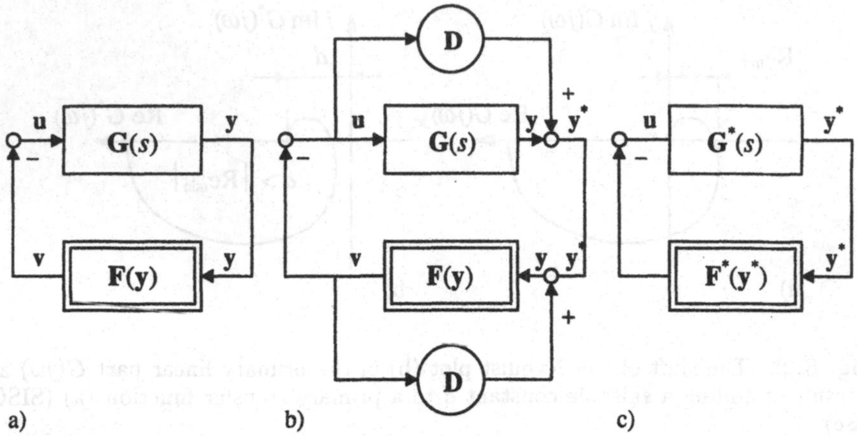

Fig. 8.24. The primary control system (a) and its secondary equivalents (b) and (c) obtained by introduction of additional degrees of freedom (represented by matrix **D**) shifting right the Nyquist plots for component transfer functions belonging to matrix $\mathbf{G}(s)$

the final choice of matrix **D** should be done after formulation of all conditions of hyperstability for both parts of the system (linear and non-linear). The choice of **D** on the basis of condition ML for only the linear part can mean that the condition MN for the non-linear part will not be fulfilled. In such a case the whole procedure of choosing **D** has to be repeated once more. Thus, during the initial stage, it is reasonable to discuss the conditions making the linear part strictly positive on the basis of suitable inequalities written in general form, where coefficients d_{ij} are not represented by numbers.

The formulation of conditions mentioned above becomes relatively easy if component transfer functions $G_{ij}(s)$ belonging to matrix transfer function $\mathbf{G}(s)$ of the primary linear part have stable poles, and stabilizing degrees of freedom represented by matrix **K** (Fig. 8.20) have not been introduced to the primary control system. In such a case the values $\mathrm{Re}_{\min}[G_{ij}(j\omega)]$ can be determined by analytic calculations or numerical analysis. There are specialized methods enabling the necessary analysis. The interesting method using curves of hyperstability (Schmitt 1996) can be treated as a good example.

If elements of matrices **K** and **D** are involved in the formulation of conditions making the linear part positive definite, then the **method of reduction of** $\mathrm{Re}(\omega)$ **to a second order sub-function** can be applied.

If the condition determining when the real part $\mathrm{Re}(\omega)$ is positive definite has the form of a second order function:

$$\mathrm{Re}(\omega) = E_2\omega^2 + E_1\omega + E_0 \geq 0\,,$$

where: $E_i = E_i(\mathbf{K}, \mathbf{D})$, then fulfillment of the following conditions I) and II):

$$\text{I)}\ E_2 > 0\,,$$
$$\text{II)}\ 4E_2E_0 - (E_1)^2 > 0\,,$$

guarantees that $\text{Re}(\omega)$ is positive definite for all ω: $[-\infty, \infty]$.

The third order function $\text{Re}(\omega)$:

$$\text{Re}(\omega) = E_3\omega^3 + E_2\omega^2 + E_1\omega + E_0\,,$$

can be reduced to a second order function, if we introduce condition $E_3 = 0$. Thus, the previous conditions I), II) and $E_3 = 0$ can be used to formulate requirements for positive definite $\text{Re}(\omega)$ of order three.

The fourth order function $\text{Re}(\omega)$:

$$\text{Re}(\omega) = E_4\omega^4 + E_3\omega^3 + E_2\omega^2 + E_1\omega + E_0\,,$$

can be reduced to a second order function, if we put $E_4 = E_3 = 0$. Putting $E_1 = E_3 = 0$ and $\omega^2 = x$, we also obtain $\text{Re}(x)$ of order two. The fifth order function $\text{Re}(\omega)$ can be reduced to a fourth order by introduction of condition $E_5 = 0$. Further conditions enabling reduction to order two can be obtained like in the previous case ($\text{Re}(\omega)$ of order four). The function $\text{Re}(\omega)$ of sixth order can be decomposed to second order sub-functions $f_1(\omega)$ and $f_2(\omega)$:

$$\begin{aligned}\text{Re}(\omega) &= E_6\omega^6 + E_5\omega^5 + E_4\omega^4 + E_3\omega^3 + E_2\omega^2 + E_1\omega + E_0 = \\ &= \omega^4(E_6\omega^2 + E_5\omega + E_4) + E_3\omega^3 + (E_2\omega^2 + E_1\omega + E_0)\end{aligned}$$

Now, putting $f_1(\omega) = E_2\omega^2 + E_1\omega + E_0$ and $f_2(\omega) = E_6\omega^2 + E_5\omega + E_4$, we can solve the problem applying the conditions for positive definite functions of order two to the following relations:

$$\begin{aligned}E_3 &= 0\,, \\ f_1(\omega) &= E_2\omega^2 + E_1\omega + E_0 > 0\,, \\ f_2(\omega) &= E_6\omega^2 + E_5\omega + E_4 > 0\,.\end{aligned}$$

Similar methods can be applied to functions $\text{Re}(\omega)$ of higher orders, although such functions are rather rarely considered for practical purposes.

The primary, standard control system and its equivalent represented by a secondary, standard form are shown in Fig. 8.25. The secondary control system is "equipped" with loops described by matrices $\mathbf{K}$ and $\mathbf{D}$. That is why conditions for hyperstability of linear and non-linear parts are usually formulated on the basis of secondary forms shown in Fig. 8.25b. The relations between the primary system and its secondary equivalent can be expressed in the form:

$$\begin{aligned}&\mathbf{G}^*(s) = [\mathbf{I} + \mathbf{G}(s)\mathbf{K}]^{-1}\mathbf{G}(s) + \mathbf{D}\,, \\ &\mathbf{F}^*(\mathbf{y}^*):\ \ \mathbf{v}^* = \mathbf{v} - \mathbf{K}\mathbf{y} = \mathbf{F}(\mathbf{y}^* + \mathbf{D}\mathbf{v}^*) - \mathbf{K}(\mathbf{y}^* + \mathbf{D}\mathbf{v}^*)\,. \qquad (8.46)\end{aligned}$$

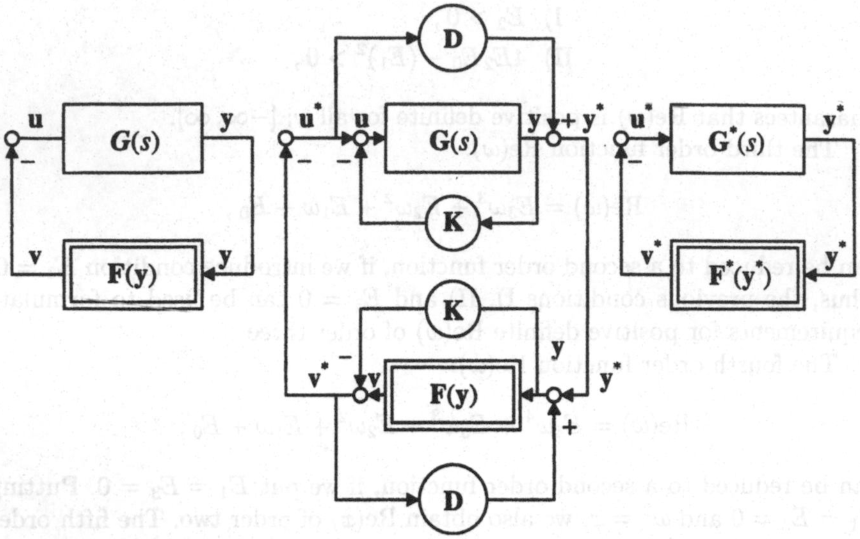

Fig. 8.25. The primary control system (a) and its equivalent, secondary form containing additional degrees of freedom **K**, **D** complete block diagram (b) and the result of its transformation to the structure of a standard control system (c)

Introduction of fictitious degrees of freedom (matrices **K** and **D**) seems a mathematical manipulation. But these degrees are of great importance and are necessary. They tie together the linear part $\mathbf{G}(s)$ and the non-linear one $\mathbf{F}(\mathbf{y})$ and ensure that the stability proof depends not on separate but on simultaneous investigations of the linear and non-linear part. The stability of a system depends on how well all elements of the system are matched – that is why both parts have to be considered together.

Explanation connected with condition MN determining hyperstability of the non-linear part

The primary form of Popov's integral inequality (8.36) is inconvenient as a tool for checking the hyperstability of the non-linear part. However, if the integrand is not negative, then (8.36) is fulfilled. It means that instead of (8.36) one can consider the secondary condition (8.47):

$$\mathbf{v}^{\mathrm{T}}(t)\mathbf{y}(t) \geq 0, \quad \forall t \geq 0\,. \tag{8.47}$$

Of course, the condition in the form (8.47) imposes additional constraints on the non-linear part (comparing with inequality (8.36)), which narrows the class of non-linear functions, but it is the price for substantial simplification of the necessary analysis. The condition (8.47) has to hold for all $t \geq 0$. If the non-linear part is a static one, then (8.47) does not depend on time.

The non-linear mapping $\mathbf{v} = \mathbf{F}(\mathbf{y})$ for a fuzzy system represents sectoral relation; i.e. the form of function $\mathbf{F}(\mathbf{y})$ depends on which sector of the input space the current vector $\mathbf{y}$ is placed in. That is why one has to formulate and check the secondary condition for the non-linear part (8.47) for each sector of input space. If dimensions of $\mathbf{v}$ and $\mathbf{y}$ are:

$$\mathbf{v} : (p, 1),$$
$$\mathbf{y} : (p, 1),$$

then condition (8.47) can be expressed in the form of sum (8.48):

$$\sum_{i=1}^{p} v_i y_i = \sum_{i=1}^{p} F_i(\mathbf{y}) y_i \geq 0, \tag{8.48}$$

where: $F_i(\mathbf{y}) = v_i$.

The secondary form of Popov's condition is usually formulated for a secondary control system containing fictitious degrees of freedom $\mathbf{K}$ and $\mathbf{D}$, therefore it can be expressed in the space $\{\mathbf{v}^*, \mathbf{y}^*\}$ of a secondary system by relation (8.49):

$$\mathbf{v}^{*\mathrm{T}} \mathbf{y}^* = \sum_{i=1}^{p} v_i^* y_i^* \geq 0. \tag{8.49}$$

The same Popov's condition expressed in the space $\{\mathbf{v}, \mathbf{y}\}$ of a primary system is given by relation (8.50):

$$[\mathbf{F}(\mathbf{y}) - \mathbf{K}\mathbf{y}]^{\mathrm{T}} \{\mathbf{y} - \mathbf{D}[\mathbf{F}(\mathbf{y}) - \mathbf{K}\mathbf{y}]\} \geq 0. \tag{8.50}$$

If matrix $\mathbf{K}$ is expressed by row vectors $\mathbf{K}_i$:

$$\mathbf{K} = \begin{bmatrix} k_{11} & \dots & k_{1p} \\ \vdots & & \vdots \\ k_{p1} & \dots & k_{pp} \end{bmatrix} = \begin{bmatrix} \mathbf{K}_1 \\ \vdots \\ \mathbf{K}_p \end{bmatrix},$$

where: $\mathbf{K}_i = [k_{i1}, \dots, k_{ip}]$, $i = 1, \dots, p$, and matrix $\mathbf{D}$ is expressed by row vectors $\mathbf{D}_i$:

$$\mathbf{D} = \begin{bmatrix} d_{11} & \dots & d_{1p} \\ \vdots & & \vdots \\ d_{p1} & \dots & d_{pp} \end{bmatrix} = \begin{bmatrix} \mathbf{D}_1 \\ \vdots \\ \mathbf{D}_p \end{bmatrix},$$

where: $\mathbf{D}_i = [d_{i1}, \dots, d_{ip}]$, $i = 1, \dots, p$, then condition for the non-linear part (8.50) obtains the form of the sum of all p outputs $v_i = F_i(\mathbf{y})$:

$$\sum_{i=1}^{p} [F_i(\mathbf{y}) - \mathbf{K}_i \mathbf{y}] \{y_i - \mathbf{D}_i[\mathbf{F}(\mathbf{y}) - \mathbf{K}\mathbf{y}]\} \geq 0. \tag{8.51}$$

The transformation of the secondary form of Popov's condition for the non-linear part to the input space of a primary system (vector $\mathbf{y}$ is the only variable in (8.51)) is very advantageous, since input space is divided into l sectors connected with the operation of a fuzzy controller, which means that condition (8.51) can be directly formulated in primary space for each sector. Thus, sectors in primary space do not have to be transferred to the secondary space of inputs $\mathbf{y}^*$.

In order to simplify the mathematical analysis the condition (8.51) can be decomposed into p conditions attached to particular outputs of the non-linear part $v_i = F_i(\mathbf{y})$ (Böhm 1995):

$$\begin{aligned} &[F_1(\mathbf{y}) - \mathbf{K}_1\mathbf{y}]\,\{y_1 - \mathbf{D}_1[\mathbf{F}(\mathbf{y}) - \mathbf{K}\mathbf{y}]\} \geq 0\,. \\ &\vdots \\ &[F_p(\mathbf{y}) - \mathbf{K}_p\mathbf{y}]\,\{y_p - \mathbf{D}_p[\mathbf{F}(\mathbf{y}) - \mathbf{K}\mathbf{y}]\} \geq 0\,. \end{aligned} \tag{8.52}$$

The particular inequality in the system (8.52) can be treated as a component of composite condition (8.51). The fulfilling of requirements insuring that all component conditions are positive definite becomes more difficult, compared to the same task for the sum (8.51) created by components (8.52). However, it is advisable to check the possibility of fulfilling the hyperstability conditions (8.52) by a suitable choice of matrices **K** and **D**. If the above attempts fail, then the next attempt of checking conditions given by the full sum (8.51) has to be made.

Both conditions, i.e. (8.51) and (8.52), are the functions of stabilizing matrix **K** as well as matrix **D** which ensures that the linear part is positive definite (its Nyquist plots are sufficiently shifted right). The **K** and **D** usually consist of a small number of elements (degrees of freedom). For example, if the linear part of a system has 2 inputs and 2 outputs, then the total number of degrees of freedom equals 8, whereas the total number of hyperstability conditions for the whole system is usually much bigger. It is the result of the existence of many operational sectors of fuzzy controllers. One among further examples shows that the total number of hyperstability conditions is equal to 70 (!). It is quite easy to understand that solving 70 inequalities by a choice of 8 variables can be a very difficult task. Sometimes, there is no solution to the problem mentioned above. So, the introduction of large numbers of fictitious degrees of freedom into a secondary system can be the only way which allows us to prove system stability.

If the number of degrees of freedom "delivered" by matrices **K** and **D** is not sufficient to prove the system stability, then consecutive, fictitious degrees of freedom r_{ii}, $i = 1, \ldots, p$, can be introduced in the form of matrices:

$$\begin{aligned} \left(\frac{1}{r_{ii}}\right) &= \operatorname{diag}\left(\frac{1}{r_{11}}, \ldots, \frac{1}{r_{pp}}\right), \\ (r_{ii}) &= \operatorname{diag}\left(r_{11}, \ldots, r_{pp}\right). \end{aligned} \tag{8.53}$$

The way of introducing additional degrees of freedom to a secondary control system is shown in Fig. 8.26.

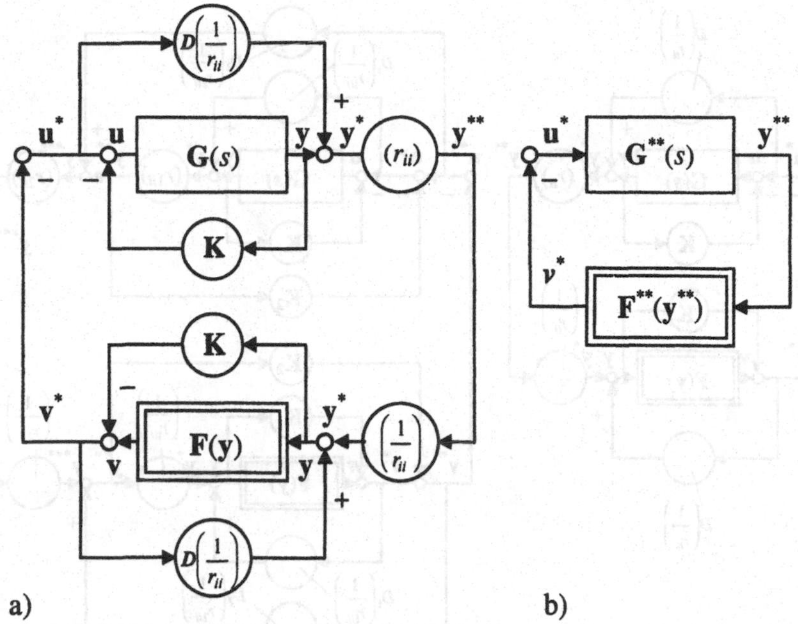

Fig. 8.26. The secondary control system containing the maximal number of fictitious degrees of freedom – the result of the first stage of expanding the linear and non-linear part: a) expanded form, b) folded form

The formulae (8.54) represent the dependence between a primary system and a secondary one:

$$\mathbf{G}^{**}(s) = (r_{ii})[\mathbf{I} + \mathbf{G}(s)\mathbf{K}]^{-1}\mathbf{G}(s) + (r_{ii})\mathbf{D}\left(\frac{1}{r_{ii}}\right), \tag{8.54}$$
$$\mathbf{F}^{**}(y^{**}) : \ \mathbf{v}^* = \mathbf{v} - \mathbf{K}\mathbf{y} =$$
$$\left[\mathbf{I} - \mathbf{K}\mathbf{D}\left(\frac{1}{r_{ii}}\right)\right]^{-1} \mathbf{F}\left[\left(\frac{1}{r_{ii}}\right)\mathbf{y}^{**} + \mathbf{D}\left(\frac{1}{r_{ii}}\right)\mathbf{v}^*\right] - \mathbf{K}\left(\frac{1}{r_{ii}}\right)\mathbf{y}^{**}.$$

If the number of operational sectors belonging to a fuzzy controller input space is very big (the number of sectors grows rapidly in two situations: for a growing number of fuzzy controller inputs and in the case of a growing number of fuzzy sets assigned to each controller input and output) then the necessity of introducing an even bigger number of fictitious degrees of freedom can occur. The same necessity can occur if one fails in proving the system hyperstability, even after "utilization" of all degrees of freedom shown in Fig. 8.26 (matrices **K**, **D**, $(1/r_{ii})$, (r_{ii})). It must not be treated as evidence confirming that the control system is unstable, since the considered **theory of stability generates** so-called **sufficient conditions of stability**.

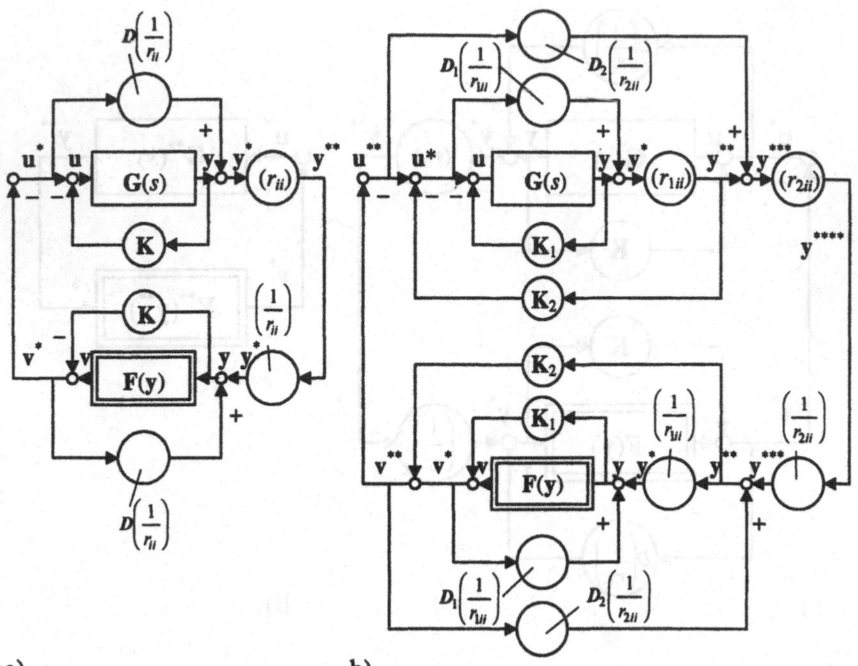

Fig. 8.27. The system "equipped" with first level of fictitious degrees of freedom (a) and its extension by introduction of second level of fictitious degrees of freedom (b)

It means that a system fulfilling stability conditions is undoubtedly stable. If stability conditions are not fulfilled then the system may be unstable or stable (a final conclusion can not be drawn). Thus, if conditions for hyperstability of a system are not fulfilled one should not neglect further attempts aimed at proof of system stability. There is still a chance for success, especially if observation of the operation of the real system seems to suggest its stability. To successfully prove the stability during the next attempt one can try expanding the space by introducing a second level of fictitious degrees of freedom see Fig. 8.27.

The readers who doubt that expanding the space by introducing a second level of degrees is an efficient method should take into account that hyperstability for most primary systems (Fig. 8.18a) can not be proved, since even initial requirements of the method (the same number of inputs and outputs of the linear part, stability of all poles representing the linear part, the system being strictly positive definite) are usually not fulfilled. Despite the above obstacles the hyperstability can be successfully proved due to the introduction of fictitious elements, which mutually reduce themselves. For comparison, one can not solve the equation:

$$x^2 + a^2 = 0\,, \tag{8.55}$$

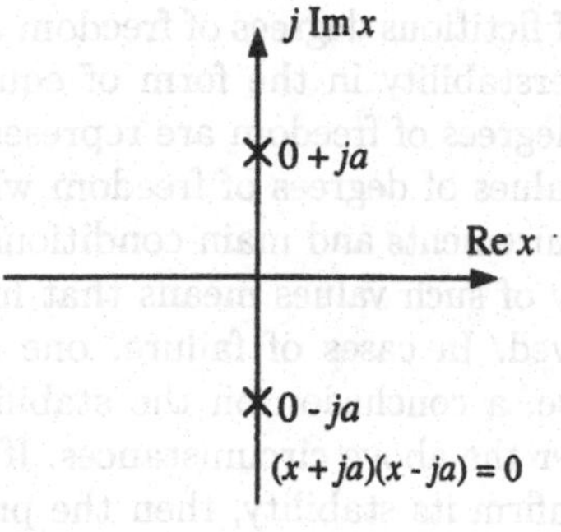

Fig. 8.28. The roots of equation $x^2 + a^2 = 0$ in complex plane

in one-dimensional space X, but a solution can be found in two-dimensional space determined by a real axis and an imaginary one, see Fig. 8.28.

If real problems seem to be apparently insoluble, then the so-called "general principle of mathematical modeling" (Bezdek 1993) advises to expand the space. The introduction of additional degrees of freedom means that the mathematical form of hyperstability conditions becomes more and more complicated. There is specialized mathematical software which can be helpful during the preparation and checking of stability conditions, since it is able to transform complicated mathematical formulas given in general form.

The additional degrees of freedom mean that the secondary linear part $\mathbf{G}^{**}(s)$ as well as the secondary non-linear part $\mathbf{F}^{**}(\mathbf{y}^{**})$ contain more variables (degrees of freedom) than the respective primary parts $\mathbf{G}(s)$ and $\mathbf{F}(\mathbf{y})$ (in fact $\mathbf{G}(s)$ and $\mathbf{F}(\mathbf{y})$) do not contain degrees of freedom at all). That is why the additional degrees of freedom allow us to form the input-output mappings in such a way that hyperstability conditions are fulfilled (the task is realizable if the real system is stable). The expansion of system space by the introduction of higher levels of degrees of freedom creates the chance for proving the hyperstability of each (hyper-)stable, real system.

Let us recapitulate: the following steps a – f determine the sequence of advised activities during investigation of hyperstability of a continuous control system in frequency domain.

a) The transformation of control system scheme to the standard form (primary system) consisting of the linear part and the non-linear part.
b) Checking if preliminary requirements for the application of hyperstability theory are fulfilled. If yes, then skip to activities described in step c), if no, skip to step d).
c) Checking of main conditions for system hyperstability. If conditions hold, then proof is finished, otherwise skip to step e).
d) Introduction of fictitious degrees of freedom and formulation of preliminary requirements given in the form of equations and inequalities containing the introduced degrees of freedom as variables. Skip to step e).

e) The introduction of fictitious degrees of freedom and formulation of main conditions for hyperstability in the form of equations and inequalities, where introduced degrees of freedom are represented as variables.

f) Looking for such values of degrees of freedom which simultaneously fulfill preliminary requirements and main conditions for hyperstability. The successful discovery of such values means that hyperstability of the system has been proved. In cases of failure, one can discontinue further attempts. Of course, a conclusion on the stability of a system can not be formulated under the above circumstances. If observations of the real system seem to confirm its stability, then the process of proving should be continued by skipping to step d).

Certain initial requirements (like, for example, controllability and observability of the linear part) can be investigated after the determination of the values of degrees of freedom fulfilling the hyperstability conditions of the secondary system.

If hyperstability is investigated for signal $w \neq 0$, then new co-ordinates (variables) have to be defined. The "0" of new co-ordinates has to represent the new point of equilibrium of the system. The exchange of variables is connected with the necessity of shifting the non-linear characteristic $F(y)$. The proper example is shown in Fig. 8.29.

The values of variables representing a state of equilibrium can be calculated on the basis of differential equations describing the operation of the system. Considering the example shown in Fig. 8.29 one can draw the conclusion that a standard system under non-zero reference signal $w \neq 0$ is equivalent to the system under reference signal $w_1 = 0$ and the shifted characteristic of the non-linear part (the origin of new co-ordinates has to be shifted to the new point of equilibrium of the system $[u_r, y_r, v_r]$).

Example 8.3.1.1. The investigation of hyperstability of a SISO fuzzy system

Let us deal once more with the system considered in Example 8.2.1. The circle criterion has been applied to check the stability of that system. The system is shown in Fig. 8.30. The fuzzy controller realizes static mapping which can be represented by a non-linear characteristic composed of segments of straight lines.

The verification of condition PL1

(The number of inputs of the linear part should be equal to the number of its outputs)
The condition is fulfilled.

The verification of condition PL2

(The linear part has to be completely controllable and observable)
The above condition will be checked later, after introduction (if necessary) of fictitious degrees of freedom.

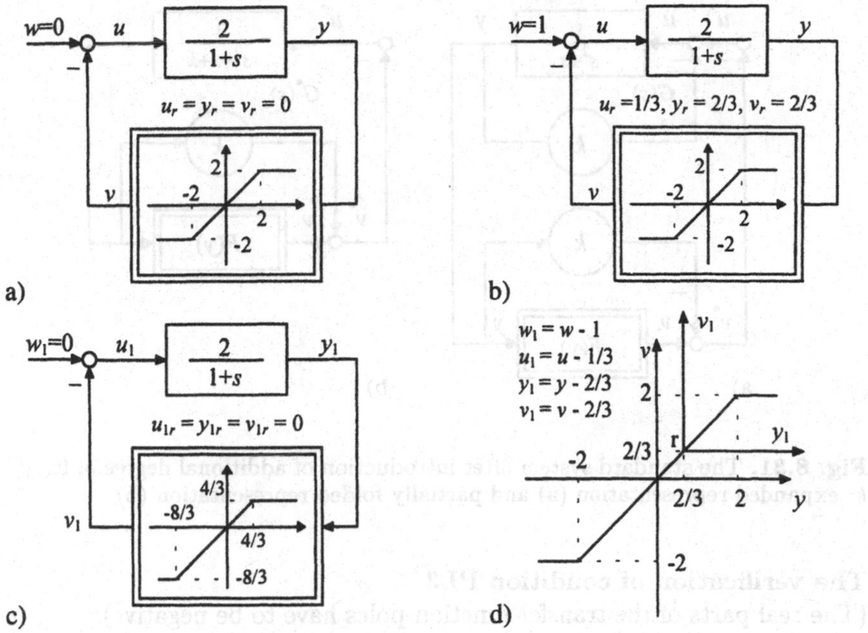

Fig. 8.29. The standard system under reference signal $w = 0$ and its state of equilibrium (a), system under signal $w = 1$ and its state of equilibrium (b), system under signal $w = 1$ after introduction of new variables and its state of equilibrium (c) and location of the new static characteristic of the non-linear part $v_1 = F(y_1)$ in relation to primary characteristic $v = F(y)$, (d)

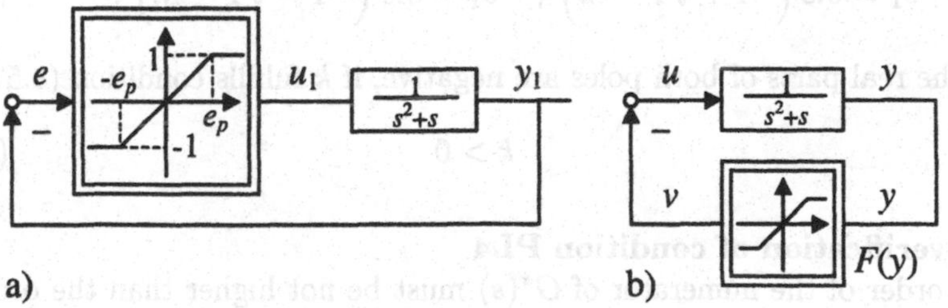

Fig. 8.30. The control system containing a static controller (a) and its transformation to the standard form (b)

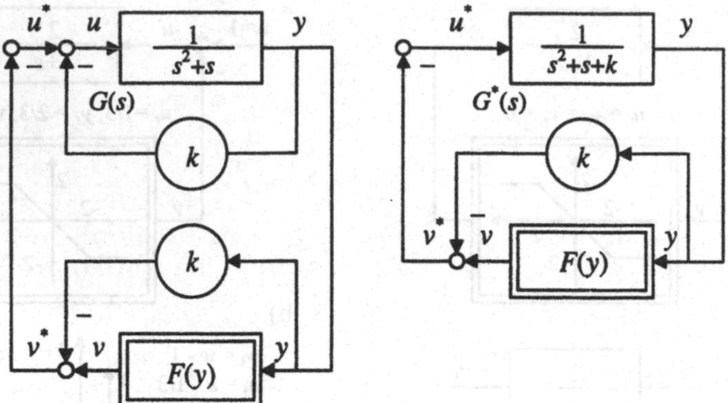

a) b)

Fig. 8.31. The standard system after introduction of additional degree of freedom k: expanded representation (a) and partially folded representation (b)

The verification of condition PL3

(The real parts of the transfer function poles have to be negative)
There are two poles: $s_1 = -1$, $s_2 = 0$. Because of the value of the second pole s_2 the condition PL3 is not fulfilled. To overcome obstructions encountered, the fictitious degree of freedom k can be introduced to the linear part as well as the non-linear part, see Fig. 8.31.

Now, the poles s_1, s_2 of the secondary linear part $G^*(s)$ obtain the values:

$$s_1 = 0.5\left(-1 + \sqrt{1-4k}\right), \quad s_1 = 0.5\left(-1 - \sqrt{1-4k}\right). \tag{8.56}$$

The real parts of both poles are negative, if k fulfills condition (8.57):

$$k > 0 \tag{8.57}$$

The verification of condition PL4

(The order of the numerator of $G^*(s)$ must be not higher than the order of its denominator)
The condition is fulfilled.

The verification of condition PN1

(The mapping $v^* = F^*(y)$ has to be unique)
It results from Fig. 8.32, that mapping represented by the "new" characteristic of the non-linear part is unique. It means that condition PN1 is fulfilled.

The verification of condition PN2

(The mapping $v^* = F^*(y)$ realized by the non-linear part has to fulfill condition $F^*(0) = 0$)
Fig. 8.32 shows that the condition is fulfilled.

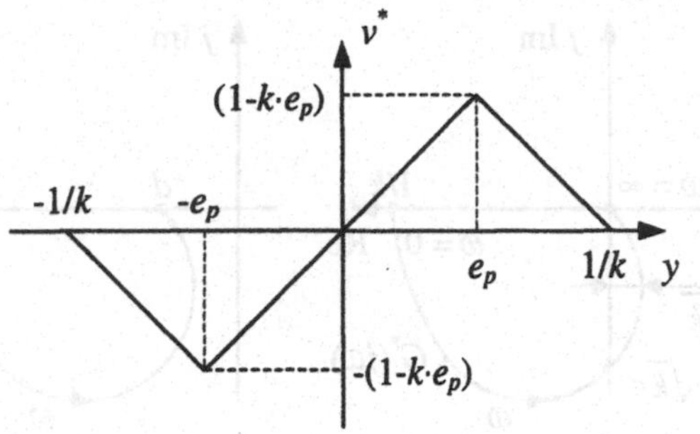

Fig. 8.32. The static characteristic of the non-linear part after introduction of fictitious degree of freedom k

The verification of main condition PL

(The linear part $G^*(s)$ has to be strictly positive real)
The above condition can be decomposed to several component requirements.

- The poles of transfer function $G^*(s)$ have to be located to the left of the imaginary axis of complex plane "s". For $k > 0$ the condition is fulfilled.
- The matrix $H(j\omega)$ of the linear part $G^*(s)$ has to be positive definite for all $\omega > 0$ and Hermitian.

$G^*(j\omega)$ and $G^*(-j\omega)$ can be expressed by (8.58) and (8.59) respectively.

$$G^*(j\omega) = \frac{k-\omega^2}{(k-\omega^2)^2+\omega^2} - \frac{j\omega}{(k-\omega^2)^2+\omega^2} = \mathrm{Re}^*(\omega) + \mathrm{Im}^*(\omega) \quad (8.58)$$

$$G^*(-j\omega) = \frac{k-\omega^2}{(k-\omega^2)^2+\omega^2} + \frac{j\omega}{(k-\omega^2)^2+\omega^2} = \mathrm{Re}^*(-\omega) + \mathrm{Im}^*(-\omega) \quad (8.59)$$

In accordance with (8.37) matrix $H^*(j\omega)$ obtains the form (8.60), which is Hermitian.

$$H^*(j\omega) = 0.5\left[G^*(j\omega) + G^{*\mathrm{T}}(-j\omega)\right] = \frac{k-\omega^2}{(k-\omega^2)^2+\omega^2} = \mathrm{Re}^*(\omega)$$
$$H^*(j\omega) = H^*(-j\omega) \quad (8.60)$$

The considered matrix will be positive definite (formula (8.39) or (8.42)), if $\mathrm{Re}^*(\omega) \geq 0$. Roughly speaking, the frequency response $G^*(j\omega)$ should be located in the right half of the complex plane. Fig. 8.33 shows that the above condition is not fulfilled.

The analysis of (8.60) leads to the statement that $\mathrm{Re}^*(\omega) \leq 0$ for $w \geq \sqrt{k}$ and the maximum of $|\mathrm{Re}^*(\omega)|$ over the considered frequency range is $|1/(1+2\sqrt{k})|$. Thus, the fictitious degree of freedom d should be added in order to fulfill the requirements under consideration – see Fig. 8.34.

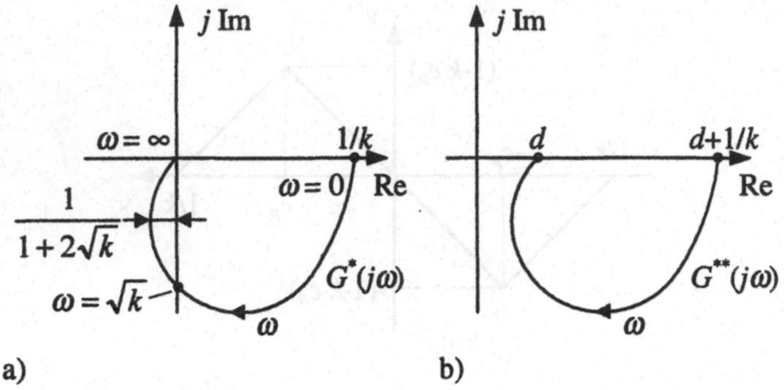

Fig. 8.33. The frequency response $G^*(j\omega)$ before introduction of the second degree of freedom d (a) and the result of introduction of the second degree of freedom (b)

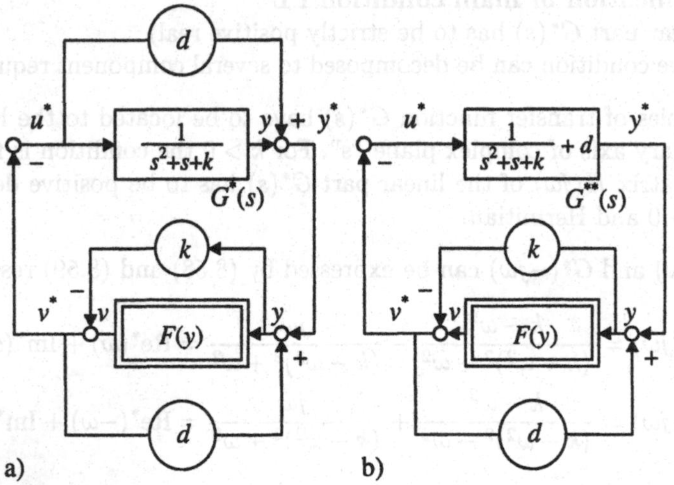

Fig. 8.34. The control system comprising two degrees of freedom k and d represented by expanded scheme (a) and partially folded scheme (b)

After adding the fictitious degree of freedom d the real part of $G^{**}(j\omega)$ can be expressed in the form (8.61):

$$\mathrm{Re}^{**} = \frac{k-\omega^2}{(k-\omega^2)^2+\omega^2} + d = \mathrm{Re}^* + d\,. \tag{8.61}$$

Since $\mathrm{Re}^*(\omega) \geq -1/(1+2\sqrt{k})$, therefore the frequency response has to be shifted right behind the imaginary axis. Thus, the minimum shift d is:

$$d \geq \frac{1}{1+2\sqrt{k}}\,. \tag{8.62}$$

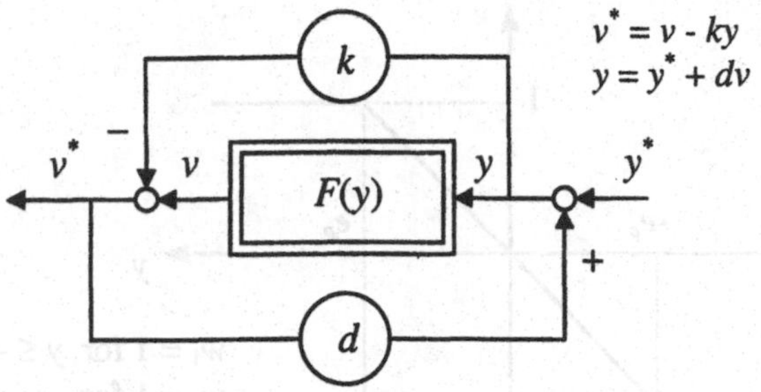

Fig. 8.35. The non-linear part comprising fictitious degrees of freedom

Let us note that minimal shift d depends on k. If k had been determined earlier (just after the analysis of stability of poles of $G^*(s)$), then the minimal value of d would have been determined in advance. But both degrees of freedom d and k have to fulfill conditions for the non-linear part as well. So, a wide space for changes of d and k can be required.

The main condition MN for the non-linear part

Here the secondary Popov's condition (8.47) obtains the form (8.63).

$$v^* y^* \geq 0 \tag{8.63}$$

The condition (8.63) becomes (8.64), after its transformation to primary space $[v, y]$.

$$v^* y^* = (v - ky)[y - d(v - ky)] \geq 0 \tag{8.64}$$

The above condition can be transformed to the form (8.65), which is called, for convenience, Popov's parabola:

$$v^* y^* = -dv^2 - (k + dk^2)y^2 + (1 + 2kd)yv \geq 0\,. \tag{8.65}$$

However, one can simplify the inequality (8.64). If we denote:

$$v - ky = A\,,$$

then (8.64) can be rewritten in the form:

$$A(y - dA) = Ay - dA^2 \geq 0\,,$$

hence:

$$Ay \geq dA^2\,.$$

Since $A^2 \geq 0$, one finally obtains (8.66).

$$d \leq \frac{y}{A}\,, \quad A \neq 0 \tag{8.66}$$

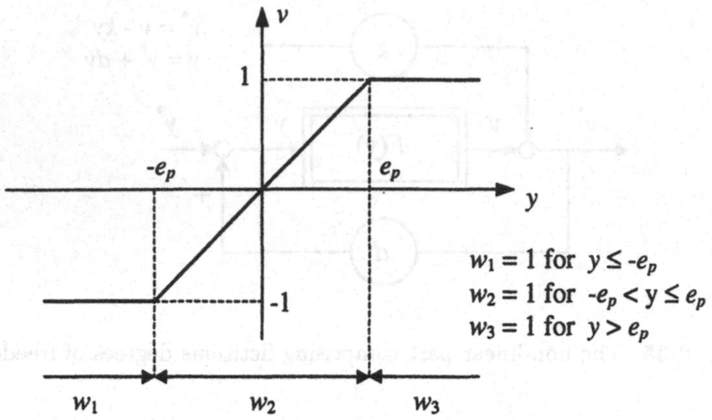

Fig. 8.36. The static characteristics of non-linear part (fuzzy controller) $F(y)$

The condition (8.66) holds for positive and negative A. It can also be decomposed to the more convenient form expressed by the two following inequalities:

$$y - dA \geq 0, \text{ for } A > 0, \qquad y - dA \leq 0, \text{ for } A < 0\,.$$

Putting primary values instead of A one obtains the transformed form of Popov's condition (8.67):

$$\begin{aligned} d &\leq \frac{y}{v - ky}\,, \\ d &\leq \frac{y}{F(y) - ky}\,. \end{aligned} \tag{8.67}$$

The analysis of system stability can be done using Popov's parabola (8.65) or the simplified form (8.67).

From (8.18), it results that the static characteristic of the non-linear part $F(y)$ can be expressed by:

$$v = F(y) = -w_1 + \frac{y}{e_p} w_2 + w_3\,, \tag{8.68}$$

where w_1, w_2, w_3 represent logical variables, which determine membership of current y-value to the one among separated sub-spaces, creating space for variable y – see Fig. 8.36. The condition (8.67) has to be fulfilled for each sector of the non-linear part $F(y)$.

For the first sector $w_1 = 1$, where $v = -1$, one obtains Popov's condition (8.65) in the form (8.69).

$$f_1(y) = -(k + dk^2)y^2 - (1 + 2dk)y - d \geq 0 \tag{8.69}$$

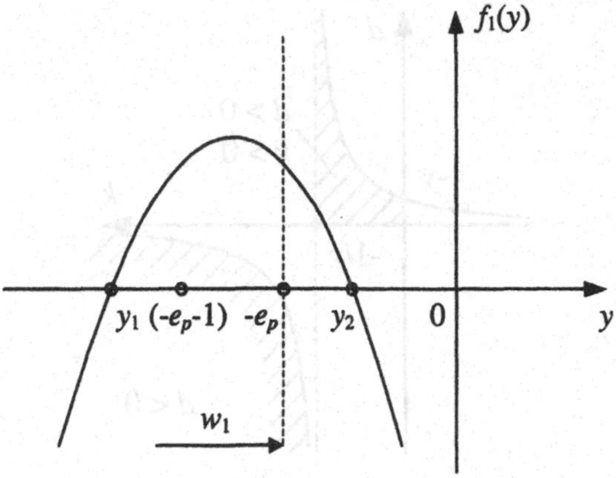

Fig. 8.37. The graphic representation of Popov's condition for the non-linear part of sector $w_1 = 1$

The graphic representation of the above condition is shown in Fig. 8.37. It results from Fig. 8.37 that the stability of a system can be proved for the following range belonging to the sector $w_1 = 1$:

$$y_1 \le y \le -e_p\,.$$

The width of the above range will increase if one finds suitable values k and d fulfilling all conditions for hyperstability. The conditions for the first sector can be worked out using Popov's parabola (Fig. 8.37) or a simplified form of Popov's condition (8.67). For the first sector $v = F(y) = -1$. Thus, the simplified Popov's condition can be expressed with (8.70):

$$d < \frac{-y}{1+ky}\,, \quad \text{for } y < 0,\ k \neq -\frac{1}{y}\,. \tag{8.70}$$

The graphic representation of that condition in the space $[k, d]$ is shown in Fig. 8.38. Fig. 8.38 clearly shows the following contradiction: our wish to prove the system stability over a wider range of variable y causes the sub-space of $[k, d]$, representing a possible choice of coefficients k and d, to decrease (shaded field in Fig. 8.38).

If the range for proof of stability in the space $[y]$ is too big it can happen that the stability of the system can not be proved at all. That is why the range for proof of stability is assumed to be as in Fig. 8.37:

$$-e_p - 1 \le y \le -e_p\,.$$

For $e_p = 1$ is:

$$-2 \le y \le -1\,. \tag{8.71}$$

Because of inequality:

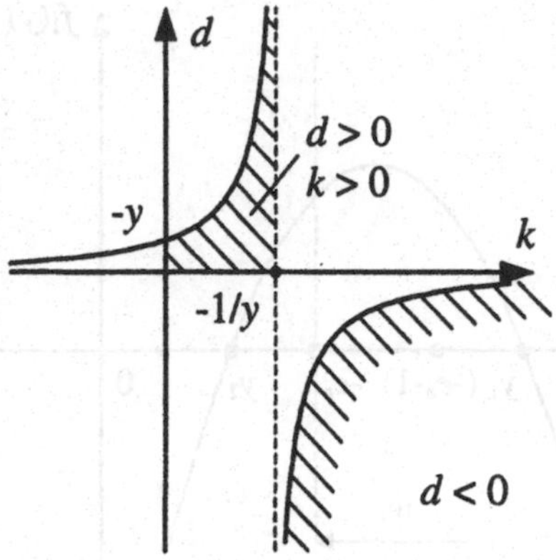

Fig. 8.38. The graphic representation of the simplified Popov's condition for the first sector of the non-linear part

$$k + dk^2 > 0\,,$$

the crown of Popov's parabola is located over its arms, see Fig. 8.37. From the relation obtained between the points creating the parabola one can conclude that fulfillment of Popov's condition in two points $y = -e_p$ and $y = -e_p - 1$ means fulfillment of Popov's condition for all points belonging to the whole assumed range (8.71). Thus the Popov's conditions over range (8.71) are:

$$d < \frac{1}{1-k}\,, \quad \text{for } y = -e_p = -1,\ k \neq 1\,, \tag{8.72}$$

$$d < \frac{2}{1-2k}\,, \quad \text{for } y = -e_p - 1 = -2,\ k \neq 0.5\,. \tag{8.73}$$

The above conditions can be represented in graphic form – see Fig. 8.39. Finally, one can draw the conclusion that both conditions (8.72) and (8.73) are fulfilled if k and d belong to the sub-space determined by inequalities (8.74):

$$0 < k < 0.5, \quad d < \frac{1}{1-k}\,. \tag{8.74}$$

For the second sector $w_2 = 1$, where $v = F(y) = y/e_p$ – see Fig. 8.36, one obtains Popov's condition (8.65) in the form (8.75), which is represented graphically in Fig. 8.40.

$$f_2(y) = \left[\left(\frac{1}{e_p} - k\right) - d\left(\frac{1}{e_p} - k\right)^2\right] y^2 \geq 0 \tag{8.75}$$

The simplified form of condition (8.75) results from (8.67):

$$d < \frac{y}{\frac{y}{e_p} - ky}\,, \quad k \neq \frac{1}{e_p}\,.$$

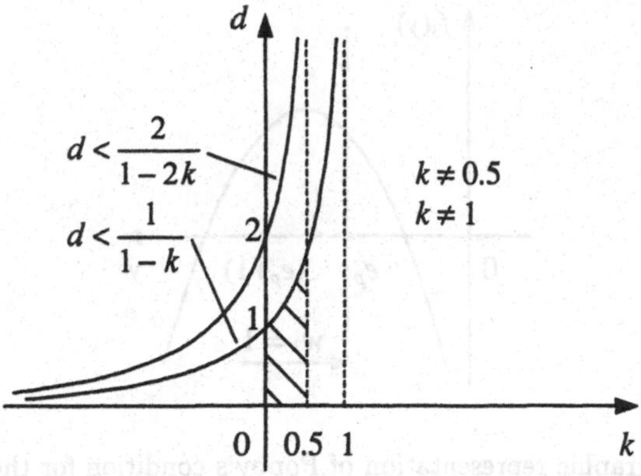

Fig. 8.39. The graphic representation of Popov's conditions for the first sector of the non-linear part

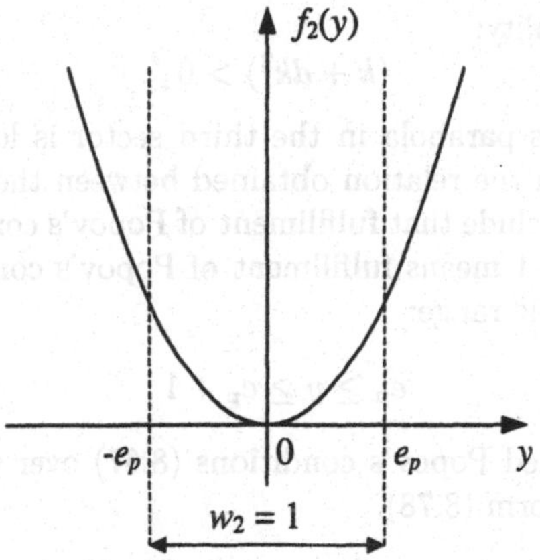

Fig. 8.40. The graphic representation of Popov's condition for the second sector of the non-linear part

For chosen value $e_p = 1$ one obtains:

$$d < \frac{1}{1-k}, \quad k \neq 1. \tag{8.76}$$

It means that the form of condition (8.76) for the second sector is identical to the form of condition (8.72) for the first sector. The condition (8.72) has been represented graphically in Fig. 8.39.

For the third sector $w_3 = 1$, where $v = F(y) = 1$, one obtains Popov's condition (8.65) in the form (8.77), which is represented graphically in Fig. 8.41.

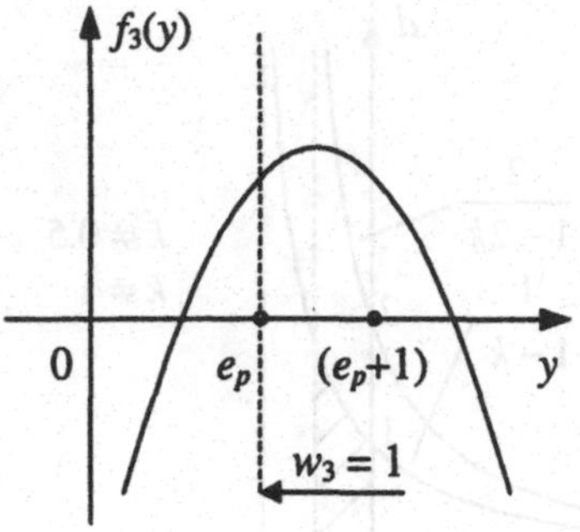

Fig. 8.41. The graphic representation of Popov's condition for the third sector of the non-linear part

$$f_3(y) = -(k + dk^2)y^2 + (1 + 2kd)y - d \geq 0 \tag{8.77}$$

Because of inequality:

$$(k + dk^2) > 0,$$

the crown of Popov's parabola in the third sector is located over its arms – see Fig. 8.41. From the relation obtained between the points creating the parabola one can conclude that fulfillment of Popov's condition in two points $y = e_p$ and $y = e_p + 1$ means fulfillment of Popov's condition for all points belonging to the whole range:

$$e_p \geq y \geq e_p + 1.$$

Thus, the simplified Popov's conditions (8.67) over the above range can be expressed in the form (8.78).

$$\begin{aligned} d &< \frac{e_p}{1 - ke_p}, \quad \text{for } y = e_p,\ k \neq \frac{1}{e_p}, \\ d &< \frac{e_p + 1}{1 - k(e_p + 1)}, \quad \text{for } y = e_p + 1,\ k \neq \frac{1}{e_p + 1}. \end{aligned} \tag{8.78}$$

If we put chosen value $e_p = 1$ to (8.78), then:

$$d < \frac{1}{1 - k}, \quad \text{for } y = 1,\ k \neq 1, \tag{8.79}$$

$$d < \frac{2}{1 - 2k}, \quad \text{for } y = 2,\ k \neq 0.5. \tag{8.80}$$

The conditions (8.79) and (8.80) for the third sector are identical to the respective conditions (8.72) and (8.73) for the first sector. The conditions (8.72) and (8.73) have been represented graphically in Fig. 8.39. All the conditions formulated for the linear part as well as the non-linear one have been collected together creating the final result (8.81):

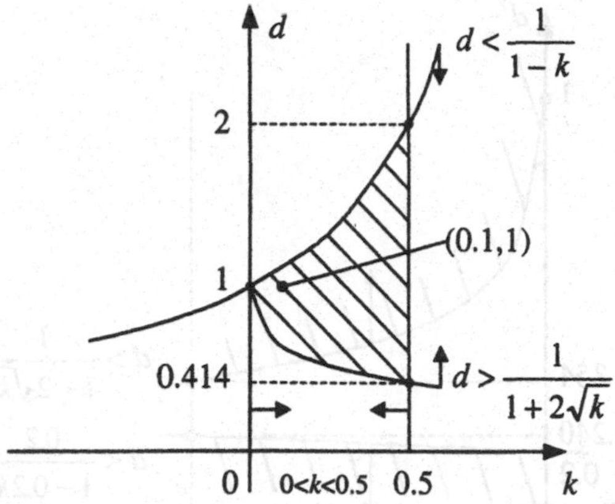

Fig. 8.42. The graphic representation of all the hyperstability conditions in respect of choice of coefficients k and d, for $e_p = 1$ (gain of controller $1/e_p = 1$)

$$k > 0, \quad d > \frac{1}{1 + 2\sqrt{k}},$$
$$0 < k < 0.5, \quad d < \frac{1}{1 - k}. \tag{8.81}$$

The graphic representation of (8.81) is shown in Fig. 8.42.

Fig. 8.42 shows that for all the conditions there is a common sub-space enabling the proper choice of k and d. For example, the choice $k = 0.1$ and $d = 1$ (Fig. 8.42) fulfills all resultant conditions given by (8.81). However, the investigation of stability for a controller having higher gain in the second sector ($e_p = 0.2$, $1/e_p = 5$) leads to the opposite conclusion, namely, there is no common sub-space enabling the choice of coefficients fulfilling all the stability conditions, see Fig. 8.43.

It can be repeated once again that the considered hyperstability conditions have to be treated as sufficient conditions. It means that the lack of a common sub-space for the choice of k and d in the case of high gain of the controller $1/e_p = 5$ does not prove system instability. According to previous advice one can try to prove system stability by adding new degrees of freedom. The investigations of stability for many degrees of freedom usually involve a lot of work, but often it is the only chance for successfully proving system stability. Proof of system stability is compulsory if industrial applications of fuzzy controllers are to be implemented.

For acceptable values of degrees of freedom k and d ($k = 0.1$, $d = 1$) one can easily check the controllability and observability of linear part $G^{**}(s)$, Fig. 8.34.

$$G^{**}(s) = G^{*}(s) + d = \frac{1}{s^2 + s + k} + d \tag{8.82}$$

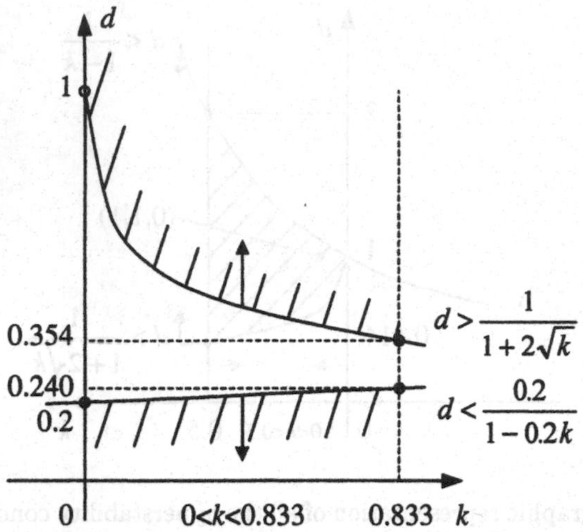

Fig. 8.43. The graphic presentation of system stability conditions in respect of the choice of degrees of freedom k and d for high gain $1/e_p = 5$ ($e_p = 0.2$) of the controller

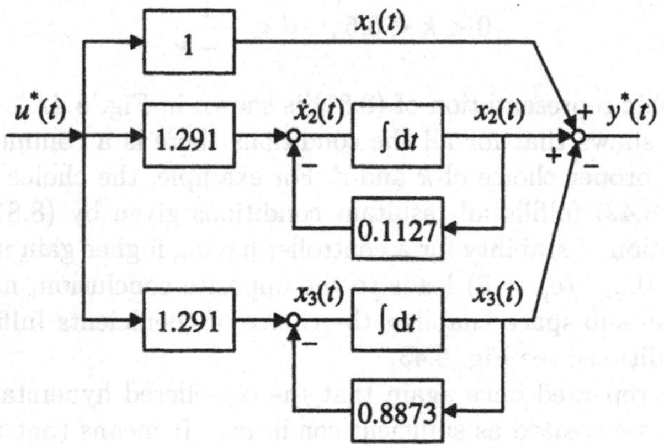

Fig. 8.44. The modal representation of linear part $G^{**}(s)$ by state variables x_1, x_2, x_3

For chosen values $k = 0.1$ and $d = 1$ the transfer function $G^{**}(s)$ obtains the form (8.83).

$$G^{**}(s) = \frac{1}{s^2 + s + 0.1} + 1 = \frac{1.291}{s + 0.1127} - \frac{1.291}{s + 0.8873} + 1 \tag{8.83}$$

The state variable representation in Fig. 8.44 shows that all state variables x_1, x_2, x_3 are controllable by input u^* and observable from the output y^* (Markowski 1985). An identical result can be obtained by application of

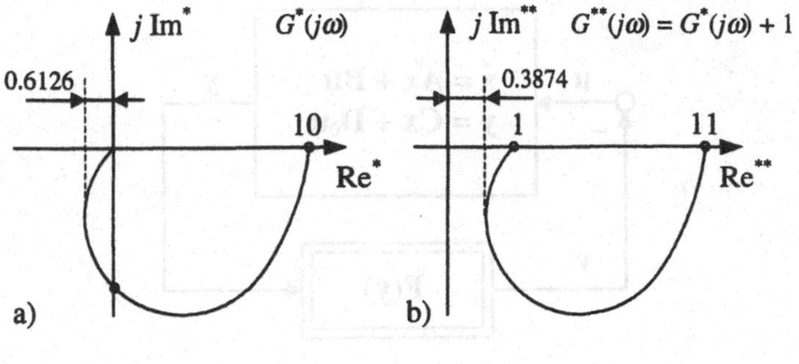

Fig. 8.45. The shift of frequency response $G^*(j\omega)$ as a result of the introduction of degree of freedom $d = 1$

strictly theoretical rules for checking the system controllability and observability, (Günther 1984).

Fig. 8.45a shows the characteristic $G^*(j\omega)$ comprising only one degree of freedom k introduced for stabilizing. Fig. 8.45b shows the characteristic $G^{**}(j\omega)$ comprising degree of freedom k and additional degree d, introduced in order to make the real part of $G^{**}(j\omega)$ positive definite.

◊ ◊ ◊

8.3.2 The time domain conditions for hyperstability of continuous, non-linear control systems containing a time-invariant non-linear part

Let state space equations of the system linear part be:

$$\begin{aligned} \dot{\mathbf{x}}(t) &= \mathbf{A}\mathbf{x}(t) + \mathbf{B}\mathbf{u}(t) \\ \mathbf{y}(t) &= \mathbf{C}\mathbf{x}(t) + \mathbf{D}_0\mathbf{u}(t) \end{aligned} \tag{8.84}$$

The standard control system (Fig. 8.46) is (asymptotically) hyperstable (Opitz 1993), if:

PL1) The number of inputs u_i of the linear part is equal to the number of its outputs y_i, which means that vectors $\mathbf{u}$ and $\mathbf{y}$ are of the same dimensions,

ML) The Kalman-Jakubowich equations of the form (8.85) have the solution containing positive definite matrix $\mathbf{P}$, any (regular) matrix $\mathbf{L}$ and any matrix $\mathbf{V}$.

$$\begin{aligned} \mathbf{A}^T\mathbf{P} + \mathbf{P}\mathbf{A} &= -\mathbf{L}\mathbf{L}^T \\ \mathbf{C} - \mathbf{B}^T\mathbf{P} &= \mathbf{V}^T\mathbf{L}^T \\ \mathbf{D}_0^T + \mathbf{D}_0 &= \mathbf{V}^T\mathbf{V} \end{aligned} \tag{8.85}$$

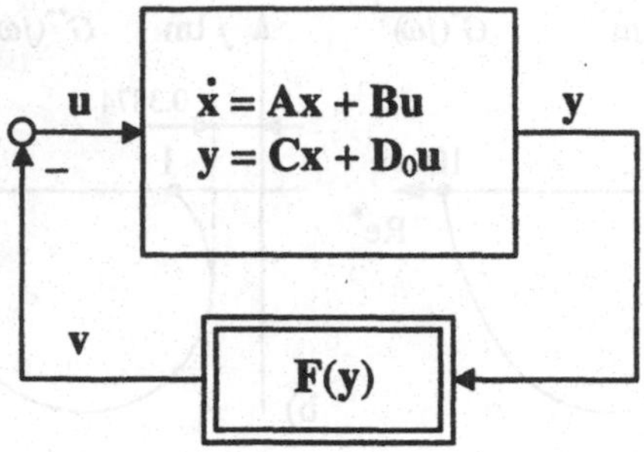

Fig. 8.46. The standard control system with a linear part described by state equations

Explanation
The square matrix is called regular (non-singular) if its rows or columns are linearly independent (Zurmühl 1964). The determinant of non-singular matrix $\mathbf{L}$ fulfills condition:

$$\det \mathbf{L} \neq 0 .$$

The conditions for the non-linear part $\mathbf{F}(\mathbf{y})$ are identical to those considered previously, where the linear part has been represented in frequency domain by matrix of transfer functions $\mathbf{G}(s)$.

To prove hyperstability using condition ML (8.85) one has to find three such matrices $\mathbf{P}$, $\mathbf{L}$, $\mathbf{V}$, which fulfill Kalman-Jakubovich equations. It usually happens that fulfillment of those equations for the primary system is not possible and fictitious degrees of freedom in the form of matrices $\mathbf{K}$, $\mathbf{D}$, (r_{ii}), $(1/r_{ii})$ have to be introduced. The method for introducing fictitious degrees of freedom is described in Section 8.3.1. The presence of additional degrees of freedom changes the primary form of state equations for the linear part (8.84) as well as the primary form of Kalman-Jakubovich equations (8.85).

It seems that analysis of hyperstability conditions in state space is a work-consuming task and transparency of the problem is rather poor. To make that analysis easier and more transparent one can replace the time-domain representation of the linear part by its frequency representation using known relation (8.86):

$$\mathbf{G}(s) = \mathbf{C}(s\mathbf{I} - \mathbf{A})^{-1}\mathbf{B} + \mathbf{D}_0 , \tag{8.86}$$

which enables further analysis according to the rules presented in Section 8.3.1.

8.3.3 The frequency domain conditions for hyperstability of discrete, non-linear control systems containing a time-invariant non-linear part

The scheme of the standard control system under consideration is shown in Fig. 8.47.

If the linear part is described by state equations (8.87):

$$\mathbf{x}_{k+1} = \mathbf{A}\mathbf{\Phi}_k + \mathbf{H}\mathbf{u}_k$$
$$\mathbf{y}_k = \mathbf{C}\mathbf{x}_k + \mathbf{D}_0\mathbf{u}_k \tag{8.87}$$

then matrix of transfer functions $\mathbf{G}(z)$ can be calculated using formula (8.88).

$$\mathbf{G}(z) = \mathbf{C}(z\mathbf{I} - \mathbf{\Phi})^{-1}\mathbf{H} + \mathbf{D}_0 \tag{8.88}$$

Several preliminary conditions must be fulfilled in order to check system hyperstability. Conditions of a similar type have been formulated as for continuous systems in Section 8.3.1.

The preliminary conditions for the linear part $\mathbf{G}(z)$

PL1) The matrix of transfer functions $\mathbf{G}(z)$ has to be a square one, which means that the number of inputs of the linear part (dimension of vector $\mathbf{u}_k$) has to be equal to the number of its outputs (dimension of vector $\mathbf{y}_k$). If the primary system does not hold the above condition, then additional signals, always equal to zero, can be introduced – see Fig. 8.48.

PL2) The linear part has to be completely controllable and observable.

PL3) All the poles of all component transfer functions $G_{ij}(z)$ belonging to $\mathbf{G}(z)$ have to be located inside the unit circle (component transfer functions have to be stable). If the primary system is not stable, then fictitious degrees of freedom k_{ij} can be introduced to the primary system. The introduction of fictitious degrees of freedom represented by matrix $\mathbf{K}$ must not change the inputs and outputs of the primary linear part. The introduction of new degrees of freedom is realized by

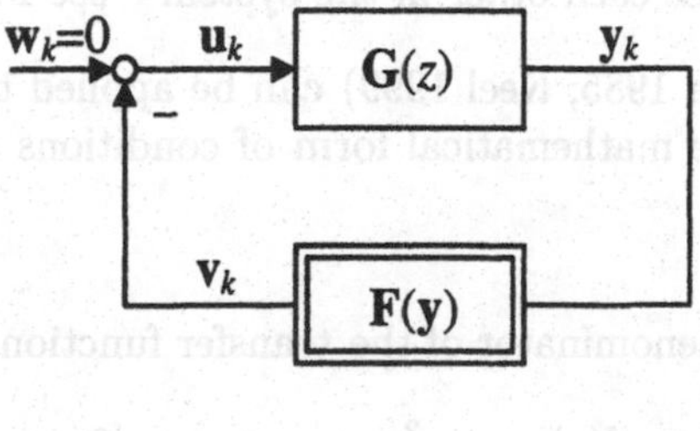

Fig. 8.47. The standard control system with a linear part represented by a matrix of discrete transfer functions $\mathbf{G}(z)$

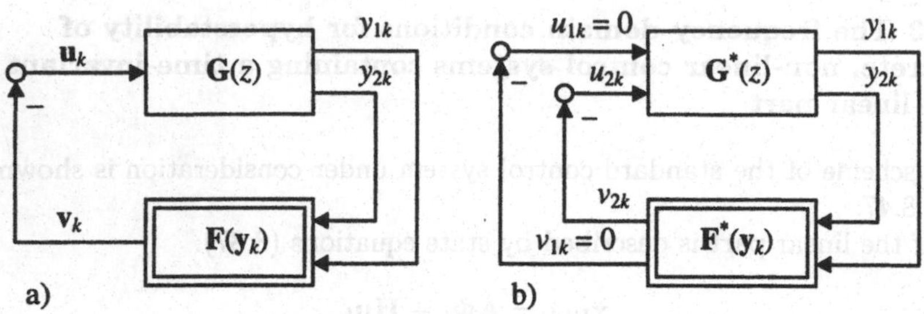

Fig. 8.48. The primary system (a) with different numbers of inputs and outputs (PL1 is not fulfilled) and its secondary equivalent (b) fulfilling PL1

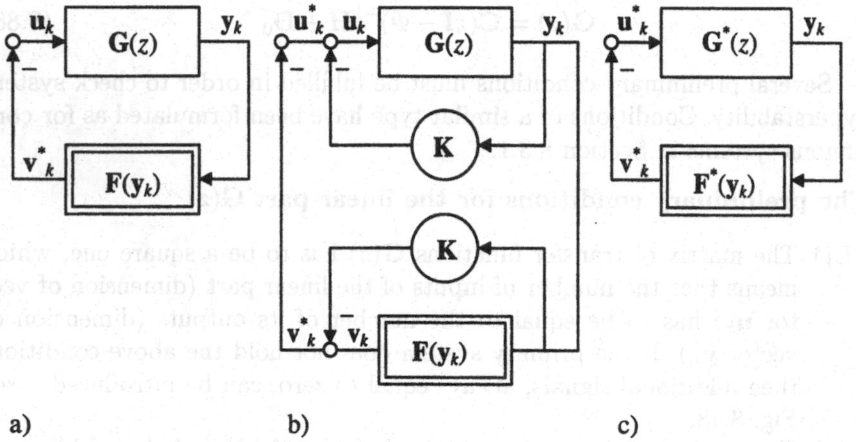

Fig. 8.49. The primary linear part $\mathbf{G}(z)$ with unstable poles (a) and the stabilizing of them by the introduction of fictitious degrees of freedom $\mathbf{K}$ (b) leading to the new, secondary form of the system (c) which is equivalent to the primary form (a)

equipping the linear part $\mathbf{G}(z)$ with feedback loops and the creation of parallel connections encircling the non-linear subsystem $\mathbf{F}(\mathbf{y}_k)$. The signals leaving the introduced parallel connections and feedback loops mutually reduce each other in the system – see Fig. 8.49.

Jury's test (Leigh 1985; Keel 1999) can be applied to check the stability of system poles. The mathematical form of conditions obtained this way is relatively simple.

Jury's test

If the second order denominator of the transfer function has the form (8.89):

$$f(z) = a_2 z^2 + a_1 z + a_0 = 0\,, \tag{8.89}$$

where $a_2 > 0$, then fulfillment of conditions (8.90) means that all the roots of function $f(z)$ belong to the interior of the unit circle.

$$\begin{aligned} f(1) &> 0 \\ f(-1) &> 0 \\ |a_0| &< a_2 \end{aligned} \tag{8.90}$$

If the third order denominator of the transfer function has the form (8.91):

$$f(z) = a_3 z^3 + a_2 z^2 + a_1 z + a_0 = 0\,, \tag{8.91}$$

where $a_3 > 0$, then fulfillment of conditions (8.92) means that all the roots of function $f(z)$ belong to the interior of the unit circle.

$$\begin{aligned} f(1) &> 0 \\ f(-1) &> 0 \\ |a_0| &< a_3 \\ |a_0^2 - a_3^2| &> |a_0 a_2 - a_1 a_3| \end{aligned} \tag{8.92}$$

The analysis of the stability of roots $f(z)$ can also be done using Hurwitz's criterion. The substitution:

$$z = \frac{1+w}{1-w}\,,$$

transforms the interior of the unit circle in complex plane Z to the right half-plane of complex plane W (Leigh 1985).

Formula (8.93) expresses the relation between the primary and the secondary form of the standard system.

$$\begin{aligned} &\mathbf{G}^*(z) = [\mathbf{I} + \mathbf{G}(z)\mathbf{K}]^{-1}\mathbf{G}(z) \\ &\mathbf{F}^*(\mathbf{y}_k): \quad \mathbf{v}_k^* = \mathbf{v} - \mathbf{K}\mathbf{y}_k = \mathbf{F}(\mathbf{y}_k) - \mathbf{K}\mathbf{y}_k \end{aligned} \tag{8.93}$$

PL4) The order of the nominator of each component transfer function $G_{ij}(s)$ belonging to the matrix of transfer functions $\mathbf{G}(z)$ has to be not higher than the order of its denominator. Real plants satisfy this requirement.

The preliminary conditions for the non-linear part $\mathbf{F}(\mathbf{y}_k)$

PN1) The operation $\mathbf{v}_k = \mathbf{F}(\mathbf{y}_k)$ has to realize unique mapping of y onto v. The non-linear elements "equipped" with memory based on the hysteresis effect (Fig. 8.50) do not satisfy the considered condition (because of multiple-valued mapping).

PN2) The mapping $\mathbf{v}_k = \mathbf{F}(\mathbf{y}_k)$ has to be of the zeroing type – property (8.94), Fig. 8.51.

$$\mathbf{F}(\mathbf{0}) = \mathbf{0} \tag{8.94}$$

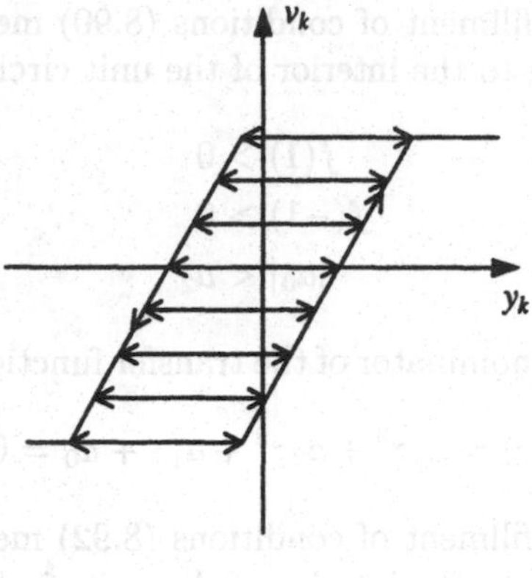

Fig. 8.50. An example of multiple-valued non-unique mapping realized by non-linear elements with hysteresis

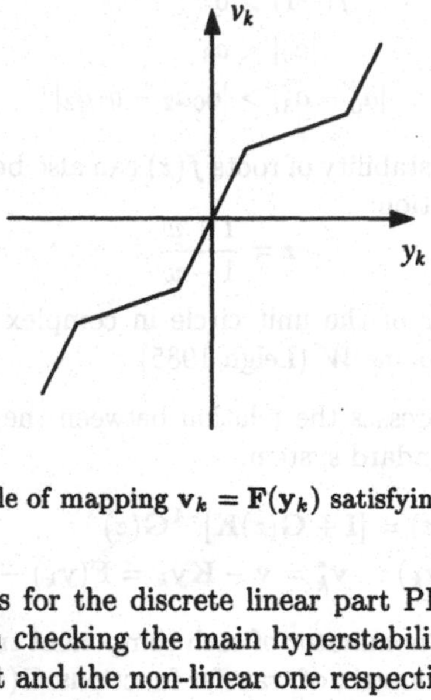

Fig. 8.51. An example of mapping $\mathbf{v}_k = \mathbf{F}(\mathbf{y}_k)$ satisfying the condition (8.94)

If initial conditions for the discrete linear part PL1 - PL4 are fulfilled, then one should begin checking the main hyperstability conditions ML and MN for the linear part and the non-linear one respectively.

The main hyperstability conditions (sufficient conditions)

The standard, discrete control system shown in Fig. 8.49a and its equivalent, secondary systems (obtained by such introductions of fictitious degrees of freedom to the primary system that inputs and outputs of its linear and non-linear parts remain unchanged and effects caused by the presence of additional degrees of freedom are mutually compensated) are asymptotically hyperstable, if:

ML) the linear part of the discrete system $\mathbf{G}(z)$ is strictly positive real,
MN) the non-linear part $\mathbf{F}(\mathbf{y}_k)$ satisfies the condition of Popov's sum for all $k_1 > 0$.

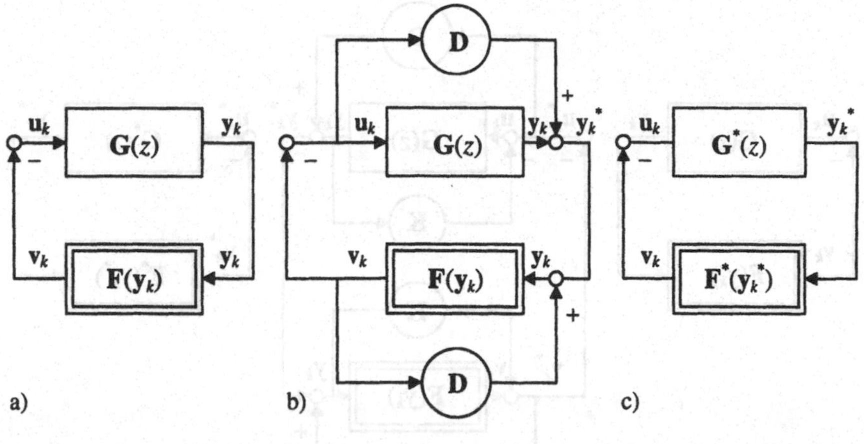

Fig. 8.52. The primary, standard control system (a) and its equivalent, secondary representation containing fictitious degrees of freedom **D**, shown in the form of a detailed block diagram (b) and a folded form (c)

$$\sum_{k=0}^{k_1} \mathbf{v}_k^{\mathrm{T}} \mathbf{y}_k \geq -\beta_0^2\,, \quad \forall k_1 > 0\,. \tag{8.95}$$

If application of condition ML to the linear part $\mathbf{G}(z)$ allows us to classify the linear part as positive real (instead of strictly positive real), then the standard system is normally stable.

Explanations connected with condition ML determining hyperstability of the discrete linear part $\mathbf{G}(z)$

The matrix of transfer functions $\mathbf{G}(z)$ is strictly positive real, if:

- poles of the all component transfer functions $G_{ij}(z)$ belong to the interior of the unit circle in complex plane Z,
- matrix (8.96) is positive definite and Hermitian for all $\omega \geq 0$.

$$\mathbf{H}(j\omega) = 0.5\left[\mathbf{G}(e^{j\omega}) + \mathbf{G}^{\mathrm{T}}(e^{-j\omega})\right] \tag{8.96}$$

The meaning of definitions used has been explained in Section 8.3.1. The matrix of transfer functions $\mathbf{G}(z)$ can not usually be classified as strictly positive real. It means that fictitious degrees of freedom represented by matrix $\mathbf{D}$ have to be introduced to the system in order to transform the linear part to a strictly positive real one – see Fig. 8.52.

The relations between a primary system and its secondary form shown in Fig. 8.52 are represented by expressions (8.97).

$$\begin{aligned} &\mathbf{G}^*(z) = \mathbf{G}(z) + \mathbf{D} \\ &\mathbf{F}^*(\mathbf{y}_k^*): \quad \mathbf{v}_k = \mathbf{F}^*(\mathbf{y}_k^*) = \mathbf{F}(\mathbf{y}_k + \mathbf{D}\mathbf{v}_k) \end{aligned} \tag{8.97}$$

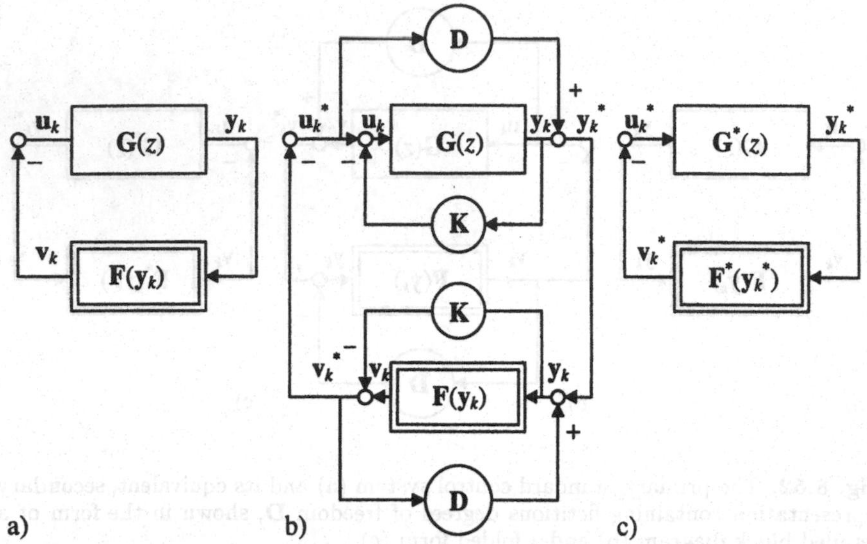

Fig. 8.53. The primary system (a) and its equivalent, secondary representation with fictitious degrees of freedom **K** and **D**, shown in the form of a detailed block diagram (b) and a folded form (c)

If a primary system is "equipped" with two matrices: the stabilizing one **K** and the matrix **D** ensuring that transfer functions G_{ij} have positive real parts (see Fig. 8.53), then relations between primary and secondary systems obtain a more complicated form (8.98).

$$\begin{aligned}&\mathbf{G}^*(z) = [\mathbf{I} + \mathbf{G}(z)\mathbf{K}]^{-1}\mathbf{G}(z) + \mathbf{D}\\&\mathbf{F}^*(\mathbf{y}_k^*):\ \mathbf{v}_k^* = \mathbf{v}_k - \mathbf{K}\mathbf{y}_k = \mathbf{F}(\mathbf{y}_k) - \mathbf{K}\mathbf{y}_k = \mathbf{F}(\mathbf{y}_k^* + \mathbf{D}\mathbf{v}_k^*) - \mathbf{K}(\mathbf{y}_k^* + \mathbf{D}\mathbf{v}_k^*)\end{aligned} \tag{8.98}$$

Explanations connected with condition MN determining hyperstability of the non-linear part

The main condition for the non-linear part of a discrete control system is represented by so-called Popov's sum (8.95). From a practical point of view Popov's sum is very inconvenient as a tool for investigations of the non-linear part. That is why the Popov's secondary condition (8.99) is usually applied to examination of the non-linear part:

$$\mathbf{v}_k^T \mathbf{y}_k \geq 0, \quad \forall k \geq 0\,. \tag{8.99}$$

However, the results obtained from (8.99) are narrower compared to results which can be theoretically obtained from (8.95). If dimensions of vectors $\mathbf{v}_k$ and $\mathbf{y}_k$ are:

$$\mathbf{v}_k : (p, 1),$$
$$\mathbf{y}_k : (p, 1),$$

then (8.99) obtains the form of sum (8.100):

$$\sum_{i=1}^{p} v_{ik} y_{ik} = \sum_{i=1}^{p} F_i(\mathbf{y}_k) y_{ik} \geq 0, \tag{8.100}$$

where: $F_i(\mathbf{y}_k) = v_{ik}$.

The secondary Popov's conditions formulated for the secondary system enlarged by fictitious degrees of freedom **D** and **K** can be written in the form (8.101):

$$\mathbf{v}_k^{*\mathrm{T}} \mathbf{y}_k^* = \sum_{i=1}^{p} v_{ik}^* y_{ik}^* \geq 0, \quad \forall k \geq 0. \tag{8.101}$$

The transformation of (8.101) to the space of a primary system $[\mathbf{v}_k, \mathbf{y}_k]$ yields the following form (8.102):

$$[\mathbf{F}(\mathbf{y}_k) - \mathbf{K}\mathbf{y}_k]^{\mathrm{T}} \{\mathbf{y}_k - \mathbf{D}[\mathbf{F}(\mathbf{y}_k) - \mathbf{K}\mathbf{y}_k]\} \geq 0, \quad \forall k \geq 0, \tag{8.102}$$

or (8.103):

$$\sum_{i=1}^{p} [F_i(\mathbf{y}_k) - \mathbf{K}_i \mathbf{y}_k] \{y_{ik} - \mathbf{D}_i[\mathbf{F}(\mathbf{y}_k) - \mathbf{K}\mathbf{y}_k]\} \geq 0, \quad \forall k \geq 0. \tag{8.103}$$

where $\mathbf{K}_i = [k_{i1}, \ldots, k_{ip}]$, $i = 1, \ldots, p$, are the rows of matrix **K**:

$$\mathbf{K} = \begin{bmatrix} k_{11} & \ldots & k_{1p} \\ \vdots & \vdots & \vdots \\ k_{p1} & \ldots & k_{pp} \end{bmatrix} = \begin{bmatrix} \mathbf{K}_1 \\ \vdots \\ \mathbf{K}_p \end{bmatrix},$$

and $\mathbf{D}_i = [d_{i1}, \ldots, d_{ip}]$, $i = 1, \ldots, p$, are the rows of matrix **D**:

$$\mathbf{D} = \begin{bmatrix} d_{11} & \ldots & d_{1p} \\ \vdots & \vdots & \vdots \\ d_{p1} & \ldots & d_{pp} \end{bmatrix} = \begin{bmatrix} \mathbf{D}_1 \\ \vdots \\ \mathbf{D}_p \end{bmatrix}.$$

The determination of **K** and **D** fulfilling all hyperstability conditions for linear and non-linear system parts means that proof of system hyperstability is successfully completed. There is much advice on how to chose **K** and **D**. The advice for a discrete system is identical to that considered in Section 8.3.1 for a continuous system.

The examples of stability analysis for SISO and MISO systems governed by neural fuzzy controllers of PD type are presented below. The investigations of hyperstability for multidimensional systems are so difficult compared

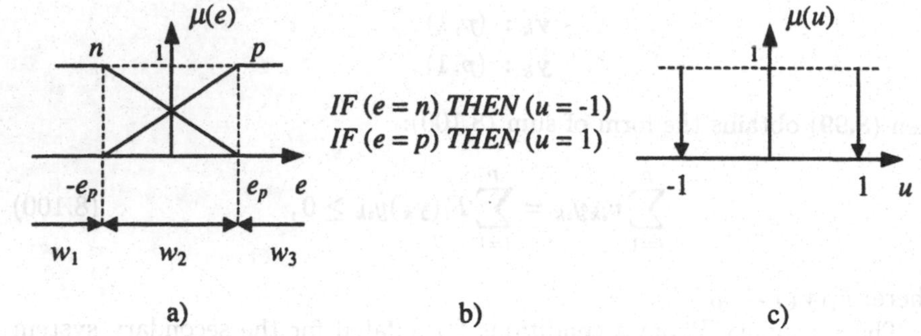

Fig. 8.54. The fuzzification (a), rule-base (b) and defuzzification (c) realized by fuzzy controller

to SISO systems that other approaches to problem solving have to be applied. The high level of difficulties encountered during exact mathematical analysis and visualization of the problem means that special methods, including numerical investigations, have to be used. The "method of characteristic points" considered in a further part of this book can be treated as a good example of such a special method as mentioned above.

Example 8.3.3.1. The investigation of hyperstability of a fuzzy control system of SISO type
Let us consider a discrete control system consisting of a continuous plant:

$$G_0(s) = \frac{1}{s \cdot (s+1)},$$

and a SISO fuzzy controller. The fuzzification element, rule base and defuzzification element of a fuzzy controller are shown in Fig. 8.54.

It results from Example 8.2.1 that input-output mapping realized by a fuzzy controller can be represented by function $F(e)$, given by (8.104) and shown graphically in Fig. 8.55,

$$u_1 = -w_1 + \frac{e}{e_p} w_2 + w_3 , \tag{8.104}$$

where: w_i – logical variables which carry information showing in which sector of controller input space the current value of input e is located. The discrete control system block diagram is shown in Fig. 8.56.

The system in Fig. 8.56 has been transformed to a standard form shown in Fig. 8.57, where the linear part is represented in s-domain by $G(s)$ or in z-domain by $G(z)$.

The discrete model of the continuous part has been obtained by applying backward difference approximation:

$$s = \frac{1 - z^{-1}}{T},$$

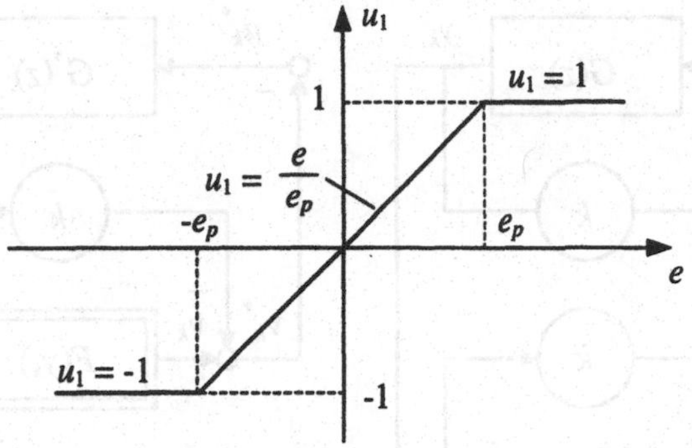

Fig. 8.55. The static characteristic of a fuzzy controller

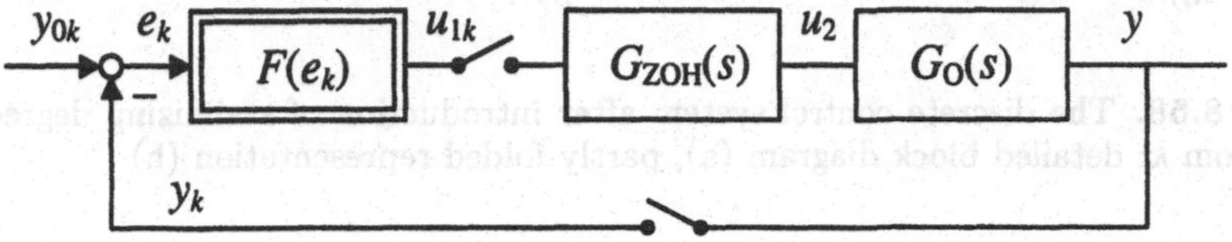

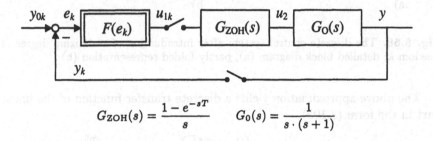

$$G_{ZOH}(s) = \frac{1 - e^{-sT}}{s} \qquad G_0(s) = \frac{1}{s \cdot (s+1)}$$

Fig. 8.56. The discrete control system consisting of fuzzy controller $F(e)$, zero-order hold element $G_{ZOH}(s)$ and continuous plant $G_0(s)$ for sample time T

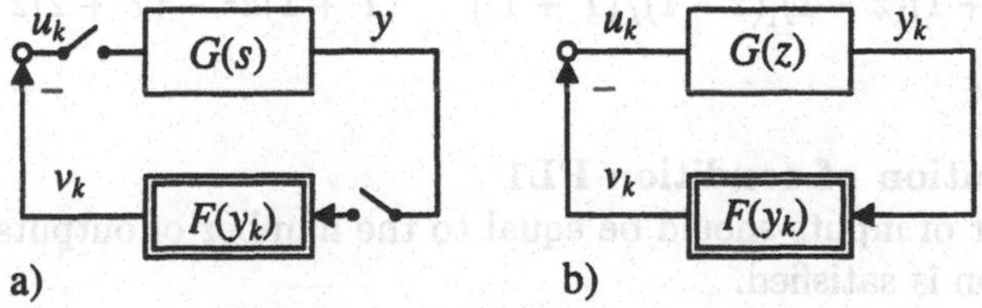

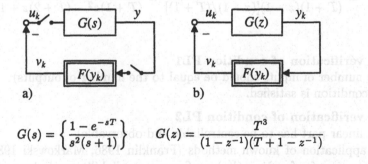

$$G(s) = \mathcal{Z}\left\{\frac{1 - e^{-sT}}{s^2(s+1)}\right\} \qquad G(z) = \frac{T^3}{(1 - z^{-1})(T + 1 - z^{-1})}$$

Fig. 8.57. The standard form of the control system considered with the linear part represented by continuous transfer function $G(s)$ (a) and by discrete transfer function $G(z)$ (b), for sample time T

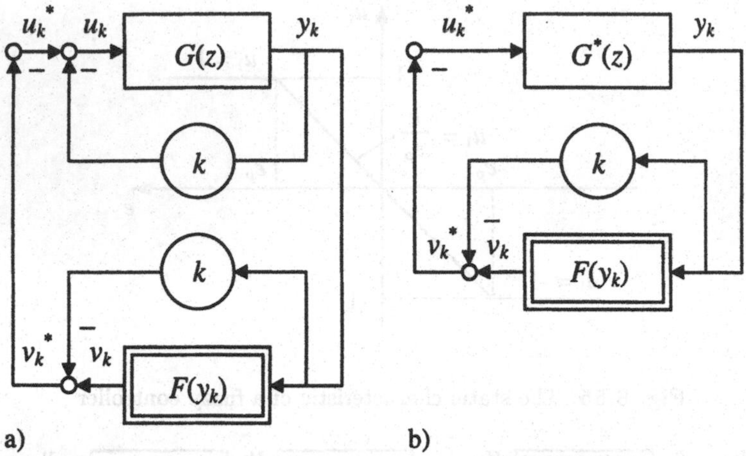

Fig. 8.58. The discrete control system after introduction of stabilizing degree of freedom k: detailed block diagram (a), partly folded representation (b)

The above approximation yields a discrete transfer function of the linear part in the form (8.105)

$$G(z) = Z\{G_{\mathrm{ZOH}}(s)G_0(s)\} = Z\left\{\frac{1-e^{-sT}}{s^2(s+1)}\right\} = \frac{T^3}{(1-z^{-1})(T+1-z^{-1})} = \\ = \frac{T^3 z^2}{(T+1)(z-1)[(z-1)/(T+1)]} = \frac{T^3 z^2}{(T+1)z^2-(T+2)z+1} \tag{8.105}$$

The verification of condition PL1
(The number of inputs should be equal to the number of outputs)
The condition is satisfied.

The verification of condition PL2
(The linear part has to be controllable and observable)
The application of known methods (Franklin 1986; Markowski 1985) to a discrete transfer function $G(z)$ confirms controllability and observability of $G(z)$ given in the form (8.105).

The verification of condition PL3
(The poles of G(z) ought to be located inside the unit circle)
The first pole $z_1 = 1$ does not belong to the interior of the unit circle. It means that stabilizing the linear part by adding a fictitious degree of freedom k (Fig. 8.58) is necessary. The second pole $z_2 = 1/(T+1)$ is always located inside the unit circle, so its stabilization is not required.

The secondary transfer function of the linear part $G^*(z)$ is determined by (8.106):

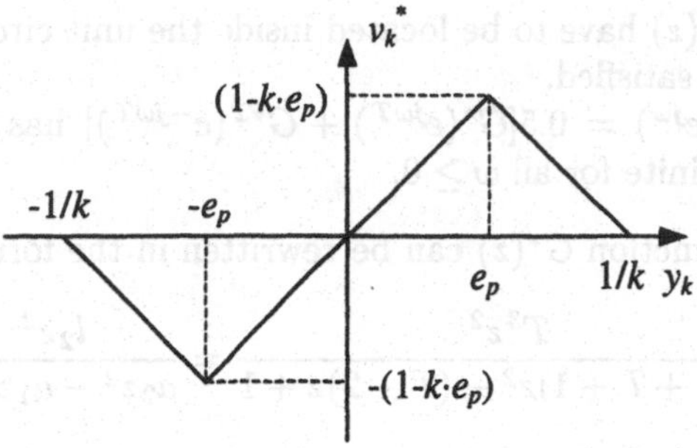

Fig. 8.59. The static characteristic of the non-linear part after introduction of a fictitious degree of freedom k

$$G^*(z) = \frac{G(z)}{1+kG(z)} = \frac{T^3z^2}{(kT^3+T+1)z^2-(T+2)z+1} \tag{8.106}$$

Applying Jury's test (8.90) one obtains conditions (8.107) for the stability of all poles of the transfer function considered.

$$\begin{aligned} f(1) &= kT^3 > 0 \\ f(-1) &= kT^3 + 2T + 4 > 0 \\ |1| &< kT^3 + T + 1 \end{aligned} \tag{8.107}$$

If:

$$k > 0 \tag{8.108}$$

then all conditions (8.107) are satisfied, which means that all poles of $G(z)$ are stable.

The verification of condition PL4
(The order of the nominator of $G^*(z)$ can not be higher than the order of the denominator)
The condition is satisfied.

The verification of condition PN1
(The function $v_k^* = F^*(y_k)$ has to represent unique mapping)
The function $v_k^* = F^*(y_k)$ is shown in Fig. 8.59. The properties of function $v_k^* = F^*(y_k)$ have been discussed in Section 8.3.1. The considered function realizes unique mapping.

The verification of condition PN2
(The mapping $v_k^* = F^*(y_k)$ has to satisfy condition $F^*(0) = 0$)
Fig. 8.59 shows that the condition is fulfilled.

The verification of main condition ML
(The linear part $G^*(z)$ has to be strictly positive real)
The above condition can be decomposed to several component conditions.

- The poles of $G^*(z)$ have to be located inside the unit circle. If $k > 0$, then the condition is satisfied.
- The matrix $H(e^{j\omega}) = 0.5[G^*(e^{j\omega T}) + G^{*\mathrm{T}}(e^{-j\omega T})]$ has to be Hermitian and positive definite for all $\omega \geq 0$.

The transfer function $G^*(z)$ can be rewritten in the form (8.109):

$$G^*(z) = \frac{T^3 z^2}{(kT^3 + T + 1)z^2 - (T+2)z + 1} = \frac{l_2 z^2}{a_2 z^2 - a_1 z + 1}, \tag{8.109}$$

where: $T = 0.1$, $l_2 = 10^{-3}$, $a_2 = 1.1 + 10^{-3}k$, $a_1 = 2.1$. Using Euler's identity:

$$e^{j\omega T} = \cos\omega T + j\sin\omega T,$$

one obtains expressions (8.110) and (8.111):

$$G^*(e^{j\omega T}) = \mathrm{Re}^* + j\mathrm{Im}^*, \tag{8.110}$$

$$G^*(e^{-j\omega T}) = \mathrm{Re}^* - j\mathrm{Im}^*, \tag{8.111}$$

where:

$$\mathrm{Re}^* = \frac{2l_2\cos^2\omega T - a_1 l_2 \cos\omega T + l_2(a_2 - 1)}{4a_2\cos^2\omega T - 2a_1(1+a_2)\cos\omega T + (a_1 - 1)^2 + a_1^2}, \tag{8.112}$$

$$\mathrm{Im}^* = \frac{-l_2\sin\omega T - (a_1 - 2\cos\omega T)}{4a_2\cos^2\omega T - 2a_1(1+a_2)\cos\omega T + (a_1 - 1)^2 + a_1^2}. \tag{8.113}$$

According to (8.37) matrix $H^*(e^{j\omega})$ can be expressed in the form (8.114):

$$H^*(e^{j\omega}) = 0.5[G^*(e^{j\omega T}) + G^{*\mathrm{T}}(e^{-j\omega T})] = \mathrm{Re}^*. \tag{8.114}$$

The matrix H^* is positive definite, if relation (8.115) is satisfied.

$$\mathrm{Re}^* = \frac{2l_2\cos^2\omega T - a_1 l_2\cos\omega T + l_2(a_2 - 1)}{4a_2\cos^2\omega T - 2a_1(1+a_2)\cos\omega T + (a_1 - 1)^2 + a_1^2} \geq 0 \tag{8.115}$$

The determination of the minimum of function $\mathrm{Re}^*(\cos\omega T)$ by using the analytical way seems to be very complicated. Furthermore, it leads to strongly non-linear conditions. Thus, the choice of coefficients k and d is not easy. There is another solution of the problem.

The denominator $M(\mathrm{Re}^*)$ can be written in the form (8.116):

$$M(\mathrm{Re}^*) = \left[2\cos^2\omega T - a_1\cos\omega T + (a_2 - 1)\right]^2 + \left[\sin\omega T(a_1 - 2\cos\omega T)\right]^2. \tag{8.116}$$

The above sum of two squares is always positive. So, if the numerator is positive too, then condition (8.115) is satisfied. It can easily be perceived that the numerator of *Re** represents a square function of variable $w = \cos\omega T$ and the properties of this function depend on values of coefficients : $a_1 = (T+2)$, $a_2 = (kT^3 + T + 1)$.

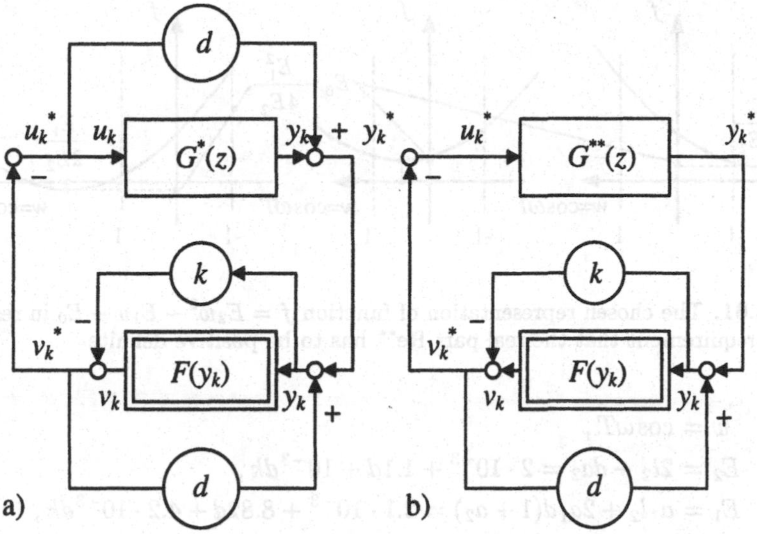

Fig. 8.60. The secondary system after introduction of fictitious degree of freedom d shown in the form of a detailed block diagram (a) and a partially folded form (b)

The value of degree of freedom k is not chosen till now. Thus, at the current stage of reasoning one can not assume the exact shape of function $Re^*(w)$ and its placement in the space. It means that values of the function may be negative. To make the function positive definite the fictitious degree of freedom d has been added, see Fig. 8.60.

Adding degree of freedom d one obtains new transfer function $G^{**}(z)$ given in the form (8.117).

$$G^{**}(z) = G^*(z) + d = \frac{l_2 z^2}{a_2 z^2 - a_1 z + 1} + d \tag{8.117}$$

The calculation of the real part of $G^{**}(z)$ yields :

$$\mathrm{Re}^{**} = \frac{2l_2 \cos^2 \omega T - a_1 l_2 \cos \omega T + l_2(a_2 - 1)}{4a_2 \cos^2 \omega T - 2a_1(1 + a_2)\cos \omega T + (a_1 - 1)^2 + a_1^2} + d \tag{8.118}$$

To make real part Re** non-negative the following condition (8.119) has to be satisfied:

$$\begin{aligned}(2l_2 + da_2)\cos^2 \omega T - [a_1 l_2 + 2a_1 d(1 + a_2)] \cos \omega T + \\ + [l_2(a_2 - 1) + d(a_2 - 1)^2 + da_1^2] \geq 0\end{aligned} \tag{8.119}$$

The notations (8.120):

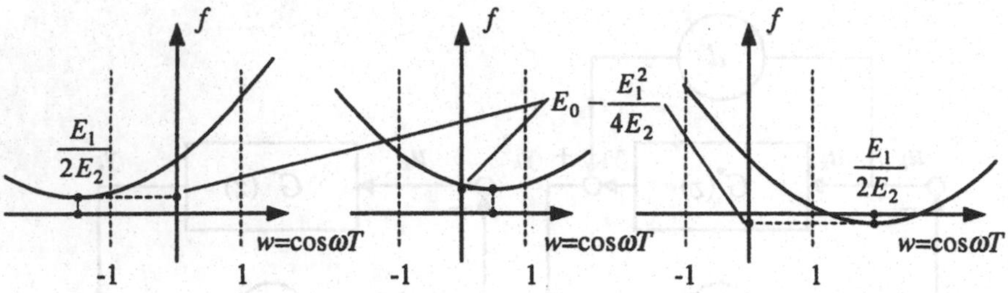

Fig. 8.61. The chosen representation of function $f = E_2w^2 - E_1w + E_0$ in respect of the requirement that the real part Re** has to be positive definite

$$\begin{aligned}
w &= \cos\omega T\,,\\
E_2 &= 2l_2 + da_2 = 2\cdot 10^{-3} + 1.1d + 10^{-3}dk\,,\\
E_1 &= a_1l_2 + 2a_1d(1+a_2) = 2.1\cdot 10^{-3} + 8.82d + 4.2\cdot 10^{-3}dk\,,\\
E_0 &= l_2(a_2-1) + d(a_2-1)^2 + da_1^2 =\\
&\quad 0.1\cdot 10^{-3} + 10^{-6}k + 4.42d + 0.2\cdot 10^{-3}kd + 10^{-6}k^2d\,, \qquad (8.120)
\end{aligned}$$

allow us to rewrite the condition (8.119) in more intelligible form (8.121).

$$f = E_2w^2 - E_1w + E_0 \geq 0 \qquad (8.121)$$

The parabola f given by (8.121) has a minimum, if coefficient $E_2 > 0$. If $k > 0$ and $d \geq 0$, then $E_2 > 0$. The determination of "polarity" of coefficients E_0, E_1, E_2 in a general case can be impossible, because necessary transformations often yield very complicated mathematical forms defining the coefficients. Furthermore, the numerical values of degrees of freedom d and k are usually involved in an analysis of coefficient polarity. The three curves representing certain classes of function f given by (8.121) are shown in Fig. 8.61. The real part Re** is positive definite, if minimum of f is positive (sufficient condition):

$$\begin{aligned}
E_0 - \frac{E_1^2}{4E_2} &\geq 0\,,\\
E_2 &> 0\,. \qquad (8.122)
\end{aligned}$$

The conditions (8.122) substantially narrow the acceptable area for choice of k and d. It can be seen in Fig. 8.61c that condition Re** ≥ 0 can also be satisfied, if the minimum of parabola f is negative. The necessary condition for f can be expressed in the form: $f(w) \geq 0$, for all possible $w = \cos\omega T$, i.e. $w : [-1, 1]$.

The acceptance of the last condition mentioned above widens the space for choice of fictitious degrees of freedom k and d, if compared to (8.122). The large area of admissible space for k and d can be meaningful, if more

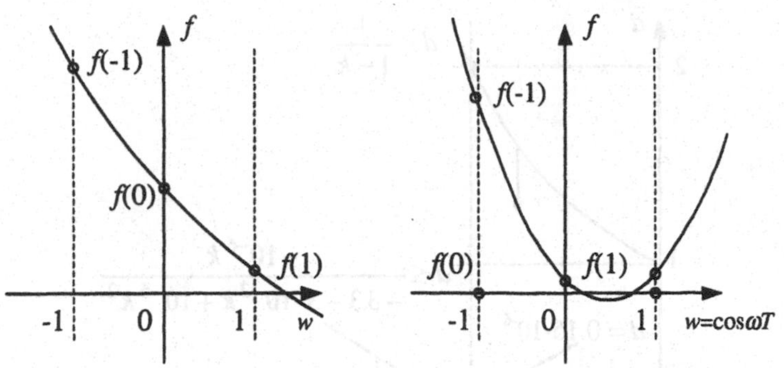

Fig. 8.62. The "advantageous" form of function $f(w)$ (a) and its "disadvantageous" form (b), where conditions for positive definite Re** are not satisfied for $w: [-1,1]$

complicated mappings $F(y)$ are considered. Further discussion reveals that the narrowed, strongly non-linear condition (8.122) can be replaced by less complicated conditions (8.123).

$$\begin{aligned} f(-1) &= E_2 + E_1 + E_0 \geq 0 \\ f(0) &= E_0 \geq 0 \\ f(1) &= E_2 - E_1 + E_0 \geq 0 \end{aligned} \tag{8.123}$$

There is a high probability that these three conditions (8.123) allow us to obtain positive definite function $f(w)$, for all $w: [-1,1]$. But it can happen that $f(w)$ is not positive definite between points $w=0$ and $w=1$ – look at the curve in Fig. 8.62b. To eliminate "disadvantageous" cases the conditions of type $f(w) \geq 0$ can be elaborated for more than 3 points. Of course, it is always possible to neglect the above advice and to check if $f(w) \geq 0$ for all $w: [-1,1]$ after the successful examination of all remaining conditions and choice of numerical values for k, d. Following the last advice, one can express conditions (8.123) in the form (8.124) depending on k and d:

$$\begin{aligned} f(-1) &= 4.2 \cdot 10^{-3} + 14.34d + 10^{-6}k + 5.4 \cdot 10^{-3}kd + 10^{-6}k^2d \geq 0 \\ f(0) &= 0.1 \cdot 10^{-3} + 4.42d + 10^{-6}k + 0.2 \cdot 10^{-3}kd + 10^{-6}k^2d \geq 0 \\ f(1) &= -3.3d + 10^{-6}k - 3 \cdot 10^{-3}kd + 10^{-6}k^2d \geq 0 \end{aligned} \tag{8.124}$$

The main conditions MN for the non-linear part

These conditions are identical to those in Example 8.3.1.1. So, for the first and third sectors they are represented by inequalities (8.125) and (8.126):

$$d < \frac{1}{1-k}, \tag{8.125}$$

$$d < \frac{2}{1-2k} \tag{8.126}$$

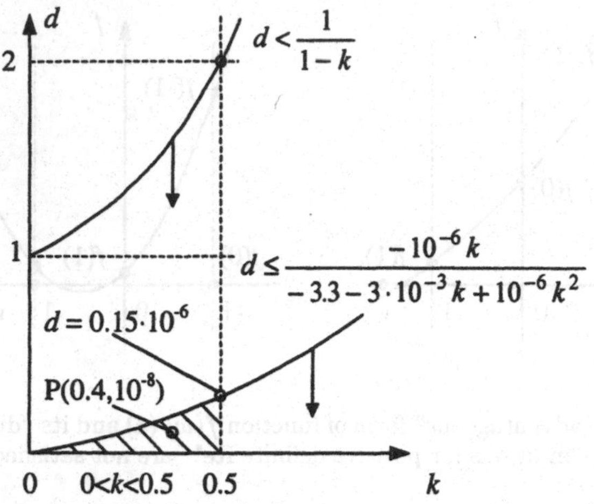

Fig. 8.63. The graphic representation of system hyperstability conditions and admissible values of fictitious degrees of freedom d and k

The main condition for the second sector of the non-linear part (see Fig. 8.58) is represented by inequality (8.125). It results from Fig. 8.39 that both conditions (8.125) and (8.126) are fulfilled if (8.125) is fulfilled. Thus this further analysis takes into consideration the condition (8.125). All conditions for the linear part (8.124) and non-linear one (8.125) are collected together in (8.127):

a) $0 < k < 0.5$
b) $4.2\cdot10^{-3} + 14.34d + 10^{-6}k + 5.4\cdot10^{-3}kd + 10^{-6}k^2d \ge 0$
c) $0.1\cdot10^{-3} + 4.42d + 10^{-6}k + 0.2\cdot10^{-3}kd + 10^{-6}k^2d \ge 0$
d) $-3.3d + 10^{-6}k - 3\cdot10^{-3}kd + 10^{-6}k^2d \ge 0$
e) $d < \frac{1}{1-k}$ (8.127)

Because all positive values k and d satisfy conditions (b) and (c), then one can neglect further analysis of the above relations. The condition (d) can be transformed to the new form (8.128). The hyperstability conditions (8.128) are presented graphically in Fig. 8.63.

$$0 < k < 0.5, \quad d > 0$$
$$d \le \frac{-10^{-6}k}{-3.3 - 3\cdot10^{-3}k + 10^{-6}k^2}$$
$$d < \frac{1}{1-k} \tag{8.128}$$

For example, coefficients $k = 0.4$ and $d = 10^{-8}$ (point P in Fig. 8.63) satisfy all conditions for the hyperstability of the system. The condition

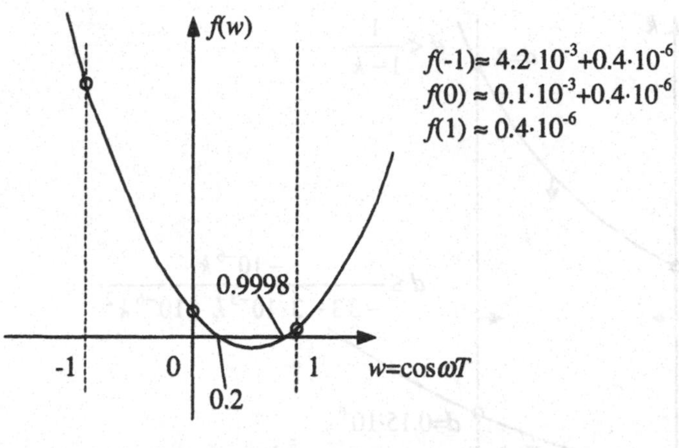

Fig. 8.64. The graphic representation of function $f = \mathrm{Re}^{**}$ shows that Re^{**} is not always positive definite (the relevant condition is transgressed between the points $w = 0.2$ and $w = 1$)

(8.121), deciding whether the real part is positive definite, is satisfied for $w = \cos\omega T = -1$, 0 and 1, but between the points $w = 0$ and $w = 1$ it is not satisfied – see Fig. 8.64.

The additional condition of (8.121) type for value w located close to the minimum of function $f(w)$ can be imposed to overcome the difficulty mentioned above. The value $w = 0.6$ seems to be a good choice – see Fig. 8.64. Hence:

$$\begin{aligned} f(0.6) &= 0.36E_2 - 0.6E_1 + E_0 \geq 0\,, \\ f(0.6) &= -0.44 \cdot 10^{-3} + 10^{-6}k + d(-0.476 - 1.96k + 10^{-6}k^2) \geq 0\,. \end{aligned} \tag{8.129}$$

The condition (8.129) can be decomposed to equivalent conditions represented by an inequality system:

$$\begin{aligned} d &\leq \frac{0.44 \cdot 10^{-3} - 10^{-6}k}{-0.476 - 1.96k + 10^{-6}k^2}\,, \\ &-0.476 - 1.96k + 10^{-6}k^2 < 0\,. \end{aligned} \tag{8.130}$$

or:

$$\begin{aligned} d &\geq \frac{0.44 \cdot 10^{-3} - 10^{-6}k}{-0.476 - 1.96k + 10^{-6}k^2}\,, \\ &-0.476 - 1.96k + 10^{-6}k^2 > 0\,. \end{aligned} \tag{8.131}$$

The condition (8.131) can not be satisfied for $0 < k < 0.5$. It means that further investigations are based on condition (8.130). The new, "supplemented" conditions are expressed by relations (8.132) and represented graphically in Fig. 8.65.

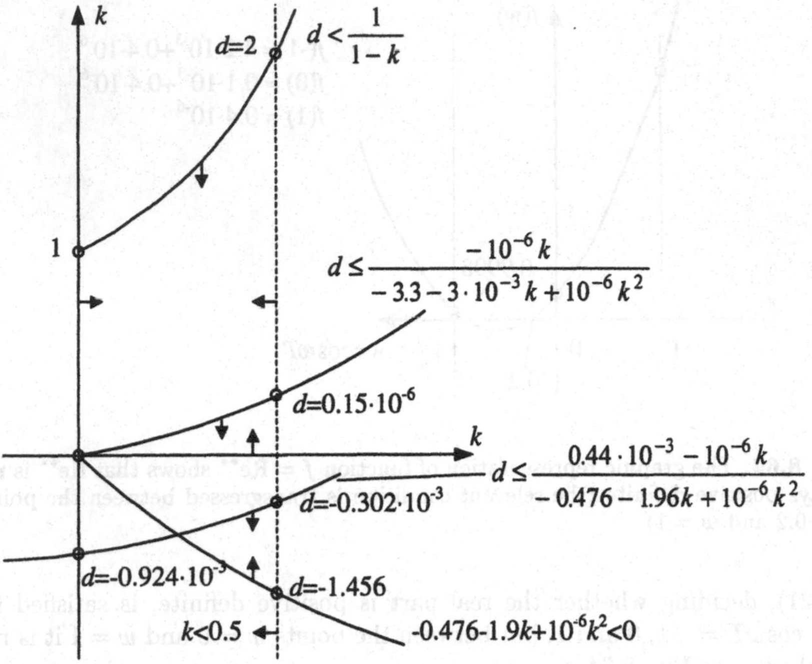

Fig. 8.65. The graphic representation of hyperstability conditions (8.132)

$$
\begin{aligned}
&0<k<0.5\,, \quad d>0\\
&d\le\frac{-10^{-6}k}{-3.3-3\cdot10^{-3}k+10^{-6}k^2}\\
&d<\frac{1}{1-k}\\
&d\le\frac{0.44\cdot10^{-3}-10^{-6}k}{-0.476-1.96k+10^{-6}k^2}\,,\\
&-0.476-1.96k+10^{-6}k^2<0\,.
\end{aligned}
\tag{8.132}
$$

There is no common sub-space $[k,d]$ where all hyperstability conditions are satisfied – see Fig. 8.65. It means that hyperstability of the system has not been proved yet. The following question arises: should one follow the advice given in Section 8.3.1 and repeat the procedure of verification after introduction of new degrees of freedom?

There is a chance of success, if the operation of a real control system is stable. If one assumes that the considered system operates in an intermediate sector, where $w_2 = 1$ (Fig. 8.54) and the errors e are small, then the fuzzy controller operation resembles the operation of a linear controller of P-type with proportional gain $k_r = 1/e_p = 1$, $(F(y_k) = 1)$.

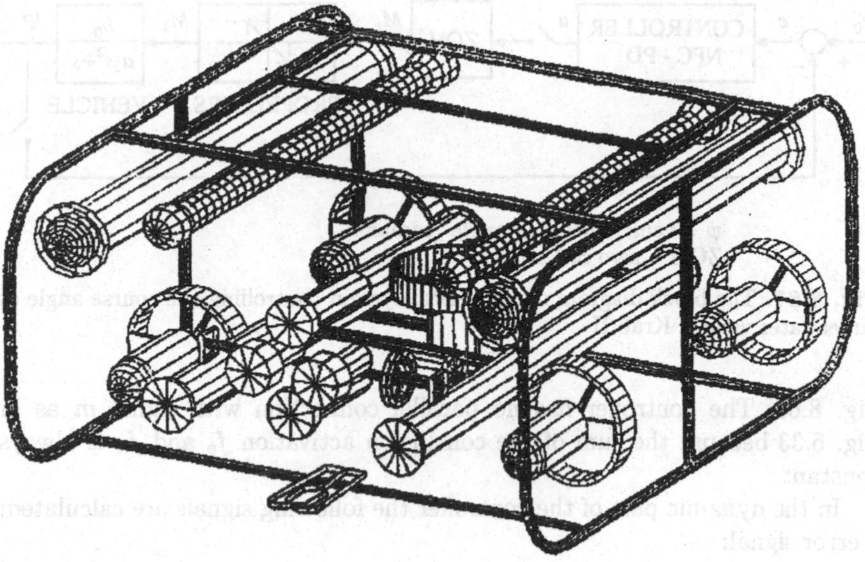

Fig. 8.66. The unmanned underwater vehicle Krab II

The known methods for investigation of discrete system stability can be used in order to check the stability of the system considered, when operating in the intermediate sector $w_2 = 1$. It results from Fig. 8.57, that the system characteristic equation , for $T = 0.1$ (s), is given by:

$$1 + F(y_k)G(z) = 0 \quad \rightarrow \quad 1.101z^2 - 2.1z + 1 = 0\,.$$

Solving the characteristic equation one obtains: $z_1 = 1.0887$ and $z_2 = 1.0113$. Both roots are located outside the unit circle. It means that the system is unstable in the second sector. The result obtained explains the reasons responsible for the failure of the hyperstability proof.

◊ ◊ ◊

The next example deals with a system controlled by a slightly more sophisticated fuzzy controller (PD). The example shows the huge increase of work involved in proving system hyperstability when there is only a slight increase in the sophistication of the fuzzy controller.

Example 8.3.3.2. The investigation of the stability of the digital system controlling the course of underwater vehicle Krab II by means of a neural fuzzy controller (NFC)

The underwater vehicle Krab II is shown in Fig. 8.66. The control system of the vehicle is described in detail in (Piegat 1996). The block diagram of the control system is shown in Fig. 8.67.

The vehicle is controlled by a self-learning neural fuzzy controller of PD-type (abbreviated to NFC). The scheme of that controller is shown in

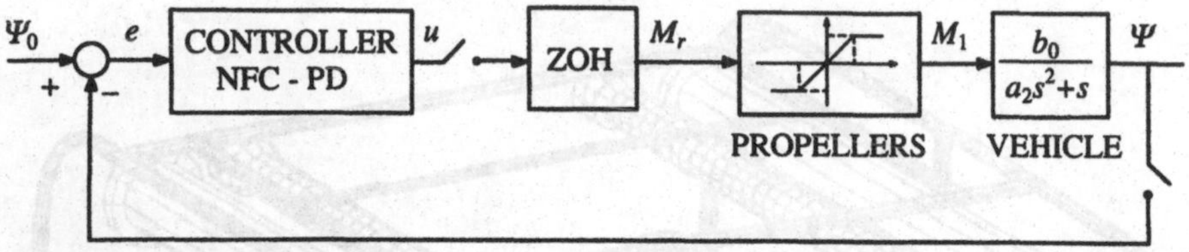

ψ – vehicle heading angle (radians)
ZOH – zero order hold

Fig. 8.67. The block diagram of the digital system controlling the course angle of underwater vehicle Krab II

Fig. 8.68. The controller has no parallel connection with signal m as in Fig. 6.33 because the sum of the conclusion activation f_p and f_n is always constant.

In the dynamic part of the controller the following signals are calculated:
– error signal:

$$e_p = e\,,$$

– integral of error signal:

$$e_I = \int_0^t e(t)\mathrm{d}t \cong \frac{T}{2}\sum_{j=1}^{k}\left[e(j) + e(j-1)\right]\,,$$

– derivative of error signal:

$$e_D = \frac{\mathrm{de}}{\mathrm{dt}} \cong \frac{1}{T}\left[e(k) - e(k-1)\right]\,,$$

where: T – sampling time.

The fuzzification, inference and defuzzification are tasks realized by the static part of the controller, whereas the signals e_P, e_I, e_D are considered as the inputs of the controller's static part. The output signal of controller M_r passes through the ZOH element and controls the propeller torque M_1 (Nm). Finally, the M_1 torque adjusts the vehicle heading angle ψ (rad). The asymmetrical saturation for propeller output signal M_1 is a typical property of propeller static characteristic – see Fig. 8.69.

The following vehicle transfer function has been identified experimentally:

$$G(s) = \frac{\psi(s)}{M_1(s)} = \frac{0.021929}{s(1+0.30372s)} = \frac{b_0}{a_2 s^2 + s}\left(\frac{\text{rad}}{\text{N}\cdot\text{m}}\right) \qquad (8.133)$$

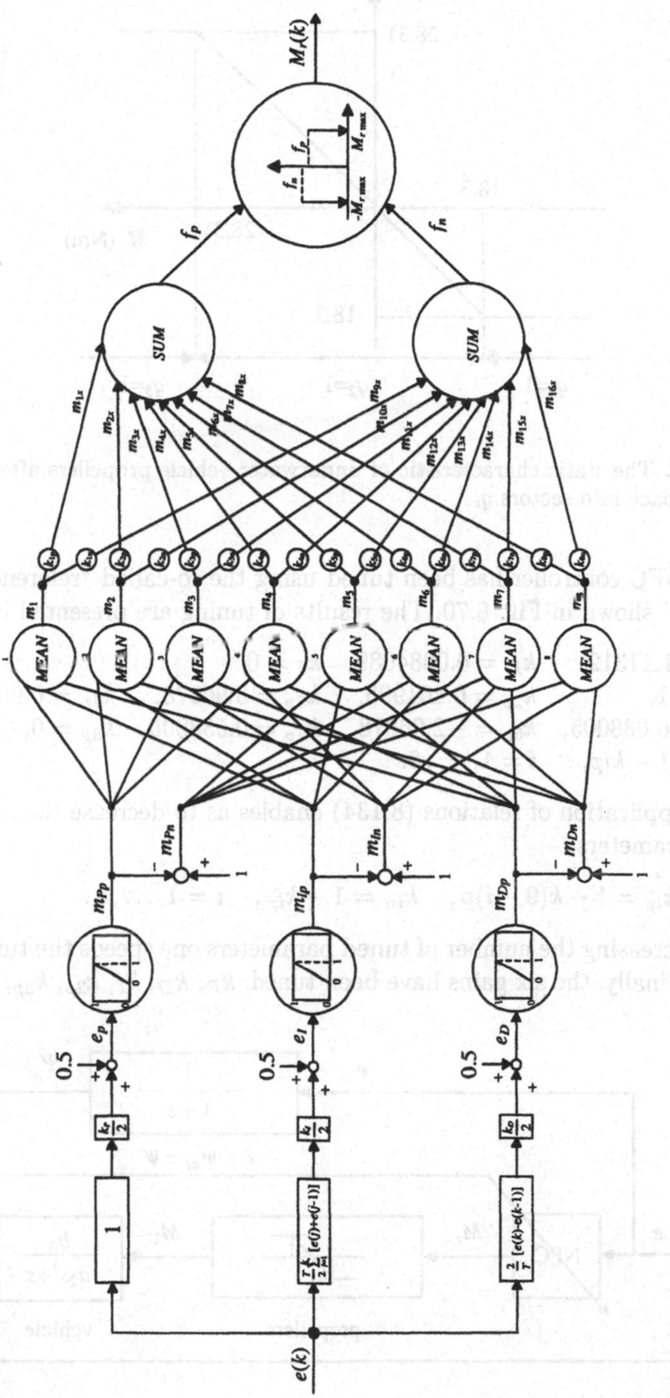

Fig. 8.68. The neural fuzzy controller of PID-type

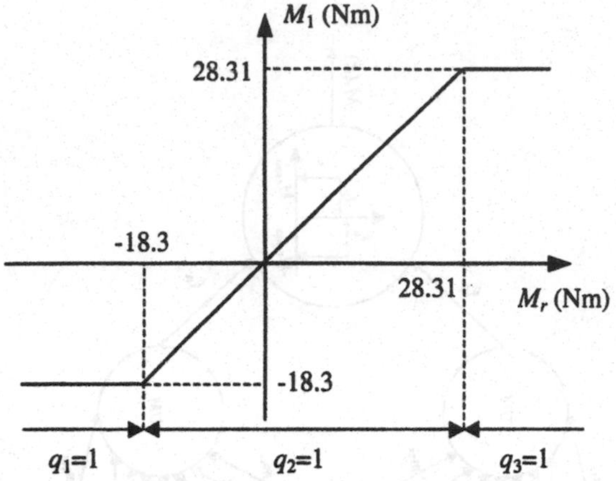

Fig. 8.69. The static characteristic of underwater vehicle propellers after division of input space into sectors q_i

The NFC controller has been tuned using the so-called "reference model structure" shown in Fig. 8.70. The results of tuning are presented below:

$$\begin{aligned}
&k_P = 1.11312, && k_D = 0.0084089, && k_I = 0 && \\
&k_{1p} = 1, && k_{2p} = 0.461995, && k_{3p} = 3.96078, && k_{4p} = 0.961995,\\
&k_{5p} = 0.038005, && k_{6p} = -2.96078, && k_{7p} = 0.538005, && k_{8p} = 0,\\
&k_{in} = 1 - kip, && i = 1, \ldots, 8. && &&
\end{aligned}$$

The application of relations (8.134) enables us to decrease the number of tuned parameters.

$$k_{ip} = 1 - k(9-i)p, \quad k_{in} = 1 - k_{ip}, \quad i = 1, \ldots, 8. \tag{8.134}$$

By decreasing the number of tuned parameters one speeds the tuning process up. Finally, the six gains have been tuned: k_P, k_D, k_I, k_{2p}, k_{3p}, k_{4p}. The

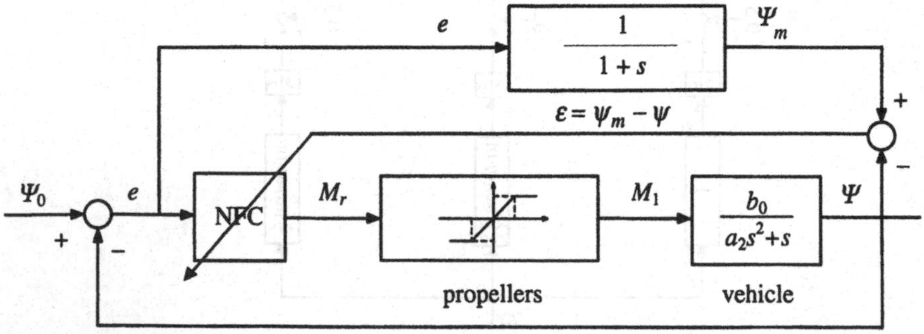

Fig. 8.70. The tuning of the NFC controller within the "reference model structure"

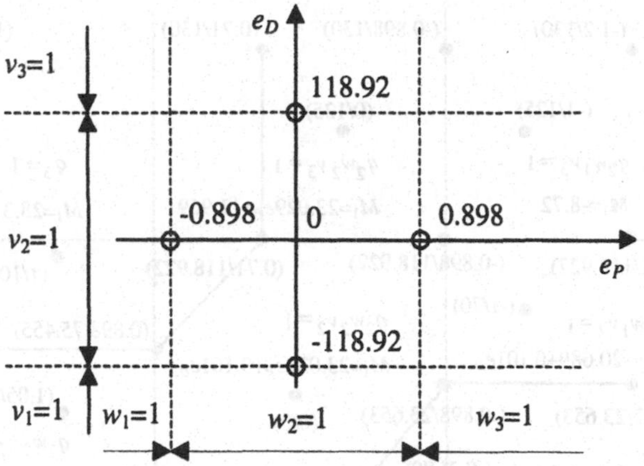

Fig. 8.71. The partition of input space of the NFC controller into sectors defined by logic variables w_i, v_j

rest of the parameters have been calculated using (8.134). The non-linear PD controller ($k_I = 0$) has been obtained as the final result of the learning process. Calculating signals in consecutive branches of the network representing the controller (Fig. 8.68) one obtains the following function realized by the NFC controller:

$$M_r = -20.689w_1 + 23.029e_P w_2 + 20.689w_3 - 1.969v_1 + 0.101e_D v_2 + 11.969v_3 \tag{8.135}$$

where: w_i and v_i are logic variables. The values of these variables indicate which of the sectors are relevant for current values of signals e_P and e_D – see Fig. 8.71.

According to the formula (8.135), each of 9 sectors of controller input space realizes its "own" control function.

The controller output signal M_r obtains its maximal value $M_r = 32.658$ in sector $w_3v_3 = 1$ and minimal value $M_r = -32.658$ in sector $w_1v_1 = 1$. It results from Fig. 8.69 that the propellers are able to generate a maximal torque only equal to 28.31 (Nm). That is why the partition of input space of the set consisting of the controller NFC and propellers has been reorganized into the final form shown in Fig. 8.72.

Fig. 8.72 shows coordinates of points characterizing particular sectors and expressions determining the output signals M_1 of the set "controller NFC + propellers", generated in respective sectors. The maximal torque generated by that driving set is 28.31 (sector $q_3 = 1$) and the minimal one is -18.30 (sector $q_1 = 1$). The values of logic variables q_i indicate that sector of input space which is suitable for the current value of the input signal controlling the propellers. In the central sector $q_2w_2v_2 = 1$, for $e_P = 0$ and $e_D = 0$, the generated signal $M_1 = 0$. The proof of stability for non-linear systems usually

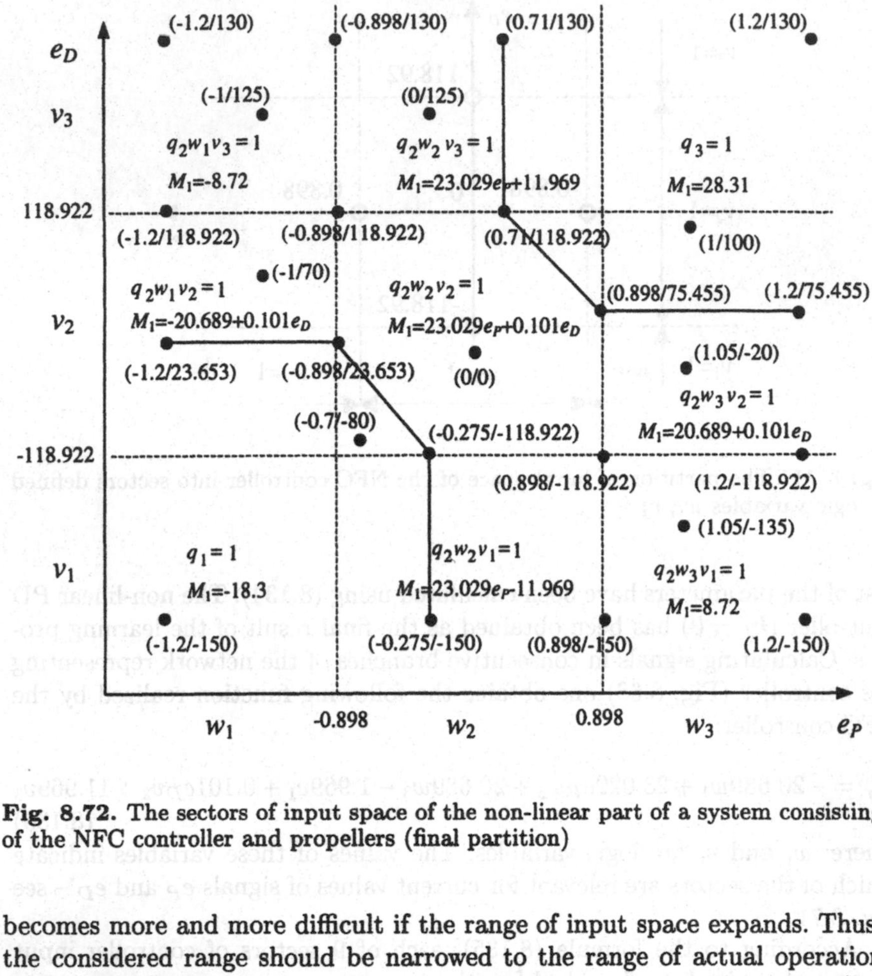

Fig. 8.72. The sectors of input space of the non-linear part of a system consisting of the NFC controller and propellers (final partition)

becomes more and more difficult if the range of input space expands. Thus, the considered range should be narrowed to the range of actual operation of the system. The ranges: $-1.2 \leq e_p \leq 1.2$ and $-150 \leq e_D \leq 130$ (see Fig. 8.72) have been chosen to avoid complications mentioned above. After successfully proving stability for the above "narrow" ranges one may enlarge the chosen ranges and attempt to obtain proof for system stability under new conditions.

The NFC controller consists of two parts (Fig 8.73): a linear one, which dynamically forms its outputs e_P, e_D and a static one $\mathbf{F}_1(e_P, e_D)$, which realizes mapping (8.135). The aggregation of elements $\mathbf{F}_1$ and F_2 into the non-linear part $\mathbf{F}$, and linear elements into the linear part, yields a quasi-standard form of control system shown in Fig. 8.74.

The surface shown in Fig. 8.75 has been obtained as a result of input/output mapping realized by the non-linear part $\mathbf{F}$ representing operation of the NFC controller and propellers. The surface is defined mainly by formula (8.135). Additionally, saturation of propeller static characteristic (see Fig. 8.69) has been taken into account.

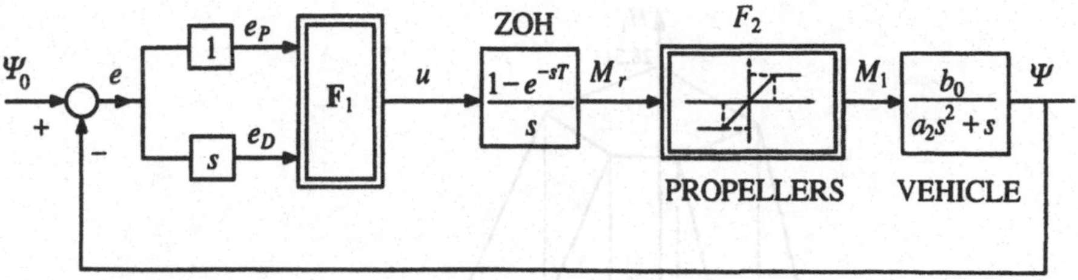

Fig. 8.73. The scheme of the underwater vehicle course control system (continuous space representation)

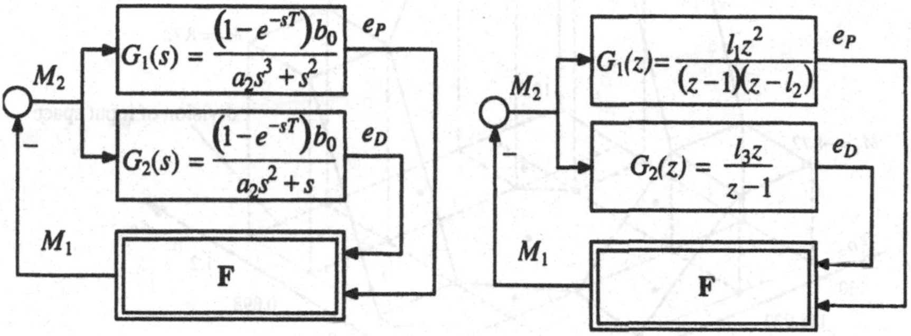

Fig. 8.74. The quasi-standard form of the control system in continuous space (a) and discrete space (b)

The surface shown in Fig. 8.75 consists of many segments. The complicated form of that surface is caused mainly by the complicated non-linear operation of NFC (number of rules, number of membership functions, etc.). There are many degrees of freedom corresponding to the control surface of the NFC controller. Thus, the control surface can be formed during the tuning process aimed at minimizing the control system performance criterion. The flat surfaces correspond to operations of classic PD controllers – see Fig. 8.76. Applying fuzzy controllers one can optimize the non-linear performance criteria much better than it can be done using classic controllers. It means that fuzzy controllers are more efficient, if quality of control processes is taken into account. The growing popularity of fuzzy controllers is closly connected with the property mentioned above.

The approximation of the linear part by discrete representation has been done using backward difference approximation (Leigh 1985), i.e. by substitution:

$$G_i(z) = G_i(s)\Big|_{\frac{1-z^{-1}}{T}}, \quad T = 0.05\,(s). \tag{8.136}$$

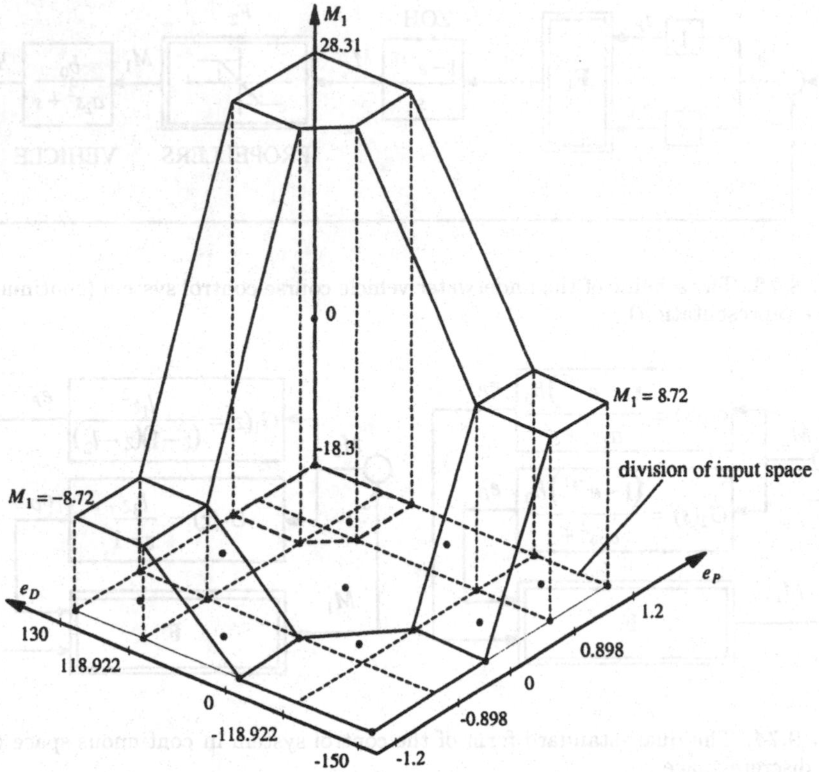

Fig. 8.75. The surface representing input-output mapping realized by the non-linear part **F** of a standard system

The discrete transfer functions approximating $G_1(z)$ and $G_2(z)$ have been obtained in the respective forms (8.137):

$$G_1(z) = \frac{7.749 \cdot 10^{-6} z^2}{(z-1)(z-0.859)} = \frac{l_1 \cdot z^2}{(z-1)(z-l_2)}$$

$$G_2(z) = \frac{1.805 \cdot 10^{-4} z}{z-1} = \frac{l_3 \cdot z}{z-1} \tag{8.137}$$

Both transfer functions (8.137) contain the pole $z_1 = 1$, which belongs to the unit circle $|z| = 1$ enclosing the system stability region. That is why two degrees of freedom k_1 and k_2 have been added in order to stabilize the system (Fig. 8.77). A single degree of freedom would be sufficient for stabilizing the system. However, the second degree of freedom has been added to make it easier to satisfy further conditions for system stability. The real parts of both transfer functions should be positive definite. That goal can be reached by the introduction of additional degrees of freedom d_1 and d_2 – see Fig. 8.77. The fictitious degrees of freedom modify transfer functions of the linear parts

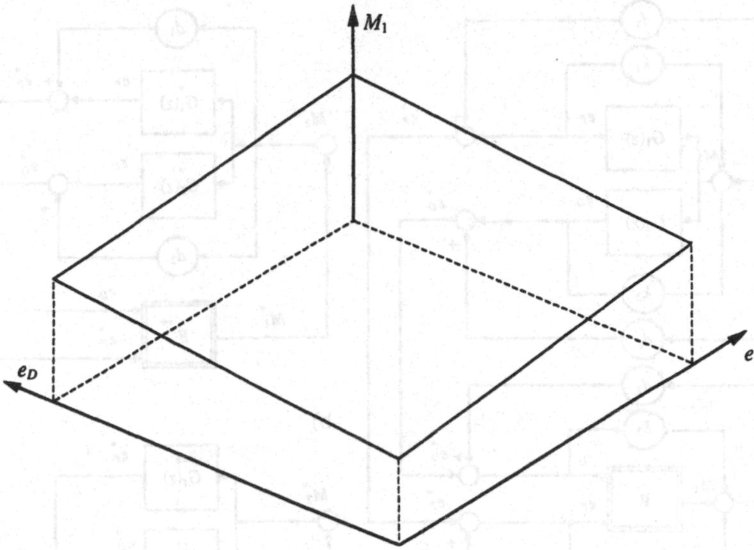

Fig. 8.76. An example of a flat control surface realized by a conventional, linear PD controller

to the forms (8.138).

$$
\begin{aligned}
G_1^{**}(z) &= \frac{e_P^*(z)}{M_2^*(z)} = \frac{l_1 z^2}{(1+k_1l_1+k_2l_3)z^2-(1+l_2+k_2l_2l_3)z+l_2} + d_1 = \\
&= G_1^*(z) + d_1 \\
G_2^{**}(z) &= \frac{e_D^*(z)}{M_2^*(z)} = \frac{l_3 z^2 - l_2 l_3 z}{(1+k_1l_1+k_2l_3)z^2-(1+l_2+k_2l_2l_3)z+l_2} + d_2 = \\
&= G_2^*(z) + d_2
\end{aligned}
\tag{8.138}
$$

The investigations of hyperstability conditions

The preliminary conditions for the linear part G(z)**

Condition PL1 (The matrix of transfer functions $\mathbf{G}^{**}(z)$ has to be square) The fictitious signals $N_{1k}^* = 0$ and $N_{2k}^* = 0$, shown in Fig. 8.78, have been added to satisfy the condition PL1. Since signal N_{2k}^* is always equal to zero, any transfer functions, for example G_1^{**}, G_2^{**}, can be "attached" to that signal. According to Fig. 8.78 one can formulate the following equation for the linear part:

$$
\underbrace{\begin{bmatrix} e_{Pk}^* \\ e_{Dk}^* \end{bmatrix}}_{\mathbf{Y}_k^*} = \underbrace{\begin{bmatrix} G_1^{**} & G_2^{**} \\ G_2^{**} & G_1^{**} \end{bmatrix}}_{\mathbf{G}^{**}(z)} \cdot \underbrace{\begin{bmatrix} M_{2k}^* \\ N_{2k}^* \end{bmatrix}}_{\mathbf{U}_k^*}
\tag{8.139}
$$

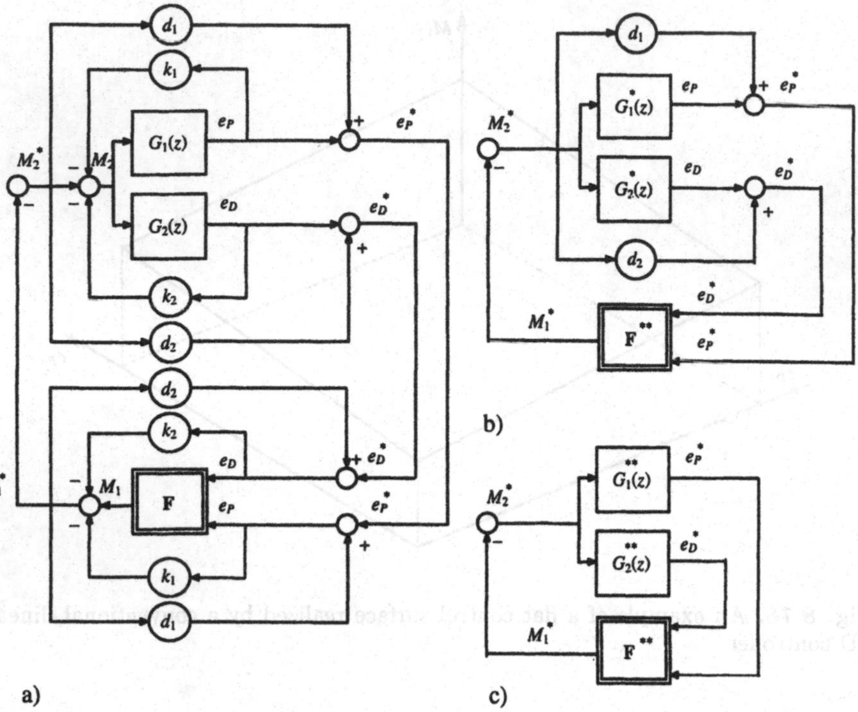

Fig. 8.77. The control system after introduction of additional degrees of freedom k_i, d_j (a) and its folded schemes (b), (c)

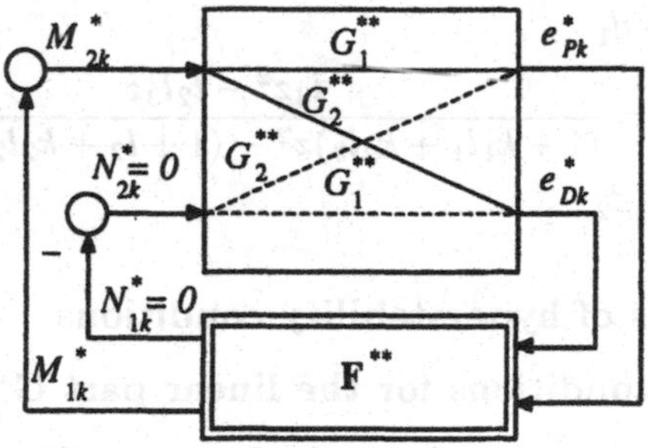

Fig. 8.78. The standard form of a control system after adding fictitious signals $N_{1k}^* = N_{2k}^* = 0$

Condition PL2 (The linear part $\mathbf{G}^{**}(z)$ has to be completely controllable and observable)
The condition will be checked last of all, after determination of values of coefficients k_i, d_j satisfying the other hyperstability conditions.

Condition PL3 (The poles of all component transfer functions $G_{ij}(z)$ have to be stable)
The requirement can be expressed in the form (8.140) resulting from Jury's test (8.90).

$$\begin{aligned} &k_1 + 3.2925k_2 > 0 \\ &k_1 + 43.292k_2 + 47968.221 > 0 \\ &k_1 + 23.292k_2 + 18240.72 > 0 \end{aligned} \tag{8.140}$$

It is important that now values k_1 or k_2 can be negative. There was not such a possibility in the case of Example 8.3.3.1, for SISO controlled plant. Now, there is a greater range of choice of k_1 and k_2. Finally, proving hyperstability can be achieved more easily.

Condition PL4 (The order of the nominator of every component transfer function $G_{ij}(z)$ can not be higher than the order of its denominator)
The condition is satisfied.

The preliminary conditions for the non-linear part $\mathbf{F}^{**}(\mathbf{Y}_k^*)$

Condition PN1 (The function $\mathbf{V}_k^* = \mathbf{F}^{**}(\mathbf{Y}_k^*)$ has to represent unique mapping)
It results from (8.135) and Fig. 8.72 that mapping realized by the primary non-linear part F can be expressed in the form (8.141):

$$M_1 = m_0 + m_P e_P + m_D e_D \,, \tag{8.141}$$

where values of coefficients m_0, m_P, m_D depend on the currently activated sector of operation of the non-linear part. For example, if $v_2 w_2 q_2 = 1$, then mapping obtains the form $M_1 = 23.029 e_P + 0.101 e_D$. The values of coefficients m_0, m_P, m_D have been listed in Fig. 8.72.

It is possible to find further equations describing the operation of the non-linear part $\mathbf{F}^{**}$. So, based on Fig. 8.77, one obtains equations (8.142).

$$\begin{aligned} e_P &= e_P^* + d_1 M_1^* \\ e_D &= e_D^* + d_2 M_1^* \\ M_1^* &= M_1 - k_1 e_P - k_2 e_D \end{aligned} \tag{8.142}$$

Finally, from (8.141) and (8.142) can be obtained:

$$\begin{aligned} M_1^* &= \frac{m_0 + (m_P - k_1)e_P^* + (m_D - k_2)e_D^*}{1 - (m_P - k_1)d_1 - (m_D - k_2)d_2} \,, \\ N_1^* &= 0 \,. \end{aligned} \tag{8.143}$$

The final result (8.143) represents unique mapping realized by the secondary non-linear part $\mathbf{F}^{**}$.

Condition PN2 (The function $\mathbf{V}_k^* = \mathbf{F}^{**}(\mathbf{Y}_k^*)$ has to map $\mathbf{Y}_k = 0$ into $\mathbf{V}_k = 0$ – zeroing condition)
If values of primary signals e_P and e_D are equal to zero, then, according to Fig. 8.72, value $M_1 = 0$ ($M_1 = 23.029e_P + 0.101e_D = 0$). Using the above information for analysis of expressions (8.142) one obtains:

$$M_1 = 0\,, \quad e_P^* = 0\,, \quad e_D^* = 0\,.$$

Because $N_1^* = 0$ independently of values e_P^* and e_D^*, then condition PN2 is satisfied.

The investigation of the main conditions for hyperstability of a system

Condition ML (The discrete linear part has to be strictly positive real)
The considered condition can be decomposed to a certain number of component conditions.

Condition ML1 (There are no unstable poles among the poles of all component transfer functions $G_{ij}(z)$)
If values k_1 and k_2 satisfy inequalities (8.140), then condition ML1 is also fulfilled.

Condition ML2 (The matrix $\mathbf{H}(j\omega) = 0.5[\mathbf{G}(e^{(j\omega)}) + \mathbf{G}^{\mathrm{T}}(e^{(-j\omega)})]$ is a positive definite, Hermitian one, for all $\omega \geq 0$)
The following expression (8.144) represents transfer function $G_1^{**}(e^{j\omega T})$:

$$G_1^{**}(e^{j\omega T}) = G_1^*(e^{j\omega T}) + d_1 = (\mathrm{Re}_1^* + d_1) + j\mathrm{Im}_1^* \tag{8.144}$$

The transfer function $G_2^{**}(e^{j\omega T})$ can be written in the form (8.145).

$$G_2^{**}(e^{j\omega T}) = G_2^*(e^{j\omega T}) + d_2 = (\mathrm{Re}_2^* + d_2) + j\mathrm{Im}_2^* \tag{8.145}$$

The elements of (8.144) and (8.145) are defined by expressions (8.146).

$$\begin{aligned}
\mathrm{Re}_1^* &= \frac{F_2^* \cos^2 \omega T - F_1^* \cos \omega T + F_0^*}{D_2 \cos^2 \omega T - D_1 \cos \omega T + D_0} \\
\mathrm{Im}_1^* &= \frac{\sin \omega T (A_1 \cos \omega T + A_2)}{D_2 \cos^2 \omega T - D_1 \cos \omega T + D_0} \\
\mathrm{Re}_2^* &= \frac{E_2^* \cos^2 \omega T - E_1^* \cos \omega T + E_0^*}{D_2 \cos^2 \omega T - D_1 \cos \omega T + D_0} \\
\mathrm{Im}_2^* &= \frac{\sin \omega T (B_1 \cos \omega T + B_2)}{D_2 \cos^2 \omega T - D_1 \cos \omega T + D_0}
\end{aligned} \tag{8.146}$$

The coefficients A_1, A_2, B_1, B_2, D_0, D_1, D_2, E_2^*, E_1^*, E_0^*, F_2^*, F_1^*, F_0^*, expressed by (8.147), are functions of degrees of freedom k_1 and k_2:

$$
\begin{aligned}
A_1 &= 3.09976 \cdot 10^{-4} \\
A_2 &= 14.40342 \cdot 10^{-6} + 12.01067 \cdot 10^{-10} k_2 \\
B_1 &= 3.09976 \cdot 10^{-4} \\
B_2 &= 3.57399 \cdot 10^{-4} + 12.01067 \cdot 10^{-10} k_1 + 5.59516 \cdot 10^{-8} k_2 \\
D_0 &= 0.01998 + 2.19084 \cdot 10^{-6} k_1 + 0.5103 \cdot 10^{-4} k_2 + 27.97587 \cdot 10^{-10} k_1 k_2 + \\
&\quad 60.0535 \cdot 10^{-12} k_1^2 + 3.25813 \cdot 10^{-8} k_2^2 \\
D_1 &= 6.90912 + 28.80528 \cdot 10^{-6} k_1 + 12.47097 \cdot 10^{-4} k_2 + \\
&\quad 24.01964 \cdot 10^{-10} k_1 k_2 + 5.59497 \cdot 10^{-8} k_2^2 \\
D_2 &= 3.43458 + 26.616 \cdot 10^{-6} k_1 + 6.19952 \cdot 10^{-4} k_2 \\
E_0^* &= 3.13582 \cdot 10^{-4} + 13.98794 \cdot 10^{-10} k_1 + 5.66026 \cdot 10^{-8} k_2 \\
E_1^* &= 6.23558 \cdot 10^{-4} + 12.01067 \cdot 10^{-10} k_1 + 5.59516 \cdot 10^{-8} k_2 \\
E_2^* &= 3.09976 \cdot 10^{-4} \\
F_0^* &= 1.09538 \cdot 10^{-6} + 60.05351 \cdot 10^{-12} k_1 + 13.98794 \cdot 10^{-10} k_2 \\
F_1^* &= 14.40342 \cdot 10^{-6} + 12.01067 \cdot 10^{-10} k_2 \\
F_2^* &= 13.308 \cdot 10^{-6}
\end{aligned} \tag{8.147}
$$

The relations (8.148) result from (8.146) and properties of trigonometrical functions ($\cos\phi = \cos(-\phi)$, $\sin\phi = -\sin(-\phi)$):

$$
\begin{aligned}
\mathrm{Re}_i^{**}(e^{j\omega T}) &= \mathrm{Re}_i^*(e^{-j\omega T}), \quad i = 1 \div 2, \\
\mathrm{Im}_i^{**}(e^{j\omega T}) &= -\mathrm{Im}_i^*(e^{-j\omega T}), \quad i = 1 \div 2.
\end{aligned} \tag{8.148}
$$

Using (8.139) and (8.148), one can calculate matrix $\mathbf{H}(j\omega)$.

$$
\begin{aligned}
\mathbf{H}(j\omega) &= 0.5\left[\mathbf{G}(e^{(j\omega)}) + \mathbf{G}^{\mathrm{T}}(e^{(-j\omega)})\right] = \\
&= 0.5\begin{bmatrix} (\mathrm{Re}_1^* + d_1 + j\mathrm{Im}_1^*) & (\mathrm{Re}_2^* + d_2 + j\mathrm{Im}_2^*) \\ (\mathrm{Re}_2^* + d_2 + j\mathrm{Im}_2^*) & (\mathrm{Re}_1^* + d_1 + j\mathrm{Im}_1^*) \end{bmatrix} + \\
&\quad 0.5\begin{bmatrix} (\mathrm{Re}_1^* + d_1 - j\mathrm{Im}_1^*) & (\mathrm{Re}_2^* + d_2 - j\mathrm{Im}_2^*) \\ (\mathrm{Re}_2^* + d_2 - j\mathrm{Im}_2^*) & (\mathrm{Re}_1^* + d_1 - j\mathrm{Im}_1^*) \end{bmatrix} = \\
&= \begin{bmatrix} (\mathrm{Re}_1^* + d_1) & (\mathrm{Re}_2^* + d_2) \\ (\mathrm{Re}_2^* + d_2) & (\mathrm{Re}_1^* + d_1) \end{bmatrix}
\end{aligned} \tag{8.149}
$$

The relations (8.148) mean that the real parts of transfer functions Re_1^* and Re_2^* are even functions of angle (ωT). Hence:

$$
\mathbf{H}(\omega T) = \mathbf{H}^{\mathrm{T}}(-\omega T) \tag{8.150}
$$

which means that matrix $\mathbf{H}(\omega T)$ is a Hermitian one.

It results from Sylvester's theorem (Kaczorek 1981), that conditions (8.151) and (8.152) have to be satisfied for positive definite matrix $\mathbf{H}(\omega T)$.

$$(\mathrm{Re}_1^* + d_1) > 0 \tag{8.151}$$
$$(\mathrm{Re}_1^* + d_1)^2 - (\mathrm{Re}_2^* + d_2)^2 > 0 \tag{8.152}$$

The inequality (8.152) can be transformed to the form (8.153):

$$[(\mathrm{Re}_1^* + d_1) + (\mathrm{Re}_2^* + d_2)]\,[(\mathrm{Re}_1^* + d_1) - (\mathrm{Re}_2^* + d_2)] > 0\,. \tag{8.153}$$

The inequality (8.153) can easily be substituted by two inequalities. Finally, the positive definite matrix $\mathbf{H}(\omega T)$ has to satisfy the three following conditions:

$$(\mathrm{Re}_1^* + d_1) = \mathrm{Re}_1^{**} > 0\,, \quad \forall\omega \geq 0\,, \tag{8.154}$$
$$(\mathrm{Re}_1^* + d_1) + (\mathrm{Re}_2^* + d_2) = \mathrm{Re}_1^{**} + \mathrm{Re}_2^{**} > 0\,, \quad \forall\omega \geq 0\,, \tag{8.155}$$
$$(\mathrm{Re}_1^* + d_1) - (\mathrm{Re}_2^* + d_2) = \mathrm{Re}_1^{**} - \mathrm{Re}_2^{**} > 0\,, \quad \forall\omega \geq 0\,. \tag{8.156}$$

Analysis of the above conditions leads to very interesting conclusions. The SISO system containing single transfer function $G(z)$, as in Example 8.3.3.1, yields a single condition of the type (8.154). The satisfied condition of the type (8.154) ensures that frequency response $G(j\omega T)$ is located in the right half-plane of the complex co-ordinates system. For a MIMO system, as in the current example, the condition (8.154) imposes the requirement that characteristic $G_1^{**}(e^{j\omega T})$ has to be located in the right half-plane and conditions (8.155), (8.156) subordinate the location of characteristic $G_2^{**}(e^{j\omega T})$ to the location of $G_1^{**}(e^{j\omega T})$. It does not mean that characteristic $G_2^{**}(e^{j\omega T})$ has to be located entirely in the right half-plane. The examples of loci of $G_1^{**}(e^{j\omega T})$ and $G_2^{**}(e^{j\omega T})$ satisfying conditions (8.154) – (8.156) are shown in Fig. 8.79.

The substitution $e^{j\omega T} = cos\omega T + j\sin\omega T$ transforms the conditions (8.154) – (8.156) to the forms depending on argument $\cos\omega T$. Further analysis of system behavior using those new forms is easier, since variability of arguments is narrowed to the range $[-1, 1]$. The numerators and denominators of the real parts of the considered transfer functions G_1^{**} and G_2^{**} are represented by parabolic functions. The denominators of both real parts are identical. Furthermore, the denominator is always positive as it is the sum of second powers of two expressions (8.157):

$$\begin{aligned}
\mathrm{Re}_1^{**} &= \frac{F_2^* \cos^2\omega T - F_1^* \cos\omega T + F_0^*}{D_2 \cos^2\omega T - D_1 \cos\omega T + D_0} = \frac{N(\mathrm{Re}_1^{**})}{D(\mathrm{Re}_1^{**})}\,,\\
\mathrm{Re}_2^{**} &= \frac{E_2^* \cos^2\omega T - E_1^* \cos\omega T + E_0^*}{D_2 \cos^2\omega T - D_1 \cos\omega T + D_0} = \frac{N(\mathrm{Re}_2^{**})}{D(\mathrm{Re}_2^{**})}\,.
\end{aligned} \tag{8.157}$$

where D_0, D_1, D_2 are determined by relations (8.147) and E_2, E_1, E_0, F_2, F_1, F_0 by expressions (8.158):

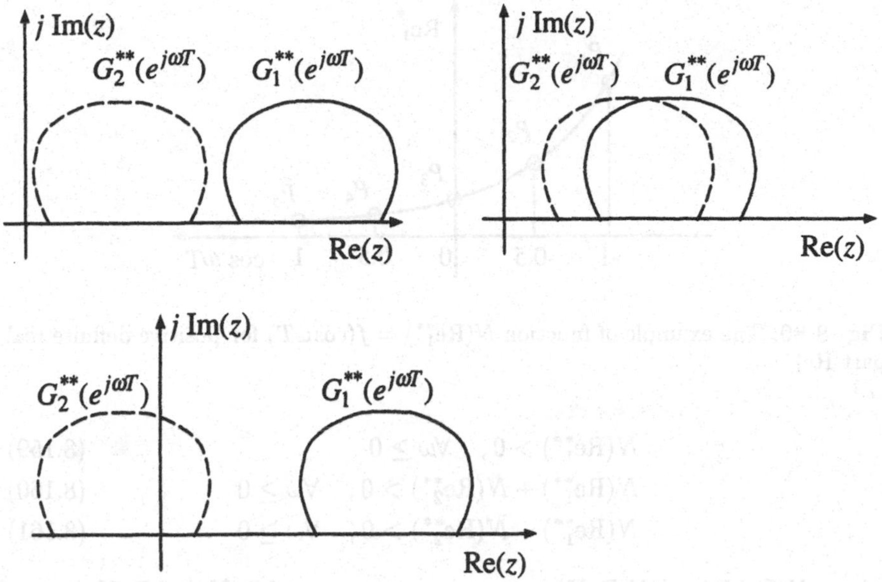

Fig. 8.79. The possible loci of $G_1^{**}(e^{j\omega T})$ in relation to the loci of $G_1^{**}(e^{j\omega T})$ for positive definite matrix $\mathbf{H}(j\omega)$ (to ensure that matrix property, $|\mathrm{Re}_2^{**}|$ has to be smaller than Re_1^{**}, for all $\omega \geq 0$)

$$
\begin{aligned}
E_0 &= 3.13582 \cdot 10^{-4} + 13.98794 \cdot 10^{-10} k_1 + 5.66026 \cdot 10^{-8} k_2 + 3.47454 d_2 + \\
&\quad 2.19084 \cdot 10^{-6} k_1 d_2 + 6.27165 \cdot 10^{-4} k_2 d_2 + 60.05351 \cdot 10^{-12} k_1^2 d_2 + \\
&\quad 5.66026 \cdot 10^{-8} k_2^2 d_2 + 27.97587 \cdot 10^{-10} k_1 k_2 d_2 \\
E_1 &= 6.23558 \cdot 10^{-4} + 12.01067 \cdot 10^{-10} k_1 + 5.59516 \cdot 10^{-8} k_2 + 6.90912 d_2 + \\
&\quad 28.80684 \cdot 10^{-6} k_1 d_2 + 12.47117 \cdot 10^{-4} k_2 d_2 + 24.02134 \cdot 10^{-10} k_1 k_2 d_2 + \\
&\quad 5.59516 \cdot 10^{-8} k_2^2 d_2 \\
E_2 &= 3.09976 \cdot 10^{-4} + 3.43458 \cdot d_2 + 26.616 \cdot 10^{-6} k_1 d_2 + 6.19952 \cdot 10^{-4} k_2 d_2 \\
F_0 &= 1.09538 \cdot 10^{-6} + 60.0535 \cdot 10^{-12} k_1 + 13.9879 \cdot 10^{-10} k_2 + 3.47454 d_1 + \\
&\quad 2.19076 \cdot 10^{-6} k_1 d_1 + 6.27165 \cdot 10^{-4} k_2 d_1 + 27.97587 \cdot 10^{-10} k_1 k_2 d_1 + \\
&\quad 60.0535110^{-12} k_1^2 d_1 + 5.66026 \cdot 10^{-8} k_2^2 d_1 \\
F_1 &= 14.4034 \cdot 10^{-6} + 12.0106 \cdot 10^{-10} k_2 + 6.90912 d_1 + 28.8068 \cdot 10^{-6} k_1 d_1 + \\
&\quad 12.47117 \cdot 10^{-4} k_2 d_1 + 24.02134 \cdot 10^{-10} k_1 k_2 d_1 + 5.59552 \cdot 10^{-8} k_2^2 d_1 \\
F_2 &= 13.308 \cdot 10^{-6} + 3.43458 d_1 + 26.616 \cdot 10^{-6} k_1 d_1 + 6.19952 \cdot 10^{-4} k_2 d_1
\end{aligned}
\tag{8.158}
$$

The identical and positive denominators of Re_1^{**} and Re_1^{**} allow us to simplify the conditions (8.154) – (8.156) to the forms (8.159) – (8.161).

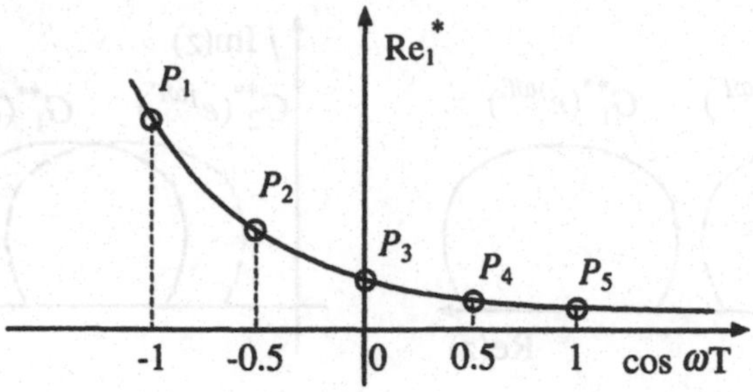

Fig. 8.80. The example of function $N(\mathrm{Re}_1^{**}) = f(\cos \omega T)$ for positive definite real part Re_1^{**}

$$N(\mathrm{Re}_1^{**}) > 0\,, \quad \forall\omega \geq 0 \tag{8.159}$$

$$N(\mathrm{Re}_1^{**}) + N(\mathrm{Re}_2^{**}) > 0\,, \quad \forall\omega \geq 0 \tag{8.160}$$

$$N(\mathrm{Re}_1^{**}) - N(\mathrm{Re}_2^{**}) > 0\,, \quad \forall\omega \geq 0 \tag{8.161}$$

where $N(\mathrm{Re}_1^{**})$ and $N(\mathrm{Re}_2^{**})$ represent numerators of Re_1^{**} and Re_2^{**} respectively.

If values of function $N(\mathrm{Re}_1^{**})$ for the argument range $[-1, 1]$ are positive, then Re_1^{**} is positive definite – see Fig. 8.80.

The fulfilling of condition (8.159) can be achieved by a suitable choice of values of coefficients F_i in (8.157). However, those coefficients are expressed by complicated functions (8.158) depending on the parameters of the primary system as well as degrees of freedom k_i, d_i. Vieta's formulas for quadratic equations could be applied to formulate the analytical representation of conditions for positive definite $N(\mathrm{Re}_1^{**})$. However, it would be a time-consuming task (to say nothing of possible mistakes). Furthermore, it can easily be predicted that the final form of the conditions obtained would be very complicated. Under such circumstances serious restrictions for the choice of values of coefficients k_i, d_i can appear. For simplification of the whole procedure one can formulate condition (8.159) for chosen points P_i located over the argument range $[-1, 1]$ – see Fig. 8.80. The points referring to arguments -1, -0.5, 0, 0.5, 1 have been chosen for further considerations.

Thus, relation (8.159) and the choice of P_i yield five conditions (8.162) – (8.166) ensuring, that $N(\mathrm{Re}_1^{**}) > 0$.

$$28.8068 \cdot 10^{-6} + 60.05351 \cdot 10^{-12} k_1 + 25.99861 \cdot 10^{-10} k_2 + d_1(13.81824 + 57.6136 \cdot 10^{-6} k_1 + 24.94234 \cdot 10^{-4} k_2 + 51.99721 \cdot 10^{-10} k_1 k_2 + 60.05351 \cdot 10^{-12} k_1^2 + 11.25578 \cdot 10^{-8} k_2^2) > 0 \tag{8.162}$$

$$46.49636 \cdot 10^{-6} + 240.21404 \cdot 10^{-12} k_1 + 79.97331 \cdot 10^{-10} k_2 + d_1(31.15098 + 93.18588 \cdot 10^{-6} k_1 + 56.22846 \cdot 10^{-4} k_2 + 159.94616 \cdot 10^{-10} k_1 k_2 + 240.21404 \cdot 10^{-12} k_1^2 + 33.83208 \cdot 10^{-8} k_2^2) > 0 \tag{8.163}$$

$$1.09538 \cdot 10^{-6} + 60.05351 \cdot 10^{-12} k_1 + 13.98794 \cdot 10^{-10} k_2 + d_1(3.47545 + 2.19076 \cdot 10^{-6} k_1 + 6.27165 \cdot 10^{-4} k_2 + 27.97587 \cdot 10^{-10} k_1 k_2 + 60.05351 \cdot 10^{-12} k_1^2 + 5.66026 \cdot 10^{-8} k_2^2) > 0 \tag{8.164}$$

$$-11.11732 \cdot 10^{-6} + 240.21404 \cdot 10^{-12} k_1 + 31.96264 \cdot 10^{-10} k_2 + d_1(3.5125 - 22.04148 \cdot 10^{-6} k_1 + 6.34378 \cdot 10^{-4} k_2 + 63.8608 \cdot 10^{-10} k_1 k_2 + 240.21404 \cdot 10^{-12} k_1^2 + 11.45 \cdot 10^{-8} k_2^2) > 0 \tag{8.165}$$

$$60.0535 \cdot 10^{-12} k_1 + 1.97727 \cdot 10^{-10} k_2 + d_1(-0.19324 \cdot 10^{-6} k_1 + 3.95453 \cdot 10^{-10} k_1 k_2 + 60.05351 \cdot 10^{-12} k_1^2 + 0.06474 \cdot 10^{-8} k_2^2) > 0 \tag{8.166}$$

A similar procedure can be applied to condition (8.160). So, relation (8.160) narrowed to points P_i, referring to arguments $\cos\omega T = -1, -0.5, 0, 0.5, 1$, can be split into five inequalities (8.167) – (8.171).

$$12.75923 \cdot 10^{-4} + 26.59914 \cdot 10^{-10} k_1 + 11.51541 \cdot 10^{-8} k_2 + (d_1 + d_2)(13.81824 + 57.6136 \cdot 10^{-6} k_1 + 24.94234 \cdot 10^{-4} k_2 + 51.99721 \cdot 10^{-10} k_1 k_2 + 60.05351 \cdot 10^{-12} k_1^2 + 11.25578 \cdot 10^{-8} k_2^2) > 0 \tag{8.167}$$

$$28.57916 \cdot 10^{-4} + 82.37524 \cdot 10^{-10} k_1 + 34.63109 \cdot 10^{-8} k_2 + (d_1 + d_2)(31.15098 + 93.18588 \cdot 10^{-6} k_1 + 56.22846 \cdot 10^{-4} k_2 + 159.94616 \cdot 10^{-10} k_1 k_2 + 240.21404 \cdot 10^{-12} k_1^2 + 33.83208 \cdot 10^{-8} k_2^2) > 0 \tag{8.168}$$

$$3.14677 \cdot 10^{-4} + 14.59324 \cdot 10^{-10} k_1 + 5.80014 \cdot 10^{-8} k_2 + (d_1 + d_2)(3.47454 + 2.19076 \cdot 10^{-6} k_1 + 6.27165 \cdot 10^{-4} k_2 + 27.97587 \cdot 10^{-10} k_1 k_2 + 60.5351 \cdot 10^{-12} k_1^2 + 5.66026 \cdot 10^{-8} k_2^2) > 0 \tag{8.169}$$

$$3.06071 \cdot 10^{-4} + 34.33256 \cdot 10^{-10} k_1 + 11.77035 \cdot 10^{-8} k_2 + (d_1 + d_2)(3.5125 - 22.04148 \cdot 10^{-6} k_1 + 6.34378 \cdot 10^{-4} k_2 + 63.8606 \cdot 10^{-10} k_1 k_2 + 240.21404 \cdot 10^{-12} k_1^2 + 11.45 \cdot 10^{-8} k_2^2) > 0 \tag{8.170}$$

$$2.5778 \cdot 10^{-10} k_1 + 0.08487 \cdot 10^{-8} k_2 + (d_1 + d_2)(3.95453 \cdot 10^{-10} k_1 k_2 + 60.05351 \cdot 10^{-12} k_1^2 + 0.0631 \cdot 10^{-8} k_2^2) > 0 \tag{8.171}$$

The condition (8.161) can be decomposed into the following inequalities:

$$-12.18309 \cdot 10^{-4} - 25.39808 \cdot 10^{-10} k_1 - 10.99543 \cdot 10^{-8} k_2 + (d_1 - d_2)(13.81824 + 57.61368 \cdot 10^{-6} k_1 + 24.94234 \cdot 10^{-4} k_2 + 51.99721 \cdot 10^{-10} k_1 k_2 + 60.05351 \cdot 10^{-12} k_1^2 + 11.25578 \cdot 10^{-8} k_2^2) > 0 \tag{8.172}$$

$$-27.64924 \cdot 10^{-4} - 77.57096 \cdot 10^{-10} k_1 - 33.03163 \cdot 10^{-8} k_2 + (d_1 - d_2)(31.15098 + 93.18588 \cdot 10^{-6} k_1 + 56.22846 \cdot 10^{-4} k_2 + 159.94616 \cdot 10^{-10} k_1 k_2 + 240.21404 \cdot 10^{-12} k_1^2 + 33.83208 \cdot 10^{-8} k_2^2) > 0 \tag{8.173}$$

$$-3.12487 \cdot 10^{-4} - 13.38264 \cdot 10^{-10} k_1 - 5.52038 \cdot 10^{-8} k_2 + (d_1 - d_2)(3.47454 + 2.19076 \cdot 10^{-6} k_1 + 6.27165 \cdot 10^{-4} k_2 + 27.97587 \cdot 10^{-10} k_1 k_2 + 60.5351 \cdot 10^{-12} k_1^2 + 5.66026 \cdot 10^{-8} k_2^2) > 0 \tag{8.174}$$

$$-3.28305 \cdot 10^{-4} - 29.52828 \cdot 10^{-10} k_1 - 11.13109 \cdot 10^{-8} k_2 + (d_1 - d_2)(3.5125 - 22.04148 \cdot 10^{-6} k_1 + 6.34378 \cdot 10^{-4} k_2 + 63.8606 \cdot 10^{-10} k_1 k_2 + 240.21404 \cdot 10^{-12} k_1^2 + 11.45 \cdot 10^{-8} k_2^2) > 0 \tag{8.175}$$

$$-1.37674 \cdot 10^{-10} k_1 - 0.04533 \cdot 10^{-8} k_2 + (d_1 - d_2)(3.95453 \cdot 10^{-10} k_1 k_2 + 60.05351 \cdot 10^{-12} k_1^2 + 0.0631 \cdot 10^{-8} k_2^2) > 0 \tag{8.176}$$

The main conditions MN for the non-linear part
The Popov's secondary condition (8.101) now has the form:

$$\mathbf{v}_k^{*\mathrm{T}} \mathbf{y}_k^* = \sum_{i=1}^{p} v_{ik}^* y_{ik}^* \geq 0, \quad \forall k \geq 0 .$$

where vectors $\mathbf{v}_k$ and $\mathbf{y}_k$ have been defined by (8.177):

$$\mathbf{v}_k^* = \begin{bmatrix} M_{\Delta k}^* \\ M_{\Delta k}^* \end{bmatrix}, \quad \mathbf{y}_k^* = \begin{bmatrix} e_{Pk}^* \\ e_{Dk} \end{bmatrix}, \tag{8.177}$$

$\mathbf{v}_k^*$ – vector of non-linear part outputs, $\mathbf{y}_k^*$ – vector of non-linear part inputs.

The application of equation (8.141):

$$M_1 = m_0 + m_P e_P + m_D e_D ,$$

representing the general formula for output M_1 of the primary non-linear part, and equation (8.142) representing properties of the secondary non linear part $\mathbf{F}^{**}$,

$$\begin{aligned} e_{Pk} &= e_{Pk}^* + d_1 M_{1k}^* , \\ e_{Dk} &= e_{Dk}^* + d_2 M_{1k}^* , \\ M_{1k}^* &= M_{1k} - k_1 e_{Pk} - k_2 e_{Dk} , \\ N_{2k}^* &= 0 , \end{aligned}$$

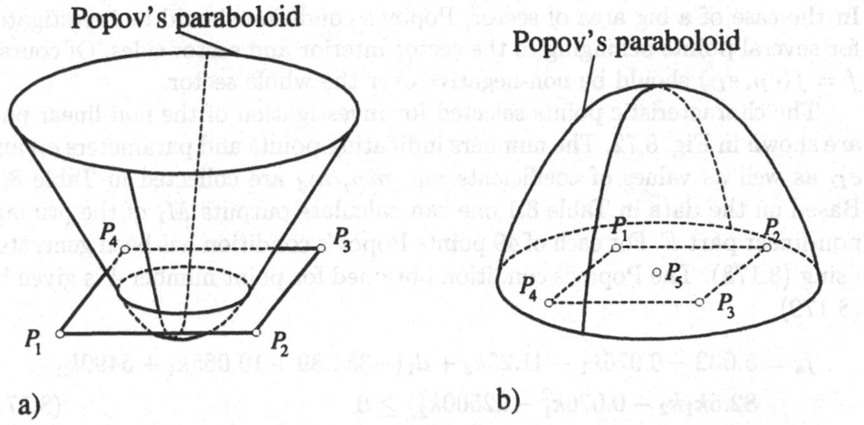

Fig. 8.81. The violation (a) and fulfillment (b) of Popov's condition over a selected sector of the non-linear part

to Popov's secondary condition (8.101) transforms it to the space $[e_P, e_D]$ of a primary system. The result of that transformations is given by (8.178).

$$\begin{aligned} f = {} & e_P(m_0 + m_P e_P + m_D e_D) - k_1 e_P^2 - k_2 e_P e_D - d_1 [m_0^2 + m_P^2 e_P^2 + \\ & m_D^2 e_D^2 + 2 m_0 m_P e_P + 2 m_0 m_D e_D + 2 m_P m_D e_P m_D + \\ & 2 k_1 e_P (m_0 + m_P e_P + m_D e_D) + 2 k_2 e_D (m_0 + m_P e_P + m_D e_D) - \\ & k_1^2 e_P^2 - k_2^2 e_D^2 - 2 k_1 k_2 e_P e_D] \geq 0 \end{aligned} \tag{8.178}$$

Fig. 8.72 shows that coefficients m_0, m_P, m_D depend on the operational sector of the non-linear part. It means that Popov's condition has to be investigated separately for every sector. Popov's condition (8.178) represents a three-dimensional surface $f = f(e_P, e_D)$. The values of coefficients influence the shape of that surface. Thus, one can obtain geometric representation of Popov's condition as a paraboloid, hyperboloid or ellipsoid. As long as values of degrees of freedom k_i, d_j are unknown, there are serious difficulties in a precise analysis of Popov's condition. We simply do not know what kind of surface is over each particular sector, what the surface locus is, if there exists a minimum or maximum of function $f = f(e_P, e_D)$ over the sector, etc.

Since exact investigation of Popov's condition over the whole sector is so difficult (it may be unrealizable), then a reasonable compromise has to be chosen. Thus, in the first step the investigations can be narrowed to a certain number of characteristic, external points and one internal point of each sector. The relevant example is shown in Fig. 8.81b.

The introduction of a condition for the internal point P_5 of each sector, see Fig. 8.81, lowers the probability of achieving the unpleasant result, resembling that shown in Fig. 8.81a (Popov's condition is satisfied for external points of sector $P_1 - P_4$ but is not satisfied for a substantial part of the sector interior).

In the case of a big area of sector, Popov's condition should be investigated for several points belonging to the sector interior and sector sides. Of course, $f = f(e_P, e_D)$ should be non-negative over the whole sector.

The characteristic points selected for investigation of the non-linear part are shown in Fig. 8.72. The numbers indicating points and parameters e_P and e_D as well as values of coefficients m_0, m_P, m_d are collected in Table 8.1. Based on the data in Table 8.1 one can calculate outputs M_1 of the primary non-linear part F. For each of 49 points Popov's condition has been generated using (8.178). The Popov's condition obtained for point number 4 is given by (8.179).

$$f_4 = 5.032 - 0.076k_1 - 41.25k_2 + d_1(-334.89 + 10.065k_1 + 5490k_2 - 82.5k_1k_2 - 0.076k_1^2 - 22500k_2^2) \geq 0 \tag{8.179}$$

Popov's condition for the remaining 48 points also obtains the form of functions of k_i, d_i. They are not published to make this section more compact.

There are 18 conditions given by inequalities (8.140) and (8.162) – (8.176) for the linear part and 49 conditions for the non-linear part. Thus, the total number of conditions for the system under consideration equals 67. Those 67 conditions have to be satisfied by the adjustment of 4 degrees of freedom (!). The above figures transparently show how important for successfully proving system hyperstability is the choice of a sufficient number of degrees of freedom.

Dealing with many local extrema is typical for non-linear conditions formulated above. A special computer program has been prepared to find the solutions $[k_1, k_2, d_1, d_2]$ satisfying all 67 conditions considered. The program algorithm has been built using the gradient method and genetic algorithms. In spite of many attempts the solution of the problem for four degrees of freedom has not been found. The unsuccessful result has led to the problem's rearrangement by the introduction of 8 degrees of freedom instead of the previous 4. The system "equipped" with 8 degrees of freedom is shown in Fig. 8.82.

Condition ML1

The application of Jury's test (8.90) (Leigh 1985) leads to the following conditions, ensuring the stability of transfer functions G_1^{**}, G_2^{**}, see Fig. 8.82 and formula (8.192).

$$\begin{aligned}
&(r_1r_2 + k_1r_2l_1 + r_1k_2l_3) - [r_1r_2(1 + l_2) + r_1k_2l_2l_3] + r_1r_2l_2 > 0\,,\\
&(r_1r_2 + k_1r_2l_1 + r_1k_2l_3) + [r_1r_2(1 + l_2) + r_1k_2l_2l_3] + r_1r_2l_2 > 0\,,\\
&(r_1r_2 + k_1r_2l_1 + r_1k_2l_3) - |r_1r_2l_2| > 0\,,
\end{aligned} \tag{8.180}$$

where : $l_1 = 7.749 \cdot 10^{-6}$, $l_2 = 0.859$, $l_3 = 1.805 \cdot 10^{-4}$.

Condition ML2

The matrix $\mathbf{H}(j\omega) = 0.5[\mathbf{G}(e^{j\omega T}) + \mathbf{G}^{\mathrm{T}}(e^{-j\omega T})]$ has to be positive definite and a Hermitian one, for all $\omega \geq 0$.

Table 8.1. The parameters of characteristic points for sectors of the non-linear part

Point no.	m_0	m_P	m_D	e_P	e_D
1	-18.3	0	0	-1.2	-150
2	-18.3	0	0	-1.2	23.653
3	-18.3	0	0	-0.898	23.653
4	-18.3	0	0	-0.275	-118.922
5	-18.3	0	0	-0.275	-150
6	-18.3	0	0	-0.7	-80
7	28.31	0	0	0.71	130
8	28.31	0	0	1.2	130
9	28.31	0	0	1.2	75.455
10	28.31	0	0	0.898	75.455
11	28.31	0	0	0.71	118.922
12	28.31	0	0	1	100
13	-8.72	0	0	-1.2	130
14	-8.72	0	0	-1.2	118.922
15	-8.72	0	0	-0.898	118.922
16	-8.72	0	0	-0.898	130
17	-8.72	0	0	-1	125
18	-20.689	0	0.101	-1.2	118.922
19	-20.689	0	0.101	-0.898	118.922
20	-20.689	0	0.101	-0.898	23.653
21	-20.689	0	0.101	-1.2	23.653
22	-20.689	0	0.101	-1	70
23	11.969	23.029	0	-0.898	130
24	11.969	23.029	0	0.71	130
25	11.969	23.029	0	0.71	118.922
26	11.969	23.029	0	-0.898	118.922
27	11.969	23.029	0	0	125
28	0	23.029	0.101	-0.898	118.922
29	0	23.029	0.101	0.71	118.922
30	0	23.029	0.101	0.898	75.455
31	0	23.029	0.101	0.898	-118.922
32	0	23.029	0.101	-0.275	-118.922
33	0	23.029	0.101	-0.898	23.653
34	0	23.029	0.101	0	0
35	-11.969	23.029	0	-0.275	-118.922
36	-11.969	23.029	0	0.898	-118.922
37	-11.969	23.029	0	0.898	-150
38	-11.969	23.029	0	-0.275	-150
39	-11.969	23.029	0	0.3	-135
40	20.689	0	0.101	0.898	75.455
41	20.689	0	0.101	1.2	75.455
42	20.689	0	0.101	1.2	-118.922
43	20.689	0	0.101	0.898	-118.922
44	20.689	0	0.101	1.05	-20
45	8.72	0	0	0.898	-118.922
46	8.72	0	0	1.2	-118.922
47	8.72	0	0	1.2	-150
48	8.72	0	0	0.898	-150
49	8.72	0	0	1.05	-135

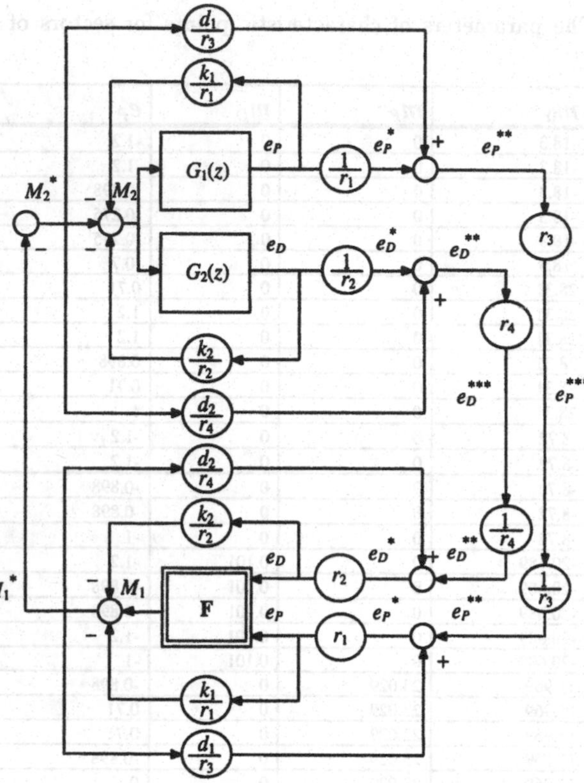

Fig. 8.82. The control system after introduction of 8 degrees of freedom (d_1, d_2, k_1, k_2, r_1, r_2, r_3, r_4)

The condition ML2 has to be satisfied for all $z_i = \cos \omega_i T$ from the range $[-1, 1]$, Fig. 8.83.

A computer program able to check condition ML2 for the measure of density Δz and range of z_i has been made. The condition ML2 can be decomposed into two conditions (8.181), referring to each point z_i.

$$\begin{aligned} &\mathrm{Re}_1^{**}(z_i) > 0 \\ &[\mathrm{Re}_1^{**}(z_i)]^2 - [\mathrm{Re}_2^{**}(z_i)]^2 > 0 \end{aligned} \qquad (8.181)$$

The computer program generated and checked both conditions for all z_i from the range assumed. The real parts of transfer functions are given by formulae (8.182):

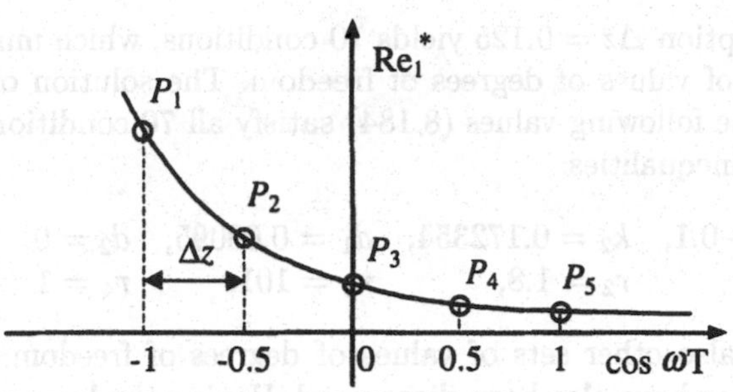

Fig. 8.83. The graphic representation of condition $\mathrm{Re}_1^{**}(z_i) > 0$ (Δz the density of investigation)

$$
\begin{aligned}
\mathrm{Re}_1^{**} &= d_1 + \frac{A_{P2}C_1}{C_1^2 + C_2^2}\,, \\
\mathrm{Re}_2^{**} &= d_2 + \frac{C_1(A_{D2} - A_{D1}z_i) + A_{D1}(1 - z_i^2)(B_1 - 2B_0 z_i)}{C_1^2 + C_2^2}\,, \\
C_1 &= B_2 - B_1 z_i + B_0(2z_i^2 - 1)\,, \\
C_2^2 &= (1 - z_i^2)(B_1 - 2B_0 z_i)^2\,, \\
A_{D2} &= r_1 r_4 l_3\,, \\
A_{D1} &= r_1 r_4 l_2 l_3\,, \\
B_2 &= (r_1 r_2 + k_1 r_2 l_1 + r_1 k_2 l_3)\,, \\
B_1 &= [r_1 r_2(1 + l_2) + r_1 k_2 l_2 l_3]\,, \\
B_0 &= r_1 r_2 l_2\,, \\
A_{P2} &= r_2 r_3 l_1\,,
\end{aligned}
\qquad (8.182)
$$

where values l_1, l_2, l_3 have been previously determined in formula (8.180). The result (8.180) means that the conditions ML2 are represented by complicated functions depending on all 8 degrees of freedom $d_1, d_2, k_1, k_2, r_1, r_2, r_3, r_4$.

Main conditions MN for the non-linear part

The investigation of the current condition means that Popov's condition (8.101), transformed to the form (8.183), is checked for all 49 points P_i from Table 8.1,

$$f(P_i) = M_1^* e_P^{***} = \frac{r_3}{r_1} e_P \cdot M_1^* - d_1 \cdot M_1^{*2} \geq 0\,, \qquad (8.183)$$

where:

$$
\begin{aligned}
M_1^* &= M_1 - \frac{k_1}{r_1} e_P - \frac{k_2}{r_2} e_D\,, \\
M_1 &= m_0 + m_P e_P + m_D e_D\,.
\end{aligned}
$$

The parameters m_0, m_P, m_D, e_P, e_D for points P_i, $i = 1, \ldots, 49$, are assembled in Table 8.1.

The assumption $\Delta z = 0.125$ yields 70 conditions, which must be satisfied by the choice of values of degrees of freedom. The solution of the problem now exists. The following values (8.184) satisfy all 70 conditions represented by non-linear inequalities:

$$\begin{array}{llll} k_1 = -0.1, & k_2 = 0.172354, & d_1 = 0.00095, & d_2 = 0, \\ r_1 = 1, & r_2 = 1.8, & r_3 = 101, & r_4 = 1. \end{array} \tag{8.184}$$

There are also other sets of values of degrees of freedom satisfying the conditions. They have also been discovered. Having the known, numeric values of degrees of freedom (8.184) one can precisely check if Popov's condition holds for any points belonging to a given sector. Previously, Popov's condition had been checked roughly, for a few characteristic points representing a given sector. Now, it is possible to detect eventual "hidden" transgressions of Popov's condition if they exist, as in Fig. 8.81a.

Further investigations show, as an example, the way of precisely checking Popov's condition for sector $q_3 = 1$, Fig. 8.72. The levels of input signals e_P and e_D assigned to that sector are given by (8.185):

$$0.71 \geq e_P \geq 1.2\,, \quad 75.455 \geq e_D \geq 130\,. \tag{8.185}$$

The output signal of the non-linear part for the considered sector takes the constant value $M_1 = 28.31$. The signal M_1^* is given by (8.186):

$$M_1^* = 28.31 + 0.1e_p - 0.14363e_D \tag{8.186}$$

Putting the values of degrees of freedom (8.184) to Popov's condition (8.183), one obtains (8.187).

$$f = M_1^* e_P^{***} = [101e_P - 0.00095M_1^*]M_1^* \geq 0 \tag{8.187}$$

By substituting (8.186) into (8.187), the (8.187) can be transformed to the form (8.188).

$$\begin{aligned} &f = (-0.02689 + 100.99999e_P + 0.00014e_D)(28.31 + 0.1e_p - 0.14363e_D) \geq 0 \\ &f = A \cdot B \geq 0 \end{aligned} \tag{8.188}$$

It is quite easy discover when the product of the two particular factors A and B representing expressions in parentheses is positive. Narrowing investigations to the rectangle in Fig. 8.72 represented by (8.185) one can determine the minimal values of A and B:

$$\begin{array}{lll} A_{\min} = 71.69366 & \text{for} & e_P = 0.71 \text{ and } e_D = 75.455\,, \\ B_{\min} = 9.70936 & \text{for} & e_P = 0.71 \text{ and } e_D = 130\,. \end{array} \tag{8.189}$$

Since A and B are positive, Popov's condition (8.187) holds for all points belonging to sector $q_3 = 1$. To check Popov's condition under more complicated circumstances, for example if one factor (A or B) takes negative values

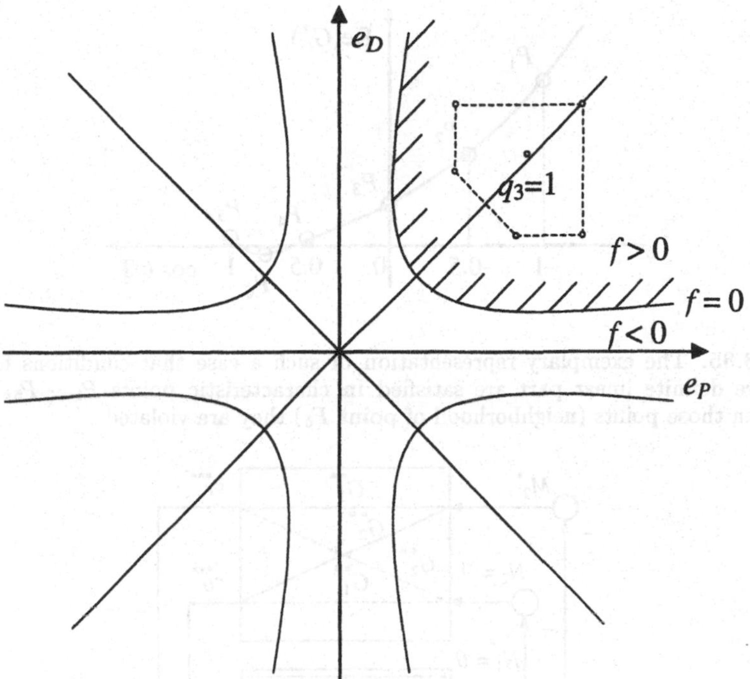

Fig. 8.84. The cross-section of plane $[e_P, e_D]$ at the point of intersection by a hyperboloid representing Popov's condition (8.190) for sector $q_3 = 1$ and ranges of positive and negative values of the condition

for e_P and e_D assigned to the investigated sector, one can apply widely known methods for the examination of functions. Popov's condition (8.188) can be transformed to the form of (8.190).

$$f = -0.76138 + 2859.30426 e_P + 0.00773 e_D - 14.5066 e_P e_D + 10.09999 e_P^2 - 0.00002 e_D^2 \geq 0 \quad (8.190)$$

The function f given by (8.190) defines a hyperboloid. Putting $f = 0$ one obtains the equation of such a curve in the plane $[e_P, e_D]$, where the positive sign of f is "switched" to a negative one or vice versa. The graphic representation of the "switching curve" (8.191) is presented in Fig. 8.84.

$$-0.76138 + 2859.30426 e_P + 0.00773 e_D - 14.5066 e_P e_D + 10.09999 e_P^2 - 0.00002 e_D^2 = 0 \quad (8.191)$$

The analytical investigation of Popov's condition (8.190) can be substituted by numerical checking. Of course, the previously chosen initial number of characteristic points (5) has to be substantially increased. So as for the nonlinear part, one has to find a precise way to check if conditions assigned to the

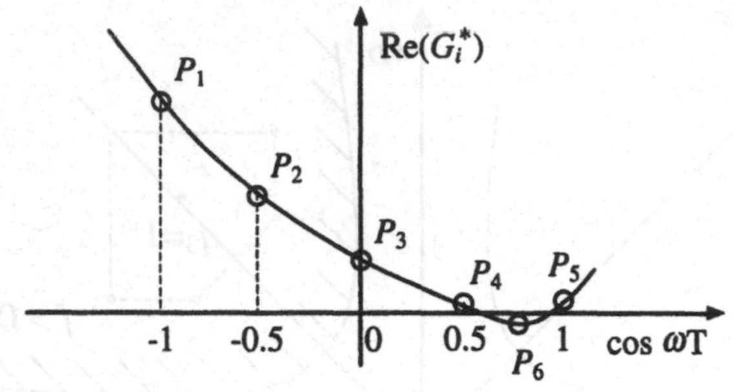

Fig. 8.85. The exemplary representation of such a case that conditions for the positive definite linear part are satisfied in characteristic points P_1 – P_5, while between those points (neighborhood of point P_6) they are violated

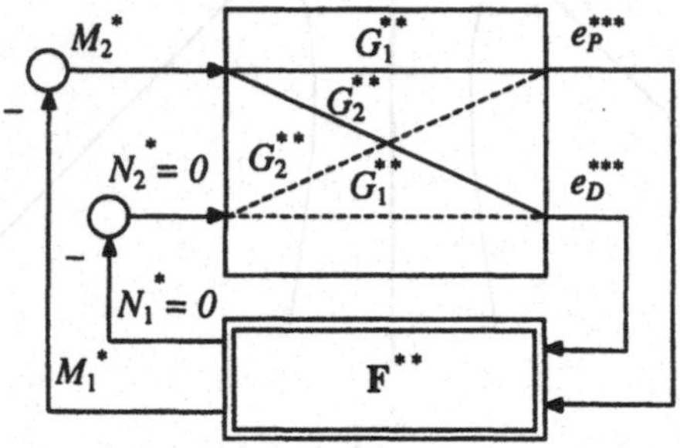

Fig. 8.86. The linear part $\mathbf{G}^{**}(z)$ of investigated system

linear part (8.159) – (8.161) hold between the chosen points P_i, $i = 1, \ldots, 5$, Fig. 8.85.

Just as for the former task, the analytic investigation can be substituted by a numeric one, if distance Δz between investigated points is sufficiently small. Numeric investigations have been carried out for numerous values Δz from the range $\Delta z < 0.125$. The investigations confirm that conditions (8.159) – (8.161) are satisfied.

Condition PL2

(The linear part has to be completely controllable and observable)

The linear part of the investigated control system, shown in Fig. 8.86, is composed of 4 component transfer functions G_i^{**}.

It results from Fig. 8.82 that particular transfer functions G_i^{**} depend on values of degrees of freedom (8.184) and can be expressed by general forms (8.192).

$$G_1^{**}(z) = \frac{e_{Pk}^{***}(z)}{M_{2k}^*} = d_1 + \frac{\frac{r_3}{r_1}G_1(z)}{1+\frac{k_1}{r_1}G_1(z)+\frac{k_2}{r_2}G_2(z)}$$

$$G_2^{**}(z) = \frac{e_{Dk}^{***}(z)}{M_{2k}^*} = d_2 + \frac{\frac{r_4}{r_2}G_2(z)}{1+\frac{k_1}{r_1}G_1(z)+\frac{k_2}{r_2}G_2(z)} \tag{8.192}$$

The primary transfer functions are determined by formulae (8.137). If values of all coefficients are put to (8.137), then further transformations yield so-called modal forms (8.193).

$$G_1^{**}(z) = 0.07925 + \frac{0.6073}{z-0.9925} - \frac{0.4618}{z-0.8655} = d_1 + \frac{c_1}{z-z_1} + \frac{c_2}{z-z_2}$$

$$G_2^{**}(z) = 0.0001 + \frac{0.6066}{z-0.9925} - \frac{0.4612}{z-0.8655} = d_2 + \frac{c_3}{z-z_1} + \frac{c_4}{z-z_2} \tag{8.193}$$

The state space representation of the linear part is shown in Fig. 8.87.

It results from Fig. 8.87, that states x_1, x_2 are available through the input u_1, while states x_3, x_4 are available through the input u_2. Hence, the controllability of states x_1, x_2 can be investigated using state equation model (8.194). The controllability of states x_3, x_4 can be investigated using state equation model (8.195).

$$\mathbf{X}_1(k+1) = \mathbf{A}\mathbf{X}_1(k) + \mathbf{B}u_1(k)$$

$$\begin{bmatrix} x_1(k+1) \\ x_2(k+1) \end{bmatrix} = \begin{bmatrix} z_1 & 0 \\ 0 & z_2 \end{bmatrix} \begin{bmatrix} x_1(k) \\ x_2(k) \end{bmatrix} + \begin{bmatrix} 1 \\ 1 \end{bmatrix} u_1(k) \tag{8.194}$$

$$\mathbf{X}_2(k+1) = \mathbf{A}\mathbf{X}_2(k) + \mathbf{B}u_2(k)$$

$$\begin{bmatrix} x_3(k+1) \\ x_4(k+1) \end{bmatrix} = \begin{bmatrix} z_1 & 0 \\ 0 & z_2 \end{bmatrix} \begin{bmatrix} x_3(k) \\ x_4(k) \end{bmatrix} + \begin{bmatrix} 1 \\ 1 \end{bmatrix} u_2(k) \tag{8.195}$$

The linear, discrete, time invariant system is controllable if and only if the matrix $\mathbf{S}$ determined by formula (8.196) has the maximum rank n, where n – order of system (Guenther 1984).

$$\mathbf{S} = \left[\mathbf{B} \vdots \mathbf{AB} \vdots \mathbf{A}^2\mathbf{B} \vdots \ldots \vdots \mathbf{A}^{n-1}\mathbf{B}\right] \tag{8.196}$$

The controllability matrices $\mathbf{S}$ (8.197) assigned to state variables x_1, x_2 and x_3, x_4 are identical, because the same matrices $\mathbf{A}$ and $\mathbf{B}$ create both models (8.194) and (8.195).

$$\mathbf{S} = \begin{bmatrix} 1 & z_1 \\ 1 & z_2 \end{bmatrix} \qquad \text{rank } \mathbf{S} = 2 \tag{8.197}$$

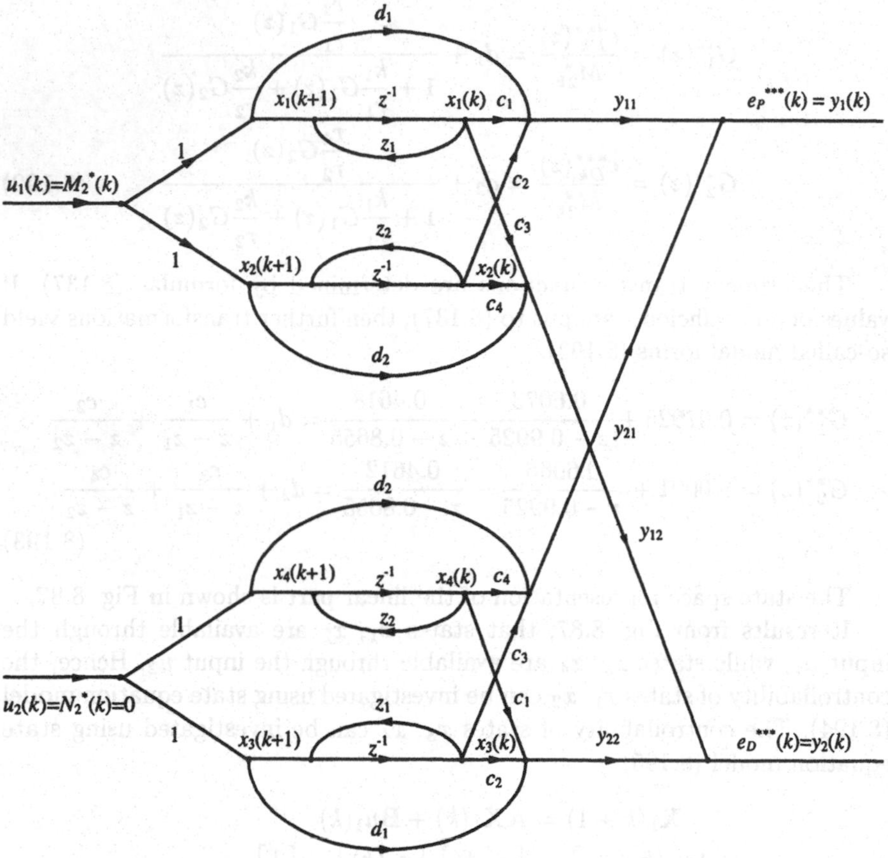

Fig. 8.87. The state space representation of the linear part $\mathbf{G}^{**}(z)$

The matrix **S** has maximal rank equal to 2 and has 2 different eigenvalues ($z_1 = 0.9925$, $z_2 = 0.8655$).

Although the controllability matrix assigned to state variables x_3, x_4 can be classified as a maximum rank matrix, in fact, the state variable x_3 and x_4 are not controlled by u_2. According to assumed rules the dummy input u_2 is always equal to zero ($u_2 = 0$) – see Fig. 8.84. However, it does not mean that controllability condition PL2 is violated. The following quotation (Opitz 1986) explains the above statement: "If all eigenvalues of a system matrix which can not be influenced have negative real parts and the matrix of transfer functions of the remaining part of a completely controllable and observable system is strictly positive real, then the system is asymptotically hyperstable". The quoted condition can be called a "softened controllability and observability condition".

The state variables x_3, x_4 are not excited by u_2, since $u_2 = 0$ (Fig. 8.87). If initial values of those state variables $x_3(0) \neq 0$, $x_4(0) \neq 0$, then after a certain transient process the variables decay to $x_3 \approx 0$, $x_4 \approx 0$ (it is "ensured" by negative eigenvalues z_1, z_2 of system matrix). Hence, the variables x_3, x_4 can only influence the linear part outputs y_1, y_2 during the initial period of system operation. Thus, the outputs y_1, y_2 depend mainly on state variables x_1 and x_2. Neglecting the transient state caused by initial conditions $x_3(0)$, $x_4(0)$ one can write the following equations:

$$y_1(k) \cong y_{11}(k) = \mathbf{C}_1\mathbf{X}_1(k) + d_1u_1(k)\,, \tag{8.198}$$
$$y_2(k) \cong y_{12}(k) = \mathbf{C}_2\mathbf{X}_1(k) + d_2u_1(k)\,, \tag{8.199}$$

where:

$$C_1 = [c_1, c_2]\,, \quad C_2 = [c_3, c_4]\,, \quad X_1^{\mathrm{T}} = [x_1, x_2]\,.$$

The state variable x_1, x_2 can be observed from both outputs y_1, y_2.

The matrix of observability $\mathbf{O}_1$ for output y_1 can be expressed in the form (8.200) (Guenther 1988). The sub-system is observable if ranks of $\mathbf{O}_1$ and state matrix $\mathbf{A}$ are identical and maximal.

$$\mathbf{O}_1 = \left[\mathbf{C}^{\mathrm{T}} \vdots \mathbf{A}^{\mathrm{T}}\mathbf{C}^{\mathrm{T}} \vdots \ldots \vdots \left(\mathbf{A}^{\mathrm{T}}\right)^{n-1}\mathbf{C}^{\mathrm{T}}\right] = \left[\begin{array}{c|c} c_1 & c_1z_1 \\ c_2 & c_2z_2 \end{array}\right] \tag{8.200}$$

The $\mathbf{O}_1$ is of rank 2, thus the sub-system represented by state variables x_1, x_2 is completely observable (except during initial period of system operation). The variables x_1, x_2 can also be observed from output y_2. The states x_3, x_4 are equal to zero or "efficiently" decay to zero (for non-zero initial conditions), so steady state values of x_3, x_4 are known.

The linear part $\mathbf{G}^{**}$, see Fig. 8.87, satisfies the "softened" observability condition. The condition PL2 of the linear part is fulfilled as the strictly positive real transfer functions G_1^{**} and G_2^{**} (8.193) are assigned to the sub-system of state variables x_1, x_2.

The presented example of investigation of hyperstability for the non-linear system of MIMO type reveals that the level of difficulties connected with proving the system stability is substantially higher if compared to linear systems. Analyzing that example one discovers how important is the choice of a proper number of fictitious degrees of freedom. For a small number of degrees of freedom the hyperstability has not been proved. Fortunately, the increase in the number of degrees of freedom has produced a good solution leading to proof. Some fuzzy-control experts maintain that the hyperstability method is a conservative one, in other words, it allows us to prove stability only for firmly stable control systems (systems having a big stability margin). The considered system can indeed be classified as globally non-linear, but locally, in individual sectors of input space (Fig. 8.72), its operation resembles the linear one. In central sector $q_2w_2v_2 = 1$ the non-linear part $\mathbf{F}$ realizes linear mapping:

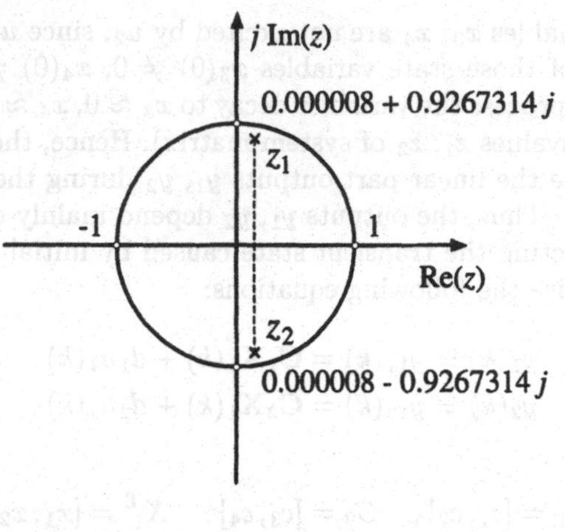

Fig. 8.88. The loci of poles z_1, z_2 for the central sector ($q_2 w_2 v_2 = 1$) of input space of the non-linear part **F**

$$M_1 = 23.029 e_P + 0.101 e_D .$$

Hence, taking into account the transfer functions of the linear part given by (8.137), see also Fig. 8.74, one obtains the following poles representing the control system (with closed control loops):

$$z_1 \cong 0.000008 + 0.926731 j ,$$
$$z_2 \cong 0.000008 - 0.926731 j .$$

The loci of those poles in Z plane is shown in Fig. 8.88.

It results from Fig. 8.88 that poles are located in the close neighborhood of the unit circle separating the stability area. Despite the critical location of poles, the stability of the system has been proved. It refutes those opinions which classify the hyperstability theory as a conservative one.

◊ ◊ ◊

References

1. Abdelnour G., Chang F. and others (1992) Design of a fuzzy controller using input and output mapping factors. IEEE Transactions on Systems, Man and Cybernetics. No. 21, pp. 952–960
2. Abel D. (1991) Fuzzy Control – eine Einführung ins Unscharfe. Automatisierungstechnik, vol. 39, No. 12, pp. 433–438
3. Ader W., Nörling A., Hollatz J. (1996) Datalyzer, a tool for fuzzy data analysis. Proceedings of the International Conference EUFIT'96, vol. 3. Aachen, Germany, pp. 1541–1545
4. Akaiwa E. (1990) Hardware and software of fuzzy logic controlled cardiac pacemaker. Proceedings of the First International Conference on Fuzzy Logic and Neural Networks. Iizuka, Japan, pp. 549–552
5. von Altrock C., Krause B., Zimmermann H.J. (1992) Advanced fuzzy logic control of a model car in extreme situations. Fuzzy Sets and Systems, vol. 48, No. 1, pp. 41–52
6. von Altrock C. (1993) Fuzzy logic. Band 1 – Technologie. R. Oldenburg Verlag GmbH, München, Germany
7. von Altrock C. (1995) Fuzzy logic. Band 3 – Werkzeuge. R. Oldenburg Verlad GmbH, München, Germany
8. Anderson B. (1968) A simplified viewpoint of hyperstability. IEEE Transactions on Automatic Control, No. 13, pp. 292–294
9. Anderson B., Bitmed R. and others (1968) Stability of adaptive systems: passivity and averaging analysis. MIT Press, Cambridge, USA
10. Angstenberger J., Walesch B. (1994) ATP-Marktanalyse: Entwicklungswerkzeuge und Spezialprozessoren für Fuzzy-Anwendungen. Automatisierungstechnische Praxis, vol. 36, No. 6, pp. 62–72
11. Aoki S., Kawachi S. (1990) Application of fuzzy control for dead-time processes in a glas melting furnace. Fuzzy Sets and Systems, vol. 38, No. 5, pp. 251–256
12. Aracil J., Garcia-Cerezo A., Ollero A. (1991) Fuzzy control of dynamical systems. Stability analysis based on the conicity criterion. Proceedings of the 4th Iternational Fuzzy Systems Association Congress, Brussels, Belgium, pp. 5–8
13. Aracil J., Gordillo F. (2000) Stability issues in fuzzy control. Physica-Verlag, Heidelberg-New York
14. Arita S., Tsutsui T. (1990) Fuzzy logic control of blood pressure through inhalational anesthesia. Proceedings of the First International Conference on Fuzzy Logic and Neural Networks. Iizuka, Japan, pp. 545–547
15. Arita S. (1991) Development of an ultrasonic cancer diagnosis system using fuzzy theory. Japanese Journal of Fuzzy Theory and Systems, vol. 3, No. 3, pp. 215–230
16. Assilian S. (1974) Artificial intelligence in the control of real dynamical systems. Ph.D. Thesis, London University

17. Åström K.J., Wittenmark B. (1989) Adaptive control. Addison-Wesley, New York, USA
18. Babuška R., Verbruggen H.B. (1995a) A new identification method for linguistic fuzzy models. Proceedings of the International Conference FUZZ-IEEE/IFES'95. Yokohama, Japan, pp. 905–912
19. Babuška R. (1995b) Fuzzy modeling a control engineering perspective. Proceedings of the International Conference FUZZ-IEEE/IFES'95. Yokohama, Japan, pp. 1897–1902
20. Babuška R., Verbruggen H.B. (1995c) Identification of composite linear models via fuzzy clustering. Proceedings of the 3rd European Control Conference. Rome, Italy, pp. 1207–1212
21. Babuška R. (1995d) Fuzzy modeling and cluster analysis toolbox for MATLAB. Proceedings of the International Conference EUFIT'95, vol. 3. Aachen, Germany, pp. 1479–1483
22. Babuška R., Sousa J., Verbruggen H.B. (1995e) Modelbased design of fuzzy control systems. Proceedings of the International Conference EUFIT'95, vol. 1. Aachen, Germany, pp. 837–841
23. Babuška R., Setnes M., Kaymak U., von Nauta Lemke H.R. (1996) Simplification of fuzzy rule bases. Proceedings of the International Conference EUFIT'96, vol. 2. Aachen, Germany, pp. 1115–1119
24. Baldwin J.F. (1995a) Fril method for soft computing, fuzzy control and classification. Proceedings of the International Conference FUZZ-IEEE/IFES'95, vol. 1. Yokohama, Japan, pp. 309–316
25. Baldwin J.F., Martin T.P. (1995b) Fuzzy modelling in an Intelligent Data Browser. Proceedings of the International Conference FUZZ-IEEE/IFES'95, vol. 4. Yokohama, Japan, pp. 1885–1890
26. Baglio S., Fortuna L., Graziani S., Muscato G. (1994) Membership function shape and the dynamic behaviour of fuzzy systems. International Journal of Adaptive Control and Signal Processing, vol. 8, pp. 369–377
27. Bartolan G., Pedrycz W. (1997) Reconstruction problem and information granularity. IEEE Transactions on Fuzzy Systems, vol. 5, No. 2, pp. 234–248
28. Bastian A. (1996) A genetic algorithm for tuning membership functions. Proceedings of the International Conference EUFIT'96, vol. 1. Aachen, Germany, pp. 494–498
29. Becker M., von Recum S. (1994) Verfahren zur bedarfsgesteuerten Abtauerkennung in Kälteanlagen unter Einsatz eines Fuzzy Entscheiders. Proceedings of the 39th Internationales Wissenschaftliches Kolloquium. Ilmenau, Germany, pp. 316–323
30. Beigy H., Eydgahi A.M. (1995) Fuzzy modelling of two-dimensional linear systems. Proceedings of the International Symposium on Methods and Models in Automation and Robotics MMAR'95, vol. 2. Międzyzdroje, Poland, pp. 775–780
31. Bellon C., Bosc P., Prade H. (1992) Fuzzy boom in Japan. International Journal of Intelligent Systems, No. 7, pp. 293–316
32. Bellman R.E. (1961) Adaptive control processes. Princetown: Princetown University Press
33. Berenji H.R., Khedkar P. (1992) Learning and tuning fuzzy logic controllers through reinforcements. IEEE Transactions on Neural Networks 1992, vol. 3, No. 5, pp. 724–740
34. Bertram T., Svaricek F. (1993) Zur Kompensation der trockenen Reibung mit Hilfe der Fuzzy-Logik. Automatisierungstechnik, vol. 41, No. 5, pp. 180–184

35. Bertram T., Svaricek F. and others (1994) Fuzzy Control. Zusammenstellung und Beschreibung Wichtiger Begriffe. Automatisierungstechnik, vol. 42, No. 7, pp. 322–326
36. Bezdek J.C. (1981) Pattern recognition with fuzzy objective function algorithms. New York: Plenum Press
37. Bezdek J.C., Ehrlich R., Full W. (1984) FCM: the fuzzy c-means clustering algorithm. Computer and Geosciences, No. 10, pp. 191–203
38. Bezdek J.C. (1993) Editorial. Fuzzy models-what are they, and why? IEEE Transactions on Fuzzy Systems, vol. 1, No. 1, pp. 1–6
39. Będzak M. (1992) Algorytmy sterowania cyfrowego bazujące na logice rozmytej i sztucznych sieciach neuronowych. Ph.D. Thesis. Technical University of Szczecin, Poland
40. Bien Z., Hwang D.H., Lee J.H., Ryu H.K. (1992) An automatic start-up and shut-down control of a drum-type boiler using fuzzy logic. Proceedings of the Second International Conference on Fuzzy Logic and Neural Networks. Iizuka, Japan, pp. 465–468
41. Bindel T., Mikut R. (1995) Entwurf, Stabilitätsanalyse und Erprobung von Fuzzy-Reglern am Beispiel einer Durchflussregelung. Automatisierungstechnik, vol. 43, No. 5, pp. 249–255
42. Bork P., Selig M., Krummen H., Schiller E. (1993) Fuzzy Control zur Optimierung der Kühlwasseraufbereitung an einer Chemie-Reaktoranlage. Automatisierungstechnische Praxis, vol. 35, No. 5, pp. 306–309
43. Bossley K.M., Brown M., Harris C.J. (1995) Neurofuzzy model construction for the modelling of non-linear processes. Proceedings of the 3rd European Control Conference. Rome, Italy, pp. 2438–2443
44. Böhm R., Krebs V. (1993) Ein Ansatz zur Stabilitätsanalyse und Synthese von Fuzzy-Regelungen. Automatisierungstechnik, vol. 41, No. 8, pp. 288–293
45. Böhm R., Bösch M. (1995) Stabilitätsanalyse von Fuzzy-Mehrgrssenregelungen mit Hilfe der Hyperstabilitätstheorie. Automatisierungstechnik, vol. 43, No. 4, pp. 181–186
46. Broel-Plater B. (1995) DSP system using fuzzy-logic technik. Proceedings of the International Conference ACS'97. Szczecin, Poland, pp. 401–406
47. Bronsztejn I.N., Siemiendiajew K.A. (1996) Matematyka. PWN, Warszawa, Poland
48. Brown M., Harris C. (1994) Neurofuzzy adaptive modelling and control. Prentice Hall, New York, USA
49. Brown M., Bossley K.M., Mills D.J., Harris C.J. (1995a) High dimensional neurofuzzy systems: overcoming the curse of dimensionality. Proceedings of the International Conference FUZZ-IEEE/IFES'95. Yokohama, Japan, pp. 2139–2146
50. Brown M., An P.E., Harris C.J. (1995b) On the condition of neurofuzzy models. Proceedings of the International Conference FUZZ-IEEE/IFES'95, vol. 2. Yokohama, Japan, pp. 663–670
51. Brown M., Ullrich T. (1997) Comparison of node insertion algorithms for Delaunay networks. Proceedings of the IMACS 2nd Mathmod Conference. Vienna, Austria, pp. 775–780
52. Bühler H. (1994) Stabilitätsuntersuchung von Fuzzy-Regelungen. GMA-Aussprachetag Fuzzy-Control, Langen BRD, VDI-Bericht Nr 1113, pp. 309–318
53. Cao S.G., Rees N.W., Feng G. (1997a) Analysis and design for a class of complex control systems. Part I: Fuzzy modelling and identification. Automatica, vol. 33, No. 6, pp. 1017–1028

54. Cao S.G., Rees N.W., Feng G. (1997b) Analysis and design for a class of complex control systems. Part II: Fuzzy controller design. Automatica, vol. 33, No. 6, pp. 1029–1039
55. Carpenter G.A., Grossberg S., Markuzon N., Reynolds J.H., Rosen D.B. (1992) Fuzzy ARTMAP: a neural network architecture for incremental supervised learning of analog multidimensional maps. IEEE Transactions on Neural Networks, vol. 3, No. 5, pp. 698–713
56. Chen H., Mitzumoto M., Ling Y.F. (1992) Automatic control of sewerege pumpstation by using fuzzy controls and neural networks. Proceedings of the Second International Conference on Fuzzy Logic and Neural Networks. Iizuka, Japan, pp. 91–94
57. Cho K.B., Wang B.H. (1995) Radial basis function based adaptive fuzzy systems. Proceedings of the International Conference FUZZ-IEEE/IFES'95, vol. 1. Yokohama, Japan, pp. 247–252
58. Cipriano A., Kamos M., Briseño H., Montoya F. (1995) Comparative analysis of two fuzzy models identification methods. Proceedings of the 3rd European Control Conference, vol. 3. Rome, Italy, pp. 1213–1218
59. Cook P.A. (1986) Nonlinear dynamical systems. London: Prentice Hall International
60. Mc Cormac S.E., Ringwood J.V. (1997) Neural and fuzzy modelling and fuzzy predictive control of a non-linear coupled multi-variable plant. Proceedings of the International Conference EUFIT'97, vol. 2. Aachen, Germany, pp. 1311–1315
61. Czogała E., Pedrycz W. (1981) On identification of fuzzy systems and its applications in control problems. Fuzzy Sets and Systems, No. 6, pp. 73–83
62. Czogała E., Pedrycz W. (1984) Identification and control problems in fuzzy systems. TIMS Studies in the Management Sciences, No. 20, pp. 437–446
63. Czogała E. (1993a) Fuzzy logic controllers versus conventional controllers. Proceedings of the 16th Seminar on Fundamentals of Electrotechnics and Circuit Theory, vol. 1. Gliwice-Ustroń, Poland, pp. 23–29
64. Czogała E. (1993b) On the modification of rule connective in fuzzy logic controllers using Zimmermann's compensatory operators. Proceedings of the International Conference EUFIT'93, vol. 2. Aachen, Germany, pp. 1329–1333
65. Czogała E. (1994) On modelling knowledge bases for fuzzy and rough fuzzy controllers using the concept of an input image. Proceedings of the International Conference EUFIT'94. Aachen, Germany
66. Culliere T., Titli A., Corrieu J.M. (1995) Neuro-fuzzy modelling of nonlinear systems for control purposes. Proceedings of the International Conference FUZZ-IEEE/IFES'95, vol. 4. Yokohama, Japan, pp. 2009–2016
67. Cytowski J. (1996) Algorytmy genetyczne. Podstawy i zastosowania. Akademicka Oficyna Wydawnicza PLJ. Warszawa, Poland
68. Daca W., Będzak M., Efner T. (1994) Fuzzy Logic für Temperaturregelung. Proceedings of the 13th IAESTED International Conference MIC. Grindelwald, Switzerland, pp. 400–401
69. Davè R.N., Krishnapuram R. (1997) Robust clustering methods: a unified view. IEEE Transactions on Fuzzy Systems, vol. 5, No. 2, pp. 270–293
70. Davis L. (1991) Handbook of genetics algorithms. New York: Van Nostrand Reinhold
71. Delgado M. (1995) Generating fuzzy rules using clustering-based approaches. Proceedings of the International Conference EUFIT'95, vol. 2. Aachen, Germany, pp. 810–814

72. Delgado M., Gomez-Skarmeta A.F., Martin F. (1997) A fuzzy clustering-based rapid prototyping for a fuzzy rule-based modeling. IEEE Transactions on Fuzzy Systems, vol. 5, No. 2, pp. 223–233
73. Dias J.M., Correia A.D. (1997) Multivariable decoupling and control by a self-organizing fuzzy system with real-time learning. Proceedings of the International Conference EUFIT'97, vol. 2. Aachen, Germany, pp. 1233–1237
74. Domański P.D., Brdyś M.A., Tatjewski P. (1997) Fuzzy logic multi-regional controllers-design and stability analysis. Proceedings of the European Control Conference EEC 97, paper TU-EG5 (679) on disc. Brussels, Belgium
75. Döbrich U., Pfeifer B.M. (1996) Technologiemarkt "Angewandte Forschung". Automatisierungstechnische Praxis, vol. 36, No. 4, pp. 80–83
76. Driankov D., Hellendoorn H., Reinfrank M. (1993) An introduction to fuzzy control. Berlin: Springer-Verlag
77. Driankov D., Hellendoorn H., Reinfrank M. (1996) Wprowadzenie do sterowania rozmytego. Wydawnictwo Naukowo-Techniczne. Warszawa, Poland
78. Dubois D., Grabisch M., Prade H. (1992) Gradual rules and the approximation of functions. Proceedings of the Second International Conference on Fuzzy Logic and Neural Networks. Iizuka, Japan, pp. 629–632
79. Dunn J.C. (1973) A fuzzy relative of the ISODATA process and its use in detecting compact well-separated clusters. Journal of Cybernetics, No. 3, pp. 32–57
80. Economou C.G. (1986) Internal model control. Extension to nonlinear systems. Ind. Eng. Chem. Process Des. Dev., vol. 21, pp. 403–411
81. Edgar C.R., Postlethwaite B.E. (1997) A fuzzy internal model controller (FIMC) for nonlinear systems. Proceedings of the International Conference EUFIT'97, vol. 2. Aachen, Germany, pp. 1217–1221
82. Eklund P., Klawonn F. (1992) Neural fuzzy logic programming. IEEE Transactions on Neural Networks, vol. 3, No. 5, pp. 815–818
83. Empacher A. (1970) Mały słownik matematyczny. Wiedza Powszechna. Warszawa, Poland
84. Feuring T. (1996) Developing algorithms for fuzzy neural networks. Proceedings of the 15th IASTED International Conference "Modelling, Identification and Control". Insbruck, Austria, pp. 133–136
85. Fijiwara Y. (1991) Image processing using fuzzy logic for video print systems. Proceedings of the International Conference IFES'91, vol. 2. Tokyo, Japan, pp. 1003–1012
86. Fiordaliso A. (1996) A prunning method for the self-structuring of fuzzy systems applied to function approximation. Proceedings of the International Conference EUFIT'96, vol. 1. Aachen, Germany, pp. 581–586
87. Fischle K., Schröder D. (1997) Stabile adaptive Fuzzy-Regelung ohne Differentiation der Regelgrösse. Automatisierungstechnik, vol. 45, No. 8, pp. 360–367
88. Fisher M., Nelles O., Füssel D. (1996) Tuning of PID-controllers for nonlinear processes based on local linear fuzzy models. Proceedings of the International Conference EUFIT'96, vol. 2. Aachen, Germany, pp. 1891–1895
89. Föllinger O. (1993) Nichtlineare Regelungen. Oldenburg Verlag. München, Germany
90. Franke H., Priber U. (1994) Echtzeit-Fuzzy-Klassifikation in Bild und Signalverarbeitung. Proc. 39 Internationales Wissenschaftliches Kolloqium, vol. 3, Ilmenau, Germany, pp. 308–315
91. Fujimoto J., Nakatani T., Yoneyama M. (1989) Speaker-independent word recoqnition using fuzzy pattern matching. Fuzzy Sets and Systems, vol. 32, No. 7, pp. 181–191

92. Fujiyoshi M., Shiraki T. (1992) A fuzzy automatic-combustion-control-system of refuse incineration plant. Proceedings of the Second International Conference on Fuzzy Logic and Neural Networks. Iizuka, Japan, pp. 469–472
93. Fukumoto S., Miyajima H., Kishida K., Nagasawa Y. (1995) A destructive learning method of fuzzy inference rules. Proceedings of the International Conference FUZZ-IEEE/IFES'95, vol. 2. Yokohama, Japan, pp. 687–694
94. Garcia C.E., Morari M. (1982) Internal model control. A unifying review and some new results. Ind. Eng. Chem. Process Des. Dev, vol. 21, pp. 308–323
95. Goldberg D.E. (1995) Algorytmy genetyczne i ich zastosowania. Wydawnictwo Naukowo-Techniczne. Warszawa, Poland
96. Gonzales A., Perez R. (1995) Structural learning of fuzzy rules from noised examples. Proceedings of the International Conference FUZZ-IEEE/IFES'95, vol. 3. Yokohama, Japan, pp. 1323–1330
97. Gorez R., Carels P. (1996) A structure and a tuning procedure for PID-like fuzzy control. Proceedings of the International Conference EUFIT'96, vol. 2. Aachen, Germany, pp. 974–979
98. Gorrini V., Salomé T., Bersini H. (1995) Self-structuring fuzzy systems for function approximation. Proceedings of the International Conference FUZZ-IEEE/IFES'95, vol. 2. Yokohama, Japan, pp. 919–926
99. Gottwald S. (1993) Fuzzy sets and fuzzy logic. Foundations and application – from a mathematical point of view. Vieweg Verlag. Braunschweig, Germany
100. Gruszecki J., Mieszkowicz-Rolka A., Rolka L. (1994) Automatyzacja procesu oceny pilotów szkolnych na symulatorze lotów. Materiały XII Krajowej Konferencji Automatyki. Gdynia, Poland, pp. 213–219
101. Gupta M., Qi J. (1991) Design of fuzzy logic controllers based on generalized T-operators. Fuzzy Sets and Systems, vol. 40, pp. 473–489
102. Gupta M., Rao D.H. (1994) On the principles of fuzzy neural networks. Fuzzy Sets and Systems, vol. 61, pp. 1–18
103. Gwiazda T.D. (1995) Algorytmy genetyczne. Wstęp do teorii. "T.D.G." Sp. Cyw. Warszawa, Poland
104. Hajek M. (1994) Optimization of fuzzy rules by using a genetic algorithm. Proceedings of the Third International Conference on Automation, Robotics and Computer Vision ICARV'94, vol. 4. Singapore, pp. 2111–2115
105. Hajek M. (1995) Multivariable fuzzy control-genetic optimization of fuzzy rules. Proceedings of the International Conference ICAUTO-95. Indore, India, pp. 783–786
106. Hakata T., Masuda J. (1990) Fuzzy control of cooling system utilizing heat storage. Proceedings of the First International Conference on Fuzzy Logic and Neural Networks. Iizuka, Japan, pp. 77–80
107. Halgamuge S., Glesner M. (1996) Fast transparent neuro-fuzzy classifiers. Proceedings of the 15th IASTED International Conference "Modelling, Identification and Control". Insbruck, Austria, pp. 407–410
108. Han C.D. (1970) Sufficient conditions for hyperstability of a class of nonlinear systems. IEEE Transactions on Automatic Control, vol. 15, pp. 705–706
109. Hanakuma Y. (1989) Ethylen plant destillation column bottom temperature control. Keisi, vol. 32, No. 8, pp. 28–39
110. Hanss M. (1994) Ein Fuzzy-Prädiktor für Bioprozesse. Proceedings of the Conference Dortmunder Fuzzy-Tage. Dortmund, Germany, pp. 206–213
111. Hanss M. (1996a) Eine Methode zur Identifikation von Fuzzy-Modellen. Automatisierungstechnik
112. Hanss M. (1996b) Design and optimization of a nonlinear fuzzy controller using fuzzy process models. Proceedings of the International Conference EUFIT'96, vol. 3. Aachen, Germany, pp. 1875–1880

113. Hathaway R.J., Bezdek J.C., Pedrycz W. (1996) A parametric model for fusing heterogeneous fuzzy data. IEEE Transactions on Fuzzy Systems, vol. 4, No. 3, pp. 270–281
114. Hauptmann W., Heesche K. (1995) A neural net topology for bidirectional fuzzy-neuro transformation. Proceedings of the International Conference FUZZ-IEEE/IFES'95, vol. 3. Yokohama, Japan, pp. 1511–1518
115. Haykin S. (1994) Neural networks – a comprehensive foundation. Macmillan College Publishing Company, Inc. USA
116. Häck M., Köhne M. (1997) Internal Model Control mit neuronalen Netzen zur Regelung eines Prozessanalysators. Automatisierungstechnik, vol. 45, No. 1, pp. 14–23
117. Heider H., Tryba V. (1994) Energiesparen durch einen adaptiven Fuzzy-Regler für Heizungsanlagen. Proceedings of the Conference Dortmunder Fuzzy-Tage. Dortmund, Germany, pp. 282–288
118. Heilbronn R. (1995) Reibkraftkompensation mittels Fuzzy-Logik. Automatisierungstechnische Praxis, vol. 37, No. 5, pp. 50–60
119. Hensel H., Holzmann H., Pfeifer B.M. (1995) Optimierung von Fuzzy-Control mit Hilfe Neuronaler Netze. Automatisierungstechnische Praxis, vol. 37, No. 11, pp. 40–48
120. Higgins C.M., Goodman R.M. (1994) Fuzzy rule-based networks for control. IEEE Transactions on Fuzzy Systems, vol. 2, No. 2, pp. 82–88
121. Hishida N. (1992) Development of the operator support system applying fuzzy algorithms for glass tube molding equipment. Proceedings of the Second International Conference on Fuzzy Logic and Neural Networks. Iizuka, Japan, pp. 1097–1100
122. Hiyama T., Sameshima T. (1991) Fuzzy logic control scheme for on-line stabilisation of multi-machine power system. Fuzzy Sets and Systems, vol. 39, pp. 181–194
123. Horikawa S.I., Furuhashi T., Uchikawa Y. (1992) On fuzzy modeling using fuzzy neural networks with the back-propagation algorithm. IEEE Transactions on Neural Networks, vol. 3, No. 5, pp. 801–806
124. Höhmann J., Nerlich H.G., Steinmeister C., Linzenkirchner K. (1993) Fuzzy Control: Regelung eines Chemiereaktors. Automatisierungstechnische Praxis, vol. 35, No. 9, pp. 514–521
125. Hsieh L.H., Groth H.C. (1994) Fuzzy Sensordatenauswertung für das automatisierte Entgraten. Proceedings of the Conference Dortmunder Fuzzy-Tage. Dortmund, Germany, pp. 173–180
126. Hung T., Sugeno M., Tong R., Yager R.R. (1995) Theoretical aspects of fuzzy control. New York: John Wiley and Sons Inc.
127. Hunt K.J., Sharbaro D., Żbikowski R., Gawthrop P.J. (1992) Neural networks for control systems – a survey. Automatica, vol. 28, No. 6, pp. 1083–1112
128. Isaka S., Sebald A. (1993) An optimization approach for fuzzy controller design. IEEE Transactions on System, Man and Cybernetics, pp. 1469–1473
129. Isermann R. (1996a) On fuzzy logic applications for automatic control, supervision and fault diagnosis. Proceedings of the International Conference EUFIT'96, vol. 2. Aachen, Germany, pp. 738–753
130. Isermann R. (1996b) Zur Anwendung der Fuzzy-logic in der Regelungstechnik. Automatisierungstechnische Praxis, vol. 38, No. 11, pp. 24–35
131. Ishibuchi H., Tanaka H., Fukuoka N. (1989) Discriminant analysis of fuzzy data and its application to a gas sensor system. Japanese Journal of Fuzzy Theory and Systems, vol. 1, No. 1, pp. 27–46

132. Ishibuchi H., Fujioka R., Tanaka H. (1993) Neural networks that learn from fuzzy if-then rules. IEEE Transactions on Fuzzy Systems, vol. 1, No. 2, pp. 85–97
133. Ishibuchi H., Murata T., Trksen I.B. (1995a) A genetic-algorithm-based approach to the selection of linguistic classification rules. Proceedings of the International Conference EUFIT'95, vol. 3. Aachen, Germany, pp. 1415–1419
134. Ishibuchi H., Nozaki K., Yamamoto N., Tanaka H. (1995b) Selecting fuzzy if-then rules for classification problems using genetic algorithms. IEEE Transactions on Fuzzy Systems, vol. 3, No. 3, pp. 260–270
135. Iwasaki T., Morita A. (1990) Fuzzy auto-tuning for PID controller with model classification. Proceedings of the International Conference NAFIPS'90. Toronto, USA, pp. 90–93
136. Jacoby E., Zimmermann C., Bessai H. (1994) Fuzzy Control in der Diabestherapie. Proceedings of the Conference Dortmunder Fuzzy-Tage. Dortmund, Germany, pp. 214–223
137. Jantzen J. (1997) A robustness study of fuzzy control rules. Proceedings of the International Conference EUFIT'97, vol. 2. Aachen, Germany, pp. 1223–1227
138. Jaszczak S., Pluciński M., Piegat A. (1996) Identyfikacja parametrów nieliniowego napędu systemu dynamicznego. Materiały Konferencji Naukowo-Technicznej "Współczesne problemy w budowie i eksploatacji maszyn" Szczecin, Poland, pp. 103–112
139. Jäkel J. (1997) Fuzzy model identification based on a genetic algorithm and optimization techniques. Proceedings of the International Conference EUFIT'97, vol. 2. Aachen, Germany, pp. 774–778
140. Kacprzyk J. (1977) Control of a nonfuzzy systems in a fuzzy enviroment with a fuzzy termination time. Systems Science, No. 3, pp. 320–334
141. Kacprzyk J. (1978) A branch-and-bound algorithm for the multistage control of a nonfuzzy system in a fuzzy enviroment. Control and Cybernetics, No. 7, pp. 51–64
142. Kacprzyk J. (1986) Zbiory rozmyte w analizie systemowej. Państwowe Wydawnictwo Naukowe. Warszawa, Poland
143. Kacprzyk J.,Fedrizzi M. (1992) Fuzzy regression analysis. Warszawa: Omnitech Press, Heidelberg: Physica-Verlag
144. Kacprzyk J. (1996) Multistage control under fuzziness using genetic algorithms. Control and Cybernetics, vol. 25, No. 6, pp. 1181–1216
145. Kacprzyk J. (1997) Multistage fuzzy control. New York: John Wiley and Sons, Inc.
146. Kaczorek T. (1981) Teoria sterowania. Tom 2. Państwowe Wydawnictwo Naukowe. Warszawa, Poland
147. Kageyama S. (1990) Blood glucose control by a fuzzy control system. Proceedings of the First International Conference on Fuzzy Logic and Neural Networks. Iizuka, Japan, pp. 557–560
148. Kahlert J., Frank H. (1994) Fuzzy-Logik und Fuzzy-Control. Vieweg Verlag. Braunschweig, Germany
149. Kahlert J. (1995) Fuzzy Control für Ingenieure. Vieweg Verlag. Braunschweig, Germany
150. Kandel A., Langholz G. (1994) Fuzzy Control Systems. London: CRC Press
151. Kang G., Lee W. (1995) Design of fuzzy state controllers and observers. Proceedings of the International Conference FUZZ-IEEE/IFES'95, vol. 3. Yokohama, Japan, pp. 1355–1360
152. Katebi S.D. (1995) Fuzzy rules generation: a learning process. Proceedings of the International Conference ICAUTO-95, vol. 1. Indore, India, pp. 63–66

153. Kaufmann A., Gupta M. (1985) Introduction to fuzzy arithmetic-theory and applications. New York: Van Nostrand Reinhold
154. Kawai H. (1990) Engine control system. Proceedings of the First International Conference on Fuzzy Logic and Neural Networks. Iizuka, Japan, pp. 929–937
155. Keel L.H., Bhattacharyya S.P. (1999) A new proof of the Jury test. Automatica 35, pp 251–258
156. Khalil H.R. (1992) Nonlinear systems. New York: Macmillan Publishing Company
157. Kiendl H. (1992) Stabilitätsanalyse von mehrschleifigen Fuzzy-Regelungssystemen mit Hilfe der Methode der harmonischen Balance. Proceedings of the International Conference 2 Workshop "Fuzzy Control", Dortmund, Germany, pp. 315–321
158. Kiendl H., Rügger J.J. (1993) Verfahren zum Entwurf und Stabilitätsnachweis von Regelungssystemen mit Fuzzy-Reglern. Automatisierungstechnik, vol. 41, No. 5, pp. 138–144
159. Kiendl H. (1997) Fuzzy Control methodenorientiert. Oldenburg Verlag. München, Germany
160. Kiriakidis K., Tzes A. (1995) Application of implicit self-tuning fuzzy control to nonlinear systems. Proceedings of the International Conference FUZZ-IEEE/IFES'95, vol. 3. Yokohama, Japan, pp. 1419–1426
161. Kinnebrock W. (1994) Optimierung mit genetischen und selektiven Algorithmen. Oldenburg Verlag. München, Germany
162. Kitamura T. (1991) Design of intelligent support system for artificial heart control. Japanese Journal of Fuzzy Theory and Systems, vol. 3, No. 3, pp. 231–240
163. Knappe H. (1994) Nichtlineare Regelungstechnik und Fuzzy-Control. Expert Verlag. Renningen-Malmsheim, Germany
164. Koch M., Kuhn T., Wernstedt J. (1993) Ein neues Entwurfskonzept für Fuzzy-Regelungen. Automatisierungstechnik, vol. 41, No. 5, pp. 152–158
165. Koch M., Kuhn T., Wernstedt J. (1994) Methods for optimal design of fuzzy controllers. Proc. 39 Internationales Wissenschaftliches Kolloqium, vol. 3. Ilmenau, Germany, pp. 275–282
166. Koch M., Kuhn T., Wernstedt J. (1996) Fuzzy Control. Oldenburg Verlag. München, Germany
167. Kolios G., Aichele P., Nieken U., Eigenberger G. (1994) Regelung eines instationär betriebenen Festbettreaktors mit Fuzzy-Kontrollregeln. Proceedings of the Conference Dortmunder Fuzzy-Tage. Dortmund, Germany, pp. 429–436
168. Korbicz J., Obuchowicz A., Uciński D. (1994) Sztuczne sieci neuronowe. Podstawy i zastosowania. Akademicka Oficyna Wydawnicza PLJ. Warszawa, Poland
169. Korzeń M. (2000) Conversion of a neural model of a MISO system into a fuzzy model with the method of "important" points of the input/output mapping surface. Diploma thesis. Institute of Control Engineering, Technical University of Szczecin
170. Kosko B. (1992) Neural networks and fuzzy systems. Englewood Cliffs: Prentice Hall Inc.
171. Kouatli I., Jones B. (1991) An improved design procedure for fuzzy control systems. International Journal Mach. Tools Manufact., vol. 31, pp. 107–122
172. Kóczy L.T., Hirota K. (1993) Interpolative reasoning with insufficient evidence in sparse rule bases. Information Sciences, No. 71, pp. 169–201
173. Krieger H.J., Kratsch T., Kuhn H., Wächter H. (1994) Das Thüringen-Fuzzy-Modul der Modularen Kommunikativen Steuerung MKS 16. Proc. 39 Internationales Wissenschaftliches Kolloqium, vol. 3. Ilmenau, Germany, pp. 332–339

174. Krone A., Bäck T., Teuber P. (1996a) Evolutionäres Suchkonzept zum Aufstellen signifikanter Fuzzy-Regeln. Automatisierungstechnik, vol. 44, No. 8, pp. 405–411
175. Krone A. (1996b) Advanced rule reduction concepts for optimising efficiency of knowledge extraction. Proceedings of the International Conference EUFIT'96, vol. 2. Aachen, Germany, pp. 919–923
176. Krone A., Teuber P. (1996c) Applying WINROSA for automatic generation of fuzzy rule bases. Proceedings of the International Conference EUFIT'96, vol. 2. Aachen, Germany, pp. 929–932
177. Kruse R., Gebhard J., Klawonn F. (1994) Foundations of Fuzzy Systems. New York: John Wiley and Sons
178. Kuhn T., Wernstedt J. (1994) SOFCON-Eine Strategie zum optimalen Entwurf von Fuzzy-Regelungen. Automatisierungstechnik, vol. 42, No. 3, pp. 91–99
179. Kuhn T., Wernstedt J. (1996) Robust design of fuzzy control. Proceedings of the International Conference EUFIT'96, vol. 2. Aachen, Germany, pp. 970–973
180. Kwon T.M., Zervakis M.E. (1994) A self-organizing KNN fuzzy controller and its neural network structure. International Journal of Adaptive Control and Signal Processing, vol. 8, pp. 407–431
181. Lacrose V., Titli A. (1997) Multivariable fuzzy control using bi-dimensional fuzzy sets. Application to a mixing tank. Proceedings of the International Conference EUFIT'97, vol. 2. Aachen, Germany, pp. 1259–1263
182. Langari R., Wang L. (1995) Fuzzy models, modular networks, and hybrid learning. Proceedings of the International Conference FUZZ-IEEE/IFES'95, vol. 3. Yokohama, Japan, pp. 1291–1308
183. Lee K.C., Min S.S., Song J.W., Cho K.B. (1992) An adaptive fuzzy current controller with neural network for field-oriented controlled induction machine. Proceedings of the Second International Conference on Fuzzy Logic and Neural Networks. Iizuka, Japan, pp. 449–452
184. Leichtfried I., Heiss M. (1995) Ein kennfeldorientiertes Konzept für Fuzzy-Regler. Automatisierungstechnik, vol. 43, No. 1, pp. 31–40
185. Leigh J.R. (1985) Applied digital control. London: Prentice Hall
186. Lewis F.L., Liu K. (1996) Towards a paradigm for fuzzy logic control. Automatica, vol. 32, pp. 167–181
187. Li Y., Yonezawa Y. (1991) Stability analysis of a fuzzy control system by the hyperstability theorem. Japanese Journal of Fuzzy Theory and Systems, vol. 3, No. 2, pp. 209–214
188. Li H.X., Gatland H.B. (1995) Enhanced methods of fuzzy logic control. Proceedings of the International Conference FUZZ-IEEE/IFES'95, vol. 1. Yokohama, Japan, pp. 331–336
189. Liao S.Y., Wang H.Q., Liu W.Y. (1999) Functional dependences with null values, fuzzy values, and crisp values. IEEE Transactions on Fuzzy Systems, vol. 7, No. 1, pp. 97–103
190. Lichtenberg M. (1994) Generierung kennliniengleicher Fuzzy-Regler aus den Parametern konventioneller (PID-) Regler. Automatisierungstechnik, vol. 42, No. 12, pp. 540–546
191. Lin C.T., Lee G. (1991) Neural-network-based fuzzy logic control and decision system. IEEE Transactions on Computers, vol. 40, No. 12, pp. 1320–1336
192. Lin Y., Cunningham G.A. (1995) A new approach to fuzzy-neural system modeling. IEEE Transactions on Fuzzy Systems, vol. 3, No. 2, pp. 190–198
193. Lin C.I., Lin C.T. (1996) Reinforcement learning for an ART-based fuzzy adaptive learning control network. IEEE Transactions on Neural Networks, vol. 7, No. 7, pp. 709–731

194. Lin Y., Cunningham G.A., Coggeshall S.V., Jones R.D. (1998) Nonlinear system input structure identification: two stage fuzzy curves and surfaces. IEEE Transactions on Systems, Man and Cybernetics – Part A: Systems and Humans, vol. 28, No. 5
195. Locher G., Bakshi B., Stephanopoulos G., Schügerl K. (1996a) Ein Einsatz zur automatischen Umwandlung von Rohdaten in Regeln. Teil 1: Prozesstrends, Wavelettransformation und Klassifizierungsbaüme. Automatisierungstechnik, vol. 44, No. 2, pp. 61–70
196. Locher G., Bakshi B. (1996b) Ein Einsatz zur automatischen Umwandlung von Rohdaten in Regeln. Teil 2: Eine Fallstudie. Automatisierungstechnik, vol. 44, No. 3, pp. 138–145
197. Lofti A., Howarth M., Thomas P.D. (1996) Non-interactive model for fuzzy rule-based systems. Proceedings of the International Conference EUFIT'96, vol. 1. Aachen, Germany, pp. 597–601
198. Lopez A.S. (1996) Tuning of a multivariable fuzzy logic controller using optimization techniques. Proceedings of the International Conference EUFIT'96, vol. 2. Aachen, Germany, pp. 965–969
199. Lopez A.S., Lafont J.C. (1997) Tuning of a decentralized multivariable fuzzy controller. Proceedings of the International Conference EUFIT'96, vol. 2. Aachen, Germany, pp. 1254–1258
200. Ławryńczuk M., Tatjewski P. (2000) Neural inverse modelling for disturbance compensation in nonlinear plant control. Proceedings of Sixth International Conference on Methods and Models in Automation and Robotics MMAR'2000 vol. 2. Międzyzdroje, Poland, pp. 721–726
201. Maeda M., Murakami S. (1991) Stability analysis of fuzzy control systems using phase planes. Japanese Journal of Fuzzy Theory and Systems, vol. 3, No. 2, pp. 149–160
202. Magdalena L., Monasterio F. (1995) Evolutionary-based learning applied to fuzzy controllers. Proceedings of the International Conference FUZZ-IEEE/IFES'95, vol. 3. Yokohama, Japan, pp. 1111–1118
203. Malki H.A., Li H., Chen G. (1994) New design and stability analysis of fuzzy proportional-derivative control systems. Transactions on Fuzzy Systems, vol. 2, No. 4, pp. 245–254
204. Mamdani E.H. (1974) Applications of fuzzy algorithms for control of simple dynamic plant. Proceedings IEEE, No. 121 (12), pp. 1585–1588
205. Mamdani E.H. (1977) Application of fuzzy logic to approximate reasoning using linguistic synthesis. IEEE Transactions on Computers, vol. C-26, No. 12, pp. 1182–1181
206. Marin J.P., Titli A. (1995) Necessary and sufficient conditions for quadratic stability of a class of Takagi-Sugeno fuzzy systems. Proceedings of the International Conference EUFIT'95, vol. 2. Aachen, Germany, pp. 786–790
207. Masters T. (1996) Sieci neuronowe w praktyce. Wydawnictwo Naukowo-Techniczne. Warszawa, Poland
208. Männle M., Richard A., Dörsam T. (1996) Identification of rule-based fuzzy models using the RPROP optimization technique. Proceedings of the International Conference EUFIT'96, vol. 1. Aachen, Germany, pp. 587–591
209. Michalewicz Z. (1996) Genetic algorithms + data structures = evolution programs. Springer Verlag, 3rd edition.
210. Mitchell M. (1996) An introduction to genetic algorithms. MIT Press. Cambridge, MA
211. Moody J., Darken C. (1989) Fast learning in networks of locally-tuned processing units. Neural Computation, vol. 1, No. 2, pp. 281–294
212. Morari M., Zafirion E. (1989) Robust process control. New York: Prentice Hall

213. Murakami S. (1989) Weld-line tracking control of arc welding robot using fuzzy logic controller. Fuzzy Sets and Systems 1989, vol. 32, pp. 221–237
214. Murata T., Ishibuchi H. (1995) Adjusting membership functions of fuzzy classification rules by genetic algorithms. Proceedings of the International Conference FUZZ-IEEE/IFES'95, vol. 4. Yokohama, Japan, pp. 1819–1824
215. Narazaki H., Ralescu A.L. (1993) An improved synthesis method for multilayered neural networks using qualitative knowledge. IEEE Transactions on Fuzzy Systems, vol. 1, No. 2, pp. 125–137
216. Narazaki H., Shigaki I., Watanabe T. (1995) A method for extracting approximate rules from neural networks. Proceedings of the International Conference FUZZ-IEEE/IFES'95, vol. 4. Yokohama, Japan, pp. 1865–1870
217. Narenda K.S. (1990a) Neural networks for control. Cambridge: MIT Press
218. Narenda K.S., Parthasarathy K. (1990b) Identification and control for dynamic system using neural networks. IEEE Transactions on Neural Networks, vol. 1, No. 1, pp. 4–27
219. Nelles O. (1996) FUREGA – fuzzy rule extraction by a genetic algorithm. Proceedings of the International Conference EUFIT'96, vol. 1. Aachen, Germany, pp. 489–493
220. Nelles O., Ernst S., Isermann R. (1997) Neuronale Netze zur Identifikation nichtlinearer, dynamischer Systeme: Ein Überblick. Automatisierungstechnik, vol. 45, No. 6, pp. 251–261
221. Nelles O., Hecker O., Iserman R. (1998) Automatische Strukturselektion fuer Fuzzy-Modelle zur Identifikation nichtlinearer dynamischer Prozesse (Subset selection for nonlinear system identification with fuzzy models). Automatisierungstechnik, vol. 46, No. 6, pp. 302–312
222. Nelles O., Fink A., Babuška R., Setnes M. (1999) Comparison of two construction algorithms for Takagi-Sugeno fuzzy models. Proceedings of 7th European Congress on Intelligent Techniques and Soft Computing – on CD-ROM. Aachen, Germany
223. Neumerkel D., Lohnert F. (1992) Anwendungsstand Künstlicher Neuronaler Netze in der Automatisierungstechnik. Automatisierungstechnische Praxis, vol. 34, No. 11, pp. 640–645
224. Nguyen T.H., Sugeno M., Tong R., Yager R.R. (1995) Theoretical aspects of fuzzy control. New York: John Wiley and Sons, Inc.
225. Nobre F.S. (1995) Genetic-neuro-fuzzy systems: a promising fusion. Proceedings of the International Conference FUZZ-IEEE/IFES'95, vol. 1. Yokohama, Japan, pp. 259–266
226. Noisser R. (1994) Beurteilung der Stabilität und der Stabilitätsreserve von Fuzzy-Regelungen mittels L_2-Stabilitätskriterium. Proceedings of the Conference Dortmunder Fuzzy-Tage. Dortmund, Germany, pp. 322–329
227. Nomura H., Hayashi I., Wakami N. (1994) A self-tuning method of fuzzy reasoning by genetic algorithm. Chapter in the book "Fuzzy Control Systems" by Kandel A., Langkolz G. London: CRC Press, pp. 337–354
228. Ohki M., Shikata T., Moriyama T., Ohkita M. (1997) A genetic algorithm with modified bit-selection probability for optimizing the fuzzy reasoning. Proceedings of the International Conference EUFIT'97, vol. 1. Aachen, Germany, pp. 694–698
229. Ohnishi T. (1991) A self-learning fuzzy control system for an urban refuse incineratoin plant. Japanese Journal of Fuzzy Theory and Systems, vol. 3, No. 2, pp. 187–200
230. Opitz H.P. (1986) Die Hyperstabilitätstheorie – eine systematische Methode zur Analyse und Synthese nichtlinearer Systeme. Automatisierungstechnik, vol. 34, No. 6, pp. 221–230

231. Opitz H.P. (1993) Fuzzy Control-Stabilität von Fuzzy-Regelungen. Automatisierungstechnik, vol. 41, No. 8, pp. A21–A24
232. Osowski S. (1996) Sieci neuronowe. Wydawnictwo Naukowo-Techniczne. Warszawa, Poland
233. Otto P. (1995) Fuzzy modelling of nonlinear dynamics systems by inductive learned rules. Proceedings of the International Conference EUFIT'95, vol. 2. Aachen, Germany, pp. 858–864
234. Pałęga A. (1996) Neuro-fuzzy-PID regulator kursu pojazdu podwodnego z fuzyfikacją potrójną. Praca dyplomowa. Technical University of Szczecin, Poland
235. Park M.K., Ji S.H., Kim M.J., Park M. (1995) A new identification method for a fuzzy model. Proceedings of the International Conference FUZZ-IEEE/IFES'95, vol. 4. Yokohama, Japan, pp. 2159–2164
236. Pedrycz W. (1984) An identification algorithm in fuzzy relational systems. Fuzzy Sets and Systems, No. 13, pp.153–167
237. Pedrycz W. (1993) Fuzzy control and fuzzy systems. New York: John Willey and Sons
238. Pedrycz W. (1995) Fuzzy control engineering: reality and challenqes. Proceedings of the International Conference FUZZ-IEEE/IFES'95, vol. 1. Yokohama, Japan, pp. 437–446
239. Pedrycz W. (1996) Fuzzy multimodels. IEEE Transactions on Fuzzy Systems, vol. 4, No. 2, pp. 139–148
240. Pedrycz W., Reformat M. (1997) Rule-based modeling of nonlinear relationships. IEEE Transactions on Fuzzy Systems, vol. 5, No. 2, pp. 256–269
241. Pfeiffer B.M. (1995a) Symbolische Analyse und Synthese von Fuzzy-Reglern durch Transformationen zwischen unscharfer und scharfer Darstellung. Automatisierungstechnik, vol. 43, No. 11, pp. 514–124
242. Pfeiffer B.M. (1995b) Imitation of human operators by "neuro-fuzzy" structures. Proceedings of the International Conference EUFIT'95, vol. 2. Aachen, Germany, pp. 804–809
243. Pfeiffer B.M. (1996) 5.Workshop "Fuzzy Control". Automatisierungstechnik, vol. 44, No. 3, pp. 141–142
244. Piegat A., Baccari M. (1995a) Shortcomings of the control with fuzzy controllers. Proceedings of the International Symposium on Methods and Models in Automation and Robotics MMAR'95, vol. 2. Międzyzdroje, Poland, pp. 759–762
245. Piegat A., Pluciński M., Skórski W. (1995b) Technical system and control algorithms of underwater vehicle Krab II. In the book "Marine Technology and Transportation", editors Graczyk T., Jastrzębski T., Brebbia C.A., Southampton: Computational Mechanics Publications
246. Piegat A., Pluciński M. (1995c) Application of the radial basis-function in modelling and identification of linear and nonlinear systems. Proceedings of the XII International Conference on Systems Science, vol. 1. Wrocław, Poland, pp. 266–274
247. Piegat A., Jaszczak S., Pluciński M. (1996) Selflearning neuro-fuzzy PID controller without simplifications. Proceedings of the International Symposium on Methods and Models in Automation and Robotics MMAR'96 vol. 3. Międzyzdroje, Poland, pp. 1195–1200
248. Piegat A., Pluciński M. (1997a) Fast-learning neuro-fuzzy PID-controller with minimal number of fuzzy regions. Applied Mathematics and Computer Science, vol. 7, No. 1, pp. 171–184.

249. Piegat A. (1997b) Stability checking of a real digital control system with a neuro-fuzzy PD-controller. Proceedings of the European Control Conference ECC 97. Brussels, Belgium, paper No. TU-AG4 (141) on disc
250. Piegat A. (1997c) Extrapolation truth. Proceedings of the International Conference EUFIT'97, vol. 1. Aachen, Germany, pp. 324–329
251. Piegat A. (1997d) Hyperstability of fuzzy-control systems and degrees of freedom. Proceedings of the International Conference EUFIT'97, vol. 2. Aachen, Germany, pp. 1446–1450
252. Piegat A. (1997e) Rule base reduction in fuzzy models. Proceedings of the International Conference ACS'97. Szczecin, Poland, pp. 395–400
253. Piegat A. (1998a) Regulator rozmyty. Patent A1(21)315 399. Biuletyn Urzędu Patentowego, No. 3(629), p. 64
254. Piegat A., Pluciński M. (1998b) Fuzzy internal model control of an underwater vehicle. Proceedings of the International Symposium on Methods and Models in Automation and Robotics MMAR'98 vol. 2. Międzyzdroje, Poland, pp. 691–695
255. Piegat A., Pluciński M. (1998c) Fuzzy inverse model control (InvMC) of an underwater vehicle. Proceedings of the International Conference EUFIT'98, vol. 2. Aachen, Germany, pp. 834–838
256. Piegat A., Pluciński M. (1998d) Modeling of nonlinear systems with application of delinearized Delaunay nets. Proceedings of the International Conference SCM. Zakopane, Poland, on disc
257. Piegat A. (1998e) Nonregular nonlinear sector modelling. Applied Mathematics and Computer Science, vol. 8, No 3, pp 101–123
258. Piegat A., Pluciński M. (2000) Firm evaluation with 2-dimensional projection method. Proceedings of the Fifth International Conference "Neural Networks and Soft Computing". Zakopane, Poland, pp. 361–367
259. Piegat A., Kraszewski P., Stolcman S. (2000) Conception of geometric-neural modeling and its employment to determination of the optimal setting path of the controlled pitch propeller ship propulsion system. Proceedings of the International Symposium on Methods and Models in Automation and Robotics MMAR'2000 vol. 1. Międzyzdroje, Poland, pp. 447–453
260. Piliński M. (1997) Sterowniki rozmyte z wykorzystaniem sieci neuronowych. Ph.D. Thesis. Technical University of Częstochowa, Poland
261. Pluciński M. (1996) Adaptacyjny układ sterowania kursem bezzałogowego pojazdu podwodnego, wykorzystujący rozmytą bazę wiedzy o obiekcie. Ph.D. Thesis. Technical University of Szczecin, Poland
262. Pluciński M. (1997) Adaptive control system with a fuzzy data base. Proceedings of the International Symposium on Methods and Models in Automation and Robotics MMAR'97 vol. 2. Międzyzdroje, Poland, pp. 809–814
263. Popov V.M. (1963) The solution of a new stability problem for controlled systems. Automatic and Remote Control, vol. 24, pp. 1–23
264. Popov V.M. (1973) Hyperstability of control systems. Berlin: Springer Verlag
265. Praca zbiorowa (1971) Poradnik inżyniera-matematyka. Wydawnictwo Naukowo-Techniczne. Warszawa, Poland
266. Preuss H.P. (1992) Fuzzy Control-heuristische Regelung mittels unscharfer Logik. Automatisierungstechnische Praxis, vol. 34, No. 4, pp. 176–184
267. Preuss H.P., Tresp V. (1994a) Neuro-Fuzzy. Automatisierungstechnische Praxis, vol. 36, No. 5, pp. 10–24
268. Preuss H.P. (1994b) Methoden der nichtlinearen Modellierung-vom Interpolationspolynom zum Neuronalen Netz. Automatisierungstechnik, vol. 42, No. 10, pp. 449–457

269. Preuss H.P., Ockel J. (1995) Lernverfahren für Fuzzy-Regler. Automatisierungstechnische Praxis, vol. 37, No. 7, pp. 10–20
270. Preut K.H., Braun H., Höhfeld M. (1995) Optimierung von Neuro-Fuzzy Netzwerken mit evolutionären Strategien. Proceedings of the International Conference Fuzzy-Neuro-Systeme'95. Darmstad, Germany, pp. 365–372
271. Pułaczewski J. (1997) Fuzzy PID digital control algorithm using Takagi-Sugeno models. Proceedings of the International Symposium on Methods and Models in Automation and Robotics MMAR'97 vol. 2. Międzyzdroje, Poland, pp. 815–818
272. Pütz A., Weber R. (1996) Automatic adjustment of weights in fuzzy rule bases. Proceedings of the International Conference EUFIT'96, vol. 2. Aachen, Germany, pp. 933–937
273. Rehfeld D., Schmitz T. (1994) Schweissprozessanalyse und Qualitätssicherung mit Fuzzy-Logik. Proceedings of the Conference Dortmunder Fuzzy-Tage. Dortmund, Germany, pp. 189–197
274. Roffel B., Chin P.A. (1991) Fuzzy control of a polymerisation reactor. Hydrocarbon Processing, No. 6, pp. 47–50
275. Rovatti R., Guerieri R. (1996) Fuzzy sets of rules for system identification. Transactions on Fuzzy Systems, vol. 4, No. 2, pp. 89–101
276. Rumelhart A., Mc Clelland I.L. (1986) Parallel Distributed Processing. Cambridge: MIT Press
277. Rumpf O. (1994) Anwendung der Methode der konvexen Zerlegung zur Stabilitätsanalyse dynamischer Systeme mit neuronalen Komponenten. Proceedings of the 39th Internationales Wissenschaftliches Kolloquim, vol. 3. Ilmenau, Germany, pp. 367–374
278. Rutkowska D. (1996) On application of genetic algorithm to fuzzy neural network learning. Proceedings of the Second Conference on Neural Networks and Their Applications. Szczyrk, Poland, pp. 420–425
279. Rutkowska D., Piliński M., Rutkowski L. (1997) Sieci neuronowe, algorytmy genetyczne i systemy rozmyte. Wydawnictwo Naukowe PWN. Warszawa-Łódź, Poland
280. Rzepka A. (2000) Visualization program of the tuning process of multidimensional, symmetric, rotable RBF-neurons with infinite support. Diploma thesis. Faculty of Computer Science and Information Systems, Technical University of Szczecin
281. Saito Y., Ishida T. (1990) Fuzzy PID hybrid control – an application to burner control. Proceedings of the First International Conference on Fuzzy Logic and Neural Networks, vol. 1. Iizuka, Japan, pp. 65–69
282. Sasaki T., Akiyama T. (1988) Traffic control process of expressway by fuzzy logic. Fuzzy Sets and Systems, vol. 26, pp. 165–178
283. Schmitt G., Günther S. (1996) Das Hyperstabilitätskurven-Verfahren als graphisches Frequenzbereichskriterium zur Stabilitätsprüfung nichtlinearer Mehrgrössenregelkreise. Automatisierungstechnik, vol. 44, No. 6, pp. 281–288
284. Serac A., Roth H., Gardus V. (1996) Adaptive fuzzy controller for a MIMO system. Proceedings of the International Conference EUFIT'96, vol. 2. Aachen, Germany, pp. 901–905
285. Sheel T., Kiendl H. (1995) Stability analysis of fuzzy and other nonlinear systems using integral Lyapunov functions. Proceedings of the International Conference EUFIT'95, Aachen, Germany, pp. 765–770
286. Shmilovici A., Maimon O. (1996) Best fuzzy rule selection with ortogonal matching pursuit. Proceedings of the International Conference EUFIT'96, vol. 1. Aachen, Germany, pp. 592–596

287. Simpson P.K. (1992) Fuzzy min-max neural networks – part 1: classification. IEEE Transactions on Fuzzy Systems, vol. 1, No. 1, pp. 32–45.
288. Singh S. (1992) Stability analysis of discrete fuzzy control systems. Proceedings of the First IEEE Conference on Fuzzy Systems, pp. 527–534
289. Sommer H.J., Hahn H. (1993) Ein einfaches Verfahren zum Test der Bibo-Stabilität bei Systemen mit Polynomialen Nichtlinearitäten und Fuzzy-Komponenten. Proceedings of the 3rd Workshop "Fuzzy Control" des GMA-UA 1.4.2., pp. 13–26
290. Sommer H.J., Hahn H. (1994) Ein einfaches Verfahren zum Test der BIBO-Stabilität bei Systemen mit Polynomialen Nichtlinearitäten und Fuzzy-Komponenten. GMA-Aussprachetag Fuzzy-Control, Langen, VDI-Bericht Nr 1113, pp. 13–26
291. Soria L.A. (1996) Tuning of a multivariable fuzzy logic controller using optimization techniques. Proceedings of the International Conference EUFIT'96, vol. 2. Aachen, Germany, pp. 965–969
292. Sousa J., Babuška R., Verbruggen H.B. (1995) Adaptive fuzzy model-based control. Proceedings of the International Conference EUFIT'95, vol. 2. Aachen, Germany, pp. 865–869
293. Stolcman S., Piegat A., Szcześniak J. (1999) Effectiveness investigation of the RBF-neural network applied to modeling of the ship propulsion systems with controllable pitch propeller (CCP). Polish Maritime Research, vol. 6, No. 2(20), pp. 11–16
294. Su M.C., Kao C.J., Liu K.M., Liu C.Y. (1995) Rule extraction using a novel class of fuzzy degraded hyperellipsoidal composite neural networks. Proceedings of the International Conference FUZZ-IEEE/IFES'95, vol. 1. Yokohama, Japan, pp. 233–238
295. Sugeno M., Yasukawa T. (1993) A fuzzy-logic-based approach to qualitative modeling. IEEE Transactions on Fuzzy Systems, vol. 1, No. 1, pp. 7–31
296. Sugiura T. (1991) Fuzzy control of a rate-adaptive cardiac pacemaker with multiple indicators. Japanese Journal of Fuzzy Theory and Systems, vol. 3, No. 3, pp. 241–249
297. Suykens J.A., De Moor B., Vandewalle J. (1995) Stability criteria for neural control systems. Proceedings of the 3rd European Control Conference. Rome, Italy, pp. 2766–2771
298. Szwedko J. (2000) Program for significance evaluation of inputs of multidimensional systems. Diploma thesis. Faculty of Computer Science and Information Systems, Technical University of Szczecin
299. Tadeusiewicz R. (1993) Sieci neuronowe. Akademicka Oficyna Wydawnicza R.M. Warszawa, Poland
300. Takagi T., Sugeno M. (1985) Fuzzy identification of systems and its applications to modeling and control. IEEE Transactions on Systems, Man and Cybernetics, vol. 15, No. 1, pp. 116–132
301. Takahashi M. (1990) Biomedical applications of fuzzy logic controllers. Proceedings of the First International Conference on Fuzzy Logic and Neural Networks, vol. 1. Iizuka, Japan, pp. 553–556
302. Tan S., Yu Y. (1995) Fuzzy modeling: an adaptive approach. Proceedings of the International Conference FUZZ-IEEE/IFES'95, vol. 2. Yokohama, Japan, pp. 889–896
303. Tanaka K., Sugeno M. (1990) Stability analysis of fuzzy systems using Lyapunov's direct method. Proceedings of the International Conference NAFIPS'90, vol. 1. Toronto, USA, pp. 133–136
304. Tanaka K., Sugeno M. (1992) Stability analysis and design of fuzzy control systems. Fuzzy Sets and Systems, vol. 45, pp. 135–156.

305. Tilli T. (1991) Fuzzy-Logik. Franzis Verlag. München, Germany
306. Tobi T., Hanafusa T. (1991) A practical application of fuzzy control for an air-conditioning system. International Journal of Approximate Reasoning, No. 5, pp. 331–348
307. Tomera M., Morawski L. (1996) Neural-network-based fuzzy logic marine autopilot. Proceedings of the International Symposium on Methods and Models in Automation and Robotics MMAR'96 vol. 3. Międzyzdroje, Poland, pp. 1207–1212
308. Ullrich T. (1997a) Datengetriebene Modellierung nichtlinearer Strecken mit Delaunay-Netzen. Automatisierungstechnik, vol. 45, No. 5, pp. 236–245
309. Ullrich T., Brown M. (1997b) Delaunay-based nonlinear internal model control. Proceedings of the International Conference ACS'97. Szczecin, Poland, pp. 387–394
310. Ullrich T., Tolle H. (1997c) Delaunay-based local model networks for nonlinear system identification. Proceedings of the IASTED International Conference on Applied Modeling and Simulation. Banff, Canada, pp. 1–12
311. Ullrich T. (1998) Untersuchungen zur effizienten interpolierenden Speicherung von nichtlinearen Prozessmodellen und Vorsteuerungen. Shaker Verlag, Aachen
312. Voit F. (1994) Fuzzy Control versus konventionelle Regelung am Beispiel der Metro Mailand. Automatisierungstechnik, vol. 42, No. 9, pp. 400–410
313. Wagner S. (1997) A specialized evolutionary method for the creation of fuzzy controllers. Proceedings of the International Conference EUFIT'97, vol. 1. Aachen, Germany, pp. 699–704
314. Wakabayashi T. (1995) A method for constructing of system models by fuzzy flexible interpretive stuctural modeling. Proceedings of the International Conference FUZZ-IEEE/IFES'95, vol. 2. Yokohama, Japan, pp. 913–918
315. Wang L.X. (1994a) Adaptive fuzzy systems and control, design and stability analysis. Englewood Cliffs: Prentice Hall
316. Wang L.X. (1994b) A supervisory controller for fuzzy control systems that guarantees stability. IEEE Transactions on Automatic Control, vol. 39, No. 9, pp. 1845–1847
317. Wang L.X. (1995a) Design and analysis of fuzzy identifiers of nonlinear dynamic systems. IEEE Transactions on Automatic Control, vol. 40, No. 1, pp. 11–23
318. Wang L.X., Jordan J.R. (1995b) The robust step performance of PID and fuzzy logic controlled SISO systems. Proceedings of the International Conference FUZZ-IEEE/IFES'95, vol. 1. Yokohama, Japan, pp. 325–330
319. Wang H.O., Tanaka K., Griffin M. (1996) An approach to fuzzy control of nonlinear system: stability and design issues. IEEE Transaction on Fuzzy Systems, vol. 4, No. 1, pp. 14–23
320. Watanabe T. (1990) AI and fuzzy-based tunel ventilation control system. Proceedings of the First International Conference on Fuzzy Logic and Neural Networks, vol. 1. Iizuka, Japan, pp. 71–75
321. Watanabe T., Ichihashi H. (1991) Iterative fuzzy modelling using membership function of degree n and its application to crane control. Japanese Journal on Fuzzy Theory and Systems, vol. 3, No. 2, pp. 173–186
322. Weinmann A. (1991) Uncertain models and robust control. Springer Verlag, Wien
323. Werntges H.W. (1993) Partitions of unity improve neural function approximation. Proceedings of the International Conference on Neural Networks, vol. 2. San Francisco, USA, pp. 914–918

324. Wu Q., Böning D., Schnieder E. (1994) Realisierung von Fuzzy-Reglern mit Hilfe von Relationsmatrizen. Automatisierungstechnik, vol. 42, No. 4, pp. 162–169
325. Yager R., Filev D. (1994) Essentials of fuzzy modeling and control. New York: John Wiley and Sons
326. Yager R., Filev D. (1995) Podstawy modelowania i sterowania rozmytego. Wydawnictwo Naukowo-Techniczne. Warszawa, Poland
327. Yamashita R., Yamakawa T. (1992) Application of fuzzy control to a localized cleanroom. Proceedings of the Second International Conference on Fuzzy Logic and Neural Networks. Iizuka, Japan, pp. 1101–1102
328. Yao C.C., Kuo Y.H. (1995) A fuzzy neural network model with three-layered stucture. Proceedings of the International Conference FUZZ-IEEE/IFES'95, vol. 3. Yokohama, Japan, pp. 1503–1510
329. Ying H. (1993) The simplest fuzzy controllers using different inference methods are different nonlinear proportional-integral controllers with variable gains. Automatica, vol. 29, No. 6, pp. 1579–1589
330. Ying H. (1994) Practical design of nonlinear fuzzy controllers with stability analysis for regulating processes with unknown mathematical models. Automatica, vol. 30, No. 7, pp. 1185–1195
331. Yu C., Cao Z., Kandel A. (1990) Application of fuzzy reasoning to the control of an activated sludge plant. Fuzzy Sets and Systems, vol. 38, No. 1, pp. 1–14
332. Zadeh L.A. (1965) Fuzzy sets. Information and Control, vol. 8, pp. 338–353
333. Zadeh L.A. (1973) Outline of a new approach to the analysis of complex systems and decision processes. IEEE Transactions on Systems, Man and Cybernetics, vol. 3, pp. 28–44
334. Zadeh L.A. (1979) Fuzzy sets and information granularity. In Advances in Fuzzy Set Theory and Applications. Editors: Gupta M., Ragade R., Yager R., Eds. Amsterdam: North-Holland, pp. 3–18
335. Zadeh L.A. (1996) Fuzzy logic = computing with words. IEEE Transactions on Fuzzy Systems, vol. 4, No. 2, pp. 103–111
336. Zell A. (1994) Simulation neuronaler Netze. Addison-Wesley (Deutschland) GmbH
337. Zengh X.J., Singh M.G. (1996) Decomposition property of fuzzy systems and its applications. IEEE Transactions on Fuzzy Systems, vol. 4, No. 2, pp. 149–163
338. Zhou J., Eklund P. (1995) Some remarks on learning strategies for parameter identification in rule based systems. Proceedings of the International Conference EUFIT'95, vol. 3. Aachen, Germany, pp. 1911–1916
339. Zimmermann H.J., Thole U. (1979) On the suitability of minimum and product operators for the intersection of fuzzy sets. Fuzzy Sets and Systems, No. 2, pp. 167–181
340. Zimmermann H.J. (1987) Fuzzy sets, decision making, and expert systems. London: Kluwer Academic Publishers
341. Zimmermann H.J. (1994a) Fuzzy set theory and its applications. London: Kluwer Academic Publishers
342. Zimmermann H.J., von Altrock C. (1994b) Fuzzy logic. Band 2. Anwendungen. Oldenburg Verlag. München, Germany
343. Zurmühl R. (1964) Matrizen und ihre technischen Anwendungen. Berlin: Springer Verlag
344. Żuchowski A., Papliński J.P. (1998) The simplification of linear models of dynamics by decomposition to zeros and poles. Proceedings of the International Symposium on Methods and Models in Automation and Robotics MMAR'98 vol. 1. Międzyzdroje, Poland, pp. 225–230

345. Žižka J. (1996) Learning control rules for Takagi-Sugeno fuzzy controllers using genetic algorithms. Proceedings of the International Conference EUFIT'96, vol. 2. Aachen, Germany, pp. 960-964

Index

accumulation 172
adaptive
– fuzzy control 599
– InvMC control 599
aggregation of premises 165
algebraic
– sum 129
– sum operator 131
anti-windup element 502
arithmetic mean-operator 124
associativity of sets 113, 127, 129
asymmetrical Gaussian function 43
asymptotic hyperstability 626

bag 20
base model 422
basic range 15
bifurcation method 613
bounded
– difference 121
– sum 129
– *SUM-MIN* inference 182
– *SUM-PROD* inference 182

c-means method 476
cardinal number of a fuzzy set 26
Cartesian product 137
Center of Gravity method 187
Center of Sums method 192
chromosome 488
circle stability criterion 618
classical
– control system structure 500
– implication 149
– relation 137, 139
closest neighborhood 411
cluster validity 480
clustering 455
commutativity of sets 112, 120, 127, 129
compensatory
– operator 132, 134
– principle 134
complete
– partition of the universe of discourse 209
– rule base 218
completeness of a fuzzy model 208
concentration operator 28
consistent rule base 218
continuity of the rule base 219
continuous
– membership function 17
– relation 149
– space 15
contrast
– decreasing operator 30
– intensification operator 29
controllability 701
controller with a hysteresis 497
convex fuzzy set 27
core of a fuzzy set 22
critical point of a membership function 41
crossover 403, 489
curse of dimensionality 335, 439
cylindrical extension 142

defuzzification 159, 184
– mechanism 159
Delaunay networks 337
denormalization of a model 245
dilatation operator 28
direct
– structure of identification 546
– version of a fuzzy PID controller 503
discontinuity of a defuzzification method 186
discrete
– membership function 17
– relation 149
– space 15
domain of discourse 15

drastic
- product 121
- sum 129
Dubois-intersection operator 121
dynamics of a man as a controller 516

Einstein
- product 121
- sum 129
empty fuzzy set 18
error
- back-propagation 377
- residuum 425
exclusive contradiction 113
expert in fuzzy modeling 367
exponential rotatable elliptic function 437
extension of the universe of discourse 440
extrapolation truth 263, 568, 580

fictitious degree of freedom 639, 645
filtering of measurement samples 410, 443
finite-support Gaussian function 42
First of Maxima method 186
first-order extrapolation 255
fitness function 488
full-dimensional clustering method 466
fuzzification 157
fuzzy
- bag 20
- extrapolation 251
- implication 152
- linguistic model 367
- logic 2
- number arithmetic 60
- numbers 13
- PID controller 502
- relation 140
- set 11, 15
- set theory 2
- zero 88, 100

Gaussian function 40, 385, 424
general control system 500
generalized
- mean-operator 124
- Modus Ponens 161
generalizing rule 490
genes 488
genetic algorithms 400, 488
geometric mean-operator 124
global fuzzy models 316
grade of membership 16
granularity of information 2
granule of information 1

Hamacher
- product 121
- sum 129
Hamacher-intersection operator 121
Hamacher-union operator 130
"hard *MIN*" operator 114
"hard" signum 114
harmonic
- mean-operator 124
- membership function 46
Height method 197
height of a fuzzy set 22
horizontal representation of a fuzzy set 23
hyper-rectangular input space partition 335
hyper-triangular input space partition 335
hyperstability 626
- conditions 630
- in the ordinary sense 626
hysteresis 325, 630

idempotency 113, 127
identification of models of dynamic plants 522
identity 113
IMC structure 561, 595
implication operators 153
incompleteness of a fuzzy model 209
inconsistency of the rules 219
incremental version of a fuzzy PID controller 503
inference 158, 160, 180
- mechanism 158
inner membership function 49
input/output stability 612
insufficiently grounded decision 219
internal model control system 561
interpolation truth 263
intersection of sets 112
intuitive membership functions 37
inverse fuzzy number 63
InvMC structure 583, 595

jump model 537
Jury's test 658

k-means method 464

Kalman-Jakubowich equations 655
kinds of membership functions 34

L-R representation of a fuzzy number 63, 72
Last of Maxima method 186
learning rate coefficient 378
left membership function 48
lexical uncertainty 56
linguistic
- Mamdani model 290
- modifiers of fuzzy sets 27
- term-set 14
- value 13
- variable 13
linguistically complete rule base 212
local
- fuzzy models 316
- model method 241
logical
- connectives 111
- product 112
- product of sets 112
- sum of sets 126
low-pass filter 565, 572
Lyapunov's method 611

Mamdani model 281
MAX-MIN inference 175
MAX-operator 128, 129
MAX-PROD inference 175
maximum absolute error method 420
mean square deviation 322
membership function 1, 16, 34
mental model 366
method of maximum absolute error points 421
Middle of Maxima method 184
MIMO 601
MIN-operator 114, 121
MISO system 66
modal value of a fuzzy set 31
model
- invertibility 525
- structure code 492
modified conclusion membership function 169
Modus Ponens tautology 161
momentum 378
monotonicity of sets 120, 129
multi-dimensional membership functions 428
multi-linear function 289
multimodels 323
multivariable fuzzy control 601
mutation 403, 489

non-convex fuzzy set 27
non-fuzzy bag 20
non-grid inputs' space partition 229
non-invertible model 550
non-jump model 537
non-linear mapping 522
non-linguistic Mamdani model 290
non-parameterized
- s-norm operators 129
- t-norm operators 121
non-rectangular input space partition 333
normal fuzzy set 19
normalization of a model 245
normalized RBF network 386
numerically complete rule base 212

observability 703
opposite fuzzy number 62
ordinary RBF network 386

parameterized
- mean intersection operator 124
- s-norm operators 130
- t-norm operators 121
partial model inversion 550
partition
- grid of the input's space 224
- of unity condition 32
perceptron networks 375
perfect control 563
perfect inversion 583
- 525
- of dynamic models 526
- of static models 525
performance criteria of control systems 508
PID controller 500
polynomial membership functions 47
Popov condition 660, 662, 692
population 488
power
- of a fuzzy set 16, 26
- of the numerical universe of discourse 15
pre-controller 586
preliminary conditions of hyperstability 628, 630
premise evaluation 162
primary control system 634

principle of incompatibility 6
PROD-operator 121
product 121
projection 148, 432
proper plant 528, 530

quality evaluation 11
quick inverse 531

radial basis functions 384
RBF neural networks 384
realizable inverse 531
rectangular input space partition 331
reduction
– of a fuzzy sets number 231
– of the rule base 229
redundancy of the rule base 222
reference model 555
relational model 311
relative power of a fuzzy set 26
resulting membership function of the rule-base conclusion 172
right membership function 48
rule base 158

s-norms 128
Schwab axioms 38
secondary control system 634, 639
self-organizing fuzzy model 363
self-tuning fuzzy model 363
sensitivity of a defuzzification method 186
set absorption 113, 127
sigmoid
– function 375
– membership function 44
significant model inputs 406
simplified
– denormalization 247
– normalization 245
single dimension clustering method 455
singleton 1, 22, 197
SISO System 60
"soft *MIN*" operator 115
"soft" signum 115
space of discourse 14
specialized structures for identification 549
stabilization of an unstable plant 564
standard SISO control system 618
static controller 495
stochastic uncertainty 56
strictly proper plant 528, 530
structure chromosome 492
structure with the reference model of disturbance rejection 591
sub-*MIN* operator 120
subnormal fuzzy set 19
super-*MAX*-operators 130
supervisory controller 614
support
– of a fuzzy set 22
– points of a model surface 224
symmetrical Gaussian function 40

t-norms 120
Takagi-Sugeno model 301
tautology 161
three-position action controller 500
transparency of a model 228
Tustin's method 500
two-position action controller 496
type 1 fuzzy set 52
type 2 fuzzy set 54

undefined sign fuzzy number 92
underwater vehicle propellers 576
union
– of sets 126
– with the complementary set 127
– with the empty set 127
universal fuzzy set 18
universe
– of discourse 14
– of discourse of a model 224

verbal model 367
vertical representation of a fuzzy set 23

weighted mean method 443

Yager-intersection operator 121
Yager-union operator 130

Zadeh's compositional rule of inference 168
zero-order extrapolation 255